Ralf Riedel (Editor)

Handbook of Ceramic Hard Materials

Related titles from WILEY-VCH

M. Swain (Ed.)
Structure and Properties of Ceramics
ISBN 3-527-26824-3

R. J. Brook (Ed.)
Processing of Ceramics
Part I: ISBN 3-527-26830-8
Part II: ISBN 3-527-29356-6

J. Bill, F. Wakai, F. Aldinger
Precursor-Derived Ceramics
ISBN 3-527-29814-2

Ralf Riedel (Editor)

Handbook of Ceramic Hard Materials

WILEY-VCH

Weinheim · New York · Chichester · Brisbane · Singapore · Toronto

Editor:
Prof. Dr. Ralf Riedel
Fachgebiet Disperse Feststoffe
Fachbereich Materialwissenschaft
Technische Universität Darmstadt
Petersenstraße 23
64287 Darmstadt
Germany

This book was carefully produced. Nevertheless, authors, editor and publisher do not warrant the information contained therein to be free of errors. Readers are advised to keep in mind that statements, data, illustrations, procedural details or other items may inadvertently be inaccurate.

Library of Congress Card No. Applied for.

A catalogue record for this book is available from the British Library.

Deutsche Bibliothek Cataloguing-in-Publication Data:
A catalogue record for this publication is available from Die Deutsche Bibliothek
    ISBN 3-527-29972-6

© WILEY-VCH Verlag GmbH, D-69469 Weinheim (Federal Republic of Germany), 2000

Printed on acid-free and chlorine-free paper.

All rights reserved (including those of translation in other languages). No part of this book may be reproduced in any form – by photoprinting, microfilm, or any other means – nor transmitted or translated into machine language without written permission from the publishers. Registered names, trademarks, etc. used in this book, even when not specifically maked as such, are not to be considered unprotected by law.
Composition: Alden Bookset, Oxford
Printing: betz-druck, Darmstadt
Bookbinding: Buchbinderei Osswald, Neustadt/Wstr.
Printed in the Federal Republic of Germany

This book is dedicate to

Ute, Vincent, Lorenz and Marlene

Preface

With increasing demand for improved efficiency of engines, plants and production processes, ceramics have gained great importance as structural engineering materials in recent years. Within the group of the so called advanced materials, carbon in form of diamond or diamond-like structures, carbides, nitrides and borides have reached an outstanding position due to their excellent hardness and thermo-chemical and thermo-mechanical properties. The distinct covalent bonding of the aforementioned structures positively influences their hardness and their tribological behavior. Moreover, a series of oxides such as stishovite, a high pressure modification of silica, or boron sub-oxides have been recently discovered to exhibit high hardness apart from the well known alumina.

There is presently much effort in basic science and applied research to work on novel ceramic hard materials denoted as super- or ultra-hard materials that can compete with the hardness of conventional diamond. Aim and scope of the research in this field is to develop hard materials with superior mechanical and chemical properties and with similar hardness. Moreover, calculations of properties of hypothetical carbon nitrides like C_3N_4 indicated that there might be compounds exhibiting even higher hardness values than that of diamond. The low-temperature synthesis of diamond and cubic boron nitride on the one hand as well as the successful research on new carbon nitrides on the other hand have caused an enormous impact around the world on both the basic science and the technological development of these novel ultra-hard materials.

With the present book we wish to review comprehensively and concisely the state of the art concerning the structure, synthesis, processing, properties and applications of ceramic hard materials in general. In particular, the synthesis, modeling and properties of novel hard materials like binary carbon nitrides, ternary boron carbonitrides and others are also addressed. It is the aim of this reference book not only to reflect the state of the art and to give a sound review of the literature, but to delineate the underlying concepts and bearing of this interdisciplinary field. With the present edition we wish to show that the field of hard materials research and development has to be recognized into the wider context of chemistry, physics as well as materials science and engineering.

The book is organized in two volumes and three parts, covering the structure and properties of ceramic hard materials (Volume 1, Part I), synthesis and processing (Volume 1, Part II) as well as the typical fields of applications (Volume 2, Part III).

Volume 1 starts with an introduction into novel ultra hard ceramics including diamond and diamond-like carbon, carbon nitrides and silicon nitrides as well as boron containing carbides, nitrides and carbonitrides. Here we wish to recognize the great fundamental and technological challenge of developing new superhard materials which can compete with the hardest counterparts such as diamond and cubic boron nitride.

In dealing with properties, the first Chapter in Part I is then devoted to the structure of crystalline and amorphous ceramic hard materials. The structural features are responsible in particular for the intrinsic materials properties such as melting point and hardness. It has been found that in many cases the hardness of a crystalline substance correlates with its melting point. Therefore, detailed knowledge of the 3dimensional arrangement of the atoms is required to understand the materials behavior under certain conditions. More details of the individual crystal structures with respect to a 3dimensional view can be found on our hard materials homepage under the web address '**www.hardmaterials.de**'. Phase transitions and materials synthesis under high pressure in laser heated diamond cells is the topic of the continuing Chapter. The materials behavior under high pressure and temperature is of fundamental interest for the synthesis of hard materials since many of the ultra-hard substances like diamond, cubic boron nitride or stishovite are formed naturally or synthetically under these harsh conditions. The next three Chapters are concerned with the mechanical behavior and corrosion of ceramic hard materials and their relation to microstructure. This correlation is an important feature since hardness is not only governed by the intrinsic atomic structure of the respective material but also to a great extend by its polycrystalline nature. Therefore, the grain morphology and grain boundary chemistry play a decisive role in the materials response under environmental or mechanical load. In the following Chapter transition metal carbides, nitrides and carbonitrides are discussed with a focus on their structure and bonding, thermodynamic behavior as well as on their physical and mechanical properties. Part I is then completed by two Chapters which deal with the theoretical design of novel sp^2-bonded carbon allotropes and novel superhard materials based on carbon and silicon nitrides. These Chapters tribute to the fact that with proceeding computerization the number of calculated novel solid structures that led to the prediction of new materials with hardness comparable to or exceeding that of diamond has increased enormously in recent years.

Part II continues with the synthesis and processing of ceramic hard materials. Since the conventional powder technological synthesis and processing of ceramics has been treated in a large number of published review articles here we concentrate on novel synthetic routes that provide ceramic hard materials. Consequently, six Chapters report on i) directed metal oxidation, ii) self-propagating high temperature synthesis, iii) hydrothermal synthesis of diamond, chemical vapor deposition of diamond (iv) and cubic boron nitride (v) films and finally vi) the polymer to ceramic transformation. All these processes are particularly suitable for the formation of refractories with high hardness. Part II is then closed by a Chapter on nano structured superhard materials. In the course of this work high hardness is achieved by microstructural control rather than by the synthesis of a distinct crystal structure.

In Volume 2 ceramic hard materials are highlighted in the light of their applications. Chapter 1 of Part III concisely reviews the history of diamond and diamond-like super abrasive tools while Chapter 2 and 3 are concerned with the application of chemical vapor deposited diamond and diamond-like carbon films. These sections include the synthesis of optical grade CVD diamond windows and discuss their physical and mechanical properties. The most important and wide-spread ceramic hard materials are based on alumina. Chapter 4 reports on the processing developments to increase the hardness of alumina based ceramics for grinding and cutting applications. Silicon carbide and silicon nitride materials are the most technologically important non-oxide compounds and have gained great significance in the field of cutting ceramics and are treated in Chapters 5 and 6. Boron-based ceramics are a further group of either established or candidate materials with extreme hardness. Therefore, Chapter 7 deals with boron carbide or transition metal borides like titanium diboride and their distinct properties and applications. In Chapter 8, classical hard metals comprised of tungsten carbide as the hard phase and cobalt as the binder phase are discussed. Volume 2 is finally completed by a data base (Chapter 9) containing approximately 130 hard materials including carbides, nitrides, borides, silicides and oxides. The data base references the crystal structure, physical properties like melting point and density, mechanical properties (Youngs modulus, micro hardness) and oxidation resistance of the respective compounds. Future developments of novel hard materials such as the recently discovered intermetallic phase $AlMgB_{14}$ will be updated on our internet homepage **'www.hardmaterials.de'**.

In closing these introductory remarks, I would like to emphasize that the special chance to place a summary of the outstanding expertise on the field of present hard materials research and development would not have been possible without the great enthusiasm and commitment of all the colleagues who contributed in the writing of this two volume set. I am grateful for their enormous efforts in compiling a fascinating series of articles imparting depth insight into the individual fields of modern hard materials research. Finally, I wish to thank the Wiley-VCH Editors Peter Gregory and Jörn Ritterbusch for encouraging me in the preparation of this book and for their continuous support throughout the editorial process.

Ralf Riedel
March 2000
Darmstadt

Foreword

One of the clearest hierarchies in materials science and engineering is provided by the property of hardness. There are, of course, many properties where remarkable differences exist between groups of materials. An example is provided by electrical conductivity where a ratio of 10^{18} can be readily found; with electrical conductivity, however, the different materials do not come into direct competitive opposition. In the case of hardness, the very value of this property lies in the ability of one material to demonstrate a higher place in the hierarchy than another; the one material is used in effect to overpower the other.

The existence of this hierarchy, which has been long recognised in the traditional measurement scale for the property, has direct relation to applications. In any use of materials it is important to be able to shape them to be fit for purpose; where the shaping process involves some type of machining, as it most commonly does, then the property of hardness becomes the unambiguous figure of merit.

It is for these reasons that there has been long standing and productive interest in hard materials, in their design, in their fabrication, in their use, and in the underlying science and engineering. It is thoroughly in keeping with this tradition of research relevant to application that the present book brings together a set of authoritative reviews of the progress which has been made.

The organisation of the book is a direct reflection of the logic which has been used in developing hard materials. One of the great attractions of the subject has been the close link that exists between hardness on the one hand and the bonding and structure of the material on the other. The link between these two has proved to be one of the best foundations on which to base materials development. The link is a central theme in the first part of the book where fine examples are given of the rich contribution which has been made and which continues to be made by fundamental studies of bonding and structure to materials performance.

It has long been recognised that the very aspect of their extreme resistance to deformation would make it a particular challenge to manufacture hard materials in reliable and cost-effective ways. It is here that the materials community has shown itself to be imaginative and forward looking in seeking innovative fabrication routes. These are well presented in the second part of the book where specific attention is given to the paths which can be used to assemble materials of precisely defined form without sacrifice of their characteristic mechanical resilience.

The most striking aspect of hard materials, however, is the direct link to applications. This link has brought an unusual degree of purpose to materials development which has enjoyed the benefits of being conducted in full recognition of the target to be reached. It has also meant that the progress made in research can be rapidly evaluated since the testing procedures relate so directly to the end use. The third part of the book accordingly gives close accounts of the performance of the different classes of hard materials in the applications context.

The contributors to this text are to be congratulated on bringing their many disciplines to bear on this central theme. Materials science is well known to undergo fashions as materials are developed and discarded and indeed as sectors of application grow and decline. The one requirement which will remain is that the forming and shaping of materials will always be necessary whatever the eventual sector of application. We can accordingly be confident that the long history of hardness studies, not least in the last two hundred years from the carbon tool steels, to high speed steels, to stellite, to tungsten carbide, to cermets, to ceramics, and now to diamond, boron, nitride and other special systems, will be continued with informed imagination and with creative innovation. The present book is a splendid platform on which to base such future development.

Richard Brook
January 2000
Oxford, UK

Contents

List of Contributors   XXVII
List of Symbols   XXXIII
List of Abbreviations   XXXIX

Introduction: Novel Ultrahard Materials
A. Zerr and R. Riedel

Introduction   XLV
Hard Materials   XLVI
Hardness   XLVII
Carbon-based Hard Materials   L
Diamond   LII
Diamond-like and Amorphous Carbon   LV
Novel Hypothetical Three-dimensional Carbon Phases   LVI
Fullerenes   LIX
Carbon Nitride (C_3N_4)   LIX
Boron-based Hard Materials   LXIV
Boron Nitrides   LXIV
Boron-rich Boron Nitrides   LXVII
Nitrogen-rich Boron Nitride   LXVIII
Boron Carbonitrides ($B_xC_yN_z$)   LXVIII
Boron Suboxides   LXXI
Silicon-based Materials   LXXI
Concluding Remarks   LXXII
Acknowledgement   LXXIII
References   LXXIII

Part I   **Structures and Properties**

1   **Structural Chemistry of Hard Materials**
W. Jeitschko, R. Pöttgen, and R.-D. Hoffmann

1.1	Introduction 3	
1.2	Diamond and Diamond-Related Structures 5	
1.2.1	The Crystal Structure of Diamond 5	
1.2.2	The Isoelectronic Compounds c-BN and SiC 6	
1.3	Crystal Chemistry of Borides and Boron Carbides 8	
1.4	The Structures of Transition Metal Carbides 12	
1.5	Silicides and Silicide Carbides of Transition Metals 20	

1.6	Nitrides 23	
1.6.1	Nitrides of Main Group Elements 24	
1.6.2	Transition Metal Nitrides 25	
1.6.3	Perspectives: Nitridosilicates 29	
1.7	Oxide Ceramics 30	
1.7.1	Hard Ceramics of Main Group Elements 30	
1.7.2	Transition Metal Oxides 32	
1.8	Amorphous Hard Materials 36	
	References 37	
2	**Phase Transitions and Material Synthesis using the CO_2-Laser Heating Technique in a Diamond Cell**	
	A. Zerr, G. Serghiou, and R. Boehler	
2.1	Introduction 41	
2.2	Technique of CO_2-Laser Heating in a Diamond Anvil Cell 42	
2.2.1	Sample Assemblage in a Diamond Anvil Cell 42	
2.2.2	Pressure Conditions in the Sample Volume 43	
2.2.3	Experimental Set-up for CO_2-Laser Heating in a Diamond Anvil Cell 44	
2.2.4	Temperature Determination 45	
2.2.5	Temperature Stabilization 45	
2.2.6	Radial Temperature Gradients 48	
2.2.7	Raman and Fluorescence Spectroscopic Analysis of Samples in a Diamond Anvil Cell 48	
2.3	Determination of Melting Temperatures at High Pressures 49	
2.3.1	Melting of Cubic BN at 10 GPa 49	
2.3.2	Melting Temperatures of Materials Relevant to the Earth's Lower Mantle 51	
2.4	Phase Diagrams, Decomposition Reactions, and Stability of Solids at High Pressures and Temperatures 54	
2.4.1	Coesite–Stishovite Phase Boundary 55	
2.4.2	High Pressure and Temperature Phase Diagram and Decomposition Reactions in a Ternary System 56	
2.4.3	Stability of a Perovskite Oxide with Respect to its Component Oxides 59	
2.5	CO_2-laser Heating Experiments on Organic Compounds 60	
2.6	Conclusion 62	
	Acknowledgments 62	
	References 62	
3	**Mechanical Properties and their Relation to Microstructure**	
	D. Sherman and D. Brandon	
3.1	Introduction 66	
3.1.1	Applications and Engineering Requirements 66	

3.1.2	Bulk Components	68
3.1.3	Coatings	70
3.1.4	Engineering Requirements	70
3.2	Principal Mechanical Properties	71
3.2.1	Elastic Modulus	71
3.2.2	Strength	72
3.2.3	Fracture Toughness	74
3.2.4	Hardness	79
3.3	Mechanical Testing of Hard Materials	81
3.3.1	Elastic Modulus	81
3.3.2	Fracture Strength	81
3.3.3	Fracture Toughness	83
3.3.4	Hardness	84
3.3.5	Indentation Toughness	86
3.3.6	Erosion, Wear and Scratch Tests	89
3.4	Microstructural Parameters and Mechanical Properties	91
3.5	Failure Mechanisms	94
3.5.1	Creep Behavior	94
3.5.2	Mechanical Fatigue	95
3.5.3	Ballistic Properties	97
3.6	Conclusions	98
	References	99

4   Nanostructured Superhard Materials
S. Veprěk

4.1	Introduction	104
4.2	Concept for the Design of Superhard Materials	109
4.2.1	Nanocrystalline Materials	110
4.2.2	Heterostructures	114
4.3	Preparation and Properties of Superhard Nanocrystalline Composites	116
4.3.1	Preparation	116
4.3.2	Properties of the ncM_nN/aSi_3N_4 Composites	119
4.3.3	Other Superhard Nanocomposites and the General Validity of the Design Principle	124
4.4	Discussion of the Possible Origin of the Hardness and Stability of the Nanostructure	128
4.4	Conclusions	133
	Acknowledgments	134
	References	134

5   Corrosion of Hard Materials
K. G. Nickel and Y. G. Gogotsi

5.1	Introduction	140

5.2	Corrosive Media 140
5.3	Corrosion Modes 141
5.3.1	Active and Passive Corrosion 141
5.3.2	Homogeneity and Location of Attack: Internal, External and Localized Corrosion 141
5.4	Corrosion Kinetics 142
5.4.1	Physical Boundary Conditions 142
5.4.2	Active Corrosion Kinetics 143
5.4.3	Basic Passive Corrosion Kinetics 145
5.4.4	Kinetic Breaks 147
5.4.5	Complex Kinetics 148
5.5	Corrosion Measurement 150
5.5.1	Experimental Methods 150
5.5.2	Corrosion Data 151
5.6	Materials 154
5.6.1	Diamond and Diamond-like Carbons 154
5.6.2	Carbides 155
5.6.3	Nitrides 166
5.6.4	Carbonitrides 173
5.6.5	Titanium Diboride 176
	References 177

6	**Interrelations Between the Influences of Indentation Size, Surface State, Grain Size, Grain-Boundary Deformation, and Temperature on the Hardness of Ceramics** *A. Krell*
6.1	Introduction 183
6.2	The Assessment of Residual Porosity and Flaw Populations: A Prerequisite for any Hardness Investigation 184
6.3	Theoretical Considerations 185
6.3.1	The Role of the Lattice and of Grain Boundaries in the Inelastic Deformation at an Indentation Site in Sintered Hard Materials 185
6.3.2	Quantitative Understanding the Load Effect on the Hardness: Theoretical Considerations Compared with Single Crystal Data 188
6.4	Influences of the Grain Size and the State of the Surface 191
6.4.1	The Grain Size Influence on the Load Effect of the Hardness: Modeling Experimental Results 191
6.4.2	The Effect of the Grain Size and the Surface State in Ceramics when Recorded by Different Measuring Approaches 193
6.5	Comparing the Grain Size Effect and the Indentation Size Effect: The Role of Grain Boundaries at Room Temperature 195
6.6	The Effects of Temperature on the Hardness of Ceramics 198
6.7	Summary 199
	References 201

7		**Transition Metal Carbides, Nitrides, and Carbonitrides**
		W. Lengauer
7.1		Introduction 202
7.2		General Features of Structure and Bonding 205
7.2.1		General Structural Features 205
7.2.2		General Features of Bonding 206
7.3		Preparation 207
7.4		Characterization 210
7.4.1		Chemical Analysis 210
7.4.2		Physical Microanalysis 211
7.5		Thermodynamics 212
7.5.1		Stability of Carbides 212
7.5.2		Nitrogen Partial Pressure of Nitrides 212
7.5.3		Phase Equilibria of Important Carbide Systems 213
7.5.4		Transition Metal–Nitrogen Systems and Structure of Phases 216
7.5.5		Carbonitride Systems 221
7.6		Properties of Important Transition Metal Carbides, Nitrides, and Carbonitrides 224
7.6.1		Melting Points 224
7.6.2		Color 224
7.6.3		Thermal and Electrical Conductivities 225
7.6.4		Thermal Expansion 228
7.6.5		Diffusivities 229
7.6.6		Elastic Properties 231
7.6.7		Microhardness 234
7.7		Industrial Applications 238
7.7.1		Cemented Carbides and Carbonitrides 238
7.7.2		Deposited Layers 241
7.7.3		Diffusion Layers 246
		Acknowledgments 248
		References 248
8		**New Superhard Materials: Carbon and Silicon Nitrides**
		J. E. Lowther
8.1		Introduction 253
8.2		Modeling Procedures 254
8.2.1		Semi-empirical Approaches 254
8.2.2		Tight-binding Schemes 255
8.2.3		*Ab initio* Pseudopotential Approach 256
8.2.4		Transition Pressures and Relative Stability 256
8.3		Carbon Nitride 257
8.3.1		Crystalline Structures 258
8.3.2		Graphitic Structures 259
8.3.3		Amorphous Structures 261

8.3.4	Relative Stability	263
8.4	Silicon Carbon Nitride	264
8.4.1	βSiC_2N_4	265
8.4.2	Near-cubic Forms of SiC_2N_4	266
8.4.3	Relative Stability	268
8.5	Conclusions	268
	Acknowledgements	269
	References	269

9 **Effective Doping in Novel sp² Bonded Carbon Allotropes**
G. Jungnickel, P. K. Sitch, T. Frauenheim, C. R. Cousins, C. D. Latham, B. R. Eggen, and M. I. Heggie

9.1	Introduction	271
9.2	Lattice Description	274
9.3	Computational Methods	276
9.4	Static Properties	278
9.5	Electronic Properties	279
9.6	Conclusions	282
	Acknowledgments	283
	References	283

Part II **Synthesis and Processing**

1 **Directed Metal Oxidation**
V. Jayaram and D. Brandon

1.1	Historical Background	289
1.2	Oxidation and Oxide Formation	290
1.2.1	Initial Oxidation	291
1.3	Related Ceramic Processing Routes	293
1.4	Directed Metal Oxidation Incubation	295
1.5	Directed Metal Oxidation Growth	300
1.5.1	Introduction	300
1.5.2	Directed Metal Oxidation Composites from Al–Mg Alloys	300
1.5.3	Directed Metal Oxidation Growth from other Aluminum Alloys	304
1.5.4	Microstructural Scale	305
1.5.5	Growth into Particulate Preforms	307
1.5.6	Growth into Fibrous Preforms	309
1.6	Mechanical Properties	310
1.6.1	Elastic Modulus	310
1.6.2	Strength and Toughness	311
1.6.3	Thermal Shock	313
1.6.4	High Temperature Strength	313
1.6.5	Wear Properties	314
1.6.6	Mechanical Properties of Fiber-reinforced DMO Composites	314

1.7	Corrosion of Directed Metal Oxidation Composites	316
1.8	Other Properties 316	
1.9	Applications 316	
1.9.1	Wear Resistant Components 317	
1.9.2	Ceramic Composite Armor 317	
1.9.3	Thermal Barriers and Heat Sinks 318	
	References 318	

2 Self-Propagating High-Temperature Synthesis of Hard Materials
Z. A. Munir and U. Anselmi-Tamburini

2.1	Introduction 322
2.2	Mechanistic Characterization of the Process 327
2.3	Effect of Experimental Parameters 331
2.3	Synthesis of Dense Materials 342
2.4	Synthesis by Field-Activated Self-propagating High-temperature Synthesis 348
2.6	Selected Recent Examples of Synthesis of Hard Materials 356
	Acknowledgment 368
	References 368

3 Hydrothermal Synthesis of Diamond
K. G. Nickel, T. Kraft, and Y. G. Gogotsi

3.1	Introduction 374
3.2	Evidence from Nature 376
3.3	Hydrothermal Synthesis 377
3.3.1	C–H–O System 377
3.3.2	Hydrothermal Treatment of SiC 382
3.4	Outlook 387
	Acknowledgments 387
	References 387

4 Chemical Vapor Deposition of Diamond Films
C.-P. Klages

4.1	Introduction 390
4.2	Preparation Methods for Diamond Films 391
4.2.1	Hot-filament Chemical Vapor Deposition 392
4.2.2	Microwave-plasma-based Methods 397
4.2.3	Preparation of Special Forms: Textured and Heteroepitaxial Films 400
4.3	Thermochemistry and Mechanism of Chemical Vapor Deposition Diamond Growth 407
4.3.1	Transformation of Graphite to Diamond at Low Pressures 407

4.3.2	Reactive Species in Diamond Chemical Vapor Deposition, the Role of CH_3 408	
4.4	Properties and Applications of Chemical Vapor Deposited Diamond 410	
4.4.1	Diamond Coated Cutting Tools 411	
4.4.2	Thermal Conductivity of Chemical Vapor Deposited Diamond: Thermal Management Applications 412	
4.4.3	Electrical Properties and Electronic Applications 413	
4.4.4	Electrochemical Use of Chemical Vapor Deposited Diamond 415	
4.5	Summary 417	
	References 417	
5	**Vapor Phase Deposition of Cubic Boron Nitride Films** *K. Bewilogua and F. Richter*	
5.1	Introduction 420	
5.2	Empirical Results 421	
5.2.1	Deposition Methods 421	
5.2.2	Morphology and Structure of cBN Films 423	
5.2.3	Film Adhesion 427	
5.3	Models of cBN Formation 427	
5.4	Sputter Deposition of cBN Films 429	
5.4.1	Sputter Deposition with Conducting Targets 430	
5.4.2	Deposition by d.c. Magnetron Sputter with a Hot Boron Target 431	
5.5	Discrimination between Nucleation and Growth Phase 433	
5.5.1	Detection of hBN–cBN Transition 433	
5.5.2	RF Magnetron Sputtering 435	
5.6	Properties of cBN Films 440	
5.6.1	Mechanical and Tribological Properties 440	
5.6.2	Optical Properties 440	
5.6.3	Electrical Properties 441	
5.6.4	Other Properties 441	
5.7	Summary and Outlook 442	
	References 442	
6	**Polymer to Ceramic Transformation: Processing of Ceramic Bodies and Thin Films** *G. D. Sorarù and P. Colombo*	
6.1	Introduction 446	
6.2	Processing of Monolithic Components 450	
6.3	Preparation and Characterization of SiAlOC Ceramic Bodies by Pyrolysis in Inert Atmosphere 452	
6.3.1	Experimental Procedure 452	

6.4	Results 453	
6.4.1	Characterization of the Pre-ceramic Precursors 453	
6.4.2	Characterization of the Pre-ceramic Components 454	
6.4.3	Characterization of the Ceramic Components 455	
6.4.4	Mechanical Characterization at High Temperature 457	
6.5	Discussion 458	
6.6	Preparation and Characterization of SiAlON Ceramics by Pyrolysis in Reactive Atmosphere 460	
6.6.1	Experimental 460	
6.7	Results and Discussion 460	
6.8	Processing of Thin Ceramic Films 463	
6.9	Experimental 463	
6.10	Results and Discussion 464	
6.10.1	Conventional Conversion Process: Annealing in Controlled Atmosphere 464	
6.10.2	Nonconventional Conversion Process: Ion Irradiation 467	
6.11	Conclusions 472	
	Acknowledgments 473	
	References 473	

Part III   Materials and Applications

1   Diamond Materials and their Applications
Edited by R. J. Caveney

1.1	Superabrasive tools: A Brief Introduction 479	
1.1.1	Introduction 479	
1.1.2	Early History 479	
1.1.3	Synthetic Diamond 481	
1.1.4	Cubic Boron Nitride 482	
1.1.5	Polycrystalline Diamond and Cubic Boron Nitride 482	
1.1.6	Chemical Vapor Deposited Diamond 484	
1.1.7	Outline of Chapter 485	
1.2	The Crystallization of Diamond 485	
1.2.1	The Carbon Phase Diagram 485	
1.2.2	Diamond Crystallization at High Pressure 487	
1.2.3	High Pressure Apparatus 490	
1.2.4	The Synthesis of Particulate Diamond Abrasives 491	
1.2.5	Growth of Large Synthetic Diamonds 496	
1.2.6	Novel Diamond Synthesis Routes 504	
1.2.7	Cubic Boron Nitride Crystallization 510	
1.3	Polycrystalline Diamond and Cubic Boron Nitride 512	
1.3.1	Natural Polycrystalline Diamond 512	
1.3.2	Synthetic Polycrystalline Diamond 512	
1.3.3	Mechanisms involved in Polycrystalline Diamond Manufacturing Process 513	

1.3.4	Polycrystalline Cubic Boron Nitride	518
1.4	New Ultrahard Materials	521
1.4.1	Introduction	521
1.4.2	Hardness	521
1.4.3	C_3N_4	523
1.4.4	Boron Rich Nitride	526
1.4.5	Boron Carbonitrides	526
1.4.6	Boron Suboxides	526
1.4.7	Stishovite	526
1.5	Industrial Applications of Diamond and cBN	527
1.5.1	Introduction	527
1.5.2	Abrasive Application	528
1.5.3	Machining of Stone and Concrete	540
1.5.4	Applications of Polycrystalline Ultra-hard Materials	548
1.5.5	Applications of Single Crystal Diamond	559
	Acknowledgments	566
	References	566

2 Applications of Diamond Synthesized by Chemical Vapor Deposition
R. S. Sussmann

2.1	Introduction	573
2.2	Properties of Chemical Vapor Deposited Diamond	574
2.2.1	Material Grades	574
2.2.2	Optical Properties	576
2.2.3	Strength of Chemical Vapor Deposited Diamond	580
2.2.4	The Young Modulus	581
2.2.5	Thermal Conductivity	582
2.2.6	Dielectric Properties	583
2.3	Optical Applications	583
2.3.1	Chemical Vapor Deposited Diamond for Passive Infrared Windows in Aggressive Environments	584
2.3.2	Windows for High-power Infrared Lasers	589
2.4	Windows for High Power Gyrotron Tubes	597
2.4.1	Window Requirements	598
2.4.2	The Development of Chemical Vapor Deposited Diamond Gyrotron Windows	599
2.5	Thermal Management of Laser Diode Arrays	606
2.5.1	Laser Diode Arrays: General Issues	607
2.5.2	Modelling of Submount Heat Resistance	607
2.5.3	Flatness of Submount	610
2.5.4	Thermal Stress	610
2.6	Cutting Tools, Dressers and Wear Parts	611
2.6.1	Cutting Tools Trends	611
2.6.2	Cutting Tool Application of Chemical Vapor Deposited Diamond	612

2.6.3	Chemical Vapor Deposited Diamond Dressers	616
2.6.4	Chemical Vapor Deposited Diamond Wear Parts	617
	References 619	

3 Diamond-like Carbon Films
C.-P. Klages and K. Bewilogua

3.1	Introduction 623	
3.2	Preparation Methods for Diamond-like Carbon Films	623
3.2.1	Hydrogenated Amorphous Carbon (a-C:H) 623	
3.2.2.	Hydrogen Free Amorphous Carbon (ta-C) 627	
3.2.3	Metal-containing Amorphous Hydrocarbon 629	
3.3	Microstructure and Bonding of Diamond-like Carbon	630
3.3.1	Amorphous Carbon and Hydrogenated Amorphous Carbon	630
3.3.2	Metal-containing Amorphous Carbon Films 634	
3.4.	Physical Properties of DLC Films 637	
3.4.1	Electrical and Optical Properties 637	
3.4.2	Mechanical Properties 639	
3.5.	Applications of DLC Films 640	
3.5.1	Adhesion of DLC Films 640	
3.5.2	Tribology of DLC Coatings 642	
3.5.3	Tribological Applications 644	
3.5.4	Other Applications 644	
	References 645	

4 Ceramics Based on Alumina: Increasing the Hardness for Tool Applications
A. Krell

4.1	Recent Trends in the Application of Ceramic Tool Materials 648
4.2	Technological Essentials for Producing Hard and Strong Tool Ceramics 650
4.2.1	Typical Defects in Ceramics Tool Materials: The State of The Art 651
4.2.2	Recent Trends in Ceramic Technologies Related to Tool Ceramics 653
4.3	Tool Materials with Undefined Cutting Edge: Sintered Grinding Materials 658
4.3.1	Technical Demands for Grinding Materials 660
4.3.2	Advanced Commercial Products: Sol/gel-derived Corundum 661
4.3.3	Sintered Alumina Grits Produced by Powder Processing Approaches 665
4.4	New Trend for Cutting Hard Workpieces: Submicrometer Cutting Ceramics for Tools with Defined Cutting Edge 666
4.4.1	Demands for Cutting Materials Used for Turning Hard Workpieces 667

4.4.2	Carbide Reinforced Composite Ceramics Based on Al_2O_3 669
4.4.3	Single Phase Sintered Corundum 670
4.4.4	Comparative Cutting Studies with Submicrometer Ceramics: Al_2O_3 and Composites Reinforced with Ti(C,N) and Ti(C,O) 670
4.5	Summary 680
	References 681

5 Silicon Carbide Based Hard Materials
K. A. Schwetz

5.1	Introduction 683
5.1.1	History 683
5.1.2	Natural Occurrence [7] 684
5.2	Structure and Phase Relations of SiC 685
5.3	Production of SiC 688
5.3.1	The Acheson/ESK Process 688
5.3.2	Other Production Methods 691
5.3.3	Dense SiC Shapes 699
5.4	Properties of Silicon Carbide 719
5.4.1	Physical Properties 719
5.4.2	Chemical Properties 720
5.4.3	Tribological Properties 723
5.5	Quality Control 734
5.6	Toxicology and Occupational Health 736
5.7	Uses of Silicon Carbide 736
	Acknowledgments 740
	References 740

6 Silicon Nitride Based Hard Materials
M. Herrmann, H. Klemm, Chr. Schubert

6.1	Introduction 749
6.2	Crystal Structure and Properties of the Si_3N_4 Modifications 753
6.3	Densification 755
6.4	Microstructural Development 758
6.4.1	Microstructural development of β-Si_3N_4 materials 758
6.4.2	Microstructural development of α'-SiALON materials 768
6.5	Properties of Si_3N_4 Materials 771
6.5.1	Mechanical properties at room temperatures 771
6.5.2	High-temperature properties of silicon nitride materials 777
6.5.3	Wear resistance of Si_3N_4 materials 782
6.5.4	Corrosion resistance of Si_3N_4 786
6.6	Conclusions/Further potential of silicon nitride materials 792
	Acknowledgements 795
	References 795

7 Boride-Based Hard Materials
R. Telle, L. S. Sigl, and K. Takagi

- 7.1 Introduction 802
- 7.2 Chemical Bonding and Crystal Chemistry of Borides 803
- 7.2.1 Chemical Bonding of Borides 803
- 7.2.2 The Crystal Structure of Borides 804
- 7.3 Phase Systems 812
- 7.3.1 Binary Phase Diagrams of Technically Important Systems 813
- 7.3.2 Ternary and Higher Order Systems 818
- 7.4 Boron Carbide Ceramics 837
- 7.4.1 Preparation of Boron Carbide 837
- 7.4.2 Sintering of Boron Carbide 839
- 7.4.3 Properties of Boron Carbide 851
- 7.4.4 Chemical Properties and Oxidation of Boron Carbide 855
- 7.4.5 Boron Carbide–Based Composites 857
- 7.5 Transition Metal Boride Ceramics 874
- 7.5.1 Preparation of Transition Metal Borides 875
- 7.5.2 Densification of Transition Metal Borides 876
- 7.5.3 Properties of Transition Metal Borides Ceramics 878
- 7.6 Multiphase Hard Materials Based on Carbide–Nitride–Boride–Silicide Composites 888
- 7.7 Boride–Zirconia Composites 888
- 7.8 Cemented Borides 895
- 7.8.1 Boron Carbide-Based Cermets 895
- 7.8.2 Titanium Diboride-Based Cermets 897
- 7.8.3 Cemented Ternary Borides 919
- 7.8.4 Potentials and Applications 927
- 7.9 Future Prospects and Fields of Application 933
  References 936

8 The Hardness of Tungsten Carbide–Cobalt Hardmetal 946
S. Luyckx 946

- 8.1 Introduction 946
- 8.2 The Hardness of the Two Component Phases 947
- 8.2.1 The Hardness of Tungsten Carbide 947
- 8.2.2 The Hardness of Cobalt 948
- 8.3 Factors Affecting the Hardness of WC–Co Hardmetal 950
- 8.3.1 Cobalt Content and Tungsten Carbide Grain Size 950
- 8.3.2 Grain Size Distribution and Cobalt Mean Free Path 952
- 8.3.3 Binder Composition and Carbon Content 952
- 8.3.4 Porosity 953
- 8.3.5 Effect of Temperature 953
- 8.4 Relationships between Hardness and Other Hardmetal Properties 960

8.4.1	Relationship between Hardness and Toughness	962
8.4.2	Relationship between Hardness and Abrasive Wear Resistance	962
8.5	Conclusions	963
	Acknowledgments	963
	References	964

9   Data Collection of Properties of Hard Materials
G. Berg, C. Friedrich, E. Broszeit, and C. Berger

9.1	Introduction	965
9.2	Profile of Properties	965
9.3	Organization and Contents of the Data Collection	966
	Acknowledgement	967
	References	991

Index   997

List of Contributors

U. Anselmi-Tamburini
Dipartimento di Chimica Fisica
Università di Pavia
27100 Pavia
Italy

M. W. Bailey
De Beers Industrial Diamond
Division Pty Ltd
Diamond Research Lab
PO Box 1770
Southdale 2135
South Africa

G. Berg
Fachgebiet und Institut für
Werkstoffkunde der TU Darmstadt
und Staatliche
Materialprüfungsanhalt
Grafenstrasse 2
D-64283 Darmstadt
Germany

C. Berger
Fachgebiet und Institut für
Werkstoffkunde der TU Darmstadt
und Staatliche
Materialprüfungsanhalt
Grafenstrasse 2
D-64283 Darmstadt
Germany

K. Bewilogua
Fraunhofer Institut für Schicht und
Oberflächentechnik (IST)
Bienroder Weg 54 E
D-38108 Braunschweig
Germany

R. Böhler
Max-Planck-Institute for Chemistry
Saarstrasse 23
D-55020 Mainz
Germany

D. Brandon
Department of Materials Engineering
Technion – Israel Institute of
Technology
Haifa 32000
Israel

J. R. Brandon
De Beers Industrial Diamond
Division Pty Ltd
Diamond Research Lab
PO Box 1770
Southdale 2135
South Africa

E. Broszeit
Fachgebiet und Institut für
Werkstoffkunde der TU Darmstadt
und Staatliche
Materialprüfungsanhalt
Grafenstrasse 2
D-64283 Darmstadt
Germany

R. C. Burns
De Beers Industrial Diamond
Division Pty Ltd
Diamond Research Lab
PO Box 1770
Southdale 2135
South Africa

R. J. Caveney
De Beers Industrial Diamond
Division Pty Ltd
Diamond Research Lab
PO Box 1770
Southdale 2135
South Africa

S. E. Coe
De Beers Industrial Diamond
Division Pty Ltd
Diamond Research Lab
PO Box 1770
Southdale 2135
South Africa

J. L. Collins
De Beers Industrial Diamond
Division Pty Ltd
Diamond Research Lab
PO Box 1770
Southdale 2135
South Africa

P. Colombo
Università di Bologna
Dipartimento di Chimica Applicata e
Scienza dei Materiali
viale Risorgimento 2
I-40136 Bologna
Italy

M. W. Cook
De Beers Industrial Diamond
Division Pty Ltd
Diamond Research Lab
PO Box 1770
Southdale 2135
South Africa

C. R. Cousins
Department of Physics
University of Exeter
Stocker Road
Exeter EX4 4QL
UK

G. J. Davies
De Beers Industrial Diamond
Division Pty Ltd
Diamond Research Lab
PO Box 1770
Southdale 2135
South Africa

B. R. Eggen
School of Chemistry, Physics and
Environmental Sciences
University of Sussex
Falmer
Brighton BN1 9QJ
UK

D. Fister
HC Starck Gmbh
Kraftwerkweg 3
D-79725 Laufenburg
Germany

T. Frauenheim
Fachbereich Physik
Universität/Gesamthochschule
Paderborn
D-33095 Paderborn
Germany

C. Friedrich
Fachgebiet und Institut für
Werkstoffkunde der TU Darmstadt
und Staatliche
Materialprüfungsanhalt
Grafenstrasse 2
D-64283 Darmstadt
Germany

Y. G. Gogotsi
Institut für Angewandte Mineralogie
Universität Tübingen
Wilhelmstrasse 56
D-72074 Tübingen
Germany

J. O. Hansen
De Beers Industrial Diamond
Division Pty Ltd
Diamond Research Lab
PO Box 1770
Southdale 2135
South Africa

M. Hoffmann
Fakultät für Maschinenbau
Institut für Werkstoffkunde II
Universität Karlsruhe
Kaiserstrasse 12
Postfach 6980
D-76128 Karlsruhe
Germany

R. D. Hoffmann
Westfälische Wilhelms-Universität
Münster
Anorganisch-Chemisches Institut
Wilhelm-Klemm-Strasse 8
D-48149 Münster
Germany

M. I. Heggie
School of Chemistry, Physics and
Environmental Sciences
University of Sussex
Falmer
Brighton BN1 9QJ
UK

V. Jayaram
Department of Metallurgy
Indian Institute of Science
Bangalore
India

W. Jeitschko
Westfälische Wilhelms-Universität
Münster
Anorganisch-Chemisches Institut
Wilhelm-Klemm-Strasse 8
D-48149 Münster
Germany

G. Jungnickel
Fachbereich Physik
Universität/Gesamthochschule
Paderborn
D- 33095 Paderborn
Germany

C. P. Klages
Fraunhofer Institut für Schicht und
Oberflächentechnik (IST)
Bienroder Weg 54 E
D-38108 Braunschweig
Germany

T. Kraft
Institut für Angewandte Mineralogie
Universität Tübingen
Wilhelmstrasse 56
D-72074 Tübingen
Germany

A. Krell
Fraunhofer Institute for Ceramic
Technologies and Sintered Materials
Winterbergstrasse 28
D-01277 Dresden
Germany

C. D. Latham
Department of Physics
University of Exeter
Stocker Road
Exeter EX4 4QL
UK

W. Lengauer
Institute for Chemical Technology of
Inorganic Materials
Vienna University of Technology
Getreidemarkt 9/161
A-1060 Vienna
Austria

J. E. Lowther
Department of Physics
University of Witwatersrand
Johannesburg
South Africa

S. Luyckx
School of Process and Materials
Engineering
University of the Witwatersrand
Johannesburg 2050
South Africa

Z. A. Munir
Facility for Advanced Combustion
Synthesis
Department of Chemical Engineering
and Materials Science
University of California
Davis CA 95616
USA

K. G. Nickel
Universität Tübingen
Applied Mineralogy
Wilhelmstrasse 56
D-72074 Tübingen
Germany

S. Ozbayraktar
De Beers Industrial Diamond
Division Pty Ltd
Diamond Research Lab
PO Box 1770
Southdale 2135
South Africa

C. S. J. Pickles
De Beers Industrial Diamond
Division Pty Ltd
Diamond Research Lab
PO Box 1770
Southdale 2135
South Africa

R. Pöttgen
Westfälische Wilhelms-Universität
Münster
Anorganisch-Chemisches Institut
Wilhelm-Klemm-Strasse 8
D-48149 Münster
Germany

F. Richter
Technische Universität Chemnitz-Zwickau
Institut für Physik
D-09107 Chemnitz
Germany

R. Riedel
Fachbereich Materialwissenschaft
Technical University of Darmstadt
Petersenstrasse 23
D-64287 Darmstadt
Germany

K. A. Schwetz
Advanced Ceramics Lab
Elektroschmelzwerk Gmbh
Max-Schaidhauf-Strasse 25
D-87437 Kempten
Germany

P. K. Sen
De Beers Industrial Diamond
Division Pty Ltd
Diamond Research Lab
PO Box 1770
Southdale 2135
South Africa

G. Serghiou
Max-Planck-Institute for Chemistry
Saarstrasse 23
Mainz
Germany

List of Contributors   **XXXI**

D. Sherman
Department of Materials Engineering
Technion – Israel Institute of
Technology
Haifa 32000
Israel

M. Sibanda
De Beers Industrial Diamond
Division Pty Ltd
Diamond Research Lab
PO Box 1770
Southdale 2135
South Africa

I. Sigalas
De Beers Industrial Diamond
Division Pty Ltd
Diamond Research Lab
PO Box 1770
Southdale 2135
South Africa

P. K. Sitch
Fachbereich Physik
Universität/Gesamthochschule
Paderborn
D- 33095 Paderbron
Germany

G. D. Sorarù
Università di Trento
Dipartimiento di Ingegneria dei
Materiali
Via Mesiano 77
I-38050 Trento
Italy

R. S. Sussmann
De Beers Industrial Diamond
Division Pty Ltd
Diamond Research Lab
PO Box 1770
Southdale 2135
South Africa

K. Takagi
Toyo Kohan Co. Ltd.
Tokyo
Japan

R. Telle
Institut für Gesteinshüttenkunde
RWTH Aaachen
Mauerstraße 5
D-52056 Aachen
Germany

S. Veprek
Institute for Chemistry of Inorganic
Materials
Technical University Munich
Lichtenbergstrasse 4
D-85747 Garching b. Munich
Germany

C. J. H. Wort
De Beers Industrial Diamond
Division Pty Ltd
Diamond Research Lab
PO Box 1770
Southdale 2135
South Africa

A. Zerr
Fachgebiet Disperse Feststoffe
Technical University of Darmstadt
Petersenstrasse 23
D-64287 Darmstadt
Germany

List of Symbols

α	absorption coefficient	
α	atomic attraction constant	
α	growth parameter	
α	power absorption coefficient	
α, β, χ	polytypes or phases	
β	atomic repulsion constant	
β	geometrical factor	
γ	rake angle	degrees
γ_i	secondary ion yield	
$\gamma(\mathbf{n})$	orientational surface energy	
γ_s	surface energy	J
Γ	width of X-ray reflection	
δ	microplastic deformability	
$\tan \delta$	dielectric loss factor	
Δc	concentratation difference	
ΔG^0_{298}	Gibbs free energy	J
ΔS_{int}	interfacial entropy	
ΔT	temperature change	K or °C
Δx	change of size or mass	
ε	elastic strain	
ε	emissivity	
$\dot{\varepsilon}$	strain rate	
η	degree of conversion to nitride	
η	fraction of reaction completed	
θ	angle	
θ	constant relating tensile strength and hardness	
2θ	X-ray scattering angle	
κ	entering angle	degrees
κ	thermal conductivity	$J\ cm^{-2}\ s^{-1}\ K^{-1}$
κ_1	thermal conductivity of rectants	
κ_2	thermal conductivity of products	
λ	empirical parameter relating bulk modulus and bond length	
λ	inclination angle	degrees
λ	layer thickness, mean free path	m
λ	polarity of bond	
λ	thermal conductivity	
λ	wavelength	m
λ	X-ray wavelength	
Λ	mean free path	m

List of Symbols

Symbol	Description	Units
μ	C/Ti ratio	
μ	coefficient of friction	
ν	Poisson ratio	
π	complementary energy	J
ρ	density	kg m^{-3}
ρ	dislocation density	
ρ	resistivity	
ρ_m	theoretical density of product	
σ	conductivity	
σ	electrical conductivity	
σ	stress	Pa
σ_0	median failure stress	Pa
σ_b	fracture strength in bending, modulus of rupture	
σ_{ij}	local stress field	
σ_s	Stefan–Boltzmann constant	J cm^{-2} s^{-1} K^{-4}
σ_Y	yield stress	GPa
τ	annealing time	
τ	help time	s
ϕ	angle between crack and tensile stress	
ϕ, ω, τ	ternary phases	
ϕ	azimuthal angle between polarization vector and substrate direction	
ϕ	constraint factor	
χ	electron affinity	
ω	wear coefficient	
$\langle x \rangle_\tau$	average diffusion distance in time τ	
□	vacancy in crystal structure	
a	indent size, half length of diagonal	m
a	crack length	m
a	depth of cut	mm
a_0	equilibrium bond distance	m
a_{cr}	critical flaw size	m
a_i	depth of lateral crack on erosion	m
a, b	heat capacity coefficients	
a, b, c	crystal unit cell parameters lattice constants	nm
A	area	m^2
A	contact area	m^2
A	material constant	
ABAB, ABCABC	stacking sequences	
A,B,H,HA	stacking positions	
b	Burgers vector	
B	bulk modulus	GPa
B	designation of a dissolved species in, e. g., a liquid	
(B$_4$C)	designation of a non-stociometric compound (solid solution)	

List of Symbols   XXXV

c	radius of radial crack	m
c	velocity of light in vacuo	m s^{-1}
c_i	interfacial concentration	
c_i	radius of lateral crack on erosion	m
C	proportionality constant	
C	specific heat per unit volume	
C_i	concentration of impurities	
C_p	heat capacity	J g^{-1} K^{-1}
d	bond length	Å
d	degree of dilution	
d	diameter of Brinell impression	m
d	diameter of Vickers impression	m
d	grain size	m
d	height of beam	m
d	layer thickness	
d	spacing of powder diffraction rings	rad
d_p	pore diameter	
D	diameter of Brinell indenter	m
D	diameter of median crack	m
D	diffusivity	
D	size of particles	
D_0	diffusion factor	
D_H, D_C	diffusion coefficients at high and low temperature	
e	unit electron charge	
E	activation energy	J
E	binding energy	J
E	Young's modulus	
E_0	theoretical Young's modulus	
E^*	$E/(1-\nu^2)$	
E_C	composite potential energy	J
E_F	Fermi energy	
E_g	band gap	J
E_i	ion energy	J
E_{imax}	maximum ion energy	J
E_P	potential energy	J
f	feed rate	mm min^{-1}
f	frequency	Hz
f	volume fraction	
f	Weibull safety factor	Hz
f_I	volume fraction of I	
F	force	N
F	statistical failure probability	
F_i	ion flux	
F_L	external work (linear elasticity)	J
(g)	vapor phase	
G	shear modulus	

List of Symbols

G	strain energy release rate	
G_{IC}	toughness, fracture energy, work of fracture	N m^{-1}
h	convective heat transfer coefficient	J cm^{-2} s^{-1} K^{-1}
h	indentation depth	
h	Planck constant	
H	hardness	
H	enthalpy	
H_{298}	enthalpy of formation at 298 K	
H_B	Brinell hardness	GPa
H_M	Meyer hardness	GPa
H_K	Knoop hardness	GPa
$H_{plastic}$	plastic hardness	
H_V	Vickers hardness	GPa
H_x	average applied compressive stress in hardness test GPa	
i, a	fluxes of impinging ions and deposited atoms	
I	rank order of test result	
I_{BB}	intensity of black body radiation	
I_i	peak height	
I_{RB}	intensity of real body radiation	
j_i	current density	
J	mass flux	
k	Boltzmann constant	
k	constant for layer growth	
k_0	reaction constant	
k_{log}	logarithmic rate constant	
k_L	linear rate constant	
k_p	parabolic rate constant	m^{2}s^{-1} or kg^{2}m^{-4}s^{-1}
k_r	reaction coefficient	
K	proportionality constant	
K	stress intensity factor	
K_{IC}, K_{ICO}	fracture toughness	Pa m$^{-1/2}$
l	diffusion path length	m
l	length of sample	m
l	long diameter of Knoop impression	m
l	span of beam	m
L	defined load	kg
L_{max}	maximum load	
m	Weibull modulus	
M	mode of deformation (I, II, or III)	
M	metal	
M	Mohs hardness	
M	weight	kg
M_i	molecular mass of impurity	
n	number	
n	order of reaction	

n	Paris exponent for fatigue	
n	refractive index	
n	stress exponent	
N	number of tests	
N_c	average coordination number	
$N_{con}(r)$	no. of constraints for coordination number r	
$\langle N_C \rangle$	average coordination number	
p	grain size exponent	
p	momentum	
p	porosity	
P	fixed load	N
P	gas pressure	
P	porosity	%
P	pressure	GPa
P_0, P_h	ambient and high pressure	Pa
P_c	confining pressure in powder	
P_N	pressure of nitrogen	Pa
$P(O_2)$	partial pressure of O_2	Pa
q	Porod scattering vector	
q	scattering vector	
Q	activation energy	J
Q	heat of reaction	J g^{-1}
Q	resonance factor	
Q_t	heat of transport	J
R	radius	
r	distance from crack tip	m
r	interatomic distance	m
r	radius of curvature	
r_0	equilibrium interatomic distance	m
R	gas constant	
R	film growth rate	
R_a	average roughness	
R_{pl}	radius of plastic zone	m
(s)	solid phase	
S	elastic recovery	%
S	stoichiometric ratio	
S^0_{298K}	entropy	J mol^{-1} K^{-1}
T	thickness of window	
t	time	s
t_c	Ch I 4 ?	
t_d	delay time	s
t_w	time for wave propagation	s
T	temperature	K or °C
T	ternary phase	
T_0	ambient temperature	K or °C
T_1, T_2	temperature limits	K or °C

List of Symbols

T_{ad}	adiabatic combustion temperature	K or °C
T_c, P_c	critical temperature and pressure	
T_e	eutectic temperature	
T_m	absolute melting temperature	K or °C
T_s	substrate temperature	K or °C
u	displacement	m
u	stoichiometry factor	
u	wave velocity	
U	internal energy	J
U_b	substrate bias voltage	
U_E	elastic strain energy	J
U_p	plasma potential	V
U_S	surface free energy	J
v	average velocity	
v	machining velocity	m min^{-1}
v	volume	m^3
v	wavenumber	cm^{-1}
v_L	longitudinal velocity of sound	m s^{-1}
v_p	volume of pores	
V	volume	m^3
V	volume fraction	
V	symbol for vacancy in a chemical formula	
V_B	applied substrate bias voltage	V
V_i	volume lost in erosion impact	m^3
V_m	molar volume of metal	
$V(\mathbf{n})$	orientational growth rate	
V_P, V_R	molar volumes of product and reactants	
w	sample thickness	m
W	RF power	W
x	carbon-to-metal ratio	
x	coordinate	
x	layer thickness	m
(1)	liquid phase	
[100]	lattice directions	
(100)	Miller indices	

List of Abbreviations

3PB	three-point bend
4PB	four-point bend
ACC	amorphous covalent ceramics
AES	Auger electron spectroscopy
AFM	atomic force microscope
APW	augmented plane wave
AR	antireflection
ASEA	Swedish company
ASTM	American Society for Testing and Materials
b.c.c.	body-centered cubic
b.c.t.	body-centered tetragonal
BB	black body
BET	Brunauer–Emmett–Teller method for determining porosity
3C, 4H, 6R	polytype notations of SiC (C: cubic, H: hexagonal, R: rhombohedral)
CAD	cathodic arc deposition
CALPHAD	calculation of phase diagrams model
cBN	cubic boron nitride
CCD	charge coupled device
CED	cutting edge displacement
CMC	ceramic matrix composite
COOP	crystal orbital overlap population
CN	coordination number
CVD	chemical vapor deposition
CVI	chemical vapor infiltrated
CW	continuous wave
d.c.	direct current
DAC	diamond anvil technique
DCC	direct coagulation casting
DF–TB	density-functional tight-binding method
DH	methyldiethoxysilane
DIN	Deutsche Industrie Norm
dlC	diamond-like carbon
DLC	diamond-like carbon
DMO	directed metal oxidation
DOS	density of states
DTA	differential thermal analysis
ECH	electron cyclotron heating
ECR	electron cyclotron resonance

EDAX	energy-dispersive analysis of X-rays
EELS	electron energy loss spectroscopy
EP	electroplated
EPMA	electron probe microanalysis
EPR	electron paramagnetic resonance
ERD	elastic recoil detection
ERDA	elastic recoil detection analysis
ESCA	electron spectroscopy for chemical analysis
ESK	Elektroschmelzwerk Kempten
EXAFS	extended X-ray absorption fine structure
f.c.c.	face-centered cubic
FEPA	Federation Europeen des Fabricants de Produits Abrasifs
FTIR	Fourier transform infrared
FWHM	full width at half maximum
FZK	Forschungzentrum Karlsruhe
GA-XRD	glancing angle XRD
gC	glassy carbon
GEC	General Electric Company, USA
GFRP	glass-fiber reinforced plastic
GGA	generalized gradient approximation
h	hexagonal
hBN	hexagonal boron nitride
HF-CVD	hot filament CVD
HIP	hot isostatic pressing
HK	Knoop hardness
HOMO	highest occupied molecular orbital
HOPG	highly ordered pyrolytic graphite
HPHT	high-pressure high-temperature
HPL	high-pressure laminate
HPMS	high-pressure microwave source
HR	Rockwell hardness
HR-TEM	high-resolution TEM
HSS	high-speed steel
HV	Vickers hardness
IBAD	ion-beam assisted deposition
ICSD	inorganic crystal structure database
ICDD	international center for diffraction data
IED	ion energy distribution
IR	infrared
ISE	indentation size effect
ISO	International Standards Organization
ITER	international thermonuclear experimental reactor
JAERI	Japan atomic energy research institute
JFM	Johnson figure of merit
JIS	Japanese Standards
KFM	Keyes figure of merit

LAS	lithium aluminosilicate
LDA	local density approximation
LDA	laser diode array
LIDT	laser induced damage threshold
LPI	liquid polymer infiltration
LPSSiC	liquid-phase sintered SiC
LPSSS	low-pressure solid-state source
LRO	long-range order
LSF	line spread function
LSI	liquid silicon infiltration
LWIR	longer wavelength infrared
MAK	Maximal zulässige Arbeitsplatz Konzentration
MAS-NMR	magic angle spinning NMR
Me–DLC	metal–DLC hybrid
MMC	metal matrix composite
MOR	modulus of rupture
MOSFET	metal-oxide silicon field effect transistor
MS	mass spectroscopy
MSIB	mass-selected ion beam
MTES	methyltriethoxysilane
MTF	modulation transfer function
MW	microwave
MWP-CVD	microwave plasma CVD
NASA	National Aeronautics and Space Administration (USA)
ncTiO$_2$	nanocrystalline titania
NDE	nondestructive evaluation
Nd-YAG	neodymium–yttrium-aluminum-garnet laser
NEA	negative electron affinity
NICALON	branded Si–C–O composite fiber from Nippon Carbon
NIRIM	National Institute for Research in Inorganic Materials (Japan)
NMR	nuclear magnetic resonance
NRA	nuclear reaction analysis
ORNL	Oak Ridge National Laboratory
PA-CVD	plasma assisted CVD
PAlC	polyaluminocarbosilane
PCS	polycarbosilane
p.p.m.	parts per million
PBC	periodic bond chain
PC	potential cycling
pcBN	polycrystalline boron nitride
pcD	polycrystalline diamond
PCS	polycarbosilane
PCT	Patent Cooperation Treaty
PTC	polytitanocarbosilane
PTES	phenyltriethoxysilane
PTFE	polytetrafluoroethylene

PP	polymer pyrolisis
PVD	physical vapor deposition
r	rhombohedral
RBAO	reaction bonded aluminum oxide
RBM	reaction bonded mullite
rBN	rhombohedral boron nitride
RBS	Rutherford back-scattering
RBSN	reaction bonded silicon nitride
RF	radio frequency
RSF	reduced spatial frequency
RSSC	reaction sintered silicon carbide
s.c.	simple cubic
s.c.cm.	standard cubic centimeters
SAD	small angle diffraction
SAXS	small-angle X-ray scattering
SCS	Textron process SiC fibers with C core and C surface
SEM	scanning electron microscope
SERR	strain energy release rate
SHS	self-propagating high-temperature synthesis
SIALON	Si–Al–O–N (silicon aluminum oxynitride) fiber
SiCAlON	SiC–AlN–Al_2OC composite fiber
Si–DLC	Si–DLC hybrid
SIF	stress intensity factor
SIMS	secondary ion mass spectrometry
SNMS	secondary neutron mass spectrometry
SP	sintered powder
SRO	short-range order
STM	scanning tunnel microscopy
ta-C	hydrogen-free amorphous carbon
taC	tetrahedral amorphous carbon
TD	theoretical density
TCNE	tetracyanoethylene
TEM	transmission electron microscope
TGA	thermogravimetric analysis
TH	triethoxysilane
TRS	transverse rupture strength (= MOR)
TZP	tetragonal zirconia polycrystals
UHP	ultahigh purity
UPS	ultraviolet photoelectron spectroscopy
UV	ultaviolet
VAMAS	Versailles Agreement on Materials and Standards
VC	vapor phase formation and condensation process
VEC	valence electron concentration
VLS	vapor–liquid–solid process
VS	vapor–solid reaction
wBN	wurtzitic boron nitride

XANES	X-ray absorption near edge structure
XPS	X-ray photoelectron spectroscopy
XRD	X-ray diffraction
YAG	yttrium aluminium garnet, yttrium aluminate, $Y_2Al_5O_{12}$
YLF	yttrium-lithium-fluorite

Part III

Materials and Applications

1 Diamond Materials and their Applications
Edited by I. Sigalas and R. J. Caveney

1.1 Superabrasive tools: A Brief Introduction
M. W. Bailey

1.1.1 Introduction

Tooling systems containing diamond and cubic boron nitride materials are used today across the breadth of modern industry, which could not function effectively without them. Their range of application covers petroleum exploration, mining, stone and concrete sawing, the grinding of tungsten carbides, glass, and ceramics, the machining of a wide range of engineering materials including aluminum alloys, modern composite materials, hard ferrous materials, and also in such unlikely fields as the cutting of frozen foods. Most applications take advantage of the extremely high hardness and wear resistance of these so-called 'ultrahard' or 'superabrasive' materials. However, more recently, the outstanding thermal and optical properties of diamond are being more widely exploited, giving rise to further (and new) areas of application.

1.1.2 Early History

Diamond, as a material, has been recognized as having unique properties for several thousands of years and references of it being used as an industrial tool can be traced back to at least 300 BC. At this time, whole stones appeared to have been used for engraving and, by 150 BC, India, where the early diamond deposits were discovered, had established an export trade with China for such diamond-tipped engraving tools for use in cutting very hard jade stone (Fig. 1).

No major change in the use of diamond as an industrial tool occurred for some considerable time. Reference was made by Pliny in his 'The History of the World,' published in AD 77, to splinters of diamond being used as engraving tools. Other industrial applications developed very slowly until the 18th and 19th centuries. In AD 1751, Diderot describes in his 'Encyclopaedia' a method of drilling rock using fragments of diamond held in the lower end of a diamond rod. This rod, a precursor to modern rock drills, was raised and dropped by an operator who also rotated it with his hand (Fig. 2). By the mid 1800s, such rock drills were machine powered and a report in 1852 claimed that a hole, 5 cm in diameter, had been drilled in granite to a depth of 37 cm in 87 min. With modern diamond technology, this can now be achieved in a matter of seconds.

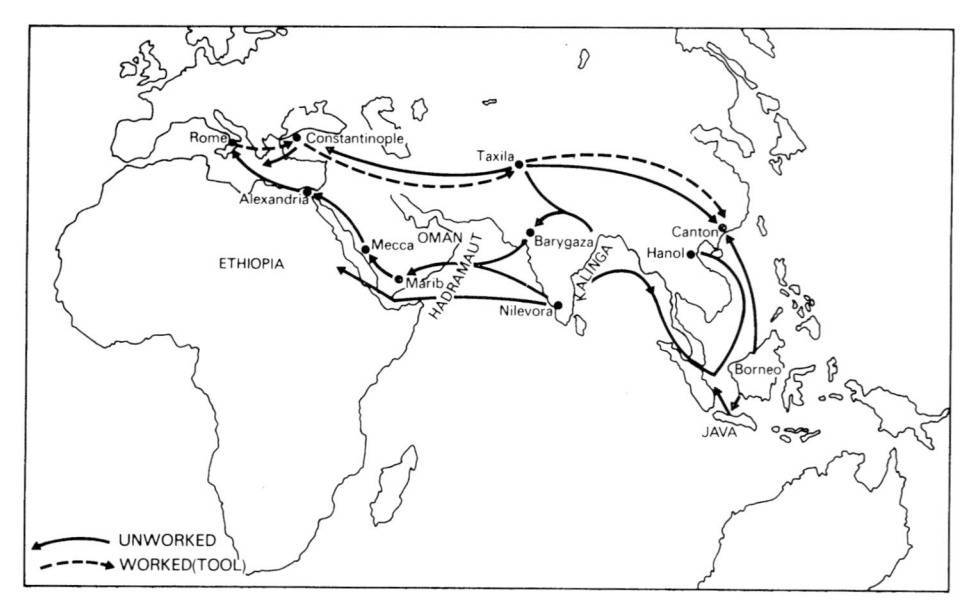

Figure 1. Diamond trade routes from the 1st century BC to the 3rd century AD.

Other developments were taking place during this period. A diamond tipped lathe tool was used by J. Ramsden in 1773, the beginning of precision turning. The use of diamond as a wire drawing die was patented in England in 1819. By the end of the 19th century, large saw blades with diamond set around their

Figure 2. The first diamond-tipped 'percussion' drill (Diderot).

periphery were in use in France for cutting stone. The cutting of marble and limestone used during the construction of large buildings in Paris in the early 1900s utilized diamond-containing saw blades up to 2–3 m in diameter.

In 1842, the first mention was made of a diamond tool utilizing relatively small pieces of diamond when Pritchard reported grinding and polishing microscope lenses using a tool consisting of diamond grit hammered into the surface of a cast iron tool. The first patent for a diamond wheel was granted to A. L. Caverdon in France, in 1878, when he produced a wheel containing fine diamond particles held onto a metal wheel form possibly by jamming them into indentations made in the surface of the wheel. This method was described by Carl Zeiss in 1906.

Most modern diamond tools, including the majority of saw blades and grinding wheels, consist of small particles of diamond held in a matrix or 'bond' which wears down during use to expose new diamond particles, which carry out the cutting. This is in contrast to the very early tools which used relatively large single stones or slivers from single stones. A patent for such a bond system was granted to the Western Electric Company in 1927. During the early 1930s, in order to grind the then-new material, cemented tungsten carbide, which had been invented by Krupp in Germany in 1928, grinding wheels containing diamond held in a resin bond were developed and several patents were granted between 1932 and 1933.

1.1.3 Synthetic Diamond

The commercial availability of synthetic diamond has been a major contributor to the dramatic advances which have occurred during the past 50 years in both diamond tool technology and also in the range of applications in which they are used. In the period between 1953 and 1958, three independent teams of scientists from ASEA in Sweden, General Electric in the United States and De Beers in South Africa were successful in producing synthetic diamond by converting graphite to diamond using high-pressure, high-temperature technology. Prior to this, the limited supply of natural diamond impeded development of new diamond tools.

The commercial availability of synthetic diamond offered two advantages. First, there was potentially unlimited availability of industrial diamond compared to the limited volumes of suitable natural material and, second, it offered the opportunity of engineering material to have specific properties suited to particular industrial applications.

Of the diamond used for industrial purposes today, some 90% is synthetic, and 10% natural. The major use of diamond grit is in the machining (sawing and drilling) of stone and concrete (Fig. 3). Finer sizes of diamond are used widely for the grinding, sawing, and polishing of glass, ceramics, tungsten carbides, and a very wide range of other mainly nonferrous engineering and industrial materials.

Specialist applications, for example fine high precision turning (Fig. 4) of nonferrous and precious materials, drawing of fine wire, and the dressing of conventional abrasive wheels have, for several decades, used natural diamond. During the 1980s, the technology to produce, economically, large synthetic single diamond crystals

Figure 3. Diamond drilling a 100 mm hole for a new sea wall drainage system.

was developed. Whole synthetic crystals or cut pieces of the order of several millimeters in size are used for these high precision industrial applications.

1.1.4 Cubic Boron Nitride

Following the successful commercial synthesis of diamond in the 1950s, the second hardest material known, cubic boron nitride, cBN, was introduced to the market in the 1960s and is complementary to diamond. The iron, and its alloying elements, in ferrous materials has a tendency to react chemically with diamond under machining conditions and this can reduce the efficiency of the tool. cBN, however, although not as hard as diamond, does not react chemically with iron and is therefore particularly well suited to machining hard ferrous materials.

1.1.5 Polycrystalline Diamond and Cubic Boron Nitride

The next notable material development was the introduction to the market in the early 1970s of polycrystalline diamond (pcD). This made available to industry

Figure 4. Natural diamond cutting tool machining a contact lens.

relatively large pieces of diamond, albeit polycrystalline rather than monocrystalline, at an economical cost. Such materials can be produced with varying mechanical properties and are used in a wide variety of cutting applications and also as a wear resistant surface. As a cutting tool material, it is used predominantly to machine nonferrous abrasive materials including drilling of rock in oil and gas exploration, and in the engineering and manufacturing industries for the machining of aluminum alloys, reinforced plastics, new wear resistant lightweight composite materials, wood, and wood composites (Figs 5 and 6). pcD is also used in noncutting applications and one of its original and important applications is as a die for drawing wire. More recently, the use of pcD as a wear resistant component in thrust bearings, work rests, and high precision gauging equipment has become more widespread.

The development of polycrystalline cBN (pcBN) in the mid-1970s made it economically possible to machine fully hardened ferrous workpieces on a lathe or milling machine when previously the only method of machining such workpieces was by grinding, normally with silicon carbide or aluminum oxide (conventional) abrasives. Huge increases in productivity can be achieved by changing to turning (or milling) with pcBN. In addition, environmental improvements can also be achieved, since in many cases pcBN does not need coolant.

Figure 5. A pcD-tipped cutter machines oak panels for kitchen furniture.

1.1.6 Chemical Vapor Deposited Diamond

The development in the late 1980s of an entirely new method of diamond synthesis using chemical vapor deposition (CVD), as an alternative to ultra-high pressure and temperature systems, has caused a considerable stir in the industry.

Figure 6. pcD machining a cast metal-matrix composite automotive brake motor in Al 20% SiC_p.

With this technique, it is possible to deposit layers of diamond onto a substrate from a hot carbon-containing plasma, enabling relatively large areas of diamond to be produced (at the time of writing, the order of $10\,000\,\text{mm}^2$). Traditional diamond tools utilize the extreme hardness of diamond (and cBN) but, with this new technology, which is still developing, the exceptional optical and thermal properties can be applied. Diamond has a very high thermal conductivity (up to six times that of copper at room temperature) and is transparent to visible and infrared radiation, opening the door to a new varied range of industrial applications where these properties can be exploited in areas such as thermal management of electronics and windows for high technology infrared equipment, in addition to its use as an alternative material for use in the more traditional diamond tool areas.

1.1.7 Outline of Chapter

Section 1.2 of this Chapter reviews the crystallization of diamond and cubic boron nitride using high pressure, high temperature techniques. Chapter 4 of Part II deals with the subject of chemical vapor deposition of diamond at low pressure.

The high pressure sintering of diamond and cubic boron nitride to form superhard composites is described in Section 1.3, a review of attempts to produce other hard materials is given in Section 1.4.

Section 1.5 summarizes the many and varied applications of diamond and cBN materials. The applications of diamond grown using the CVD technique are discussed in Chapter 2 of Part III.

1.2 The Crystallization of Diamond

G. J. Davies

1.2.1 The Carbon Phase Diagram

It was established that diamond is an allotrope of carbon by Tennant in 1797 [1]. This led to many attempts to crystailize diamond using various carbonaceous starting materials, but it was not until about a century and a half later that successful synthesis was proven, as referred to in Section 1.1.3. The first clear success was by a Swedish group at the ASEA Company in February 1953 [2]. This was followed by the General Electric Company in December 1954 [3]. The early attempts and the subsequent successes are well reviewed [4–6].

Natural diamond has been used by man since at least biblical times, not only as a gem but also, due to its extreme properties relative to other materials, as an abrasive and even as a medicine in crushed form, 'a panacea for all ills'. It is not intended here to enter the debate concerning the crystallization of natural diamond. Some general

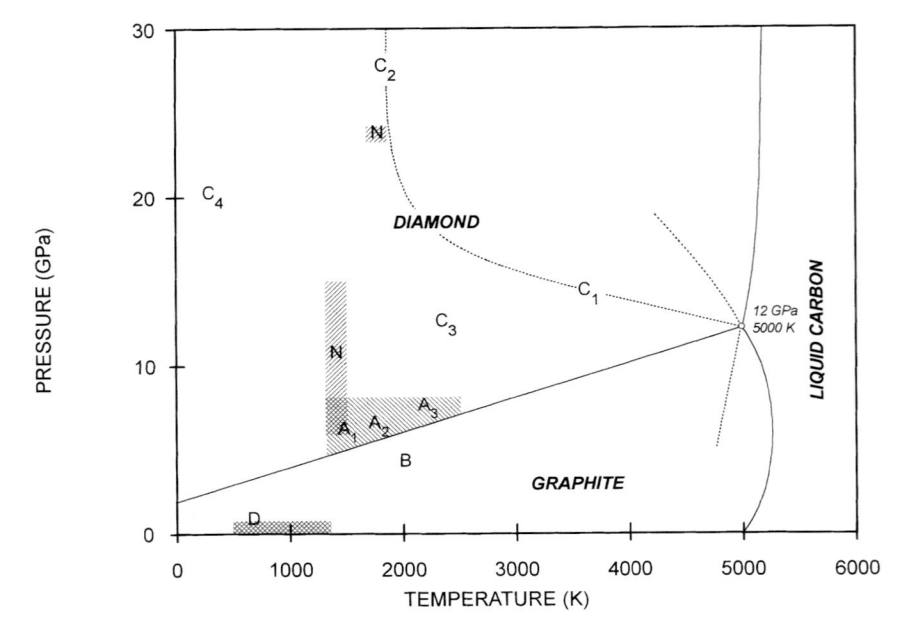

Figure 7. The phase diagram of carbon adapted from [9].
A: region of solvent/catalyst-based synthesis of diamond from graphite using:
  A_1: commercially used transition metal solvent/catalyst [16]
  A_2: nonconventional metal solvents: Cu, Zn, Ge [17–20]
  A_3: nonmetallic solvents, mainly metal salts, silicates [22–28]
B: the synthesis of diamond from phenolic resin and cobalt [30]
C: region of direct conversion of hexagonal graphite to either hexagonal or cubic diamond, where the dotted line C_{1-2}, represents the solid-solid transformation line of graphite or hexagonal diamond to cubic diamond through fast pressure/temperature cycles [9]
C_3: conversion of hydrocarbons to cubic diamond [33]
C_4: transformation of C_{60} fullerenes into polycrystalline diamond [35]
D: low pressure metastable growth of diamond [12, 13, 36–38]
N: regions of natural diamond genesis [7, 8, 21, 29].

natural diamond genesis information is to be found elsewhere [7, 8]. The conditions under which natural diamonds are formed will, however, be touched upon when the phase transformation diagram for carbon is discussed later in this section on crystallization.

To gain a balanced general view of the many ways in which man has crystallized diamond, the thermodynamically stable and metastable phases of elemental carbon and reaction dynamics between them, over obtainable pressures and temperatures, should be considered. The most up to date phase and transformation diagrams are to be found in a review by Bundy *et al.* [9]. Figure 7 is an adaptation of the pressure–temperature phase transition diagram taken from this review article, together with further information gleaned from the extensive diamond- and carbon-related scientific literature.

The dominant thermodynamically stable forms of carbon are graphite, diamond, liquid, and vapor. The phase diagram presented by Bundy [9] and adapted in Fig. 7

has some new notable features over many of the earlier diagrams. The slope of the diamond melting line is positive, and molten carbon is said to be metallic with little evidence for transformations from electrically conducting to nonconducting forms. For consideration of diamond crystallization, one of the most important features is the diamond–graphite equilibrium line which stretches from 1.7 GPa at 0 K to 12 GPa at 5000 K, the diamond–graphite–liquid carbon triple point. The activation energies required to initiate transformations from one form of carbon to another are very largely due to the high cohesive energy of carbon in its crystalline structures. Davies and Evans [10] showed that the activation energy for the graphitization of the {110} surface of pure diamond in vacuo is 728 ± 50 kJ mol^{-1}, which is about twice the bond energy of the C–C bond in aliphatic hydrocarbons (3.7 eV per σ C–C bond). This is also about the same energy required for the vaporization of graphite. These very high activation energies result in very large areas of pressure and temperature over which a metastable form of carbon can subsist. Probably the most important example of this is the indefinite existence of diamond at room temperature and pressure even though, at these conditions, it is very 'deep' within the graphite stability field. To overcome the activation energy for conversion to graphite in vacuo, diamond must be heated to a temperature of about 2000 K before graphitization becomes rapid [10]. In addition to the allotropes of carbon, there are polytypes of carbon, that is structural variations based upon lattice plane stacking difference. Important examples are hexagonal diamond, which occurs in methods of very rapid and hot direct transformation into diamond, and rhombohedral graphite, produced by cold working normal hexagonal graphite crystals. When the nanoscale groupings of pure carbon atoms are considered, other solid forms with very distinctive properties are known and are probably best described as 'metastable' over the whole range of pressures and temperatures in Fig. 7. These are amorphous forms such as carbon black and glassy carbon, condensate of very long linear carbon molecules collectively called carbynes and, of course, the now very extensive variations of fullerenes (C60, C70, etc.) and the derived nanotubes [11].

Importantly for diamond crystallization, all these forms of carbon, including graphite, very stubbornly persist at pressures far into the diamond stable region. The practical synthesis of diamond could thus be considered as the search for means of finding practical convenient kinetic pathways to avoid the high activation barriers between the multifarious nondiamond carbon structures and diamond. The two very general approaches are to use solvents for carbon and to precipitate diamond primarily in the diamond stable part of the phase diagram or to 'quench' carbonaceous species from very high energy states such as vapors, liquid, or plasmas. Some of the more recently reported means could well be considered to be combinations of these [12, 13]. Some of the most important diamond crystallization techniques in the recent literature will now be itemized using Fig. 7 as a guide.

1.2.2 Diamond Crystallization at High Pressure

The shaded region in Fig. 7 labeled A_{1-3}, with pressure from about 4.5 GPa to 9 GPa and temperature from about 1475 K to 2475 K, gives the conditions under which the bulk of the reported solution grown, diamond stable region, crystallization of diamond has taken place.

The graphite–diamond equilibrium expression in this region, following Berman and Simon [14] is

$$\Delta G_T^P = \Delta H_T^O - T\Delta S_{T+O}^O \int^P \Delta V \, \mathrm{d}P = 0 \qquad (1)$$

It has stood the test of time and experimental verification [15].

A_1 is where most of the commercial diamond abrasive products and high pressure polycrystalline diamond products are manufactured using alloys of transition metals, mainly iron–nickel, cobalt–iron and to a lesser extent nickel–manganese for the grits and larger individual crystals, and mainly cobalt for the polycrystalline products [16]. In this reference, the general transition metal solvent work up to the end of the 1980s is covered. In principle many of the other transition metals and alloys can be used but a large number of them require very difficult to attain and control temperatures and pressures dependent upon melting points, carbide stabilities, and carbon solubilities.

The position A_2, somewhat higher in pressure and temperature than A_1, covers the region where the higher melting and or lower carbon solubility metals and alloys can be used. Diamond synthesis from graphite has been demonstrated using copper, zinc, and germanium [17, 18]. Magnesium was shown to be a workable solvent/catalyst [19] at about 7.7 GPa and 2275 K. Later, by the addition of 50–60% copper to magnesium, diamond synthesis conditions were reduced to 6 GPa and 1825 K [20].

The A_3 position in Fig. 7 is the general condition under which inorganic compounds have been used to aid in the conversion of graphite to diamond. The more recent work in this area has mainly been carried out at the National Institute for Research in Inorganic Materials, Tsukuba, Japan. This work was probably stimulated by the possibility that volatiles such as CO_2, H_2O, CH_4, and O_2 may play an important role in natural diamond formation [21]. Carbonates [22], hydroxides [23] and sulphates [23, 24] with alkali and alkali earth cations have proven to work at 7.7 GPa and temperatures in the range 1825–2475 K. The diamonds formed were aggregates with particles up to 20 μm in size with normal faceting. In order to clarify the possible role of CO_2, silver carbonate has been investigated compared to silver metal [25]. This compound aided diamond formation above about 2075 K and 7.7 GPa, whereas pure silver did not act as a solvent/catalyst under these conditions. Further, volatile-rich silicate melts of kimberlite composition (kimberlite is the igneous rock usually associated with natural diamond in diamond bearing pipes) and SiO_2–H_2O combinations aided graphite conversion to fine diamond at 7.7 GPa and 20750–2475 K [26, 27]. The need for the presence of volatiles when using silicates to aid diamond formation has been recently confirmed [28]. In this report, anhydrous magnesium silicate melts corresponding to the composition of the minerals forsterite (Mg_2SiO_4) and enstatite ($MgSiO_3$) were compared with hydrated versions of these compositions. Fine diamonds of euhedral shape were formed only in the hydrated magnesium silicate graphite systems above about 2075 K at 7.7 GPa. No diamond was detected in the anhydrous systems up to

2175 K at this pressure even after treatments lasting 8 h or more. These findings are consistent with the current theories of natural diamond genesis.

The two regions both marked N in Fig. 7 are intended to illustrate the conditions under which natural diamond formed in the earth's crust, probably over geological time scales (about 3.3 billion years). There are two environments linked to natural diamond formation, the peridotitic (depicted by the lower pressure N region on Fig. 7) and the eclogitic (at higher pressures and about 200 K hotter, also marked N).

The pressure given by the N positions are estimated from the accepted correspondence of 0.1 GPa pressure to 3 km depth into the earth crust. The lower pressure limit of N is given by the boundary zone between the upper region lithosphere and the mobile asthenosphere at about 150 km in the earth's mantle.

It is believed that natural diamonds, crystallized from CO_2 or CH_4 volatiles below the lithosphere and after existing for great periods of time at these depths, are brought to the surface by subsequent eruptions of magma. For more detailed discussions of the geology, mineral inclusions in natural diamond and general matters related to diamond genesis, the reader is referred to Harris [7], Kesson [8], Haggerty [29], and Meyer [21].

The position marked B at 4 GPa and temperatures between 1775 and 2075 K, gives the condition under which well faceted diamonds of about 200 μm diameter have been crystallized using specially treated phenolic resins as the source of carbon and molten cobalt as solvent [30]. The notable point of this particular crystallization is that it occurred well into the graphite stable region of Fig. 7 and is thus an example of the 'metastable growth of diamond'. An explanation for this may be found in consideration of the relative solubilities of phenolic resins and diamond in molten cobalt allowing dissolution of one metastable form and precipitation of another, namely diamond. This is an example where rules governing the transitions between metastable states, such as the Ostwald and Ostwald–Volmer rules, can be applied [31].

The curved dashed line C_1–C_2 in Fig. 7 depicts the proposed boundary between the formation of the hexagonal (below the line) or cubic (above the line) polytypes of diamond, after explosive shock compression or fast heating, mainly using laser heated diamond anvil cells, of graphite structures in the diamond stable region, described by Bundy *et al.* [9]. This reference is a detailed review article and is a major source of reference for work concerned with the direct conversion of graphite and other carbon structures to diamond. The diamonds that can be retrieved from these various methods, either of the hexagonal or cubic form, are always small and crystallographically defective. This is because of the very short times at high temperature and pressure, and the need to satisfy fast thermal quench requirements. Despite this, diamond powders have been commercially produced by the Du Pont Company of the USA and others using explosive shock of metal containing small crystals of graphite as a carbon source [32].

The direct conversion of hydrocarbons to diamond has also been demonstrated. C_3 is the region where organic compounds such as anthracene, camphene, fluorene, pyrene, sucrose, polyethylene, adamantane, and paraffin wax have been converted to very fine diamonds under high pressure (about 1μm in size) (Wentorf [33]). Aliphatic compounds seem to work but compounds with aromatic rings and/or large amounts of nitrogen only formed graphite of excellent crystallinity under high pressure pyrolysis and did not form diamond.

There is a growing body of literature describing the crystallization of diamond from fullerenes as a starting material. One of the latest [34] claims that C_{60} can be transformed to diamond at 5 GPa and 1675 K using a solvent/catalyst without forming graphite first. The generation of 'bulk' polycrystalline diamond at room temperature has been reported by the crushing of C_{60} at about 20 GPa [35], (position C_4 in Fig. 7). The conditions used in the polycrystalline diamond (cobalt-matrix) anvil cell were deliberately nonhydrostatic with respect to pressure. It was suggested, from the high efficiency of conversion and fast kinetics involved, that, if cheap bulk C_{60} became available at some time in the future, a feasible method for the commercial production of polycrystalline diamond might result.

The metastable crystallization of diamond in the appropriate pressure range of 0–0.5 GPa and temperature range 675–1300 K has now been accomplished using a 'galaxy' of techniques in region D in Fig. 7. These include chemical vapor deposition from carbonaceous gases, mainly methane, hydrogen mixtures using many varied means of generating plasma [36, 37] (see also Chapter 4 of Part II), the use of a combination of lasers and carbon dioxide as source (the so called QQC deposition process) [13], hydrothermal synthesis in the C–H–O and C–H–O-halogen systems [38], low pressure solid source processes (the so called LPSSS methods) [12], and many more. The literature in this general area of metastable diamond growth has now become extensive, as discussed in Section 1.2.6.

From a commercial point of view, the production of diamond abrasives is, at this time, dominated by the high-pressure, high-temperature, transition metal solvent/catalyst techniques where the crystallization takes place in the diamond stable region.

1.2.3 High Pressure Apparatus

Early attempts at generating high pressure used opposed anvils, frequently referred as 'Bridgman anviis' after the father of high pressure, Professor P.W. Bridgman [39]. A constraint of this system was the limited volume of material to which the high pressure could be applied. This problem was addressed by forming recesses in the two opposing anvils, to form the so-called) 'toroidal' device used extensively in Russia [40]. However, most commercial growth of diamond using metallic solvent/catalysts has been carried out in the belt/girdle and multi-anvil devices described below.

1.2.3.1 Belt/Girdle Devices

The pressure limit of the piston and cylinder apparatus [39] was extended to allow commercial production of synthetic diamonds by designing the cylinder to accommodate pistons shaped as truncated cones.

If the die is reinforced by prestressing with concentric rings with interference fits, the device can sustain a pressure of about 6 GPa through use of an appropriate gasket (talc, pyrophyllite, and mixtures thereof) at the die/anvil interface.

The gasket allows an even distribution of stress in the die taper and along the tapered anvil cone. In this way, both the die and anvil are supported by the resulting compressive stresses.

The essential difference between the belt and girdle devices is the shape of the die bore/taper and the matching anvils. For the belt apparatus, the die bore is curved continuously from center to outer surface whereas, for the girdle, the die bore is straight and proceeds out on both sides in a linear progression.

1.2.3.2 Multi-anvil Devices

The first multi-anvil press was invented by Von Platen [41]. It comprised six anvils in a cubic arrangement. The benefit of a muiti-anvil device is the possibility of multi-staging, which allows the yield strength of a compressed component to be increased. Russian workers [42] have made extensive use of a multi-staged split sphere device. Pressure is applied to two sets of anvils. The outer arrangement contains eight anvils which form an octahedral-shaped cavity, and an inner set of six anvils is placed to form a cubic-shaped central cavity that contains the high pressure capsule in which the diamonds are grown.

1.2.4 The Synthesis of Particulate Diamond Abrasives

Crystal growers in most of the world's research and commercial organizations [16, 43, 44] have efficiently harnessed the solvent/catalyst growth of diamond to produce both abrasive grits and large single crystals. Scientists have developed mechanisms to explain the formation of diamond. These have been described by many workers and could involve the formation of a C^+ ion or the formation of an intermediate carbide which decomposes at synthesis conditions. However, the more accepted explanation of growth is based on supersaturation. For example, if the well known Ni–C equilibrium phase diagram, Fig. 8 [45], at 5.7 GPa is studied, an explanation of the formation of diamond under isothermal (thin film growth) or temperature gradient conditions can be obtained. Both diamond growth techniques are possible since the solubility of diamond in the molten solvent/catalyst, nickel, is less than the solubility of graphite.

This section will focus upon the crystallization principles used to produce commercially synthetic diamond 'grit' abrasives. The size of particles discussed here are within an envelope of about 850–200 μm for grits used mainly in stone and concrete sawing (saw sizes) and from about 200–50 μm used for grinding applications (wheel sizes).

The solvent/catalyst crystallization method generally uses graphitic carbons as the source material at constant high temperature and high pressure. The general pressure and temperature conditions used are all within the diamond stable region, at position A_1 in Fig. 7. The lowest conditions are about 5 GPa and 1500 K, and the highest about 5.8 GPa and 1775 K. These conditions are determined by the transition metal alloy systems together with the crystal nucleation and growth conditions all necessary to crystallize the diamond structures required for specific

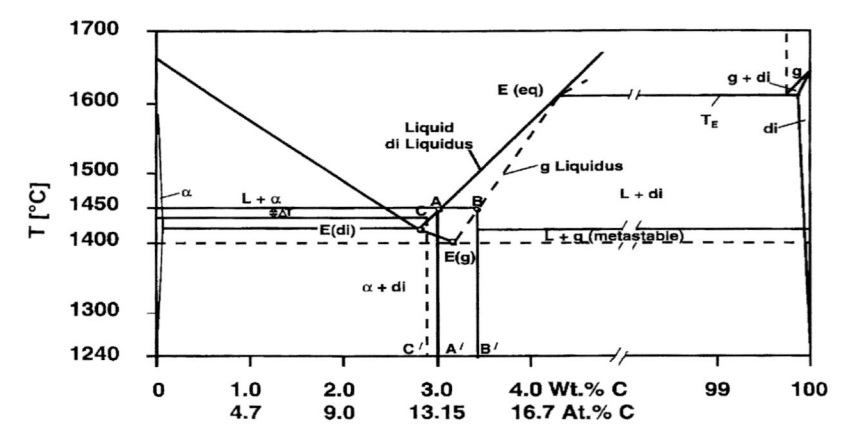

Figure 8. Ni–C phase diagram at 5.7 GPa [45].

product types. The diamond industry has spent many years and significant amounts of money developing and refining the large scale high pressure equipment appropriate to generate and maintain economically the required crystallization conditions. The high pressure equipment used by the two largest manufacturing companies, namely De Beers Industrial Diamond Division and the General Electric Corporation are proprietary, highly evolved designs based upon the belt and/or girdle apparatus described in Section 1.2.3.1. The Chinese producers mainly use cubic multi-anvil high pressure designs, and the Russians and Ukrainians use various modifications of the toroidal equipment.

Possibly the first question a crystal grower must ask when investigating or developing a crystal growth method is, 'how is an appropriate driving force for nucleation and growth generated and controlled'?

The central part of Fig. 8 is considered as representative of a general transition metal/carbon phase diagram at a pertinent pressure (5.7 GPa) and temperature range. The driving force for crystallization of diamond using 'ideal' graphite as source carbon would be given by the difference in solubility between graphite and diamond (the supersaturation) at the chosen conditions. For example, in Fig. 8 at 1450°C and 5.7 GPa this would be the solubility difference AB, (A'B'). Note that this supersaturation for diamond crystallization (let it be Δc) would decrease as temperature is increased and would essentially be zero at the diamond/graphite equilibrium temperature, $E_{(eq.)}$ in Fig. 8. The slopes of the diamond/metal liquidus line and graphite/metal liquidus line at constant temperature will change with pressure. The difference between these solubility lines, Δc, will thus also change with pressure and similarly become zero at the diamond/graphite equilibrium condition. Thus the driving force for crystallization, Δc, is a function of the 'distance' from the diamond graphite equilibrium line in pressure/temperature space, Fig. 9. The supersaturation, Δc, can of course be related to the nucleation rate and growth rate of the diamond in the classical way as shown in Fig. 10.

Note the extremely strong dependence of nucleation rate upon supersaturation and that there is generally a critical level of supersaturation, Δc_{crit}, which must be

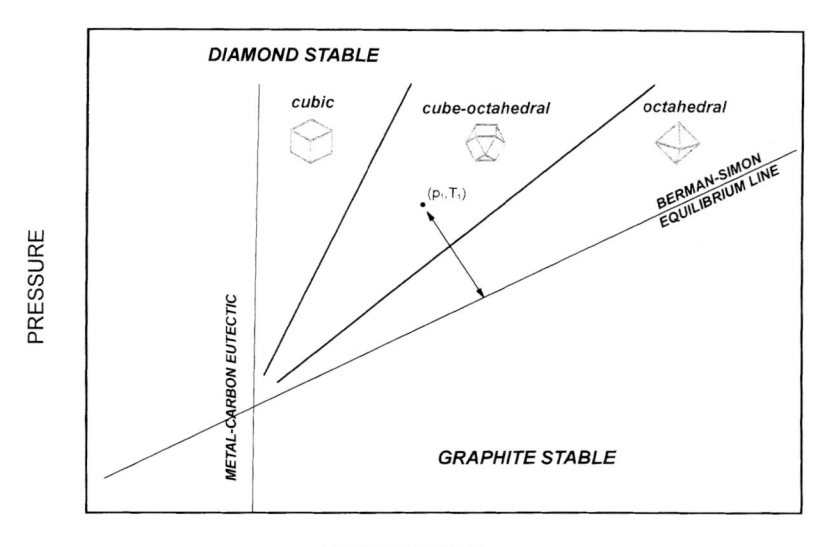

Figure 9. Pressure–temperature diagram for diamond synthesis showing typical conditions (P_1, T_1). The supersaturation Δc is a function of $P_1 T_1$ giving rise to different crystal morphology at different conditions.

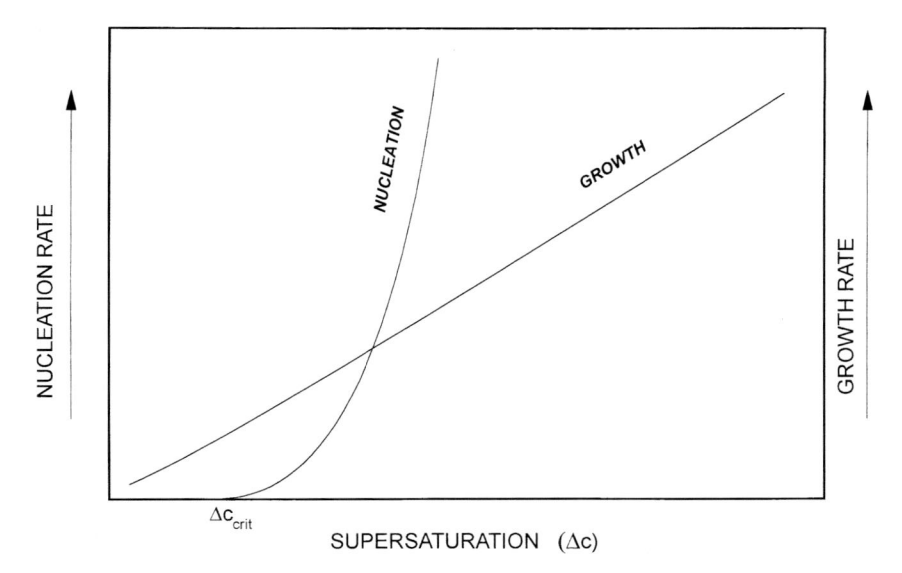

Figure 10. The classical relationships of the rates of nucleation and growth to supersaturation, Δc, for crystallization from a solution.

exceeded before the probability for nucleation becomes large enough for an observable nucleation rate. The growth rate is an increasing function of supersaturation. The relative growth rates of different crystallographic growth zones is also a function of Δc, hence the morphology shift with pressure, temperature condition as depicted in Fig. 9.

The relationships between solubility, supersaturation, pressure, and temperature depend upon the details of the microscopic mechanisms operative in the context of the detailed active impurity chemistry and the specific geometric relationship set up between the source carbon, molten metal solution, and nucleation site, and or growing diamond. For an introduction into these extensive subjects the reader is referred to Burns and Davies [16] where the issues such as nitrogen and boron active impurity chemistries, diamond surface reconstruction, morphology and solvent/catalyst are discussed.

An important conclusion to be drawn from the above fundamental explanation is that the structure of the source carbon as it pertains to solubility in the chosen molten metal is very important. Will and Graf [46] compare six graphite types using an iron–nickel solvent and show a clear dependence of rate of diamond formation and diamond size upon the starting crystallinity of the graphites. Moreover these authors include consideration of the influence of gaseous impurities in the graphites and the cluster theories of Sunagawa [47] concerning the carbon species in solution in the metal. As the source carbon became more crystalline, the pressure needed for diamond nucleation increased by up to 0.5 GPa. This could be interpreted as the need to increase the pressure in order to exceed the critical supersaturation for spontaneous nucleation, Δc_{crit}, owing to the much lower solubility of the recrystallized and thus more crystalline graphite.

Another striking example of the importance of the structure of source carbon leading to crystallization consequences via the supersaturation driving force is described in Wang et al. [48]. Here, the pressure needed for spontaneous nucleation was shown to increase with time as the graphitic source material recrystallized in the high pressure reaction chamber.

In addition to the explained (at least in general terms) dependence of crystallization on the structure of the source carbon, the character of the transport of solute from dissolving source to growing diamond crystals needs to be considered.

When a diamond synthesis reaction chamber from a process for making grits is broken open after completion of the crystallization cycle, it is observed that each grown crystal is separated by a metal alloy film from the undissolved carbon source material [16]. The metal film, which of course was molten during the growth of the diamonds, is usually about 50–200 μm in thickness. Figure 11 is a schematic drawing of this growth geometry. The carbon source material is dissolved at the molten metal interface, D, and the resultant carbon species in solution transported by diffusion to the growing diamond surface. The dimensions of these molten metal films are such that it is unlikely that convection plays any significant role in carbon transport for this type of crystallization. The growth of diamond grits of good morphology occurs through lateral propagation of steps over the surface. The detailed kinetics of surface adsorption, surface diffusion and incorporation of carbon species into the step structures needs to be studied to understand the

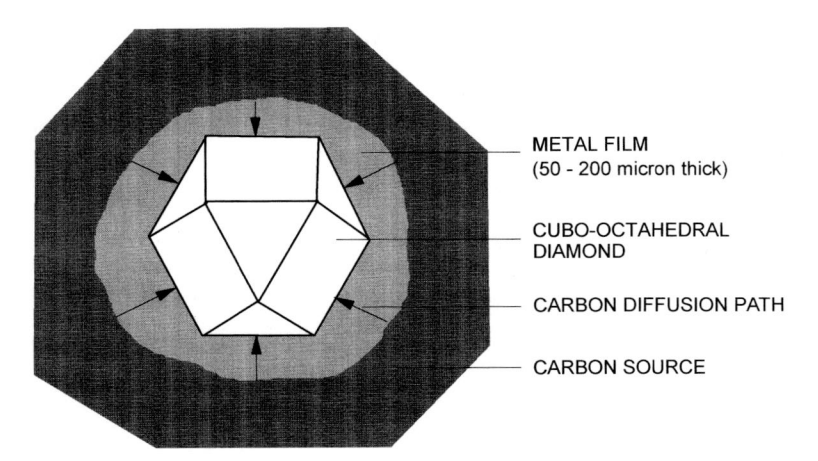

Figure 11. Schematic growth environment of a synthetic diamond particle.

microscopic mechanisms of growth. The different chemical and reconstructed structures of the lowest energy crystal facets of diamond give rise to different growth rate responses to the flux of carbon species resulting in a growth zone history and particular final morphology for each crystal. The morphology of diamond grits may be related to the pressure and temperature conditions of growth through the relative growth rate response of the different crystallographic surfaces to the flux set up by the supersaturation, Δc, produced by these pressure and temperature conditions, as in Fig. 10 [16]. Thus a morphology map may be drawn on such diagrams showing the regions of pressure and temperature of common morphology [16]. The difficulty of investigating the detailed growth mechanisms is extreme due to the 'inaccessibility' of the growing crystal in practical high pressure, high temperature equipment.

The crystal defect types and distributions which characterize any real crystal, diamond abrasive grits being no exception, may also be discussed in terms of the driving force for crystallization, and the impurities in solution and/or suspension in the solvent and the geometry/transport considerations of the growth environment. Examples of this are the oriented metal solvent inclusions described by Wakatsuki et al. [49]. These inclusions occur as an array, along certain growth zone boundaries leading to certain edges of the crystals. The most important crystal defects for diamond abrasives are those which may affect their behavior in applications and also those which give rise to their appearance, such as color and clarity. A discussion of the characterization of synthetic diamond abrasives in terms of strain, impurity content, both as inclusions and lattice defects, and crush strengths is given by McCormick et al. [50]. These authors consider the macroscopic structure such as shape, external morphology, cracks, and large inclusions together with the microscopic structure such as small inclusions, dislocations, and some nitrogen related lattice defects. Using strain sensitive techniques such as Raman spectroscopy, photoluminescence, and birefringence, together with information from extensive literature, they arrive at hierarchies of the macroscopic and microscopic structures and properties that affect the strengths of the diamond particles. Not surprisingly, the most important macroscopic structures of the diamond particles turn out to be size, shape, crack, and inclusion content. It was also demonstrated that particular nitrogen related lattice defects such as the H3 and N–V center affect the strength of the crystals. For a comprehensive discussion of the observed impurity lattice defect centres such as H3 and N–V, the reader is referred to Davies [51]. An important observation in [50] was that some of the microscopic defects may well strengthen the diamond by impeding dislocation motion and possibly acting as crack arresters.

The very complex nature of the macroscopic and microscopic structures as they affect strength and the behavior of diamond abrasive/particles still requires extensive work to elucidate fully. However, from a crystallization point of view, gaining control over the crystallization behavior is the key to the production of optimal diamond abrasives. This, of course, may be achieved by choice and manipulation of the pressure/temperature conditions, source carbon structure and solvent/catalyst metal type, leading to control over nucleation and growth rates.

1.2.5 Growth of Large Synthetic Diamonds

R. C. Burns and M. Sibanda

Large diamonds are best grown using the temperature gradient technique, where diamond is used as the source of carbon. In this technique, as described by Wentorf [52], the driving force for reconstitution, under diamond stable conditions, is provided by the higher solubility of diamond in a hot zone of the solvent/catalyst and the consequent crystallization of diamond in a cooler zone at constant pressure (Fig. 9).

The design of the high pressure cell also influences the rate at which this carbon transport mechanism proceeds. A schematic diagram of a high pressure cell used to grow reconstituted diamond is shown in Fig. 12. A useful axial temperature gradient is established by manipulating the distance between the diamond source and seed pad, which (at constant cell pressure) determines the growth rate and hence, quality and size of the diamonds grown on the seeds in a fixed time. The small, well formed diamond seeds, which act as a template for new growth, are mechanically attached to the seed pad. For a reaction cell pressure of about 5.5 GPa, the temperature at the diamond source is about 1450°C. The solvent/catalyst bath is agitated by convection and soon saturates with carbon. The solvent/catalyst in the vicinity of the seeds, at a temperature of 1420°C, is therefore supersaturated with respect to diamond, providing the driving force for diamond growth on the seeds, which are selected to be about 0.5 mm in size [53, 54].

The nucleated diamond is restrained by the seed pad, resulting in truncated cubo-octahedral growth, whereas the orientation of the grown diamonds depends on the orientation of the seed. The habit of the reconstituted diamond depends largely on the solvent/catalyst, internal pressure and temperature of the reaction cell.

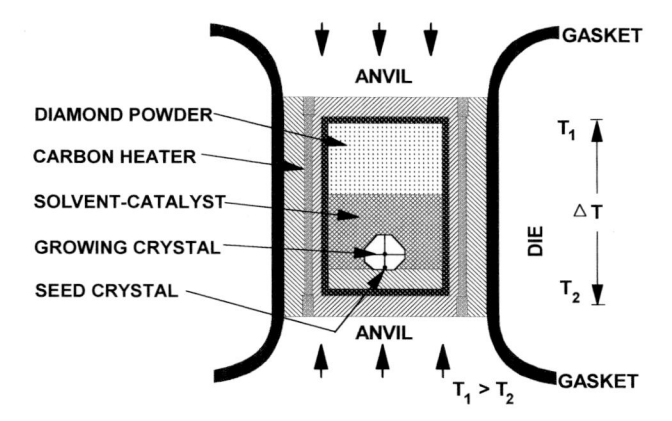

Figure 12. Schematic diagram of a high pressure cell used in the reconstitution method.

Diamonds of a cubic habit tend to be formed at a lower growth temperature than those of an octahedral habit [47] (refer to Fig. 9).

The chosen spacing between diamond seeds is determined by the intended size of the reconstituted diamond. This spacing also affects the growth rate. In practice it is found, for a fixed temperature gradient, that the smaller the spacing between seeds, the slower the growth rate. Figure 13 shows a selection of large synthetic diamonds (0.5 to 1.8 ct in weight) grown by De Beers Diamond Research Laboratory from seeds using the temperature gradient technique.

Crystal perfection is determined by a reasonably low growth rate which is achieved by optimizing, in combination, the space between seeds, the axial temperature gradient and the solvent/catalyst under diamond synthesis conditions.

Figure 13. Diamonds grown by De Beers Diamond Research Laboratory from seeds using the temperature gradient technique.

1.2.5.1 Solvent/Catalysts for the Growth of Large Synthetics

1.2.5.1.1 Type Ib Diamonds Effective solvent/catalysts for high pressure, high temperature diamond synthesis (thin-film and temperature gradient techniques) are derived from metals and alloys of Group VIII of the periodic table and metals such as chromium, manganese, tantalum and niobium [55–57]. The effectiveness of transition metals and their alloys depends on a number of factors which include: their melting points, the degree to which they influence the supersaturation of the molten solution and also the stability of possible metal carbides at synthesis conditions [58].

Most commercial solvent/catalyst alloy systems make use of a combination of metals [16, 59–61] (e.g. iron–nickel–chromium, iron–nickel, iron–cobalt, nickel–manganese–cobalt, or manganese–nickel–copper), which enable lower temperatures and pressures to be used. Large synthetic diamonds of different colors can be grown using such combinations with or without introducing additives or getter elements, the most common color being yellow (type Ib). The amount of nitrogen in synthetic diamonds depends on the composition of solvent/catalyst used, synthesis temperature and pressure and also the growth rate [16]. In this regard, the solubility of nitrogen in the solvent/catalyst and how it is influenced by other elements is also an important consideration [62]. The higher the solubility, the more readily nitrogen will be retained by the solvent/catalyst rather than be incorporated into the growing crystal. This is the case for iron relative to cobalt [63]. Furthermore, the solubility also increases with temperature. This is consistent with observations of very pale crystals being grown with pure iron at relatively high temperatures [43, 64].

The choice of solvent/catalyst alloy for synthetic diamond growth not only has an influence on nitrogen content, but also on other diamond characteristics such as morphology, color, and inclusion content. The dependence of some of these diamond characteristics on solvent/catalyst typically used for research and commercial production by institutes and companies such as De Beers, Sumitomo, General Electric, and NIRIM, amongst others, are summarized in Table 1.

1.2.5.1.2 Type IIa and IIb Diamonds Aluminum, titanium, niobium, and zirconium are used to remove nitrogen, in the growth of near colorless synthetic diamonds (type IIa) [62, 66, 67]. These nitrogen-getter elements have a high affinity for nitrogen and contribute to nitrogen removal by forming stable nitrides. As mentioned earlier in the chapter, there is a relationship between nitrogen in the grown diamond and nitrogen solubility in the solvent/catalyst.

The dependence of nitrogen content on the alloy composition is shown in Fig. 14, where it can be seen that an increase in cobalt content (which is accompanied by decreased nitrogen solubility in the iron-cobalt solvent/catalyst alloy) results in increased nitrogen concentration in the grown diamond [67].

Nitrogen concentration in the crystals decrease with increasing amounts of aluminum or titanium added to the solvent/catalyst (Fig. 15). Even when large amounts of aluminum are added, it is difficult to achieve purity levels of less than 0.1 p.p.m. nitrogen. This is to be expected as the aluminum nitrides are believed to readily decompose, whilst titanium nitrides are not readily decomposed, making titanium

Table 1. Summary of large synthetic diamonds grown for research purposes by some of the leading producers of synthetic diamonds

Solvent	Max. size (ct)	Growth time (h)	Growth rate (mg h^{-1})	Color	Nitrogen content (p.p.m.)	Morphology
Co–Ti [62]	3.4	300	2.3	near colorless	<0.01–2	{111}, {100} > {113}, {115}, {110}
Fe–Al [65]	4.6	500	1.8	near colorless	0.4–11	{111} > {113} > {100} > {110}
Fe–Co [62]	25	1000	5	yellow	100–300	{111}, {100} ≫ {110}, {113}
Fe–Al–B (0.02 wt % B) [65]	5.1	760	1.3	blue	–	{111}, {110}, {100}, {113}, {115}
Ni [43]	–	–	–	yellow	50–100	{111}, {110} ≫ {110}, {113}
Ni–Fe	3.5	–	–	yellow	50–400	{111}, {100}
Fe–Co–Ti [65]	4	–	–	near colorless	<0.01	{111}, {100} > {113}, {115}, {110}

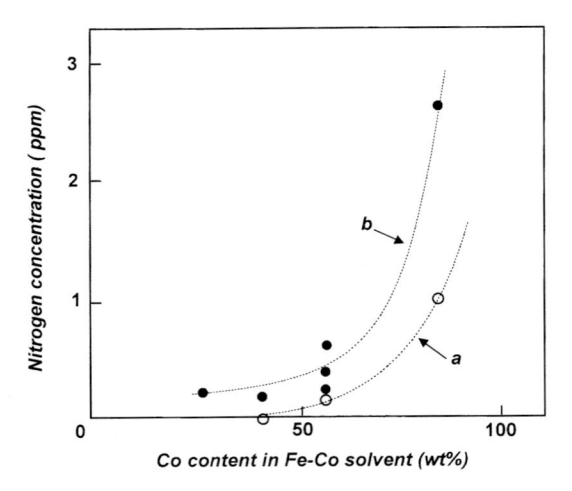

Figure 14. Nitrogen concentration in diamond versus the cobalt content of *iron + cobalt* solvent. The diamonds were grown with 1.5 weight-% titanium addition using (a) high-purity graphite and (b) synthetic diamond powder [67].

a more efficient nitrogen-getter under conventional synthesis conditions. At higher temperatures, however, aluminum becomes a more efficient nitrogen-getter than titanium as shown by diamond synthesis experiments at 1550°C where the use of titanium yielded yellowish and heavily included crystals [62].

Boron is often added in the form of a stable metal boride to gettered reaction volumes, to generate blue semiconducting diamonds (type IIb) [62].

1.2.5.2 Purity of Large Synthetic Diamonds

Synthetic diamonds invariably contain

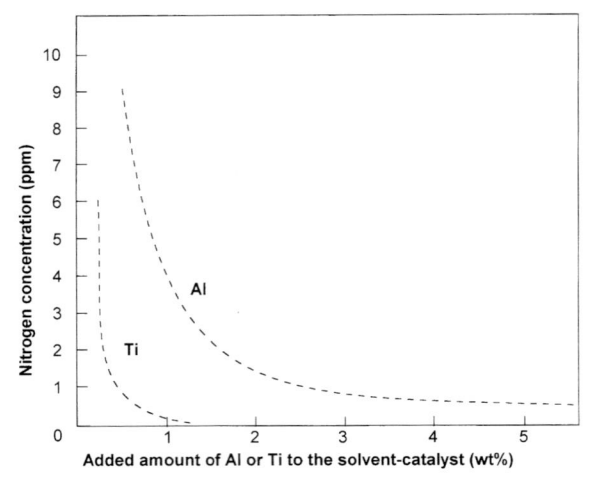

Figure 15. Nitrogen concentration of diamond grown with Al or Ti added as a nitrogen getter [67].

impurities. These impurities fall into two classes. There are impurities on an atomic scale (point defects) and these include nitrogen, boron, and metallic elements such as nickel. The second class of inclusions are present in the form of metallic or nonmetallic particles, both microscopic and macroscopic. The occurrence of these inclusions is controlled by crystal growth kinetics.

1.2.5.2.1 Inclusions Solvent/catalyst metal inclusions can occur in large synthetic diamonds. The number, size, and shape of these inclusions depends to a large extent on the growth rate and the synthesis temperatures and pressures. Extensive research in this area shows that the uptake of metallic inclusions occurs predominantly during the early stages of growth when the carbon flux density is very large, particularly when a small seed crystal is used [68]. One suggestion is the use of a two-stage growth method [63, 69] in which a small recess is made on the seed pad and the seed is located at the bottom, such that it does not protrude above the recess. The net result is that carbon supply to the seed is limited as the seed grows within the recess during the initial stages of growth. Once the seed has grown out of the recess, it has an enlarged surface which can accept more carbon without compromising quality. The uptake of inclusions during the early stages of a growth can, therefore, be minimized by paying special attention to the depth and diameter of the seed recess.

The use of larger seed crystals tends to increase the growth rate. The coupled effect of growth rate and seed crystal size [70] on the uptake of inclusions is summarized in Fig. 16 where it is evident that the size of the seed crystal is an important parameter in the growth of high purity diamonds. In general, the population and size of inclusions increases with an increase in growth rate. There are many different types of inclusions that occur and most of these have been comprehensively dealt with elsewhere [44]. One of these types of inclusions results from hoppered growth when there is a high supersaturation [71], as shown in Fig. 17. Low growth rates and stable growth conditions are therefore important for the growth of high purity large synthetic diamonds.

The choice of solvent/catalyst alloy which incorporates a getter is important in controlling the quality of grown diamond. In the growth of near-colorless diamonds, the tendency for the getter elements to form stable carbides can be a problem. When titanium is used, a very stable carbide is formed whilst, with aluminum, the Al_4C_3 carbide formed is readily decomposed at diamond synthesis conditions. Inclusions therefore occur more readily with titanium than with aluminum gettered systems. At high levels of titanium or aluminum, inclusion content can be a problem. When titanium is used, many fine particles of TiC are formed [72, 73] and included in the growing diamond crystal in addition to the solvent/catalyst metals (Fig. 18). It has been suggested [73] that the addition of copper to the reaction volume has the effect of suppressing the formation of titanium carbide. Copper is thought to form a cation in the molten solvent/catalyst bath which can decompose the carbide.

1.2.5.2.2 Point Defects Nitrogen and boron are ubiquitous atomic-scale impurities in synthetic diamonds. The free energy of formation of nitrides and borides in the solvent/catalyst melt will determine the extent to which these impurities are

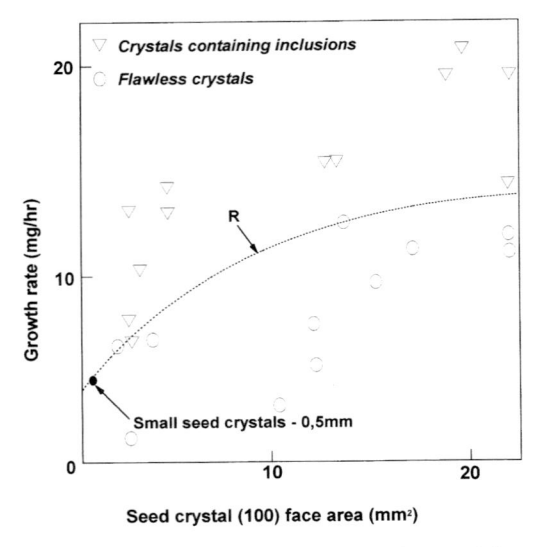

Figure 16. Dependence of critical growth rate of diamond crystals on {100} faces of seed crystals, on varying seed size [70].

available for incorporation into ie growing crystal.

As mentioned in Section 1.2.5.1.1, unless nitrogen–getters are used during synthesis, nitrogen is responsible for the highest concentration of impurity-related point defects in synthetic diamond. This impurity is incorporated principally as isolated substitutional atoms which give rise to a strong optical absorption below 500 nm, leading to a yellow or yellow-brown color. Such diamonds are classified as

Figure 17. Cross section of a cobalt-titanium grown crystal with inclusions resulting from hoppered growth.

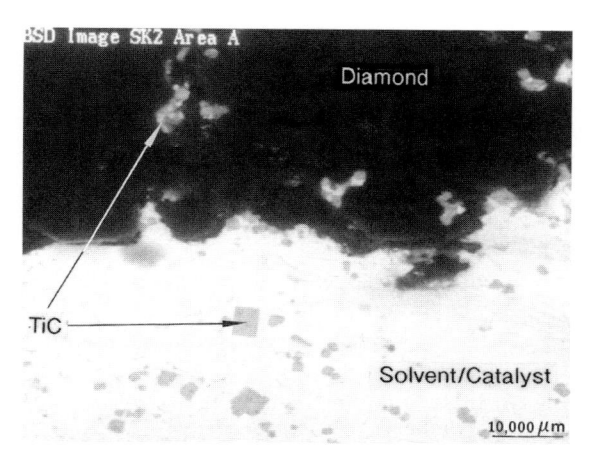

Figure 18. Micrograph of cross-section of a carbon, solvent/catalyst mixture Co–Fe–Ti showing large TiC crystallites and some TiC particles that have been trapped at the diamond-solvent/catalyst interface.

type Ib. Synthetic diamonds tend to take up impurities at different concentrations for each of the types of growth sector (regions of the crystal with a common crystallographic growth plane). Under standard synthesis conditions, nitrogen concentrations are highest in {111} growth sectors, typically 100–300 p.p.m. for iron–cobalt grown crystals [62]. Generally, the average nitrogen concentration in {100} sectors is about half that for {111} sectors [73, 74]. However, at low growth temperatures, cube growth predominates and the nitrogen concentration in {100} sectors exceeds that in {111} sectors [75]. The reduced solubility of the solvent/catalyst for nitrogen at low temperatures has led to nitrogen concentrations up to about 1000 p.p.m. for the iron–cobalt system. Nitrogen concentrations are substantially less in the minor growth sectors ({110}, {113} and occasionally {115}) compared with {111} and {100}, with the {110} sectors containing about 1 p.p.m. If nitrogen getters are used to reduce the amount of nitrogen available to the growing diamond, as described in Section 1.2.5.1.2, the residual nitrogen concentration tends to be approximately proportionally similar to the ungettered case.

Blue, semiconducting type IIb diamonds are synthesized by adding both boron, and a nitrogen getter to the reaction volume. The total boron acceptor concentration is highest for {111} sectors, somewhat less for {110} sectors and substantially less for the remaining types of growth sector [74–76]. The presence of uncompensated boron acceptors gives rise to a blue coloration. The strength of the blue color, or indeed any residual yellow color, depends upon the difference between the boron acceptor and nitrogen donor concentrations for any growth sector.

When nickel or cobalt is used as the solvent/catalyst, these elements are incorporated in the diamond lattice as optically active point-defects, but only in {111} growth sectors [77, 78].

Frequently, when large, high-quality synthetic diamonds are discussed, the question of their potential use as gemstones is raised. At the time of writing, none of the major manufacturers of synthetic diamonds markets material for gem use. However, small numbers of cut and polished synthetic diamonds, mostly produced in the former Soviet Union using the so-called BARS equipment, have appeared in the gem trade in recent years. The majority have been yellow in color but a few near-colorless synthetic diamonds have also appeared. As uncut crystals, it is easy to distinguish synthetic diamonds from natural diamonds by their distinctive morphology and the presence of a seed. But cut and polished synthetic diamonds are also readily identifiable as such by standard gemmological techniques such as microscopy of inclusions and of features on polished faces (due to sector-dependent differential hardness) and zonation of color or fluorescence (due to sector-dependent differential uptake of optically active impurities) [79–84]. Near-colorless and boron-doped, blue in color, synthetic diamonds tend to exhibit strong, long-lived phosphorescence. To enable large numbers of diamonds to be screened rapidly for the presence of synthetic diamonds, instruments have been developed. The first of these instruments detects an optical absorption line at 415 nm present in the majority of natural diamonds. Those diamonds not exhibiting this feature may be examined using a second instrument which, by exciting near-surface fluorescence with very short wavelength ultraviolet light, produces a fluorescence pattern from which the growth structure of the diamond may be inferred. These patterns are very different for synthetic as compared to natural diamond [85].

1.2.6 Novel Diamond Synthesis Routes

J. O. Hansen

1.2.6.1 Introduction: Thermodynamically Stable and Metastable Processes

The two major innovations in diamond technology in this century must be the reproducible high pressure process with metal solvent/catalyst, and the low pressure sustained growth from methane-hydrogen plasma reviewed by Kamo *et al*. [86] which we call CVD. There may be some argument that the work of Angus *et al*. [87] or the patents of Eversole [88] predate and anticipate metastable diamond synthesis by CVD, but the authors did not achieve sustained growth. They did demonstrate that the decomposition of methane or carbon monoxide can produce active carbon atoms or hydrocarbon fragments, which in a suitable temperature range and in the presence of diamond surfaces to act as nuclei, will permit overgrowth of a new diamond phase. The appearance of graphite nuclei leads to a cessation in diamond growth which suggests that the success of the 'metastable' growth, without plasma methods, depends upon an energy barrier to nucleation of graphite. In the now mature CVD technology, graphite nucleation is inhibited by the continuous bombardment of the growing surface by atomic hydrogen, which has a greater tendency to etch graphite than diamond [89]. Even so, it appears that certain 'tricks of the trade' are required to initiate nucleation of diamond on a nondiamond substrate because of a substantial energetic barrier.

The graphite–diamond equilibrium line in Fig. 7 is based on a thermodynamic analysis by Berman and Simon [14], which has been reviewed recently [90] in the light of new data, particularly the direct electrochemical determination of the free energy difference for the diamond to graphite transformation [91]. From the equilibrium expression of Berman and Simon, Eq. (1) in Section 1.2.2, it may be possible to include additional terms which offer an explanation for some cases of novel non-CVD synthesis, or to expand the existing terms such as ΔH^0, the enthalpy change at zero pressure.

1.2.6.2 Increase in Enthalpy: Synthesis from Nongraphitic Carbons

Onodera *et al.* [30] and Brannon and McCollum [92] have published evidence of diamond growth from nongraphitic carbons at pressures below the phase-equilibrium line at 2–4 GPa and 1300–1900°C. They suggest as explanation an increased chemical potential (free energy) of the nongraphitic phase with respect to diamond, or the catalytic action of the hydrogen associated with the cokes, and not present in graphite, by analogy with diamond CVD. The second explanation seems unlikely, since hydrogen has been shown to be a deleterious impurity in metal solvent/catalyst synthesis from graphites [93, 94].

The enthalpy of nongraphitic carbon (furan coke) has been measured at low temperature [95, 96] and found to be higher than both graphite and diamond, by between 0.4 and 2.8 kJ mol^{-1}. This would have the effect of eliminating or reversing the sign of the first term in Eq. (1) and so decreasing the pressure requirement for diamond synthesis, perhaps as far as point B in Fig. 7. As Vereschagin *et al.* [96] point out, there is an even greater driving force for graphite crystallization. Even in synthesis from synthetic graphites of high graphitization index ($d_{002} = 0.337$ nm), there is a competitive process of recrystallization of graphite as coarse single crystal flakes [56]. See also Fig. 19.

The increased enthalpy of the nongraphitic carbon follows from a higher internal energy because of lower total bond energy or equivalently, a higher concentration of defects by comparison with single-crystal graphite. A similar argument has been used by Bar-Yam and Moustakas [97] for the case of CVD synthesis. High vacancy concentrations in graphite and diamond can cause diamond to become the stable phase because the formation energy of vacancies in diamond is lower than that in graphite.

The possible success of a synthesis route using nongraphitic carbon would seem to depend upon the suppression of graphite nucleation and purity factors, since growth of high quality diamond requires low levels of impurities. This is more difficult to achieve in a carbon which has not been graphitized.

1.2.6.3 Interfacial Energy: Growth in the Colloid Size Range

Ideal diamond has an exceptionally high surface energy because of the high density of strong bonds which must be destroyed in order to generate unit area of surface. It has been estimated at 5000–9000 mJ m^{-2} depending upon the crystal face [98]. Graphite has a lower average surface energy but there is extreme anisotropy, with the basal plane having an energy of 150 mJ m^{-2} and the prism planes over 4000 mJ m^{-2} [99]. This explains the tendency for graphite to crystallise as large thin flakes, so reducing the contribution of the edge surfaces to the total surface

Figure 19. Micrograph of a section of a diamond synthesis capsule in polarized light at 200× magnification. Recrystallization of coarse graphite flakes of high crystallinity is a process which competes with diamond growth even in the diamond stable region. The tabular shape suggests that surface energy is a significant driving force for recrystallisation even in this size range.

(interfacial) energy.

In practice, these solid–vacuum interfacial energies are not likely to be seen, since the diamond surface and the graphite prism-face surfaces will reconstruct or react with adatoms such as hydrogen and oxygen to reduce their surface energy. Under atmospheric conditions, diamond is terminated by a range of hydride and oxide groups and is a low energy surface [100].

During high pressure solvent/catalyst synthesis, the interfacial energies of metal melt–diamond and metal melt–graphite have values lying between the metal surface tension, $1800\,mJ\,m^{-2}$, and the solid–vacuum surface energy. Some values have been estimated from contact angle data [101] although the complexities of contact angle work in the diamond stable region are formidable.

By including a surface energy (interfacial energy) term in the equilibrium expression it is possible to make some interesting predictions.

$$\Delta G_T^P = \Delta H_T^O - T\Delta S_{T+O}^O \int \Delta V_T\,dP + \Delta(\gamma A) = 0 \qquad (2)$$

where γ is the interfacial (surface) energy of any phase and A is its molar area, which is not an intensive variable like pressure or temperature but varies with both the size of the particles and the density of the phase, diamond having a lower molar area for the same particle size than graphite.

For instance, using this approach, Nuth [102] has predicted that diamonds (in vacuo) are more stable than graphite in the size range 1–5 nm, a result which was supported by Badziag et al. [103] using a different approach.

There is experimental support for this in the finding that nanometre diamonds are abundant in primitive meteorites and possibly in the interstellar dust clouds [104]. Similarly, growth of graphite in the diamond-stable region during high pressure, high temperature synthesis, and the corollary, an overpressure for nucleation of diamond, are explainable using the additional interfacial energy term without the need for kinetic arguments [105, 106]. Once the growing particles are in the size range of microns, the molar surface area becomes too small for this term to have any effect and the classical phase equilibrium line is applicable.

There is another way of looking at the same effect, since a surface energy per unit area is equivalent to a line force per unit length, or surface tension, hence there is an excess pressure in a spherical droplet given by the Laplace equation:

$$\Delta P = \frac{2\gamma}{r} \quad (3)$$

where γ is the surface energy of any phase and r is the radius of curvature. For diamond with sigma at $5\,\mathrm{J\,m^{-2}}$ and r at 1 nm, the excess pressure in a nanometre sized sphere is 5 GPa, which would be sufficient to stabilize the diamond phase to at least 1600 K.

This could explain processes such as laser pyrolysis of hydrocarbons [107], where small quantities of nanometre sized diamond have been detected in the carbonaceous products. A recent result suggests that it is even possible to generate and stabilize a diamond core in the centre of a graphitic 'onion' a giant fullerene, by electron irradiation and annealing of the outer graphite layers. In this process, 'the sputtering-induced loss of atoms and the closure of the shells around vacancies lead to extreme surface tension, which is finally responsible for the spherical shape of the onions' [108].

This publication is a good example of the development of analytical techniques such as high resolution electron microscopy lattice imaging [108] and micrometre-area Raman spectroscopy (microRaman) [109] which permit the researcher to determine that diamond is present in crystallites of submicrometre, perhaps even nanometre size. Early scientists like Moissan [110] and Hannay [111] performed experiments in the pressure and temperature ranges now being cited as feasible for hydrothermal synthesis of diamond [38, 112, 113]. Had they been equipped to use the analytical techniques now available, they may have been proven correct in their claims, but in a particle size regime much finer than they were able to inspect.

1.2.6.4 Non-equilibrium Thermodynamics: Superequilibrium Atomic Hydrogen

The process of CVD of diamond may be explained by kinetic arguments [89], or alternatively by the nonequilibrium thermodynamic coupling theorem [114].

In the latter case, it is the presence of a superequilibrium concentration of atomic hydrogen and the coupling of the process of recombination of hydrogen atoms with the processes of diamond growth which allow the diamond deposition reaction to occur in a direction contrary to that prescribed by conventional thermodynamics. Diamond synthesis processes which are characterized by continuous generation of atomic hydrogen or halogens would then fall into the class of nonequilibrium processes. The generation of atomic hydrogen could be by means of microwave or discharge plasma, flame or even a hot filament. This topic is covered more fully in Chapter 4 of Part II of this book (C.-P. Klages), but some special cases will be listed here.

Table 2. Classification of low-pressure synthesis methods

Nominally metastable thermodynamic explanation is possible	Truly metastable: explanations in terms of nonequilibrium thermodynamics, atomic H	Undetermined: perhaps controversial
HPHT: nongraphitic carbon [30], [92]	LPSSS (low pressure solid state source) as by Roy et al. [116, 117]	Triple laser process using CO_2 (QQC process) [13]
Hydrothermal synthesis – nonflow [113, 38]	Hydrothermal flow synthesis with flame [116, 117]	PVD using ion beams – reviewed by Prins [123]
Laser in absence of hydrogen [105, 115]	Laser in presence of hydrogen [118]	Precipitation from nickel–sodium hydroxide melts [124]
Diamonds in interstellar space [102–104]	Fluidized bed deposition on particles from acetylene flame [119]	
Electron irradiation of carbon 'onions'; nanometre scale HPHT [108]	Discharge electrolysis from water:organics mixtures [120–122]	

1.2.6.5 Classification of Low-pressure Methods

Using this criterion, it is possible to separate those methods which are nominally metastable, but for which there is an explanation in terms of the additional free energy components discussed in Sections 1.2.6.1 to 1.2.6.3, from those which are truly metastable. There is a third category, undetermined, where the authors have insufficient knowledge of the process to make a judgement. Perhaps these are truly different in their mechanism.

1.2.6.6 Extension of High-pressure Methods

In Section 1.2.3 we reviewed the design of high pressure systems for diamond synthesis. The major development in recent decades has been an almost thousand-fold increase in working volumes. The pressures achieved using cemented carbide technology have, however, not increased significantly. The advent of the diamond-anvil-cell DAC [125] has allowed researchers to attain pressures of thousands of kilobars, whilst simultaneously collecting spectroscopic data on phase changes. This is possible only because of the transparency of the diamond anvils. Most research on ultrahigh pressure has been done with DACs. There might be two reasons for industrial interest in pressures of 10 GPa. The first is the synthesis of diamond from nonmetallic solvents [24, 126] such as carbonates at pressures of 7.7 GPa and 2000°C. This might offer different and novel types of diamond to those presently commercially made.

The second is the possibility of direct conversion, solvent free, of graphite to diamond. Table 3 lists reported conditions for the direct conversion.

The direct conversion is dependent upon the crystallinity of the starting carbon [127], and the product, when analysed by XRD, appears to be hexagonal diamond.

Figure 20 shows the martensitic transformation, by compression in the c-direction and in-plane buckling, of two crystalline forms of graphite. Hexagonal graphite,

1.2 The Crystallization of Diamond 509

Table 3. Direct conversion of graphite to diamond

Authors	Apparatus	Date	Pressure and Temp	Comments
Bundy and Kaspers [127]	Modified belt	1967	13 GPa and 1000°C	Hexagonal diamond
Boehler [128]	DAC	1994	10 GPa and 1700°C	
Xu and Huang [129]	DAC	1994	6 GPa and 1700°C	Raman at 1333 cm^{-1}

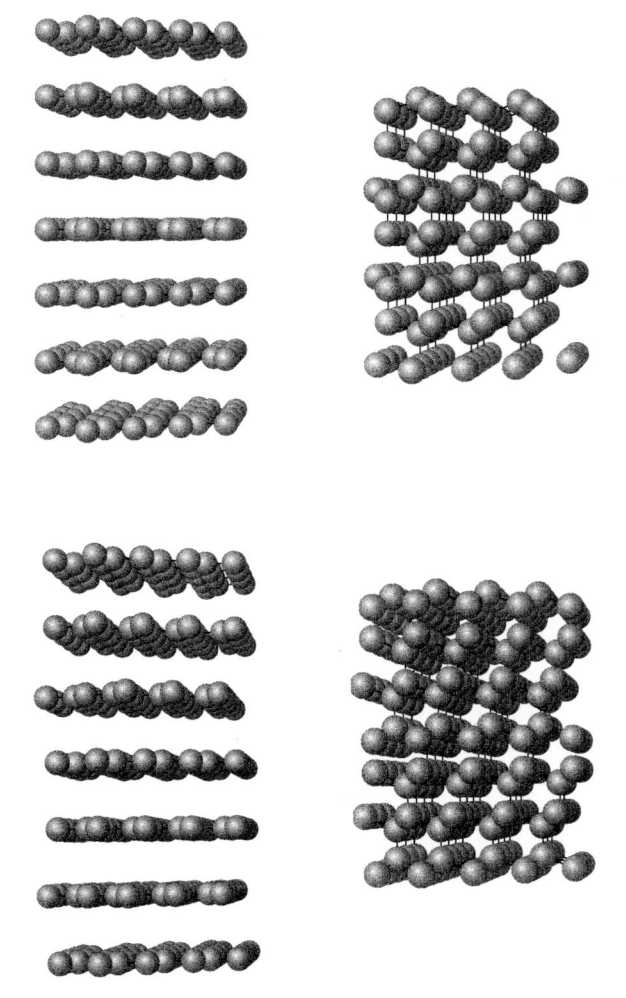

Figure 20. Computer-generated perspectives of the martensitic transformation processes, hexagonal graphite to hexagonal diamond, and rhombohedral graphite to cubic diamond. The structure on the right is generated by compacting the structure on the left in the vertical direction and puckering those atoms that find a neighbor in an adjacent plane so as to encourage the formation of the new sp^3 bonds in the vertical direction [132]. Software designed by Pawel Wzietek, Massy, France.

which has layers repeating in the order ABAB..., transforms to hexagonal diamond, or lonsdaleite. This is a metastable form of diamond and is never seen in diamond synthesized from a solvent/catalyst. Higher pressures and temperatures are required for the conversion of hexagonal graphite to cubic diamond, above line C_1 in Fig. 7, which suggests that the mechanism requires melting of carbon [130]. Rhombohedral graphite has a layer sequence ABCABC... (Fig. 20) and can convert to cubic diamond by a martensitic process of compression and buckling. Rhombohedral graphite is metastable towards hexagonal graphite and rare except in some natural crystals. Shear forces in high pressure or shock synthesis processes could cause rapid conversion of small domains of hexagonal graphite to the rhombohedral form, after which the direct conversion to cubic diamond may proceed. It is unlikely that this could happen over domains much larger than the 100 nm crystallite size seen currently in shock-synthesized diamond [131].

1.2.6.7 Summary and Conclusions

There have been many innovations leading to growth of nanometre sized crystals of diamond. A few processes, like the LPSSS and Me-C-H [117] appear to offer the possibility of combining the advantages of low-pressures with the commercial need for a volume-based rather than a surface-based process, and of growing crystals of micron rather than nanometre size. Direct conversion of graphite to diamond at macroscopic sizes will depend upon the synthesis and stabilisation of suitable graphite or mixed sp^2/sp^3 precursors.

1.2.7 Cubic Boron Nitride Crystallization

G. J. Davies

Cubic boron nitride (cBN) is presently the second hardest material known to man after diamond. The compound boron nitride is unknown in nature. The combination of atoms of boron and nitrogen in stoichiometric arrangements is such that B–N is isoelectronic with C–C. It is not surprising that, by analogy with carbon structures, similar crystalline structures of boron nitride occur. By analogy with graphite, hexagonal boron nitride is a soft, slippery material with an anisotropic, hexagonal structure made up of sheets containing interlinked six-membered rings ($3 \times$ B–N). There are some differences however, notably, the π-orbital overlap of graphite is absent, so hBN is an electrical insulator and is white in color. For kinetic reasons, all conventional chemical synthetic routes for making boron nitride result in the hexagonal structure. Again, by analogy with carbon, boron nitride can take up the diamond structure with alternating atoms being boron and nitrogen. This is the common zinc blende or sphalerite structure. This phase of boron nitride has a hardness about half that of diamond; and a bulk modulus of about 367 GPa compared to the 435 GPa of diamond.

The value of cBN as an abrasive lies in its much higher oxidation stability compared to that of diamond (1200°C compared to 600°C) and in its reduced chemical

1.2 The Crystallization of Diamond 511

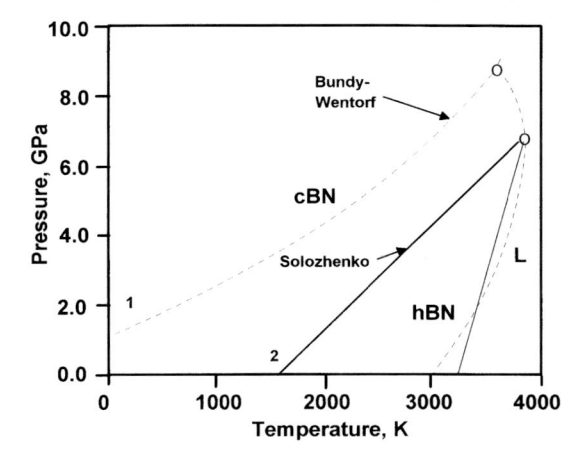

Figure 21. Phase diagram of boron nitride showing the latest stability fields suggested by Solozhenko [134] compared to the earlier Bundy–Wentorf diagram [138].

interaction with ferrous materials when compared to diamond. It is, therefore, the abrasive of first choice for the grinding of hard, ferrous materials.

When the formation of cBN was first reported [133], the pressure and temperature phase diagram was believed to be very similar in form and stability field positions to that of carbon, with the equilibrium line between the hexagonal and cubic phases intercepting the pressure axis when temperature is at absolute zero [134]. The most recent phase diagram for boron nitride is that of Solozhenko, shown in Fig. 21 [135]. This diagram was initially calculated and subsequently partially experimentally substantiated. The hexagonal-cubic-liquid triple point is at about 7 GPa and 3800 K. The hexagonal-cubic equilibrium line intercepts the temperature axis at just above 1500 K. However, the exact temperature of the intercept is not well-established [136], although it is now generally agreed that cBN is the stable form of boron nitride at room temperature and pressure. This is where boron nitride differs significantly from diamond. It does not transform to the hexagonal structure until temperatures of at least 1200 °C are exceeded.

cBN is synthesized using the same general technology as that used for commercial diamond synthesis. A mixture of hBN as source and various catalyst/solvents are subjected to pressures in the range 4.5–5.5 GPa and temperatures between about 1500 K and 2000 K. The elements initially investigated and demonstrated to function as solvent/catalysts were the alkali and alkali earth metals, as well as antimony, tin, and lead [137]. Subsequently, the nitrides of lithium, calcium. and magnesium were shown to be more effective. Single crystals of cBN, in the size ranges appropriate to grinding applications, both black and amber in color, are now readily synthesized by several companies on a commercial basis.

The recent developments in the boron nitride phase diagram, particularly with respect to the room temperature and pressure thermodynamic stability of cBN, have stimulated work aimed at producing cBN over a wider range of conditions [136]. In conjunction with this, it is now known that a wide range of nitrogenous compounds are capable of aiding the transformation of the hexagonal to the cubic form [138].

1.3 Polycrystalline Diamond and Cubic Boron Nitride

S. Ozbayraktar

1.3.1 Natural Polycrystalline Diamond

Although the first natural polycrystalline diamond was discovered in the 19th century, it was only investigated on a scientific basis during the 1970s, and was classified into two broad types: Carbonado and Ballas [139].

Carbonado is 'a porous, randomly polycrystalline diamond aggregate' [140]. Ballas on the other hand is defined as 'polycrystalline diamond of oriented globular growth'. Its crystallites have $\langle 110 \rangle$ directions oriented radially.

Experiments so far have clearly indicated that these natural forms of sintered diamond are truly polycrystalline ceramics, exhibiting transgranular fracture as a result of diamond to diamond bonding.

1.3.2 Synthetic Polycrystalline Diamond

In 1958, Hall [141] discussed the desirability of preparing a 'cemented diamond composition analogous to WC' and hinted that experiments to produce polycrystalline diamonds were underway. But it was not until 1970, when he reported details of his procedures [142], that he established experimentally practical pressure and temperature fields where pure diamond powder can be sintered within times ranging from several days down to about one second. He mentions hard refractory materials like borides, carbides, nitrides and oxides as suitable binders.

In December of the same year, Stromberg and Stephens [143] published a paper titled 'Sintering of Diamond at 1800–1900°C and 60–65 kBar'. Their manuscript was originally dated 17 December 1969. In their work, natural diamond powders of sizes 0.1–10 µm were carefully cleaned in several solvents and then heated at 500°C in a vacuum of 10^{-7} torr in order to remove adsorbed gases. The cleaned particles were then loaded, under argon, into a similarly cleaned and outgassed tantalum crucible which was sealed by electron beam welding. The capsule was heated under pressure and temperature for 1 h. They used indium as a 'cushion' to protect the diamond specimen on unloading. Small amounts (less than 1wt%) of boron, silicon, or beryllium were found to aid the sintering process. Most of the sintered specimens gave hardness values of over 7000 kg mm^{-2}, with a maximum of 8800 kg mm^{-2}.

In 1971, Katzman and Libby, from the University of California at Los Angeles [144], described the formation of sintered diamond compacts in the presence of 20 vol% of cobalt binder. The sintering was carried out for about 20 min at 6.2 GPa in a piston and cylinder type high pressure-high temperature apparatus, and at a temperature (1590°C) which was above the cobalt-diamond eutectic, 1570°C at 62 kBar. They claimed their results suggested that the cobalt cleans the diamond surfaces of any adsorbed gases that might prohibit or retard grain growth or fusion, thereby making unnecessary the surface cleaning and degassing procedures described by Stromberg and Stephens. They also claim that the optimum cobalt content was 20% by volume. An attempt to substitute cobalt with nickel under identical sintering conditions also led to a somewhat softer material. Compacts made with diamond grit sizes of 1–5 µm and 0–2 µm were harder than those made with grit sizes of 10–20 µm.

Wentorf and Rocco in 1973 [145] described *in situ* diamond sintering on top of a layer of cemented tungsten carbide (WC). The idea here was that any solvent/catalyst metal (cobalt) which was present in the metal carbide would act as a source for liquid phase sintering. The sintering method involves placing in a cylindrical mold, a 0.5–1 mm layer of diamond powder on top of a thicker layer (10 mm) of WC–10wt% cobalt powder. The mold is then subjected to high pressures (>5 GPa) and temperatures (1400–1600°C) for at least 10 min to 1 h. Under these conditions, the cobalt melts and liquid phase sintering of both the carbide and the diamond occur.

Veraschagin *et al.* [146], in 1975 showed that, if the metallic binder was initially in the form of a disc (rather than a powder), better packing densities were achieved and the molten binder would thus be found mainly in the voids rather than in between diamond grains, thus resulting in strong and uniform diamond-to-diamond bonding throughout the compact.

1.3.3 Mechanisms involved in Polycrystalline Diamond Manufacturing Process

The pcD manufacturing process can be roughly divided into three stages: cold compaction, hot compaction, and liquid phase sintering of the diamond compact.

1.3.3.1 Cold Compaction

Kolomiitsev [147] and Uehara [148] studied the compaction behavior of diamond compacts under high pressures. Kolomiitsev showed that there are basically three processes taking place during cold compaction, see Fig. 22: particle rearrangement, crushing of diamond particles, and filling of voids by crushed particles.

At a given pressure, coarser particles crush more than finer particles. Their average particle size changes drastically and the particle size distribution becomes truly bimodal. At a given pressure, porosity of coarser powders is lower than that of finer powders. The change in porosity with pressure is bigger at lower pressures than at higher pressures for a given starting particle size. Pore size of coarser powders is

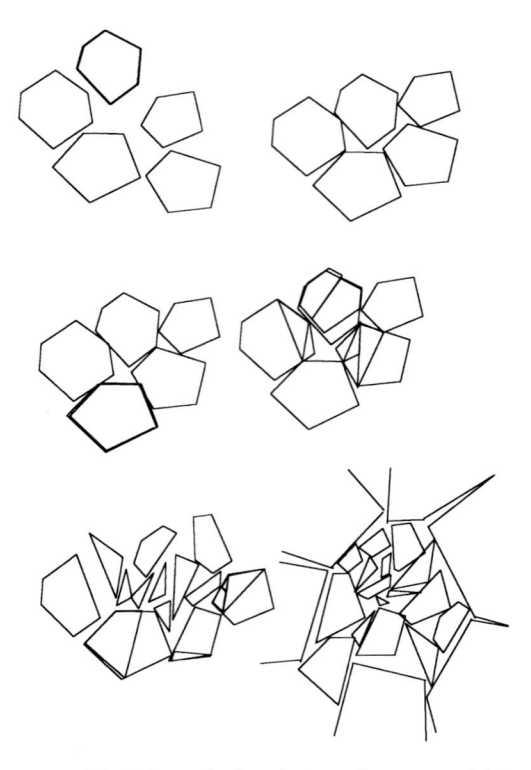

Figure 22. Schematic description of processes taking place during cold compaction.

larger at a given pressure. But the difference between coarse and fine grains gets smaller as pressure is increased.

Crushing of diamond particles during this stage dramatically shifts the particle size distribution.

The powder, which has a lower number of sharp edges per particle, is more stable to compression forces than the one with many spallings, sharp edges, etc. Since their weakest planes are the planes where they contain inclusions; the density, size and shape of the inclusions may be a critical factor in the crushing of diamond particles during the cold compaction stage.

1.3.3.2 Hot Compaction

We describe hot compaction as the stage where temperature is applied together with pressure but the temperatures are not high enough to start infiltration/liquid phase sintering processes.

During hot compaction the following processes are believed to be taking place, as shown in Fig. 23: graphitization of the diamond surfaces facing the voids, plastic deformation of diamond grains, and densification of the compact (shrinkage of pores).

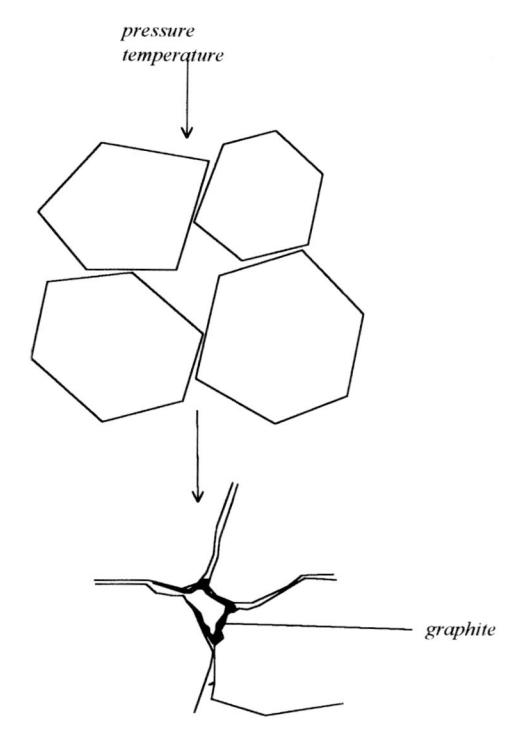

Figure 23. Deformation of diamond grains under high pressure and temperature.

Once the heat is turned on at high pressures, the densification of the diamond compact proceeds mainly by further crushing and rearrangement of the crushed particles up to 700°C [149]. After 700°C, densification proceeds with plastic deformation [150–152]. The extent of plastic deformation is a function of temperature and pressure. At higher temperatures, the diamond grains become more round, their sharp edges disappear, they start to deform in the zones of contact with each other. Pores get smaller, even diminish at a rate initially determined by their size at the end of the cold compaction stage. Pore size reduction rate is faster for coarser pores than for smaller pores [149]. But the time required for a pore to disappear is constant at a given pressure and temperature to a first approximation, regardless of size [153, 154].

Also at higher temperatures, graphitization of the diamond surfaces, which are not in contact with another diamond surface, takes place. The amount of graphitization is a function of the temperature at a given pressure [150]. The degree of graphitization in diamond compacts can be related to the specific resistivity of the compacts since graphite is a relatively conductive material compared to diamond which is an insulator. That is, the more graphitization takes place, the more conductive the material becomes.

At even higher temperatures, bulk graphitization of diamond grains starts to take place and the hardness and wear properties of pcD produced (binderless) start to degrade [150]. At a given temperature, if the pressure is increased, graphitization seems to decrease and the properties improve. One can conclude from these observations that, at a given plastic deformation, the degree of graphitization determines the change in hardness and wear resistance. But when both plastic deformation and graphitization are taking place, properties are more sensitive to plastic deformation until bulk graphitization starts taking place. Finer particles tend to graphitize more than coarser particles at a given pressure and temperature. This is most probably due to their larger surface area in contact with the pores.

1.3.3.3. Surface Chemistry Effects

Surface chemistry of the starting diamond particles has an effect on the graphitization kinetics of the free surfaces of diamond particles. Surface adsorbed oxygen gas reacts with diamond and speeds up the process of graphitization on these surfaces [149]. This may be due to the fact that the activation energy for the process of oxidation is lower than the activation energy for the process of graphitization without participation of the gas phase. Moreover, such graphitization may begin at lower temperatures.

1.3.3.4 Liquid Phase Sintering

There are a number of binders used for pcD sintering. The most commonly used ones are iron group materials which are known to be good solvent/catalysts for diamond synthesis, namely cobalt, iron, nickel, and manganese or various combinations of these.

Utilization of binder materials for the sintering of diamond compacts has threefold benefits: decreasing of sintering temperatures and pressure, cleaning diamond particle surfaces of graphite, and electron discharge cuttability for tool making.

There are various methods of adding binder material into diamond compact: mixing it with the diamond powder, coating diamond particles with it, infiltrating from a disc of binder metal, and infiltrating from a substrate containing binder metal.

We will concentrate here on the mechanism of infiltration from a substrate because it is the most commonly employed method in the industry.

There are basically four stages in the infiltration of diamond compacts [155, 156].

During first stage, temperature is just sufficient to cause melting of $Co-W-C$ at the interface due to excessive amounts of carbon which cause the eutectic temperature to drop. The amount that is melted during this stage immediately infiltrates into diamond due to the very high pressure gradient between the porous body and the solid interface. However, this stage is very short and it is rapidly followed by the second stage.

The second stage corresponds to melting of the binder in the bulk of the substrate. As the temperature increases, it reaches the eutectic temperature of binder in the substrate and binder becomes liquid and starts infiltrating into diamond compact again under the same pressure gradient which is roughly equal to the capsule internal pressure.

As the pores in the diamond layer become filled, the magnitude of the pressure gradient is decreased. Accordingly, the rate of mass transfer is decreased, although its direction is maintained. As this pressure gradient decreases, the effect of the pressure gradient in the reaction volume of the capsule, which was found to be about 1 GPa in the axial direction by Russian researchers in their own capsule [157], starts dominating the infiltration gradient. Whatever the direction of the pressure gradient is, this stage of mass transfer is completed when there is no pressure gradient at the interface between diamond layer and the substrate and when the pores are completely filled. Thus, the rate of mass transfer approximates zero and the binder content in the diamond layer becomes a maximum.

The graphitization of the diamond surface, described in Section 1.3.3.2, is important in terms of infiltration of the solvent/catalyst binder. Metals such as cobalt dissolve graphite more readily than diamond at the sintering pressures (Fig. 9b). Hence, the more graphitization that has taken place, the more cobalt is expected to be in the final material.

A characteristic feature of the fourth mass transfer stage is a change in its direction: it has been observed [155–157] that the quantity of liquid phase in the diamond layer is decreased and increased in the substrate. The beginning of mass transfer indicates that a pressure gradient arises in the system, directed from the diamond layer to the substrate. The origin of this phenomenon was explained by Shulzhenko et al. [155, 156] in terms of temperature gradients and their changes resulting from the surface tension of the molten binder–diamond interface.

Also, this effect can be explained by the fact that, as the sintering of diamond compact proceeds, the area of the contacts between the diamond grains increases. This accelerates shrinkage and the densification process. This in turn according to Shulzhenko et al. squeezes the excess cobalt, which cannot be accommodated in the ever shrinking pores and cavities in between diamond grains, out of the diamond layer. One destination of this stage of mass transfer is the diamond/substrate interface.

As a result of the change in direction of binder mass transfer in the fourth stage, the binder content of the diamond layer is decreased by an amount of the order of couple of percent [155, 158]. The binder content of the diamond layer can be controlled by lowering the sintering temperature and extending the sintering time.

Almost complete exclusion of cobalt from large areas where two grains have grown together is observed by TEM studies on polycrystalline diamonds sintered with cobalt binder [159]. No thin, residual layer of cobalt was detected at the grain boundaries. The thickness of regrown diamonds is usually much lower than the original diamond grains it is precipitated onto, and it joins with the regrown diamond precipitated onto a neighboring diamond grain forming a high angle grain boundary. So it can be said that only the surfaces of the diamonds in contact with a pore just before infiltration takes place are potential candidates for reprecipitation of new dislocation free diamond. The areas next to the diamond to diamond contact points or surfaces are the preferred locations for diamond reprecipitation due to lesser surface energy requirements.

1.3.3.5 Conclusions

During cold compaction of diamond powders, coarser powders densify more and have larger size pores than finer powders at a given pressure.

During hot compaction, the densification of the diamond compact proceeds mainly by crushing and rearrangement of the crushed particles up to 700°C. Above that temperature, the densification proceeds with plastic deformation. The extent of plastic deformation is a function of temperature and pressure. Pore size reduction rate is faster for coarser pores than for smaller pores.

The degree of plastic deformation and of graphitization taking place during hot compaction seems to influence the wear resistance and hardness of the sintered diamond compacts. When both plastic deformation and graphitization are taking place, properties are more sensitive to plastic deformation until bulk graphitization of diamond grains starts.

Infiltration rate of the metallic binder from the substrate is influenced by pressure gradients created by the applied external load, pressure gradients inside the capsule and the pore size of the diamond compact at the onset of infiltration, temperature, and the substrate binder mean free path and chemistry.

1.3.4 Polycrystalline Cubic Boron Nitride

As referred to in Section 1.2.7, cubic boron nitride (cBN) was first synthesized in 1957 [133].

Polycrystalline cubic boron nitride is preferable over single crystal cBN because it can be manufactured in bigger sizes and has higher fracture toughness. However, like its counterpart pcD, it can only be sintered at very high pressures and temperatures with the aid of binders because of its strong covalent bonds.

pcBN, having a hardness second only to pcD in terms of polycrystalline materials, and being relatively less reactive with ferrous metals, is the cutting tool material of choice for a wide range of applications such as finish and interrupted cutting of hardened steel, gray, and ductile cast iron machining and finds a wide range of applications in the production of automotive parts. These applications will be discussed further in Section 1.5.4.

Commercially, pcBN tool materials are produced using sintering additives and/or binding materials at pressures of 4–6 GPa and at temperatures of 1200–1500°C. The sintered pcBN has a higher thermal stability than pcD.

The first polycrystalline cBN material was sintered by Wentorf and Rocco [160] in 1971 on a cemented carbide substrate using alloys of nickel, cobalt and iron with aluminum as binders. cBN to cBN bonding is believed to be achieved by a liquid phase sintering process involving these binder alloys.

In February, 1980, Sumitomo from Japan filed the patent 'Sintered compact for a machining tool and a method of producing the compact' [161]. This patent basically covers any compact with 10–80 vol% cBN and a balance of binder material that can comprise any carbides, nitrides, borides, or silicides of metals of groups IVa, Va, or VIa. Specifically mentioned are titanium, zirconium, hafnium, vanadium, niobium,

Table 4. Properties of the pcBN products and other cutting tool materials.

	DBC 50*	Amborite*	SYNDITE 010	WC	Al_2O_3 + TiC	Sialon
Density [g cm^{-3}]	4.297	3.41	4.127	14.7	4.28	3.20
Compressive strength [GPa]	3.55	2.73	4.74	4.50	4.50	3.5
Fracture toughness [MPa.m$^{1/2}$]	3.64	6.36	8.39	10.80	3.31	5
Knoop hardness [GPa]	27.5	31.5	50	13.0	17.0	13
Young modulus [GPa]	607	653	993	620	370	300
Modulus of rigidity [GPa]	258	288	453	258	160	117
Bulk modulus [GPa]	315	297	412	375	232	227
Poisson's ratio	0.178	0.13	0.102	0.22	0.22	0.28
Thermal expansion coefficient [$10^{-6}K^{-1}$]	4.7	4.9	3.8	5	7.8	3.2
Thermal conductivity [W.m^{-1}K^{-1}]	44	100	120	100	16.7	20–25

tantalum, chromium, molybdenum, and tungsten (which are, in fact, the elements of groups IVb, Vb and VIb).

In Sumitomo's patent, high heat conductivity together with high hardness are mentioned as paramount properties when selecting a suitable binder material. A rapid rise and fall in temperature in the tool during interrupted cutting often leads to the formation of cracks. Therefore, a binder material that couples high thermal stability and hardness with high thermal conductivity at high temperatures is suitable for this kind of application.

The manufacturing process (especially the cold and hot compaction stages) of pcBN is very similar to that of pcD, summarized in Section 1.3.3. However, the types of binders used and liquid phase sintering mechanisms can be quite different. In pcBN production, 'reactive sintering' plays a major role in terms of driving the densification process, whereas in pcD, dissolution and precipitation and/or adhesion/coalescence are the main driving mechanisms. A typical example for pcBN would be the following reaction [162].

$$3Al + 2BN \rightarrow AlB_2 + 2AlN \qquad (4)$$

where the reaction products AlB_2 and AlN act as binders [163]. However, the chemical reaction to achieve these products is the driving force for sintering this particular pcBN product.

The properties of the pcBN products are not only determined by the final microstructure, but also by the phases formed during sintering. The pcBN properties are listed in Table 4 and the properties of the main phases are given in Table 5.

A direct conversion process can produce translucent pcBN which has almost theoretical density and very high thermal conductivities. In 1972 and later in 1974, Wakatsuki et al. [164, 165] sintered a polycrystalline cBN directly from hBN at pressures as low as 55 kBar and temperatures of 1100–1400°C. Corrigan [166] observed that thermal conductivity of directly converted compacts increases with increasing grain size. Fukunaga and Akaishi [167] used a small amount of Mg_3BN_3 to promote conversion from hBN to pcBN at pressures of 5–6 GPa and at 1500°C.

Table 5. Properties of the main phases found in pcBN products after sintering.

	cBN	AlN	TiB2	TiN	TiC	Al$_2$O$_3$	Diamond (for reference)
Crystal sructure	cubic	hexagonal	hexagonal	cubic	cubic	hexagonal	cubic
Density [g cm^{-3}]	3.48	3.26	4.52	5.22–5.44	4.92	3.96	3.51
Melting point [°C]	2700	2300	3197	2930	3065	2050	–
Knoop hardness [GPa]	47	12	26.5	19	28–35	23	57–104
Young modulus [GPa]	700–800	318	434–540	370	450	427	1141
Thermal expansion coefficient [10^{-6} K^{-1}]	3.2	3.9	8.1	8	8.6	8.1	1.5–4.8
Thermal conductivity [W.m^{-1} K^{-1}]	150–700	200	80	25	33–43	14	500–2000

1.4 New Ultrahard Materials

I. Sigalas

1.4.1 Introduction

Diamond's combination of properties make it a unique material. Although hardness is its primary characteristic, thermal conductivity, compressive strength, refractive index, spectral transmittance, and chemical stability are either the highest or among the highest found in nature.

This combination of properties can be traced to the same structural characteristics of diamond that give rise to its high hardness. It is therefore reasonable to expect that other ultrahard materials would also exhibit such a suite of properties. This would make them also desirable for a number of industrial applications.

New ultrahard materials might possess new attributes other than a higher hardness that would make them more attractive than diamond in some cases. A different chemical composition would give rise to different interaction of cutting tool and workpiece as is the case for cBN, as would a different crystal structure and ultrahard particle morphology.

The search for new ultrahard materials is motivated largely by these considerations, but also from purely economic ones. It might be possible to synthesize such materials through routes cheaper than those involved in the case of diamond, thus accessing new applications and new markets. In this paragraph we briefly review the basic ideas behind the search for new ultrahard materials, as well as the latest developments in the search for some of the 'identified' cases.

1.4.2 Hardness

Hardness is a measure of a material's ability to resist elastic and plastic deformation. The hardness of non-ideal material is determined by the intrinsic stiffness of the material, as well as by the nature of its defects, be they point defects, dislocations, or macroscopic defects such as microcracks etc. For ideal systems, the hardness of a material will scale with its bulk modulus.

Figure 24 shows the Knoop hardness as a function of the bulk modulus for a number of representative materials [168]. Table 6 shows the hardness as well as the bulk modulus of a number of representative materials.

The bulk modulus of a solid can be calculated by means of *ab initio* calculations. A review of this field can be found elsewhere [169]. Such calculations require extensive computer time. Liu and Cohen in 1989 [170] proposed an empirical formula for the bulk modulus of solids with the zinc blende structure:

$$B = (19.71 - 2.20\lambda)/d^{3.5} \tag{5}$$

where B is the bulk modulus in GPa, d is the bond length in Ångstroms, and λ is a measure of the ionicity of the compound. For purely covalent compounds (group

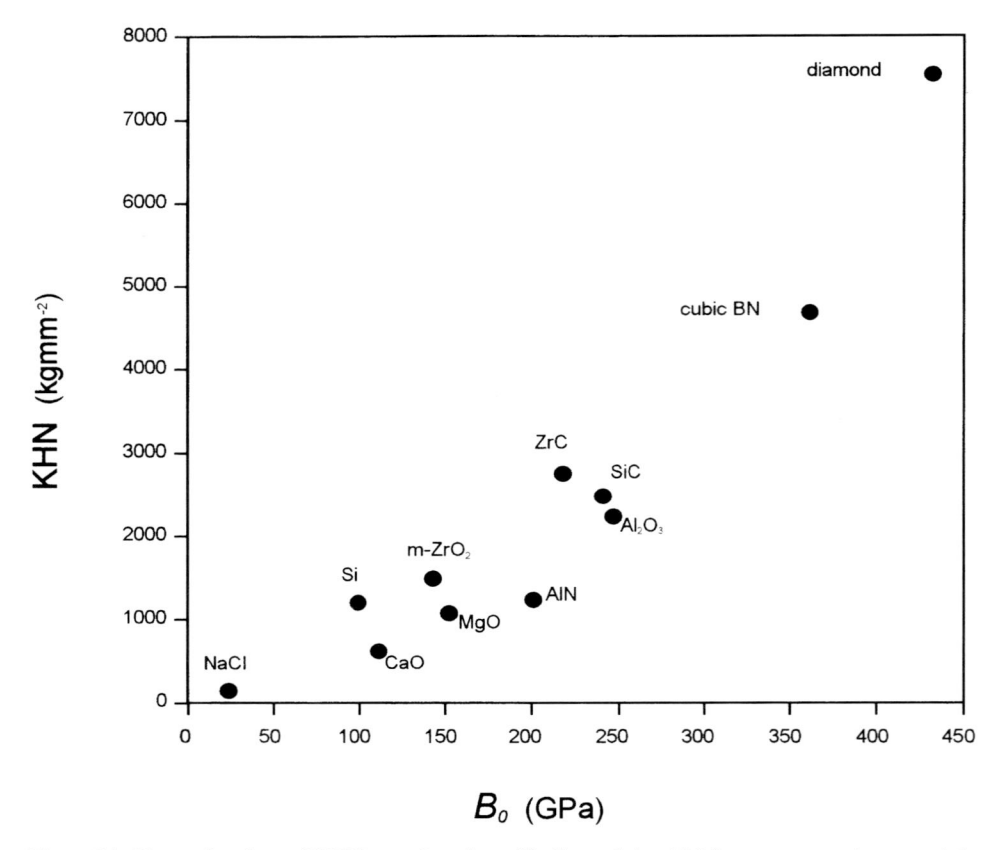

Figure 24. Knoop hardness (KHN) as a function of bulk modulus (B_0) for representative materials.

IV), $\lambda = 0$ while, for compounds of groups III–V and II–VI, λ is equal to 1 and 2 respectively. This expression can be modified to allow for nontetrahedrally coordinated compounds as follows [171, 172]:

$$B = (N_c/4)(1971 - 220\lambda)d^{-3.5} \qquad (6)$$

Where N_C is the average coordination number.

From Eq. (6) it follows that in order for the bulk modulus to be high, the crystal must have: a large coordination number; a high degree of covalency; and short interatomic distances, meaning small atoms. Although a high coordination

Table 6. Bulk moduli and microhardness for light covalent ceramics

Material	B (GPa)	H (GPa)	H actual GPa [228]
C (diamond)	443	77	75–100
Si_3N_4	220	35	17
SiC	211	33	26

number may be attainable in metallic materials, where it can be as high as 12, metallic bonds are much weaker than covalent bonds which can allow a maximum coordination number of 4 in naturally occurring materials. Because of these considerations the search for new materials is confined to tetrahedrally coordinated compounds with high degree of covalency and with small atoms.

On the basis of this model, it was suggested that, for $\lambda = 0.5$ and $d = 1.47$–1.49, a hypothetical tetrahedral compound between carbon and nitrogen would have a bulk modulus of 461–483 GPa, which would exceed that of diamond (443 GPa).

However, not all potentially hard or ultrahard materials can be found within the two top periods of the periodic table, nor are they confined to tetrahedral coordination. A number of sp^2 coordinated structures have been proposed which, mainly due to their high density, would posses a high bulk modulus and would therefore potentially exhibit high hardness [173–175]. Superdiamond structures have been proposed by Diedrich et al. [176] and Alberts et al. [177], based on polymerized derivatives of tetraethylmethane $C(C_2H_4)_4$ and the carbon skeleton of allene $CH_2=C=CH_2$ respectively. Leger [168] investigated the feasibility of generating ultrahard materials by producing the high pressure phases of a number of oxides. Such phases would have a higher degree of coordination and bonds with a higher degree of covalency than their ambient pressure allotropes.

1.4.3 C_3N_4

In 1989, Liu and Cohen reported local density approximation pseudopotential calculations on a hypothetical carbon nitride phase, βC_3N_4 which suggested that it could be metastable under ambient conditions [170, 178] and would have a short C–N bond length (0.147 nm), low ionicity ($\approx 7\%$) and a bulk modulus higher than that of diamond. Figure 25 shows the crystal structure of hexagonal βC_3N_4 derived from βSi_3N_4 [179].

Alternative structures have been proposed by Liu and Wentzcovitch [180], Wentzcovitch and Martins [181] and Wentzcovitch [182], proposing a cubic zinc blende structure with one carbon vacancy per unit cell, and a structure resembling graphitic CN with one carbon vacancy per four nitrogen sites. This study was based on an *ab initio* molecular dynamics scheme with a variable cell shape algorithm.

More recently Teter and Henley [183], using first principles pseudopotential total energy techniques, predicted a cubic form of C_3N_4 with a zero pressure bulk modulus of 495 GPa, higher than that of diamond. The same authors predicted that αC_3N_4 and graphite-structure C_3N_4 are energetically more favorable than βC_3N_4. Figures 26 and 27 show the structures of the proposed cubic C_3N_4 and αC_3N_4 compounds respectively.

The above theoretical predictions have generated a great deal of experimental activity aimed at synthesising a new ultrahard compound.

Several attempts have been made to synthesize thin film C_3N_4, be it cubic or hexagonal. These include reactive magnetron sputtering [184–187], laser ablation [188] ion beam assisted deposition (IBAD) [189], plasma [191, 192] and plasma-enhanced chemical vapour deposition [190].

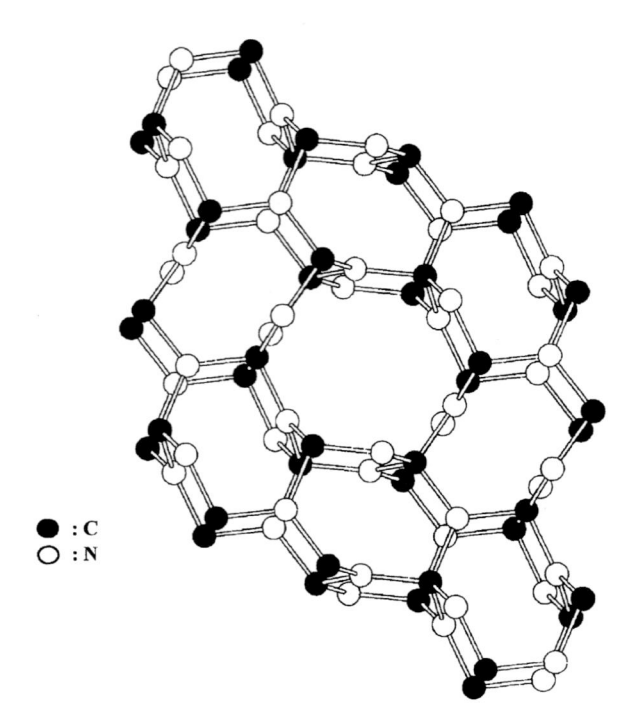

Figure 25. Hexagonal crystal structure of βC_3N_4 derived from βSi_3N_4. Two types of sp$^{2'}$ bonded N atoms are present. The first type is in a trigonal planar configuration with the trigonal plane perpendicular to the c-axis. The second type forms, together with the carbon atoms, the tunnel at the center [179].

In most efforts, the films produced were either amorphous, or their nitrogen content was less than the target of 57%. D. Li et al. [184] using a magnetron sputtering system produced crystalline carbon nitride/titanium nitride composite coatings with a reported hardness of 55 GPa.

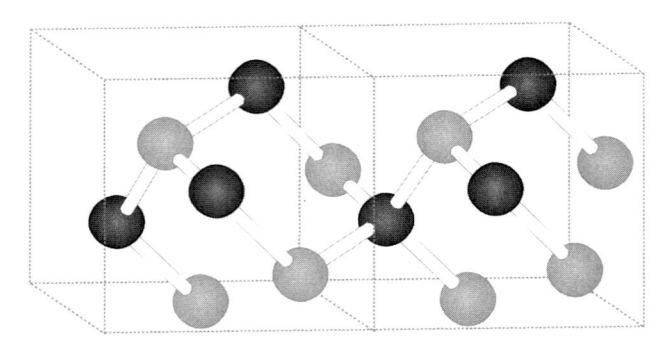

Figure 26. Two adjacent unit cells of cubic C_3N_4 carbon atoms are shown schematically as the black atoms [192].

β-C₃N₄

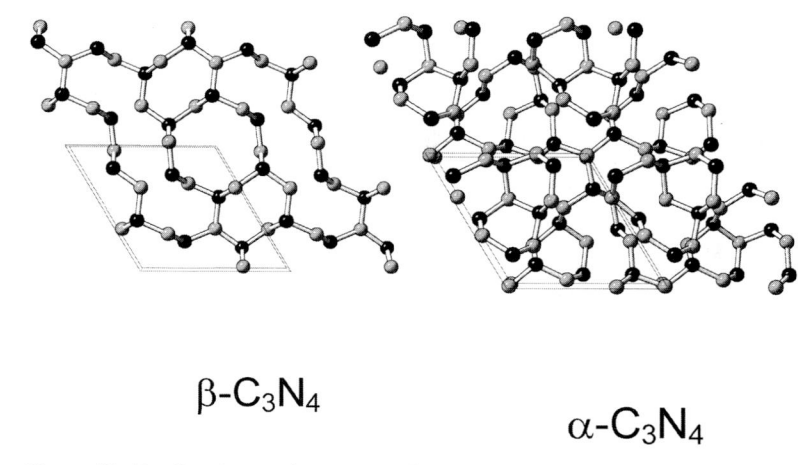

α-C₃N₄

Figure 27. Predicted crystal structure of $\alpha C_3 N_4$ [192], compared with that of $\beta C_3 N_4$.

Zhang et al. [194] at Harvard were able to synthesize an amorphous CN thin film with composition C_2N. This material had exceptional hardness [195–197]. In some cases, tiny crystallises of covalent C–N were observed in the amorphous films [196–202]. Bhusari et al. in February 1997 [193] were able to synthesise large crystals (750 µm) of Si-containing carbon nitride consisting of a predominantly C–N network, by microwave CVD.

More recent efforts involving nitrogen ion beam assisted deposition [203, 206, 211], hot filament CVD with or without rf plasma [204, 205, 209], or microwave plasma enhanced CVD [213, 214], with bias assistance [207, 208, 212] were able to produce thin films containing C_3N_4, some with the $\beta C_3 N_4$ structure and some with the cubic or α-structure.

Yan et al. [205] using hot filament CVD were able to synthesize fully crystalline films combining both $\alpha C_3 N_4$ and $\beta C_3 N_4$.

Xu et al. [210] using a reactive magnetron plasma source claim to have deposited polycrystalline $\beta C_3 N_4$ films with crystallites as large as 20 µm.

Xiao-Ming et al. [211] using argon arc nitrogen-assisted bombardment claim to have deposited completely polycrystalline $\beta C_3 N_4$ films. They quote a measured hardness of 52.6 GPa.

Similar results were obtained by Wu et al. [212] who used RF plasma enhanced CVD.

Finally Chen et al. [213], using microwave plasma enhanced CVD, claim to have grown crystals larger than 10 µm, and propose that the incorporation of silicon from the substrate into the C–N structure promotes $\beta C_3 N_4$ crystal growth.

Efforts to synthesize C_3N_4 at high pressures have been much less extensive than those made to synthesize this material in the thin film form. A number of trials [214] indicate that it may be possible, through the use of pressure to incorporate nitrogen into sp^3-bonded carbon nitride.

1.4.4 Boron Rich Nitride

B_3N, B_5N and B_4N have been synthesized by CVD methods [215–217]. The B_4N structure, with a crystal structure analogous to that of rhombohedral B_4C, may exhibit high hardness. No reported hardness values are available at present.

1.4.5 Boron Carbonitrides

In addition to binary compounds made with elements from the 2d period, ternary compounds may also exhibit higher thermal stability that diamond [218]. Such compounds can be considered as solid solutions of carbon in boron nitride. Efforts to synthesize such compounds have been made both at low and at high pressures.

Ternary BC_xN_y films were first produced in 1972 by Badzian et al., using a CVD process [219]. Montasser et al. [220] in 1984 produced films of hardness in the range 4–33 GPa. Loeffler et al. [221] studied the influence of substrate temperature on the crystallinity of BC_xN_y films deposited by PA CVD. Hegermann et al. [222] investigated the influence of carrier gases nitrogen, argon and helium and the influence on the applied power density (about 1–4 W cm^{-2}) of the film deposited by the PA CVD process.

1.4.6 Boron Suboxides

Boron suboxides with hardness comparable to that of diamond were first reported by A. R. Badzian in 1988 [223]. These suboxides were produced by reacting B_2O_3 with boron at 1600–2000°C and 7 kbar. Subsequent melting under argon gave a sintered compound with an oxygen content of 4–5% and a microhardness of 60 GPa.

The introduction of oxygen reduces the electron deficiency of the β-rhombohedral structure of elemental boron. Filling these voids with atoms results in a cross linking of the boron icosahedra and can contribute to a significant increase of the rigidity of the structure.

Recently McMillan et al. [224] reported the synthesis of B_6O icosahedra at pressures of 5–6 GPa. Itoh et al. [225] reported the synthesis of B_6O powder by reacting B_2O_3 with boron at 1350–1400°C and the subsequent sintering of that powder with B_4C or cBN at 3–7 GPa and 1500–1800°C for 10–30 min.

Although the hardness of the resulting compacts exceeded 40 GPa, the fracture toughness did not exceed 1.5 MPa m$^{1/2}$.

The synthesis of B_6O was patented initially in 1992 [226, 227], but no boron suboxide products have appeared in the market as yet.

1.4.7 Stishovite

The search for new ultrahard materials has also been extended into nonboron based oxides. The expectation was that high pressure allotropes would exhibit a high coordination of the metallic ion, as well as more ionic bonding, thus hopefully possessing higher stiffness and, therefore, higher hardness.

Leger *et al.* [166] in 1994 measured the bulk modulus of the high pressure phases of HfO_2 and RuO_2, which they obtained at pressures above 42 and 12 GPa. The resulting bulk moduli place the two compounds above diamond in the case of HfO_2 and between cBN and diamond for RuO_2.

In 1996 Leger *et al.* [228] reported the synthesis of stishovite from α quartz at 20 GPa and 1100°C and measured a hardness of 33 GPa for the polycrystalline compact produced. As the samples obtained were very small, a maximum load of 1.9 N during hardness testing was possible. Synthesis of this material in larger quantities has not yet been possible.

1.5 Industrial Applications of Diamond and cBN

M. W. Bailey

1.5.1 Introduction

As discussed earlier, the modern industrial diamond business began to expand significantly following the commercial availability of synthetic diamond in the late 1950s. Before then, only natural diamond produced as a by-product of gem diamond mining was available and was, by its nature, limited in volume. Demand was increasing to the point where new application development was a largely self-defeating exercise since, the more successful they were in terms of using large volumes of diamond, the less feasible they became in terms of total world supply. With the introduction of synthetic diamond, this situation changed dramatically on two important fronts. Firstly, the only constraint remaining on the development of new applications was the cost and technology of producing a suitable product; and secondly the constraints on product characteristics imposed by the nature of a natural raw material were removed: diamond could now be specifically engineered for its intended use.

Natural and synthetic diamond are complementary, not competing products. While today a minor player, natural industrial diamond still plays an important role in a variety of special applications where one or other of its particular characteristics will make it the preferred material (refer to Section 1.5.5 for some examples).

Industrial applications of diamond have developed over the years as a result of developments by tool makers, machinery manufacturers and the advent of new materials. In addition, new diamond (and cBN) products have evolved, either as a result of technological advances in synthesis or in response to the requirements of a new application, and it is this multi-partnership relationship within the industry which has resulted in the dramatic growth and diversification since the early 1960s.

1.5.2 Abrasive Application

1.5.2.1 Bonded Tools

One remarkable property of diamond is its hardness and associated resistance to abrasion and it is this which has led to its use as an industrial abrasive. Cutting and polishing stone, glass, and other hard materials were among its early areas of application. Diamond has one limitation in respect of its application, this being its reaction with iron at high temperatures (causing a reversion to graphite and hence high rates of wear), which in general can make it uneconomic in the machining of the ferrous materials by comparison with other conventional abrasives such as aluminum oxide and silicon carbide. Cubic boron nitride (cBN) does not react in this way and hence, although only having 50% of the hardness of diamond, it is still substantially harder than the conventional abrasives and this makes it suitable as a high performance abrasive for use on ferrous workpieces. Figure 28 summarizes the major areas of application for diamond and cBN abrasives.

The advantages of using diamond and cBN abrasives are due mainly to their high hardness and abrasion resistance which, when applied correctly, result in longer tool lives, higher levels of productivity, closer tolerances, and many other advantages.

The majority of diamond and cBN abrasive is used in so-called 'bonded tools', although much polishing and fine finishing is carried out using the abrasive as a loose powder or in a slurry when mixed with an oil or other carrier medium. Diamond and cBN abrasives are normally used in particle sizes of approximately 1 mm down to less than 0.1 μm, and their sizing is subject to international standards, the most widely used being the FEPA standard [229].

In a bonded tool, the abrasive particles are held in a bond or matrix material, usually distributed randomly but in a controlled concentration. The main functions of the bond are:

− to hold the abrasive particles
− to form a molded profile of the required shape which is usually attached to a hub or carrier

Diamond → **Non-ferrous materials:**
Glass - flat, decorative, optical
Engineering ceramics
Cemented tungsten carbides
Reinforced plastics
Natural stone
Semiconductor materials

Cubic Boron Nitride (cBN) → **Ferrous materials:**
Hard ferrous components

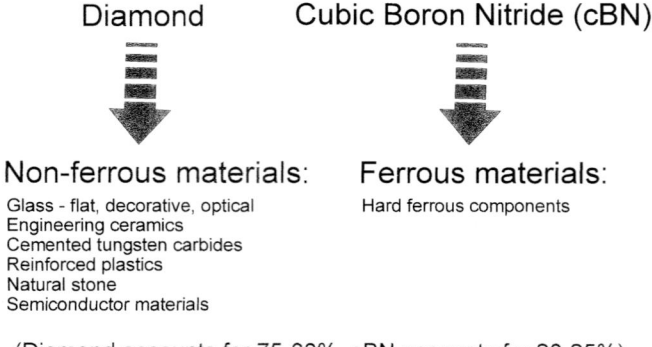

(Diamond accounts for 75-80%, cBN accounts for 20-25%)

Figure 28. Main application areas for diamond and cBN abrasives.

Figure 29. Basic principle of bonded tools.

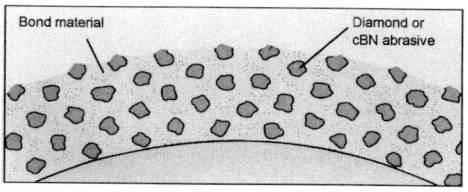

Layers of diamond particles in a resin, metal or vitrified bonded tool

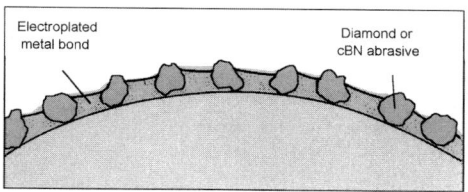

Single layer of diamond particles in an electroplated tool

– to release worn abrasive particles as they reach the end of their useful lives, thus exposing new ones.

It is therefore a general requirement of a bonded tool that the abrasive and the bond material wear at similar rates.

The most common types of bond used are as follows:

– resin, normally thermosetting phenolic or polyimide resins
– vitrified, based on glass frits
– metal, normally sintered alloys
– electroplated, single layer or a few layers of abrasive attached to a hub by an electro-deposited metal layer, normally nickel.

Figure 29 shows the concept diagramatically. Figure 30 shows a selection of bonded tools including saw blades and grinding wheels.

In the case of the first three bond types, since the rate of bond erosion and rate of wear of the abrasive have to be matched if optimum performance is to be obtained, the selection of the abrasive which has the correct characteristics of size, impact strength, and fracture mode matching the characteristics of the bond is very important. Workpiece material plays a major role in this selection process but also the type of machining operation, the machining conditions and the primary requirement of the operation are important. Maximum tool life, high rates of material removal or high quality of surface finish are common requirements but normally a compromise between one or more is required. Figure 31 shows some of the issues which affect the final machining performance. In the case of the fourth bond type, electroplated (or EP), the abrasive particles are embedded in a metal layer to something less than 50% of their total dimension as shown schematically in Fig. 29.

The abrasive particles wear down to the level of the supporting EP layer, at which point the tool is at the end of its life. With this type of tool, the choice of abrasive

Figure 30. A selection of bonded tools.

depends primarily upon the workpiece material, the machining conditions and the primary role of the machining operation (e.g. to produce a high surface finish or high material removal rates). The role of the EP bond layer is to support the abrasive and not wear significantly during the life of the tool. When new, EP tools have a high degree of protrusion of the abrasive particles above the bond layer (much higher than the other types of bond) and hence can machine at high material removal rates while generating low cutting forces. The term often used to describe such a

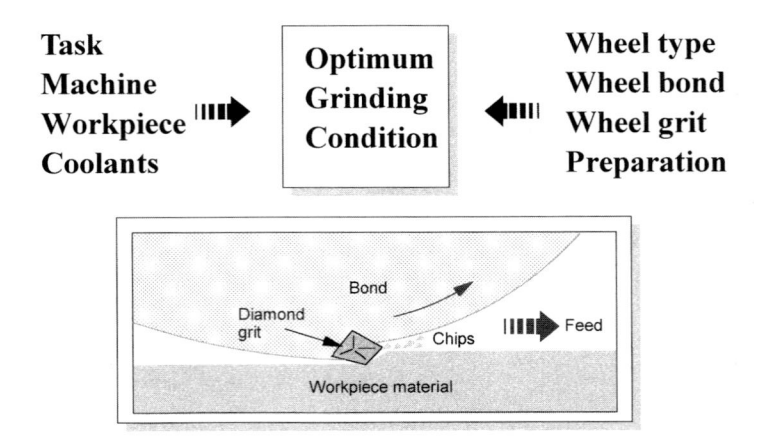

Figure 31. Main parameters affecting tool performance.

- General purpose bond
- Simple to manufacture
- Relatively cheap
- Easier to use than vitrified or metal (more forgiving)
- Easy to dress
- Profile holding not as good as vitrified or metal

Typical applications:
General purpose for all types of material
Sintered carbide, Non-oxide ceramics
Cermets, Natural stone (polishing)

Resin bond diamond grinding the teeth of a cemented carbide cutter

Figure 32. Main attributes of a resin bond.

characteristic is 'free-cutting'. As the abrasive particles become worn, so this free cutting character is gradually reduced and the cutting forces increase until the end of the useful life of the tool is reached.

1.5.2.2 Grinding with Diamond and Cubic Boron Nitride Abrasives

1.5.2.2.1 Introduction Grinding with diamond and cBN abrasives is employed by a very wide range of industries and application areas. In dentistry, small EP diamond drills are widely used, spectacle lenses and crystal glass items are machined using diamond, large diamond wheels are used to produce high volumes of cemented tungsten carbide cutting tool inserts to precise tolerances and cBN wheels are widely used in the automotive and bearing industries to produce high volumes of precision ground components.

In grinding operations, all four of the main bond types are used. As a general guide they may be considered to have specific advantages and the choice is often dependent upon which characteristic is the more important. Figures 32–35 show a

- Very high wear resistance
- Good profile holding
- Difficult to dress
- Manufacturing technology more advanced than resin

Typical applicatons:

Decorative, flat, optical glass
Aluminium oxide

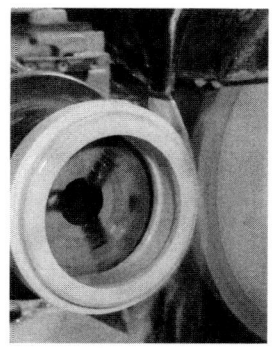

Metal bond diamond grinding a high alumina ceramic seal

Figure 33. Main attributes of a metal bond.

- Single layer of Ni-electroplated abrasive
- Cheap to manufacture
- Excellent for profile grinding
- Limited life
- High stock removal possible

Typical applications:

Green carbide, Aluminium oxide
Ferrites, Natural stone (profiling)
Precious stone, Dental
Reinforced plastics,
Semi-conductors
Profile grinding hard ferrous (cBN)

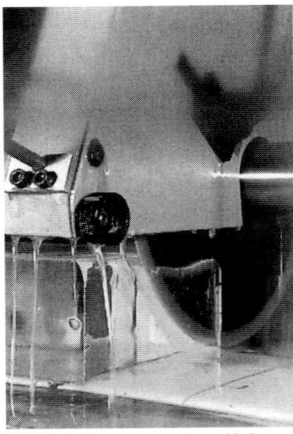

Electroplated diamond saw blade slicing a block of optical glass

Figure 34. Main attributes of an electroplated bond.

typical application for each of the four bond types and some of their particular advantages. The characteristics of the abrasive used can be selected from a wide range available; size, resistance to impact and fracture characteristics being among the more important and Fig. 36 summarizes the main differences and their effects.

Grinding operations can take place with the application of a coolant or lubricant (wet grinding) or with no such application (dry grinding).

1.5.2.2.2 Grinding with Diamond Abrasives In grinding applications, the main types of bond used are metal and resin bonds.

In resin bond wheels, low strength friable diamond types are normally used, a typical diamond being shown in Fig. 37. Since the resins normally used for the manufacture of diamond tools have relatively low strength and are also sensitive to high

- Hard and abrasion resistant
- Up to 30% porosity improves chip clearance independent of grit size
- Holds its edge/profile well
- Longer life than resin
- Easily dressed
- More difficult to manufacture than resin

Vitrified bond diamond grinding of a tungsten carbide roll

Typical applications:

Non-oxide ceramics
Hardened steels (cBN)

Figure 35. Main attributes of a vitrified bond.

Stronger, blocky grits -
withstand higher forces -
greater exposure -
higher rates of stock removal -
lower degree of surface finish -
suited to more ductile,
long chipping workpiece materials

Weaker, friable grits -
withstand lower forces -
less exposure -
lower rates of stock removal -
higher degree of surface finish -
suited to hard, brittle,
short chipping workpiece materials

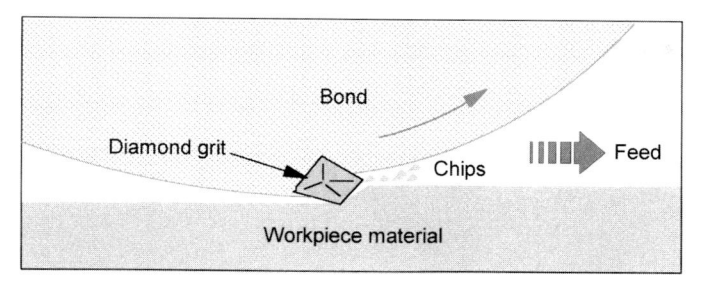

Figure 36. Basic effects of abrasive characteristics.

temperatures, a thick metal cladding of about 10 µm of nickel or sometimes copper is applied to the diamond particle such that typically it will comprise 50–60% of the total particle weight (Fig. 37). This cladding functions in two main ways (Figs 38 and 39), firstly by increasing the surface area of the particle and hence aiding its retention by the bond, and second by acting as a heat sink to protect the bond from thermal damage due to the high interfacial temperatures generated between the abrasive particle and the workpiece, particularly under heavy grinding conditions. These temperatures, although of short duration, can reach in excess of 800°C as the particle passes through the arc of cut.

Figure 37. A typical friable diamond abrasive PDA321 used (normally in metal clad form) in resin bonded grinding wheels.

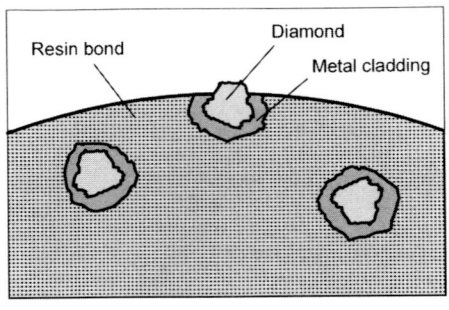

Figure 38. The role of metal cladding.

Resin bond tools are used to machine a wide variety of non-ferrous, usually abrasive, workpiece materials. A very small number of ferrous materials can be ground economically with diamond, for example, some carbon rich cast irons which, providing machining temperatures are maintained relatively low, do not chemically react with the diamond, and other difficult-to-machine materials such as some grades of stainless steel which have a tendency to work harden during machining. The major application area for resin bond diamond wheels is the grinding of cemented tungsten carbide (which was also its initial application area in the 1940s).

The list below gives the main areas of application in descending order of relative size:

– grinding cemented tungsten carbide
– grinding ceramics and cermets
– grinding semiconductors
– grinding polycrystalline diamond (pcD)
– machining stone/glass
– others

Although the machining of cemented tungsten carbide represents approximately half the resin bond tools used, other areas are growing, notably semiconductors, ceramics and cermets, pcD and the machining of stone.

The bond of an abrasive tool performs various tasks:

◆ Holds the abrasive particles

◆ Forms a moulded part or segment
(attached to a carrier -blank or hub- by brazing/welding/ adhesive fixing/sintering or electroplating methods)

◆ Release worn diamond at a controlled rate
- wear rates of abrasive particles and bond must match for a given workpiece and machining parameters

Figure 39. The role of the bond.

Figure 40. A typical high strength diamond abrasive PDA999 used in grinding wheels.

Metal bond diamond wheels are predominantly used in the machining of glass including windows for the automotive and construction industries, decorative glass such as lead crystal and also optical and electronic components. The diamond types most commonly used in metal bond tools are relatively high in impact strength with regular crystal shapes. Figure 40 shows a typical high strength diamond abrasive. Sintered metal bonds commonly based on bronze are much more abrasion resistant than resin bonds, and therefore better suited to machining workpiece material which produce highly abrasive swarf, glass being one example. The list below summarizes the main application areas for metal bond diamond wheels.

– grinding glass
– machining refractory materials
– machining semi-precious stones
– grinding pcD
– grinding ceramics
– honing (mostly ferrous materials)

Electroplated diamond wheels are used in applications where free-cutting characteristics are required. Fibreglass and other composite materials are in general highly abrasive although not particularly hard. The high abrasion resistance of diamond makes it the ideal tool material for machining these types of material and the very open texture and free-cutting nature of EP tools mean that these materials can be machined quickly. Figure 41 shows an example.

Vitrified bond diamond wheels are most commonly used to machine polycrystalline diamond. pcD is difficult to machine and often has to be machined to close tolerances when being used as a cutting tool material as shown in Fig. 42. Both metal and resin bond grinding wheels are also used, but vitrified bond wheels are becoming the most popular, since they provide a good compromise between relatively good wheel life, and hence ease of maintaining dimensional tolerance on the pcD workpiece, and the rate at which the pcD can be ground. Another advantage of vitrified

Figure 41. An EP diamond wheel cutting glass.

520mm diameter SYNDITE
CTC002-tipped saw blade for wood

Flank grinding a PCD-tipped saw

Figure 42. A circular saw blade tipped with pcD.

bond systems is that a degree of porosity can be built into the bond. Resin and metal bonds are effectively 100% dense and clearance for the swarf produced during grinding is provided by the protrusion of the abrasive particle above the surface of the bond. The pores, or voids, in porous vitrified bonds provide extra clearance volume for the removal of swarf out of the grinding zone and also the transport of coolant into it.

In general, all diamond grinding wheels operate at average peripheral speeds in the range 20–$30\,\text{ms}^{-1}$. If a material is being ground dry, without coolant, the wheel speeds tend to be lower, and also the difficult materials such as cermets and pcD tend to be ground at lower speeds to avoid the excess generation of heat. A small number of specialized applications use much higher wheel speeds, one example being the sawing of glass tubes which, providing the machine is specially designed for the purpose, with particular attention to the coolant application, can be cut much more quickly with a wheel rotating at speeds up to 60–$80\,\text{ms}^{-1}$.

1.5.2.2.3 Grinding with Cubic Boron Nitride Abrasives Cubic boron nitride abrasives are used to machine hard ferrous materials. As with diamond, all four of the main bond types are used and, during the early days of its use, in the 1970s and early 1980s, resin bond tools were the majority. Today, vitrified bond tools are the most commonly used type, particularly in high volume production operations, with resin, electroplated, and metal bonds failing in second, third and fourth place respectively. A reason for the growth in the use of vitrified bond cBN tools is that this type of bond system offers a good compromise between high material removal rates, ease of use, and low wear rate resulting in the ability to hold tight tolerances.

The growth in the use of cBN since the early 1970s has been due to its penetration into volume production operations, for example in the automotive and aerospace industries. Prior to that time, cBN wheels were used primarily in the toolroom for sharpening and re-sharpening high speed steel drills, milling cutters and other tooling and also in small wheels for precision jig grinding (Fig. 43).

The economics of cBN grinding have been improved by the adoption of creep feed grinding techniques and also the use of higher wheel peripheral speeds. Both of these developments required the development of suitable grinding machines.

Creep feed (or deep) grinding techniques were originally developed using conventional wheels and the principle of the technique is shown in Fig. 44. In 'normal' or reciprocating grinding mode, a shallow depth of cut is used together with a relatively high table speed to achieve the required material removal rate. Under these conditions, the arc of cut, and hence the area of wheel in contact with the workpiece at any one time, is very small. In creep feed grinding, a significantly deeper cut is made at a slow table speed. Under these conditions, the arc of cut is much longer and hence the number of particles instantaneously in contact with the workpiece is much greater. The specific rate of material removal is a function of the depth of cut and the feed rate of the table. In general, when comparing the two modes of grinding, and assuming equal material removal rates, the main differences for creep feed grinding are that

- the number of particles in contact is higher
- the load on each particle is lower

Figure 43. Dry grinding with cBN.

- the total normal load in the machine is greater (due to the greater number of particles)
- the life of the grinding wheel is longer
- surface finish produced is normally improved.

Reciprocating Grinding

DOC = 5 - 50 Micron
Table speed = 5 - 15 m/min

Creep Feed Grinding

DOC = 5 - 25 millimetres
Table speed = 0.1 - 1 m/min

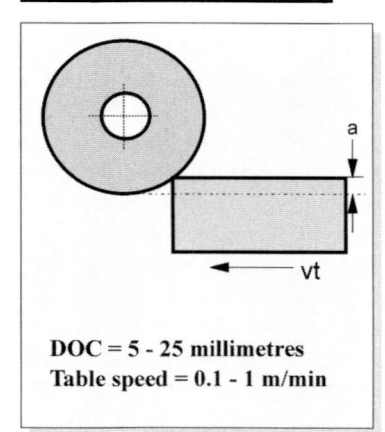

Figure 44. Principle of creep feed and reciprocating grinding.

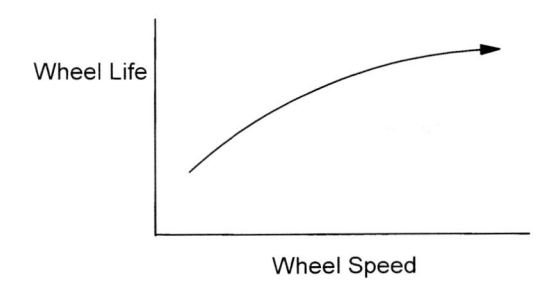

Figure 45. Effect of wheel speed on grinding wheel performance.

A second development in cBN grinding is the adoption of higher wheel peripheral speeds. In the early days of grinding with cBN, wheel speeds similar to those used for diamond wheels were used. It was found however that increasing the wheel peripheral speed improved the life of the grinding wheel providing coolant was used (Fig. 45). Initially speeds up to 60 ms^{-1} were thought to be optimum but today, wheel speeds in the range 60–150 ms^{-1} are common in industry. This has been made possible by improvements in the mechanical design of both the grinding wheels and the machines on which they are used since, at high speeds, safety becomes a major issue which has to be addressed by suitable speed certification for the wheels and safety guarding of the machine. At various academic institutions, much higher wheel speeds have been used, and wheel life has continued to improve. In one such series of tests in Germany, speeds of 500 ms^{-1} were run in one series of experiments.

As a general rule in any grinding operation, increasing the rate of material removal results in a decrease in wheel life. Hence, there is usually a balance in the cost of the time taken to grind and the cost of the wear on the wheel (Fig. 46). By adopting either, or both, the technologies of creep feed grinding techniques and high wheel speeds, higher material removal rates (and hence rates of production) can be achieved whilst retaining, or even improving, the wheel life and hence direct machining cost due to wheel wear.

The field of application for cBN grinding is almost entirely that of hard ferrous metal machining, although some superalloys are also ground using cBN. The list

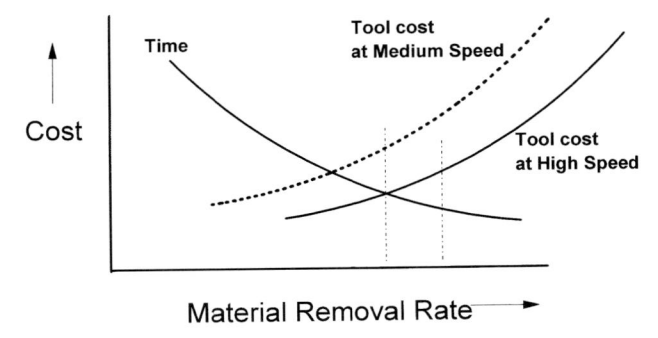

Figure 46. Trend of cost *versus* removal rate.

540   *1 Diamond Materials and their Applications*

600mm dia.resin bond wheel containing ABN360
Grit size B126 (120/140)
Material removed 5-20 micron

Figure 47. Grinding automotive piston rings with cBN.

below gives the main application areas:

- automotive and aerospace material machining
- tool and cutter grinding
- gear grinding
- machining of bearings
- general grinding of hardened steel components
- honing of hardened steel

The largest area of application is in the automotive and aerospace industries, grinding of camshafts, crankshafts and other precision engine parts such as turbine blades and injector nozzles being examples (Fig. 47). Precision grinding of gear tooth profiles is another area where cBN, either in vitrified or EP tools, is used (Fig. 48). The original application area for cBN, tool and cutter grinding, is still a cBN grinding application, the grinding of high speed steel drills and milling cutters, both during manufacture and re-sharpening during use being examples (Fig. 49).

In addition to faster metal removal rates and better tolerances compared to grinding methods using conventional abrasives, cBN wheels have the ability to produce lower workpiece temperatures during grinding, which reduces the potential for thermal damage. This can be of significant importance when critical components are being machined.

1.5.3 Machining of Stone and Concrete

The largest application area for diamond abrasives in terms of volume consumed is the sawing, drilling and surfacing of natural stone and concrete. This application area grew dramatically during the 1970s and 1980s, due to a number of influences. Industrial diamond became more widely available and also became available in an

Figure 48. Electroplated gear grinding wheel.

ever increasing range of size and also other characteristics such as strength. Developments by tool manufacturers and machine builders improved the economics of machining hard natural stone types such as the granites, making their processing more economical. There was also a movement towards the use of granite from marble in the construction industry. One of the reasons for this was that it was recognized that granite was far more durable than marble in respect of both wear if used on floors, and also more resistant to chemical attack and discoloration in urban environments if used as a cladding on the exterior of prestige buildings. Its use, therefore, became more widespread and the quantity of diamond abrasives used to process it has continued to increase.

Figure 49. Flute grinding high speed steel.

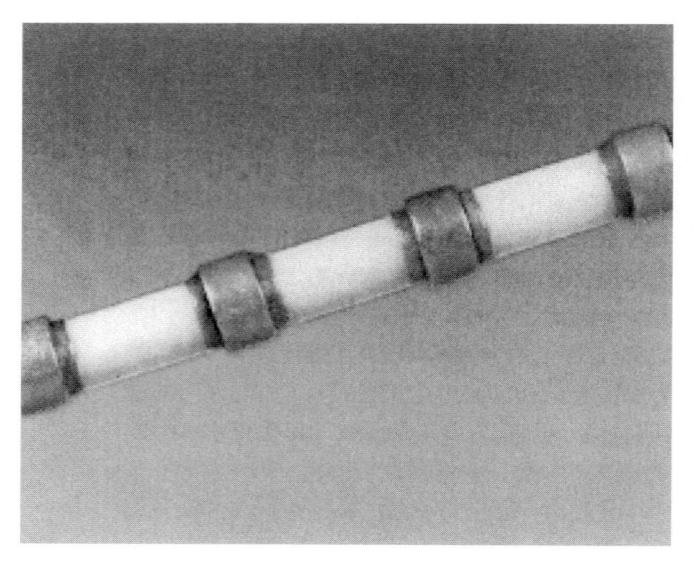

Figure 50. Section of a diamond wire saw.

The types of diamond abrasive used for cutting stone and concrete are normally those with high impact strength and of a relatively coarse size.

The normal production route for stone products includes:

- extraction from the quarry in the form of large blocks
- squaring and stabbing the rough blocks in the stone yard
- cutting the blocks into either cladding or tiles
- surfacing and polishing the finished items.

In all these operations, diamond tools are widely used. For quarrying, diamond wire sawing is established as the standard method of extraction for soft rock (e.g. marble) and the technology for hard rock is being established. Development of diamond wire began in the early 1970s but it came into common use only in the late 1980s. It consists of a high tensile steel wire fitted with beads containing the diamond abrasive, typically 40 beads per metre. The overall diameter of the beads is 6–12 mm (Fig. 50). Two intersecting holes drilled in the quarry face and the wire, which can be up to 500 m long, is fed through them, joined to form an endless loop and then driven at speeds of 20–50 ms^{-1} while tension is applied, thus cutting a slot through the rock and enabling a block to be cut (Fig. 51). The resulting block suffers much less damage compared with using explosive techniques and this, in turn, leads to less waste in subsequent processing. A major problem encountered with early designs was erosion of the wire by the fine particles of stone produced by the sawing, leading to breakage. This is now alleviated by protecting the wire with rubber or plastic sleeving bonded to the wire between the beads.

When quarrying granite, a relatively low wire speed is typically used, 20–30 ms^{-1} and, for a medium grade of material, a block measuring 30 m long × 10 m wide × 5 m deep can be separated from the surrounding material in less than 12 h.

Figure 51. Wire sawing in a stone quarry.

However, because the block produced is of a regular size, up to a 40% increase in usable material is possible with commensurate similar improvements in the economics.

Following extraction from the quarry, the large blocks are either sawn into thick slabs for use in the monument industry or they are squared for subsequent production into tiles or stone cladding. Diamond wire saws are used for these operations together with large circular sawing machines utilizing diamond saw blades of up to 4 m diameter (Fig. 52). The advantages of using diamond tools compared to machines utilising silicon carbide abrasives are increased speed of production, improved accuracy of cut and much reduced pollution.

Tile production is a major application for diamond saw blades. The sawn blocks are machined on multi-blade machines using up to 32 blades, 1.6 m in diameter on one spindle (Fig. 53). These machines typically cut the granite blocks into a series of slabs 12 mm thick, with a thickness tolerance of ±0.5 mm normally required. A horizontal blade then traverses the block to separate the slabs produced and these

Figure 52. Block sawing of granite.

Figure 53. Multi-blade sawing of granite to produce tiles.

slabs in turn are ground to final thickness and sawn to final size, both operations using diamond tools. The final stage in the production of tiles is polishing which is normally achieved using automatic multistage machines (Fig. 54). In this final process, many of the stages are now completed using diamond tools, the early

Figure 54. Polishing natural stone tiles.

Figure 55. Resin bond polishing head.

stages in polishing using metal bond tools with high strength diamond to achieve the high material removal rate required and the later stages using resin bond tools incorporating lower strength, more friable, grades of diamond particle. Figure 55 shows a selection of polishing tools.

Another major application area for diamond abrasive tools is in the construction, refurbishment and demolition industries. In many instances, concrete, masonry and also roadways need to be cut or drilled and diamond tools provide an economic solution, their advantages being speed, accuracy and also reduced damage to the surrounding structure compared to other methods using percussive hammers or thermal lances. Circular saw blades are used in a wide variety of operations and metal bond tools are most commonly used. On roadways, both asphalt and concrete, diamond saws are used to cut trenches during repairs and also to provide slots to embed sensor wires for traffic management systems (e.g. traffic signals) (Fig. 56). Concrete roadways, runways and pedestrian walkways can, under wet conditions, become slippery and diamond blades, up to 100 spaced along a spindle, are used to cut a series of parallel grooves in the concrete to aid water drainage and improve skid resistance and hence safety in critical areas (Fig. 57).

Saw blades are used to cut openings in walls and floors during both building and refurbishment and can also be used during controlled demolition. The advantage of a diamond saw, compared to a saw of conventional abrasive, is its ability to cut both concrete, masonry and steel reinforcement (Fig. 58).

The majority of all diamond sawing of concrete is carried out with use of water as a coolant. This has the combined effect of controlling the temperatures at the cutting edge and also removing the swarf produced by the saw. Some modern blades are designed to run dry (Fig. 59). These are now widely used on construction sites for cutting tiles and the minor correction work which needs to be carried out.

Wire saw machines, similar in principle to those used for cutting stone, are also used widely for large demolition operations. When using a circular saw, there is a limit on the depth of cut which can be made due to the presence of the drive spindle

Figure 56. Roadway sawing.

Figure 57. Roadway grooving to improve skid resistance.

Figure 58. Sawing in the construction industry.

Figure 59. Dry sawing of concrete.

Figure 60. Wire sawing of heavily reinforced concrete.

in the centre of the blade. This is a problem if deep sections have to be cut. The wire saw does not have any such limitation and concrete, heavily reinforced with steel as used in large structures such as road bridges, can be cut with modern wire saw machines. Figure 60 shows such an operation.

Drilling stone with diamond bits is now widely used and concrete, both plain and steel reinforced, brickwork and other masonry have been drilled for many years using diamond core bits (Fig. 61). These tools can range in size from approximately 20 mm to in excess of a meter. They are all metal bonded tools and were until recently designed to operate with water as a coolant. In some areas, the use of water can be inconvenient and a relatively recent development has been the design of core bits which do not need to be cooled with water. These dry drill bits are widely used on construction sites to produce accurate holes in walls for electrical, plumbing and other services (Fig. 62).

1.5.4 Applications of Polycrystalline Ultra-hard Materials

1.5.4.1 Introduction

The polycrystalline derivatives of diamond and cBN, normally referred to as pcD and pcBN respectively, provide engineers with materials which have many of the properties of diamond and cBN, notably hardness and abrasion resistance, but in the form of relatively large isotropic pieces, usually in the form of flat discs. These materials have a wide variety of uses, both as a defined edge cutting tool

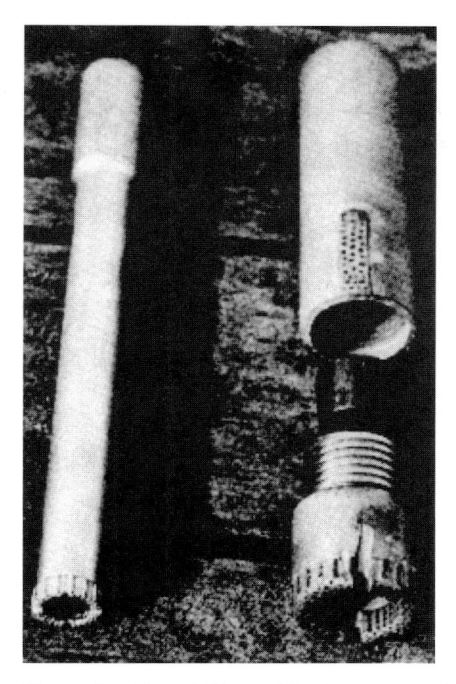

Figure 61. Core drills used in the construction industry.

element and also as a wear resistant material. As a cutting tool material, the general rules which apply to abrasives also apply to pcD and pcBN.

– pcD is for use in non-ferrous applications.
– pcBN is for use on ferrous workpieces.

Figure 62. Dry drilling of masonry.

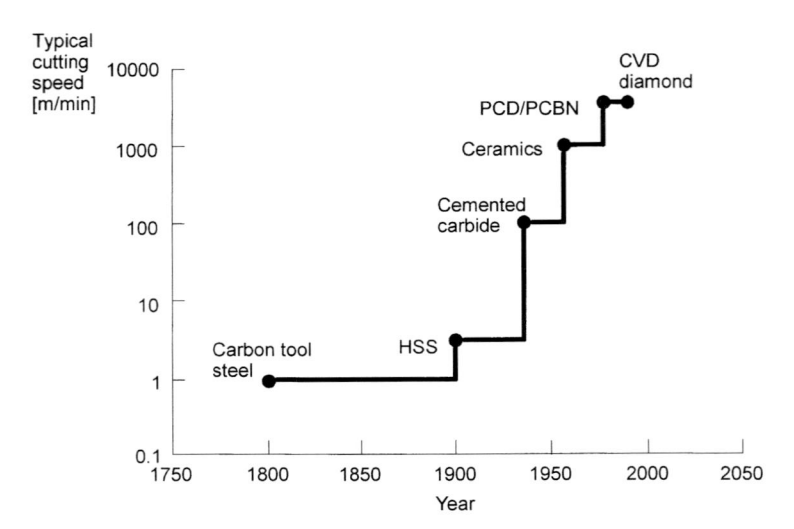

Figure 63. Development of cutting tool materials.

1.5.4.2 Cutting Tools of Polycrystalline Diamond

In most successful applications of pcD cutting tools, the advantages over more conventional tool materials, including sintered tungsten carbide and ceramics, are due to their high hardness and abrasion resistance. This in turn results in longer tool life giving rise to closer tolerances and the ability to work efficiently at high cutting speeds and hence give higher levels of productivity. In addition, workpiece materials which are difficult to machine can be machined efficiently with pcD. Continuous material's development over the years has played a very significant role in the advances made in machining technology, resulting in higher levels of productivity and, in particular, closer tolerances over long machine runs which are now required by modern manufacturing industries. Figure 63 shows the achievable cutting speeds which have been made possible by advances in cutting tool material technology. Figure 64 shows schematically how current materials relate to each other in terms of two important characteristics, toughness and abrasion resistance/hot hardness. The ideal cutting tool material would feature very high toughness with equally high wear resistance. Although, today, such a material does not exist, it is the objective of many development programmes (see Section 1.4).

The main application areas for pcD are the machining of:

– non-ferrous metals, e.g. aluminum alloys, magnesium alloys, brass, copper
– wood and wood composites
– fibre-glass and carbon fibre composites
– plastics and rubber
– mineral materials

The automotive and aerospace industries provide a large number of applications for pcD, particularly with the movement to more widespread use of advanced alloys

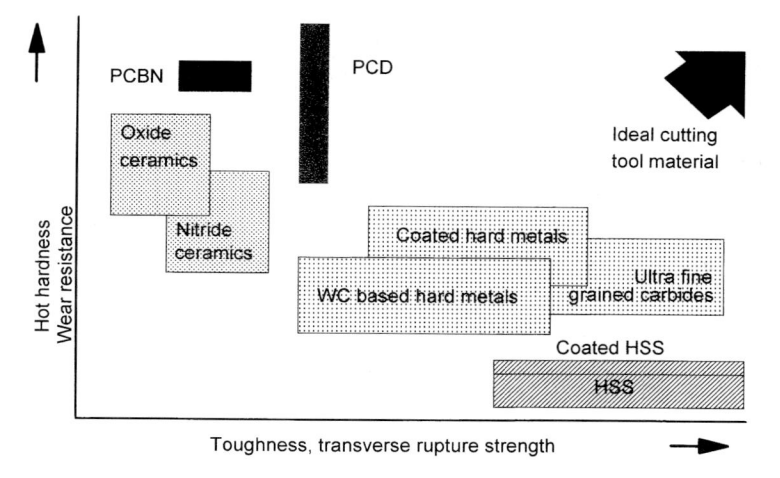

Figure 64. Properties of cutting tool materials.

and reinforced plastics. The main application areas in the automotive industry are summarized in Fig. 65.

An example of an engine component commonly machined with pcD is an aluminum alloy engine block (Fig. 66) manufactured from an silicon-silicon alloy

- **Engine components**
- **Gearbox components**
- **Transmission components**
- **Braking systems**

→ Aluminium alloys MMC's

- **Interior fittings**
- **Body panels**

→ Reinforced plastics

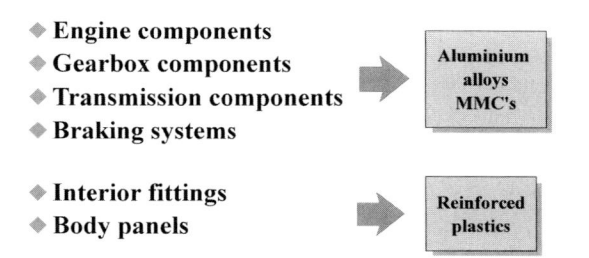

Figure 65. pcD applications in the automotive industry.

Figure 66. Rough milling Al-9% Si engine blocks with pcD.

Figure 67. Face milling Al-12% Si gearboxes with pcD.

which, dependent upon the alloy, can be very abrasive. Similar classes of material are used in the manufacture of gearbox casings and these can also be machined more efficiently with pcD than with any other cutting tool materials (Fig. 67). The cylinder bores of aluminum-silicon alloy engine blocks are also typically machined with pcD. Figure 68 shows such a component from a V-12 engine being finish bored with pcD.

Figure 68. Finish boring of cylinder blocks with pcD.

Figure 69. SYNDITE 025 machining 20% SiC reinforced Al MMC brake discs.

A material originally used in racing cars and motorcycles is a metal matrix composite (MMC) made of aluminum reinforced with silicon carbide particles. Its light weight and high strength made it ideal for use as a disc brake rotor and also other components. The presence of the silicon carbide makes it a difficult material to machine, pcD being the best cutting tool material. Figure 69 shows a typical MMC machining operation on a brake disc rotor. Machining of reinforced plastics is also a major application for pcD tooling. Figure 70 shows part of a glass reinforced plastic (GFRP) truck body section which is machined with a pcD router.

A relatively recent application area for pcD has been in the woodworking industry. Wood composites such as medium density fibreboard (MDF), chipboard, laminated boards with plastic coatings, for use in furniture and on floorings, and also composites, used in wall claddings and in ceilings, all tend to be highly abrasive

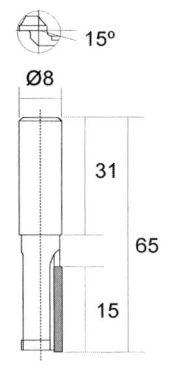

Figure 70. pcD milling cutter machining a GFRP truck bumper.

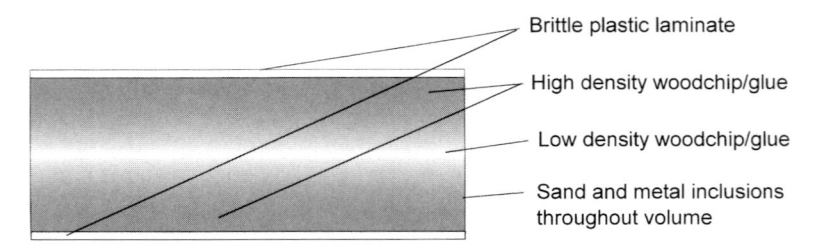

Figure 71. Schematic cross-section through typical abrasive particle board.

when being machined. Figure 71 shows a typical cross-section through a wood laminate. Saw blades, routers and other profiling tools now use pcD tips for high volume production. Figure 72 shows a profiled pcD tipped cutter being used to produce furniture panels in MDF, a commonly used material for this application and Fig. 73 shows typical wear rate results obtained when machining such materials and compares high speed steel, cemented tungsten carbide and pcD in terms of tool wear rate as a function of distance machined.

1.5.4.3 Drilling with Polycrystalline Diamond

The major advantage of using pcD in a rock drilling application is that pcD drill bits are designed to remove rock by shearing rather than by crushing or grinding, as is the case with most other types of drill. Rock removal by shearing is very efficient, requiring only 15–20% of the energy required by crushing and grinding. This, in

Figure 72. pcD is typically used on medium density fibreboard (MDF), chipboard, and abrasive hardwoods.

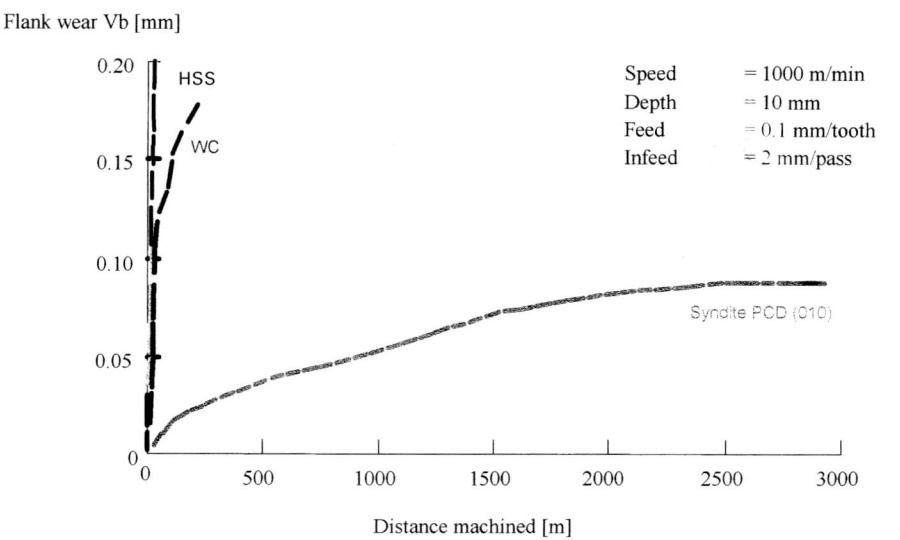

Figure 73. Wear rates of HSS (high speed steel), WC (tungsten carbide) and pcD routers machining chipboard.

turn, means that pcD drill bits have the potential to drill much faster than conventional bits and, in addition, the high abrasion resistance of pcD results in the cutters remaining sharp and hence extending considerably the life of the drill bit.

Drilling for oil and gas is a highly specialized subject: pcD cutters were introduced to this application area in the early 1980s and, since that time, the design of the drill bits and the pcD cutting elements themselves have resulted in ever-increasing performance.

A factor which limited the early pcD oil and gas exploration bits was premature failure, either of the tungsten carbide supporting the pcD or of the braze attaching the cutter to the bit body. Today, most of these difficulties have been overcome, and pcD bits offer high penetration rates and long bit lives when drilling medium hard abrasive rock formations.

pcD drill bits are also used in mining, although the volume is considerably less than that in the oil and gas exploration industry. One example is in methane drainage, where holes are drilled into the rock surrounding the coal seam and the methane pumped out to prevent it seeping into the mining area and hence causing a fire hazard. Also, in underground mining, holes have to be drilled in large quantities for roof bolts to be inserted to provide roof support, and blast holes have to be drilled into which explosives are packed. Figure 74 shows a selection of typical pcD mining bits.

1.5.4.4 Non-cutting Application of Polycrystalline Diamond

pcD, being hard and abrasion resistant, is potentially ideal for use as a wear resistant material. One traditional non-cutting application is in wire drawing, where pcD is

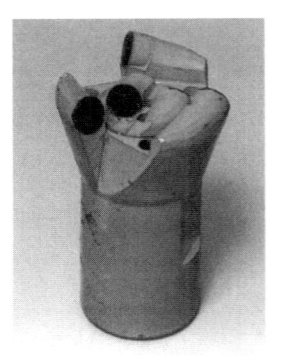

**2.5 inch PCD
3-wing
blast hole bit**

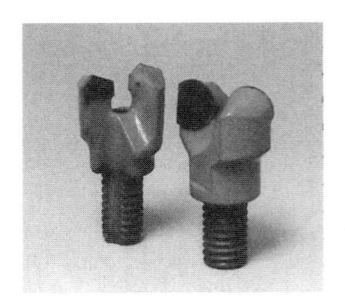

**1 inch PCD
2-wing
roof bolt bits
with different
cutter designs**

Figure 74. Selection of mining bits.

used as a die to reduce fine wire (Fig. 75). pcD wire drawing dies are most widely used to draw non-ferrous wires such as copper where the life, compared to cemented tungsten carbide, is up to 1000 times longer.

A more recent application field has been the use of pcD as a wear part, such as a bearing, a workpiece support rest in a machining operation, a nozzle for abrasive fluids and in high precision gauges. Figure 76 shows a pcD used as a workpiece support rest for an automotive engine component and Fig. 77 shows precision automatic gauging equipment fitted with pcD tipped measuring fingers. In such cases, the high wear resistance of pcD leads to improved accuracy being achieved.

Hot drawing 600-700°C
10 die string 0.49-0.168mm
Die life 50x WC
Better surface finish

Figure 75. Drawing wire with diamond dies.

Figure 76. SYNDITE pcD wear parts support a crankshaft during machining.

1.5.4.5 Machining Ferrous Materials with Polycrystalline Boron Nitride

pcBN is used almost exclusively for machining hard ferrous materials. As described earlier, cBN and pcBN do not react with iron at high temperatures in the same way as diamond, and hence can be used to machine hard ferrous metals. To machine a hard ferrous material effectively, the temperature of the workpiece in the zone immediately ahead of the contact between it and the cutting tool (the shear zone) has to be increased such that the workpiece becomes softer and hence easier to machine. Figure 78 shows the hardness/temperature curve for a typical hard ferrous material. By selecting the correct machining conditions of cutting speed and tool geometry, these relatively high temperatures can be achieved, and because pcBN retains its

Figure 77. Measuring diamond honing heads with a pcD-tipped micrometer.

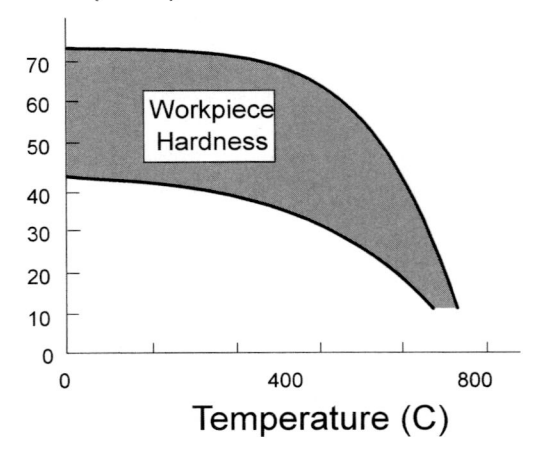

Figure 78. Workpiece material hardness as a function of temperature.

hardness at these high temperatures, efficient machining can be conducted. The high temperatures generated are limited to the cutting zone only, providing the machining conditions are selected correctly. The bulk temperature of the workpiece remains low and coolant is only required in a minority of cases. pcBN tool materials fall largely into two classes:

– high cBN content (>90% by volume cBN) in a ceramic binder
– low cBN content (<60% by volume cBN) in a ceramic binder

High cBN content materials are used in rough machining applications, whereas lower cBN content tools tend to perform better in finishing applications where shallow depths of cut are being taken and a good surface finish is required.

Abrasion resistant cast-irons, for example Ni-hard, are used to manufacture pumps for pumping slurry and gravel in the quarrying, mining and associated heavy industries. These materials are difficult to machine with conventional tooling materials but can be machined easily with a high cBN content tool material such an Amborite. Steel rolling mill-rolls, which are normally ground to final size, can also be machined with pcBN, the advantage being higher production rates and easier swarf disposal since most rough machining operations using pcBN are conducted dry, without coolant (Fig. 79). pcBN can also be used in milling, and Fig. 80 shows an example of a milling operation on a machine tool slideway.

In the automotive industry, high content pcBN is used to machine a wide range of cast iron components such as brake discs (Fig. 81) and cast iron cylinder blocks (Fig. 82).

Using low-content pcBN tools for finishing previous operations often replaces a grinding operation. In many cases, a hardened steel component was finish ground because its hardness meant that it was not possible to machine it in any other way. Many such components may now be turned to final tolerance and surface

**Ni-HARD Pump Bodies (left)
High Chromium Iron Rolls (right)**

Figure 79. Typical applications of high cBN-content pcBN tools.

finish using suitable pcBN tooling. Figure 83 shows a gear component being finished in this way.

Low content pcBN can also be used in fine milling. Figure 84 shows an automotive gearbox component being finish milled with a low content pcBN (DBC50).

1.5.5 Applications of Single Crystal Diamond

Natural diamond crystals were the original diamond tools used hundreds of years ago for cutting and engraving purposes, because the outstanding properties of

Figure 80. AMBORITE milling hardened meehanite slideways.

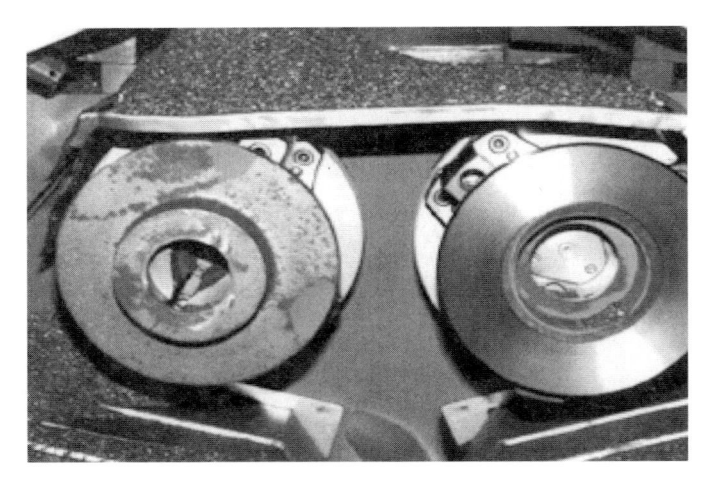

Figure 81. Machining cast iron brake discs with AMBORITE.

hardness and abrasion resistance were recognized, but it was only possible to use relatively large pieces of diamond, several millimeters in size. More recently, other outstanding properties have been recognized, in particular the high degree of optical transparency of type II diamonds and the very high thermal conductivity of (the relatively rare) type IIA diamond. In recent years, the technology to synthesize

Figure 82. AMBORITE machining a six-cylinder cast iron engine block.

Figure 83. Finish machining a case-hardened automotive gear with DBC50.

economically synthetic single crystal diamond of several millimeters in size has been developed and this has contributed to the availability of relatively large diamonds for industrial use.

Single diamond crystals are currently used in a wide variety of industrial applications, albeit some highly specialized, and these range from rock drilling to sophisticated applications in micro-surgery and electronics.

One advantage of synthetic crystals is that they can be synthesized and processed to specific required sizes and also crystallographic orientations chosen according to the application (Fig. 85).

Figure 84. DBC50 finish milling a transmission component (60HRc).

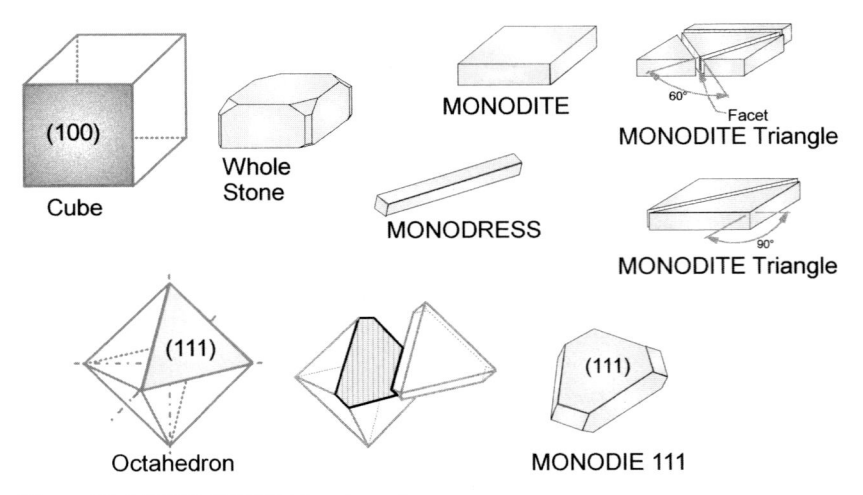

Figure 85. MONOCRYSTAL orientation.

1.5.5.1 Diamond Truing and Dressing Tools

Diamond tools are employed for truing and dressing conventional abrasive wheels, like aluminum oxide and silicon carbide wheels. Truing ensures that newly installed wheels run concentrically on the machine spindle, whilst dressing restores the desired surface topography and texture to grinding wheels that have become dull, glazed, or loaded with grinding debris. The standard types of truing and dressing tools available include single-point, multi-point, chisel, rotary, and indexing tools (Fig. 86). This wide variety caters for all types of conventional abrasive wheel and all types of truing and dressing operations, such as those required to produce both straight and profiled wheel faces, those that have steps, radii and shoulders, and for plunge forming etc. Rotary dressers are used for truing and dressing vitrified-bond cubic boron nitride wheels. Custom-made tools are also produced for specialized applications.

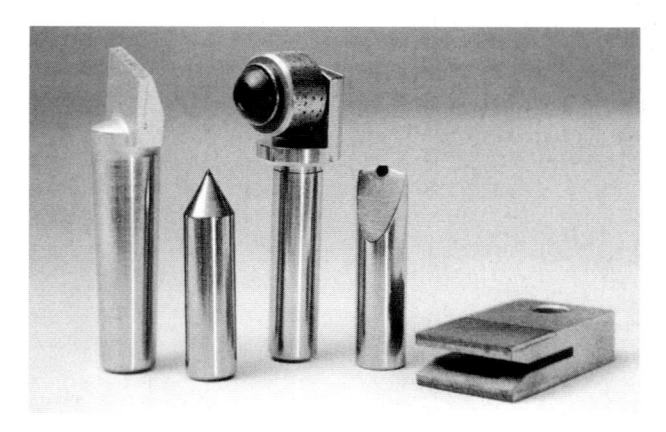

Figure 86. Selection of diamond dressers.

1.5.5.2 Diamond Cutting Tools

Diamond cutting-tool applications are many and varied, ranging from rock drilling to the finish machining of nonferrous metals and plastics. In rock drilling with surface-set diamond drill bits, the structural strength, fracture resistance, hardness and wear resistance of diamond are most in evidence because rock is removed by crushing rather than shearing.

Its crystal structure and extreme hardness also make it possible for diamond to accept and retain cutting edges which are flawless and unbroken over their entire length when viewed at high magnification.

Diamond tools are used for turning, milling and engraving non-ferrous metals, including precious metals, non-metallic materials, abrasive composites etc. (Fig. 87). Typical of applications for which diamond has been successfully used are the turning of aluminum alloy photocopier cylinders and computer memory discs, copper printing rolls, non-ferrous metal optical components and plastic contact lenses, the fly-cutting of multi-facetted scanner mirrors, the production of precious metal jewellery and the restoration of crazed plastic aircraft windows to their original transparency. In such operations, diamond tools give reduced downtime and improved quality, thus making it possible to achieve greater economy in comparison with other types of cutting tool. In many cases, the superior finish obtained with diamond machining means that no subsequent polishing or buffing operations are required, thus improving the economics even further.

Figure 87. Typical diamond cutting tools.

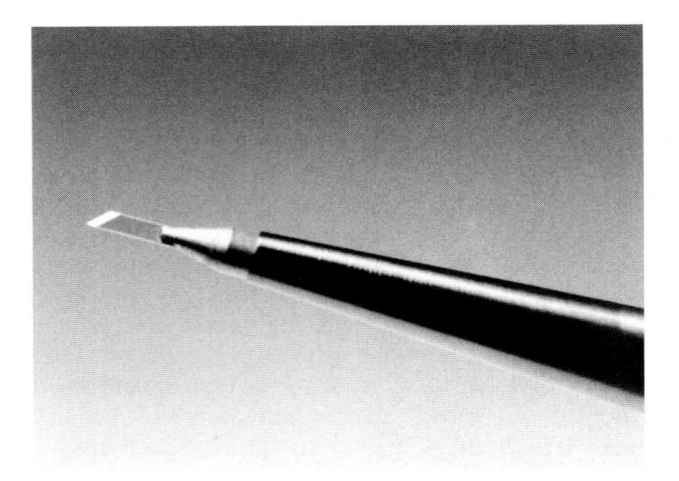

Figure 88. Surgical scalpel for micro-surgery.

In the medical field, diamond blades are used in surgical scalpels (Fig. 88) and in microtome knives (Fig. 89) to prepare tissue samples for microscopic examination. Diamond knives are also used for cleaving fibre optics.

1.5.5.3 Non-Cutting Applications of Diamond Tools

In comparison with the use of diamond as a cutting tool material, the consumption of diamond for non-cutting applications is small. Nevertheless, high-quality-single crystal industrial diamond is often the only suitable material for specific tasks, due to its outstanding mechanical, optical, electrical, chemical or thermal properties.

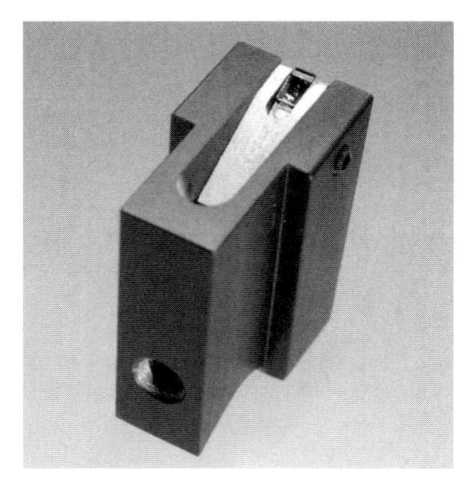

Figure 89. Diamond microtome knife.

Figure 90. Ultra-high pressure diamond anvils.

The types of equipment in which diamond crystals are used include anvils for high-pressure research (in excess of a million atmospheres or Megabar) (Fig. 90), heatsinks, bearings for precision chronometers and other ultra-sensitive electronic meters, styli for audio equipment and surface measuring instruments, hardness-testing indenters (Fig. 91) and distance stops on machine tools.

In addition, diamond is used for metal-forming applications, such as burnishing, engine turning, and the ruling of diffraction gratings, where little or no material is actually removed. The most important of such forming operations is wire drawing. In top-quality wire-drawing dies, the ultimate wear resistance and surface polish rely on the particular qualities of diamond crystals. Such dies enable large quantities of

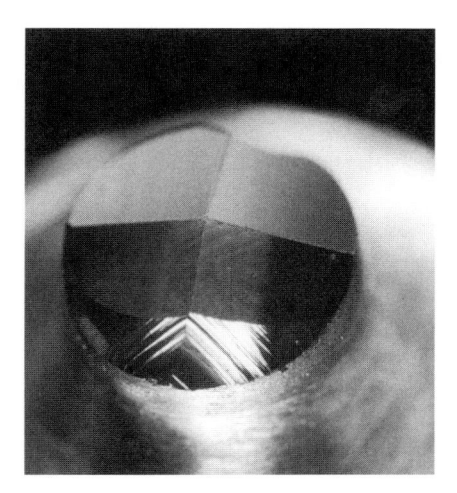

Figure 91. Diamond hardness indentor.

Figure 92. Diamond wire drawing dies.

very fine wire (down to 10 μm in diameter) to be produced with the required precision (Fig. 92).

Acknowledgments

The authors are indebted to Dr I. Sigalas and Dr R. J. Caveney for the overall editing of the article, and to Mrs Chancellor Teffo for her very competent secretarial help.

References

1. S. Tennant, *Phil Trans. R. Soc.* 1797, **97**, 591.
2. H. Liander and E. Lundblad, *Arkiv Kemi*, 1969, **16**, 139–149.
3. F. P. Bundy, H. T. Hall, H. M. Strong, and R. H. Wentorf, *Nature*, 1955, **176** (No. 4471), 51–55.
4. F. P. Bundy, *Nature* (London), 1973, **241**, 116–118.
5. G. Davies, *Diamond*, Adam Hilger, Bristol, 1984, Chapter 4.
6. H. M. Strong, *Am. J. Phys.* 1989, **57**, 794–802.
7. J. W. Harris, Diamond geology, in *The Properties of Natural and Synthetic Diamond*, J. E. Field (Ed.), Academic Press, London, 1992, pp. 342–393.
8. S. E. Kesson and A. E. Ringwood, *Chemical Geology*, 1989, **78**, 97–118.

9. F. P. Bundy, W. A. Bassett, M. S. Weather, R. J. Hemley, H. K. Mao, and A. F. Goncharov, *Carbon*, 1996, **34**, 141–153.
10. G. J. Davies and T. Evans, *Proc. Ro. Soc. Lond.*, 1972, **A328**, 413–427.
11. M. L. H. Green, M. J. Rosseinsky, and E. S. C. Tsang, *2nd International Interdisciplinary Colloquium on the Science and Technology of the Fullerenes*, Keble College and University of Oxford Museum, Oxford, Abstracts, Elsevier Science, Oxford, 1996, p. 186.
12. R. Roy, in *Advances in New Diamond Science and Technology*, S. Saito (eds), MYU, Tokyo, 1994, pp. 17–22.
13. P. Mistry, M. C. Turchan, S. Liu, G. O. Granse, T. Baurmann, and M. G. Shara, *Innovations in Materials Research*, 1996, **1**(2), 193–207.
14. R. Berman and F. Simon, *Zeitschrift fur Elektrochemie*, 1955, **59**, 333–337.
15. C. S. Kennedy and G. C. Kennedy, *J. Geophys. Res.*, 1976, **81**(14), 2467–2469.
16. R. C. Burns and G. J. Davies, Chapter 10 in *The Properties of Natural and Synthetic Diamond*, J. E. Field (Ed.), Academic Press, London, 1992.
17. H. Kanda, M. Akaishi, and S. Yamaoka, *Appl. Phys. Lett.* 1994, **65**, 784.
18. S. K. Singhal and H. Kanda, *J. Cryst. Growth*, 1995, **154**, 297.
19. N. V. Novikov and A. A. Shul'zhenko, in *Science and Technology of New Diamond*, S. Saito, O. Kukunaga, and M. Yoshikawa (Eds), KTK Sci, Tokyo, 1990, p. 339.
20. A. V. Andreyev and H. Kanda, *Diamond Relat. Mater.*, 1997, **6**, 28–32.
21. H. O. A. Meyer, *Inclusions in Diamond in Mantle Xenoliths*, P. H. Nixon (Ed.), John Wiley and Sons, 1987, pp. 501–522.
22. M. Akaishi, H. Kanda, and S. Yamaoka, *J. Cryst. Growth*, 1990, **104**, 578.
23. M. Akaishi, H. Kanda, and S. Yamaoka, *Jap. Appl. Phys.*, 1990, **29**, L1172.
24. M. Akaishi, *Diamond Relat. Mater.*, 1993, **2**, 183–189.
25. L. Sun, M. Akaishi, S. Yamaoka, H. Yamaoka, and S. Nakano, Special Issue of The Review of High Pressure Science and Technology, *AIRAPT-16* 1997, **6**, 36.
26. M. Arima, K. Makayama, M. Akaishi, S. Yamaoka, and H. Kanda, *Geology*, 1993, **21**, 968.
27. M. Akaishi, *Proceedings of 3rd NIRIM International Symposium on Advanced Materials*, Tusukuba, Japan, Mar 4–8, 1996, 75.
28. H. Yamada, M. Akaishi, and S. Yamaoka, Special Issue of The Review of High Pressure Science and Technology, *AI RAPT-16*, 1997, **6**, 35.
29. S. E. Haggerty, *Nature*, 1986, **320**, 34–37.
30. A. Onodera, K. Terashima, T. Urushihara, K. Suito, H. Sumiya, and S. Satoh, *J. Mater. Sci.*, 1997, **32**, 4309–4318.
31. A. Barti, S. Bohr, R. Haubner, and B. Lux, *Int. J. Refract. Metal Hard Mater.*, 1996, **14**, 145–147.
32. G. R. Cowan, B. W. Dunnington, and A. H. Holtzman, U. S. Patent, USP34010119, Du Pont de Nemours USA, 1965.
33. R. H. Wentorf Jr., *J. Phys. Chem.*, 1965, **69**, 3063–3069.
34. Y. Ma et al., *Appl. Phys. Lett.*, 1994, **65**, 822.
35. M. N. Rugueiro, P. Monceau, and J. L. Hodeau, *Nature*, 1992, **355**, 237–239.
36. Y. Muranaka, H. Yamashita, and H. Miyadera, *Diamond Rel. Mater.*, 1994, **3**, 313–318.
37. H. E. Spear and J. P. Dismukes (eds), Parts II and III in, *Synthetic Diamond, Emerging CVD Science and Technology*, John Wiley and Sons, 1994, pp. 21–304.
38. R. Roy, D. Ravichandran, P. Ravindranathan, and A. Badzian, *J. Mater. Res.*, 1996, **11**(5), 1164–1168.
39. P. W. Bridgman, *Phys. Rev.* 1935, **48**, 825–832.
40. L. F. Vereschagin, A. A. Semerchan, N. N. Kuzin, and Y. A. Sadkov, *Dokl. Akad. Nauk SSR*, 1968, **183**, 565–567.
41. B. von Platen, in *Modern Very High Pressure Techniques*, R. H. Wentorf (Ed.), Butterworths, Washington, DC, 1962, pp. 234–239.
42. Yu. N. Pal'yonov, Yu. I. Malinvosky, Yu. M. Borzdov, A. F. Khokhryakov, A. I. Chepurov, A., Godovikov, and N. V. Sobolev, *Doklady Akademii Nauk SSSR, Earth Science Section* 1990, **315**, 1221–1224.
43. H. Kanda and O. Fukunaga, in *Advances in Earth and Planetary Science, High Pressure Research in Geophysics, Vol 12*, S. Akimoto and M. H. Manghnani (Eds), Riedel, Dordrecht, 1982, pp. 525–535.

44. Yu. N Pal'yonov, A. F. Khokhryakov, Yu. M. Borzdov, A. G. Sokol, V. A. Gusev, G. M. Rydov, and N. V. Sobolev, *Geol. Geofiz*, 1997, **38**, 882–906.
45. H. M. Strong and R. E. Hanneman, *J. Chem. Phys.*, 1967, **46**, 3668–3676.
46. G. Will and G. Graf, *High Press. Res.*, 1994, **12**, 17–27.
47. I. Sunagawa, Morphology of natural and synthetic diamond crystals, in *Material Science of the Earth's Interior*, I. Sunagawa (Ed.), Tokyo Terra Scientific Publishing Co, 1984, pp. 303–330.
48. Y. Wang, R. Tanakabe, and M. Wakatsuki, *Proc. 15th AIRAPT Int Conf. on High Pressure Science and Technology*, Warsaw, Sept 11–15, 1995, pp. 235–237.
49. M. Wakatsuki and K. Takaro, K. lnone, Inclusions in synthesised crystals of diamond, in *8th AIRAPT Conf. Proc., High Pressure in Research and Industry*, Uppsala, pp. 369–372.
50. T. L. McCormick, W. E. Jackson, and R. J. Nemanich, *J. Mater. Res.*, 1997, **12**(1), 253–263.
51. G. Davies, *Chemistry and Physics of Carbon*, Vol. 11, P. L. Walker (Ed.), Marcel Dekker, New York, 1982, pp. 70–71.
52. R. H. Wentorf, *J. Phys. Chem.* 1971, **75**, 1833–1837.
53. H. M. Strong and R. E. Tuft, US Pat. 4073380, 1978.
54. Y. Suji, *et al.*, Jpn. Pat 2631/1986, 165–167.
55. F. P. Bundy, H. M. Strong, and R. H. Wentorf Jr, *Chemistry of Physics of Carbon, Vol. 10*, 1973, 213–272.
56. M. Wakatsuki, in *Materials Science of the Earth's Interior* I. Sunagawa (Ed.), Terra Scientific Publishing Company, Tokyo, 1984, pp. 351–374.
57. Yu. A Kocherzhinskii and O. G. Kulik, *Powder Metall Metal Ceramics*, 1996, **35**(7–8), 470–483.
58. H. Kanda, in *Advances in New Diamond Science and Technology*, S. Saito, N. Fujimori, O. Fukunaga, M. Kamo, K. Kobashi, and M. Yoshikawa (Eds), MYU, Tokyo, 1994, pp. 507–512.
59. M. Wakatsuki, *Jpn J. Appl. Phys.* 1966, **5**, 337.
60. T. Sugano, N. Ohashi, T. Tsurumi, and O. Fukunaga, *Diamond Relat. Mater.* 1996, **5**, 29–33.
61. H. M. Strong and R. M. Chrenko, *J. Phys. Chem.* 1971, **75**(12), 1838–1843.
62. R. C. Burns, S. Kessier, M. Sibanda, C. M. Welbourn, and D. L. Welch, *Proceedings of the 3rd NIRIM International Symposium on Advanced Materials*, 1996, pp. 105–111.
63. H. Kanda, T. Ohsawa, O. Fukunaga, and I. Sunagawa, *J. Cryst. Outgrowth*, 1989, **94**. 115–124.
64. P. Cannonand and E. T. Conlin, *J. Am. Chem. Soc.* 1964, **86**, 4540–4544.
65. R. C. Burns, J. O. Hansen, M. Sibanda, R. A. Spits, C. M. Welbourn, and D. L. Welch, *Diamond Rel. Mater.*, 1999, **8**, 1433–1437.
66. H. Kanda and S. C. Lawson, *Indust. Diamond Rev.* 1995, **2**, 56–61.
67. H. Sumiya and S. Satoh, *Diamond Rel. Mater.*, 1996, **5**, 1359–1365.
68. W. Li, H. Kagi, and M. Wakatsuki, *Trans. Mater. Res. Soc. Jpn.*, 1994, **14B**, 1451–1454.
69. M. Wakatsuki, W. Li, Y. Gohda, and L. Y. Ding, *Diamond Rel. Mater.*, 1996, **5**, 56–64.
70. S. Yazu, *J. Soc. Mater. Sci. Jpn.*, 1993, **42**(476), 588–592.
71. H. Kanda, *New Diamond*, 1990, 58–62.
72. R. A. Chapman, Private Communication, 1996, De Beers Diamond Research Laboratory, South Africa.
73. S. Satoh and H. Sumiya, *Jpn. Mater. Res.*, 1995, **15**, 183–195.
74. R. C. Burns, V. Cvetkovic, C. N. Dodge, D. J. F. Evans, M. L. T. Rooney, and C. M. Welbourn, *J. Cryst. Growth*, 1990, **104**, 257–279.
75. S. Satoh, H. Sumiya, K. Tsuji, and S. Yazu, in *Science and Technology of New Diamond*, S. Saito, O. Fukunaga, and M. Yoshikawa (Eds), KTK Scientific Publishers/Terra Scientific Publishing Co., Tokyo, 1990, pp. 351–355.
76. H. Kanda, T. Ohsawa, and O. Fukunaga, in *Abstracts 2nd Meetinq of Diamond*, Tokyo, 1987 pp. 23–24.
77. M. L. T. Rooney, *J. Cryst. Growth*, 1992, **116**, 15–21.
78. A. T. Collins, H. Kanda, and R. C. Burns, *Phil. Mag. B*, 1990, **61**(5), 797–810.
79. S. L. Lawson, H. Kanda, K. Watanabe, I. Kiflawi, and Y. Sato, *J. Appl. Phys.*, 1996. **79**(8), 1–10.
80. J. I. Koivula and C. W. Fryer, *Gems and Gemology*, 1984, **20**, 146–158.

81. J. E. Shigley, E. Fritsch, C. M. Stockton, J. I. Koivula, C. W. Fryer, and R. E. Kane, *Gems and Gemology*, 1986, **22**, 192–208.
82. J. E. Shigley, E. Fritsch, C. M. Stockton, J. I. Koivula, C. W. Fryer, R. E. Kane, D. R. Hargett, and C. W. Welch, *Gems and Gemology*, 1987, **23**, 187–206.
83. M. L. T. Rooney, C. M. Welbourn, J. E. Shigley, E. Fritsch, and I. Reinitz, *Gems and Gemology*, 1993, **29**, 38–45.
84a. J. E. Shigley and E. Fritsch, *J. Cryst. Growth*, 1993, **128**, 425–428.
84b. J. E. Shigley, E. Fritsch, I. Reinitz, and T. M. Moses, *Gems and Gemology*, 1995, **31**, 256–264.
85. C. M. Welbourn, M. Cooper, and P. M. Spear, *Gems and Gemology*, 1996, **32**(3), 3 156–169.
86. M. Kamo, Y. Sato, S. Matsumoto, and N. Setaka, *J. Cryst. Growth*, 1983, **62**(3), 642–644.
87. J. C. Angus, H. A. Will, and H. S. Stanko, *J. Appl. Phys.*, 1968, **39**, 2915.
88. W. G. Eversole, US Patent 3, 030, 187 and US Patent 3, 030, 188 1962.
89. K. V. Ravi, *Mater. Sci. Eng.*, 1993 **B19**, 203–227.
90. R. Berman, *Solid State Commun.*, 1996, **99**(1), 35–37.
91. K. T. Jacob, *Solid State Commun.*, 1995, **94**(9), 763–765.
92. C. J. Brannon and S. L. McCollum, in *New Diamond Science and Technology*, R. Messier *et al.* (Eds), Materials Research Society, Pittsburgh, PA, USA, 1991, p. 117.
93. A. Tsuzuki, S.-I. Hirano, and S. Naka, *J. Mater. Sci.*, 1985, **20**, 2260–2264.
94. H. Uchikawa, H. Hagiwara, and K Nakamura, in *Science and Technology of New Diamond*, S. Saito *et al.* (Eds), Terra Scientific Publishing Co., 1990, 227.
95. Hawtin, J. B. Lewis, N. Moul, and R. H. Phillips, *Roy Soc. Phil. Trans. Ser A.*, 1966, **261**(116), 67–95.
96. L. F. Vereschagin, E. N. Yakoviev, L. M. Buchnev, and B. K. Dymov, *Teplofizika Vysokikh Temperatur*, 1977, **15**(2), 316–321.
97. Y. Bar-Yam and T. D. Moustakas, *Nature*, 1989, **32**, 786–787.
98. A. A. Abramzon and A. A. Novozhenets, *Russ. J. Phys. Chem.*, 1992, **66**(8), 1220–1222.
99. J. Abrahamson, *Carbon*, 1973, **11**, 337–362.
100. J. O. Hansen, R. G. Copperthwaite, T. E. Derry, and J. M. Pratt, *J. Colloid Interface Sci.*, 1989, **130**(2), 347–358.
101. R. A. Munson, *Carbon*, 1967, **5**, 471–474.
102. J. A. Nuth, *Astrophys. Space Sci.*, 1987, **139**(1), 103–109.
103. P. Badziag, W. S. Verwoerd, W. P. Ellis, and N. R. Greiner, *Nature*, 1990, **343**(6255), 244–245.
104. E. Anders and E. Zimmer, *M eteoritics*, 1993, **28**, 490–514.
105. A. N. Nesterov, Y. I. Merezhko, and V. V. Chernikov, *Russ. J. Phys. Chem.*, 1983, **57**(2), 307–308.
106. D. V. Fedoseev, *Colloid J. USSR*, 1978, **40**(2), 341–342.
107. D. V. Fedoseev, B. V. Deryagin I. G. Varshavskaya, and A. V. Laverentev, *Prog. Surf. Sci.*, 1994, **45**, 84–87.
108. F. Banhart and P. M. Ajayan, *Nature*, 1996, **382**, 433–435.
109. B. Wei, J. Zhang, J. Liang W. Liu Z Gao, and D. Wu, *J. Mater. Sci. Lett.*, 1997, **16**, 402–403.
110. H. Moissan, *C. R. Acad. Sci. Paris*, 1894, **118**, 320–325.
111. J. B. Hannay, *Nature*, 1880, **22**, 255–256.
112. Y. G. Gogotsi K. G. Nickel, and P. Kofstad, *J. Mater. Chem.* 1995, **5**(12), 2313–2314.
113. M. A. Cappeli, US Patent 5,417,953, 1995.
114. J-Y Wang, Y-Z Wan, D-W Zhang, and Z-J Liu, *J. Mater. Res.*, 1997, **12**(12), 3250–3253.
115. F. Davanloo, E. M. Juengerman, D. R. Jander, T. J. Lee, and C. B Collins, *J. Mater. Res.*, 1990, **5**(11), 2398–2404.
116. R. Roy, H. S. Dewan, and P. Ravindranathan, *Mater. Res. Bull.*, . 1993, **28**, 861–866.
117. R. Roy, K. A. Cherian, J. P Cheng, A. Badzian, C. Langlade, H. Dewan, and W. Drawl, *lnnov. Mater. Res.*, 1996, **1**(1), 165–87.
118. P. A. Molian and A. Waschek, *J. Mater. Sci.*, 1993, **28**, 1773.
119. M. Horio, A. Saito, K. Unou, H. Nakazono, N. Shibuya, S. Shima, and A. Kosaka, *Chem. Eng. Sci.*, 1996, **51**(11), 3033–3038.
120. Y. Namba, *J. Vac. Sci. Technol. A.*, 1992, **10**(5), 3368–3370.
121. M. Takaya, Japanese Patent Hei 6-25896, 1994.
122. T. Suzuki, T.]shihara, T. Yamazaki, and S. Wada, *Jpn J. Appl. Phys.*, 1997, **36** part 2, No. 4B, L504–L506.

123. J. F. Prins, *Diamond Rel. Mater.*, 1993, **2**, 646–655.
124. K. U. Cherian, M. Komath, S. K. Kulkarni, and A. Ray, *Diamond Rel. Mater.*, 1994, **4**, 20–25.
125. A. Jayaraman, *Sci. Am.*, 1984, **250**, 42–54.
126. Y. A. Litvin and L. T. Chudinovskikh, *Trans. Russian Acad. Earth Sci.*, 1997, **355A**(6), 908–911.
127. F. P. Bundy and J. S. Kaspers, *J. Chem. Phys.*, 1967, **46**(9), 3437–3446.
128. R. Boehier, Private communication, Jun 1994, Max Planck Inst, Mainz Germany.
129. J-A Xu and E. Huang, *Rev. Sci. Instrum.*, 1994, **65**(1), 204–207.
130. J. Kleiman, R. B. Heimann, D. Hawken, and N. M. Salansky, *J. Appl. Phys. (USA)*, 1984, **56**(5), 5 1440–1454.
131. O. Bergmann and N. F. Bailey, in *High Pressure Explosive Processing of Ceramics*, R. A. Graham and A. B. Sawaoka (Eds), Trans Tech duPont Wilmington, USA, 1987.
132. S. Fahy, S. G. Louie, and M. L. Cohen, *Phys. Rev. B*, 1986, **34**(2), 191–199.
133. R. H. Wentorf, Jr., *J. Chem. Phys.*, 1957, **4**, 956.
134. R. H. Wentorf, Jr., *New Diamond Sci. Technol.*, 1990, 1029–1037.
135. V. L. Solozhenko, *J. Hard Materials*, 1995, **6**, 51–65.
136. N. V. Novikov and V. L. Solozhenko, *J. Chem. Vap. Dep.*, 1996, **4**, 240–252.
137. R. H. Wentorf, Jr., *J. Chem. Phys.*, 1961, **34**(3), 809–812.
138. V. L. Solozhenko, *Adv. Mater. '96: Proceedings of the Third NIRM International Symposium*, Tsukuba, Japan, 1996, pp. 119–124.
139. F. P. Bundy and R. H. Wentorf, *J. Chem. Phys.*, 1963, **38**, 1144–1149.
140. Y. Moriyoshi, M. Kamo, N. Setaka, and Y. Sato, *J. Mater. Sci.*, 1983, **18**(1), 217–224.
141. H. T. Hall, *Rev. Sci. Instrum.*, 1958, **29**(4), 267.
142. H. T. Hall, *Science*, 1970, **169**, 868–869.
143. H. D. Stromberg and D. R. Stephens, *J. Am. Ceram. Soc.*, 1970, **49**(12), 1030–1032.
144. H. Katzman and W. F. Libby, *Science*, 1971, **172**, June 1971, 1132–34.
145. R. H. Wentorf, W. A. Rocco, Patent SAP 7315038, 1973.
146. L. F. Veraschagin, A. A. Semerchan, Patent UKP 1382080, 1975.
147. A. I. Kolomiitsev and V. E. Smirnov, *Almazyi Sverkhtverdye Materialy*, 1980, **7**, 4–5.
148. K. Uehara and S. Yamaya, *Science and Technology of New Diamond*, S. Saito (Ed.), Tokyo, KTK Sci Publ., 1990, pp. 203–209.
149. A. A Shulzhenko, V. G. Gargin, V. A. Shishkin, and A. A. Bochechka, Ukrainian Academy of Sciences, Institute for Superhard Materials, printed in *Kiev Naukova Dunko*, 1989, Chapter 4, 93–173.
150. Yu A. Kocherzhinskii, A. A. Shulzhenko, and V. A. Shishkin, *Vliyanie Vysok Davieniya Na Structuru I Svoistva Materialov*, 1983, 34–40.
151. O. A. Voronov and A. A. Kaurov, *Soviet J. of Superhard Materials*, 1994, **16**(1), 6–10.
152. S. H Robertson, PhD Thesis, University of Reading, 1984.
153. D. V. Fedoseev, G. A. Sokolina, and E. N. Yakoviev, *Sov. Phys. J.*, 1985, **30**(5), 425–427.
154. D. V. Fedoseev, G. A. Sokolina, E. N. Yakoviev, and A. V. Lavrentev, *Sov. Powder Metall. Ceram. Mater.*, 1984, **23**(9), 683–686.
155. A. A. Shulzhenko, S. A. Bozhko, I. A. Ignatusha, and A. N. Vashchenko, *Sov. J. Superhard Mater.*, 1990, **12**(5), 14–19.
156. A. A. Shulzhenko, S. A. Bozhko, A. I. Ignatusha, and A. N. Vashchenko, *Sov. J. Superhard Mater.*, 1988, **10**(5), 13–17.
157. S. A. Bozhko, A. I. Ignatusha, V. G. Delevi, and A. A. Budyak, *Sov. J. Superhard Mater.*, 1989, **11**(5), 32–35.
158. M. Akaishi, S. Yamaoka, J. Tanaka, T. Ohsawa, and O. Fukunaga, *J. Am. Ceram. Soc.*, 1987, **70**(10), C237-C239.
159. J. C. Walmsley and A. R. Lang, *J. Mater. Sci.*, 1988, **23**(5), 1829–1834.
160. R. H. Wentorf, Jr. and W. A. Rocco, U. S. Patent 3 767 371, 1973.
161. Sumitomo Electric Industries, Ltd., US Patent 4 334 063, 1982.
162. A. A. Shulzhenko, S. A. Bozhko, A. N. Sokolov, A. Petrusha, N. P. Bezhenar, A. I. Ignatusha, *Cubic Boron Nitride Synthesis, Sintering and Properties*, The Ukrainian Academy of Sciences, Institute for Superhard Materials, Kiev Naukova Dumka, 1993, 254.
163. P. N. Tomlinson, R. J. Wedlake, *Ind. Diamantenrundschau*, 1983, **7**(4), 234–241.
164. M. Wakatsuki, K. Ichinose, and T. Aoki, *Mater. Res. Bull.*, 1972, **7**, 999–1004.

165. M. Wakatsuki and K. Ichinose, *4th International Conference on High Pressure, Kyoto*, 1974, 441–445.
166. F. R. Corrigan, *6th AIRAPT Conference on High Pressure Science and Technology,*' 1979, **1**, 994–999.
167. O. Fukunaga and M. Akaishi, *High Press. Res.*, 1990, **5**, 911–913.
168. J. M. Leger, J. Haines, and B. Blanzat, *J. Mater. Sci. Lett.*, 1994, **13**, 1688–1690.
169. M. C. Payne, M. T. Teter, D. C. Alan, T. A. Arias, and J. D. Ioannopoulos, *Rev. Mod. Phys.*, 1992, **64**, 1045.
170. A. Y. Liu and M. L. Cohen, *Science*, 1989, **245**, 1981.
171. M. L. Cohen, *J. Hard Mater.*, 1991, **2**, 13.
172. P. K. Lou, M. L. Cohen, and G. Martinez, *Am. Phys. Soc.*, 1987, **35**, 9190–9194.
173. R. Hoffmann, T. Hughbanks, M. Kerttesz, and P. H. Bird, *J. Am. Chem. Soc.*, 1983, **105**, 4831.
174. M. A. Tamor and K. C. Hass, *J. Mater. Res.*, 1990, **5**, 2273.
175. A. Y. Liu, M. L. Cohen, K. C. Hass and M. A. Tamor, *Phys. Rev. B*, 1991, **43**, 6742.
176. F. Diedrich and Y. Rubin, *Angew. Chem. Int. Ed. Engl.*, 1992, **31**, 1101.
177. H. Alberts, T. Sekine, H. Kanda, Y. Bando, and K. Hojou, *J. Mater. Sci. Lett.*, 1990, **9**, 1376.
178. A. Y. Liu and M. L. Cohen, *Phys. Rev. B.*, 1990, **41**, 10727–10734.
179. J. V. Badding, *Adv. Mater.*, 1997, **9**, 877–886.
180. A. Y. Liu and R. M. Wentzcovitch, *Phys. Rev.*, 1994, **1350**, 10362.
181. R. M. Wentzcovitch and J. L. Martins, *Solid State Commun.*, 1991, **78**, 831.
182. R. M. Wentzcovitch, *Phys. Rev.*, 1991, **B44**, 2358.
183. D. M. Teter and R. J. Henley, *Science*, 1996, **271**, 53–55.
184. D. Li, S. Lopez, Y. W. Chung, M. S. Wong, and W. D. Sproul, *J. Vac. Sci. Technol.*, 1995, **13**, 1063.
185. T. A. Yeh, C. L. Lin, J. M. Sivertsen, and J. H. Judy, *IEEE Trans. Laga.*, 1991, **27**, 5163.
186. J. J. Cuomo, P. A. Leary, D. Yu, W. Reuter, and M. Frisch, *J. Vac. Sci. Technol.*, 1979, **16**, 299.
187. H. Sjöström, W. Lanford, B. Hjövarso, K. Xing, and J. E. Sundgren, *J. Mater. Res.*, 1996, **11**, 981.
188. C. Niu, Y. Z. Lu, and C. M. Lieber, *Science*, 1993, **261**, 334.
189. K. Ogata, J. F. D. Chubaci, and F. Fujimoto, *J. Appl. Phys.*, 1994, **76**, 3791.
190. J. Schwan, D. Dworschak, K. Jung, and H. Erhardt, *Diamond Rel. Mater.*, 1994, **3**, 1034.
191. T. Y. Yen and C. P. Chon, *Appl. Phys. Lett.*, 1995, **67**, 2801.
192. Y. Guo and W. Goddard III, *Chem. Phys. Lett.*, 1995, **237**, 72–76; J. E. Lowther, *Phys. Rev. B*, 1998, **57**, 5724–5727.
193. D. M. Bhusari, C. K. Chen, K. H. Chen, T. J. Chuang, L. C. Chen, and M. C. Liu, *J. Mater. Res.*, 1997, **12**, 322–325.
194. Z. J. Zhang, S. Fan, J. Huang, and C. M. Lieber, *Appl. Phys. Lett.*, 1996, **68**(19), 2639–2641.
195. F. Fujimoto and K. Ogata, *Jpn. J. Appl. Phys.*, 1993, **32**, L. 420.
196. D. Liu, Y. W. Chung, M. S. Wong, and W. D. Sproul, *J. Appl. Phys.*, 1993 **74**, 219.
197. Z. M. Ran, Y. C. Du, Z. F. Ying, Y. X. Qiu, X. X. Xiong, J. D. Wu, and F. M. Li, *Appl. Phys. Lett.*, 1994, **65**, 1361.
198. C. Niu, Y. Z. Lu, and C. M. Lieber, *Science*, 1993, **261**, 334.
199. S. Kumai and T. L. Tansley, *Solid State Commun.*, 1993, **88**, 803.
200. D. Norton, K. J. Boyd, A/H. Al-Bayari, S. S. Todorov, and J. W. Rabalais, *Phys. Rev. Lett.*, 1994, **73**, 118.
201. K. M. Yu, M. L. Cohen, E. E. Hailer, W. L. Hausen, A. Y. Liu, and I. C. Wu, *Phys. Rev.*, 1994, **B49**, 5034.
202. J. P. Riviera, D. Texier, J. Delafond, M. Jaouen, E. L. Mathe, and J. Chanmond, *Mater. Lett.*, 1995, **22**, 115.
203. X. W. Su, H. W. Song, F. Z. Cui, W. Z. Li, and H. D. Li, *Surf. Coat. Technol.*, 1996, **84**, 388–391.
204. Y. Zhang, H. Li, and Q. Xue, *Mater. Res. Soc.*, 1996, 317–322.
205. C. Yan, Liping Guo, and E. G. Wang, *Phil. Mag. Lett.*, 1997, **75**, 155–162.
206. K. Yanamoto, Y. Koga, K. Yase, S. Fuziwara, and M. Kubota, *Jpn. J. Appl. Phys.*, 1997, **36**, L230–L233.
207. E. G. Wang, Chen Yan, and Guo Liping, *Physica Scripta*, 1997, **T69**, 108–114.

208. M. J. Yacaman, J. J. M. Gil, and M. Sarikaya, *Mater. Chem. Phys.*, 1997, **47**, 109–117.
209. D. W. Wu, W. Fan, H. X. Guo, M. B. He, X. O. Meng, and X. J. Fan, *Solid State Commun.*, 1997, **103**, 193–196.
210. S. Xu, Han-Shi Ki, S. Lee, and Yin-Au Li, *IEEE*, 1997, 213–214.
211. Xiao-Ming He, Li Shu, Wen-Zhi Li, and Heng-De-Li, *J. Mater. Res.*, 1997, **12**, 1595–1602.
212. Wu Dawei, Fu Dejan, Guo Huaixi, Zhang Zhihong, Meng Xianquan, and Fan Xiangiun, *Phys. Rev. B*, 1997, **56**, 4949–4954.
213. Chen, L. L., D. M. Bhusari, C. Y. Yang, K. H. Chen, T. J. Chuang, M. C. Lin, C. K. Chen, and Y. F. Huang, *Thin Solid Films*, 1997, **303**, 66–75.
214. J. Badding, *Adv. Mater.*, 1997, **9**, 877.
215. K. Ploog, P. Rauh, W. Stoeger, and H. Schmidt, *J. Cryst. Growth*, 1972, **13/14**. 350.
216. K. Ploog, H. Schmidt, E. Amberger, G. Will, and K. H. Kossobutzki, *J. Less-Common Met.*, 1972, **29**, 161.
217. H. Saitoh, K. Yoshida, and W. A. Yarbrough, *J. Mater. Res.*, 1993, **8**, 8.
218. L. Vel, G. Demazeau and J. Etournea, *Mat. Sci. Eng.*, 1991, **BID**, 149.
219. A. R. Badzian, T. Niemyski, and E. Ockusmik, *Proc. 3rd Int. Conf. on CVD*, F. A. Glaski (Ed.), The Electrochemical Society, Penington, NJ, 1972, p. 747.
220. K. Montasser, S. Hattori, and S. Morita, *Thin Solid Films*, 1984, **117**, 311.
221. J. Loeffier, F. Steinbach, J. Bill, J. Mayer, and F. Aidinger, *Z. Metailkol*, 1996, **87**, 170.
222. D. Hegermann R. Riedel, W. Dresier, C. Oehr, B. Schindler, and H. Brunnen, *Chem. Vap. Depos.*, 1997, **3**, 257
223. A. Badzian, *Appl. Phys. Lett.*, 1988, **53**, 2494.
224. P. F. McMillan, W. T. Petuskey, H. Hubert, K. J. Kingma, L. A. Garvie, A. Grzechnik, and A. Chizmeshya, *Adriatico Research Conference*, Miramare, Trieste, Italy, July 1997.
225. H. Itoh, I. Maekawa, R. Yanamoto, and H. Iwahara, *Rev. High Press. Sci. Technol.*, 1998, **7**, 986–988.
226. C. Ellison-Hayashi, M. Zandi, F. J. Csillag, and Shih-Yee Kuo, US Patent 5, 135, 892, 19912.
227. C. Ellison-Hayashi, M. Zandi, D. K. Shetty, P. Kuo, R. Yeckley, and F. Csillag, US Patent 5330,937, 1994.
228. J. M. Leger, J. Haines, M. Schmidt, J. P. Petitet, A. S. Pereira, and J. A. H. da Jomada, *Nature*, 1996, **383**, 401.
229. Federation Européene des Fabricants de Produits Abrasifs (FEPA), Standard for Superabrasives Grain Sizes, 2nd Edn. September, 1997.

2 Applications of Diamond Synthesized by Chemical Vapor Deposition

R. S. Sussmann

2.1 Introduction

The ability of Chemical Vapor Deposition (CVD) technology to synthesize diamond as large plates of high and controlled purity [1–5] is enabling a host of new applications in science and technology that were hitherto hampered by the practical difficulties inherent in the use of natural diamond. The basic principles of CVD for the synthesis of polycrystalline diamond have been discussed in the chapter by Klages *et al.* [6]. The aim of this chapter is to discuss a selected range of applications to illustrate how CVD diamond is emerging as a key (and in some cases unique) engineering material.

The impact that CVD diamond is starting to have on some areas of modern technology is in some way similar to the impact that the invention of the high pressure and high temperature (HPHT) techniques for the synthesis of diamond had on the abrasive industry in the early sixties [7]. At that time the industrial implementation of HPHT technology gave a renewed impetus to an existing but relatively small diamond abrasive industry. HPHT synthesised diamonds have been the foundation of what is now a relatively mature industry in which the extreme hardness and wear resistance of diamond are exploited for numerous abrasive and cutting tool applications [8].

The potential importance of diamond in technology follows from the wide range of technically desirable thermomechanical, optical, and electronic properties exhibited by diamond [9–13]. In addition to its extreme hardness and exceptional wear resistance, diamond is the stiffest known material, has a broad transmission spectral range (from the ultraviolet to the far infrared and extending to microwave frequencies), has the highest room-temperature thermal conductivity, one of the lowest thermal expansion coefficients and is radiation hard and chemically inert to all acid and base reagents.

What makes diamond so attractive, and in many cases unique, is not only the extreme value of some of the above properties such as hardness, thermal conductivity or the Young modulus, but the combination of two or more of these properties in a high performance product, as will be discussed later in this chapter.

The enormous interest that CVD diamond has generated since the mid eighties has been based on the perception that this new diamond synthesis technology could overcome all the limitations previously encountered with the availability of large size natural diamond specimens [14, 15]. As was the case in the abrasives industry, the use of natural diamond in other fields of technology has been relatively small and restricted to niche applications in optics [16], the thermal management of small electronic devices (such as laser diodes or impatt oscillator diodes) [17, 18], and others [19, 20]. For these types of applications only the relatively rare high-purity Type IIa form of natural diamond could be used. The most abundant form of natural diamond, Type Ia, contains relatively large concentrations of nitrogen which degrade properties such as optical transmission [21], thermal conductivity [22], and electronic transport [23]. Another major difficulty in the use of natural diamond is the stringent limitation to the size of specimens that are practically available. Most natural Type IIa windows sold commercially are below 5 mm in diameter and typical heat sinks are smaller than 1×1 mm [16, 18]. The cost of natural single crystal windows of sizes in excess of 19 mm, if available, would be outside the budget range of most practical applications [16]. The largest and the most spectacular natural diamond product ever made was a window, 18.2 mm in diameter and 2.8 mm thick, used for the main pressure cell of the pioneer Venus probe launched on August 1978 [24]. Purity and size are also serious limitations in the use of HPHT synthetic diamond in these types of applications because of the high intrinsic single substitutional nitrogen concentration and the practical maximum size of typical HPHT diamond crystals [7].

It is within this perspective that efforts in recent years have been directed to the development of a CVD diamond technology that allows the routine manufacture of large plates of diamond of high and controlled purity. Some of this effort has met with considerable success [1–5, 25]. In the following sections of this chapter we will briefly review the properties of various types of CVD diamond materials that have been developed to address specific applications. We will illustrate through a few selected examples how the current ability to manufacture CVD diamond (including large area plates and three-dimensional shapes) as a reliable and robust engineering material is enabling the use of diamond in an increasing range of technically demanding applications.

2.2 Properties of Chemical Vapor Deposited Diamond

2.2.1 Material Grades

For most practical applications CVD diamond is synthesized as a polycrystalline material. Many of its properties can be expected to be strongly influenced by its grain structure, including factors such as grain size, preferred orientation, intragranular purity, microstress, etc. All of these factors depend strongly on the synthesis technique and process conditions [7] and although there are generic features which are common to all CVD diamond materials, substantial variations in properties such as tensile strength, optical transparency, thermal conductivity and others can be found in material from different origins. It is not yet possible, for instance, to define a generic optical grade of CVD diamond because optically transparent CVD diamond specimens prepared in different laboratories are known to exhibit

2.2 Properties of Chemical Vapor Deposited Diamond

Figure 1. Optical grade CVD diamond window, 100 mm diameter, 0.7 mm thick.

different values of fracture strength, intrinsic bulk absorption, and total forward scatter [26, 27].

The majority of the properties described in this section refer to the type of CVD diamond that is currently available as a commercial product sold under the trade name of DIAFILM [28]. With sufficient control of synthesis conditions, the extra degree of freedom conferred by the polycrystalline structure has been used to formulate different 'grades' of CVD diamond in which properties are optimized for specific applications. Optical grades, for instance, exhibit the highest optical transparency and can be made routinely in discs up to 100 mm in diameter. In contrast mechanical grades are dark in color but have a higher mechanical strength and can be made in plates up to 160 mm in diameter. Various thermal grades have been developed for thermal management applications which offer a compromise between good thermal conductivity and cost. Most of the applications currently pursued with CVD diamond are relatively new and in certain cases the quality of the corresponding grade is subject to continuous development in which factors such as cost and performance are gradually optimized.

Figure 1 shows an example of a 100 mm diameter, 0.7 mm thick, optical grade CVD diamond window which has been polished on both sides to an optical finish. The as-grown grain morphologies of optical (2 mm thick) and mechanical (0.5 mm thick) grade samples are shown in Fig. 2. Figure 3 shows the Raman spectrum of an optical grade CVD diamond sample measured by a Renishaw instrument with an excitation wavelength of 633 nm. The main Raman line is centered at $1332\,\mathrm{cm}^{-1}$ and the full-width-half-maximum (FWHM) line width is approximately $3.0\,\mathrm{cm}^{-1}$, which compares well with typical values of $2.8\,\mathrm{cm}^{-1}$ to $3.1\,\mathrm{cm}^{-1}$ observed in high-quality single-crystal Type IIa natural diamond specimens measured with the same instrument.

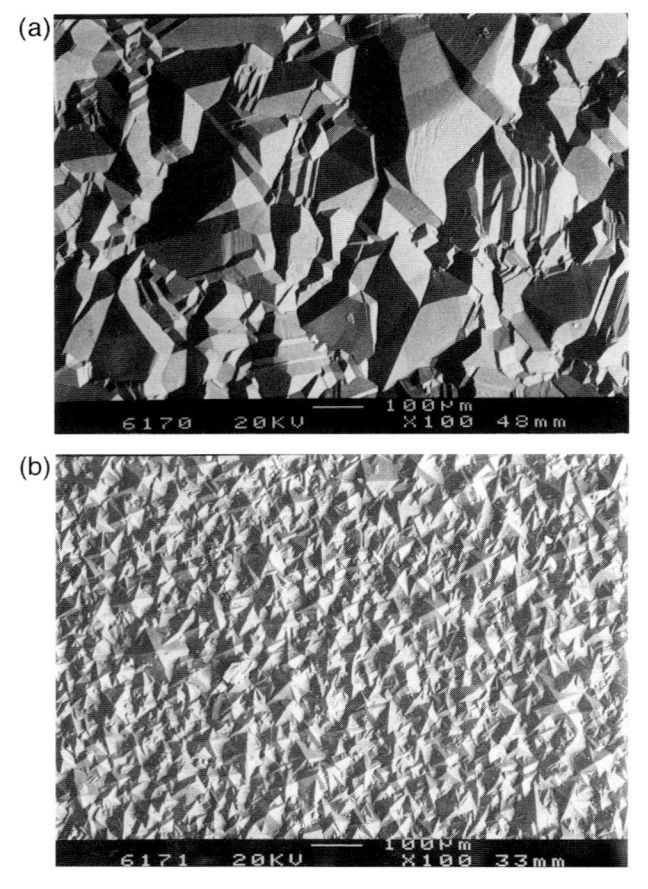

Figure 2. (a) SEM of the as-grown morphology of an optical grade DIAFILM sample 2 mm thick. (b) SEM of the as-grown morphology of a mechanical grade DIAFILM layer 0.5 mm thick.

2.2.2 Optical Properties

For a sample of optical grade CVD diamond 1 mm thick in which both faces have been polished to an optical finish the optical transmission from the fundamental cut-off at 220 nm to the far infrared (IR) part of the spectrum is shown in Figs 4 and 5. For comparison, the transmission of a high-quality single-crystal natural Type IIa sample is also shown. For wavelengths beyond the fundamental cut-off at 220 nm the transmission rises to reach values close to the theoretical reflection limited value of approximately 71% assuming a refractive index of approximately 2.4 [29, 30]. At shorter wavelengths there is a reduction in transmission relative to the Type IIa sample, which can be partly attributed to scattering and, close to the fundamental edge, to a bulk absorption mechanism [31].

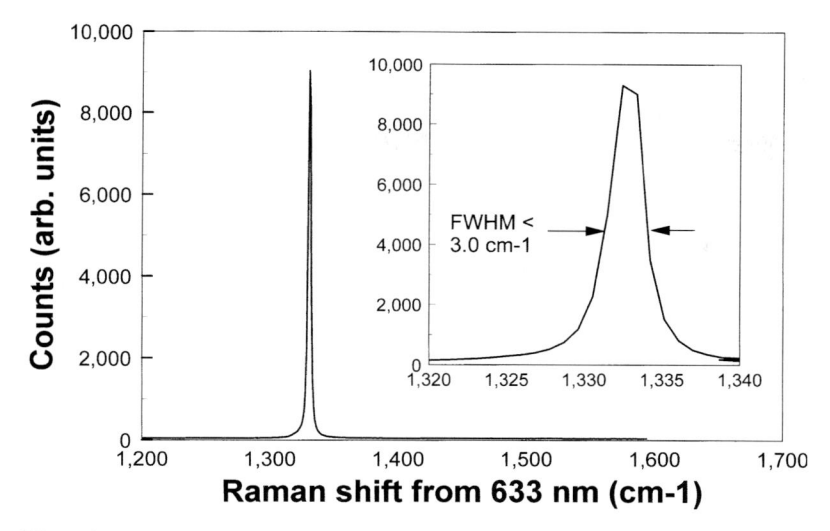

Figure 3. Raman spectrum of an optical grade CVD diamond specimen measured with a Renishaw instrument with an excitation wavelength of 633 nm.

Beyond the intrinsic, multi-phonon absorption bands (between 4000 cm^{-1} and 1500 cm^{-1}) [30], the transmission of the CVD diamond specimen reaches the maximum reflectivity limited value (Fig. 5) of 71.4% (using a refractive index value of 2.375 [32, 3]) and is identical, within experimental accuracy, to that of high-quality single crystal Type IIa samples measured with the same instrument.

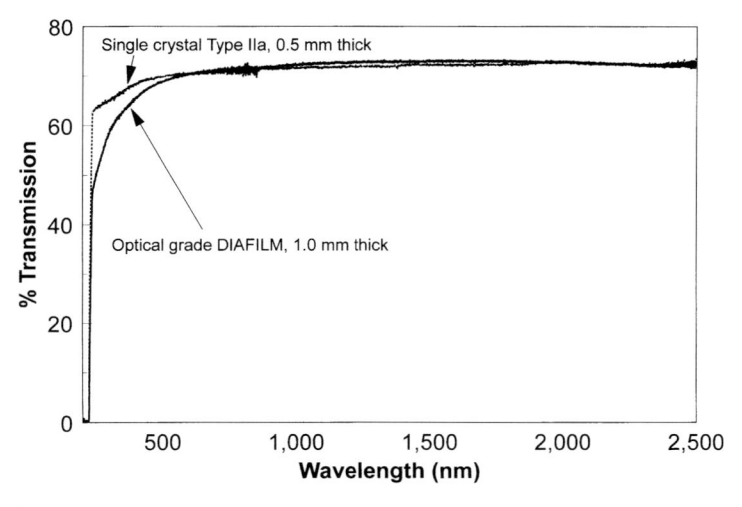

Figure 4. Transmission spectrum for an optical grade CVD diamond window 1.0 mm thick in the UV-visible-near IR spectral range compared to that of a high quality natural Type IIa window 0.5 mm thick.

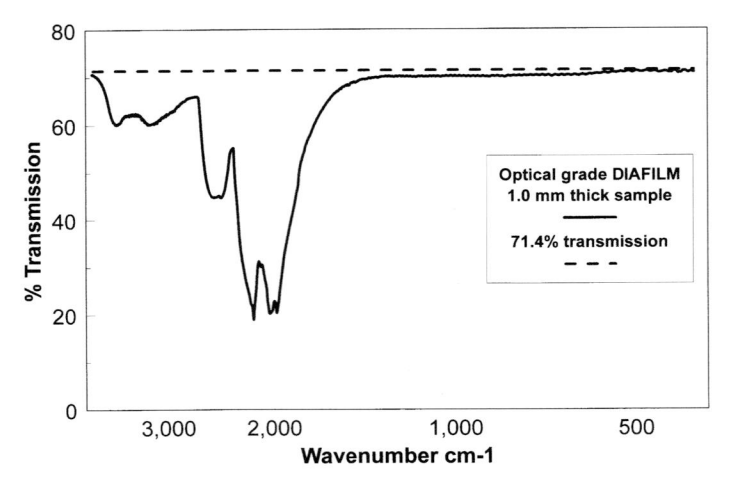

Figure 5. Transmission spectrum for an optical grade CVD diamond window 1.0 mm thick in the infrared spectral range.

Transmission measurements have been recently reported [33] extending the range of the spectrum of Fig. 5 to much longer IR wavelengths (to $20\,\text{cm}^{-1}$ or $500\,\mu\text{m}$) showing that there are no measurable absorption features within that range.

The intrinsic absorption at $10.6\,\mu\text{m}$ wavelength of optical grade samples measured by calorimetry is in the range $0.10\text{--}.029\,\text{cm}^{-1}$ [27, 34]. It is thought that some of these values may be affected by surface absorption. This aspect is currently being investigated. As will be discussed below, values of intrinsic absorption below $0.1\,\text{cm}^{-1}$ are acceptable for most practical applications, due to the very high thermal conductivity of diamond, coupled with the low change in refractive index with temperature.

The low angle scattering of optical grade CVD diamond at $\approx 10\,\mu\text{m}$ wavelength has been assessed by measurements of modulation transfer function (MTF) using samples of optical grade CVD diamond with accurately polished surfaces to evaluate one aspect of their performance as imaging windows [4, 35]. The MTF is the best parameter available for defining image quality. A modulation transfer function curve effectively shows the contrast in the image as the spatial frequency of that image changes (so that the curve will always drop to zero at a high enough spatial frequency). The MTF of the components was determined from the width of the image of a slit source. Any component which is placed in the light path which has a poor optical performance (whether because of form shape errors or scatter) will increase the broadening of the image of the slit source (the line spread function, LSF).

An infrared interferometer was used to determine the LSF due to the form shape error and this can be subtracted from the total broadening as measured in the infrared MTF instrument. What remains is the broadening due to the scatter alone.

Figure 6 shows a comparison of the LSF for the system with and without a sample showing that the CVD diamond optical window introduces only a very small broadening. The computer converts these data into an MTF trace such as

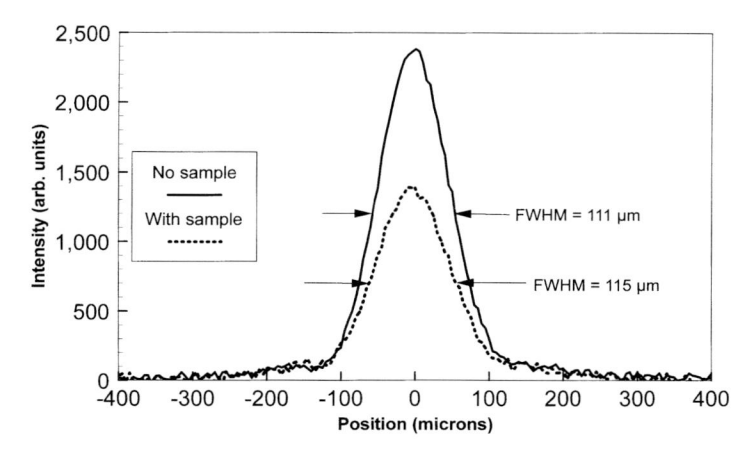

Figure 6. Comparison of the Line Spread Function measured with and without a sample in the system.

that in Fig. 7 (MTF vs RSF, reduced spatial frequency) which also shows the MTF degradation caused by form shape errors and the MTF of the system with no window in place (the diffraction limited curve). The effect of scatter is therefore reflected in the difference between the lower and middle curves. The conclusion is that optical grade CVD diamond shows minimal scatter and degradation in MTF

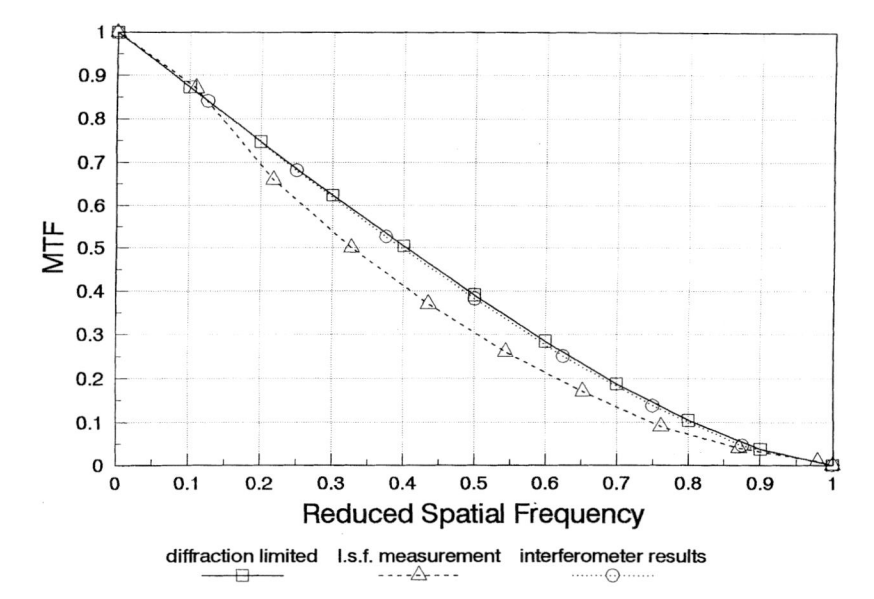

Figure 7. Modulation Transfer Function (MTF) measured with a CVD diamond window (lower curve), the MTF degradation caused by form shape errors (middle curve) and the MTF of the system with no window in place showing the diffraction limited case (upper curve).

in the far infrared and is therefore a good candidate for windows required in imaging systems. Such an application will be discussed in 2.3.1.

2.2.3 Strength of Chemical Vapor Deposited Diamond

C. S. J. Pickles

The fracture stress of CVD diamond has been evaluated using the three-point bend geometry [15]. The details of these measurements and its relative merits have been discussed elsewhere [4, 35].

The fracture stress is plotted against thickness in Figs 8 and 9 for the growth (coarse grain) and nucleation (fine grain) surfaces respectively, for the optical and mechanical grades of material. All of the samples in this study were in the as-grown state. These strength data suggest several conclusions.

First, the measured strength depends strongly on thickness. There is an increase in grain size on the growth side with thickness and it is speculated that the size of the critical flaw is limited by the grain size close to the tested face, as discussed previously [4, 25, 35]. The variation of the stress level through the thickness is another factor which may influence the thickness dependence of the strength. A more precise analysis has been performed in which the experimental data seem to be in good agreement with the model, as reported elsewhere [35].

Second, the reproducibility of the strength values is high. If the data from samples of thickness between 0.4 and 1.4 mm are assessed, they give the Weibull moduli [15] shown in Table 1 (a higher value indicates a more reproducible strength, c.f. sapphire = 2.1, zinc sulphide = 5.4, silicon carbide = 10). These can be corrected to allow for the thickness effect [25, 35]. The corrected values for the growth surface are also given in the table.

The observations suggest that CVD diamond has bulk (rather than surface)

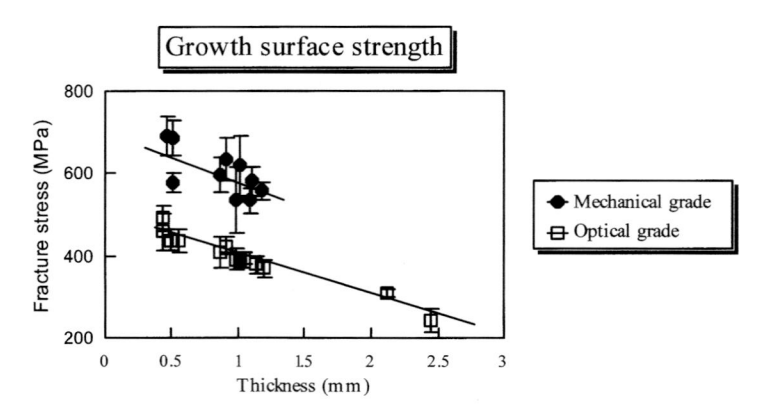

Figure 8. Fracture stress as a function of thickness measured with the growth face in tension.

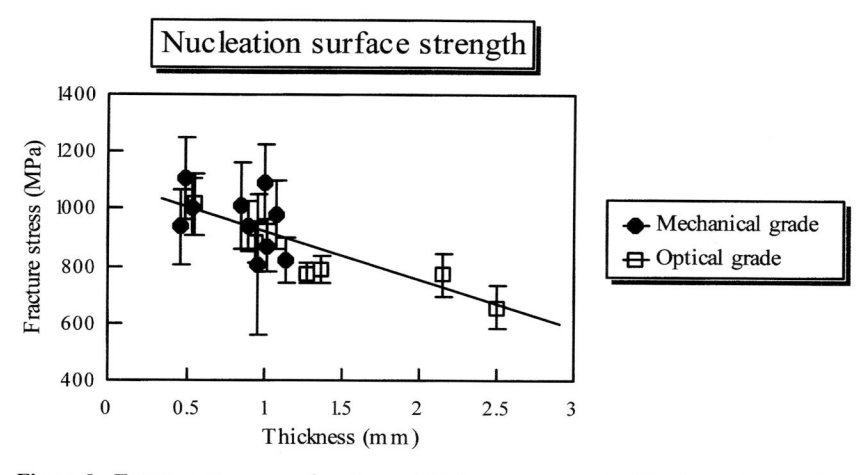

Figure 9. Fracture stress as a function of thickness measured with the nucleation face in tension.

strength controlling flaws [26] which are related to the size of the crystal grains.

2.2.4 The Young Modulus

Early measurements of the Young modulus on relatively thin specimens (up to 300 µm) gave values in the range 986–1079 GPa [2, 36]. More recent results have been obtained on optical and mechanical grade specimens as a function of temperature in the range 20–800°C [37]. For temperature values below 700°C (which is the onset of oxidation in diamond) the Young modulus is relatively insensitive to temperature as shown in Fig. 10. The average value for the Young modulus in optical grade material at room temperature is 1133 GPa and that of mechanical grades is 1166 GPa. Results published previously using samples of a different origin [38] show that values between 242 and 539 GPa may be exhibited if the CVD diamond contains a high concentration of defects such as micron-size voids or microcracks. This illustrates the great variability in material properties that can be found in CVD diamond of different origins and the need for CVD diamond suppliers to provide precise material specifications to end-users.

Table 1. The Weibull moduli calculated from the data in Figs 8 and 9

CVD diamond type	Growth surface	Growth surface (corrected for thickness variation)	Nucleation surface
Mechanical grade	9.8	11.6	6.5
Optical grade	11.6	23.1	11

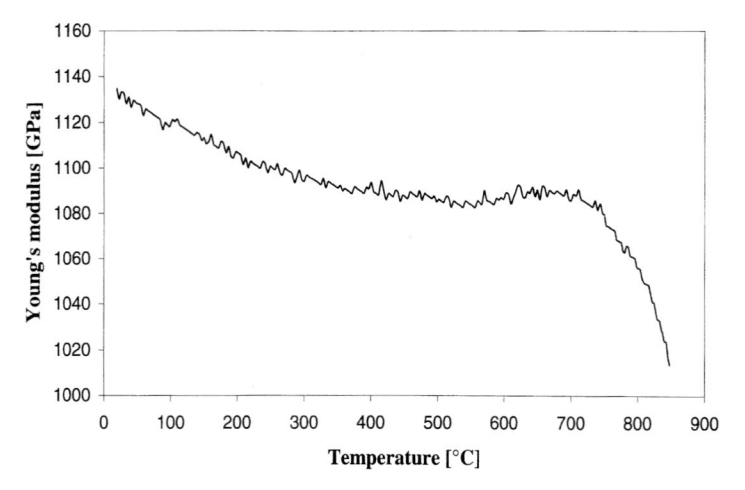

Figure 10. Young modulus of an Optical Grade CVD diamond specimen in the 20°C to 800°C temperature range [37].

2.2.5 Thermal Conductivity

The thermal diffusivity of CVD diamond has been measured over the temperature range 200–425 K using the thermal flash technique which measures the thermal diffusivity perpendicular to the plane of the layer (through-the-plane) [1, 3].

Figure 11 shows the temperature dependence of the thermal conductivity for different grades of CVD diamond: optical grade, and three grades of thermal

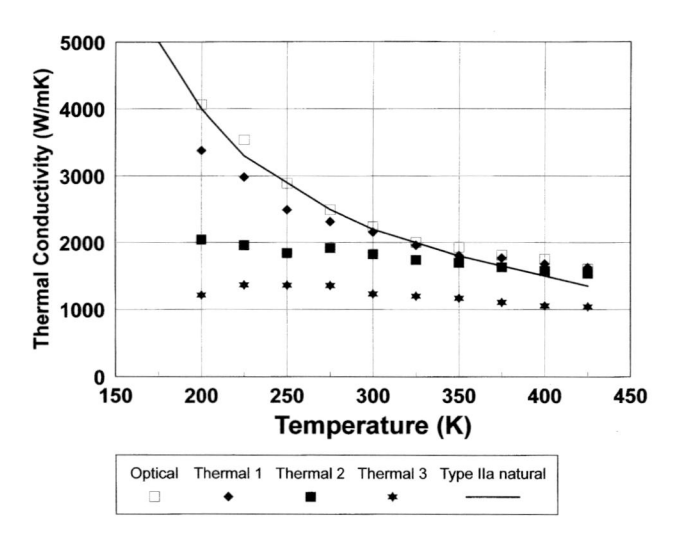

Figure 11. Temperature dependence of the thermal conductivity for different grades of DIAFILM CVD diamond: optical grade and three thermal grades.

DIAFILM specimens. These results are compared with literature values of natural Type IIa single crystal diamond [39]. The difference between the thermal conductivity of optical and thermal grade specimens becomes more appreciable at lower temperatures. The optical grade specimen seems to follow the Type IIa trend reaching a value of 4000 W mK^{-1} at 200 K, probably the highest reported so far for polycrystalline CVD diamond at that temperature. For the 'thermal-3' grade, which shows the lowest conductivity and is of dark color, the temperature dependence is not very pronounced, in agreement with previous publications [40, 41]. The 'thermal-1' grade has a thermal conductivity better than 1800 W mK^{-1} at room temperature which is adequate for most demanding applications, as discussed in 2.5.2. The ability to define different grades of thermal material is an example of the cost-performance optimisation that is possible with CVD diamond.

2.2.6 Dielectric Properties

At microwave frequencies in the range 72–145 GHz, the critical parameters for high-power transmission are the dielectric characteristics of the window material: the dielectric loss factor tan δ and the permittivity ε'_r (or the refractive index $n = \varepsilon'^{1/2}_r$) because they affect power absorption and reflection [42]. The dielectric loss factor tan δ in low loss samples is usually measured as the decrease in the Q factor of a resonant cavity [43]. Low dielectric loss materials find application as the output windows of high-power microwave tubes. A specific case is that of windows for Gyrotron tubes operating in the 70–170 GHz frequency region with output powers in excess of 1 MW, as will be discussed later.

Values of the dielectric loss of CVD diamond have been measured over the past 3 years as a suitable material grade for dielectric window applications was being developed [5]. For open resonant cavity measurements, samples are usually required to be of at least 30 mm in diameter and of thickness in excess of 0.87 mm depending on the measurement frequency and the accuracy required. For recent CVD diamond, values of tan δ below 10^{-5} have been achieved. A specific example is a window 100 mm in diameter and 1.6 mm thick which exhibited a tan δ value of 0.6 ($\pm$ 0.2) 10^{-5}. This is the lowest value so far reported for CVD diamond and would enable the material to be used as output windows in Gyrotron tubes of powers in excess of 2 MW as discussed in 2.4.

2.3 Optical Applications

In this section, applications will be discussed which illustrate the versatility and advantages of CVD diamond as an infrared and multi-spectral window material. We describe the use of CVD diamond optical elements including CVD diamond domes and flat plates as windows for IR seekers or imaging systems in high-speed flight or other mechanically aggressive environments. Then we describe the use of CVD diamond windows for the transmission of high-power IR laser beams.

2.3.1 Chemical Vapor Deposited Diamond for Passive Infrared Windows in Aggressive Environments

C. J. H. Wort

The extreme hardness and abrasion resistance of diamond makes this material an ideal choice for applications in which the optical components are exposed to aggressive environments. The use of diamond for optical elements for infrared seeking missiles and other military uses constitute probably the best examples of this type of application as discussed in more detail in this section. There are, however, a host of other applications in industry outside the military field in which the attributes of diamond optics are of great advantage. These include windows used in the monitoring of chemical reactions, or in the analysis of fluids which contain abrasive components.

Originally conceived during the 1940s to allow armed forces to operate effectively under the cover of darkness, infrared thermal imaging fulfilled a need that could not be met with radar, acoustic, or visible sensors [44]. Thermal imaging relies on sensing the heat emitted by a body, by virtue of its temperature alone; thus thermal imaging is purely passive in its operation. The wavelength of the emitted light is temperature dependent and the IR detector is optimized for maximum sensitivity over a limited wavelength (and hence temperature) range.

Infrared detector materials are delicate and require protection from the environment by a transparent window. In many military applications this environment can be extremely harsh, such as the environment encountered by a high-speed, heat-seeking missile or a tank window exposed to a sandstorm. The damage caused by a high velocity rain drop impact onto a zinc sulphide window [45] is clearly illustrated in Fig. 12. The erosion/impact damage seriously affects the component strength and also the ability to form an acceptable image behind the window.

Owing to the absence of strong absorption by water or carbon dioxide molecules in the atmosphere, two wavebands are basically transparent and are generally used by IR imaging equipment. The mid waveband is between 3 and 5 µm wavelength and the long waveband is between 8 and 14 µm wavelength. For a body at 300 K there is considerably more emission per unit area (termed excitance) in the 8–14 µm waveband than the 3–5 µm waveband (the excitance is about thirty times more). For a body at 600 K, the levels of excitance in each waveband are similar but at 2000 K the 3–5 µm excitance is a factor of about nine times higher [15]. This is one of the main reasons why the 3–5 µm operating waveband has been favored for heat-seeking missiles designed to follow the hot air trail left by a jet engine. The other main reason why the mid waveband seekers are currently preferred is that there are several durable window or dome materials that are transparent in that spectral range, such as sapphire or magnesium oxide [15] and that are currently available on a commercial basis. Figure 13 shows typical heat-seeking missiles fitted with 3–5 µm wavelength transparent domes.

In contrast, the longer wavelength infrared (LWIR) band is desirable for imaging cooler bodies (such as the aircraft itself rather than just the hot engines, which can be 'masked') and although LWIR detector materials and missile seekers have been

2.3 Optical Applications 585

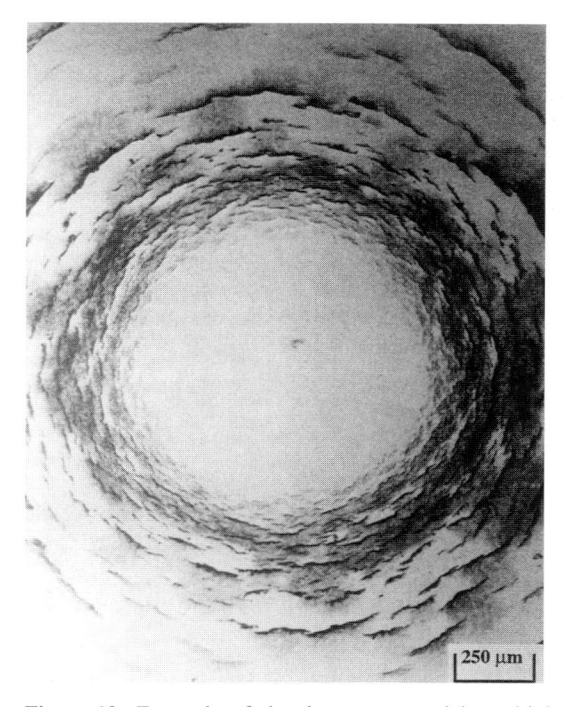

Figure 12. Example of the damage caused by a high velocity water drop onto a zinc sulphide window (from [45]).

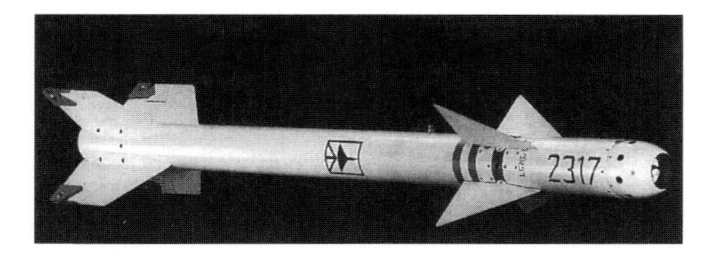

Figure 13. Examples of air-to-air heat-seeking missiles fitted with 3 to 5 μm wavelength transparent domes.

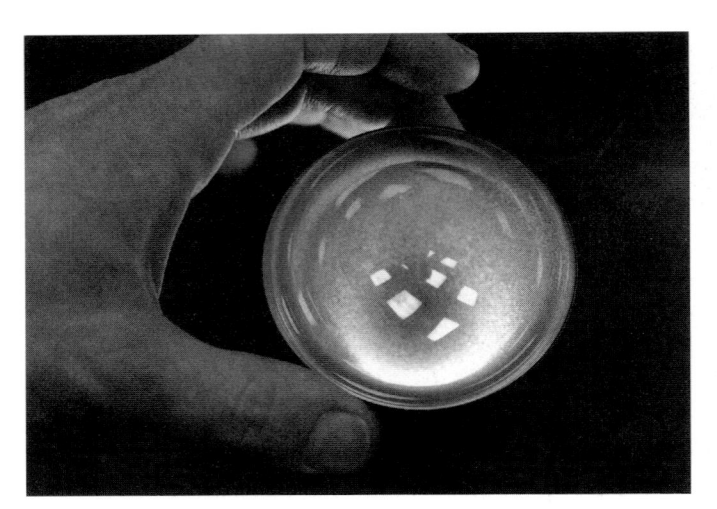

Figure 14. Example of a CVD diamond hemispherical dome, 70 mm in diameter and over 1.0 mm thick that has been semi-processed on both surfaces.

available for some time, they have not been exploited due to the lack of erosion resistant windows and domes that transmit in the 8–14 μm waveband [14]. Materials such as germanium and zinc sulphide, for instance, are transparent in the longer wavelength IR band but even with the best available protective coatings they are not able to withstand the rain erosion encountered during air carriage, let alone a missile flight scenario.

Diamond has been recognized to be an ideally suited material for this application [15, 14, 25] because of its good IR transmission and its extreme mechanical properties. It was only with the introduction of CVD diamond technology and its potential for synthesizing diamond films over large areas that the IR seeker community started to consider the practical possibility of using diamond for this application. One approach has been the use of CVD diamond as a protective coating for other softer IR materials such as ZnS or Ge [46]. However the successful development of thick-film or bulk CVD diamond [2f, 13] motivated active development projects on the synthesis of free-standing CVD diamond domes. This work has met with considerable success [25, 47] resulting, for instance, in the synthesis of CVD diamond hemispherical dome geometries 70 mm diameter, ≥1 mm thick that have been semi-processed on both surfaces as shown in Fig. 14. For imaging applications the form shape needs to be very precise and the radii of the two surfaces have been consistently machined to better than ±1 μm, which is technically difficult to achieve in such a hard material.

Figure 15 shows the thermal image of a commercial aircraft through a CVD diamond dome fitted (such as that shown in Fig. 14) to an 8–14 μm thermal imaging camera. There is a small loss of image resolution due to imperfections in both the surface finish (causing scattering) and the form shape of the dome being tested; however, the engines are clearly visible as being hotter than the bulk of the plane, and the plane shape is clearly identified.

Figure 15. Thermal image of a commercial aircraft through a CVD diamond dome such as that shown in Fig. 14. The engines are clearly visible as being hotter than the main body and the shape of the aircraft is clearly identified.

Table 2 compares IR window properties of CVD diamond with other, 8–14 μm wavelength transparent materials showing the exceptional optical and thermo-mechanical properties available with CVD diamond. As in other applications, it is the combination of several of these properties within one window material that makes diamond so special. One example is the remarkable sand erosion resistance of CVD diamond coupled to the high transparency in several spectral regions. The latter property may be important for future military imaging systems since there are considerable advantages in being able to 'see' simultaneously in several different spectral regions through a common window. When compared with multi-spectral ZnS (which is the only other truly, multi-spectral material for visible, LWIR, and radar frequencies) it can be appreciated that diamond is not only more transparent but also more resistant to sand erosion by close to a factor of a million times! The extreme sand erosion resistant properties of diamond are graphically illustrated in Fig. 16 which compares the reduction in transmission, as a function of sand erosion time, for CVD diamond and sapphire (which is the most erosion resistant of the multi-spectral, 3–5 μm waveband materials). A potential future application for a multi-spectral window would be on a tank, where both the IR sensing and laser range finding could be performed through the same window as used by the tank driver.

Thermal shock of the window (or dome) is an important consideration for the design engineers of future missile systems because of the high thermal gradients generated during rapid in-flight acceleration, and is currently limiting the performance of existing heat-seeking missiles. The combination of a high thermal conductivity coupled with a low thermal expansion coefficient render diamond virtually immune to thermal shock [15]. Using a thermal shock figure of merit as defined in note 3 of Table 2 diamond is approximately 200 times better than sapphire, which is

Table 2 Property comparison between high quality CVD diamond and alternative window materials that transmit in the long wave (8–14 μm) thermal imaging region

Property	CVD Diamond (high optical quality)	ZnS (FLIR grade)	ZnS (Multi-spectral)	Germanium
Vicker's hardness (GPa)	81 ± 18	2.3	1.5	8.5
Fracture toughness (MPa.m$^{1/2}$)	>6	0.81	1	0.57
Young modulus (GPa)	1000–1100	74.5	74.5	103
Poisson ratio	0.1 (assumed)	0.29	0.28	0.28
Tensile strength (MPa)	350–800	103	60	90
Rain impact DTV (m s^{-1}) 2 mm drop size	525 (>1.25 mm thick)	170	165	175
Sand erosion (mg kg^{-1}) @ 80 m s^{-1} C25/52 sand	0.18	19,000 (2)	156,750 (2)	12,382 (2)
Sand erosion (mg kg^{-1}) @ 34 m/s C25/52 sand	Negligible	950	7,840	1,730
Thermal conductivity @ 300 K (W mK)	1800–2200	19	27	59
Thermal conductivity @ 750 K (W mK)	1,200	–	–	–
Thermal expansion coeff. (p.p.m./K) at 273 K	0.9	6.6	6.3	6.1
Thermal expansion coeff. (p.p.m./K) at 473 k	2	7.7	7.5	–
Thermal shock (mild) FOM ($\times 10^3$ W m^{-1}) (3)	~1000	2.6	2.1	6.1
Refractive index at 10.6 μm	2,376	2.19	2.19	4
dn/dT (K^{-1}) (at 300 K)	9.6×10^{-6}	41×10^{-6}	36×10^{-6}	400×10^{-6}
Dielectric const. D (35 GHz)	5.68 ± 0.15	8.35	8.39	~17
% increase in D at 773 K (%)	4.3	~8	~8	–
Loss tangent at 35 Hz	<0.0002	–	–	–
10.6 μm abs coeff. (cm^{-1})	0.03–0.1	0.2	0.2	0.02
Max. window temperature (K) for extended operation in air at 10.6 μm wavelength (1)	1023**	573* 873**	573* 873**	343***
Useful transmission range (μm)	0.3 to 3.0 and 7.0 to >500	7.7 to 11.2	0.4 to 11.5	1.8 to 23.0

(1) * Band edge moves towards shorter wavelength, introducing significant absorption.
 ** Mass loss due to oxidation.
 *** Free-carrier absorption makes material opaque.
(2) Calculated sand erosion mass loss values at 80 m/s for ZnS and Ge (to allow a comparison with CVD diamond) are based upon up-scaling theories (using erosion rate velocity scaling factors) from work undertaken at the Cavendish Laboratory [45].
(3) FOM = (Material strength × (1 − Poisson's Ratio) × thermal conductivity/thermal expansion coeff. × Young modulus) [15].

generally considered to be the 3–5 μm material which is most resilient to failure by thermal shock.

The above considerations illustrate the versatility of CVD diamond technology in the ability to synthesize three-dimensional shapes of high quality diamond and

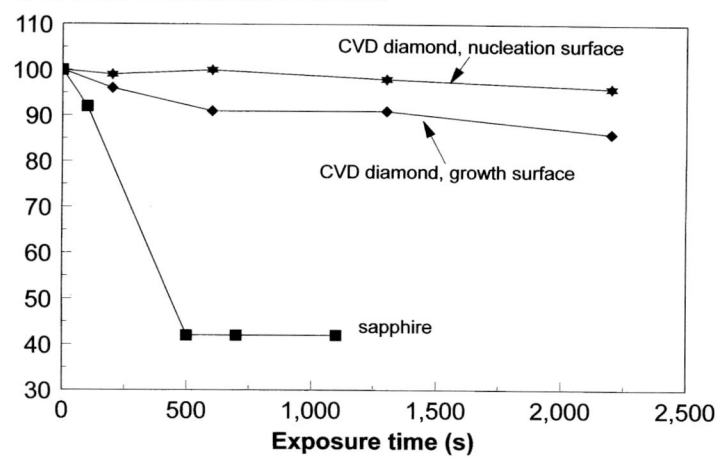

Figure 16. Reduction in transmission as a function of sand erosion time for a CVD diamond window compared to a sapphire window.

explains why CVD diamond is emerging as one of the preferred material options for applications requiring multi-spectral transmission in harsh environments.

2.3.2 Windows for High-power Infrared Lasers

Considerable progress has been achieved in recent years on improving the optical properties of CVD diamond and in making this material available in large sizes and with reproducible and consistent properties [4, 5, 35] so that its use in actual laser systems has become a practical reality [48]. This section reviews the technical issues relevant to the performance of windows for high power lasers comparing the specific case of CVD diamond and ZnSe for the transmission of CO_2 laser beams.

A major requirement for windows used in the transmission of high power beams is the reduction of wave-front distortion which may result in defocusing of the beam and loss of power density. There are two major sources of wave-front distortion: thermal effects and geometrical or form-shape aberrations [49].

2.3.2.1 Thermal Effects
S. E. Coe

A window can absorb heat from the laser beam by intrinsic bulk absorption or at the surfaces (through intrinsic surface absorption mechanisms or at the anti-reflection (AR) coatings). The absorbed heat induces radial and axial temperature gradients which in turn result in two distinct forms of optical distortion:

Table 3. Comparison of properties between CVD diamond and ZnSe

Property at 300 K	Diamond (optical-grade DIAFILM)	ZnSe
Thermal conductivity (W mK^{-1})	1800–2100	16
Thermal expansion (p.p.m. K^{-1})	0.9	7.6
Thermo-refraction (dn/dT) (T^{-1})	1×10^{-5}	5.8×10^{-5}
Fracture strength (MPa)	400–600	55
Weibull modulus	11–23	5
Young modulus (GPa)	>1050	70
Typical thickness (mm) used for a 25 mm diameter window	0.7–1.0	6
Bulk absorption coefficient (cm^{-1})	0.1–0.03	$\approx$0.0005
Total absorption (AR/AR coated, 0.1% absorption per surface) (%)	0.9	0.23

(i) The thermo-refractive effect is the result of changes in refractive index with temperature and is described by the thermo-refractive coefficient dn/dT, where n is the refractive index of the window material.

(ii) Thermo-elastic effects which are a caused by bulging (or other form-shape distortions) or by changes in the optical constants by thermal stress gradients through the photo-elastic effect.

Materials which have relatively small thermal gradients and do not change geometry with temperature will show the lowest values of thermal distortion. Critical parameters of relevance in these effects are thermal conductivity, thermal expansion coefficient, thermo-refractive constant, bulk absorption coefficient, absorption at the AR coatings and the photo-elastic constants. The total wave-front distortion can be described by the variation in phase (or optical-path-length) of a transmitted beam with radial position. For the thermo-refractive effect, for instance, the phase change will be (dn/dT) $\times t \times \Delta T$ (where t is the thickness of the window and ΔT is the temperature change) and is thus smaller for a thinner window. The minimum acceptable thickness of a window will be determined largely by the ability to support vacuum or pressure without breaking or distorting [50]. Critical parameters in these cases are the fracture strength and the Young modulus respectively.

The current standard material for optical elements in CO_2 lasers is ZnSe [27] because of its very low intrinsic absorption at 10.6 µm [14, 15, 34]. Table 3 compares the critical parameters of CVD diamond and ZnSe showing that CVD diamond, for the reasons discussed above, has the potential to handle much greater beam powers. This has been long recognized [51] and the thermal effects in diamond laser windows have been theoretically modelled in previous work [27, 51–53]. Some of the earlier results [51–53] were derived before reliable data were available on the properties of polycrystalline diamond and were therefore very speculative. The following is an up-date of the predicted thermal effects in optical grade CVD diamond and ZnSe windows where some of the earlier calculations are revisited.

The analysis presented here assumes that bulging and thermo-elastic effects in either diamond and ZnSe can be neglected [27, 52–54]. The assumed configuration for each case is that of a circular window, 25 mm in diameter, which is edge

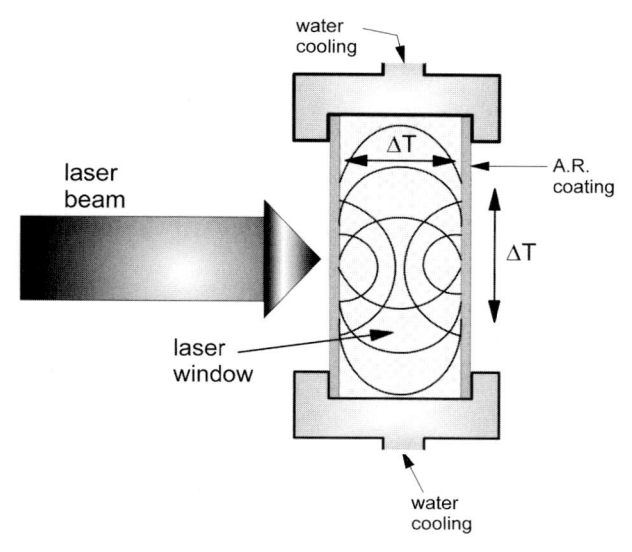

Figure 17. Schematic of the axially symmetrical temperature gradients generated in an edge-cooled window fitted with AR coatings when traversed by a laser beam.

cooled at a fixed temperature resulting in an axially symmetrical temperature profile as shown in Fig. 17. An AR coating with a total absorption at 10.6 μm of 0.1% per surface is assumed for both windows. In the case of ZnSe, the absorption at the AR coating dominates, leading to the highest temperature being at the surface.

Because of the difference in mechanical strength and Young modulus (see Table 3), it is expected that the thickness of CVD diamond windows could be made thinner than those of ZnSe. Using the criterion that the thickness should be sufficient to prevent fracture, the thickness-to-diameter ratio must be [35, 50, 55]:

$$t/D > 0.5(kfP/S)^{1/2} \quad (1)$$

where:

- k = constant determined by the boundary conditions and dependent on the Poisson's ratio
- t = sample thickness
- D = sample diameter
- P = pressure differential
- f = safety factor (depends on the value of the Weibull modulus)
- S = fracture strength.

Using the criterion of acceptable deformation imposes the condition [50, 52]:

$$t/D > 1.01\{(n-1)(P/E)^2(D/\lambda)\}^{1/2} \quad (2)$$

where E is the Young modulus.

Using Eq (1), the ratio of the CVD diamond-to-ZnSe thickness (for the same diameter) is:

$$t_{CVDD}/t_{ZnSe} = \{(k_{CVDD}f_{CVDD}S_{ZnSe})/(k_{ZnSe}f_{ZnSe}S_{CVDD})\}^{1/2} \qquad (3)$$

Using the parameters listed in Table 3, assuming that the k parameter is similar in both materials and using the same safety factor for ZnSe and CVD diamond results in $t_{CVDD}/t_{ZnSe} \approx 1/3$. Equations (1) and (2) predict a minimum t/D ratio of 0.017 for CVD diamond (using a safety factor of 4 and a fracture strength value of 400 MPa) or a thickness above 0.4 mm for a 25 mm diameter window. In practice, it is found that a better flatness and overall shape stability can be achieved if the thickness is close to 1.0 mm. For these calculations it will be assumed that the CVD diamond window is 1.0 mm thick and the ZnSe 6.0 mm, which is the typical thickness used for ZnSe windows in the industry [27]. The higher thickness used for ZnSe may be partly justified by the need to use a higher safety factor due to the lower value of its Weibull modulus [35].

The absolute bulk absorption of optical grade CVD diamond windows has been found to be in the range 0.1–0.03 cm^{-1} [27]. A value of 0.05 cm^{-1} has been assumed for this calculation.

The analysis has been done for a continuous wave, 5 kW incident beam that is assumed to have a Gaussian profile with a 14 mm diameter at 14% ($1/e^2$) of maximum intensity. For this level of power the temperature increases are relatively low and the temperature dependence of parameters such as dn/dT [54] have been ignored, but the temperature dependence of the thermal conductivity for optical grade CVD diamond (Fig. 11), although not very pronounced, has been taken into account.

The analysis solves the fundamental heat propagation equations using a two-dimensional computational model in which values of the radial and axial variations of the temperature and the refractive index are calculated. The calculated temperature gradients for the two cases are shown in Fig. 18. Figure 19 shows the radial dependence of the average (across the thickness) temperature. The temperature in CVD diamond shows a maximum variation (center to edge) of 3.4°C compared to 23.4°C for the ZnSe case. The temperature variations over the effective diameter of the beam are 1.7°C and 12.4°C for the CVD diamond and ZnSe windows respectively.

The phase or optical path length variation $\Delta\Phi = (dn \times t)$ is a function of radial position in the ZnSe and CVD diamond windows and is shown in Fig. 20. The variation for CVD diamond over the effective beam width is over a factor of 250 smaller. Three factors combine to yield this large difference: the smaller temperature rise in CVD diamond due to the higher thermal conductivity (ΔT a factor of more than 7 smaller in CVD diamond), the smaller value of dn/dt for diamond (a factor of 5.8), and the thickness of the window, which for CVD diamond is only 1/6th of that in ZnSe. For an example similar to that shown in Fig. 18 the thermal lensing effect in the ZnSe window was found to be equivalent to a lens of 3–5 m focal length, whereas for the CVD diamond window the effect was negligible [27, 34].

As a general rule it can be assumed that the lensing effect can be neglected if the difference in $\Delta\Phi$ is smaller than $\lambda/10$ [51], or 1 μm for a CO_2 laser beam. As shown in

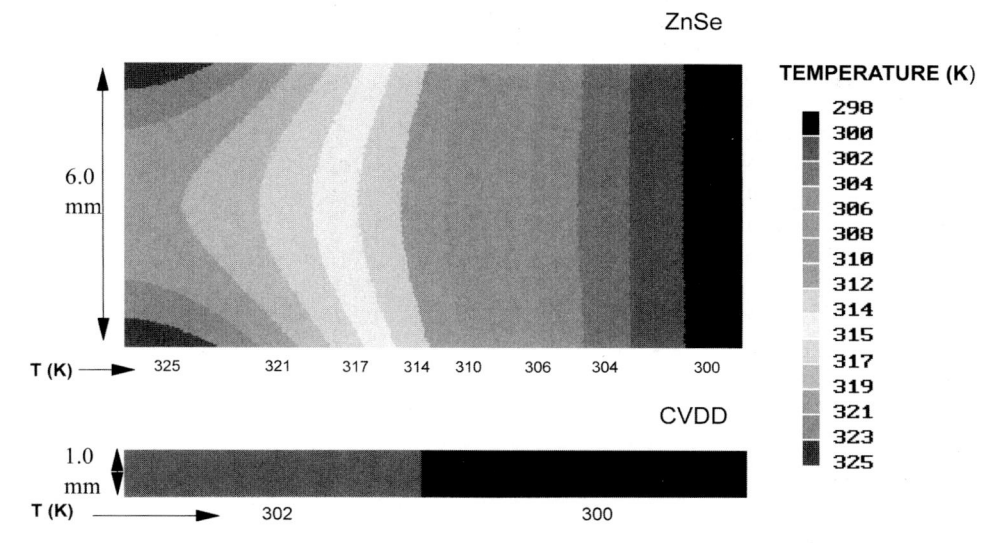

Figure 18. Calculated temperature gradients in a CVD diamond and a ZnSe windows, edge cooled and with AR coatings.

Fig. 20, $\Delta\Phi$ is approximately 8 μm for ZnSe but under 0.04 μm for CVD diamond. Recent results have been reported using electronic speckle pattern interferometry (ESPI) [56] using a CO_2 laser of power up to 1 kW on a CVD diamond and a ZnSe windows of 25 mm aperture and AR coated on both sides. The ZnSe window starts to deform at moderately low power levels (50 W). In the CVD diamond window discernible distortions start at powers over 700 W. These

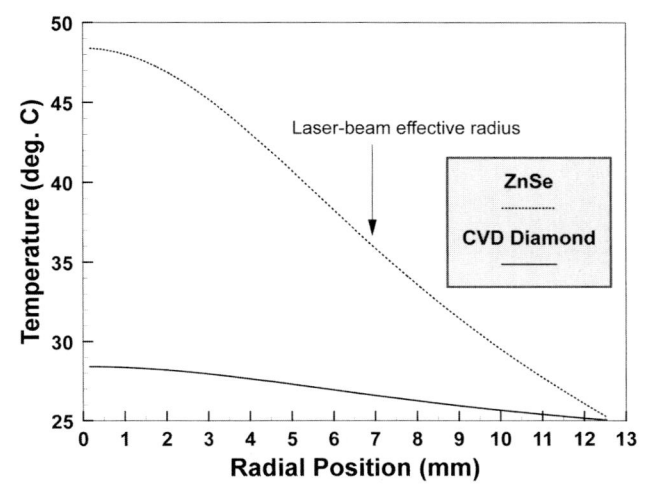

Figure 19. Radial dependence of the average (across the thickness) temperature for the CVD diamond and the ZnSe windows derived from the calculations shown in Fig. 18.

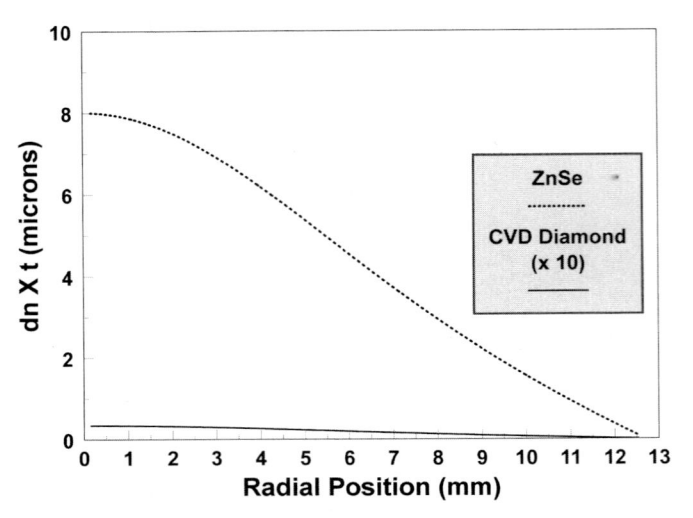

Figure 20. Radial variation of the optical path length $\Delta\Phi = dn \times t$ in the CVD diamond and the ZnSe windows derived from the temperature profiles shown in Fig. 19.

measurements predict a deformation in the ZnSe window for a beam power of 5 kW of approximately 5 μm, in rough agreement with our previous calculation. The equivalent deformation measured in that experiment for a 1 kW beam was at least one order of magnitude lower in the CVD diamond window than in ZnSe.

Using parameters similar to the ones used in the above modeling, Wild [54] estimated that a CVD diamond window will introduce an optical distortion of 1 μm for a laser beam of at least 160 kW and at that level of incident power, the generated thermal stress in the window should not exceed 100 MPa, a factor of at least four below the fracture strength of optical grade CVD diamond <1 mm thick [25, 35, 54]. From this perspective it could be concluded that the power handling capability of an edge cooled CVD diamond window 1.0 mm thick is of at least 160 kW before serious thermal degradation is observed.

However, this may be an oversimplification because it ignores possible laser damage effects. There is evidence [57] that for CW beams:

(i) the laser induced damage threshold (LIDT) is a decreasing function of total power, in fact the power damage threshold scales with diameter rather than with area, and
(ii) LIDT can be considerably reduced due to surface imperfections such as pits or digs.

The first is a heating effect and it can be expected that, because of its high thermal conductivity, diamond will outperform most other materials. For ZnSe, LIDT values of $3\,\text{kW}\,\text{mm}^{-1}$ have been measured [54]. Since temperature excursions in diamond are at least a factor of seven lower than in ZnSe (Fig. 19) a thermal LIDT of at least $20\,\text{kW}\,\text{mm}^{-1}$ can be expected for CVD diamond (probably much larger because of the higher mechanical and chemical integrity of diamond compared

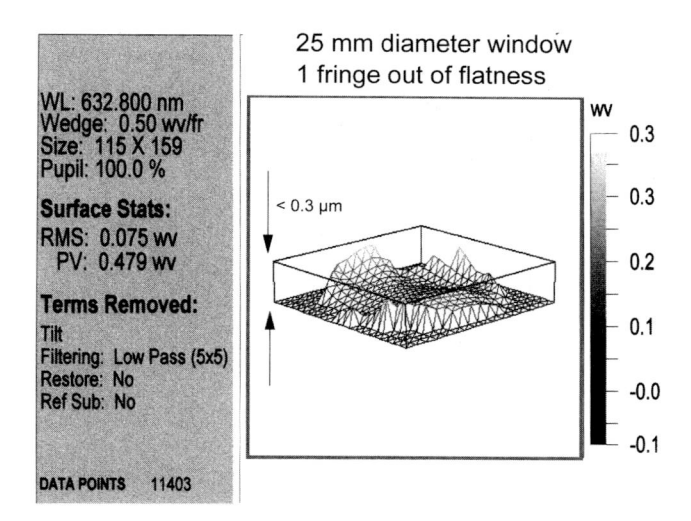

Figure 21. Interferometric surface profile of a 25 mm CVD diamond window showing a flatness better than one fringe (633 nm illumination).

with ZnSe). For a 160 kW beam, this implies a minimum beam diameter of 8 mm, considerably smaller than actual beam sizes in lasers of this power level. What has not yet been investigated is the potential decrease in LIDT that could be introduced by surface and bulk features typical of polycrystalline CVD diamond [2, 35]. For pulsed lasers, considerably higher LIDT values have been reported [58]. Clearly, more experimental work is required before the true power handling capability of CVD diamond is truly understood.

2.3.2.2 Geometrical Distortions: Window Flatness
C. S. J. Pickles

Beam distortion can be generated when the form shape of the surfaces of a window deviate from perfect flatness. The flatness of optical elements as measured by a interferometer is expressed as the number of fringes (twice the wavelength of the light beam probe) shown in the interference pattern of the optical element placed against a perfect flat surface [59]. Depending on the application, an acceptable number of fringes is specified. Applications in laser optics often require a flatness of < 4 fringes and sometimes < 1 fringe using visible illumination (typically 633 nm wavelength). Figure 21 shows the interferometer pattern of the surface of a CVD diamond Window which has been polished to a flatness better than 1 fringe (633 nm illumination). This is typical of what can be achieved presently for CVD diamond showing that current technology is sufficient to address applications such as exit windows, where a flatness of 1 to 3 fringes is considered adequate. Better results have been achieved using techniques currently under development as shown in Fig. 22 where a flatness of 0.5 fringe has been achieved in a window 48 mm in diameter. This

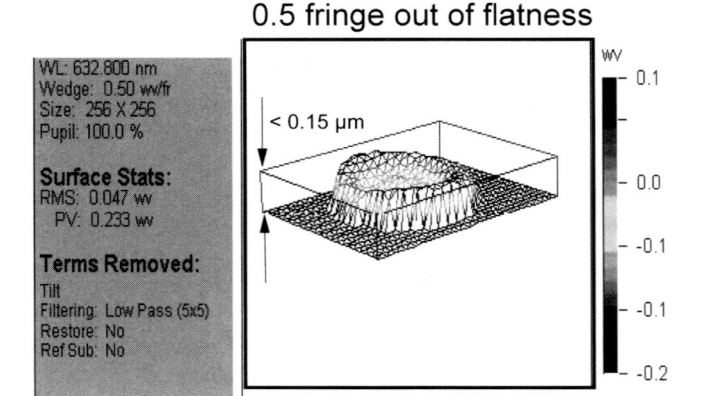

Figure 22. Interferometric surface profile of a CVD diamond window showing that by the use of an improved process, a flatness of 0.5 fringe can be obtained over a 48 mm diameter.

suggests that CVD diamond will be able to address a wider range of applications as an optical element in laser systems.

The ability to process CVD diamond to controlled form shapes has also allowed the production of samples with wedge angles up to 1 degree, as required to eliminate interference fringes for broad band IR transmission, as is the case in synchrotron beam lines [4, 33].

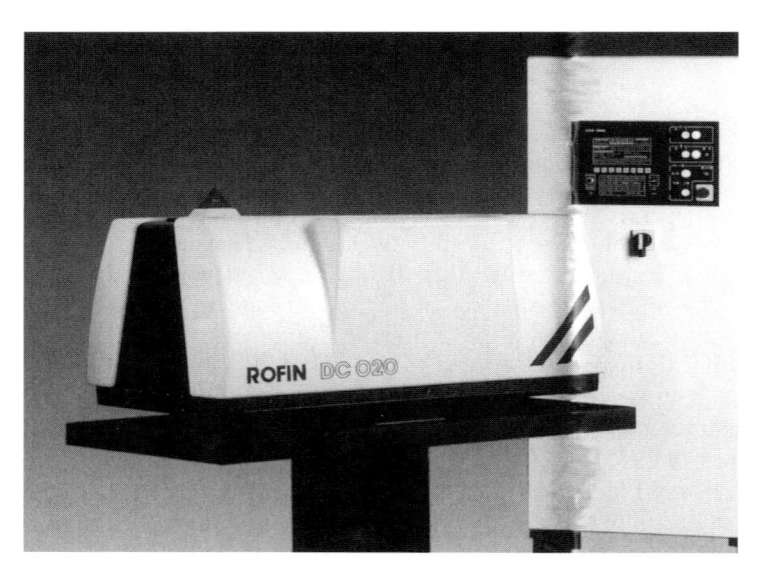

Figure 23. The Rofin Sinar 'Diffusion cooled' DC-series CO_2 slab laser with output power up to 3500 W [60].

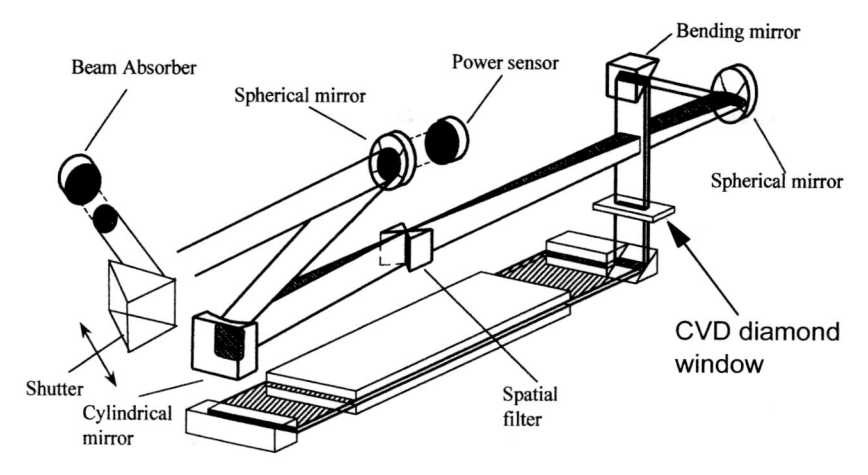

Figure 24. Schematic diagram showing the location of the CVD diamond window in the 'Diffusion cooled' Rofin Sinar DC-series CO_2 slab laser [60].

2.3.2.3 Scattering

Scattering in the infrared has been discussed in 2.2.2 and by Union *et al.* [34] concluding that, at $\approx 10\,\mu m$, a good quality optical grade CVD diamond window does not show a level of scattering that is likely to affect its infrared imaging properties and is capable of transmitting a diffraction limited image. For systems operating at wavelengths in the UV-visible spectral range, scattering effects may be more pronounced, although recent measurements indicate that a good quality CVD diamond window could exhibit values of scattering at 633 nm similar to that of a single crystal Type IIa specimen [54].

CVD diamond windows are now being used in commercial high power CO_2 lasers. An example is the Rofin Sinar Diffusion-cooled CO_2 slab laser with an output power of 3500 W [48] shown in Fig. 23. In this laser, the CVD diamond window is used to seal the slab cavity, as shown schematically in Fig. 24.

2.4 Windows for High Power Gyrotron Tubes

Measurements of dielectric loss in high quality CVD diamond at millimetre wave microwave frequencies, reported in 1993 [2], furnished for the first time clear evidence that the loss tangent of CVD diamond could be comparable or lower than conventional dielectrics such as sapphire or boron nitride. This initial data and subsequent first dedicated dielectric property studies [61] intensified the considerable amount of interest that had already existed in the nuclear fusion community for CVD diamond as a high power window material [62] especially for the development of Gyrotron tubes [63] for the first 'International Thermonuclear Experimental Reactor' (ITER). This section will discuss why, once other advantages of diamond are taken into account (high thermal conductivity, radiation hardness, the relative insensitivity of the dielectric properties with temperature), CVD diamond is being considered not only as an alternative option but probably as the only practical solution for high power Gyrotron tube and Torus windows in ITER.

2.4.1 Window Requirements

It is believed that electron cyclotron heating (ECH) at frequencies in the range 70–170 GHz is one of the major candidates for heating, current drive, and start-up of plasmas in thermonuclear reactors such as the ITER Tokamak or the Stellarator W7-X [64]. The preferred option to supply this source of microwave power is Gyrotron tubes [63] which are currently being developed with an output CW power of at least 1 MW [65]. Figure 25 shows an example of a 1.3 MW, 140 GHz Gyrotron tube being developed at the Forschungzentrum Karlsruhe (FZK) [66]. A key component in these tubes is the exit window which should be able to carry high output powers.

Figure 25. Example of a 1.3 MW, 140 GHz Gyrotron tube built at the Forschungzentrum Karlsruhe [66].

Additionally, the barrier windows in the Plasma Vacuum Torus should be able to withstand high doses of radiation without degradation.

The critical parameters for high-power windows are the dielectric characteristics of the window material: the dielectric loss-factor $\tan \delta$ and the permittivity ε_r' (or the refractive index $n = \varepsilon_r'^{1/2}$) because they affect power absorption and reflection. The power absorption coefficient α is related to $\tan \delta$ by [43]

$$\alpha = 2\pi n \tan \delta / \lambda$$

where λ is the wavelength of the microwave radiation in free space.

To minimize power reflection at a given frequency the thickness t of the window needs to be an integer number of half dielectric wavelengths [64]

$$t = N\lambda/2n$$

where N is an integer. This requirement can be relaxed if a Brewster angle window is used, but the penalty is that a much larger diameter window is necessary [67].

The requirement of CW operation puts extremely high demands on the material properties of the dielectric vacuum barrier window that serves as both the primary tritium containment boundary and as the output window of the Gyrotron. The window options available to Gyrotron manufacturers have been discussed in detail by Thumm et al. [64]. From this analysis it is appreciated that all other options apart from diamond require sophisticated and, in some cases, impractical cooling systems involving, for instance, cryogenically edge-cooled windows, liquid-surface-cooled double-disc windows or distributed windows in which segments of dielectric are metal bonded to microchannel cooled metal ribs. Another shortcoming of some of the materials considered, such as sapphire or gold doped silicon, is the pronounced temperature dependence of the loss factor and refractive index, which may lead to thermal runaway and/or detuning of the window resulting in high levels of reflected power.

This background explains the keen interest in diamond as an alternative material. Theoretical thermal modeling results such as that illustrated in Fig. 26 [66] show that a simple edge cooled diamond window approximately 100 mm in diameter and with a value of loss tangent below 10^{-4}, should be able to withstand powers in excess of 1 MW with a temperature rise at the centre of the window below 240°C, which is far below the temperature (600°C) at which diamond starts to degrade by oxidation, although thermal stress may be a limiting factor at these temperatures. This relatively low temperature excursion is a direct consequence of the high thermal conductivity of diamond. Table 4 compares properties of diamond relevant to this application with those of other competitive materials.

2.4.2 The Development of Chemical Vapor Deposited Diamond Gyrotron Windows

J. R. Brandon

The detailed requirements for the use of diamond as a window material in high power Gyrotron tubes can be listed as follows.

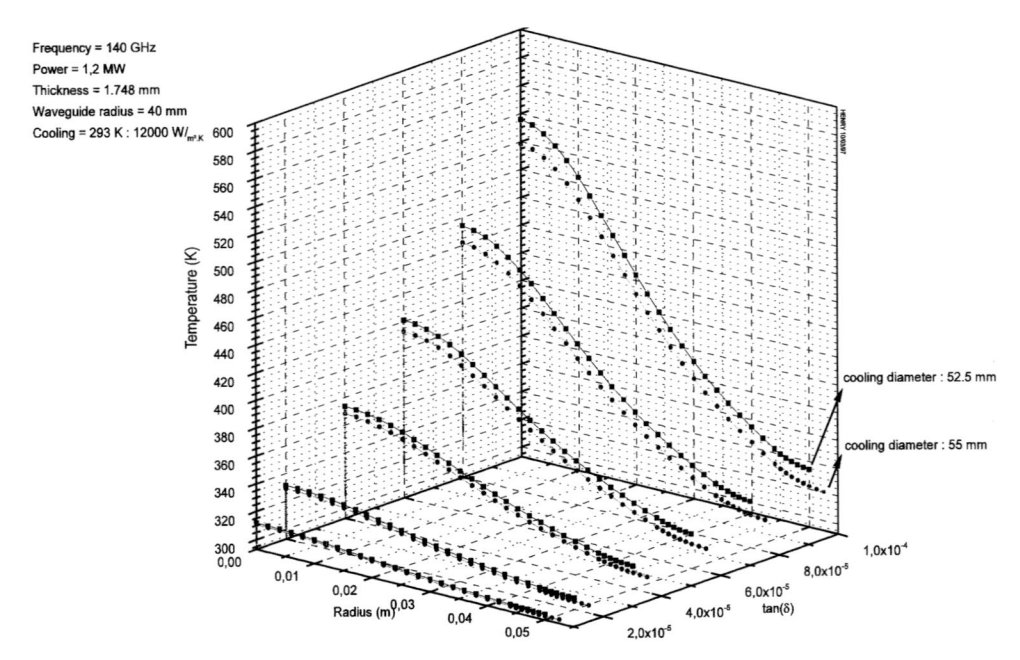

Figure 26. Results from a theoretical thermal model showing the predicted temperature rise at the center of an edge cooled diamond window for a 1 MW incident microwave beam. Results are shown for different values of dielectric loss and for two different cooling diameters assuming an exposed aperture of 100 mm diameter [66].

(i) A reproducible value of loss tangent below 10^{-4} (from modeling results as in Fig. 26) achieved in discs of at least 100 mm in diameter and up to 2.2 mm thick with reasonable radial and axial uniformity. The thickness requirement is dictated by the need to minimize power reflection as discussed above and by the need to support vacuum with a reasonable safety factor.
(ii) The variation with temperature of the dielectric constants (tan δ and n) needs to be very small in the temperature range of interest (room temperature to approximately 200°C) to minimize the risk of thermal runaway or the detuning of the window for minimum power reflection.
(iii) The diamond windows need to be bonded to vacuum-tight and high-temperature bakeable metal flanges for mounting to the Gyrotron tube and reactor ports.
(iv) The dielectric properties need to be relatively insensitive to radiation damage. This requirement is critical for the tritium barrier window that will be exposed to a radiation field of both neutrons and gamma radiation [68].

The following summarizes some of the progress achieved to date partly as part of a collaboration between De Beers Industrial Diamonds and the FZK-ITP, that show convincingly that CVD diamond is indeed a very attractive option for this application.

Table 4. Properties of candidate Gyrotron window materials at room temperature.

Property	Diamond CVD – p.c.	Sapphire s.c.	Silicon Au-doped s.c.	BN CVD – p.c.
Thermal conductivity (W mK^{-1})	>1800	40	150	50
Fracture strength (MPa)	400–600	410	3000* (170)	80
Density (g cm^{-3})	3.5	4	2.3	2.3
Specific heat capacity (J gK^{-1})	0.5	0.8	0.7	0.8
Young modulus (GPa)	>1050	385	190	70
Thermal expansion coefficient (10^{-6} K^{-1})	1.1	5.5	2.5	3
Permittivity (at 145 GHz)	5.67	9.4	11.7	4.7
Dielectric loss (at 145 GHz) tan δ (10^{-5})	0.8–10	20	0.35	115
Temperature dependence of dielectric properties	small	steep	steep	

p.c.: polycrystalline, s.c.: single crystal
* The fracture strength of conventional single crystal silicon is 170 MPa [36]. The value of 3000 MPa has been reported after a special surface treatment [76].

2.4.2.1 Dielectric Loss

Table 5 shows the dielectric loss values that have been achieved for different types of CVD diamond windows. The four samples DB 1 to DB 4, of diameters between 30 and 40 mm, were the result of a development effort to achieve the lowest value of dielectric loss in specimens of the required thickness. The results achieved demonstrate that values of tan δ below 10^{-5} are possible in specimens of thickness in excess of 1.7 mm. These values of loss are close to the sensitivity of the measuring technique for specimens in this thickness range. When the four samples were stacked together, the more accurate loss value measured was tan δ = $1(\pm 0.3) \times 10^{-5}$ [69].

Samples DB 5, DB 6, DB 7 and DB 8 are full size (96–120 mm diameter) windows, synthesized under different conditions.

Sample DB 5 (named 'Super FZK' elsewhere [70]) is 100 mm diameter and demonstrates that the low values of dielectric loss measured in small diameter

Table 5. Dielectric loss of CVD diamond window samples.

Sample	Diameter (mm)	Thickness (mm)	Dielectric loss tan δ (10^{-5}) (at 145 GHz)
DB – 1	50	1.8	0.8
DB – 2	30	1.79	0.8
DB – 3	30	1.74	0.8
DB – 4	30	1.72	0.8
DB – 1 to 4 (stacked)		7.06	1.0
DB – 5 'Super FZK' [70]	100	1.6	0.6 (±0.2) (70 to 370 K)
DB – 6 'Star of FZK' [70]	120	2.2	2 (70 to 370 K)
DB – 7	100	1.85	2
DB – 8	96	2.23	13–15

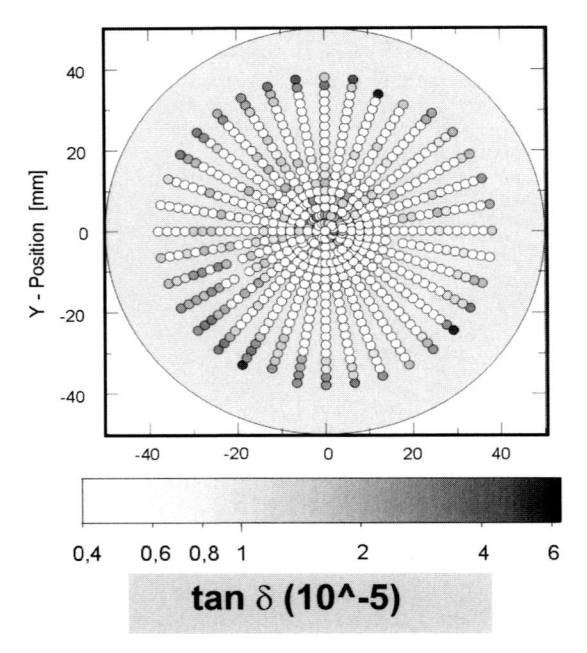

Figure 27. Spatial distribution of the dielectric loss in a 100 mm diameter CVD diamond window 1.6 mm thick measured with a spatial resolution of 6 mm. The isolated points of higher loss at the edge may be caused by scattering losses. Over most of the area the measured dielectric loss is equal or below tan $\delta = 10^{-5}$ which is close to the sensitivity of the measuring technique. This may account for some of the measured isolated spurious higher values of loss.

specimens (DB 1 to 4) can also be achieved in full size windows. The spatial uniformity of the dielectric loss in this window is shown in Fig. 27 which shows a map of tan δ measured with a beam diameter of 6 mm [69]. It can be seen that over most of the active aperture of the window (a central area of approximately 80 mm diameter) tan δ is below 10^{-5} and close to the sensitivity of the measuring technique, which may account for spurious isolated higher values of loss. The temperature dependence of this window over the 70 to 370 K temperature range indicates that the loss is largely independent of temperature over this range and that the average tan δ exhibits a value of 0.6 ($\pm$ 0.2) 10^{-5} [70], which is the lowest loss tangent so far reported for a CVD diamond window.

Sample DB 6 is probably the largest Gyrotron window ever made (named 'Star of FZK') and exhibits a loss of 2×10^{-5} over the same temperature range as DB 5 [70].

The window DB 8, which has a relatively high loss tangent of over 13×10^{-5}, was the first CVD diamond window to be mounted in a high power Gyrotron tube (shown in Fig. 28) at the Japan Atomic Energy Research Institute (JAERI) [71–73]. This window was tested in 1998 using the 170 GHz JAERI/Toshiba Gyrotron at output powers of 0.5 MW for 8 s with a water cooled edge showing a temperature increase at the center of not more than 130°C, consistent with theoretical modelling

Figure 28. The first high-power Gyrotron (JAERI/Toshiba, 170 GHz) fitted with a CVD diamond window. Measurements using beam powers up to 1 MW for 10 seconds have already been performed.

[73]. The second window for the JAERI Gyrotron (DB7) has a lower loss tangent of approximately 2×10^{-5}. This window was tested at the end of 1999 using also the 170 GHz Gyrotron for powers up to 1 MW for 10 s. The temperature increase at the center was less than 30°C, also in very good agreement with theoretical modelling [73]. Because diamond reaches almost steady state for times in excess of 3 s the above experiments show convincingly that CVD diamond windows are able to transmit millimeter wave powers of at least 1 MW continuous wave. At the moment of going to press, in excess of 20 full size CVD diamond windows have been delivered to members of the international nuclear fusion community. These exhibit routinely values of loss tangent of $2-3 \times 10^{-5}$.

It is not clear yet if the low values of dielectric loss of samples DB 1 to DB 7 represent a fundamental limit of a loss mechanism in diamond or if they are still affected by impurities or defects in the polycrystalline CVD diamond structure.

From measurements in previous samples [74], there is clear evidence that, in samples of relatively low quality (tan δ greater than 70×10^{-5}), darker regions of the sample have higher values of loss. Even in the highest quality polycrystalline CVD diamond, such as that of the sample shown in Fig. 27, it is possible to observe localized features related to the grain structure of the layer. It can be speculated that these features may be a contributing factor to the dielectric loss of polycrystalline CVD diamond. However, from a practical point of view, this is probably not of great concern in the foreseeable future since it has been estimated that diamond windows with values of tan δ below 2×10^{-5} (a factor of two above that of the window shown in Fig. 27) should have a power handling capability of at least 2.5 MW [64].

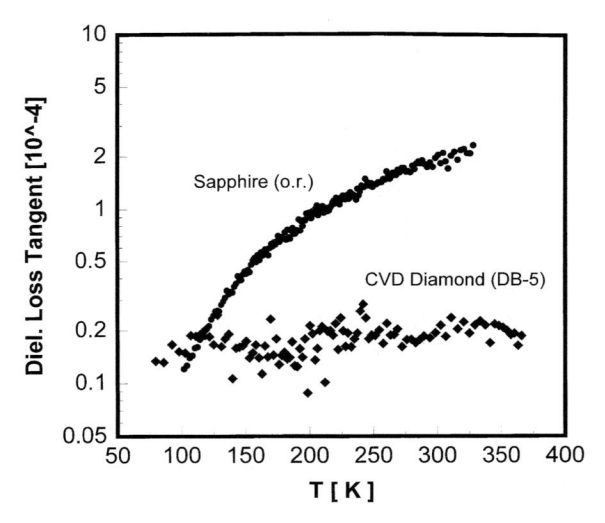

Figure 29. Temperature dependence of tan δ for a CVD diamond sample measured at 145 GHz compared with sapphire [75].

2.4.2.2 Temperature Dependence of Dielectric Constants

Preliminary measurements of the temperature dependence of the dielectric properties of polycrystalline CVD diamond are available. Figures 29 and 30 show the temperature dependence of tan δ and ε'_r measured at 145 GHz for a CVD diamond sample compared to sapphire [75]. In this temperature range (100–370 K), the dielec-

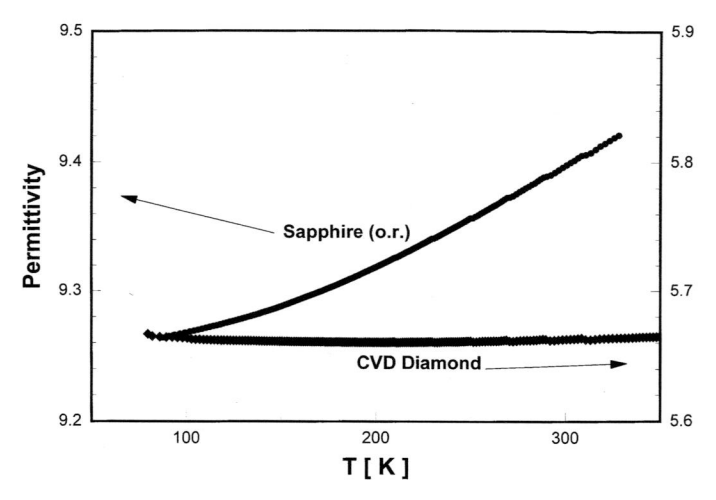

Figure 30. Temperature dependence of the dielectric constant of a CVD diamond sample at 145 GHz compared with sapphire [75].

Figure 31. Examples of CVD diamond windows mounted onto vacuum flanges. The two at either end show 100 mm diameter windows mounted to conventional CONFLAT vacuum flanges. The one in the middle illustrates a double flange configuration designed specifically for Gyrotron tube assemblies. All these vacuum assemblies have been tested to be vacuum tight to better than $10-9\,\mathrm{mbar\,l\,s^{-1}}$ after thermal cycling up to 450°C.

tric constants of diamond are largely insensitive to temperature, in contrast to sapphire. In 1999 the FZK team extended the measurements of dielectric loss to temperatures up to 700 K [75] showing that in a high quality CVD diamond window the loss tangent at 700 K is not more than 5×10^{-5} from a value at 300 K of 3×10^{-5}. The results described in 2.4.2.1 indicate that even with transmitted powers of up to 1 MW the temperature increase is likely to be considerably less than 100°C and therefore there is no major risk of thermal runaway.

2.4.2.3 Effects of Radiation

Measurements of the effect of radiation on the dielectric properties of CVD diamond have been reported using neutron irradiation experiments with fast neutron fluences up to at least $10^{20}\,\mathrm{n\,m^{-2}}$ (energy >0.1 MeV) [65]. Differences before and after irradiation were found to be more pronounced at lower frequencies. At 145 GHz specimens that started with values of tan δ of 2×10^{-5} maintained these levels (or even showed a decrease in loss). Further experiments to extend radiation fluencies to $10^{21}\,\mathrm{n\,m^{-2}}$ are in progress [68].

2.4.2.4 Mounting Techniques

A metal-to-diamond bonding technique has been developed to attach CVD diamond windows to vacuum flanges or other assemblies. This is required to mount the windows to the Gyrotron tubes or reactor ports via suitable waveguide sections. Figure 31 shows examples of three 100 mm diameter CVD diamond windows mounted to two different types of flanges. The two at either end of the picture are

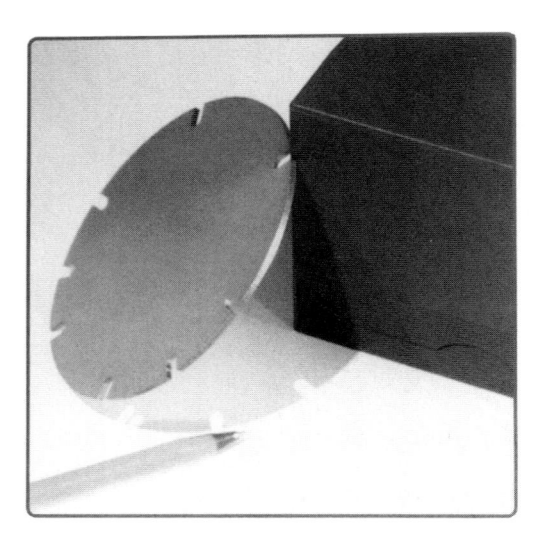

Figure 32. Window disc with slots to reduce hoop stresses and to facilitate bonding alignment.

test mechanical grade specimens mounted to conventional CONFLAT vacuum flanges. The one in the middle illustrates a double flange configuration designed specifically for mounting on to a Gyrotron tube allowing for water cooling of an exposed section of the edge of the window. This bonding technique is based on an aluminum alloy and has been demonstrated to withstand thermal cycling at temperatures up to 450°C without degradation of the vacuum integrity (measured at better than 10^{-9} mbar l s^{-1} [64]. Figure 32 shows a new development in which slots are cut at the edge of the window to reduce hoop stresses and to facilitate bonding alignment.

The results described above indicate that CVD diamond windows are very close to meeting all the requirements listed at the beginning of this section, and promise to be the preferred option for high power Gyrotron windows for ITER and other experiments.

2.5 Thermal Management of Laser Diode Arrays

Natural Type IIa single crystal diamond heatsinks have been used for the thermal management of microwave and laser diode devices for well over a decade [17, 18]. The restriction in the size of available natural diamond has, however, limited the use of diamond to the heat management of small, discrete devices such as single junction laser diodes. The commercial availability of large CVD diamond plates has opened a host of new possible applications in which diamond can be used in the heat management of larger electronic and optoelectronic devices [1, 77, 78].

2.5.1 Laser Diode Arrays: General Issues

A specific example is the use of CVD diamond to improve the heat dissipation in laser diode arrays (LDA) [78, 79]. In these devices a set of laser diode junctions are monolithically integrated on a single chip. A typical LDA bar is 0.1 mm in height, the resonator length of each junction between 0.3 and 2 mm, the junction width between 0.1 to over 0.2 mm and the total length of the array between 5 and 15 mm to accommodate 5–75 single emitters side by side along the length of the bar [80]. During operation of a LDA, more than half of the electrical input power is dissipated in heat which needs to be efficiently removed in order to reduce the junction temperature and thus prolong the life of the device. Typical operating junction temperatures are 50–55°C and raising this temperature by only 10°C results in a reduction by a factor of two in lifetime [79].

The thermal resistance of a diode laser bar depends on the area size of the active junctions. Mounting the diode laser bar on a high thermal conductivity material serves to increase the effective area of the junction before contacting the water cooled, copper heat-sink. These submounts are therefore heat-spreaders, rather than heat-sinks [1]. A typical LDA requires a heat-spreader submount over 10 mm in length, which is well within the current capability of CVD diamond technology but would have been almost impossible to implement by the use of natural diamond. The other advantage of using electrically insulating materials, such as CVD diamond, is the possibility to address electrically single emitters of an array individually for applications such as marking and printing [80]. The efficiency of a heat-spreader will be a function of its intrinsic thermal conductivity and of its actual dimensions and is measured by the decrease in the total thermal resistance of the package, equivalent to the total reduction in temperature of the device junction for a given dissipated power.

There are a number of issues that need to be addressed before CVD diamond can be used effectively as a heat-spreader submount.

(i) Optimization of submount geometry and thermal conductivity.
(ii) The flatness of the submount, to ensure uniform thermal contact.
(iii) Metallization issues such as thickness and thermal stability.
(iv) Thermal stress and expansion mismatch.

2.5.2 Modelling of Submount Heat Resistance

S. E. Coe

Numerical simulations of the junction temperature increase of a laser diode array have been performed for a package configuration as illustrated in Fig. 33. Calculations have been performed with and without a CVD diamond heat-spreader. The decrease in thermal resistance has been calculated for different values of the heat-spreader thickness, depth and thermal conductivity. The simulations have been performed for an array of laser diodes which are individually 200 µm wide

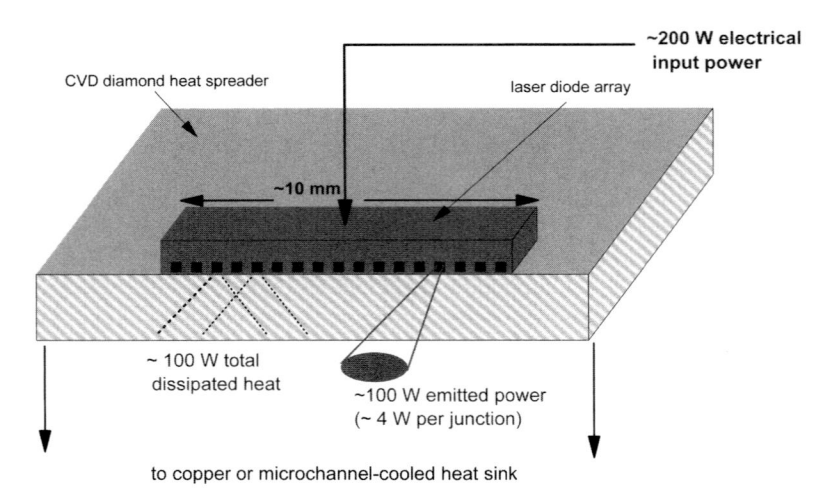

Figure 33. Package configuration analysed in the work consisting of the laser diode array mounted on a CVD diamond spreader which is mounted on a copper heat-sink block.

and separated by a spacing of 200 µm. The length of the heat-spreader has been assumed to be constant at 11 mm. This is a three-dimensional calculation which simultaneously takes into account heat dissipation parallel and perpendicular to the direction of the LDA bar, in contrast with previous two-dimensional calculations reported in the literature [79].

Figures 34 and 35 show that the thermal resistance decreases as a function of heat-spreader thickness and depth. Heat dissipation through the rear (or depth) of the heat spreader is important, because in an array of diodes there is thermal cross-talk between junctions in which the temperature of one junction is affected by that of adjacent junctions. Figure 34 shows that heat-spreader depths between 2 and 3 mm are required for a thickness of 300 µm. Also, because of thermal cross-talk, the

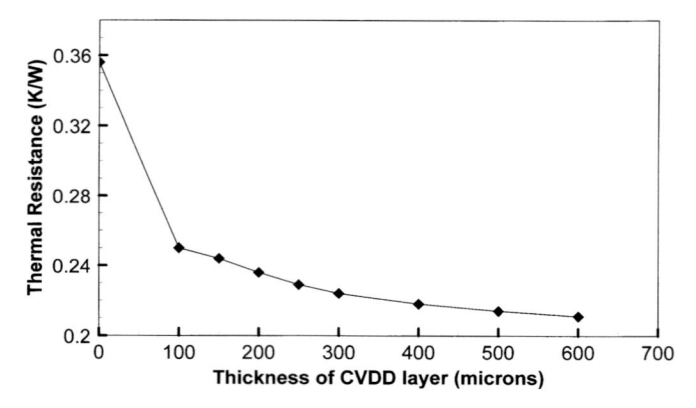

Figure 34. Thermal resistance decrease as a function of the thickness of the CVD diamond heat spreader for the configuration shown in Fig. 33.

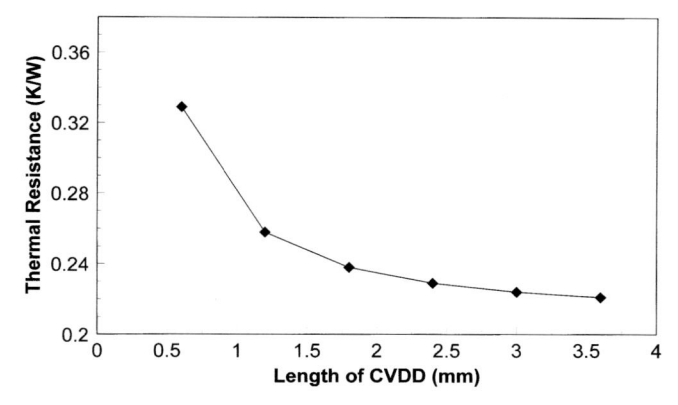

Figure 35. Thermal resistance decrease as a function of the depth of the CVD diamond heat spreader for the configuration shown in Fig. 33.

influence of heat spreader thickness is not as pronounced as would have been expected in a single diode junction and very little gain is achieved for a thickness above 300 μm, as shown in Fig. 35.

Figure 36 shows the decrease in thermal resistance as a function of the thermal conductivity of the spreader, showing that, for a total dissipated power of 90 W (or 3.6 W per junction), a reduction close to 5°C is obtained by increasing the thermal conductivity from 1000 W m^{-1} K^{-1} to values close to 2000 W m^{-1} K^{-1}. This reduction in temperature is significant in view of the increase in life of a factor of two expected for a decrease on 10°C, as mentioned above. As discussed in 2.2.5, values of thermal conductivity of 1000 W m^{-1} K^{-1} are typical of the lower thermal grade whereas the higher grades exhibit a thermal conductivity close to 2000 W m^{-1} K^{-1} (Fig. 33). The result of Fig. 36 shows, therefore, that for this application it is advantageous to use a higher thermal conductivity CVD diamond grade.

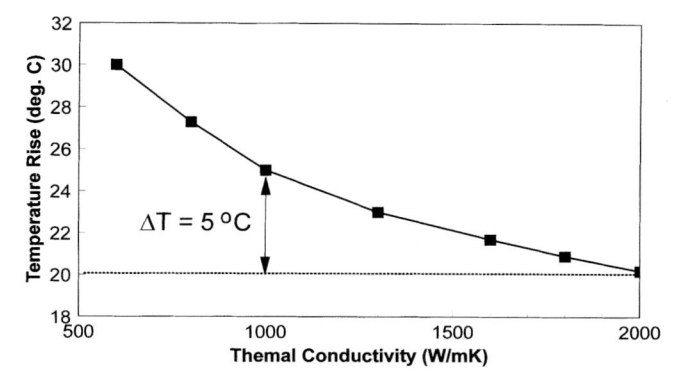

Figure 36. Decrease of the thermal resistance as a function of the thermal conductivity of the CVD diamond heat spreader.

These results indicate that an ideal CVD diamond heat-spreader would reduce the thermal resistance of a 25 element array by approximately 36% from 0.36 K W^{-1} to 0.23 K W^{-1}. Experimental results [79] show that, in a similar array, the measured reduction in thermal resistance is only 22%. This may be a consequence of the extra thermal resistance introduced by the metallization. If a metallization thermal resistance value of 0.12 K W^{-1} is added, the results from the above calculations agree well with experimental results. Using a novel microchannel heat-spreader made of CVD diamond reductions in thermal resistance up to 75% have been predicted [81].

2.5.3 Flatness of Submount

The analysis discussed above is an ideal case in which the effects of intermetalic layers is ignored. If the heat spreader is not very flat there will be differences in metal layer thickness between different sections of the array, leading to a non-uniform temperature distribution. This effect is thought to account for differences in thermal resistance observed between different heat-spreaders [79]. It is estimated [80] that the required flatness should be better than 1 μm cm^{-1}.

As discussed in 2.3.2.2, it is possible to process CVD diamond plates to accuracy close to one visible fringe over 25 mm. This is equivalent to a deviation from flatness of less than 0.13 μm cm^{-1}, well within the requirements for an efficient LDA heat spreader.

2.5.4 Thermal Stress

Due to the substantial difference between the thermal expansion coefficient of CVD diamond (1 p.p.m. K^{-1}) and laser diode material (6 p.p.m. K^{-1} for GaAs), thermal

Figure 37. Example (courtesy of Jenoptik [83]) of a laser diode array bar which has been pre-cracked to reduce effects of thermal stress.

stress can seriously affect the reliability and life of these devices [79, 80, 82]. A novel approach has been proposed called stress induced dicing (SID) where the tensile stress in the laser diode bar after soldering and during cooling down is relaxed by cracking the bar at defined lines between emitters [83]. Figure 37 shows an example (courtesy of Jenoptik) of such a laser diode bar mounted on a pre-metallized CVD diamond spreader using a Au:Sn (80:20) solder alloy.

2.6 Cutting Tools, Dressers and Wear Parts

J. L. Collins, M. W. Cook and P. K. Sen

2.6.1 Cutting Tools Trends

The use of nonferrous materials and composites in modern industry, particularly automotive, is substantial and becoming increasingly important as the demand for stronger, lighter, more wear resistant components increases. Highly abrasive, high performance materials, such as fibre reinforced plastics, high SiAl alloys, metal matrix composites (MMCs) and even abrasive wood composites are steadily increasing in demand

Also, there is significant development within the machining industry to increase machining speed and eliminate or reduce the amounts of cutting fluid used for environmental and cost reasons. Components are being manufactured to continually higher levels of precision and consistency.

These trends in workpiece material, manufacturing philosophy, and, product quality have led to a strong demand for polycrystalline diamond (pcD) when machining nonferrous materials, and polycrystalline cubic boron nitride (pcBN) when machining ferrous materials. These materials are increasingly providing the best economic solution to high volume, precision manufacturing problems. More recently, CVD diamond has become available and is currently being evaluated and developed in cutting tool type applications. CVD diamond cutting tool materials fall into two categories – a thin film conformal coating less than 30 μm in thickness or a thick film solid diamond usually around 0.5 mm in thickness. The application of thin film CVD diamond is highly substrate sensitive. The adhesion to tungsten carbide, particularly three dimensional tools like taps and drills, remains a persistent technical obstacle to overcome. CVD thick film can be used free-standing, or has to be brazed directly to a tool body or a substrate that is in turn brazed again to a tool body.

The strong acceptance of pcD in machining non-ferrous materials is based on its properties being highly suited to the requirements of modern day manufacturing. CVD diamond has similar properties and it is anticipated that it will gradually become integrated into diamond based cutting tool material selection and application.

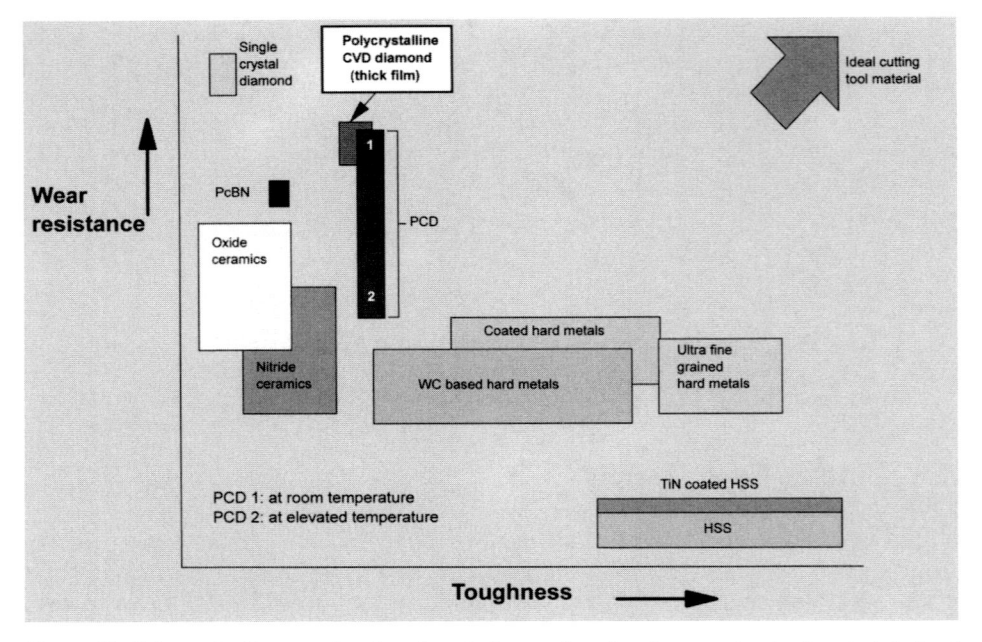

Figure 38. Schematic diagram showing the relative merits of various cutting tool materials for the machining of different materials. It can be seen that CVD diamond occupies a specific niche in the machining of materials requiring exceptional wear resistance.

2.6.1.1 Polycrystalline and Chemical Vapor Deposited Diamond

An indication of the relative position of various cutting tool materials with respect to their wear resistance and fracture toughness is shown in Fig. 38. Diamond is by far the most wear resistant material, particularly in pure single crystal form, but because single crystal diamond is a brittle material, it does not have the average toughness of the polycrystalline forms – both pcD (cobalt-containing) and CVD diamond (which contains no metal phases).

CVD thick film diamond complements pcD throughout a wide range of general machining applications. Its main advantage over pcD is its greater thermochemical stability. Potential disadvantages have been its relative brittleness, and its lack of electrical conductivity. But recent advances have resulted in the development of an electrically conductive grade of CVD diamond material, demonstrating that the use of spark erosion cutting and processing techniques (as are used extensively in the diamond tool making industry, and in particular for woodworking tools) could be also adopted for CVD diamond tools.

2.6.2 Cutting Tool Application of Chemical Vapor Deposited Diamond

The following case studies are typical examples of where CVD diamond's performance is increasing the range of diamond cutting tool materials available to industry.

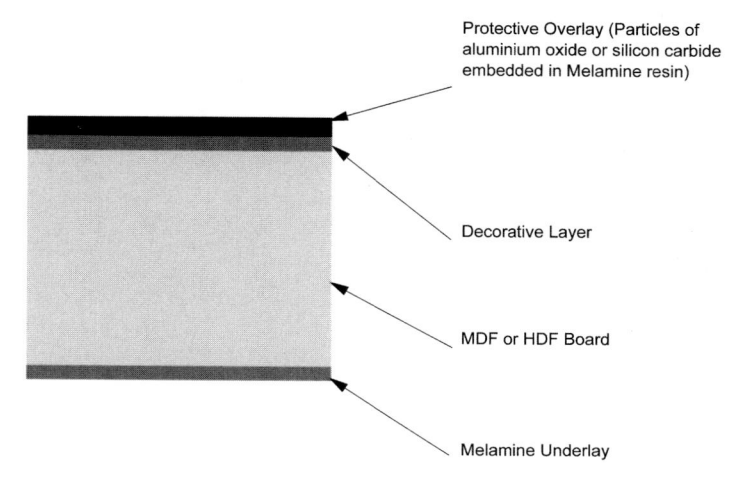

Figure 39. Section through typical HPL wood flooring panel.

2.6.2.1 High Speed Finish Machining of High Pressure Laminate Wood Flooring

During the 1990s, the wood working industry has seen a phenomenal growth of the use of PCD cutting tools in sawing, milling, and profiling of high and medium density fibre board, chipboard, particle board – all known to be abrasive materials. In the last five years, high pressure laminate (HPL) – a generic wood product, with its decorative aluminum oxide coated surface for high wear resistance, has emerged as a flooring material. Figure 39 shows the makeup of a HPL wood flooring material. In this application, CVD thick film diamond has recently been tested alongside pcD and single crystal diamond tools for the finish machining of the highly abrasive coated layer of HPL wood flooring material, with very encouraging results [84] as described below.

2.6.2.1.1 Machining Parameters

Material machined : HPL boards Al-oxide layer
Cutter diameter : 200 mm
No. of teeth : 8
Spindle speed : 6000 r.p.m.
Cutting speed : 63 m s^{-1}
Feed rate : 53 m min^{-1}
Feed per tooth : 1.1 mm

For CVD diamond materials, the tool edges were prepared by mechanical grinding using a standard cutter grinder and diamond wheels. The cutting edge of synthetic single crystal diamond was prepared by scaife polishing, and the pcD tools' edges were prepared by the standard Electric Discharge Grinding (EDG) method.

The criterion by which tool life is measured is chipping of the board edge. When chips of an 'unacceptable' size are produced, then the tool is changed. 'Unacceptable' chip size is determined visually by the machine operator.

2.6.2.1.2 Test Results Figure 40 shows, for each test, the average linear length machined by the diamond tools, before the tools were taken out on the basis of visible edge chipping of the floor panels.

The edge wear of CVD tools appears to have been more influenced by grain pull out and erosion at the grain boundaries than wear.

All the materials evaluated in these tests are capable of machining HPL flooring, however, it is quite clear that single crystal diamond gives the longest tool life and CVD diamond gives a marginal advantage compared with PCD. Both synthetic single crystal diamond and CVD diamond materials, while providing high tool life, need to be evaluated in the greater cost benefit equation for specific workpiece materials, the requirements of individual HPL flooring products and the manufacturing environment pertaining.

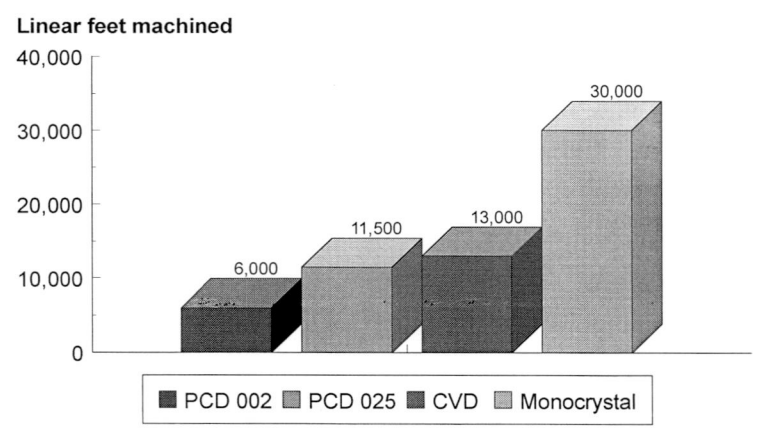

Linear feet machined

Tool	Linear feet machined
PCD 002	6,000
PCD 025	11,500
CVD	13,000
Monocrystal	30,000

Figure 40. Average linear length machined by different tool materials.

2.6.2.2 Dry Machining of SiCp Al Metal Matrix Composite

MMCs are typically alloys of aluminum reinforced with ceramic particles (usually silicon carbide). This composite material combination has the lightness of aluminum but the strength and temperature resistance of cast iron. Consequently, MMCs have tremendous potential in the automotive and aerospace industries for the replacement of cast iron to considerably reduce weight.

MMCs, although readily machinable, are highly abrasive and will dull the cutting edge of conventional tools in a matter of seconds. The performance of CVD diamond thick and thin film and the extremities of pcD grain sizes of 2 and 25 µm have been compared when machining an MMC. The machining conditions and parameters are detailed below.

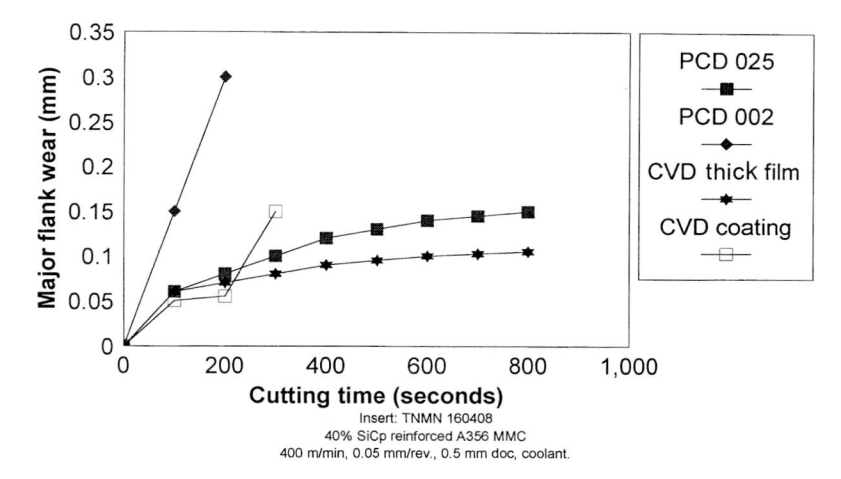

Figure 41. Flank Wear of CVD Diamond and PCD Machining MMC.

2.6.2.2.1 Machining Conditions/Parameters

Material Machined : 40% SiC$_P$ Al (A356) – MMC
Insert Style : TNMN 160408
Cutting Speed : 400 m min^{-1}
Feed Rate : 0.05 mm rev.$^{-1}$
Depth of Cut : 0.5 mm
No Coolant

2.6.2.2.2 Test Results Figure 41 shows the flank wear on the respective tool relative to the machining time. In the case of the CVD thin coated tool, initially the rate of tool wear is low but as soon as the 'thin' coating is worn through the flank wear increases rapidly. The lowest rate of tool wear is observed on the CVD thick film (0.5 mm) insert. Although testing is limited, the CVD thick film flank wear indicates that the edge has 'bedded in' and steady state wear has been established. This suggests that the ultimate tool life could be quite considerable.

The machining of MMCs, particularly high volume fraction, >35%, generates high temperature and involve having to efficiently shear hard silicon carbide particles in the cut zone. The high thermal conductivity/resistance to thermal degradation and abrasion resistance of CVD are particularly suitable for machining high volume fraction MMCs.

Table 6 summarizes the relative merits of the three major diamond cutting tool materials – pcD, CVD diamond, and synthetic single crystal – as current technology stands, from the perspective of tool performance.

It is clear that the various diamond cutting tool formats are largely complementary and, with the development of electrically conductive CVD diamond, the potential application areas of the machining of highly abrasive workpiece materials becomes a reality.

Table 6. The relative merits of different diamond cutting tool materials: tool performance.

	PCD	CVD diamond (thick film)	Single crystal
Toughness	✓✓✓	✓✓	✓
Abrasion resistance	✓✓	✓✓	✓✓✓
Thermo-chemical resistance	✓	✓✓✓	✓✓✓

2.6.3 Chemical Vapor Deposited Diamond Dressers

Diamond is extensively used for the dressing of precision grinding wheels. The grinding process using conventional, diamond and cBN abrasives is extensively used in the finishing of ferrous, tungsten carbide and ceramic components. Here, the geometrical tolerance, the surface finish of the wheel and its condition is critical and the dresser's function is to achieve and maintain these characteristics.

CVD diamond has recently been available in a range of dresser log formats and has shown promising results. CVD diamond potentially has a number of advantages compared to other diamond dresser material formats. The polycrystalline nature of CVD makes it insensitive to orientation. It also has extremely uniform wear characteristics. This means that dresser manufacture is more straight forward and the wear of the dresser is more consistent. The possibility of extracting long sections of CVD from a large disc means that products like blade dressers can be made with continuous diamond pieces compared to numerous discrete stones or pieces that are required to make up the working length of the dresser. Also CVD diamond has increased toughness, meaning that chip-free grinding of the dresser point is easier to achieve.

CVD diamond has been compared to single crystal synthetic diamond in single point dressing tests. The conditions and results of these tests are described below.

2.6.3.1 Dresser Test Conditions

Wheel Type: Aluminium oxide WA46 KV (300 mm O/D)
Wheel Speed: 2300 r.p.m.
In-Feed: 0.02 mm wheel diameter^{-1}
Feed: 4 mm s^{-1}
Flood Coolant

2.6.3.2 Test Results

From Fig. 42 it can be seen that CVD diamond exhibits almost identical rates of wear compared to single crystal synthetic diamond. Effectively this means that CVD diamond offers a range of geometric and orientation advantages in dresser design without having to compromise on the wear resistance of the dresser.

Dresser Wear (mm)

Dressing Parameters
Wheel   : Al-Oxide, WA46KV (300 mm dia. x 50 mm wide)
Speed   : 2300 rpm.
In-Feed : 0.020 mm / wheel dia.
Feed   : 4 mm / sec.
Coolant : Flood coolant.

CVD Diamond
MONODRESS

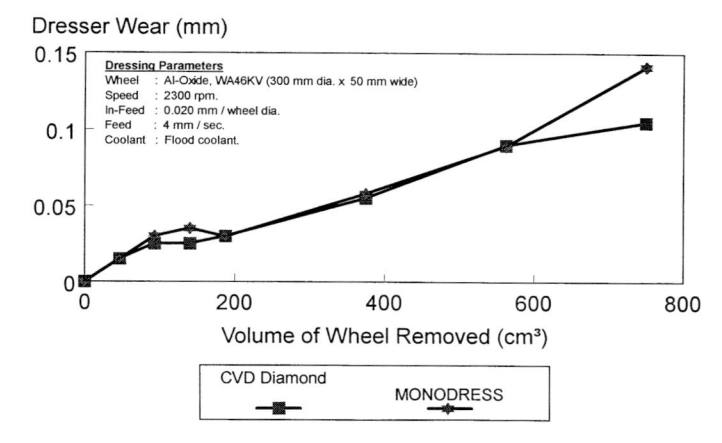

Figure 42. CVD diamond and Monodress single point dresser wear relative to volume of grinding wheel removed.

2.6.4 Chemical Vapor Deposited Diamond Wear Parts

The extreme abrasion and erosion resistance of polycrystalline CVD diamond combined with chemical inertness, make it a perfect candidate for use in applications such as wear parts, solid-state sensors, actuators and tools [85, 86]. The combination of diamond's hardness, high Young modulus, very low coefficient of friction (less than 0.1 in air) and low wear rates (less than $10^{-15}\,\mathrm{m^3\,N^{-1}\,m^{-1}}$) under extreme loading conditions [87, 88] potentially give diamond great tribological advantage over most other materials: The advent of polycrystalline CVD diamond deposition technology means that these tribological advantages can be exploited in applications such as large bearings or seals. These can be fabricated from either CVD diamond coated parts or from thick layers of free standing CVD diamond. Examples of these applications areas are pumps, valves and pipelines used in the extremely harsh tribological conditions seen in oil fields, often in remote locations [89].

Limited data is available on the wear rate of CVD diamond [90, 91]. A recent study has been conducted in order to investigate the basic tribological properties of bulk and thin-film CVD diamond. The sliding wear characteristics of a number CVD diamond materials have been evaluated using a modified Denison T62 abrasion [92] pin-on-disc test rig, operated in dry and lubricated sliding wear conditions, and compared with a typical hardened steel which is often used for wear part components.

Discs and pins of each material were made for testing; the pins all had chamfered edges to prevent edge-galling and were made square to minimise any misalignment thus reducing the contact area. Each material was tested first in dry sliding wear and then in lubricated sliding wear and only tested against components made of the same material.

The test parameters were identical in each case, with a contact stress of 15 MPa and sliding speed of 150 m min^{-1}. The tests were performed for 48 h where possible (to steady state or until sliding speed diminished through increased friction).

The tests all began at nominal room temperature of 20°C and, where required, the lubricant used was a typical heavy white mineral oil; excess lubricant was removed by a repeated solvent wash prior to weighing the pin to determine the specific wear rate.

The temperature rise associated with both the dry and lubricated wear tests did not exceed 90°C. Such a temperature rise can be associated with the low coefficients of friction (and carrier mechanisms during the lubricated tests). Such increases are not sufficient to noticeably alter the properties of the CVD diamond or to breakdown the mineral oil lubricant.

The coefficient of friction determinations have been made at specific times into the sliding wear tests. These have an associated uncertainty of approximately ±10%, predominantly from the (unsteady) change in friction coefficient with time as the sample surface is modified by wear and vibration/noise associated with the friction monitoring transducer.

It is worthwhile noting that there are several sources of error in the determination of the specific wear rate which give approximately an order of magnitude uncertainty in the quoted values, in particular, the mass loss through wear of the pins is minuscule – a consequence of the excellent abrasion resistance of CVD diamond – and such a low loss is easily affected by any residual lubricant.

As can be seen from the data presented in Table 7 the surface finish of the CVD diamond has a considerable effect on both the wear rate and coefficient of friction.

Table 7. Wear test results of CVD diamond and hardened steel

Material	Surface finish – R_a (µm)	Dry or lubricated	Specific wear rate ($\times 10^{-18}$ m^3 N^{-1} m^{-1})	Friction coefficient (initially)	Friction coefficient (at 24 hours)
Hardened steel	0.07	Dry	N/A	Galled at ~15 s	N/A
Hardened steel	0.07	Lubricated	1	0.08	0.08
CVDD (bulk) (Nucleation face)	0.06	Dry	986	0.115	<24 hour test
CVDD (bulk) (Nucleation face)	0.1	Lubricated	0.4	0.038	0.049
CVDD (bulk) (Growth face)	0.06	Dry	835	0.15	<24 hour test
CVDD (bulk) (Growth face)	0.1	Lubricated	0.7	0.018	0.018
CVDD (thin film) (As-grown face)	1.52	Dry	611	0.124	0.246
CVDD (thin film) (As-grown face)	1.52	Lubricated	27	0.11	0.08
CVDD (thin film) (Lapped face)	0.23	Dry	758	0.116	0.196
CVDD (thin film) (Lapped face)	0.23	Lubricated	84	0.11	0.05

However, whereas the hardened steel could not be run without a lubricating layer, the CVD diamond does not require lubrication or might be self-lubricating.

It is clear that there is a great potential for very low wear rate and low coefficient of friction CVD diamond materials prepared with the required surface finish for wear part applications, where component lifetime is determined by sliding wear in low lubricant or lubricant-free applications.

References

1. R. S. Sussmann, DIAFILM, A New Material for Optics and Electronics, *Indust. Diamond Rev.*, 1993, **53**, 555, pp. 63–72.
2. R. S. Sussmann, J. R. Brandon, G .A. Scarsbrook, C. G. Sweeney, T. J.-Valentine, A. J. Whitehead, and C. J. H. Wort, *Diamond Rel. Mater.*, 1994, **3**, 303–312.
3. C. J. H. Wort, C. G. $weeney, M. A. Cooper, G. A. Scarsbrook, and R. S. Sussmann, *Diamond Rel. Mater.*, 1994, **3**, 1158–1167.
4. R. S. Sussmann, C. S. J. Pickles, J. R. Brandon, C. J. H. Wort, S. E. Coe, A. Wasenczuk, C. N. Dodge, A. C. Beale, A. J. Krehan, P. Dore, A. Nueara, and P. Calvani, *Nuovo Cimento, Section D*, 1998, **12** (April), 503–526.
5. R. S. Sussmann, J. R. Brandon, S. E. Coe, C. S. J. Pickles, C. G. Sweeney, A. Wasenczuk, C. J. H. Wort, and C. N. Dodge, CVD Diamond, A. New Engineering Material for Thermal Dielectric and Optical Applications, *Proceedings of the Industrial Diamond Association of America meeting on Ultrahard Materials*, Windsor, Ontario 28–30 May 1998, and *Indust. Diamond Rev.*, 1998, **58**(578), 69–77.
6. C. P. Klages, Chemical Vapor Deposition of Diamond Films, in *Handbook of Ceramic Hard Materials*, R. Riedel (Ed.), Wiley-VCH, Weinheim, 2000, pp. 390–419.
7. G. J. Davies, Section 2 of Diamond Materials and their Applications, in *Handbook of Ceramic Hard Materials*, R. Riedel (Ed.), Wiley-VCH, Weinheim, 2000, pp. 485–504.
8. M. W. Bailey, Section 1 of Diamond Materials and their Applications, in *Handbook of Ceramic Hard Materials*, R. Riedel (Ed.), Wiley-VCH, Weinheim, 2000, pp. 479–485.
9. R. Berman Chapter 1, J. E. Field Chapter 9, D. Tabor Chapter 10, A. T. Collins Chapter 3, C. D. Clark, E. W. Mitchell and D. J. Parson Chapter 2, and C. A. Brooks Chapter 12, in *The Properties of Diamond*, J. E. Field (Ed.), Academic Press, London, New York, San Francisco, 1979.
10. C. D. Clark, A. T. Collins, and G. S. Woods Chapter 2, R. Berman Chapter 7. J. E. Field Chapter 12, C. A. Brookes Chapter 13, D. Tabor and J. E. Field Chapter 14, J. Wilks and E. M. Wilks Chapter 15, and J. E. Field Chapter 18, in *The Properties of Natural and Synthetic Diamond*, J. E. Field (Ed.), Academic Press London, San Diego, New York, Boston, Sydney, Tokyo and Toronto, 1992.
11. J. E. Field in *Proceedings of the NATO Advanced Study Institute on Diamond and Diamond-like Films and Coatings 22 July 1990*, R. E. Clausing, L. L. Hardon, J. C. Angus, and P. Koidl (Eds), Plenum Press New York, London, 1991, pp. 17.
12. Chapters 1, 3, 5, 7 and 8. in *Handbook of Industrial Diamonds and Diamond Films* M. A. Prelas, G. Popovici, and L. K. Bigelow (Eds), Marcel Dekker Inc. New York, Basel, Hong Kong, 1998.
13. Chapters 1, 2, and 5. in *Properties and Growth of Diamond*, G. Davies (Ed.), EMIS Data Review Services, INSPEC IEE, London, 1994.
14. J. Save, in *Infrared Optical Materials and Their Antireflection Coatings*, Adam Hilger, Bristol, Boston, 1985, p. 119.
15. D. C. Harris in *Infrared Windows and Dome Materials*, Tutorial Texts in Optical Engineering, Vol. TT10 – SPIE Optical Engineering Press, Bellingham, WA (USA), 1992, Chapter 5.

16. M Seal and W. J. P. van Enckevort, *Proceedings of SPIE – Diamond Optics*, 1988, **969**, 144–152.
17. M Seal in *Industrial Diamond Review*, P. A. Daniel (Ed.), 1971, **31** (Nov), 464–469.
18. J. Doting and J. Molenaar, *Proceedings of the Fourth Annual IEEE Semiconductor Thermal and Temperature Measurement Symposium, San Diego Califomia, February*. IEEE CH2530–418810000–0113, 1988, pp. 113–117.
19. R. J. Caveney in Chapter 20 of Ref. [9].
20. M. Seal in Chapter 16 of Ref. [10].
21. G. S. Woods in Chapter 3 of Ref. [13].
22. J. W. Vandersande in Chapter 3. of Ref. [13].
23. A. T. Collins in Chapter 9. of Ref. [13]; B. A. Fox, L. Harstell, D. M. Malta, H. A. Wynands, G. Tessmer, and D. L. Preifus in *Proceedings of the Materials Research Society Symposium*, 1996, **416**, 319–329.
24. M. Seal, *Indust. Diamond Rev.*, 1979, **39**(461), 115–118.
25. J. A. Savage, C. J. H. Wort, C. S. J. Pickles, R. S. Sussmann, C. G. Sweeney, M. R. McClymont, J. R. Brandon, C. N. Dodge and A. C. Beale. *SPIE Proceedings of the Window and Dome Technologies and Materials Symposium, Orlando, Florida, April*, 1997, **3060**, 144–159.
26. D. C. Harris, *Development of Chemical Vapour Deposited Diamond for Infrared Optical Applications, Status Report and Summary of Properties, Report No NAWCWPNS TP 8210*, Naval Air Warfare Center Weapons Division, China Lake, CA (USA), 1994.
27. M. Massari, P. Union, G. A. Scarsbrook, R. S. Sussmann, and P. Muys, *Proceedings of the SPIE Laser-Induced Damage in Optical Mateilals, Boulder, Colorado, November*, 1995, *SPIE*, **2714**, 177–184.
28. DIAFILM is a. trade mark of De Beers Industrial Diamond Division.
29. R F. Potter in *Handbook of Optical Constants of Solids*, E. D. Palik (Ed.), Academic Press Inc. Orlando. San Diego, New York, London, Toronto, Montreal, Sydney, Tokyo, 1985, Chapter 2.
30. M. E. Thomas and W. J. Tropf, *SPIE*, 1994, **2286**, 1144–151.
31. V. Vorlicek, J. Rosa, M. Vanacek, M. Nesiadek, and L. M. Stals, *Proceedings of the 7th European Conference on Diamond, Diamond-like and Related Materials*, Tours France, September 1996, *Diamond Rel. Mater.*, 1997, **6**, 704–707.
32. D. F. Edwards and H. R. Philipp, in *Handbook of Optical Constants of Solids*, E. D. Palik (Ed.), Academic Press Inc. Orlando, San Diego, New York, London, Toronto, Montreal, Sydney, Tokyo, 1985, subpart 3, pp. 668.
33. P. Dore, A. Nucara, D. Cannavo, G. De Marzi, P. Calvani, A. Marcelli, R. S. Sussmann, A. J. Whitehead, C. N. Dodge, A. J. Krehan, and H. J. Peters, *J. Appl. Optics*, 1998, **37** (July), no. 21.
34. P. Union, P. Muys, D. Vyncke, B. Depuydt, and P. Boone, *SPIE*, 1995, **2870**, 521–527.
35. C. S. J. Pickles, J. R. Brandon, S. E. Coe, and R. S. Sussmann, *Proceedings of Symposium IV, Diamond Films, 9th Cimtec – World Forum on New Materials, Florence, Italy, June 1998*, P Vincenzini (Ed.), Faenza, Italy, 1999, pp. 435–454.
36. T. J. Valentine, A. J. Whitehead, R. S. Sussmann, C. J. H. Wort, and G. A. Scarsbrook, *Diamond Rel. Mater.*, 1994, **3**, 1168–1172.
37. F. Szücs, University of Berlin and California Institute of Technology, Pasadena, California. Private communication, 1998, see also: F. Szücs, M. Werner, R. S. Sussmann, C. S. J. Pickles, and H. J. Fecht, *J. App. Phys.*, 1999, **86** (December) no. 11, 6010–6017.
38. F. Szücs, C. H. Moelle, S. Klose, H. J. Fecht, S. Fassbender, N. Meyendorf, and M. Werner, in *Proceedings of the International Conference on Micro Materials, April 1997 (MICRO-MA '97)* B. Michel and T. Winkler (Eds), 1997, pp. 569–572.
39. J. W. V. Vandersande in Chapter 1 of Ref. [13].
40. J. E. Graebner, M. E. Reiss, L. Seibles, T. M. Hartnett, R. P. Miller, and C. J. Robinson, *Phys. Rev. B*, 1994, **50**(6), 3702.
41. E. Wörmer, C. Wild, W. Muller-Sebert, R. Cocher, and P. Koidl, *Diamond Rel. Materi.*, 1996, **5**, 688–692.
42. See for instance P. A. Rizzi, Chapter 2.6 in *Microwave Engineering*, Prentice-Hall International, Englewood Cliffs, NJ (USA), 1988, p. 41.
43. A. C. Lynch and R. N. Clarke, *IEE Proceedings A*, 1992, **139**, 221.
44. A. P. O'Leary in *Electro-Optics, Special Report on Thermal Imaging Systems*, 2nd edn, Jane's Information Group, Antony Rowe Ltd, Chippenham, Wiltshire, UK, 1994.

45. E. J. Coad, *The Response of CVD Diamond and other Brittle Materials to Multiple Liquid Impacts*, PhD Thesis, University of Cambridge, 1996.
46. M. D. Hudson, C. J. Brierley, A. J. Miller, and A. E. Wilson in the *Proceedings of the Window and Dome Technologies and Materials Symposium, Orlando, Florida*, April 1997, *SPIE*, **3060**, 196–202.
47. C. J. H. Wort, J. R. Brandon, B. S. C. Dorn, J. A. Savage, R. S. Sussmann, and A. J. Whitehead in *Applications of Diamond Films and Related Materials: Third International Conference*, A. Feldman, Y. Tzeng, W. Yarbrough, M. Yoshikawa and M. Murokawo (Eds), NIST Special Publications, 1995, **885**, 569–572.
48. Rofin Sinar Technologies, Annual Report, 1997, pp 6. 7.
49. C. A. Klein, *Optical Engineering*, 1990, **29**(4), 343–350.
50. M. Sparks and M. Cottis, *J. App. Phys.*, 1973, **44**, 787–794.
51. S. Singer, Diamond optics, A. Feldman and S. Holly (Eds), *SPIE*, 1988, **969**, 168–177.
52. C. A. Klein in *Laser Induced Damage in Optical Materials*, 1991, *SPIE*, **1624**, 475–492.
53. C. A. Klein, *Diamond and Related Materials*, 1993, **2**, 1024–1032.
54. C. Wild, Chapter 10.3 in *Low-Pressure Synthetic Diamond*, B. Dischier and C. Wild (Eds), Springer Verlag, Heidelberg, 1998.
55. S. P. Timoshenko and J. N. Goodier, Chapter 12 in *Theory of Elasticity*, 3rd edn, 1970, McGraw-Hill, NY, pp. 385.
56. B. Depuydt, P. Boone, P. Union, and P. Muys in *Proceedings of the 4th International Worshop on Laser Beam and Optics Characterisation, Munich, June 1997*, A. Giesen and M. Morin (Ed.), 1997, 657–663.
57. N. Ellis and D. E. Greening *Diamond and Related Materials*, 1993, **2**, 572–581.
58. R. S. Sussmann, G. A. Scarsbrook, C. J. H. Wort, and R. M. Wood. *Diamond Rel. Mater.*, 1994, **3**, 1173–1177.
59. H.-K. Karow, *Fabrication Methods for Precision Optics*, John Wiley and Sons, 1993, p. 620.
60. Tim Holt, Rofin Sinar 1998, Private communication.
61. R. Heidinger, *Digest 19th International Conference on Infrared and Millimeter Waves, Lausanne (CH)*, SPIR, 1991, **1576**, 441.
62. L. Redcliffe, *Digest 16th International Conference on Infrared and Millimeter Waves, Sindon (J)*, JSAP Cat No AP 941228, p. 277.
63. M. Thumm, *State of the art of high power Gyro-devices and free-electron Masers, Wissenschaftliche Berichte FZKA 5564*, Forschungszentrum Karlsruhe GmbH, Karlsruhe 1995, (ISSN 0947–8620).
64. M. Thumm, J. of Infrared and Millimeter Wave, 1998, **19**, No. 1, pp. 3–14.
65. M. Thumm, *Fusion and Engineering Design*, 1995, **30**, 139–170.
66. M. Thumm, FZK Institute, 1997, Private communication.
67. 0. Braz, G. Dammertz M. Küjntze, and M. Thumm, *International Journal of Infrared and Millimeter Waves*, 1997, **18**(8), 1465–1477.
68. R. Heidinger, A. Ibarra, and J. Molia, *J. Nuclear Materials*, 1998, **258–263B**, part B, pp. 1882–1826.
69. R. Spörl, R. Schwab, R. Heidinger, and V. Parshin, *CVD Diamond for High Power Gyrotrons: Characterisation of Dielectic Properties*, ITG-Pachbericht, Berlin, April 1998.
70. R. Heidinger, R. Spörl, M. Thumm, J. R. Brandon, R. S. Sussmann, and C. N. Dodge, Programme Committee, 23rd International Conference on Infrared and Millimeter Waves, Colchester, Essex, UK, 'CVD Diamond Windows for High Power Gyrotrons', September 1998, pp. 223–225.
71. O. Braz, A. Kasugai, K. Sakamoto, K. Takahashi, M. Tsunekoa, T. Imai, and M. Thumm, *Int. J. Infrared Millimeter Waves*, 1997, **18**(8), 1495–1503.
72. A. Kasugai, K. Sakamoto, K. Takahashi, M. Tsuneoka, T. Kariya, T. Imai, O. Braz, M. Thumm, J. R. Brandon, R. S. Sussmann, A. Beal, and D. C. Ballington, *Rev. Sci. Instrum*, 1998, **68**(5).
73. K. Sakamoto, Japan Atomic Energy Research Institute, Private communication.
74. R. Heidinger, R. Schwab, R. Spörl, and M. Thumm, *22nd International Conference on Infrared and Millimeter Waves*, Wintergreen, Virginia, July 1997, P. Freund (Ed.).
75. R. Heidinger, FZK Institute, Private communication.

76. S. M. Hu, Critical stress in silicon brittle fracture, and effect of ion implantation and other surface treatments, *J. Appl. Phys.*, May 1992, **53**(5), 3576–3580.
77. R. C. Eden, in *Handbook of Industrial Diamonds and Diamond Films*, M. A. Prelas, G. Popovici, and L. K. Bigelow (Eds), Marcel Dekker Inc, New York, Basel, Hong Kong, 1998, Chapter 30, 1073–1102.
78. J. Bonhaus, D. Lorenzen, E. Kaulfersch, A. Denisenko, A. Zaitsev, S. Kwong, K. Man, H. Reichi, and W. R. Fahrner, *Proceedings of the International Conference on Micro Materials, Micro Mat '97, Berlin*, April 1997, B. Michel and T. Winkler (Eds), 1997, 505–508.
79. D Lorenzen, S. Heineman, J. Bonhaus, *Proceedings of the International Conference on Micro Materials, Micro Mat '97, Berlin,* April **1997** (Ed.: B. Michel and T. Winkler), **1997**, 513–516.
80. D. Lorenzen, Jenoptik Laserdiode, 1998, Private communication.
81. K. E. Goodson, K. Kurabayashi. and R. F. W. Pease, *IEEE Transactions on Components Packaging and Manufacturing Technology*, 1997, Part B. – Advanced Packaging 20-1, 104–109.
82. S. Weiss, E. Zakel, and H. Reichi, *IEEE Transactions on Components Packaging and Manufacturing Technology*, 1996, Part A. Vol. **19-1**, 46–53.
83. Patent assigned to Jenoptik Laserdiode, DE 196 44 941 C1 (D. Lorenzen, 1998. Private communication).
84. P. K. Sen, M. W. Cook, and R. Achilles, To be published: *Proceedings of the Industrial Diamond Association of America Meeting on Ultrahard Materials*, Windsor, Ontario 28–30 May, 1998.
85. S. J. Bull and A. Matthews, *Diamond for Wear and Erosion Applications, Diamond and Related Materials*, 1992, **1**, 1049–1064.
86. B. Lux and R. Haubner, *Phil. Trans. R. Soc. Lond.* A, 1993, **342**, 297–311.
87. T. Le Huu, H. Zaidi, and D. Paulmier, *Wear*, 1977, 203–204, 442–446.
88. M. N. Gardos in *Protective Coatings and Thin Films*, Y. Pauleau and P. B. Barna (Eds), Kluwer Academic Publishers, The Netheriands 1997, 185–196.
89. D. Cooper, F. A. Davies, and R. J. K. Wood, *J. Phys. D: Appl. Phys.*, 1992, **25**, A195-A204.
90. A. Edemir and G. R. Fenske, *Proceedings of the 51st Annual meeting of the STLE, Cincinnati – Ohio*, May 1996, 1–8.
91. R. A. Hay and J. M. Galimbarti, *Handbook of Industrial Diamonds and Diamond Films*, M. A. Prelas, G. Popovici, and L. K. Bigelow (Eds), Marcel Dekker Inc, New York, Basel, Hong Kong, 1998, Chapter 33, 1135–1147.
92. M. J. Neale (Ed.), *Tribology Handbook*, Butterworths, London, 1973 (2nd Edition 1994).

3 Diamond-like Carbon Films

C.-P. Klages and K. Bewilogua

3.1 Introduction

The term 'diamond-like carbon' was coined in the early 70s in order to denote amorphous carbon films, prepared by physical vapor deposition techniques, which were expected to approach genuine diamond, as far as density and mechanical properties are concerned. Today the expression (frequently abbreviated to 'DLC') is used for a fairly wide range of amorphous films containing carbon atoms in different states of hybridization and possibly also further elements such as hydrogen, silicon, or metals ('Si–DLC', 'Me–DLC'). There are presently no generally accepted criteria for the use of the term DLC or for a clear distinction between 'diamond-like' on the one hand and 'graphite-like' or 'polymer-like' on the other. This review will therefore pragmatically emphasize those kinds of amorphous carbon-based films which have found the greatest interest from a scientific or an applicational point of view. Aside from hydrogen-free amorphous carbon (a-C) and especially highly tetrahedrally coordinated amorphous carbon (ta-C), deserving the denotation 'diamond-like' most of all, hydrogenated films (a-C:H) and metal-containing hydrogenated films (MeC:H) will be treated in more detail, owing to the technical importance for mechanical-tribological applications which these materials have gained. Section 3.2 will present the deposition methods which have been used or are being used for the preparation of DLC films. Microstructures and properties will be treated in sections 3.3 and 3.4 and the final section will deal with applicational aspects of DLC coatings.

3.2 Preparation Methods for Diamond-like Carbon Films

3.2.1 Hydrogenated Amorphous Carbon (a-C:H)

In 1976 Holland and Ojha [1] reported the deposition of a-C:H films using a glow discharge in hydrocarbon gases excited by a radio frequency (RF, 13.56 MHz) powered substrate electrode. The counter-electrode was the grounded wall of the deposition chamber. Because this technique was relatively simple in operation the work of Holland and Ojha can be seen as the starting point of intensive research and development efforts aiming at practical applications of DLC films. Hydrogenated amorphous carbon films have relatively low electrical conductivities, typically less than $10^{-6}\,\Omega^{-1}\,cm^{-1}$. The operation of the substrate with RF allows deposition of a-C:H films with thickness of several micrometers on conducting as well as on insulating substrates without surface charging effects. On the other hand, attempts to excite the substrate electrode in the glow discharge only with DC (direct current) power were reported (e.g. by Whitmell and Williamson [2] and later by Grill and Patel [3]). There are indications that these DC deposition processes are sustained by secondary electrons generated at masks or shield surrounding the substrates [4, 5]. These secondary electrons compensate the surface charges. But at least for thicker films the charging of the films becomes an essential problem if their resistivity is higher than about 10^8 Ωcm. This can be explained by a simple estimation of the voltage drop over the film (see below). Today such two-electrode processes with only DC excitation have nearly no importance. However, there are intensive efforts to find alternatives to the RF processes, because this technique has only a limited potential for economic up-scaling of the coated parts dimensions. One alternative is the substitution of RF in the MHz range by lower frequencies, especially medium frequencies up to 100 kHz. In 1986 Catherine and Couderc [6] revealed that films prepared with 13.56 MHz or with 50 kHz, respectively, had nearly the same properties. At present the medium frequency technique is interesting above all from the viewpoints of up-scaling and costs.

Following the work of Holland and Ojha several other methods for preparation of a-C:H were developed, for example: sputter deposition [7, 8], ion plating [8, 9], cascade arc plasma jet [10], and simultaneous operation of a RF and a microwave (2.45 GHz) discharge [11].

Important parameters of all a-C:H deposition processes are the ion energy, commonly expressed by the negative substrate bias voltage U_b, the power density at the substrate, the pressure of the used hydrocarbon gas or vapor, the ion and neutral flux densities, and the substrate temperature. In contrast to the diamond film deposition the substrates must not be heated additionally. Therefore it is an essential advantage of DLC films that they can be deposited on temperature-sensitive parts. In Table 1 typical parameter ranges of different a-C:H deposition processes are summarized. Besides the technique, the type of precursors, gas pressures and bias voltage, the power density at the substrate and the linear growth rate are considered. Data on particle flux densities can be found only in few papers. All the mentioned parameters influence structure and properties of the films. It should be noted that before starting the deposition process, the substrates must be cleaned by sputter etching with noble gas ions in order to achieve a good adhesion.

The following discussion of characteristic features of a-C:H deposition processes will be concentrated on the RF and the ion plating method.

Figure 1 shows a scheme of the RF glow discharge deposition method. The RF power (the mostly applied frequency is 13.56 MHz) is capacitively coupled to the substrate electrode, the counterelectrode commonly is the wall of the vacuum chamber having ground potential. In the frequency range of some MHz only the electrons, but not the ions, can follow the RF voltage. Due to the large differences in the mobilities of electrons and ions a high negative DC self-bias U_b will be generated at the substrate electrode if the area of the powered electrode is markedly smaller than that of the grounded electrode (asymmetric system, see e.g. [12]). In

Table 1. Summary of a-C:H deposition parameters for different methods.

Deposition technique	Precursor C_xH_y	Gas pressure (Pa)	$-U_b$(V)	Power density (W cm^{-2})	Rate (μm h^{-1})	Remarks	References
RF plasma deposition	CH_4 C_6H_6–He	5–53 8–27		0.1–0.8 0.2–0.8	≤3	RF 13.56 MHz water cooled electrodes	[6]
MF plasma deposition	CH_4 C_6H_6–He	4–80 7–80		0.2–0.6 0.2–0.6	≤2	MF 50 kHz water cooled substrates	[6]
DC plasma deposition	C_2H_2 C_6H_{12}	≤10	500 800	0.05–0.2	≤5	$T_s = 180°C$	[3]
Microwave + RF plasma	CH_4–Ar	10	0–800		3–4	$T_s = 180$–220°C	[11]
Ion plating	C_6H_6	≈0.01	100–3000	0.2	≈3	Water cooled substrates	[8, 5]

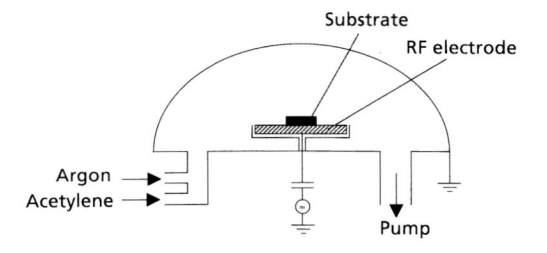

Process pressure: 1 to 10 Pa

Self-bias: -50 to -800 V

Power density: 0.05 to 0.5 W cm^{-2}

Current density: ~ 0.6 mA cm^{-2}

Deposition rate: ~ 1 μm h^{-1}

Figure 1. Schematic drawing of the radio frequency (RF) plasma technique for deposition of a-C:H films with some typical process parameters.

practice this condition is usually fulfilled. The self-bias drops across the plasma sheath in front of the negatively biases substrate electrode down to the plasma potential U_p. The maximum ion energy E_{imax} of the ions arriving at the substrate surface is $e(U_p - U_b)$, where e is the unit electron charge. In asymmetric systems the mean plasma potential is a few volts positive. Therefore U_b is a sufficiently good measure for the maximum ion energy. However, the ions can lose energy if their mean free path in the plasma sheath is smaller than the thickness of the sheath. For the pressure range typical for RF deposition of a-C:H (≈ 10 Pa) Bubenzer et al. [13] derived the relation $E_i \approx U_b/P^{1/2}$. Catherine and Couderc [6] confirmed this relation for the RF process. The self-bias U_b varies with the discharge parameters according to $U_b \sim (W/P)^{1/2}$ where W is the RF power and P the pressure of the process gas [6]. For the film growth rate the relation $R \approx U_b P$ was found experimentally [13]. The deposition rate also depends on the type of precursor C_xH_y used. The highest rates can be realized with benzene (C_6H_6) and cyclohexane (C_6H_{12}), the lowest with methane. One possible explanation is that the rate decreases with increasing ionization potential of the source gas [14].

The negatively self-biased substrate surface is bombarded with ions which are generated in the discharge ($C_xH_y^+$ and noble gas ions if noble gases are added). However, not only the $C_xH_y^+$ ions introduce carbon into the film. Comparing the calculated mass deposition rates of major ions (CH_3^+ for CH_4 gas and $C_6H_6^+$ as well as $C_6H_5^+$ for C_6H_6 precursor) with the experimentally determined deposition rates Catherine and Couderc [6] found clear discrepancies in the correlations between rate and RF power and between rate and pressure. In order to overcome these, the adsorption of activated species and their incorporation into the growing film must be taken into account additionally to explain the experimental data.

Concerning the details of a-C:H deposition mechanisms there is still a lot of uncertainties. A comprehensive discussion of the situation in this field was given by Möller [15].

Besides pressure and power the substrate temperature can influence the growth rate of a-C:H films. Commonly the rate decreases with increasing temperature [15, 16], owing to a declining contribution of adsorbed molecules to the film growth rate [15]. Thus the observed temperature-dependence of the growth rates is an additional support for the adsorption model.

In the RF glow discharge process the plasma generation and the film deposition will be realized by the same electrode. The advantage of the ion plating method is that plasma generation and film deposition are separated. Some typical process parameters for this method are noted in Table 1. The plasma can be generated by fast electrons (≈ 100 eV) from a hot cathode-anode arrangement [8, 9] or by microwaves [17]. Positive ions from the plasma then are accelerated to the negatively biased substrate. This is done with DC voltage only. A simple approximation shows that the use of DC becomes problematic for film resistivities $> 10^8$ Ωcm. The voltage drop over the film is estimated as $\Delta U = \rho d_f j_i$ where ρ is the resistivity of the film, d_f the film thickness and j_i the current density at the substrate. For typical values $d_f = 3 \mu$m, $j_i = 0.1$ mA cm^{-2} and $\rho = 10^8$ Ωcm [8] follows $\Delta U = 3$ V. For substrate voltages of 1000 V and more such a voltage drop is negligible [8, 5]. For higher resistivities or for insulating substrates this deposition method becomes unsuitable.

If benzene vapor is used, about 70% of the ions striking the substrate consist of $C_6H_x^+$ ($x \leq 6$). As in the case of RF plasma deposition not only ions contribute to the film growth. The contribution of the hydrocarbon ions to the film deposition is not more than 50%, further contributions are due to adsorbed and polymerized species [5].

The properties of the deposited a-C:H films, to be discussed in sections 3.4 and 3.5, strongly depend on the chosen process parameters, especially on pressure, power density, bias voltage and temperature. Therefore the deposition conditions must be carefully adjusted for preparation of films on parts or tools to be utilized in industrial practice. Today both RF plasma and ion plating deposition methods are in use for research and development and for prototype production in the industry.

3.2.2. Hydrogen Free Amorphous Carbon (ta-C)

There is a considerable interest in hydrogen free amorphous carbon films with high sp^3 contents (high contents of tetrahedrally bonded carbon atoms), because – compared to a-C:H – such films have properties much closer to diamond. For the preparation of ta-C films a source of energetic carbon species is needed, either with or without mass filtered carbon ions. Already in 1971 Aisenberg and Chabot [18] deposited carbon ions of 40–100 eV separated from a special source creating argon and carbon ions by sputtering a graphite target. In their pioneer work they introduced the designation diamond-like carbon for the deposited material. This was the first case of preparation of hydrogen free DLC. At present several different

methods to prepare such films are known [19, 20]:
(i) Ion beam deposition (with or without mass selection).
(ii) Sputtering from a graphite target.
(iii) Arc evaporation from graphite cathodes (with or without filtered beams).
(iv) Laser ablation from a graphite target.
(v) Ion-beam assisted deposition (IBAD: simultaneous bombardment of the growing film with noble gas ions).

Mass selected ion beam deposition is a special method with well defined conditions concerning ion masses and energies. Hofsäss et al. [21] for example used $^{12}C^+$ ions, in the most cases with deposition energies of 100 eV. This method is interesting for basic research, but seems to have no perspectives for industrial uses.

From the viewpoint of practical applications especially the sputter and arc evaporation techniques are interesting. Using the sputter deposition from graphite targets with argon and an argon-hydrogen mixture, respectively, Savvides and Window [22, 23] prepared amorphous carbon films with high contents of sp^3 bonded carbon. The sp^3/sp^2 ratio in the films increased with increasing energy per carbon atom. The increase of the sp^3 content in the case of hydrogen-containing sputter gas was explained by a preferred reaction of atomic hydrogen with sp^2 bonded carbon as known from diamond film deposition.

The vacuum arc evaporation technique seems to have a high potential for industrial applications. Arc plasmas are characterized by a high degree of ionization which is especially suitable to obtain high sp^3 contents in amorphous carbon films. A disadvantage of arc evaporation is the occurrence of graphitic microparticles. These particles can be avoided by filtering the particle beam [24]. However, such filter systems reduce the deposition rates drastically. As an alternative method which allows to control the arc process and to reduce the microparticle emission the laser arc technique was developed by Scheibe et al. [25]. Recently this technique was presented as a new industrial technology for high rate deposition of hard amorphous carbon [26].

In laser ablation systems a laser beam strikes on a graphite target surface and produces a flux of carbon material from the target to the substrate where the film grows (for a review see [27]). The emitted particles have a broad range of energies and can be utilized to prepare both hydrogen free as well as hydrogenated amorphous carbon films.

Ion beam assisted deposition techniques are characterized by a generation of carbon from a target by evaporation, laser ablation or sputtering [28] and a simultaneous bombardment of the growing film with argon ions. The low energy ion bombardment (up to some 100 eV) leads to dramatic changes in the film structure and the creation of dense carbon with high concentration of sp^3 bonds.

Several models are presently under discussion to explain the formation of highly tetrahedrally coordinated carbon ta-C. According to one of these theories the ta-C deposition can be described as a subplantation process [19, 29]. The principle of this process is that carbon ions penetrate the first atomic layers and enter interstitial positions, thereby increasing the local density and inducing local compressive stress. The densification can cause a change of the hybridization from sp^2 to sp^3. This process occurs in an energy range which obviously depends on the preparation method used [19]. Considering the experimental data, high sp^3 fractions in the a-C films (>70%) were found in the energy range between 20 and 600 eV. For too low energies the ions can not penetrate, will be stuck at the surface and form sp^2-carbon. For too high energies the ion dissipates the energy in a 'thermal spike' and relaxation of the densified regions occurs. At present several methods to prepare ta-C films are known, but mostly on a laboratory level. Therefore the deposition technique is still not suitable for broad technological applications as they are already realized in different fields for a-C:H or MeC:H coatings.

3.2.3 Metal-containing Amorphous Hydrocarbon

An essential problem to be solved before introducing coatings to practice is a good and reliable adhesion. The very promising properties of a-C:H films for a long time could not be utilized because of their high compressive stress. For thickness of more than 1–2 µm the adhesion, especially on steel, was insufficient. This problem is still more pronounced for ta-C films. Today there are possibilities to improve the adhesion markedly, for example by interlayer systems (for more details see Section 3.5).

Another way to overcome the disadvantage of bad adhesion is the incorporation of metals into the a-C:H films (MeC:H). Such films were first prepared by Dimigen and coworkers in 1983 using a reactive RF sputter technique [30, 31]. A large number of metals, either forming carbides (like Ti, Ta, and W) or not (Au or Cu), were investigated [32]. For applications in tribology the carbide-forming metals are preferable. MeC:H films with suitably adjusted metal contents turned out to have considerably lower compressive stress than a-C:H, but nearly the same excellent friction behavior. On the other hand the wear resistance of MeC:H coatings is lower than that of a-C:H (see Section 3.5.2). As a function of metal-content, minimum wear rates were measured for metal to carbon ratios of 0.1–0.2 [33]. With respect to industrial requirements, reactive DC magnetron sputter techniques were developed for the MeC:H preparation [34, 35]. Magnetron sputtering has a great potential owing to larger deposition rates about one order of magnitude higher than with the RF method and the excellent possibilities for up scaling of the coating devices. An industrial DC magnetron batch coater for the WC:H deposition was presented by Hofmann et al. [36]. Metals (mostly Ti, W) or metal carbides (WC) will be used as target materials. The carbon content can be controlled by the reactive gas content in the sputter gas (e.g. Ar + C_2H_2). Like in the a-C:H deposition the substrates must not be heated additionally. A substrate bias is not necessary to deposit MeC:H films, but a moderate bias voltage ($U_b = -100$ to -200 V DC) clearly improves the coating quality. Other typical process parameters to prepare MeC:H coatings with metal to carbon ratios of 0.1 to 0.2 are:

- power density at the target: 10 W cm^{-2}.
- acetylene concentration (ratio of gas flows [C_2H_2]/[C_2H_2 + Ar]): 30–40%
- total gas pressure: 0.3–0.8 Pa

- substrate sputter cleaning: Ar ions ($U_b = -600$ V DC)
- substrate current density: >1 mA cm^{-2} (unbalanced magnetron mode)
- substrate holder temperature: <200°C
- time for deposition process: about 100 min

Concerning the mechanism of MeC:H sputter deposition two components must be considered [35, 37].

(i) metal and carbon are sputtered from the target in a Ar-hydrocarbon gas mixture. The carbon material consists of loosely bonded hydrogenated carbon which will be deposited and resputtered continuously at the target surface. Such a mechanism can explain the very high deposition rates (more than 10 μm h^{-1} on substrates stationary in front of the target) reached with the DC magnetron [35].
(ii) Hydrogenated carbon is deposited directly from the plasma surrounding the substrate (a-C:H process similar to that discussed in Section 3.2.1). In the case of magnetron sputter deposition this contribution is small ($\approx 10\%$).

Today MeC:H coatings are widely used in industrial practice, for example for machine parts improved with wear resistant surfaces exhibiting low friction. On the other hand the wear resistance of MeC:H has still not reached the level of a-C:H. Therefore a further improvement of the MeC:H quality as well as a higher productivity and reliability of a-C:H and ta-C coatings will be necessary.

3.3 Microstructure and Bonding of Diamond-like Carbon

3.3.1 Amorphous Carbon and Hydrogenated Amorphous Carbon

Among the Group IV elements carbon takes an exceptional position owing to its ability to exist in bonding configurations formed by different hybridizations of its s and p atomic orbitals [38]: The tetrahedral sp^3 hybridization with four σ bonds to neighboring atoms is realized in the diamond allotrope, graphite is formed by trigonally sp^2-bonded carbon atoms, while polyacetylene (C$_2$)$_n$ is an example of a compound with sp-hybridized carbon with two σ bonds arranged to form a straight line. The description of DLC structures – even if based solely on carbon (a-C), without any hydrogen or metal incorporation – is complicated by the fact, that all kinds of hybridizations will in general be present in the amorphous matrix which may also contain nanocrystalline carbon phases. Additionally, there can be a chemical ordering, such as by clustering of sp^2-bonded atoms to longer chains or planar graphite-like segments within a predominantly sp^3-bonded matrix [39]. In hydrogen-containing DLC films as grown by plasma CVD or reactive PVD methods in hydrogen atmospheres, different amounts of H atoms can be bonded to the sp, sp^2 and sp^3 bonded carbon atoms, resp.

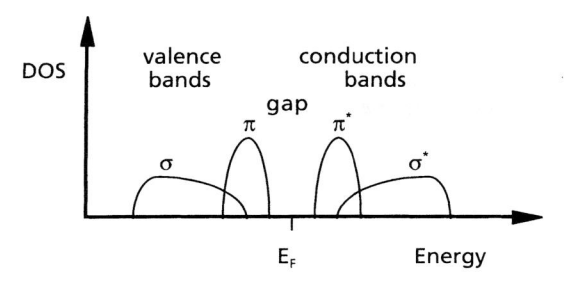

Figure 2. Schematic drawing of electron density (DOS) of states of amorphous carbon phases. E_F = Fermi energy (after [14]).

The most important methods which have been in use to reveal the structure and bonding of a-C and a-C:H films have been summarized in review articles [39, 40]. In order to understand these methods and the electrical and optical properties of DLC, a closer look to its electronic structure is necessary. Figure 2 is a simplified schematic diagram of the electron density-of-states (DOS) of amorphous carbon, as it is frequently shown in the literature (e.g. in [14]). A pair of σ and σ^* bonding and antibonding states is due to the bonds forming the skeleton of the a-C or a-C:H network. Owing to the weaker π bonds, the π/π^* couple is located more closely to the Fermi energy E_F. (Strictly speaking, the nomenclature of σ and π molecular orbitals would require the presence of a mirror plane within a molecule [38]. Nevertheless, this simplified concept is of certain value for a discussion of bonding in DLC.)

The hydrogen content in a-C:H films can be determined by a number of methods, including classical chemical combustion analysis, nuclear reaction analysis (NRA), elastic recoil detection (ERD), nuclear magnetic resonance (NMR) spectroscopy, or an analysis of C–H band intensities in infrared vibrational absorption spectroscopy (IR). The most reliable information about the relative amounts of sp^2- and sp^3-bonded carbon (sp-bonded C atoms usually do not play an important role) comes from solid state ^{13}C-NMR spectroscopy [41] and electron energy loss spectroscopy (EELS) [42–44] (see also the references to other original papers in [39] and [40]). The results of Weissmantel et al. already pointed to the existence of a special carbon structure in films deposited either by dual beam sputtering or by condensation of species from a low pressure ionization system operated with benzene. Fink et al. used RF plasma deposition of a-C:H from benzene and measured the EELS spectra for samples grown at different substrate bias voltages before and after annealing up to 1000°C. The $\sigma + \pi$ plasmon energy positions and intensities are significantly dependent on the sp^3 and sp^2 fractions. The concentration of carbon in sp^3 configuration was found to be roughly two third in as-grown films and increased after annealing. The hydrogen content declined strongly from 60 to 25 at-%, when the bias voltage was raised from 200 to 1200 V, accompanied by a significant density increase (1.49 to 1.75 g cm^{-3}). A strong dependence of the sp^2 carbon and the hydrogen content, in a-C:H films grown by RF or DC plasma and ion beam deposition, was also noted in the NMR measurements performed by Kaplan et al. [41].

A compilation of the sp^3 percentages and H contents of typical a-C and a-C:H films was given by Robertson [45]. Table 2, containing also density, hardness, and

Table 2. Structural characteristics and properties of various kinds of carbon, data taken from an earlier compilation [45].

	% sp^3	at-% H	Density, d (g cm^{-3})	Hardness, H (GPa)	Gap (eV)
Diamond	100	0	3.515	100	5.5
Graphite	0	0	2.267	–	−0.04
Glassy carbon	≈0	–	1.3–1.55	2–3	0.01
Polyethylene	100	67	0.92	0.01	6
a-C, evaporated	1	–	1.9–2.0	2–5	0.4–0.7
a-C, sputtered	2–5	–	1.9–2.4	11–15	0.4–0.7
a-C, MSIB (ta-C)	90 ± 5	<9	3.0	30–130	0.5–1.5
a-C:H, hard	30–60	10–40	1.6–2.2	10–20	0.8–1.7
a-C:H, soft	50–80	40–65	0.9–1.6	<5	1.6–4

MSIB = mass selected ion beam deposition.

band gap energy values, compares these special forms of carbon and hydrogenated carbon to a few 'conventional' materials. It is important to note, that a 100% sp^3 content, accompanied with a band gap energy in excess of 5 eV, is found in diamond as well as in polyethylene, representing the extremes of the hardness and density scales. In DLC, a high amount of tetrahedral bonding is present in very soft, polymer-like a-C:H films with high H content and low density as well as in amorphous a-C films, grown by mass selected ion beam deposition (MSIB), which have a low or zero H content and densities as well as hardness values approaching those of diamond. The latter, ta-C (tetrahedral amorphous carbon) films, which can also be grown by filtered vacuum arc deposition [46], have in recent years gained considerable scientific interest. Several models have been proposed to explain the formation of these very dense amorphous carbon phases [39], [47]. A prerequisite to obtain maximum density and sp^3 content is a suitable average kinetic energy of carbon ions in the 20–600 eV region. If the energies deposited per carbon atom during film growth are much smaller, e.g. during evaporation of graphite ($E \approx 0.1$ eV) or plasma deposition of an a-C:H film from a hydrocarbon at small substrate bias voltages, soft films of low density and hardness will be deposited, consisting predominantly of sp^2-bonded carbon (in a-C) and strongly hydrogenated sp^3-bonded carbon (in a-C:H).

Among the methods mentioned above, IR spectroscopy is by far the most inexpensive and – if the analysis of absorption band intensities is done with great care – a powerful tool for structural investigations of hydogenated amorphous carbon films. The interrelation of deposition parameters and structural data of Table 3 which were reported by Dischler [48], are partly based on IR spectroscopic data (namely the ratios of sp^n-hybridized carbon atoms and ratios of CH$_m$-bonded segments) and can be considered as typical for plasma deposition of a-C:H. Growth of a dense, hard phase with high refractive index is favored by high ion energies of carbon-containg species deposited on the substrate, which are achieved under conditions of high bias voltages and relatively low pressures. In case of the softer a-C:H film, a large amount of hydrogen is present in methyl groups which do not contribute to the rigidity of the film. For the harder sample, methyl groups are virtually

Table 3. Deposition conditions and structural properties of hard and soft, polymer-like a-C:H films (after [48]).

	a-C:H (hard)	a-C:H (soft)
Frequency (MHz)	2.3	2.3
Self bias voltage (kV)	1.0	0.1
Benzene pressure (Pa)	3.2	6.1
Estimated kinetic ion energy (eV)	100	10
Deposition temperature (°C)	50	25
$sp^3:sp^2:sp$ (%)	68:30:2	53:45:2
$CH_3:CH_2:CH$ (%)	0:40:60	25:60:15
Optical gap (eV)	1.3	3.0
Refractive index	2.0	1.65
Density (g cm^{-3})	1.65	1.3

absent and all sp^3-bonded carbon atoms make at least two bonds to neighboring carbon atoms, thus contributing to the cross-linking of the network.

Quantitative investigations of the connectivity in a-C:H random networks were introduced by Robertson [49] and Angus [50], based on the pioneering work on random covalent networks of Phillips, Thorpe and Döhler et al. [51–53]. In considerations of this kind, the number of translational degrees of freedom of atoms ($=3$ for one atom in three-dimensional space), forming the random network, are balanced out against the number of constraints N_{con} on their movement imposed by bonds to neighboring partners. It can be shown that in three-dimensional networks whose bonding is dominated by directed valence bonds to nearest neighbors the number of constraints for one atom is

$$N_{con}(r) = r^2/2 \text{ for } r \leq 2 \text{ and } N_{con}(r) = 5r/2 - 3 \text{ for } r \geq 2 \qquad (1)$$

r is the coordination number. As long as there are less constraints than degrees of freedom in a random network, atoms can still move in order to minimize strain energy, the network is 'floppy'. As soon as the degrees of freedom are exhausted by an increased number of bonds, the network will be fully constrained, additional cross-linking will then increase the strain energy rapidly. For a monatomic network this happens at an average coordination number $r_{av} = 2.4$. For a completely constrained network consisting of sp^2- and sp^3-bonded carbon atoms as well as hydrogen atoms with molar ratios x_3, x_4 and x_1, Angus [50] deduced an equation for the sp^3/sp^2 ratio (x_4/x_3) as a function of the hydrogen content x_1:

$$x_4/x_3 = (8x_1 - 3)/(8 - 13x_1) \qquad (2)$$

In order to be composed exclusively from sp^3 carbon atoms (corresponding to a zero denominator in Eq. (2)) and still be not overconstrained, a network should, according to Eq. (2), contain a hydrogen atom fraction x_1 of $8/13 = 0.615$ [54]. Modification have to be introduced, however, in order to take account of the clustering of sp^2-bonded carbon [55]. In early papers on DLC structure and properties, extended clustering of the sp^2-bonded carbon atoms to graphite-like aromatic planes was invoked in order to explain, among other properties, the small band gaps of a-C and a-C:H. For a band gap of, say, 1.2 eV, typical for hard a-C:H, clusters involving up to 25 benzene rings are required according to this assumption [40]. Modern atomistic simulations of the detailed electronic structure of a-C and a-C:H, however, point towards much smaller, mainly chainlike clusters, containing only 2–10 sp^2-bonded atoms. The small band gap is a result of the distortion of these structural elements, as discussed in more detail in [55]. In any case, the fact that ta-C films can be prepared, with sp^3 carbon fractions in excess of 80% and virtually no hydrogen, demonstrates, that the range of accessible sp^3 contents extends far beyond the limits suggested by the considerations of network connectivity.

The highly cross-linked, dense microstructure of amorphous carbon films, grown under the assistance of energetic particles, becomes also evident from diffusion and permeation experiments. As already pointed out by Angus and Jansen, the fact, that even after several years a virtually unchanged content of Ar atoms, entrapped during plasma or low-energy ion beam deposition from Ar-containing gas phases, can be detected in a-C:H films, shows that the Ar diffusion coefficient, at least in certain portions of the film, must be lower than 10^{-18} cm^2 s^{-1}. In conventional polymers, the corresponding figures are in the 10^{-9} to 10^{-6} cm^2 s^{-1} region [54]. Using depth profiling of deuterium by SIMS after exposition of a-C:H samples, grown by RF plasma deposition from acetylene, to D_2O vapor (160 h at 85°C and 80–85% relative humidity), Klages et al. [56] showed, that a virtually hermetical sealing against humidity can be achieved: In films of 120 nm to 1.9 μm thickness, deposited with a substrate bias of at least −50 V, D incorporation could be detected only within the outer 20 to 50 nm thick regions. Only in a 150 nm thick film grown at a bias voltage of −10 V, a diffusion of D through the whole film thickness was observed. Conventional organic polymers and coatings as used for corrosion protection, on the other hand, are easily permeable to water, oxygen and ions.

3.3.2 Metal-containing Amorphous Carbon Films

Structural investigations of metal-containing amorphous hydrogenated carbon films, prepared by co-evaporation or cosputtering of a polymer and a metal, plasma polymerization of a metal-containing monomer, or reactive sputtering of a metal or metal carbide in a hydrocarbon atmosphere, have generally pointed to a granular structure of the deposits, containing the metal or (in case of carbide formers) a metal carbide as nanocrystalline segregates in an amorphous matrix. A review of the relevant literature until 1989 was published by Klages and Memming [32].

Structural information came from transmission electron micrographs TEM and mainly from X-ray diffraction (XRD) measurements, revealing the crystallographic phases formed and (by application of the Scherrer equation $D = \lambda/\Gamma \cos\theta$) allowing an estimation of the average size of the particles D from the X-ray wavelength λ and the width Γ of a reflection on the 2θ scale of scattering angles. Reactively sputtered MeC:H films containing not too high concentrations of refractory metals Me = Ti, V, W, Mo, Nb were found to contain carbide particles MeC with sizes between about 1 and 10 nm typically. For iron-containing films the presence of the carbide Fe_5C_2 in superparamagnetic segregated nanocrystals was inferred from magnetic and Mößbauer spectroscopic measurements.

The microstructure of the amorphous a-C:H matrix surrounding the metal or metal carbide particles in MeC:H, especially its sp^3/sp^2 ratio, is still largely unknown. The application of IR absorption spectroscopy, for example, is possible only for very small metal contents, owing to the rapidly increasing optical absorption induced by the metal. A dominance of hydrogen bonding to sp^3 carbon atoms was indicated, for example, in Ta-C:H films containing 4 at-% Ta. The systematic shift of the G-line in Raman spectra of W-C:H towards larger wavenumbers (1528 ... 1564 cm^{-1}) with increasing W contents (7 ... 25 at-%) was used by Schiffman to argue in favor of a decreasing sp^3 content [57]. The hydrogen concentration generally decreases with growing metal content: In Ta-C:H, a H/C ratio of about 0.48 is found by Rutherford backscattering analysis (RBS) for a virtually Ta-free film grown from a methane-containing atmosphere, declining to 0.22 in films with a Ta/C atomic ratio in excess of 0.66 [32]. Using secondary ion mass spectrometry, similar results have been obtained for other metals, too [57, 58].

In recent years the application of small angle X-ray scattering (SAXS) working with synchrotron radiation has turned out to be valuable tool for an analysis of the nanostructure of MeC:H films [57, 59–61]. SAXS allows an investigations of structural inhomogeneities of solids, resulting in a modulation of the local electron density, on length scales in the 1–100 nm region. Figure 3 shows SAXS curves as obtained from measurements on reactively sputtered WC:H films with a varying content of W [57]. In all cases investigated so far in detail (Me = Au, Pt, W, Fe) the scattering intensity at large values of the scattering vector q (Porod region) (typically $q > 3.0\,nm^{-1}$) decreases as q^{-4}, indicating a relatively sharply defined, smooth surface of the particles. In the medium q region (Guinier region, typically $10 < q\,nm < 3$), a maximum or shoulder is usually observed, permitting a rough estimate of the inter-particle distance $d \approx 2\pi/q_{max}$. The frequently observed increase of intensity towards the lowest q values used points to the existence of a relatively small number of particles having a much larger diameter and therefore a significant

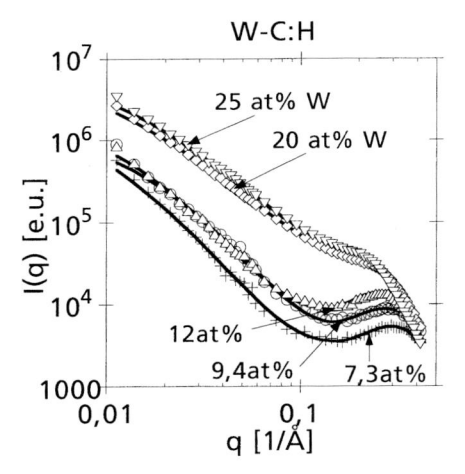

Figure 3. Small angle X-ray scattering (SAXS) curves for WC:H films with differing metal contents. Continuous lines = simulation, see text. (From [57] by courtesy of K. I. Schiffmann).

volume fraction, among a majority of smaller ones. The continuous lines in Figure 3 are simulated curves, based on a model of hard homogeneous spheres in a homogeneous matrix. In order to represent the particle sizes, a combination of two log-normal distributions was used, one main distribution representing the majority of particles with a relatively small width, plus an additional broad background distribution representing few, much larger particles. Particle radii and distances ranging from 0.6 to 0.8 nm and 2 to 3 nm, resp. could be deduced from these investigations for W-C:H films with W contents between 7 and 25 at-%.

Metal and carbide particle distances and radii in MeC:H films have also been measured by applying scanning tunnel microscopy (STM) imaging. In order to take into account tip convolution effects, leading to an appearant particle enlargement and to particle hiding, statistical methods were proposed by Schiffmann *et al.*, using an off-line tip radius determination by analysing appearant particle radii and a Monte Carlo model to compute fractions of hidden particles. A very good agreement was generally observed for radii and distances as measured by STM, TEM, SAXS and (for the radii only) XRD [57, 60].

The anisotropies in two-dimensional SAXS measurements as observed by Fryda [59] using sample tilting with respect to the incoming synchrotron radiation beam are, according to TEM micrographs of film cross sections [57], not due to an elongated shape of the segregated particles, but possibly to a chain-like ordering along the growth direction. The particles themselves are usually nearly spherical and do not exhibit a pronounced shape anisotropy along the growth direction. The average distance d of the particles, at a given volume fraction directly related to their average size, correlates quite closely with the melting point T_m of the corresponding bulk metal. The plot against T_m^{-1} in Figure 4 is based on the evaluation of the position of the maximum or shoulder in the Guinier region of SAXS scattering curves for a number of MeC:H films [59] plus the data for Au from [57]. Qualitatively, an increasing mobility of the atoms of lower melting metals

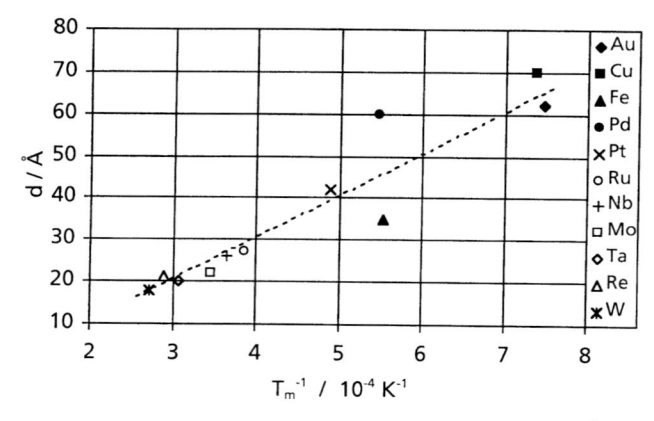

Figure 4. Plot of the average center of mass distances d (in Å) of segregated metal and carbide particles, respectively, in MeC:H films against the reciprocal melting points $1/T_m$ of the corresponding bulk metals, according to data taken from [59] and, for Au, [57].

during film growth and segregation of the metal or metal carbide phases can be made responsible for their increasing sizes and distances.

3.4 Physical Properties of DLC Films

3.4.1 Electrical and Optical Properties

The room temperature electrical conductivity of amorphous carbon varies over more than 15 decades from around 1 to $10^{-16} \, \Omega^{-1} \, cm^{-1}$, depending on the growth methods and growth parameters used [49, 62]. A fairly exhaustive review, including data, mostly measured over a range of temperatures, for samples prepared by evaporation, sputtering, ionized carbon and hydrocarbon deposition, laser evaporation, arc discharge deposition, and plasma deposition has been published by Frauenheim et al. [63]. The temperature dependence of conductivity has frequently been discussed in terms of Eq. (3) [64, 65]:

$$\sigma(T) = \sigma_0 \exp\{-(T_0/T)^n\}; \; n = 1 \ldots 1/4 \qquad (3)$$

The value of n depends on the conduction mechanism: $n = 1$ represents conduction by activation to extended states as in typical crystalline semiconductors, or by hopping (i.e., phonon-assisted quantum-mechanical tunneling) between localized states on nearest-neighbor sites, frequently found in highly doped crystalline and in amorphous semiconductors at sufficiently high temperatures. For decreasing temperatures, it will become increasingly difficult to find a nearest-neighbor site with a suitable energy separation and it becomes more favorable for the electrons to make larger hops to more distant sites, in order to find a site within a suitable energy distance: The geometry of the most probable path will depend on the temperature, the value of n becoming smaller than 1. Mott's famous $T^{-1/4}$ law will be obtained as a limiting case for a constant DOS around the Fermi energy. In nonhydrogenated amorphous carbon films with small sp^3 content, the transition to $n = 1$ behavior will happen only at relatively high temperatures, $n < 1$ has therefore been observed up to 300–500 K. For hydrogenated films, on the other hand, $n = 1$ prevails down to about 200–300 K temperatures and only at lower temperatures a transition to variable range hopping occurs [63].

Meyerson and Smith were the first to grow and investigate a-C:H films doped with boron and phosphorus, by additions of diborane or phosphine to the hydrocarbon gas used for deposition [66]. With a dopant concentration of 10%, the room temperature conductivity increased by about five orders of magnitude to $10^{-7} \, \Omega^{-1} \, cm^{-1}$. However, as shown by Thiele et al., this effect ist not a true p- or n-type doping in the conventional sense (i.e. a shift of the Fermi level towards the valence or conduction band) but an increase of localized states near the Fermi level, supporting a hopping conduction mechanism [67].

The mechanism of electrical conductivity in highly sp^3-bonded, virtually hydrogen-free ta-C, having the highest band gap among the dense amorphous carbon phases, and the question of its dopability is presently under discussion. Data published for undoped ta-C exhibit a large spread in room temperature conductivity, from 10^{-8} to $10^{-12} \Omega^{-1}$ cm^{-1}, even for quite similar sp^3 fractions [39]. Amaratunga *et al.* were the first to grow ta-C (filtered arc deposition) with additions of N, P, and B [68]. The conclusions drawn in this paper and in subsequent publications from this group have recently been critically reviewed by Hofsäss [69] on the basis of their own experimental results. According to his investigations, the conduction in undoped ta-C is mainly due to hopping in band-tails states and (at higher electrical fields) Poole–Frenkel excitation of electrons from traps into the conduction band. Doping with N or B leads to an increase of conductivity by 3–4 orders of magnitude at room temperature, presumably due to a generation of localized states related to three-fold coordinated B or N sites, contributing to hopping as well as Poole-Frenkel conduction. This view is supported by completely symmetrical *I–V* curves measured for layered diode-like structures containing N- and B-doped material.

Electrical properties of MeC:H films, deposited by reactive sputtering from metal or metal carbide targets in hydrocarbon atmosphere, have been described elsewhere [32]. In case of Ta as a typical refractory, carbide-forming metal, the electrical behavior can be divided into three categories, depending on the metal content:

- a 'dielectric' region with thermally activated conductivity according to Eq. (3) at less than about 13 at-% Ta,
- a transition region with metallic behavior (i.e. residual conductivity at vanishing temperature) but still a positive or zero slope of log(σ)–T curves and $\sigma > 10^2\ \Omega^{-1}$ cm^{-1} at medium Ta contents (14 and 22 at-% Ta) and
- a metallic region with with negatively sloped curves log(σ)–T and $\sigma > 10^3\ \Omega^{-1}$ cm^{-1} for 44 and 94 at-% Ta.

In contrast to Au- and Cu-C:H, for example, showing a percolation-type sudden increase of room temperature conductivity at a medium metal content of several 10 vol-%, the $\sigma(RT)$ values in Ta-C:H films tend to increase continuously by about six orders of magnitude right from the lowest concentrations in nearly metal free films up to less than 10 at-% Ta. It is only at low temperatures, that an insulator/metal transition is observable as a four orders of magnitude jump in σ(3 K) between 11.5 and 14.3 at-%. A detailed discussion of the electrical properties of MeC:H films is given elsewhere [70].

Similar to the electrical conductivities optical properties like absorption coefficients and refractive indices are critically dependent on the preparation conditions used to grow a-C or a-C:H. Except for absorption coefficients below about 10^{-4} cm^{-1}, the photon energy E dependence of the absorption coefficient α can generally be described by the Tauc relation (4) [71]:

$$(E\alpha)^{1/2} = G(E - E_0) \tag{4}$$

The slope G varies relatively weakly among different kinds of amorphous carbon films, the optical gap, however, depends quite strongly on the H-content and sp^3/sp^2 ratio of the films, owing to the fact, that sp^2-related π states (see Fig. 2) are mainly responsible for absorptions in and near to the visible region. Typical values for a-C and a-C:H are shown in Table 2. While an upper limit of 1.5 eV is still quoted in this survey for ta-C, prepared by mass selected ion beam deposition, considerably higher values approaching or even exceeding 3 eV seem now feasible according to the review [39]. Thus, the very hard and wear resistant hydrogen-free ta-C films might gain importance for optical applications in the visible region.

Refractive indices for ta-C films are typically close to the value of diamond ($n = 2.4$) [39]. Incorporation of hydrogen lowers n to values between about 1.8 and 2.3 in a-C:H. In plasma deposition of these films, hydrocarbon pressure P and substrate bias voltage U_B are the most important growth parameters allowing a tuning of n. Dense films with high refractive indices are favored by low P and high U_B. To a first approximation, holding for n up to 2, constant refractive indices can be achieved for a constant ratio U_B/P [62].

3.4.2 Mechanical Properties

For tribological applications of diamondlike carbon films, hardness, H, and stiffness, measured as the Young modulus, E, are key physical properties. The hardness of a material is its resistance against permanent geometric deformation under mechanical loading. In indentation testing, it is equivalent or proportional to the average pressure under a sharp indenter and can be calculated from the applied pressure divided through the (projected) contact area between the indenter and the material tested. For hardness measurements on thin films of less than 1 to several μm thickness, conventional microhardness measurements with microsocopic evaluation are difficult due to the small sizes of the indents. Therefore nanoindentation techniques working with a simultaneous registration of force and indentation depth with resolutions of a few μN and nm, resp. have been introduced [72]. Aside from the hardness, Young modulus can be estimated, using these methods, from the slope of the unloading curve [73]. Jiang *et al.* applied the method for an investigation of a-C:H films grown by RF plasma deposition from methane and acetylene [74, 75]. Hardness values between 5 and 20 GPa were typically found, correlating closely with Young moduli of the films, $H \approx 0.1\ E$. In MeC:H coatings, the metal content is the prime factor determining H and E [76, 35]: For WC:H and Ta-C:H, H increases from around 20 GPa for films with virtually zero or only a few vol-% of MeC precipitation to about 40 GPa at about 50 vol-%. Over the same range of metal contents E grows from 100 to 200 GPa ($H/E \approx 0.2$).

Even higher H and E values, approaching those of diamond ($\approx$100–120 and 1100 GPa) are characteristic for hydrogen-free, highly tetrahedrally coordinated ta-C films as to be obtained by ion beam or vacuum arc deposition, owing to the high average coordination numbers $r_{av} \geq 3.6$ in these networks [39].

Based on their characteristic chemical bonding and microstructure, as well as mechanical properties, a gross division of technically important materials into the three categories metals, polymers and ceramics is frequently made. It is tempting to compare the spectrum of properties of DLC films to those of these main material groups. In [33] this has already been done for the narrower class of tribologically interesting MeC:H films with low metal contents up to about 20 at-%. However, the essential results basically hold also true for DLC coatings in general. According to these considerations, the exceptional mechanical-tribological properties of DLC can be attributed to

- hardness values ($\approx$10–100 GPa) similar or even exceeding those of ceramic inorganic materials (10–20 GPa) and by far larger than for metals (<8 GPa),
- elastic moduli ($\approx$100–1000 GPa) similar or exceeding metals and ceramics (<600 GPa), combined with
- elasticities as measured by H/E ratios (0.1–0.2) which are throughout larger than for metals (<0.03) and ceramics (<0.06) and approach the lower limit of a range which is more typical for organic polymers (0.2–0.5), and
- for hydrogenated film surfaces, surface energies (30–50 × 10^{-3} N/m) in a region typical for polymers and much less than for metals (300–2000 × 10^{-3} N/m) and ceramics (200–1300 × 10^{-3} N/m) (see [33] for references to the sources used).

Friction coefficients, being not a physical property in the strict sense, but depending on the contact partners and many other external parameters, will be reported in the next section. The combination of low friction coefficients, typically ≤ 0.2, with mechanical and surface properties which are partly polymer-like (elasticity, surface energy) and partly ceramic- or metal-like (hardness, Young modulus) qualitatively explains the outstanding position of DLC films as low-friction, highly wear-resistant coatings on which most of their present applications are based [77].

3.5 Applications of DLC Films

Applications of DLC films in several fields like optics, magnetic disks, tribology and machine elements or biomedical purposes have been reported in the literature. A review on this topic was given by Matthews and Eskildsen [78].

The dominating field of application at present is that of tribological coatings on machine parts and tools. This shall be discussed in more detail. However, for all applications of the coatings a sufficient adhesion is required. Therefore this question will be treated first in the next section.

3.5.1 Adhesion of DLC Films

The high compressive stresses in hard a-C:H and ta-C films (up to 10 GPa, depending on the deposition conditions, especially the bias voltage) [79] frequently lead to film delamination and thus limit the attainable film thickness. For many years thicknesses of more than 1–2 µm could not be realized on technically relevant substrates like steel, hardmetal or glass. Considerable improvements were achieved by applying intermediate layers consisting of aSi:H [80], metals (Al, Ti, Ni) [81, 82], or multilayers (Ti–TiCN–TiC) [83], for example. For the characterization of

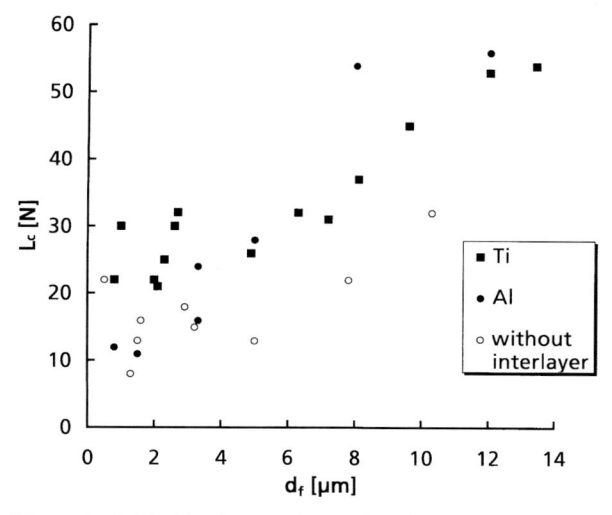

Figure 5. Critical load L_c against a-C:H film thickness d_f for deposition with Al- and Ti- interlayers or without interlayers, respectively, on polished steel substrates.

adhesion the critical load, L_c, or the crack pattern caused by a Rockwell indentor is frequently used. The critical load is a measure determined by scratch tests [84], utilizing a diamond tip moving over the film with continuously increasing normal load. L_c is the load where an intensive acoustic emission occurs, the friction coefficient between tip and sample drastically increases or a failure of the film can be observed by a microscope. All three criterions are in use. Therefore a comparison of data from different laboratories is difficult. Figure 5 shows the effect of aluminum and titanium interlayers on the critical load as a function of the a-C:H film thickness d_f. L_c was determined with the second and third criteria. The values determined with both methods were nearly identical. These films were prepared by RF plasma deposition from acetylene after magnetron sputtering of the metal interlayers (thickness about 0.5 µm) onto polished steel substrates in the same coating machine. In particular in the range $d_f > 3$ µm a pronounced improvement of the adhesion was found. The hardness of these coatings was about 25 GPa. Even films thicker than 10 µm having excellent adhesion can be achieved by this method. It should be noted that the effect of interlayers is still more pronounced if substrates with rougher surfaces, like ground substrates are to be coated.

However, the reliable and reproducible preparation of well-adhering a-C:H films is commonly not a simple task and requires a good control of several process parameters. Traces of oxygen in the films, to give just one example, can cause a drastic deterioration of the film adhesion.

On the other hand, there are very reliable industrial processes for deposition of MeC:H coatings. MeC:H films have lower stresses than a-C:H (<1 GPa) [34, 85] Therefore their tendency to delaminate is considerably reduced. Furthermore, intermediate layers can be prepared very easily using non-reactive sputtering from

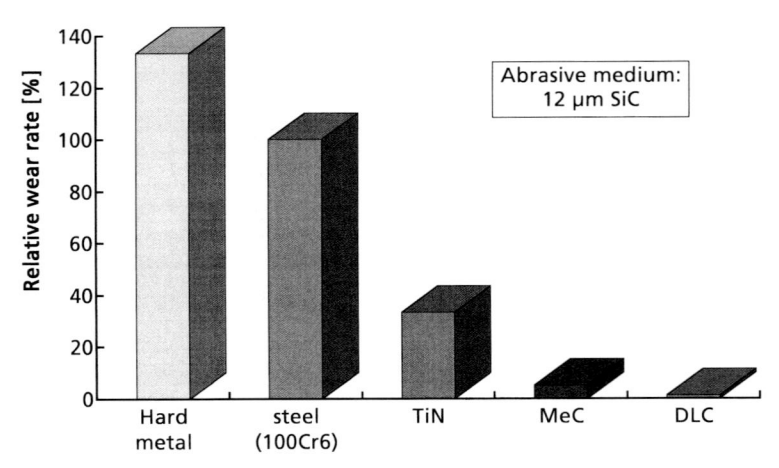

Figure 6. Abrasive wear of DLC coatings compared to other coating and substrate materials. As abrasive medium SiC with a particle dimension of 12 µm was used.

the metal target (see Section 3.2.3.), before the MeC:H deposition process is started.

3.5.2 Tribology of DLC Coatings

After satisfying solutions for the adhesion problem had been developed, DLC has become very interesting for tribological applications owing to the very favorable combination of low abrasive and adhesive wear with low friction coefficients. The following discussion concerns typical data on wear and friction of DLC films, but has to be taken into account that these properties can be influenced by the deposition conditions like bias voltage, gas pressure etc. Figure 6 shows the rates of abrasive wear for different types of hard coatings and reference materials. MeC:H and still more pronounced a-C:H exhibit excellent values. In Fig. 7 adhesive wear rates, measured in a pin on disc experiment, are presented. As pin a ball of bearing steel (100Cr6) was used. Figure 7a shows the results for MeC:H and TiN coated discs as well as for a steel disc, Fig. 7b the results for the 100Cr6 counterpart (pin). In the case of MeC:H the extremely low wear of the counterpart is an evidence of the low material transfer between steel ball and coating. It should be mentioned that the wear rates of MeC:H films depend on the metal content. However, for different metals [32, 85] and both for RF and DC magnetron sputter deposition the minimum values of abrasive wear were determined to be in the concentration range of 0.1 to 0.2 (metal to carbon ratio).

Friction coefficients, µ, against steel (steel 100Cr6) are shown in Fig. 8. While the friction coefficient of titanium nitride strongly increases with growing load due to a material transfer of steel onto the coating, the µ values of WC:H (Me/C ratios in the above mentioned range) even decrease with increasing load. The friction coefficients of metal-free a-C:H are very close to those of MeC:H. Very low friction coefficients (<0.2 against 100Cr6) were also determined for laser-arc

3.5 Applications of DLC Films 643

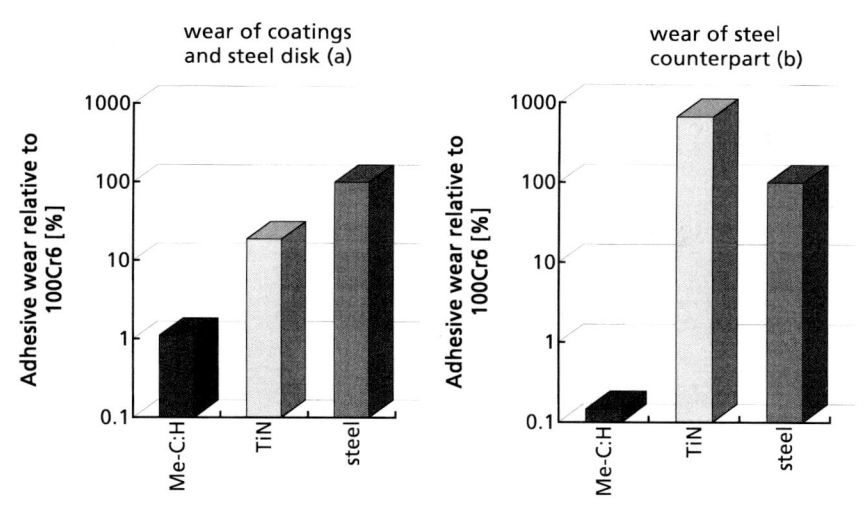

Figure 7. Adhesive wear measured in a ball on disk arrangement: (a) MeC:H and TiN coated discs and a disc of ball bearing steel 100Cr6 and (b) wear of the counterpart (100Cr6 ball). Measurements were performed under ambient conditions.

deposited films [86].

The friction coefficients of DLC films can considerably be lowered further by incorporation of silicon. Minimum μ values were reported for films with 10–20 atom-% Si [87, 88]. Another important factor influencing the friction properties is the humidity. For a-C:H and WC:H (W/C < 0.4) films the friction coefficients

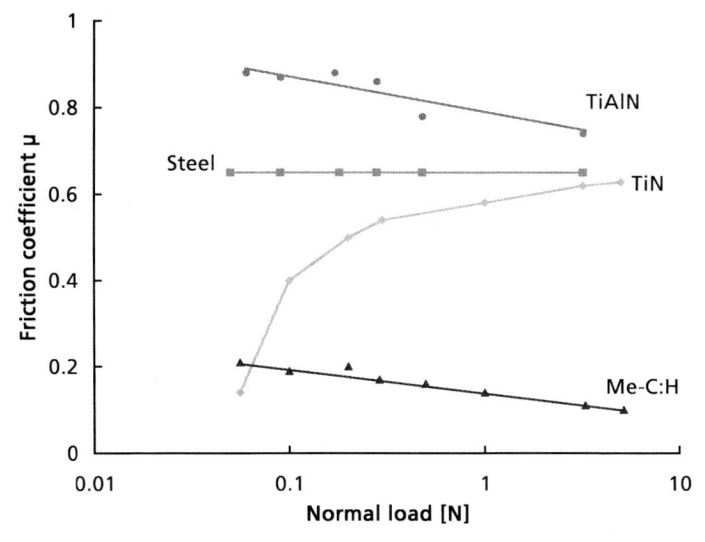

Figure 8. Friction coefficients against normal load for MeC:H, TiN, TiAlN and steel measured against ball bearing steel under ambient conditions (up to about 50% relative humidity).

drastically declined from $\mu > 0.2$ to $\mu < 0.05$ with decreasing relative humidity ($<1\%$) [89, 90]. For silicon-containing films very low μ values (<0.1) were measured even up to relative humidities of 75% [88, 91]. A possible reason for the different friction behavior could be a material transfer from the film to the steel ball (decreasing μ) or in the opposite direction (increasing μ) [90]. In the case of silicon-containing films reactions of Si with the atmosphere were assumed. The generated silicon ceramics are known to have extremely low friction coefficients in water [91]. However, there are still several open questions concerning the friction mechanisms of DLC films. It should be noted here that any published values of friction coefficients must be considered with caution because measuring conditions like humidity, loads, material of counterparts or sliding distance, which can have a strong influence on the result, can be extremely different [78].

3.5.3 Tribological Applications

DLC coatings have been applied successfully for industrially used components and tools to minimize wear and friction and in some cases to avoid the use of lubricants. For example gears, ball bearing races, valves, knives and other parts have been coated with WC:H, see [78]. Roth *et al.* [17] coated blades for fibreboard material and stamping tools with a-C:H. There is a great application potential for DLC films in the field of forming tools, such as for aluminum- and titanium-based alloys [92]. With coated tools the cost intensive and ecologically harmful employment of forming lubricants can be avoided or at least diminished.

By incorporation of certain elements the wetting behavior of DLC can be modified considerably. The surface tension of a-C:H can be decreased by additions of fluorine or silicon and oxygen, while it is enlarged by the incorporation of boron and nitrogen [81]. Low surface tension coatings have been applied successfully to reduce the adhesion of organic material on tools used for polymer fabrication. Another interesting application is the improvement of the heat transfer in steam condensation heat exchangers, thanks to the fact, that the water vapor does not form a heat-insulating water film on the coated heat exchanger surface, but it condenses in form of droplets which easily run down the vertically arranged heat exchanger, exposing a virtually water-free surface.

3.5.4 Other Applications

Due to their chemical inertness, high abrasion resistance, low friction and good biocompatibility DLC films are very promising candidates for biomedical applications. For example, femoral heads of hip protheses have been coated successfully with DLC to reduce the production of small wear particles, which can lead to reactions on a cellular level. Furthermore DLC was shown to be a potential material for artificial heart valves where the films must be non-thrombogenic and must have a long-time stability in contact with blood [93]. Mitura *et al.* [94] investigated DLC films prepared by ion beam decomposition, RF plasma and pulsed arc deposition, for coating implants. In spite of different film properties all three materials turned out as very suitable.

Ion beam deposited DLC films are used to increase the abrasion and chemical resistance of sunglass lenses, of float-glass windows in supermarket laser scanners and of infrared windows [95]. Very thin amorphous carbon films are used to protect magnetic disks and heads. Recently it was found that especially nitrogen containing carbon films (CN_x) have an excellent wear resistance (about four-fold value of a-C) [96]. One reason for this difference is presumably the different roughness of the films. Using an atomic force microscope (AFM) roughness values r.m.s. of 0.25 nm and 0.63 nm, respectively, were measured for CN_x and for a-C.

For the application of coatings, also cost and benefit aspects must be considered. It shall be noted here that Matthews and Eskildsen [78] discussed this topic in detail for four different DLC production methods taking into account prices for deposition equipments to be obtained in the UK.

References

1. L. Holland and S. M. Ojha, *Thin Solid Films* 1976, **38**, L17–L19.
2. D. S. Whitmell and R. Williamson, *Thin Solid Films* 1976, **35**, 255–261.
3. A. Grill and V. V. Patel, *Diamond Films Technol.* 1992, **4**, 219–233.
4. L. Holland, *J. Vac. Sci. Technol.* 1977, **14**, 5–15.
5. K. Bewilogua and D. Wagner, *Vacuum* 1991, **42**, 473–476.
6. Y. Catherine and P. Couderc, *Thin Solid Films* 1986, **144**, 265–280.
7. C. Weissmantel, K. Bewilogua, K. Breuer, D. Dietrich, U. Ebersbach, H.-J. Erler, B. Rau, and G. Reisse, *Thin Solid Films* 1982, **96**, 31–44.
8. C. Weissmantel, in *Thin Films from Free Atoms and Particles*, J. Klabunde (Ed.), Academic, New York, 1985, pp. 153–201.
9. T. Namba and Y. Mori, *J. Vac. Sci. Technol.* 1983, **A1**, 23–27
10. M. W. Kroesen, D. C. Schram, and M. J. F. van de Sande, *Plasma Chem. Plasma Proc.* 1990, **10**, 49–69.
11. L. Martinu, A. Raveh, A. Domingue, L. Bertrand, J. E. Klemberg-Sapieha, S. C. Gujrathi, and M. R. Wertheimer, *Thin Solid Films* 1992, **208**, 42–47.
12. B. Chapman, *Glow Discharge Processes*, Wiley, New York, 1980, Chapter 5.
13. A. Bubenzer, B. Dischler, G. Brandt, and P. Koidl, *J. Appl. Phys.* 1993, **54**, 4590–4595.
14. J. Robertson, *Surf. Coat. Technol.* 1992, **50**, 185–203.
15. W. Möller, *Appl. Phys. A* 1993, **56**, 527–546.
16. H. Kersten and G. M. W. Kroesen, *J. Vac. Sci. Technol.* 1990, **A8**, 38–42.
17. D. Roth, B. Rau, S.Roth, J. Mai, and K.-H. Dittrich, *Surf. Coat. Technol.* 1995, **74–75**, 637–641.
18. S. Aisenberg and R. W. Chabot, *J. Appl. Phys.* 1971, **42**, 2953–2958.
19. Y. Lifshitz, *Diamond Rel. Mater.* 1996, **5**, 388–400.
20. J. J. Cuomo, D. L. Pappas, J. Bruley, J. P. Doyle, and K. L. Saenger, *J. Appl. Phys.* 1991, **70**, 1706–1711.
21. H. Hofsäss, H. Binder, T. Klump, and E. Recknagel, *Diamond Rel. Mater.* 1993, **3**, 137–142.
22. N. Savvides, *Thin Solid Films* 1988, **163**, 13–32.
23. N. Savvides and B. Window, *J. Vac. Sci. Technol.* 1986, **A4**, 504–508.
24. A. Anders, S. Anders, and I. G. Brown, *Plasma Sources Sci. Technol.* 1995, **4**, 1–12.
25. H.-J. Scheibe, B. Schultrich, and D. Drescher, *Surf. Coat. Technol.* 1995, **74–75**, 813–818.

26. H.-J. Scheibe, D. Drescher, B. Schultrich, M. Falz, G. Leonhardt, and R. Wilberg, *Surf. Coat. Technol.* 1996, **85**, 209–214.
27. A. A. Voevodin and M. S. Donley, *Surf. Coat. Technol.* 1996, **82**, 199–213.
28. F. Rossi and B. Andre, *Jpn. J. Appl. Phys.* 1992, **31**, 872–879.
29. J. Robertson, *Diamond Rel. Mater.* 1993, **2**, 984–989.
30. H. Dimigen and H. Hübsch, *Philips Tech.Rev.* 1983/84, **41**, 186–197.
31. H. Dimigen and H. Hübsch, R. Memming, *Appl. Phys. Lett.* 1987, **50**, 1056–1058.
32. C.-P. Klages and R. Memming, *Mater. Sci. Forum*, 1989, **52–53**, 609–644.
33. H. Dimigen and C.-P.Klages, *Surf. Coat. Technol.* 1991, **49**, 543–547.
34. E. Bergmann and J. Vogel, *J. Vac. Sci. Technol.* 1987, **A5**, 70–74.
35. K.Bewilogua and H. Dimigen, *Surf. Coat. Technol.* 1993, **61**, 144–150.
36. D. Hofmann, H. Schuessler, K. Bewilogua, H. Hübsch and J. Lemke, *Surf. Coat. Technol.* 1995, **73**, 137–141.
37. J. T. Harnack and C. Benndorf, *Diamond Rel. Mater.* 1992, **1**, 301–306.
38. F. L. Pilar, *Elementary Quantum Chemistry*, MacGraw-Hill, New York, 1968.
39. Y. Lifshitz, in *The Physics of Diamond*, A. Paoletti and A. Tucciarone (Eds), IOS Press, Amsterdam, 1997, p. 209.
40. J. Robertson, in *Diamond and Diamond-like Films and Coatings*, R. E. Clausing et al. (Eds), Plenum, New York, 1991, p. 331.
41. S. Kaplan, F. Jansen, and M. Machonkin, *Appl. Phys. Lett.* 1985, **47**, 750–753.
42. C. Weissmantel, K. Bewilogua, D. Dietrich, H.-J. Erler, H.-J. Hinneberg, S. Klose, W. Nowick, and G. Reisse, *Thin Solid Films* 1980, **72**, 19–31.
43. S. Mühling, K. Bewilogua, and K. Breuer, *Thin Solid Films* 1990, **187**, 65–75.
44. J. Fink, Th. Müller-Heinzerling, J. Pflüger, B. Scheerer, B. Dischler, P. Koidl, A. Bubenzer, and R. E. Sah, *Phys. Rev. B* 1984, **30**, 4713–4718.
45. J. Robertson, *Diamond Rel. Mater.* 1992, **1**, 397–406.
46. D. R. McKenzie, D. Muller, B. A. Pailthorpe, Z. H. Wang, E. Kravtchinskaia, D. Segal, P. B. Lukins, P. D. Swift, P. J. Martin, G. Amaratunga, P. H. Gaskell, and A. Saeed, *Diamond Rel. Mater.* 1991, **1**, 51–59.
47. H. Hofsäss, H. Feldermann, R. Merck, M. Sebastian, and C. Ronning, *Appl. Phys. A*, 1988, **A66**, 153–181.
48. B. Dischler, in *Proceedings of E-MRS Symposium on Amorphous Hydrogenated Carbon, Symposium C, Strasbourg, France*, Les Editions de Physique, Paris, 1987, **XVII**, pp. 189–201.
49. J. Robertson, *Adv. Phys.* 1986, **35**, 317–374.
50. J. C. Angus, in *Proceedings of E-MRS Symposium on Amorphous Hydrogenated Carbon, Symposium C, Strasbourg, France*, Les Editions de Physique, Paris, 1987, **XVII**, pp. 179–188.
51. J. C. Phillips, *Phys. Rev. Lett.* 1979, **42**, 153–181.
52. M. F. Thorpe, *J. Non-Cryst. Sol.* 1983, **57**, 355–370.
53. G. H. Döhler, R. Dandoloff, and H. Bilz, *J. Non-Cryst. Sol.* 1980, **42**, 87–96.
54. J. C. Angus, F. Jansen, *J. Vac. Sci. Technol.* 1988, **A6**, 1778–1782.
55. J. Robertson, *Diamond Rel. Mater.* 1997, **6**, 212–218.
56. C.-P. Klages, A. Dietz, T. Höing, R. Thyen, A. Weber, P. Willich, *Surf. Coat. Technol.* 1996, **80**, 121–128.
57. K. I. Schiffmann, Thesis, University of Hamburg, published as *Fraunhofer-IST-Berichte aus Forschung und Entwicklung*, Fraunhofer IRB Verlag, Stuttgart, 1997, Nr. 1.
58. P. Willich and R. Bethke, in *Secondary Ion Mass Spectroscopy (SIMS X)*, A. Benninghoven, B. Hagenhoff and H. W. Werner, Wiley, New York, 1996.
59. M. Fryda, *Fortschritt-Berichte VDI, Reihe 5*, VDI Verlag, Düsseldorf, 1993, **303**, p. 20.
60. K. I. Schiffmann, M. Fryda, G. Goerigk, R. Lauer, and P. Hinze, *Ultramicroscopy* 1996, **66**, 183–192.
61. K. I. Schiffmann, M. Fryda, and G. Goerigk, *Mikrochim. Acta* 1997, **125**, 107–113.
62. J. C. Angus, P. Koidl, and S. Domitz, in *Plasma Deposited Thin Films*, J. Mort and F. Jansen (Eds), CRC, Boca Raton, FL, 1986, p. 89.
63. Th. Frauenheim, U. Stephan, K. Bewilogua, F. Jungnickel, P. Blaudeck, and E. Fromm, *Thin Solid Films* 1989, **182**, 63–78.
64. H. Overhof, P. Thomas, *Electronic Transport in Hydrogenated Amorphous Semiconductors*,

Springer Tracts in Modern Physics, Vol. 144, G. Höhler (Ed.), Springer Verlag, Berlin, 1989, p. 10.
65. R. Zallen, *The Physics of Amorphous Solids*, Wiley, New York, 1983, p. 252.
66. B. Meyerson and F. W. Smith, *Solid State Commun*. 1982, **41**, 23–27.
67. J. U. Thiele, B. Rubarth, P. Hammer, A. Helmbold, B. Kessler, K. Rohwer, and D. Meissner, *Diamond Rel. Mater*. 1994, **3**, 1103–1106.
68. G. A. J. Amaratunga, D. E. Segal, and D. R. McKenzie, *Appl. Phys. Lett*. 1991, **59**, 69–71.
69. H. Hofsäss, in *Proc. 1st Int. Specialist Meeting on Amorphous Carbon (SMAC '97)*, Amorphous Carbon: State of the Art, S.R.P. Silva *et al*., (Eds.), Singapore: World Scientific, 1998, p. 296–310.
70. H. Köberle, *PhD Thesis*, University of Hamburg, 1989.
71. J. Tauc, R. Grigorovici, and A. Vancu, *Phys. Stat. Sol*. 1966, **15**, 627–637.
72. P. Wierenga, A. J. J. Franken, *J. Appl. Phys*. 1984, **55**, 4244–4248.
73. J. L. Loubet, J. M. Georges, O. Marchesini, and G. Meille, *J. Tribol*. 1984, **106**, 43–48.
74. X. Jiang, K. Reichelt, and B. Stritzker, *J. Appl. Phys*. 1989, **66**, 5805–5808.
75. X. Jiang, K. Reichelt, and B. Stritzker, *J. Appl. Phys*. 1990, **68**, 1018–1022.
76. M. Fryda, K. Taube, and C.-P. Klages, *Vacuum* 1990, **41**, 1291–1293
77. E. Rabinowicz, *Friction and Wear of Materials*, Elsevier, Amsterdam, 1987.
78. A. Matthews and S. S. Eskildsen, *Diamond Rel. Mater*. 1994, **3**, 902–911.
79. M. Tamor, in *Proc. Third Int. Conf. on Applications of Diamond Films and Related Materials*, A. Feldman *et al*. (Eds), NIST Special Publications 885, 1995, p.691.
80. R. Butter, M. Allen, L. Chandra, A. H. Lettington, and N. Rushton, *Diamond Rel. Mater*. 1995, **4**, 857–861.
81. M. Grischke, K. Bewilogua, K. Trojan, and H. Dimigen, *Surf. Coat. Technol*. 1995, **74–75**, 739–745.
82. H. Ronkainen, J. Vihersalo, S. Varjus, R. Zilliacus, U. Ehrnsten, and P. Nenonen, *Surf. Coat. Technol*. 1997, **90**, 190–196.
83. D. P. Monaghan, D. G. Teer, P. A. Logan, I. Efeoglu, and R. D. Arnell, *Surf. Coat. Technol*. 1993, **60**, 525–530.
84. P. J. Burnett and D. S. Rickerby, *Thin Solid Films* 1988, **157**, 233–254.
85. M. Grischke, K. Bewilogua, and H. Dimigen, *Mater. Manuf. Proc*. 1993, **8**, 407–417.
86. H.-J. Scheibe and D. Klaffke, *Surf. Coat. Technol*. 1993, 57, 111–115.
87. S. Miyake, *Surf. Coat. Technol*. 1992, **54–55**, 563–569.
88. K. Oguri and T. Arai, *Thin Solid Films* 1992, **208**, 158–160.
89. K. Enke, H. Dimigen, and H. Hübsch, *Appl. Phys. Lett*. 1980, **36**, 291–292.
90. R. Memming, H. J. Tolle, and P. E. Wierenga, *Thin Solid Films* 1986, **143**, 31–41.
91. K. Oguri and T. Arai, *J. Mater. Res*. 1992, **7**, 1313–1316.
92. K. Taube, M. Grischke, and K. Bewilogua, *Surf. Coat. Technol*. 1994, **68/69**, 662–668.
93. R. S. Butter and A. H. Lettington, *Proc. Third Int. Conf. on Applications of Diamond Films and Related Materials*, A. Feldman *et al*. (Eds), NIST Special Publications 885, 1995, p.683.
94. S. Mitura, P. Niedzielski, D. Jachowicz, M. Langer, J. Marciniak, A.Stanishevsky, E. Tochitsky, P. Louda, P. Couvrat, M. Denis, and P. Lourdin, *Diamond Rel. Mater*. 1996, **5**, 1185–1188.
95. F. M. Kimock and B. J. Knapp, *Surf. Coat. Technol*. 1993, **56**, 273–279.
96. D. Li, E. Cutiongco, Y.-W. Chung, M.-S. Wong, and W. D. Sproul, *Diamond Films Technol*. 1995, **5**, 261–272.

4 Ceramics Based on Alumina: Increasing the Hardness for Tool Applications

A. Krell

4.1 Recent Trends in the Application of Ceramic Tool Materials

New trends in the development and application of hard ceramic grinding and cutting materials will be discussed in this Chapter. For both groups of tools, modern technical demands drive the development of submicrometer microstructures that exhibit significantly increased hardness and reliability. Manufacturing approaches and resulting properties will be described for both advanced single phase sintered alumina materials and for composite ceramics.

The fundamental difference between grinding and cutting tools is the existence of a geometrically well-defined cutting edge for the latter group, whereas grinding is performed by a plurality of geometrically not defined edges and tips. Most grinding grits are inorganic nonmetallic materials (diamond, cubic boron nitride cBN, fused or sintered ceramic grinding materials based on alumina Al_2O_3 or silicon carbide SiC). For cutting wheels the share of these materials in large-scale manufacturing is reported to be more or less constant at only 14–16%, compared with about 80% for hard metals [1]. Table 1 reviews the most important ceramic tool materials that are commercially available for the machining of metals, but this contribution will not discuss materials with limited applications (e.g. garnet grits for grinding wood and glass) or with super-hard characteristics (diamond or cBN).

In recent years, the intensity of research intended to find basically new solutions was rather different for two main sectors. For grinding metals there was an avalanche of patent applications in the early eighties associated with the development of the first polycrystalline (sintered) grits, and this stream of innovations continues now. Here, among the great number of new applications there is only a small percentage of 'speculative' inventions concerning materials with unsure scientific and technical prospects (e.g. an invention claiming diamond reinforced alumina without describing the sintered dense body, thus without determination of technically relevant properties). On the contrary, there is a remarkably large number of inventions which are regarded as so important that an international (European or PCT (Patent Cooperation Treaty)) application is filed without a foregoing national priority. The similarly large number of oppositions on the one hand and of reproaches for the violation of claims demonstrate that such patent applications describe commercially highly sensitive subjects. Most of the recent inventions concern the technology of materials and tools based on polycrystalline sintered corundum (α-Al_2O_3), but there are also propositions to manufacture SiC grits with a polycrystalline microstructure (where the importance of SiC grits is

Table 1. Commercially available tool materials for machining metals. HV1 (testing load 1 kg) and HV10 (10 kg load) hardness values are given for grinding and cutting materials to avoid the misuse of microhardness data

	Vickers hardness (GPa)	Fracture toughness K_{IC} (MPa m$^{1/2}$)	Bending strength (MPa)
Ceramic grinding grits	HV1		
Fused aluminas			
Brown corundum (Al_2O_3; impurities)	17.0–18.5	1.0–2.5	–
White corundum (Al_2O_3)	19.0–20.5	2.0–2.2	–
Ruby (Al_2O_3:Cr)	18.5–20.5	2.4–3.1	–
Al_2O_3/ZrO_2 eutectic	15.5–17.5	4.0–5.0	–
Sintered corundum (sol/gel)			
with nearly equiaxed crystal shape	14.0–22.5	2.9–4.1	–
with platelet-shaped crystals	18.7–19.0	3.3–3.7	–
SiC	25.5–27.5	2.5–3.0	–
Cutting tool ceramics			
Al_2O_3	HV10		
+ 4% ZrO_2	18.0–20.0	3.6–4.5	500–600
+ 15% ZrO_2	17.0–19.0	5.0–6.0	600–750
Al_2O_3 + Ti(C, N) [≈ 30%]	19.0–22.0	4.0–5.5	550–800
Al_2O_3 + SiC$_{Whisker}$ [≈ 20%]	17.0–20.0	6.0–8.0	650–950
Si_3N_4	15.0–18.0	6.5–9.0	750–1000
cBN cutting tools	27.0–40.0	3.5–6.6	600–800
Diamond (polycrystalline)	>4500	6.0–9.0	750–1000

generally less compared with alumina because of its poor ability to machine ferrous workpieces).

On the other hand, there has been some stagnation in the field of tools with geometrically defined cutting edges after the development of advanced Al_2O_3/ZrO_2 and Al_2O_3/TiC composites at the beginning and of tough Si_3N_4 cutting tools and SiC-whisker reinforced grades at the end of the eighties (high-strength tetragonal zirconia polycrystals cannot be used for machining metals). At least until 1997 it was the position of most manufacturers that they can satisfy the current requests of their customers with the available ceramics. Consequently, many laboratories are busy performing extended screening tests with commercial grades. A somewhat special position has to be attributed to efforts intended to improve the chemical resistance of Si_3N_4 by surface coatings. A very recent tendency is the increasing interest in a new class of materials with submicrometer microstructures, a trend which is discussed in Section 4.4.

The strong motivation for these developments in grinding materials is the real possibility of new inventions able to surpass the present state of the art by 50–100% (in terms of rate of material removal or maximum time of use of a tool), the chance to find materials that redouble the productivity of the industrial grinding operation and reduce substantially the time-out for the change of tools.

Significant recent changes in the market shares of the different kinds of cutting ceramics shown in Table 1 should be noted. There has been an increase in the use of ceramic cutting tools, attributed to increasing use of silicon nitride tools.

Two different reasons exist. First, the requirement for high *reliability* in automated manufacturing lines (without running a tool to the possible maximum time of use) favors the application of tool materials with a high fracture toughness (like Si_3N_4) tolerating, in the mean time, their lower hardness. Second, a shift in the ratio of machined materials from steel towards globular and vermicular cast iron also requires tougher tools like silicon nitride, especially when high feed rates are necessary. A drawback of nitride tools is their rather high chemical reactivity with iron: against steel and cast iron (like GG25), at feed rates of about 0.3 mm rev^{-1} the wear of Si_3N_4 is usually higher than for Al_2O_3-based ceramics. Hence, even for machining cast iron, nitride tools can be applied with advantage only when it is impossible to use the less tough Al_2O_3/ZrO_2 or Al_2O_3/TiC composites due to the required high feed rates. Attempts have been made to overcome this problem by TiN and/or oxide coatings, but real success was obtained only in some very special cases.

There is another trend, which has not yet altered market shares, because the new qualities are not commercially available: this is the substitution of certain grinding operations by highly precise cutting approaches, especially in machining very hard materials. It is expected that turning may cut the machining time to 1/3 compared with grinding, and is sometimes associated with significant savings of consumed energy, and even the compromise of first turning with subsequent final grinding reduces the time to 2/3 compared with an all-grinding process. Additional advantages are the less expensive recycling of turning chips compared with the waste disposal of the grinding silt and investment costs of the equipment which can be reduced by a half for precise turning [1, 2]. On the other hand, investment in new machine tools is imperative for precise cutting of hard materials (the mechanical stability of the most machine tools used today in the automotive industry is probably not sufficient for the new demands). More widespread use of cutting rather than grinding is restricted by the limited lifetime of today's cutting ceramics and the increase in workpiece roughness after rather short cutting paths. Investigations by large-scale manufacturers indicate a lack of real alternatives to cBN for turning hard workpieces with high feed rates, and drilling and milling is the field of hard metal tools, but for precise finishing Al_2O_3/TiC composites exhibit an advantage in the cost–performance ratio [2]. It is expected that such findings will stimulate the development of significantly more fine grained and harder cutting tool ceramics based on corundum. For example, it was in 1997 that a German company launched a new submicron $Al_2O_3/Ti(C,N)$ composite. Indeed, such $Al_2O_3/Ti(C,N)$ tools show the fewest problems when compared with other commercial ceramics (which, however, all have not been developed to meet the demands of precise machining of hard surfaces), but an analysis of the technical demands under this special condition points to better opportunities with other ceramics (cp. Section 4.4.4).

4.2 Technological Essentials for Producing Hard and Strong Tool Ceramics

It is beyond the scope of this paragraph to give a review of ceramic technologies, but the development of tools with an advanced *technical* performance needs an understanding of the required *basic* mechanical parameters, and of the real chances new technologies offer for their improvement. Also, it must be emphasized that the properties of different new tool materials can be compared only if the effect of their different microstructures is not obscured by processing-related imperfections: the investigation of new materials requires samples that are different in their microstructures but with similarly small frequency and size of defects. Much care has to be taken to perform this recent research (the results of which are described below) because for extremely fine-grained microstructures this demand is not easy to meet.

4.2.1 Typical Defects in Ceramics Tool Materials: The State of The Art

It is known that defects reduce the strength of brittle materials due to stress concentrations, which cannot relax by plastic deformation, as in metals. However, there is also evidence that the flaw population has a similarly strong impact on the hardness. The most promising way to develop new tool materials with improved hardness, wear resistance, and reliability is to reduce the grain size in the sintered microstructures: an approach which requires the use of increasingly fine-grained raw materials, be it within the framework of advanced powder technologies or of sol/gel or precursor approaches.

Unfortunately, it is just this trend towards extremely fine-grained, nanoscale, raw materials that often prevents the realization of the expected progress in the mechanical properties, simply because with the reduced particle sizes there is an increased risk of defects in the green microstructure.

4.2.1.1 Powder Processing

All 'dry' powder pressing technologies require the powder to be granulated before compaction, otherwise the specific volume of non-agglomerated particles in or beyond the micrometer range is so great that on compaction the decrease of volume becomes excessive, causing extended elastic relaxation (and associated cracking) on removing the load. The key is to produce granules (by spray drying or other approaches) that are large and strong enough for technical handling but soft enough to be completely destroyed on compaction to avoid relicts of the granules in the pressed and finally sintered bodies (Fig. 1). Cold isostatic pressing can significantly reduce the size and frequency of such defects (provided there is optimized slurry preparation and granulation [3]), but it is impossible to avoid them entirely.

Liquid shaping is a promising alternative. Avoiding granulation, a slurry of a dispersed powder is poured into a mold. A problem with liquid shaping is that homogenization and pouring introduces air into the slurry (the more the more fine-grained the powder). The *complete* elimination of gas from the slurry is difficult, and some porosity may appear in the green bodies that cannot be removed in solid state sintering.

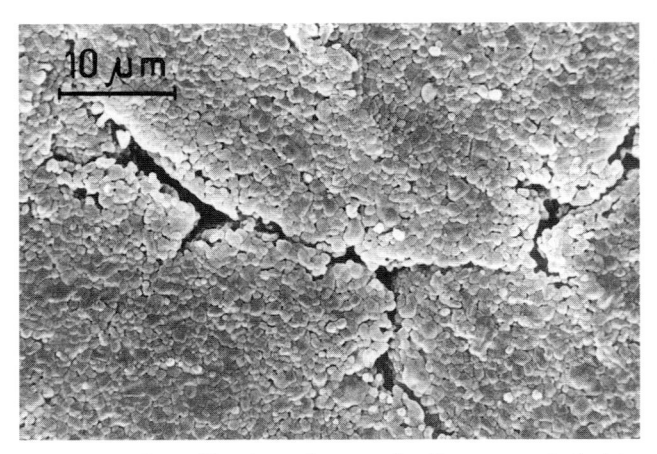

Figure 1. Relicts of hard powder granules (from spray drying) in pressureless sintered alumina ceramics (solid state sintering).

4.2.1.2 Sol-gel Technologies

One of the reasons for developing sol/gel and precursor technologies is the use of raw materials with particle sizes < 30–50 nm, enabling the production of very fine-grained microstructures. However, the raw material exists in a phase composition which is different from the final product, and a complex sequence of phase transformations takes place before or during sintering. Most of these transformations are controlled by the formation of (or pre-existence of artificially introduced) seeds. Without seeding, the transformation starts at a few isolated locations, extends from these nuclei in three dimensions, and may produce a textured microstructure with a preferential crystallographic orientation in rather large subregions and with a porosity that is hard to remove even at high sintering temperatures (promoting grain growth and destroying the benefit of using the nanoscale raw material).

But even *with* seeding the fundamental problem is that a seed is required from a phase formed in a later transformation step, and the seed particles should be as fine-grained as possible to be homogenized (at a low concentration) with small distances between them. However, seed particles are always *much larger* than the particles of the precursor or in the sol (if they were available more fine-grained, people would use them as a raw powder). As a consequence, the coarser character of the seed determines the spatial homogeneity of the transformation and of the resulting microstructure. With a typical size of seed particles between 0.05 and 0.3 µm and seed concentrations between 1 and 5%, the homogeneity cannot be better than in the micrometer range. Therefore, even advanced sol/gel derived microstructures contain defects like that in Fig. 2. On the one hand, submicrometer materials with a narrow grain size distribution are easily formed and can be rather perfect over distances of several dozens of micrometers. On the other hand, it is almost impossible to avoid small (0.5–5 µm) clusters of very small pores, the spacing of which seems to correlate with the spacing of seeds in the sol/gel process.

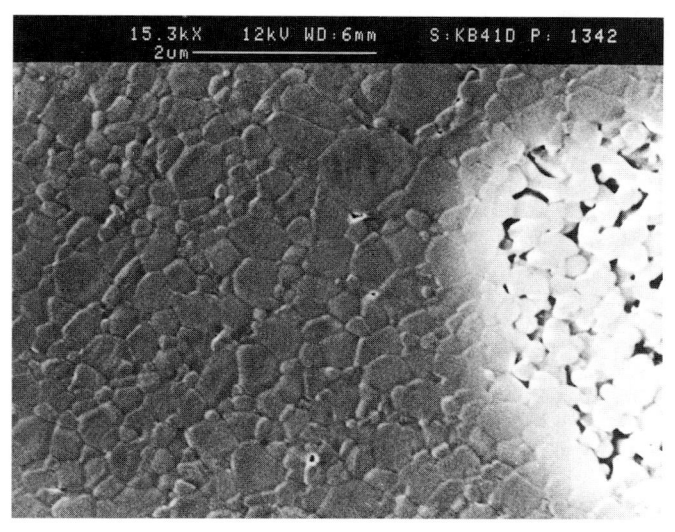

Figure 2. Microstructure with typical cluster of micropores in a submicrometer sintered alumina ceramic produced from a boehmite sol with corundum (α-Al_2O_3) seeds <0.2 µm.

Typical flaws like the small pores in Fig. 2 are too small to be seen with the optics of most hardness testers. Hence, an evasive positioning of indents is impossible, and the testing gives a hardness which will be considerably lower than the inherent hardness of the perfect submicrometer microstructure. In fact, this reduced hardness represents a 'true' technical property of the material (because the flaws will affect other hardness-related properties as well, e.g. the tribological behavior), and the real drawback comes from the limitation of assessable *properties* and not from the influence of measuring procedures.

4.2.2 Recent Trends in Ceramic Technologies Related to Tool Ceramics

4.2.2.1 Technological Background

The drawbacks of known industrial technologies have initiated a number of new developments in both powder synthesis and powder processing. Within the first 5–10 years, it is often difficult to distinguish which of the problems of a new technical approach are simply associated with the present, not optimized state of the art, and which are really characteristic of the new technology. Therefore, in the following, the discussion is focused on examples where the original reference work shows that the investigations avoided such technological problems in the fabrication of sintered parts, as far as it is generally possible within the framework of a given (optimized) approach. Of course, disadvantages which are inherent in a technology in the present state of the art can never be excluded. For this reason *different* flaw populations are characteristic of the various technical approaches and cause *processing-related property-limits* which often cannot be overcome by *materials* development [3].

As to new raw materials, the most significant step has been the development of *chemically derived* substitutes for materials of fundamental industrial importance like alumina or zirconia (most grades of which were originally manufactured from mineral sources). Since the late eighties, new corundum powders have been commercially available with fine particle sizes ($\approx 0.2\,\mu m$), which offers chances for developing extremely fine-grained microstructures with a high density. The small size and the increased curvature of the particle surface are expected to increase the driving forces of solid state diffusion on sintering, to reduce the required temperature, and to reduce grain growth. However, these *potential* improvements are often offset by the increased tendency of the fine-grained powders to agglomerate on sintering a body which contains agglomerates, densification proceeds first *inside* these agglomerates (where the particles are in closer contact), and, as a consequence, with the shrinkage of the agglomerates their boundaries will move away from each other [4]. Then, to eliminate the *sintering-related* new (and rather *coarse!*) interagglomerate porosity, much higher temperatures are required than expected for the fine-grained powder used. And, of course, the impact of the high temperatures on these fine particles is an undesired, strong grain growth (first starting inside the agglomerates). Therefore, for many years the published progress achieved with using these new raw materials was much less than hoped.

Success came from increasing understanding of the relationship between sintering temperature, obtained density, and grain size as a function of spatial homogeneity and of interfaces, similarly to the microstructural dependence of the strength and the hardness of ceramics [3, 4]. The general idea concerning the mechanism of sintering has been outlined above. The expertise starts from the simultaneous or sequential application of several dispersal approaches to bring the extremely fine-grained powders into a state that gives the right basis for the adjusted use of different shaping procedures. Successful experiments were reported with direct coagulation casting (DCC) [5] and centrifugal compaction [6] as well as with cold isostatic pressing of freeze dried granules, pressure filtration, and gelcasting [3]. Cold isostatic pressing easily gives dense sintered products with an average crystal size of $0.5\,\mu m$. Gelcasting means shaping by casting a (preferentially aqueous) slurry which contains a small percentage of a polymerizable additive. After polymerization, the soft but geometrically stable body is transferred to thermal processing. A bending strength of 700–900 MPa was reported for pressureless sintered single phase alumina bodies (which is about twice the conventional value) [3].

It is interesting to note that it is really the improved *homogeneity* of compaction of the powder particles in the 'green' bodies that is the key to smaller grain sizes with an increased hardness, wear resistance, and strength: in spite of a similar green density (similar starting porosity at the beginning of sintering) of all bodies produced by different shaping, the sequence dry (uniaxial) pressing → cold isostatic pressing → pressure filtration → gelcasting gives a dramatic decrease of the sintering temperature by about 250°C (always using the same raw powder and avoiding any doping additives). Therefore, the improved homogeneity of compacted particles in the given sequence is the one possible reason for the improved sintering behavior.

For most machining operations where really hard tools are required (e.g. grinding or cutting of hardened steel), either *more complex* strength properties than a maximum bending strength of compact bodies are required, or the macroscopic strength is generally less important owing to the applied small feed rates (as in precise cutting). Therefore, advanced principles for obtaining a high macrostrength for sintered shaped bodies manufactured from these tool ceramics [3] will not be discussed here in detail, whereas special emphasis shall be given to prospects for improved hardness and wear resistance.

4.2.2.2 New Prospects for the Hardness

It seems obvious that this increased homogeneity is also the reason for the (quantitatively evaluated [3, 7]) reduced frequency and size of defects and increased strength. Equally clear is the influence of the reduced sintering temperature on the reduced crystallite size, the increased hardness, and wear resistance. Figure 3 shows the increase in hardness obtained by decreasing crystallite sizes (under the condition of a residual porosity $\leq 1\%$): single phase alumina ceramics can be produced by pressureless sintering with a hardness that exceeds the hardness of hot pressed TiC reinforced composites with a high carbide concentration of

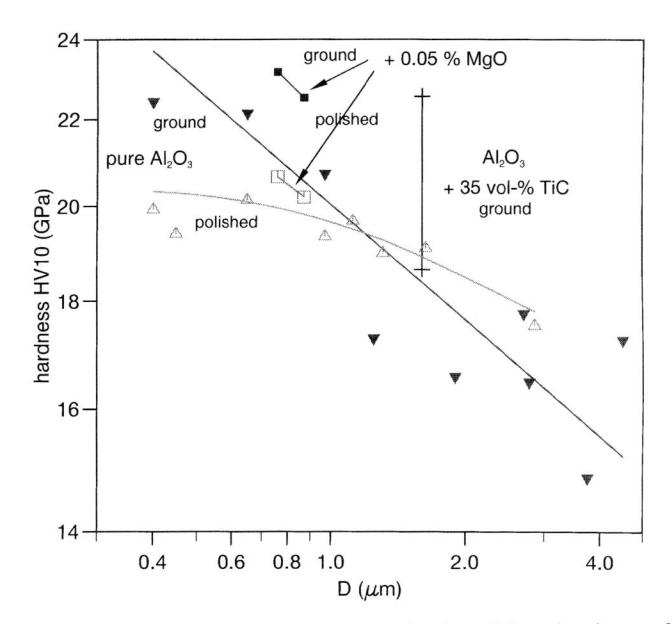

Figure 3. Effect of the crystallite size in the Vickers hardness of Al_2O_3 ceramics (HV10: testing load = 10 kg).

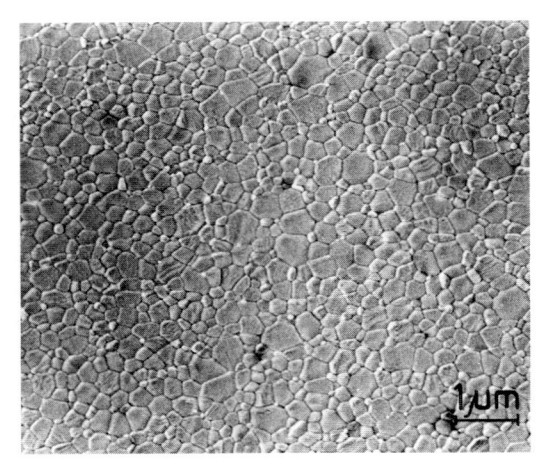

Figure 4. Microstructure of a single phase submicrometer alumina grinding material manufactured by powder processing (average crystallite size 0.4 μm, hardness HV10 (ground) = 22.4 ± 0.7 GPa).

45 vol-%; Fig. 4 shows such a submicrometer sintered alumina microstructure. Further improvement is possible with small MgO additives (note that *micro*hardness measurements give artificially higher values related simply to the smaller size of the indents; these microhardness data are, however, not representative for the technical behavior of tool ceramics).

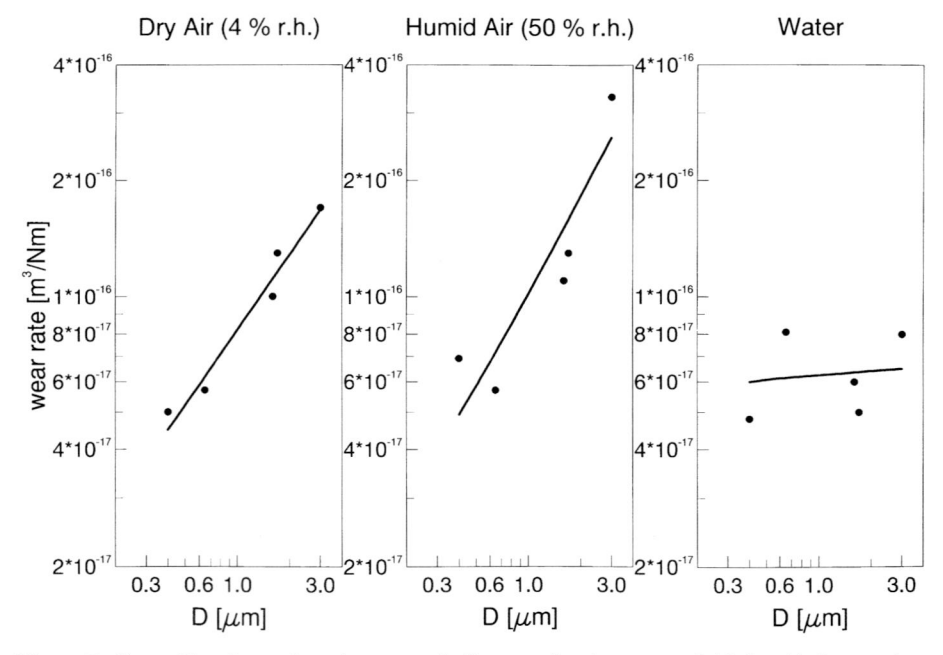

Figure 5. Crystallite size and environmental effects on fretting wear of Al_2O_3–Al_2O_3 couples.

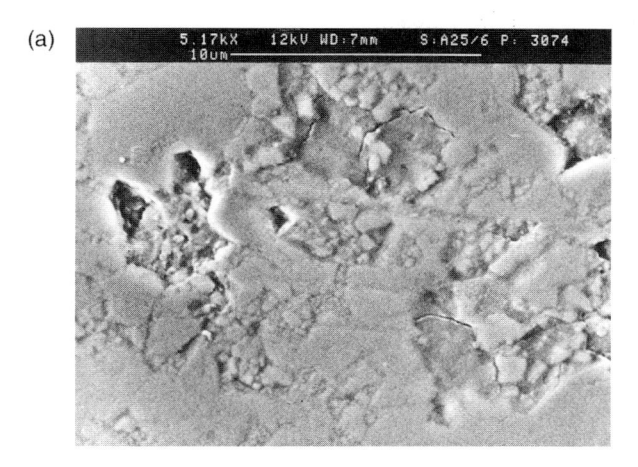

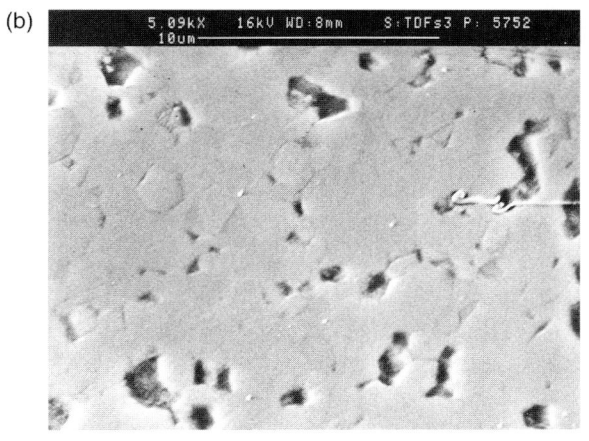

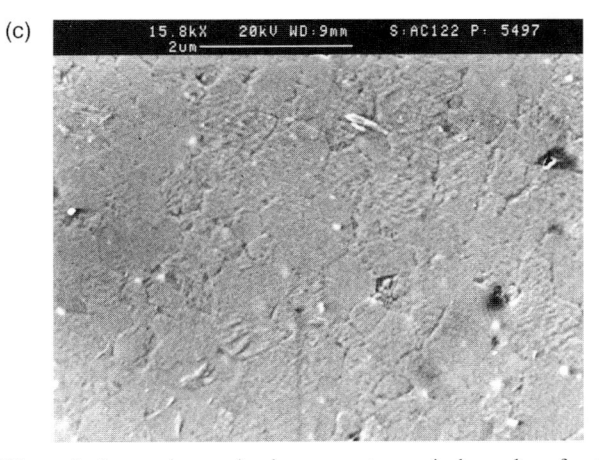

Figure 6. Increasing grain size promotes grain boundary fracture and grain pull-out (fretting wear in dry air; cf. Fig. 5). Average grain sizes: (a) 3 µm; (b) 1.6 µm; (c) 0.6 µm (bottom) [11].

When comparing data from different sources, the condition of the surface investigated has to be noted: machining introduces high dislocation densities [8], and the hardness of ground ceramic surfaces is usually higher than measured on polished surfaces, an effect that is the more significant the more fine-grained the microstructure is (Fig. 3).

4.2.2.3 A Leap in Wear Resistance

The grain size–hardness relationship in Fig. 3, published first in 1995 [9], has opened new prospects for improved tool materials. Until that time it had not been clear whether reducing the grain size of sintered alumina ceramics would cause a technically significant increase in hardness; in the USA it was accepted that the hardness increases down to crystallite sizes of 2–4 µm, whereas smaller grain sizes do not contribute to an improved hardness [10]. But even people with a different opinion did not have a chance to test it because more fine grained microstructures could not be prepared until the early nineties.

Figure 5 shows investigations of the fretting wear of alumina: depending on the environmental conditions, submicrometer ceramics may exhibit a wear rate which is an order of magnitude lower than measured for conventional Al_2O_3. It is important to understand that this improvement is *not* only a simple consequence of the increased hardness, but is also associated with reduced grain pull-out in the wear track of the submicrometer material (Fig. 6) [11].

4.3 Tool Materials with Undefined Cutting Edge: Sintered Grinding Materials

Grinding is an indispensable process in industrial machining. With the exception of special technologies like the grinding of wood or glass with garnet grits, it is generally assumed that high quality grits exhibit a high hardness. The consequences of a high hardness are rather complex, and we will show that a more than proportional increase of the grinding rate with a rising hardness of Al_2O_3 grits (Fig. 7 [12]) demonstrates only one among several positive aspects of a high hardness. Also, hard grinding materials are advantageously applied not only for grinding metals or very hard surfaces: hard SiC grits are likewise used for machining glass, some heavy metals, lacquers, chipboards, and plastics. For working steel, on the other hand, corundum grits (alumina, α-Al_2O_3) are superior to (the much harder!) SiC due to their extreme chemical stability. Alumina grits are manufactured by melting tons of raw powders in an electric arc ('fused' or 'electro-' corundum). Depending on the impurity content and additives, brown, white and Zr-corundum are different types of fused grits where each of the quasi-macroscopic grains (available in the size range 0.01–0.5 mm) has an internal structure close to that of a small single crystal. As to the goal of a high hardness, the advantage of the absence of pores is to some degree offset by the disadvantage of a microplastic deformability that is

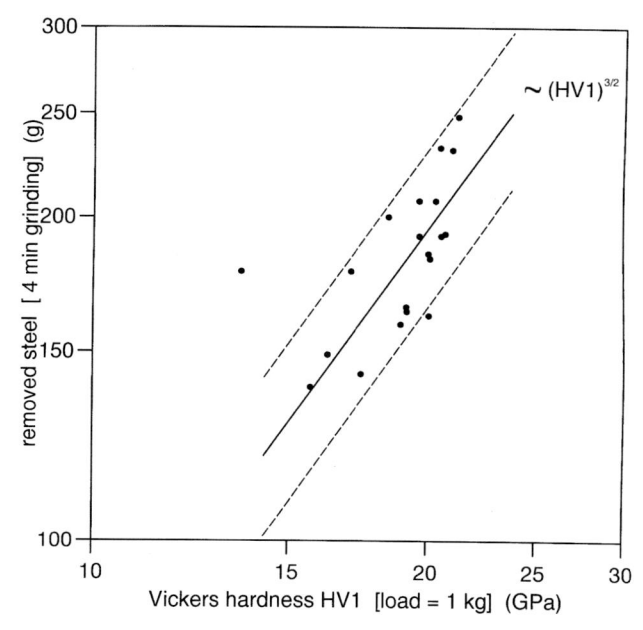

Figure 7. Influence of the hardness of the grinding grit on the rate of removed material on grinding the cross-section of a steel tube (flexible disc, different alumina grits; the results have been normalized for a unified grit toughness of $K_{IC} = 3.6\,\mathrm{MPa\,m^{1/2}}$ to demonstrate the hardness effect without the additional influence of K_{IC}) [12].

lower than in polycrystals owing to the lack of grain boundaries. Zr-corundum, in fact, contains such micro-interfaces because with higher concentration of zirconia (ZrO_2) the grits are a eutectic of finest Al_2O_3 and ZrO_2 lamellae, but as the hardness of the zirconia subregions is much lower than that of corundum the macroscopic hardness of the eutectic is not improved but even somewhat reduced. On the other hand, the eutectic structure increases the fracture toughness to twice that of grits which are free of zirconia (here, toughness is a parameter which may be understood as a strength related to a unified defect size). On machining cold rolled steel sheets, the increase of the grinding rate by about 20–30%, in spite of the reduced hardness, has to be attributed to this increase of the toughness in Zr-corundum grits.

For many applications fused grits are unsatisfactory, and efforts started at the beginning of the eighties to develop much harder (single phase) corundum grades by producing grits with a polycrystalline internal structure. This idea would affect not only the hardness, but also the fracture toughness, the microfracture behavior and the formation of chips. A difficult problem to be solved was that most of the conventionally known approaches for the manufacture of compact polycrystalline sintered ceramics cannot be used for producing irregularly shaped small grits (e.g. uniaxial pressing or cold isostatic pressing of powders). Hence entirely new procedures had to be found to obtain a 'green' density of the intermediate (not yet sintered) product that is high enough to enable sintering until a high density with less than 1–2% of residual porosity is reached. Presently approaches that associate minimum crystal sizes with a maximum density of the sintered grits are the primary target of research.

Efforts to increase the hardness by using composite materials with harder additives are much less important for grits than in the field of compact ceramics. The reason is the chemical nature of available hard materials (carbides like SiC, TiC, but also nitrides or, in a few experiments, diamond). The inability of SiC to grind ferrous materials is well known, and a similar observation had to be made for carbide reinforced composite alumina grits, in spite of their improved hardness. Further, a carbide additive in an oxide matrix does not always result in an increased hardness because the low diffusion of the carbides (dominated by covalent bonds) often gives increased porosity or (after sintering at a higher temperature) a coarser microstructure. On the other hand, highly pure (and therefore chemically very stable) sintered corundum grades have been developed recently, which after pressureless sintering exhibit a hardness similar to or even higher than the hardest hot-pressed composites with high carbide concentrations (cp. Fig. 5). All these features limit the motivation for new *composites* and increase the trend towards advanced, thermodynamically stable sintered grits on a *pure oxide* basis.

4.3.1 Technical Demands for Grinding Materials

The following discussion is focused on grinding materials for machining steel and cast iron, because these are most important industrially. Two different requirements are to increase working efficiency and to minimize undesirable interactions of the tool with the ground surface.

A frequently used criterion in patent applications is an increased rate of material removal. Another parameter is the quality of the ground surface: the tool should run for a long time with insignificant changes in the roughness of the ground workpiece. With new grinding materials that increase the grinding rate by 30–100% (see paragraph 4.3.2 and 4.3.3) or which give a roughness inside of the tolerance limit for a doubled working path, the economic benefit is even larger than given by the percentage of the reduced time per process (or per workpiece). The increasing output of the manufacturing lines increases the importance of down-times, affected also by the frequency of tool exchanges. The extended life of grinding tools gives an additional contribution to the economic efficiency of a line if, besides the reduced time in the individual grinding step per workpiece, an increasing number of workpieces can be manufactured with one tool.

In many grinding operations the development of heat is very important: undesirable changes in the microstructure and the properties of the ground material have to be avoided (e.g. in hardened steel), the introduction of residual surface stresses on machining is tolerable within given limits only. Since grinding tools can give high material removal rates at fairly low contact pressure (minimizing the process temperature), the final grinding of functional surfaces in ball bearings can hardly be substituted by other approaches today. The interest in grinding materials that exhibit a high grinding power at low process temperatures has been further promoted by the need to improve the environmental compatibility of manufacturing, a trend that is manifested by efforts to decrease or to avoid the consumption of coolants (reduced coolant concepts, dry grinding). There are manifold ways to develop such grinding materials. One approach is microstructural engineering: the shape of the crystals in the microstructure governs the fracture behavior of the grinding grain, the shape of the cutting edges and tips, and, finally, the chip formation. One of numerous ideas was the composition of grinding tools with hollow sphere-shaped bodies that, during grinding form always new (sharp) grain edges.

4.3.2 Advanced Commercial Products: Sol/gel-derived Corundum

Obviously, to develop grinding grains between 10 μm and 500 μm in size and with a *polycrystalline* internal structure is possible only if the grit is manufactured as a polycrystal with crystallite sizes much smaller than the dimension of the grinding grain, preferentially < 1 μm. However, the available corundum raw powders have particle sizes of about 0.5 μm, and until now it is impossible to create microstructures with crystallite sizes of less than about 2 μm from such powders by pressureless sintering. Success came with the application of sol/gel procedures that use chemical precursors of the final corundum phase as extremely fine-grained (usually nanoscale) raw materials. Most sol/gel approaches are expensive, but the grinding industry has benefited from the unique situation with alumina where sols can be prepared from boehmite (an aluminum monohydrate AlOOH) readily available from other chemical processes (the price is about $5 per kg or less). Unfortunately, the sequence of chemical and crystallographic transformations from the precursor to the final phase generates substantial risks in forming inhomogeneous, porous microstructures, running counter to the goal of high hardness. Patent applications suggest different doping [13] and seed [14, 15] additives intended to control these transformations. As a result it is, in fact, possible to improve the strength, and few examples offer significant advantages in the hardness compared with white fused corrundum. Nevertheless, on grinding rolled steel sheets such grits exhibit an increased power expressed, for example, by a rate of removed material which is 140–150% compared with brown corundum or 115–125% compared with the fused Al_2O_3–ZrO_2 eutectic. The comparison with fused grades in Fig. 8 includes two intermediate qualities of sol/gel derived (sintered) alumina grits together with an advanced material of the sol/gel group the grinding rate of which is increased by about 150% related to fused Al_2O_3–ZrO_2 [16].

In Fig. 8, the improved technical performance of the polycrystalline grinding materials *in spite of* their almost unchanged hardness seems to disagree with the

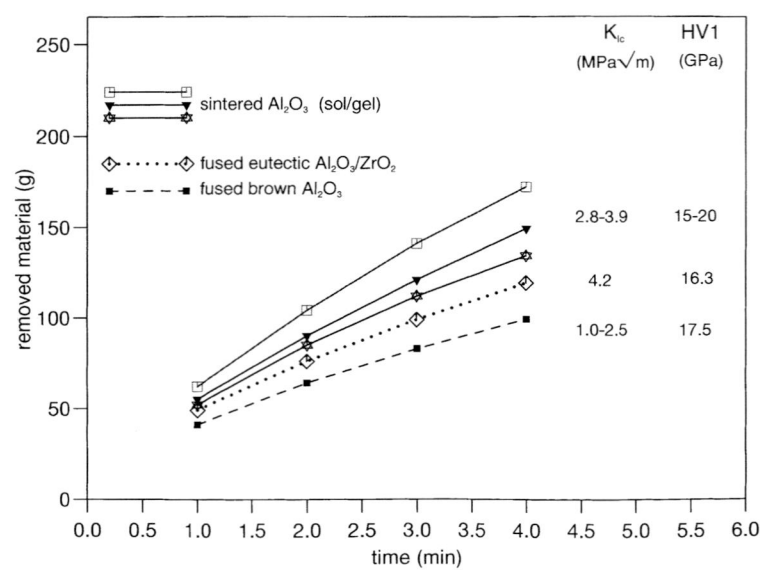

Figure 8. Comparison of the material removal rate of sol/gel derived sintered and of fused grits on the basis of Al_2O_3 (flexible fiber disc, 6000 min^{-1}, contact pressure 30 N; workpiece: steel tube [ST W 22 DIN 1543] with diameter 195 mm, wall thickness 1 mm).

concept of developing more powerful grits by increasing the *hardness* (due to a fine-grained microstructure). Obviously, there are other strong influences, and first of all strength properties, modifications of the microfracture behavior are expected to govern the results in Fig. 8. It has been recognized for a long time that the strength of ceramics depends on microstructural inhomogeneities ('defects', 'flaws') which locally increase the externally applied mechanical load, causing sudden, 'brittle' failure. What is important for grinding grains is, however, not a generally high but an adjustable (depending on the specific grinding process) strength: on the one hand, some global minimum strength under given loading conditions is imperative to run the grit for some time but, on the other hand, originally sharp edges and tips of a very strong grain would be worn and would deteriorate the grinding efficiency within less than a minute: an effect that is readily observed with very strong (e.g. carbide reinforced) ceramic grains. Therefore, the strength must be low enough to enable microfracture processes on a local microscale for re-generating new sharp edges during the grinding process.

Since it is impossible to evaluate the strength of a geometrically irregular grinding grain by standard approaches, the fracture toughness is measured by indentation procedures on mounted samples. The method derives the toughness from the length of indentation cracks formed at a given load. The analysis of a larger number of grains gives the statistical distribution of the loading capacity in different grades of grinding materials, and these statistics can be correlated with the technical

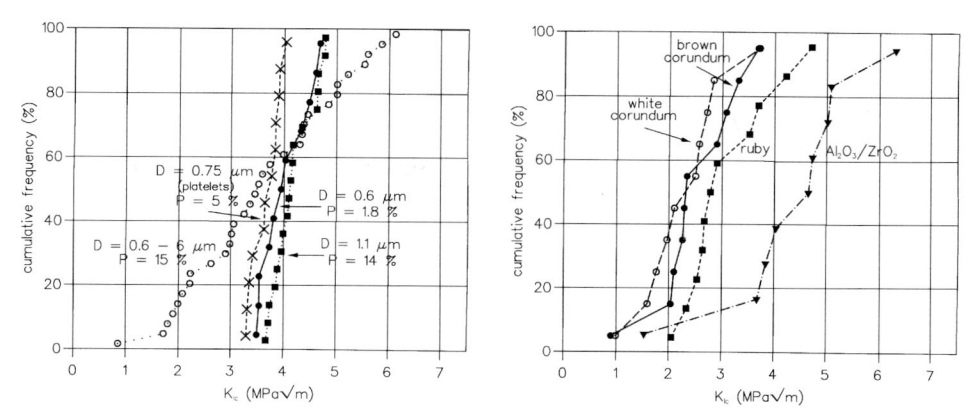

4.3 Tool Materials with Undefined Cutting Edge: Sintered Grinding Materials

Figure 9. Microstructural dependence of the fracture toughness in grinding grains on the basis of alumina prepared by sol/gel approaches (left) and in fused grits (right).

grinding performance of the grits under different conditions [16]. Figure 9 shows different toughness distributions obtained in sol/gel by producing microstructures with different porosity, crystallite size, and shape (as exemplified by Fig. 10). Obviously, the polycrystalline nature of sintered materials is a strong tool for modifications of the mechanical properties which are impossible in fused corundum grades.

From Fig. 9 it can be understood that the sol/gel derived grits owe their increased grinding efficiency (Fig. 8) first of all to their improved global stability (fracture toughness, strength). On the other hand, it is very instructive to understand the grinding performance of the Al_2O_3–ZrO_3 eutectic: with a very high toughness (Fig. 9) and with a hardness which is smaller than observed for zirconia-free fused grits but in no way inferior to the sol/gel grades (Fig. 8), it is impossible to attribute the only intermediate grinding efficiency of this material (Fig. 8) to deficits in its toughness or hardness. On the contrary, the Al_2O_3–ZrO_2 eutectic is a very clear example for the power-limiting effect of too high a strength of grinding grains, decreasing the frequency of microfracture processes which are needed to re-generate sharp cutting edges.

Of course, the concept of an adjustable strength that enables an optimum frequency of microfracture events does not mean tolerance of hardness deteriorating defects. From Fig. 2 it can be shown that within the framework of sol/gel approaches the frequency and size of such defects can be minimized but hardly avoided. In spite of the high number of patent applications devoted to improved sol/gel-derived grinding materials, it has been impossible to overcome the fundamental limit associated with this technology: the discrepancy between the unique chances to use nanoscale raw materials and the spatial heterogeneity introduced by the indispensable but always coarser seeds. With the obviously negative impact of flaws like those in Fig. 2 on the hardness and wear resistance *even if the porosity is no more than 1–2%*, it has to be expected that sol/gel derived products will not be able to exceed some limiting level of properties.

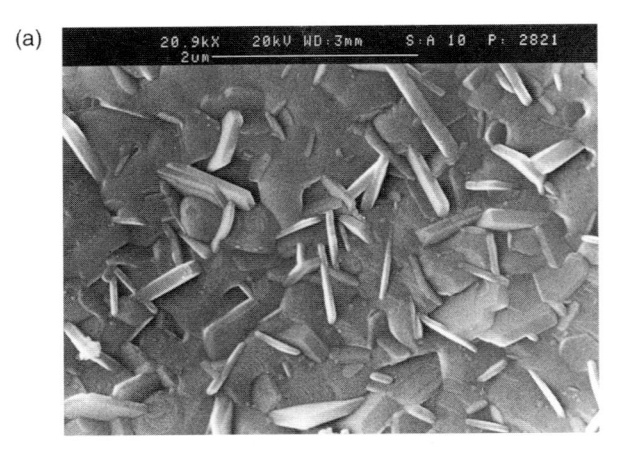

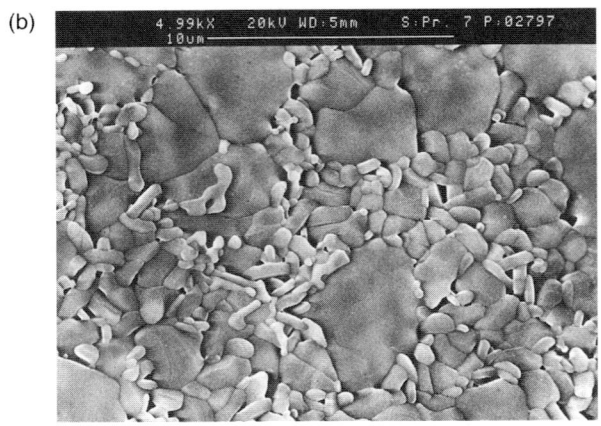

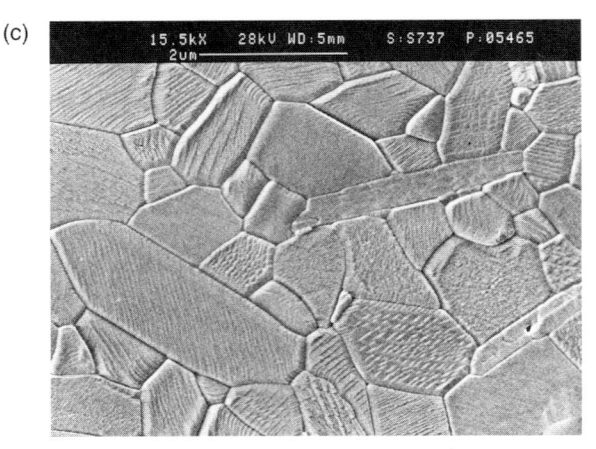

Figure 10. Microstructure of sol/gel derived sintered alumina grits: (a) platelet shaped microstructure ('Cubitron', manufactured by MMM, USA), (b) bimodal grain size distribution after seeding with transitional alumina nuclei, (c) anisotropic grain growth (Ce doping) [16].

4.3.3 Sintered Alumina Grits Produced by Powder Processing Approaches

Sol/gel-derived grinding materials probably have been the first example where a *precursor approach* was successfully used in the industry for a manufacturing volume of several *tons per year*. The *scientifically* fundamental importance of this comes from the circumstance that it was also the first example of highly dense and submicrometer ceramics with advanced mechanical properties. This progress was first achieved in the industry and only later on investigated in scientific institutes in greater detail. And again, as to the mentioned limitations of the sol/gel approach, it was the requirements of the industry which promoted a new idea: to complete the family of fused and sol/gel derived grinding ceramics by a third group which due to a generally different manufacturing process should exhibit not a stepwise improvement but another class of quality, similarly as the sol/gel grits have shown a generally improved performance compared with the fused materials. Success came with an apparent return to well-known powder processing technologies, but on a rather different level, as described in Section 4.2.

Considering that such basic influences like grain size and porosity effects on hardness are often a subject of controversy and are not completely understood, it is not a surprise that only recently has hardness been discussed as a function of defects [17]. Figure 2 shows small defects (2 µm or less) which will significantly deteriorate the hardness parameter which is effective for the grinding process. On the other hand, larger flaws like cracks or pores of about 5–20 µm do not necessarily decrease the grinding efficiency: on grinding, such flaws will become nuclei of microfracture processes (which are useful to re-generate sharp cutting edges), and these fracture processes will *eliminate the flaws* from the microstructure of the mesoscopic cutting edge (the hardness of which is determined by remaining *smaller* defects). Hence, the 'effective' (local) hardness may be higher than a value measured with the influence of coarser pores.

All microstructural understanding, however, can give rise to a new class of grinding materials only if the basic mechanical properties are adjusted to some *application-related* optimum. Therefore, for grinding as for cutting tools, investigation of technical performance under defined conditions is necessary. Figure 11 shows the results of test series performed by Hermes Abrasives, Hamburg (Germany). The most interesting result is the large progress in the grinding efficiency obtained by the *powder derived* material compared with the advanced sol/gel grits (all submicrometer microstructures); the difference is of similar magnitude to the advantage of the sol/gel grits compared with fused brown corundum. Compared with the Al_2O_3–ZrO_2 grinding grit, powder processing gives a single phase alumina grade which exhibits twice the material removal rate of the eutectic. It has to be emphasized that this improvement does not depend on the extremely small crystallite size of the microstructure (e.g. ≤ 0.5 µm, for which range Fig. 3 indicated a decreasing grain size influence in the hardness): the result in Fig. 11 relates to a powder-derived material with a grain size of about 1 µm [16] and was equally reproduced with crystallite sizes between 0.5 and 1 µm. An example of a powder-derived tool material with a submicrometer microstructure was given in Fig. 4.

With a fracture toughness of 3.3–4 $MPa^{1/2}$m for all sintered corundum microstructures *independent* sol/gel or powder processing, the additional data included in Fig. 11

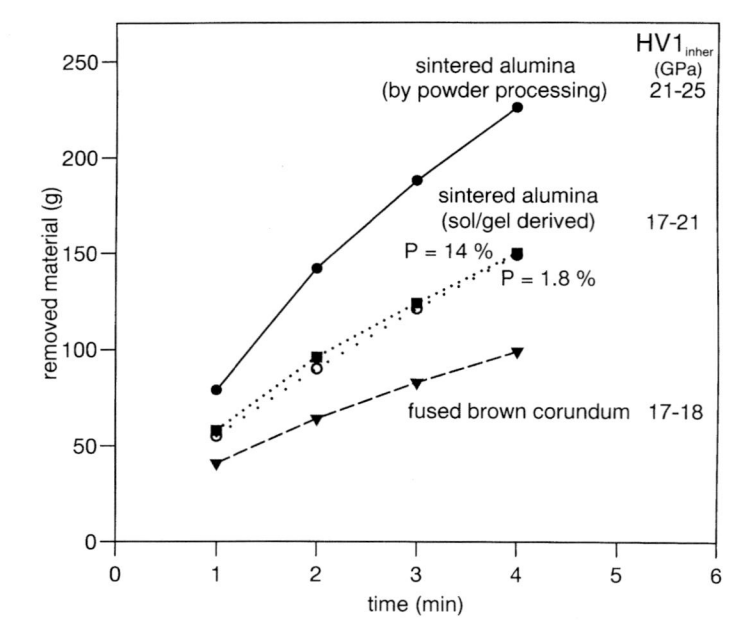

Figure 11. Comparison of the grinding performance of known (fused and sol/gel derived) grinding materials on the basis of Al_2O_3 with a new single phase corundum grit produced by a powder processing approach (same test conditions as given for Fig. 8) [16].

demonstrate the importance of the 'effective' (local) hardness on technical performance and especially for the stepwise increase of the grinding power in the ranking sequence fused corundum → sol/gel derived sintered alumina → sintered alumina by powder processing. This effective hardness (which is inherent to the microstructure) is diminished by small defects like that in Fig. 2, and with their elimination by substituting sol/gel by powder approaches the hardness increases (without decreasing the crystallite size!) and increases the grinding efficiency. With a sufficiently large number of tests, results like that given in Fig. 7 can be obtained which demonstrate a general trend of an increasing grinding power with increasing hardness (under the condition of an always *equal chemical* interaction of *all* grits with the ground metal).

4.4 New Trend for Cutting Hard Workpieces: Submicrometer Cutting Ceramics for Tools with Defined Cutting Edge

Characteristic for the late nineties was the increasing use of tough grades of silicon nitride, first of all when high feed rates are requested. With their rather low hardness, however, Si_3N_4 ceramics are outside the scope of the present discussion.

Most of the machining of *hardened* steel and *hard* cast iron, on the other hand, is done by grinding, a process that today is necessary for the final finish of functional surfaces, like in ball bearings. Nevertheless, there are several limitations of grinding which motivate the search for other methods: compared with cutting operations (whenever applicable) grinding is a time-consuming process (up to a factor of three), the equipment is more expensive, and disposal or recycling of the grinding silt is difficult from both ecological and economic points of view. It has been shown that even without a complete substitution the working time can be cut to 2/3 by combining a first turning step with final grinding [2]. If turning operations shall be used for the final finish as well, important tool life criteria are the surface roughness of the machined workpiece and the stability of the geometrical position of the cutting tip (which governs the accuracy of the produced piece). This paragraph deals with the technical demands on cutting tools which might be able to give fresh impetus to such efforts without increasing the costs (as it is the case with tools manufactured of diamond or cubic boron nitride). Therefore, the discussion will be focused on ceramics on the basis of alumina as outlined already in Section 4.1.

4.4.1 Demands for Cutting Materials Used for Turning Hard Workpieces

Table 2 illustrates the primary technical demands on cutting ceramics for machining hardened steel or hard cast iron. These demands are a consequence of the required wear resistance in dynamic contact with the abrasive carbide particles in the microstructure of the cut metals. As to more fundamental properties of the cutting plate, there is the requirement for the combination of maximum hardness and sufficiently high microscopic fracture toughness of the grain boundaries to minimize the frequency of grain pull-out at the cutting tip.

From the understanding of the influences of the microstructure on hardness and wear properties, an average grain size of substantially less than 1 µm and a low frequency of microflaws are imperative to meet these demands. On the other

Table 2. Demands on basic properties of cutting tool ceramics for machining hardened steel: x = important, xx = very important

Technical demands	On precise machining	
	Hard cast iron	Hardened steel
Wear resistance (especially against the abrasive impact of carbide particles)	xx	xx
Low thermal conductivity	x	x
Low creep rate (at high temperature)	x	xx
High oxidation stability		x
High chemical stability (preventing solid state reactions with constituents of the iron-based workpieces)		xx

hand, however, special measures may be required to balance the increasing high-temperature creep usually associated with small grain sizes. Reinforcement with whiskers (SiC whiskers are most common [18]) is one way to reduce the creep rate [19], an approach that also increases the thermal shock resistance [20]. Another way to reduce the creep rate is grain boundary engineering. Besides the prevention of impurity associated amorphous grain boundary films it was observed that a small concentration of yttria (0.1%) can reduce the creep rate of sintered alumina [21].

As it is also known from other fields of working metals, the different forms of the chips from machining cast iron and steel cause a much more intense interaction with the cutting plate for the latter material. Therefore, the demands concerning the chemical and oxidation stability are more stringent on working hardened steel than on machining hard cast iron (Table 2).

If the wear resistance is increased by decreasing the grain size of the ceramic microstructures, an additional advantage comes from a feature illustrated by Fig. 6. The reduced wear rate is associated with a decreasing amount of grain pull-out at smaller grain sizes, and the same effect can improve the quality of the ground ceramic surfaces. With a constant grinding procedure in finishing alumina ceramics (diamond wheel, grit size 40–50 µm), the roughness of a ground 0.6 µm alumina ceramic is about one half the value measured on a ground 3.6 µm microstructure [7]. Obviously, the improved quality of the ceramic cutting edge increases the chances for tool life at a small roughness of the machined hardened steel.

To prevent spontaneous microstructural changes in the workpiece and to obtain a high efficiency of the machining process, all approaches try to realize conditions with low cutting forces and a low heat input into the finished surface. Whereas this is generally true for both grinding and cutting operations, it is also accepted that the use of cutting ceramics is promoted if local temperatures are high enough for some softening of the (usually brittle) surfaces of hardened steel or hard cast iron [2]. To this end, ceramics with a low thermal conductivity are preferred. Table 3 reveals little differences with the exception of the most expensive tool materials (cBN, diamond) which are on the *less favorable* side (in fact, the high thermal conductivity of the latter gives an improved thermal shock behavior which, however, is considered as less important when small feed rates are applied on precise machining hardened steel). Much more significant differences exist for the hardness of the different ceramics (Table 1) and for their mechanical resistance at high temperatures. As to its oxidation stability and its chemically almost inert behavior in contact with iron or steel, pure sintered corundum is second to no other ceramic.

Table 3. Thermal conductivity of ceramic cutting tool materials.

Cutting tool ceramics	Thermal conductivity [W m^{-1} K^{-1}]
$Al_2O_3 + \approx 15\%$ ZrO_2	≈ 28
Al_2O_3	≈ 30
$Al_2O_3 + \approx 30\%$ Ti(C, N)	≈ 35
Si_3N_4	30–35
Cubic boron nitride (cBN)	40–120
Diamond	500–600

4.4.2 Carbide Reinforced Composite Ceramics Based on Al_2O_3

Previous investigations of precise cutting operations with hardened steel or hard cast iron were focused on black composite ceramics of Al_2O_3 with about 30% Ti(C,N). No positive results are known for grades based on Si_3N_4 or Al_2O_3/ZrO_2; probably, hardness and chemical stability of silicon nitride and of zirconia do not meet the demands of this machining procedure.

For example, a commercial Al_2O_3/Ti(C,N) tool was used for the final (precise) machining of the inner diameter of a ball bearing ring [2]. The average crystallite size of such cutting ceramics is about 1–1.5 µm. With the large body of evidence for the improvement of hardness and wear resistance by decreasing grain sizes, it is not surprising that most manufacturers have started to develop these tools with a submicrometer microstructure, but the covalent nature of the carbide bonds prevents pressureless sintering of the composites at temperatures of 1600°C or less. With TiC concentrations of 25–35%, the most fine grained of these composites exhibit average sizes of Al_2O_3 and TiC subregions of about 0.8–1 µm associated with a HV10 hardness up to 23 GPa (measured at a testing load of 10 kgf). Compared with today's commercial grades with HV10 < 21.5 GPa, this is an increase of almost 10%. Nevertheless, these tools showed only similar behavior on fine turning gray cast iron (GG25, 1000 m min^{-1}), hard cast iron (60–65 Shore, 150 m min^{-1}) and on turning a hardened steel 100CrMo6 for ball bearings (180 m min^{-1}, Fig. 12).

The latest patent applications try to obtain more fine grained microstructures outside the known compositions Al_2O_3/TiC or Al_2O_3/Ti(C,N). An alumina composite with 25% (by weight) of a mixed carbide/carbonitride consisting of about 90% Ti(C,N) and 10% WC was hot pressed at 1600°C yielding a microstructure with

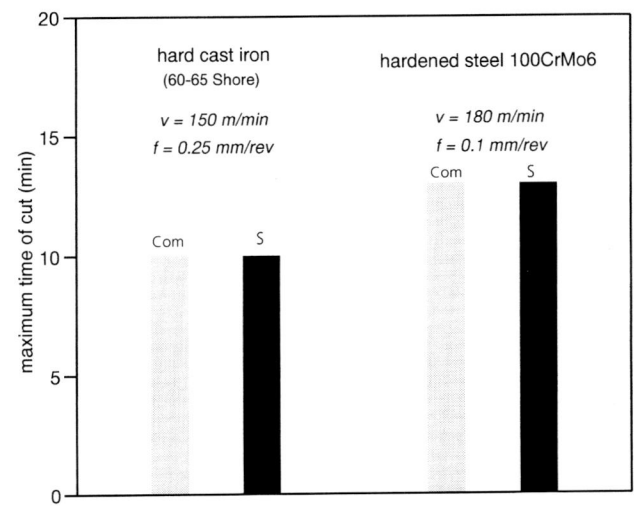

Figure 12. Turning hard workpieces with different ceramic composite tools. Com = advanced commercial Al_2O_3/Ti(C,N) grade, concentration of Ti(C,N) about 25–30 vol-%. S = laboratory grade: average sizes of Al_2O_3 and TiC subregions 0.8–1.0 µm, TiC concentration 35 vol-%.

an average grain size of 0.5 μm [22]. On turning hardened steel (X63CrMoV5) with a discontinuous cutting regime, the comparison with an unspecified commercial tool showed lifetime (or cutting path) that was three times greater. Another recent development uses oxycarbide and oxycarbonitride compositions (Ti(C,O), Ti(C,O,N)) with different stoichiometry instead of TiC and obtains crystallite sizes of about 0.6 μm (after sintering at 1700°C and hot-isostatic post densification at 1435°C); the hardness is HV10 = 22.5–23 GPa [23]. Figure 13 shows the microstructure of these materials which is much more fine grained than an advanced (also claimed submicrometer) commercial material with the same carbide concentration (on the market since 1997). As another example, Fig. 13 includes a new material with an increased concentration of 35 vol-% of the covalent phase which can be manufactured with the same small grain size in spite of the high carbide loading.

The cutting performance of these new materials was tested in a direct comparison with similarly hard but single phase (absolutely oxidation resistant) pure alumina ceramics. The results are discussed in Section 4.4.4.

4.4.3 Single Phase Sintered Corundum

From Table 2, an advanced cutting ceramic for the precise machining of hard workpieces should associate a maximum hardness with finest grain sizes (to minimize grain pull-out at the cutting edge) and high chemical and oxidation resistance. With today's knowledge, it is not clear if carbide or carbide reinforced tools will ever be able to meet the thermodynamic and chemical side of these demands: in fact, the (negative) free enthalpy of TiC and TiN is only half of the value of Si_3N_4 [24], and even the silicon nitride is unable to cut mild steel (not to mention hardened steel). On the other hand, from the data in Fig. 3 there exist pure sintered corundum ceramics with a hardness that even after pressureless sintering is in no way inferior to hot-pressed carbide reinforced composites, with a strength of 800–900 MPa [3] (after HIP up to 1300 MPa [25]) that equals or even exceeds the strength of the composites, *and* with the additional great advantage of highest chemical and oxidation resistance.

Cutting tests performed with these recently developed alumina grades are rare, some first examples are compared in Section 4.4.4 with the performance of commercial and of submicrometer composite tools.

4.4.4 Comparative Cutting Studies with Submicrometer Ceramics: Al_2O_3 and Composites Reinforced with Ti(C,N) and Ti(C,O)

4.4.4.1 Mild Conditions: Machining Cast Iron and Steel

First investigations of the cutting performance of submicrometer tools have been performed in Asia [26–28]. Unfortunately, all these tests describe only the machining of unhardened steel and cast iron. On *milling* cast iron, the submicrometer sintered Al_2O_3 showed almost no advantage and was inferior to $70Al_2O_3/30TiC$ and

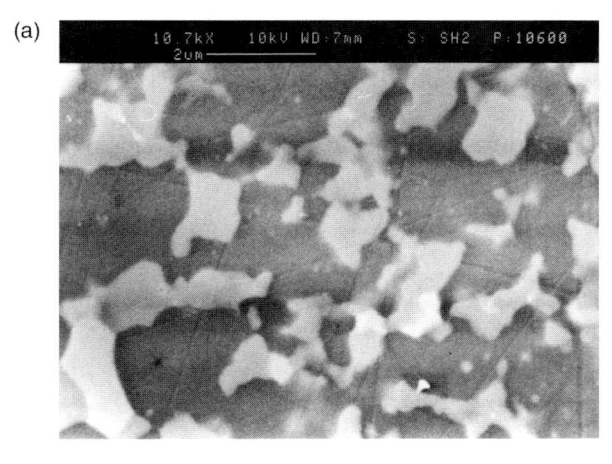

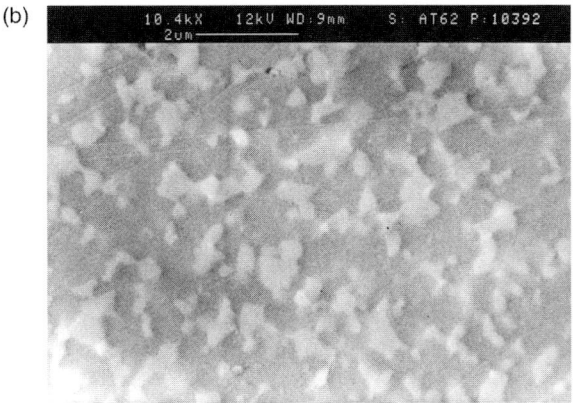

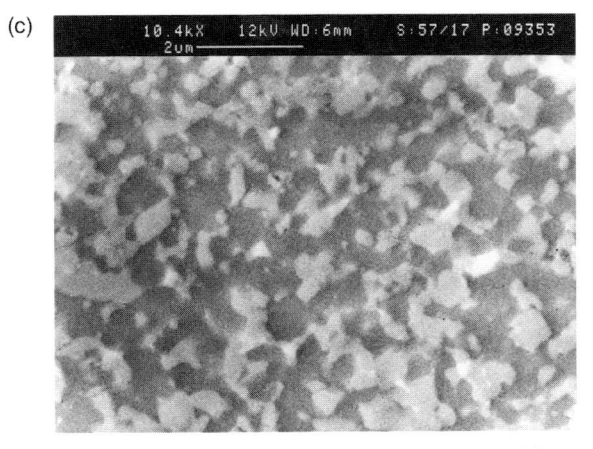

Figure 13. Latest tool composites based on Al_2O_3: (a) commercial 'submicronstructured' $Al_2O_3/$Ti(C,N) composite and new laboratory grades [23] with (b) Ti(C,O;N), (c) and with an increased concentration of about 35 vol-% Ti(C,O).

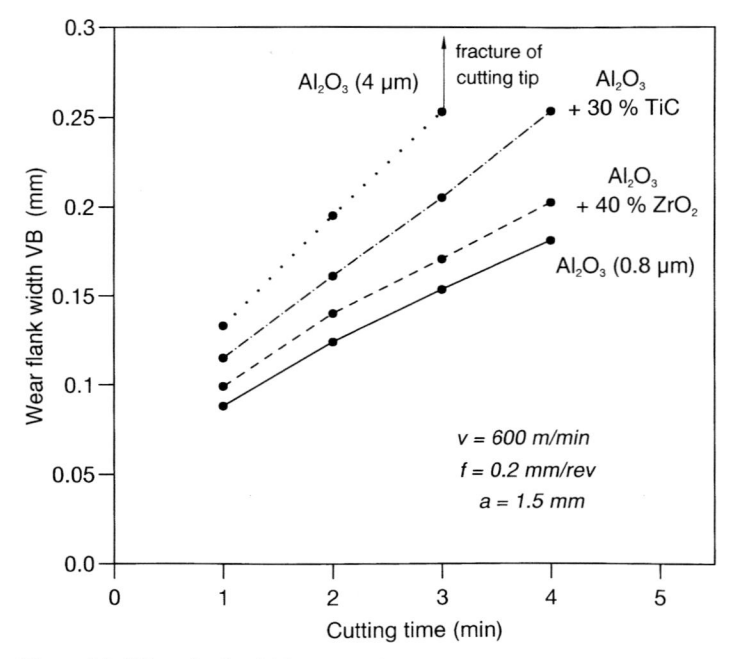

Figure 14. Wear flank width on turning steel (Japanese code S45C). G = sintered corundum with grain size 4 μm. Adapted from Takahashi et al. [26].

$60Al_2O_3/40ZrO_2$ composites, a result which is hardly surprising: in this application, the differences in the tool *hardness* are less important, and the improvement in the *toughness* of submicrometer Al_2O_3 microstructures is so small ($<4\,\text{MPa}\,\text{m}^{1/2}$ [3, 29]) that it is technically unimportant. On the contrary, on *turning* cast iron or steel (Fig. 14) the progress obtained by the submicrometer alumina was significant. It should be emphasized that an advantage of the pure alumina tool compared with carbide reinforced composites and with zirconia toughened alumina was observed just under these conditions where during continuous cutting an increased processing temperature at the cutting edge applies.

4.4.4.2 Machining Hard Cast Iron and Hardened Steel

The behavior of the tools described in the last section supports suggestions of a preference of submicrometer alumina ceramics for the precise machining of *hardened* steel and *hard* cast iron. First investigations in this field started in 1998 at the Dresden Institute of Ceramic Technologies and Sintered Materials. All turning tests were performed with a feed rate, $f = 0.1\,\text{mm}\,\text{min}^{-1}$ and a depth of cut, $a = 0.2\,\text{mm}$. The size of the cutting inserts was SNGN120412 (i.e. $12.7 \times 12.7 \times 4.76\,\text{mm}^3$ with a 1.2 mm radius) with a 20° chamfer (width 0.2 mm). The plates were positioned with a rake angle $\gamma = 6°$, an inclination angle $\lambda = -4°$, and a large entering angle $\varkappa = 45°$ chosen to maximize the selectivity of the tests (as to the tool materials)

Table 4. Properties of new tool ceramics compared with advanced commercial grades.

	Composition	Average grain size (composites: size of single phase subregions) (μm)	Hardness HV10 (Vickers, 10 kg testing load, ground surfaces) (GPa)
Composites			
SH1 (CeramTec) commercial	$Al_2O_3 + \approx 33$ vol-% Ti(C,N)	1.5	21.16 ± 0.23
'submicron'	$Al_2O_3 + \approx 33$ vol-% Ti(C,N)	<1	21.22 ± 0.54
AT60A	$Al_2O_3 + \approx 33$ vol-% Ti(C,O)	0.7	22.18 ± 0.77
AT60	$Al_2O_3 + \approx 37$ vol-% Ti(C,O)	0.67	22.45 ± 0.85
Pure alumina			
AC41	Al_2O_3	0.56	22.08 ± 0.90

by high loads at the cutting tip. A new commercial grade of an $Al_2O_3/Ti(C,N)$ tool ceramic was claimed 'submicronstructured' by the producer and introduced into the market 1997 especially for machining hard materials; it was compared here with an advanced standard grade SH1 (CeramTec AG, Plochingen, Germany) and used as a reference to evaluate the cutting power of new laboratory grades (Table 4). Both the submicrometer and the SH1 tools are composites based on Al_2O_3 with about 30% of Ti(C,N).

On the turning hardened steel at high cutting velocities > 250 m min^{-1} there is a significant input of heat not only into the formed chips but also into the cut surface of the shaft. Depending on the state of wear of the tool, cutting forces and process temperatures may increase to an extent that causes softening of the hardened steel, and the hardness was observed to drop from HRC = 57–60 to values of less than 50. Therefore, the decrease of the hardness has to be recorded as an additional parameter in such investigations, and an intermediate machining operation with a low velocity $v = 180$ m min^{-1} was required before every new run to restore the original hardness.

4.4.4.2.1 Intermediate Demands: Machining Hard Cast Iron Figures 15–17 show the results for continuous cutting chilled globular cast iron G-X300CrMo153 (Rockwell hardness HRC = 43–45). The flank wear of the tools is given in Fig. 15. The high hardness of the oxycarbide reinforced submicrometer composite reduces its wear by an amount similar to the progress of the commercial submicrometer tool compared with the standard material SH1. Much more surprising is the qualitative leap in the lower wear of the pure alumina insert which cannot be explained by its hardness: the hardness is only just intermediate between the (lower) hardness of the commercial composites and the (higher) value of AT60A. Also, the grain size of the sintered corundum material is not very different from the microstructural data given in Table 4 for the $Al_2O_3/Ti(C,O)$ composite AT60A.

Therefore, Fig. 15 suggests that reducing the grain size of TiC, Ti(C,N), or Ti(C,O)-reinforced composites may indeed increase the hardness and reduce the wear in turning hard cast iron, but the effect is limited, probably because of chemical

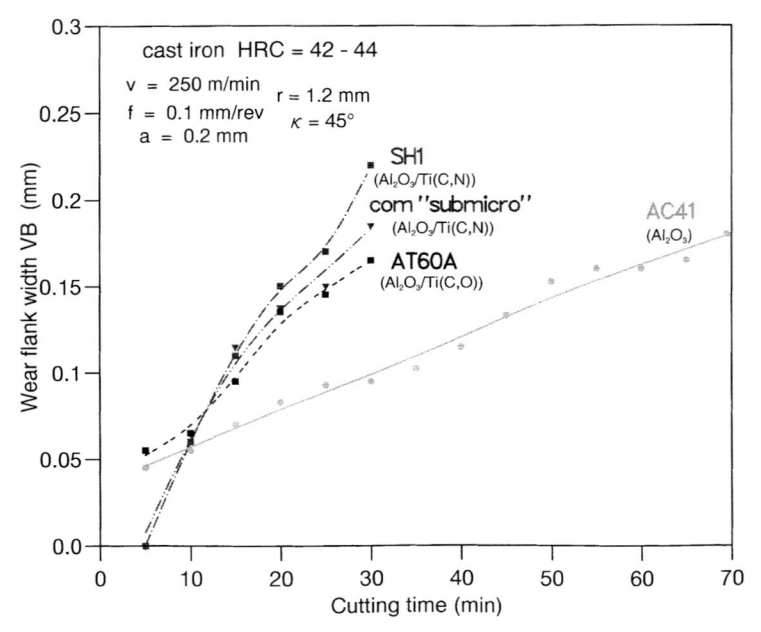

Figure 15. Flank wear width of submicrometer ceramic tools on turning hard globular cast iron (Rockwell hardness HRC = 43–45), see Table 4 for tool characterization.

interactions between the iron and the carbide phases of the tools. To eliminate this interaction by using a pure alumina tool can greatly improve the wear resistance as far as the process temperature at the cutting tip is not too high to initiate additional wear due the low high-temperature creep resistance of the submicrometer alumina microstructure.

The cutting edge displacement (CED) characterizes the obtained *precision* in turning hard materials. Figure 16 (CED) shows that complex wear processes may give a different ranking of the tool materials compared with Fig. 15 (flank wear). Again, all of the new laboratory submicrometer grades exhibit much less wear than the advanced commercial tools. However, with the criterion of the cutting edge displacement it is the ceramic with the highest hardness which shows the best performance (composite AT60A, Table 4).

Another important criterion for the quality of tools in precision turning of hard materials is the obtained roughness of the cut metal surface. The graphical presentation in Fig. 17 gives the *average* roughness R_a. A fixed ratio between the statistically averaged *maximum depth* of roughness R_{ZD} and R_a was observed in all the test (Fig. 17). The arbitrary scatter of data was rather large, and no significant difference was observed between the surfaces cut by the SH1 or by the commercial submicrometer tools. However, all other inserts show the same ranking as observed for the flank wear in Fig. 15. Improved results are provided by the oxycarbide reinforced composites (AT60), but a more significant progress results from the use of pure submicrometer alumina inserts.

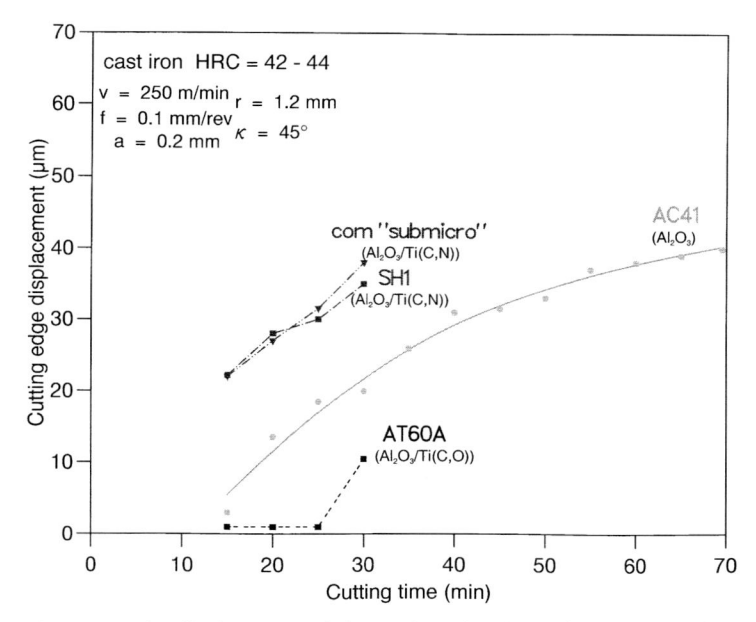

Figure 16. The displacement of the cutting edge determines the precision of the turning operation (hard cast iron). Average grain sizes of the tool microstructures [in composites: sizes of single phase subregions] are $D(SH1) = 1.5\,\mu m$, $D(AT60A) = 0.70\,\mu m$, $D(AC41) = 0.56\,\mu m$ (see Table 4).

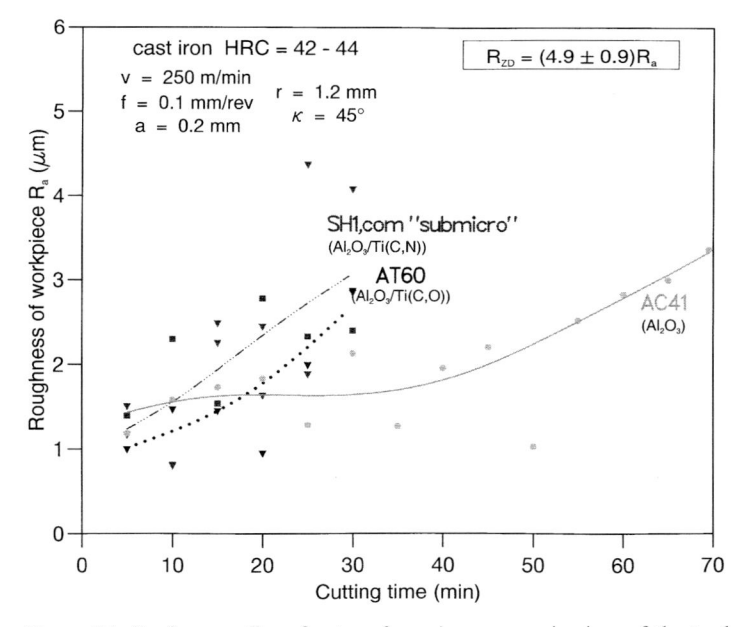

Figure 17. Surface quality of cut surface. Average grain sizes of the tool microstructures (in composites: sizes of single phase subregions) are $D(SH1) = 1.5\,\mu m$, $D(AT60) = 0.67\,\mu m$, $D(AC41) = 0.56\,\mu m$ (see Table 4).

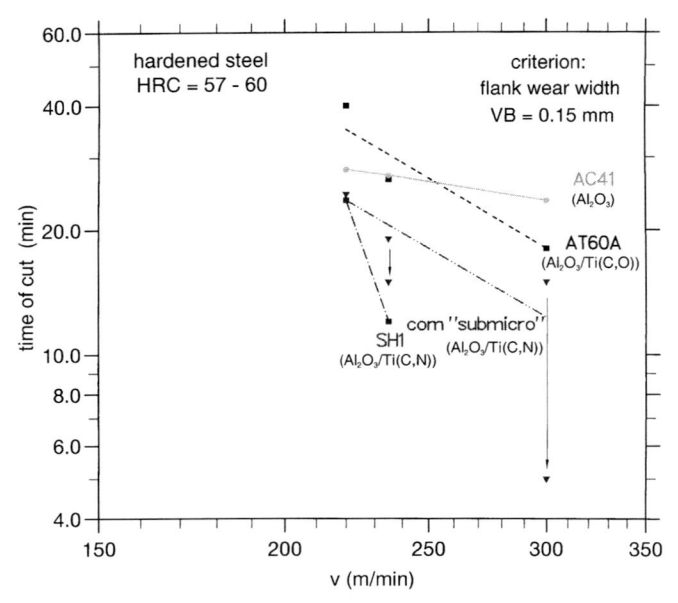

Figure 18. Life time of advanced ceramic inserts on machining hardened steel (see Table 4 for materials characterization). The arrows indicate frequently observed deviations.

Whereas turning of hard materials is usually performed with asymmetrically positioned inserts (entering angle $\varkappa = 10\text{--}30°$), it was mentioned above that $\kappa = 45°$ was chosen in these investigations to maximize the selectivity of the tests (as to the tool materials) by high loads at the cutting tip. For geometrical reasons, a lower roughness is expected for both smaller (10–30°) and larger (60–80°) entering angles. Therefore, it should be noted that in Fig. 17 the increase of R_a towards values between 2 and 4 µm is directly associated with this very special orientation of $\kappa = 45°$ and *does not* indicate a disadvantage of the tested *tool materials*. Indeed, in an additional test performed with SH1 and with the commercial sub-micrometer tool at $\kappa = 75°$ all R_a values were < 1.5 µm for the whole time of cut (i.e. up to 0.2 mm flank wear at least).

4.4.4.2.2 Extreme Conditions: Continuous and Discontinuous Cutting of Hardened Steel

Figures 18–19 give the results for continuous cutting of hardened steel 90MnCrV8. Its high hardness (HRC = 57–60) and the choice of tool materials with a low thermal conductivity cause a large input of heat into the formed metal chips. These appear rather flaming than red-glowing, an effect which is optically more spectacular on machining hardened steel with *pur corundum* inserts than using *composite* tools the thermal conductivity of which is about 15% higher (see Section 4.4.1, Table 3). On machining hardened steel with composite tools, the lower temperature gives more faceted chips.

For the reference tools (SH1 and the commercial submicrometer ceramic), the high temperature is associated with a critical influence of the cutting velocity

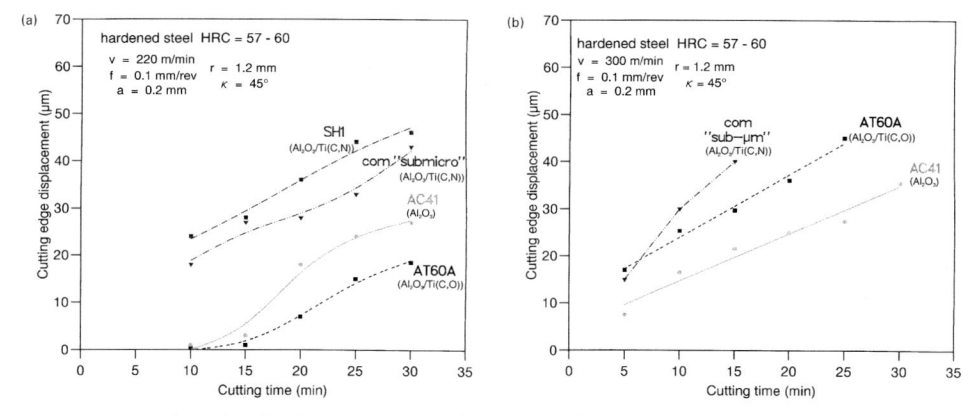

Figure 19. Cutting edge displacement on turning hardened steel with advanced ceramic tools: (a) at $v = 220 \, \text{m min}^{-1}$ (above), (b) at $300 \, \text{m min}^{-1}$ (below); see Table 4 for materials characterization.

between 200 and 250 m min^{-1}. Flank wear is small at $v = 220 \, \text{m min}^{-1}$, but already at $v = 235 \, \text{m min}^{-1}$ a greater crater-wear is observed. At $v = 235 \, \text{m min}^{-1}$, more than 50% of the tested commercial inserts exhibited a sudden, strong wear increase during the first 10–20 min associated with local or global fracture at the cutting edge, and it becomes difficult to record a continuously increasing flank wear up to VB > 0.2 mm.

Whereas the cutting edge displacement was immeasurably small during the first 5 min, it was overtaken by fracture soon later and could not be determined optically (even at $v = 220 \, \text{m min}^{-1}$ the cutting edge displacement of commercial tools was often shadowed by local features of microfracture). Hence, it was impossible to measure the cutting edge displacement of the reference tools at $v = 235 \, \text{m min}^{-1}$. Therefore, tests started with $v = 220 \, \text{m min}^{-1}$, but some of the new laboratory grades exhibited extremely few wear at this velocity, and additional experiments had to be performed at higher rates. It should be noted, however, that already at $v = 235 \, \text{m min}^{-1}$ the reference curves for SH1 and the commercial submicrometer material characterize but a minority of inserts standing this condition (about 50% failed at shorter times by cutting edge fracture). Therefore, these data (Fig. 18) overestimate the real behavior of the commercial grades at this velocity.

Whereas even at the lower velocity of $v = 220 \, \text{m min}^{-1}$ almost 50% of both commercial grades failed by fracture during the first 5–15 min of cutting hardened steel, no one such event happened with the laboratory grades (where crater-wear was also observed but to a lesser extent compared with the commercial ceramics). This behavior correlates with another surprising observation made on preparing the hardened shaft for the tests. This preparation is difficult with shafts of hard cast iron or hardened steel because it associates the high hardness of the counterpart with discontinuous cutting conditions at a changing frequency and power of impacts. Cubic boron nitride inserts (dreborid®, Lach company, Hanau/Germany) failed on machining the hard (globular) cast iron, and hard metal inserts (WC/6% Co, Vickers hardness $\approx 16 \, \text{GPa}$, $K_C \approx 9 \, \text{MPa m}^{1/2}$) were not able to cut the rough outer shell of the hardened steel shaft at velocities between 50 and 100 m min^{-1}. The same observation had to be made with the commercial ceramic composites (SH1 and the submicrometer grade fractured after $t < 2$ min). On the contrary, both the pure alumina insert AC41 and the composites AT60A and AT62 were successfully used with $v = 120$ m min^{-1}, feed rate $f = 0.3$ mm rev^{-1}, and a depth of cut $a = 1.5$ mm. These new, submicrometer cutting ceramics did not only not fracture macroscopically during one hour of severe cutting (with *one* tip) but retained microscopically nearly perfect cutting edges providing a high surface quality of the hardened shaft in the first cut ($R_a < 1$ μm for $t \leq 15$ min [32]). Surprisingly, under these severe conditions the general performance of the pure alumina ceramics was even more prospective than the behavior of the new submicrometer composites [32].

For SH1, the manufacturer's data sheet gives a strength of 600 MPa (no available data for the submicrometer grade). The bending strength of the Ti(C,O) reinforced composite AT60A is about 800 MPa [23], and a lower strength of about 650–700 MPa was observed for the pure submicrometer alumina ceramic [3]. The fracture toughness of all of the laboratory grades is 3.3–3.8 MPa m$^{1/2}$, whereas surprisingly high values of 5.5 MPa m$^{1/2}$ (SH1) and 6.6 MPa m$^{1/2}$ (submicron commercial composite) are given by manufacturer's information for these references. None of these data explain the much smaller fracture risk of the submicrometer laboratory grades on machining hardened steel. Obviously, complex wear-induced flaw-generation processes will have to be considered to understand the improved global stability of the new laboratory grade microstructures.

No differences were observed in the roughness of the hardened steel surface when different ceramics were used. At $v = 235$ m min^{-1} ($f = 0.1$ mm rev^{-1}, $a = 0.2$ mm), R_a was ≤ 1 μm for the first 10–15 min of cut for all of the tested ceramics, increasing to about 1–1.5 μm after 30 min ($\kappa = 45°$: see comment at the end of Section 4.4.4.2.1!). On the other hand, the tools showed very different flank wear and cutting edge displacements. Figure 18 gives the influence of high cutting velocities on the life of advanced cutting ceramics on machining hardened steel (flank wear criterion VB = 0.15 m min^{-1}). As already mentioned, $v = 200$–220 m min^{-1} is the upper limit of reliable use for the latest generation of commercial Al$_2$O$_3$/Ti(C,N) composites. At this velocity, best results are obtained with the new submicrometer composites with Ti(C,O) which exhibit the highest hardness of all of the tested ceramics (see Table 4). At higher cutting velocities, however, the relationship compared with the pure submicrometer alumina changes, and pure Al$_2$O$_3$ again gives the greater advantage. A similar smaller effect of the cutting velocity on Al$_2$O$_3$ inserts compared with composites was noticed on cutting hard cast iron (Fig. 15, Fig. 17).

Contrary to the new ceramics, most commercial submicrometer inserts fractured within the first 5–10 min when used at $v = 300$ m min^{-1}, and the one continuous wear curve measured up to 15 min and displayed in Fig. 19b was obtained at a softer part of the steel shaft (HRC = 52–57).

At all cutting velocities the new laboratory grades are clearly superior to advanced commercial inserts. Contrary to a sometimes different behavior of flank wear and cutting edge displacement on turning hard cast iron, both parameters were well correlated on machining hardened steel. Comparing the new submicrometer Al_2O_3/Ti(C,O) composite with submicrometer pure corundum, Fig. 19 shows a velocity-dependent different ranking of the cutting edge displacement (CED) of these tools in perfect correlation with the data in Fig. 18 (flank wear); the CED curves for the Al_2O_3/Ti(C,O) composite and for pure Al_2O_3 were observed close to each other at an intermediate velocity $v = 235\,\text{m min}^{-1}$ (as suggested by the flank wear data in Fig. 18).

Comparing this behavior on machining hardened steel with the turning of hard cast iron, again both the pure alumina tool AC41 and the laboratory grade composite AT60A exhibited less wear than the commercial cutting ceramics, but the changing ranking of AC41 and AT60A ceramics is worth some comments. It is obvious that the technical ranking of both tool ceramics will be affected by some important basic properties:

(i) For Al_2O_3/TiC composites, chemical interactions cause preferential (local) wear of the covalent carbide constituents of the microstructure even at room temperature and with almost inert conditions (dry air, fretting wear against alumina), with the consequence that at equal crystallite sizes the pure sintered corundum is more wear resistant than the composite [30, 31]. The same ranking is observed for the flank wear of ceramic tools on turning hard cast iron (Fig. 15).

(ii) On machining hardened steel with $v = 220\,\text{m min}^{-1}$, flaming chips indicate much higher process temperatures than on turning hard cast iron at $v = 250\,\text{m min}^{-1}$. Whereas pure submicrometer alumina exhibits intense creep at temperatures $> 1100°C$, it can be assumed that the continuous network of covalent crystals reduces the creep rates in composites with more than 25 vol-% of TiC, Ti(C,N) or Ti(C,O). Indeed, comparing at $v \leq 250\,\text{m min}^{-1}$ the machining of cast iron and steel, the increasing temperature increases the wear of alumina more than the wear of the best composite AT60A and changes their ranking: at a similar submicrometer grain size, AT60A exhibits the lowest flank wear and the smallest cutting edge displacement.

(iii) Of course, this advantage of the composite will be lost when at yet higher temperatures oxidation of the Ti(C,O,N) phases or more intense reactions with the steel start. With the results in Fig. 18 and Fig. 19, the commercial composites with Ti(C,N) are more susceptible to such processes than the new laboratory grades with Ti(C,O). In fact, no information about the specific nature of such reactions can be obtained from the present results, but it is obvious that on machining hardened steel an increase of the cutting speed (and of the process temperatures) causes a more intense deterioration of the wear resistance of even the best composite AT60A than observed for the pure alumina tool AC41 (Fig. 18, Fig. 19): already at $v = 235\,\text{m min}^{-1}$ both ceramics exhibit the same wear, and with $v = 300\,\text{m min}^{-1}$ the sintered corundum again shows the best performance.

At a velocity of $v = 300\,\text{m min}^{-1}$, the increasing with time wear deteriorates the cutting edge to an extent that obviously causes an increasing input of heat into the surface of the machined steel. For example, on machining the hardened steel with the submicrometer composite ceramic AT60A the hardness of the cut surface decreased from originally HRC = 59–60 to values of 57–59 (25 min) and 55–56 (30 min) [32]. This influence of high-velocity turning on the hardness of workpieces is important for different aspects in the development of new machining technologies. On the one hand, the softening of the hardened steel promotes unexpectedly high materials removal rates. In this way, turning of hardened steel with cutting ceramics that are much less expensive than cubic boron nitride (cBN) seems possible now. If, on the other hand, for the final step of precision machining a constant high hardness of the cut surface is required, critical limits exist for the cutting velocity (about $250 \, \text{m min}^{-1}$ in the investigations described above) or for the tolerable tool wear (here at $v = 300 \, \text{m min}^{-1}$: VB ≤ 0.15 mm). These technical demands are, of course, affected by the choice of the ceramic tool depending on its thermal and wear properties.

4.5 Summary

Recent developments show somewhat different trends for sintered ceramic *grinding* materials and for *cutting* tools.

(i) In the field of grinding grits, extended innovative work is directed towards a substantial further increase of the grinding efficiency (high materials removal rates) and increased tool life within narrow tolerances for the surface quality of the ground workpiece.
(ii) For cutting tools, innovation is focused on two aspects: the development of very tough and highly reliable grades for high feed rates and cutting depths (e.g. for globular and vermicular cast iron), and of extremely hard and chemically resistant microstructures for the precise machining of hardened steel and hard cast iron.

Apart from temperature dependent transformation toughening (ZrO_2 additives), high values of the fracture toughness are obtained first of all by anisotropic crystal shapes (whisker reinforcement, rod shaped crystals with high aspect ratio in Si_3N_4, platelets); generally new mechanism are not likely to be found in the near future.

Both the demands for advanced grinding tools and for the precise machining of hard cast iron and hardened steel with cutting ceramics are expected to promote the further development of powder technologies for the manufacture of highly perfect submicrometer microstructures for applications in fields where diamond or cBN tools are too expensive.

For *grinding materials*, corundum (α-Al_2O_3) ceramics are dominant at present. Here, contrary to other fields of ceramic technologies, precursor (sol/gel) approaches represent an advanced industrial state of the art. Extended developments are yet in progress, but, on the other hand, it is just the wide industrial use of these procedures which gives insight to some general limitations. New powder routes revealed such a surprisingly large increase of the observed grinding efficiency of the grits compared with sol/gel derived materials, that another leap in the quality has to be expected similar to the progress enabled ten years ago by the introduction of sol/gel approaches.

At present, submicrometer *cutting ceramics* for machining hard workpieces are developed in two classes of materials. In the industry, first tests in Europe are known for carbide reinforced composites based on Al_2O_3, whereas more fundamental consideration (Sections 4.4.1 and 4.4.3) and first laboratory results from Asia and Europe (Section 4.4.4) indicate advantages of new grades of pure corundum tools for cutting both mild steel and for machining hard cast iron/hardened steel. Besides of precision machining, first surprising results suggest that pure Al_2O_3 cutting tools may be applied for the machining of hardened steel even under conditions where at present only cubic boron nitride can be used (discontinuous cut). On turning hardened steel with pure alumina, high materials removal rates are possible if at high cutting velocities some softening of hardened steel workpieces can be tolerated in a preliminary machining step.

References

1. A. Raab and W. Zielasko, Schneidstoffeinsatz in der Großserienfertigung, Lecture at the meeting of the VDI-Division Schneidstoffanwendung, 21.11.1996, Siegmaringen.
2. G. Wellein and J. Fabry, Hartdrehen ist mehr als der Einsatz von CBN, *Werkzeug-Technik* 1996, **45**, 7–10.
3. A. Krell and P. Blank, The influence of shaping method on the grain size dependence of strength in dense submicrometre alumina, *J. Europ. Ceram. Soc.* 1996, **16**, 1189–1200.
4. F. F. Lange, Sinterability of agglomerated powders, *J. Am. Ceram. Soc.* 1984, **67**, 83–89.
5. T. J. Graule, F. H. Baader, and L. J. Gauckler, Shaping of ceramic green compacts direct from suspensions by enzyme catalyzed reactions, *cfi Ber. Dtsch. Keram. Ges.* 1994, **71**, 317–323.
6. S. Tashima, Y. Yamane, H. Kuroki, and N. Narutaki, Cutting performance of high purity alumina tools formed by a high speed centrifugal compaction process, *J. Jpn. Soc. Powd. Powd. Metall.* 1995, **42**, 1464–1468.
7. A. Krell, Fracture origin and strength in advanced pressureless sintered alumina, *J. Am. Ceram. Soc.* 1998, **81**, 1900–1906.
8. B. J. Hockey, Observations by transmission electron microscopy on the subsurface damage produced in aluminum oxide by mechanical polishing and grinding, *Proc. Br. Ceram. Soc.* 1972, **20**, 95–115.
9. A. Krell and P. Blank, Grain size dependence of hardness in dense submicrometer alumina, *J. Am. Ceram. Soc.* 1995, **78**, 1118–1120.
10. S. D. Skrovanek and R. C. Bradt, Microhardness of a fine-grain-size Al_2O_3, *J. Am. Ceram. Soc.* 1979, **62**, 215–216.
11. A. Krell, Improved hardness and hierarchic influences on wear in submicron sintered alumina, *Mater. Sci. Eng.* 1996, **A209**, 156–163.
12. A. Krell, Fortschritte in der spanenden Metallbearbeitung durch Keramiken mit Submikrometer-Gefüge, in *Hartstoffe, Hartstoffschichten, Werkzeuge, Verschleißschutz*, Proc. of the Hagen Symposium for Powder Metallurgy, 13–14 Nov. 1997, R. Ruthhardt (Ed.), Werkstoff-Informationsgesellschaft, Frankfurt, 1997, pp. 57–76.

13. H. G. Sowman and M. A. Leitheiser, Non-fused aluminium oxide-based abrasive mineral, a process for its production and abrasive products comprising the said abrasive mineral, *European Patent Application* EP-24 099 B1, Int. Cl.^{6} C09K3/14, 25.01.1984.
14. T. E. Cottringer, R. H. Van de Merwe, and R. Bauer, Abrasive material and method for preparing the same, *US Patent* 4 623 364, C09C1/68, 18.11.1986.
15. G. Bartels, G. Becker, and E. Wagner, Process for the production of a ceramic polycrystalline abrasive, *US Patent* 5 034 360, C04B35/10, 23.07.1991.
16. A. Krell, P. Blank, E. Wagner, and G. Bartels, Advances in the grinding efficiency of sintered alumina abrasives, *J. Am. Ceram. Soc.* 1996, **79**, 763–769.
17. A. Krell and P. Blank, Al_2O_3 sintering material, process for the production of said material, and use thereof, *European Patent* 756 586, Int. Cl.^{6} C04B35/111, publ. 19.10.1997.
18. N. G. L. Brandt and Z. D. Senesan, Ceramic cutting tool reinforced by whiskers, *US Patent Application* 4 867 761, Int. Class. C04B35/117, 10.09.1989.
19. A. G. Evans and B. J. Dalgleish, Creep and fracture of engineering materials and structures, in *Proc. 3rd Int. Conf. at University College*, B. Wilshire and R. W. Evans (Eds), The Institute of Metals, London, 1987, pp. 929–955.
20. T. N. Tiegs and P. F. Becher, Thermal shock behavior of an alumina-SiC whisker composite, *J. Am. Ceram. Soc.* 1987, **70**, C109–C111.
21. J. D. French, J. Zhao, M. P. Harmer, H.-M. Chan, and G. A. Miller, Creep of duplex microstructures, *J. Am. Ceram. Soc.* 1994, **77**, 2857–2865.
22. K. Friedrich, S. Lehmann, M. Fripan, D. Klotz, and H. Ziegelbauer, Mischkeramik auf Aluminiumoxidbasis, *European Patent Application* 755 904 A1, Int. Cl.^{6} C04B35/117, 29.01.1997.
23. A. Krell, L.-M. Berger, E. Langholf, and P. Blank, Hartstoffverstärkte Al_2O_3-Sinterkeramiken und Verfahren zu deren Herstellung, *German/International Patent Application* DE-196 46 344 A1, Int. Cl.^{6} C04B35/117, 14.05.1998 / WO-98/21161 A1, C04B35/117, 22.05.1998.
24. M. Fripan and U. Dworak, Keramische Hochleistungsschneidstoffe in der Guß- und Stahlzerspanung, in *Schneidstoffe* H. Kolaska (Ed.), Verlag Schmid, Freiburg, 1988, pp. 194–214.
25. J. Koike, S. Tashima, S. Wakiya, K. Maruyama, and H. Oikawa, Mechanical properties and microstructure of centrifugally compacted alumina and hot-isostatically-pressed alumina, *Mater. Sci. Eng.* 1996, **A220**, 26–34.
26. T. Takahashi, Y. Katsumura, and H. Suzuki, Cutting performance of white ceramics tools having high strength, *J. Jpn. Soc. Powd. Powd. Metall.* 1994, **41**, 33–37.
27. S. Tashima, Y. Yamane, H. Kuroki, and N. Narutaki, Cutting performance of high-purity alumina ceramic tools formed by a high-speed centrifugal compaction process, *J. Jpn. Soc. Powd. Powd. Metall.* 1995, **42**, 464–468.
28. G. K. L. Groh, L. C. Lim, M. Rahman, and S. C. Lim, Effect of grain size on wear behaviour of alumina cutting tools, *Wear* 1997, **206**, 24–32.
29. A. Muchtar and L. C. Lim, Indentation fracture toughness of high purity submicron alumina, *Acta Mater.* 1998, **46**, 1683–1690.
30. A. Krell and D. Klaffke, Effects of grain size and humidity on the fretting wear in fine-grained alumina, Al_2O_3/TiC, and zirconia, *J. Am. Ceram. Soc.* 1996, **79**, 1139–1146.
31. V. N. Koinkar and B. Bushan, Microtribiological studies of Al_2O_3, Al_2O_3/TiC, polycrystalline and single-crystal Mn–Zn ferrite, and SiC head slider materials, *Wear* 1996, **202**, 110–122.
32. A. Krell, P. Blank, L.-M. Berger, and V. Richter, Alumina tools for machining chilled cast iron, hardened steel, *Bull. Am. Ceram. Soc.* 1999, **78**(12), 65–73.

5 Silicon Carbide Based Hard Materials

K. A. Schwetz

5.1 Introduction

SiC is considered by many as the most important carbide, because of its extreme properties and the variety of present and potential commercial applications. According to Kieffer [1] silicon carbide belongs to the group of nonmetallic hard materials, that is materials whose great hardness and high-melting point result from a high fraction of covalent bonding. Superhard compounds are obviously formed by combination of the four low atomic number elements: boron, carbon, silicon, and nitrogen to form a quarternary system (see Fig. 1); carbon as diamond, boron–nitrogen as cubic boron nitride, boron–carbon as boron carbide, and silicon–carbon as silicon carbide belong to the hardest materials hitherto known.

5.1.1 History

Berzelius [2] first reported the formation of silicon carbide in 1810 and 1821. It was later rediscovered during various electro chemical experiments especially by Despretz [3], Schützenberger [4], and Moissan [5]. However it was Acheson [6], who first realized the technical importance of the carbide as a hard material and, by 1891, had managed to prepare silicon carbide on a large scale. Originally, Acheson had completely different things in mind: he wanted to produce artificial diamonds by recrystallizing graphite in an aluminum silicate melt; at a time when high-pressure and high-temperature chemistry was unknown. His efforts proved fruitless, but he was rewarded by the discovery of a far more versatile material, silicon carbide.

Believing that this was a compound of carbon and corundum, he named the new substance 'carborundum' (Fig. 2). Today, silicon carbide has become by far the most widely used nonoxide ceramic material, with an annual world production of about 700 000 tonnes. Owing to its great hardness, heat resistance, and oxidation resistance it has become firmly established as an abrasive as well as a raw material for making refractories such as firebricks, setter tiles and heating elements. Another major use of silicon carbide is as a siliconizing and carburizing agent in iron and steel metallurgy.

Self-bonded silicon carbide materials in molded form and of high SiC content in order to make full use of the excellent material properties, have been available only from the late 1960s onwards. They are being increasingly used as refractory materials as well as for structural components in mechanical engineering, and have proved highly successful for use under extreme conditions involving abrasion, corrosion, and high temperatures.

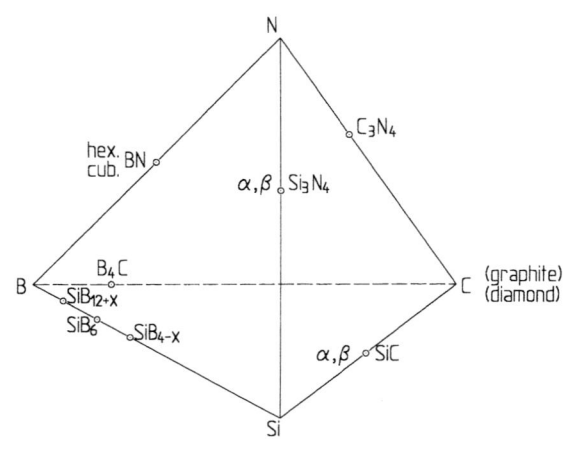

Figure 1. Nonmetallic hard materials in the system B–C–Si–N.

5.1.2 Natural Occurrence [7]

Naturally occurring silicon carbide was first discovered by Moissan [8] in a meteorite from the Diablo Canyon in Arizona, and subsequently several small deposits have been found [9–11]. Extraterrestrial occurrences of silicon carbide are even

Figure 2. Crystalline SiC grain from the Acheson process.

rarer [7]. Its existence in interstellar space has been shown by spectral measurements, and it has also been found in samples of moon rock.

5.2 Structure and Phase Relations of SiC

As shown in the phase diagram in Fig. 3, silicon carbide is the only binary phase in the silicon–carbon system, with the composition 70.05 weight-% Si, 29.95 weight-% C. Silicon carbide does not have a congruent melting point. In a closed system at a total pressure of 1 bar it decomposes at $2830 \pm 40°C$ into graphite and a silicon-rich melt (Fig. 2). This is the highest temperature at which silicon carbide crystals are formed. In an open system, silicon carbide starts to decompose at $\approx 2300°C$ with the formation of gaseous silicon and a residue of graphite. SiC and Si form a degenerate eutectic at $1413°C$ and $0.02\,\text{at}\%$ C. The solubility of carbon in liquid silicon is $13\,\text{at}\%$ C at the peritectic temperature [12].

Silicon carbide exists in several modifications being polymorphic and polytypical and crystallizing in a diamond lattice, like silicon [13].

Polycrystalline silicon carbide obtained by the Acheson process exhibits a large number of different structures (polytypes), some of which dominate. These can be classified in the cubic, hexagonal, and rhombohedral crystal systems (Table 1).

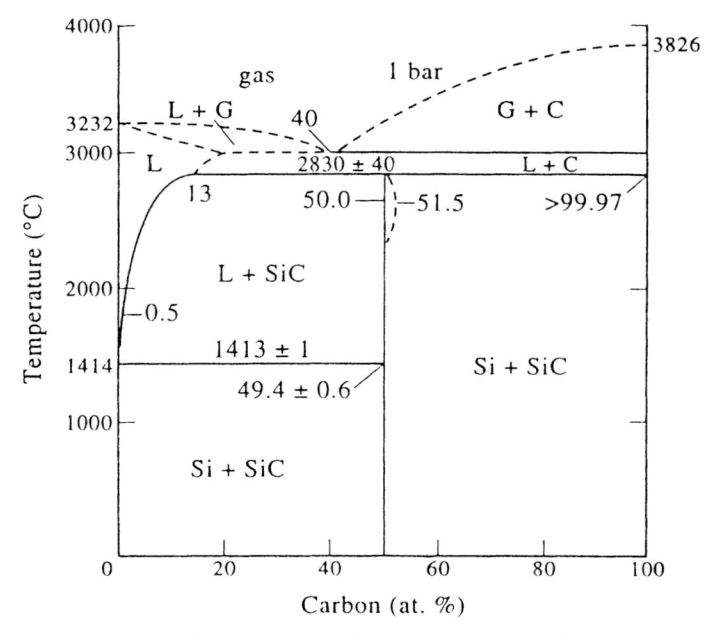

Figure 3. Phase diagram of the silicon–carbon system at 1 bar total pressure after Kleykamp and Schumacher [12].

Table 1. Dominant SiC types

Modification	Polytype
αSiC (high-temperature modification)	6H
	15R
	4H
βSiC (low-temperature modification)	3C

Since silicon carbide predominantly exists in its beta form at temperatures below 2000°C, this is referred to as the low-temperature modification. Cubic βSiC is metastable and, in accordance with Ostwald's rule, is formed initially in SiC production from silicon dioxide and carbon. βSiC (polytype 3C) can also be prepared at about 1450°C from simple mixtures of silicon and carbon or by hydrogen reduction of organosilanes at temperatures below 2000°C. Above 2000°C, only the hexagonal and rhombohedral types are stable if there are no stabilizing influences which raise the transformation temperature [14, 15]. Thus, in a nitrogen atmosphere at high pressure, βSiC is the stable form, so that above 2000°C αSiC transforms to βSiC. On the other hand, a hexagonal polytype, the wurtzite or 2H modification of SiC, has been obtained at temperatures as low as 1300–1600°C by the decomposition methyltrichlorosilane CH_3SiCl_3 [16]. This 2H polytype, which does not occur in commercial SiC, and the most common 3C, 4H, 6H and 15R structures are called short-period polytypes. Long-period polytypes are much less common and consist of blocks of short-period polytypes which are broken by regularly occurring stacking faults.

The basic element of the silicon carbide structure is the tetrahedron [17] due to sp^3 hybridization of the atomic orbitals. This tetrahedron consists of a silicon or a carbon atom at the spatial center, surrounded by four atoms of the other kind. The SiC- bond is 88% covalent. The tetrahedra are arranged in such a way that units of three silicon and three carbon atoms form angled hexagons which are arranged in parallel layers as shown in Fig. 4.

The layer sequences can repeat themselves in the cycles ABC, ABC ... (zinc blende, type 3C) or AB, AB ... (wurtzite, type 2H), according to cubic or hexagonal close packing. In addition, numerous others stack sequences are formed in the case of silicon carbide, resulting in many similar polytypes.

About 180 different polytypes are currently known. They all have the same density, 3.21 g cm^{-3}. The formation and stability of the various polytypes are not yet clearly understood. Written polytype nomenclature [18] indicates the number of layers in the repeating layer pack by a numeral, while the crystal system is denoted by the letters C, H, or R.

The amounts of the most frequently encountered polytypes 3C, 4H, 15R, and 6H can be quantitatively determined by X-ray diffraction techniques [19], which can indicate the temperature of formation of the silicon carbide since the stability of a given polytype depends on, among other things, temperature. According to Inomata et al. [20] 2H-SiC seems to be stable below 1400°C, 3C between 1400 and about 1600°C, 4H between 1600 and 2100°C, 6H above 2100°C and 15R above 2200°C.

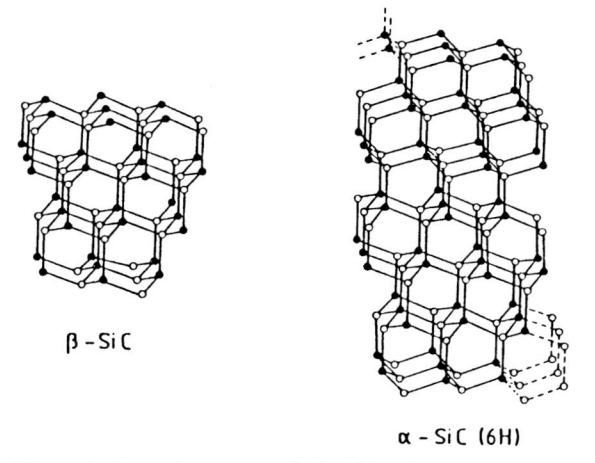

Figure 4. Crystal structure of the SiC polytypes 3C and 6H.

There is no doubt, however, that impurities also play a part in the formation of the various polytypes, together with the surrounding gas atmosphere and growth-kinetic processes. Preferential stabilization of the 4H polytype is achieved with additions of aluminum up to concentrations of ≈ 0.4 weight-%, which is thought to substitute for silicon in the silicon carbide lattice [21–24]. According to Lundquist [21] pure αSiC mainly crystallizes as 6H, with small amounts of aluminum 15R also occurs, and above 0.1 weight-% aluminum, 4H predominates (see Fig. 5). Also it is well known today that nitrogen can be incorporated into the SiC lattice, whereby it stabilizes not only the cubic βSiC form but also the wurtzite 2H polytype [14, 25]. Whereas the nitrogen content obtained in 3C-SiC by gas/solid diffusion is low, e.g. only <1 atom %, extremely high nitrogen contents of up to ≈ 14 atom % were

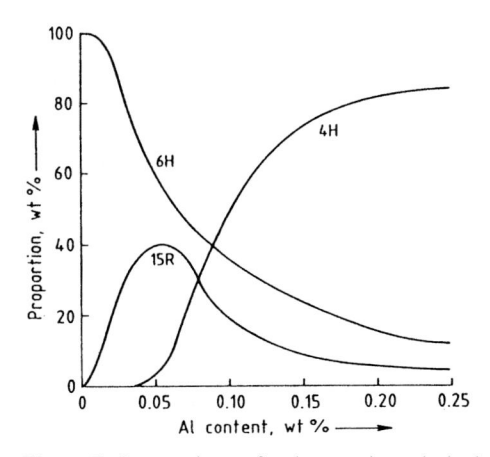

Figure 5. Proportions of polytypes in technical-grade SiC as a function of Al content.

recently found for nitrogen-stabilized 2H SiC when formed from carbothermal reduction of silicon nitride, Si_3N_4, in the presence of a liquid phase [26].

Transmission electron microscopy using high resolution lattice imaging techniques allows polytype analysis within single grains in the microstructure of dense SiC bodies [27].

5.3 Production of SiC

5.3.1 The Acheson/ESK Process

Abrasive and refractory grade silicon carbide is produced industrially according to the following reaction

$$SiO_2 + 3C = SiC + 2CO$$

from high-purity quartz sand (99.5% SiO_2) and petroleum coke, in electrically heated resistance furnaces [28]. The raw materials should have a maximum grain size of 10 mm. In addition to the reactants, sawdust is sometimes added to increase the porosity of the mix and to facilitate the venting of the carbon monoxide formed. The reaction is strongly endothermic: the heat of formation is 618 kJ mol^{-1} SiC, corresponding to 4.28 kWh kg^{-1} SiC. The process takes place at temperatures of between 1600°C and 2500°C and is far more complex than the above equation because of the intermediate reactions, some of which occur with participation of the gas phase [29], [36].

Acheson SiC furnaces are up to 25 m long, 4 m wide, and 4 m high, consuming up to 5000 kVA, and running for about 130 h. The largest furnaces operate at up to 25000 A. The furnace has a rectangular cross section and consists of two water-cooled graphite electrodes at the end and two movable side walls, whose purpose is to retain the raw material mixture. When about half of the mixture has been charged to a depth even with the electrodes, a column of tamped graphite powder or a solid graphite rod is inserted to connect the electrodes. This so-called core is then covered with additional reaction mixture. The electrodes are then connected to the transformer, and the current is switched on. The silicon carbide is formed in the hot core in the form of a polycrystalline, compact cylinder, the so-called roll. Temperatures within the furnace reach 2500°C causing the silicon carbide to recrystallize. The surrounding mixture acts as a thermal insulator. The carbon monoxide byproduct escapes through the surface of the reaction mixture and through slits in the side walls, where it burns.

After the current is switched of, the furnace is allowed to cool for several days. Then the side walls are removed, and the excess reaction mixture is carried off. The roll of silicon carbide contains not only the original resistance core, but also graphite formed by decomposition of silicon carbide. The boundary between the graphite and the SiC is very sharply defined. The innermost SiC zone is of the highest quality, as the purest and largest crystals can form in this region. Moving

(a)

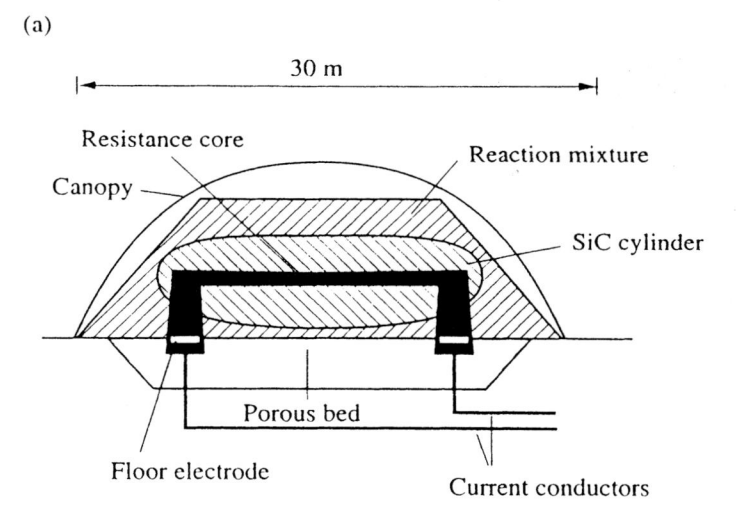

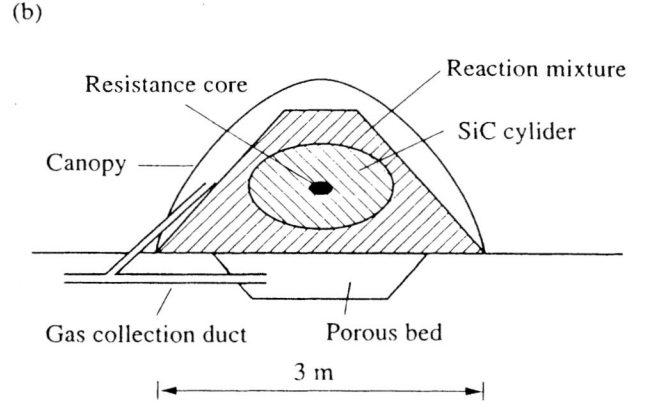

```
|←―――――――― 30 m ――――――――→|
```

Resistance core
Canopy
Reaction mixture
SiC cylinder
Porous bed
Floor electrode
Current conductors

(b)

Resistance core
Reaction mixture
SiC cylinder
Canopy
Gas collection duct
Porous bed

```
|←――― 3 m ―――→|
```

Figure 6. Silicon carbide production by the ESK process [7, 30]: (a) longitudinal and (b) radial cross-sections through the furnace.

outwards, with decreasing temperature, the crystal growth rate decreases and the SiC roll becomes more and more finely crystalline. The outer zone is composed of very fine crystals of βSiC. While in the case of the classic Acheson furnaces, with their laterally positioned electrodes and side walls, the carbon monoxide burns away on the surface, polluting the atmosphere inside and outside the furnace building, the ESK (Elektroschmelzwerk Kempten) process [30] developed in the 1970s (Fig. 6), is ecologically safe. It operates with floor electrodes, which enter the reaction mass from below a porous floor, and the carbon monoxide is collected by a gastight plastic canopy and transported through gas collecting pipelines to a power station where it is converted into electrical energy after burning. This energy is recycled and amounts to 20% of the energy necessary for the production

Figure 7. SiC roll at the ESD facility in Delfzijl, The Netherlands.

of silicon carbide. The silicon carbide cylinders inside the linear or U-shaped resistance furnaces are up to 60 m long and weigh about 300 tonnes after a run of 10 days (see Fig. 7). A comparative review of some Acheson plants has been published [31].

The lumps of silicon carbide obtained on dismantling the furnace are taken to a sorting area and broken down with hydraulic equipment and compressed air hammers. The finely crystalline outer zone (the so-called amorphous zone) is further size reduced (e.g., in a jaw crusher) and is usually returned to the furnace process by adding it to the coke/sand mixture. Alternatively, this 90–95% SiC material is a valuable raw material for the production of grain refiners for the ferrous metal industry.

The coarsely crystalline compact pieces from the inner zone of the SiC roll are sorted according to criteria such as color, structure, etc. This top-grade material forms the largest fraction of the SiC yield from a furnace charge.

Further processing of the silicon carbide into the various macro- and microgrits depends upon such factors as the required particle size distribution [32] and shape, and involves breaking, grinding, sieving, air classification, and elutriation.

For applications that require a particularly high grade of silicon carbide, such as industrial ceramics, the refractory industry, and ceramic-bound abrasives, the granular product must be washed with alkali or acid to remove adhering traces of elemental silicon, metals, metal compounds, graphite, dust, and silica.
Commercial SiC grades vary in composition from 90% to 99.5% SiC. Light green SiC is 99.8% pure, and, as the purity decreases to 99%, the color changes to dark green, and, at 98.5% to black.

Before αSiC is converted into molded parts by sintering, it must be intensively milled into the submicrometer range, that is to a mean particle size of less than 1 μm (see Fig. 8b), in attritors [33, 34] or ring-gap mills. Adherent impurities which interfere with the sintering process, such as SiO_2 or wear from milling equipment, are removed by wet chemical purification [35].

Figure 8. Silicon carbide submicron powders: (a) βSiC powder, specific surface area: $12.6\,\mathrm{m^2\,g^{-1}}$ [40] and (b) αSiC powder, $S = 15.8\,\mathrm{m^2\,g^{-1}}$ [35].

5.3.2 Other Production Methods

5.3.2.1 βSiC Powder

5.3.2.1.1 Carbothermal Reduction of Silica This process is based on the following equation

$$SiO_2 + 3C \xrightarrow{\leq 1.900°C} \beta SiC + 2CO.$$

According to the reaction mechanism proposed by Weimer [36] the reaction takes place in several steps:

$$SiO_{2(s)} + C_{(s)} \rightarrow SiO_{(g)} + CO_{(g)}$$
$$SiO_{2(s)} + CO_{(g)} \rightarrow SiO_{(g)} + CO_{2(g)}$$
$$C_{(s)} + CO_{2(g)} \rightarrow 2CO_{(g)}$$
$$SiO_{(g)} + 2C_{(s)} \rightarrow SiC_{(s)} + CO_{(g)}.$$

The initial reaction with formation of gaseous SiO at points of direct contact between C and SiO_2 requires temperatures above $\approx 1600°C$ at atmospheric pressure (solid/solid reaction). The high rate of $SiO_{(g)}$ formation suggests that the further two gas/solid reactions make the final carbon reduction of the SiO to SiC possible once the C/SiO_2 contact points are consumed. Carbon crystallite diameter was found to have a substantial influence on the rate of reaction and the size of synthesized SiC.

This process has been used in the early 1960s to fabricate pure ultrafine βSiC powders especially suited for the semiconductor and pigment industry [37, 38]. During the first step of the General Electric Process [37] a silica gel is formed by hydrolysis of silicon tetrachloride in an aqueous solution of sugar (sucrose). Upon dehydratization at 300°C the sugar pyrolyzes and an intimate amorphous carbon/amorphous silica mixture is obtained which is subsequently fired at 1800°C in an inert atmosphere to form βSiC.

Later the process was modified for the fabrication of βSiC sintering powder [39, 40], using lower reaction temperatures and as starting materials the combinations tetraethylsilicate/sugar and fumed amorphous silica/sugar, respectively. Particle size control corresponding to powder surface areas in the $10-20\,m^2\,g^{-1}$ range (see Fig. 8a) was necessary, because ultrafine powders have high oxygen contents (surface silica), and often low green densities result due to their high surface area.

Since 1981 a continuous version of the Acheson process has been commercially operated by the Superior Graphite Company [41]. A mixture of silica sand and petroleum coke is fed continuously to the furnace and heated electrothermally to a temperature of 1900°C. The run-off-furnace product is a free flowing grain generally smaller than 3 mm, comprising agglomerates of βSiC crystallites and free carbon. The grain is friable and is readily processed to make sized microgrits and sintering powder by dry milling, wet grinding, classification, and wet chemical purification (removal of 5 weight-% free carbon).

5.3.2.1.2 Synthesis from the Elements Prochazka [39] synthesized submicron βSiC powder with a surface area of $7\,m^2\,g^{-1}$ using direct reaction of high-purity silicon powder with carbon-black at temperatures between 1500 and 1650°C, but saw little densification with boron and carbon additions, suggesting that free silicon is detrimental to densification.

The *in situ* formation of silicon carbide according to the reaction

$$Si_{(l,g)} + C_{(s)} = SiC_{(s)}$$

is mainly used in the manufacture of reaction sintered SiC components. After shaping a mixture of fine-grained SiC powder and carbon and exposure at high temperatures to liquid or vapor-phase silicon, the carbon is reacted to secondary silicon carbide, which bonds together the primary SiC grains. Excess silicon (about 10 vol-%) is usually left to fill the pores, thus yielding a nonporous body (Si–SiC).

5.3.2.1.3 Deposition from the Vapor Phase Submicron βSiC can be continuously produced by decomposing gaseous or volatile compounds of silicon and carbon in inert or reducing atmospheres, at temperatures above 1400°C [42]. Particle size and morphology depend considerably on the reaction temperature and on the composition in the gas phase. Powders obtained below 1100°C are amorphous. A great variety of reactants and several methods of heating, such as d.c. arc jet plasma [43], high frequency plasma [44], laser [45], and thermal radiation are possible. However, large scale commercialization of these processes has been limited by the very high cost of production.

Small particle sizes are produced, 0.01–0.5 μm, with greater than 70% efficiency using silicon tetrachloride ($SiCl_4$) and methane (CH_4) or methyltrichlorosilane (CH_3SiCl_3) as the gaseous precursors in the presence of hydrogen as the plasma gas:

$$CH_3SiCl_{3(g)} \xrightarrow{H_2} SiC_{(s)} + HCl_{(g)}$$

$$SiCl_{4(g)} + CH_{4(g)} \xrightarrow{H_2} SiC_{(s)} + 4HCl_{(g)}.$$

If boron trichloride is used for the boron dopant together with a small excess of the carbon feedstock, sinterable SiC powders are obtained.

The laser synthesis allows laboratory scale production of high-purity SiC powder with controlled particle sizes. Using a CO_2 laser and a mixture of silane (SiH_4) and methane (CH_4) or acetylene (C_2H_2) βSiC particles in the size range 5–200 nm have been obtained. An advantage of using SiH_4, rather than $SiCl_4$, as a reactant is that it has a strong absorption band near the wavelength of the laser (10.6 μm).

Amorphous SiC powders with metal impurity levels below 60 p.p.m. total have been synthesized by thermal decompostion of a methylsilane gas stream at temperatures above 600°C according to the equation

$$CH_3SiH_3 \xrightarrow{T} SiC + 3H_2$$

in an induction heated graphite reactor [46].

5.3.2.2 SiC Whiskers

SiC whiskers are discontinuous, monocrystalline hair or ribbon-shaped fibres in the size range 0.1–5 μm in diameter and 5–100 μm in average length (see Fig. 10a).Their aspect ratio (length/diameter) ranges from 20 to several 100 and is an important figure. SiC whiskers are all grown from the gas phase on solid or liquid substrates under conditions that result in a small defect size (0.1–0.4 μm) in the whisker. Such small defect size results in very high strength, of about 9–17 GPa. Whiskers grown at temperatures below 1800°C are composed of mostly βSiC, those grown above 2000°C are of αSiC.

According to Shaffer [47] preparation methods for whiskers can be divided into three routes:

(i) vapor–liquid–solid process (VLS),
(ii) vapor phase formation and condensation (VC),
(iii) vapor–solid reaction (VS).

5.3.2.2.1 The Vapor–Liquid–Solid Process In the VLS process silicon and carbon-rich vapors (usually CH_4, SiO, or $SiCl_4$) react at 1400°C to from SiC on a liquid

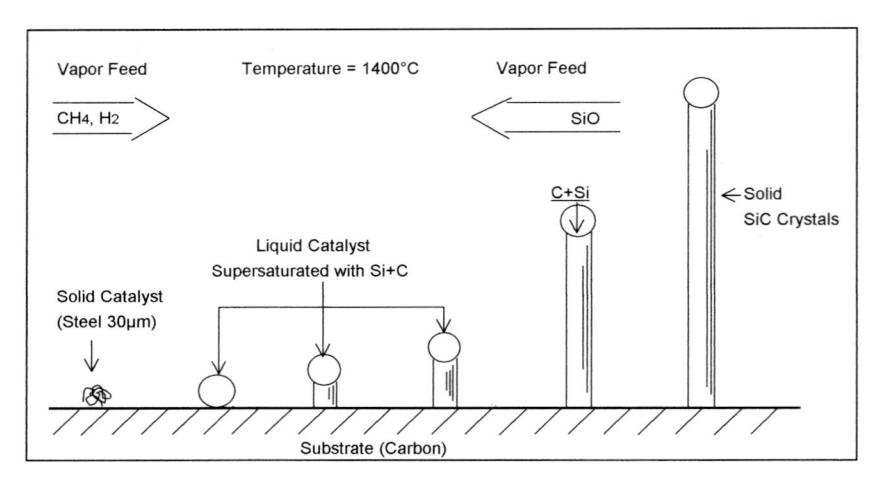

Figure 9. VLS process for SiC whisker growth (1400°C).

alloy:

$$\text{Si}_{(d)} + \text{C}_{(d)} \xrightarrow{1400°C} \text{SiC}.$$

(d) dissolved in liquid iron alloy

Microscopic particles of an alloy are distributed on a substrate (graphite) then exposed to silicon- and carbon-rich vapors. The presence of a liquid catalyst, such as a transition metal or usually an iron alloy, distinguishes this method from all other whisker preparation methods.

Although the VLS-method yields near-perfect whiskers and has been investigated since the early 1960s, it has not been developed to economical commercial practice up to now.

Figure 9 illustrates the VLS whisker growth sequence. At 1400°C the solid catalyst melts and forms the liquid catalyst ball. Carbon and silicon from the gas phase dissolve into the liquid catalyst, which soon becomes supersaturated and solid SiC precipitates at the interface with the carbon substrate. Continued dissolution of gas species into the liquid catalyst allows the whisker to grow, lifting the catalyst ball from the substrate as additional SiC precipitates. These VLS whiskers are typically larger in diameter (4–6 μm) than those formed by the vapor–solid process (VS-whiskers). They frequently grow to length of tens of millimeters.

5.3.2.2.2 Vapor Phase Formation and Condensation
To grow whiskers by this process, bulk SiC is vaporized by heating to very high temperatures (>2200°C) usually under reduced pressure. Upon cooling αSiC whiskers form on nucleation sites.

$$\text{SiC}_{bulk} \xrightarrow{\text{Sublimation}} \text{SiC}_{whisker}$$

The addition of lanthanum, yttrium, neodymium, or zirconium leads to an increase in grow rates. However, today no whiskers are commercially produced by this sublimation process.

5.3.2.2.3 Vapor–Solid Reaction At present the principal commercial method for SiC whisker production is the carbothermic reduction of low cost silica sources at temperatures of 1500–1700°C. The reaction for the formation of VS-cubic βSiC whiskers occurs in two steps:

$$SiO_2 + C \rightarrow SiO + CO \uparrow,$$

$$SiO + 2C \xrightarrow{<1800°C} \beta SiC + CO \uparrow.$$

Many different VS-processes are known according to the starting raw materials and catalysts species involved. In all cases, a catalyst is used for whisker formation to be favored versus particulate SiC formation. Commercially available VP-whisker grades are typically less than 1 μm in diameter (submicron) and up to 30 μm in length and show strength/Young modulus values of 1–5 GPa/400–500 GPa. During whisker handling and processing, rigorous protection has to be provided since smaller whiskers (<3 μm diameter) can become lodged in the lungs and represent a health hazard. An inexpensive VS-process utilized calcined rice hulls, which contain both silica and carbon, as a precursor to SiC whiskers.

The characteristics of commercial SiC whiskers as claimed by producers are given in Table 2.

βSiC whisker have been added to a wide variety of ceramics (Al_2O_3, Si_3N_4, AlN, $MoSi_2$, mullite, cordierite, and glass ceramics) in an effort to achieve increased toughness. In most cases, hot pressing has been required to achieve near-theoretical density of the matrix. A real success story in the advanced ceramic market was the introduction of a 25–30 weight-% SiC whisker reinforced hot-pressed alumina for the high-speed cutting of high nickel content alloys. Thus, by incorporation of SiC whiskers into the alumina matrix a composite with increased strength (by over 50%), fracture toughness (by 100%), increased fatigue and thermal shock resistance was obtained [48]. In this composite, whisker pull out and crack bridging have been identified as the major toughening mechanisms. Crack bridging consists in bridging the crack behind the crack tip by unfractured whiskers or fibers. The effect depends on the strength of the fiber and on its aspect ratio. Inhibiting crack

Table 2. Comparison of the properties of βSiC whisker (supplier's data).

Supplier	Tokai Carbon	Tateho	Shin Etsu	C-Axis Technology	Millenium Materials Inc.	ART
Phase	β	β	β	β	β	β
Length (μm)	20–50	5–200	3–20		30–200	
Diameter (μm)	0.3–1.0	0.05–1.5	0.15–0.25	0.3–1.0	1–3	0.5–1.7
Aspect ratio		20–200	20–50	20–50		5–20
Density (g cm^{-3})	3.2	3.18		3.21	3.21	3.21

Table 3. Morphological characteristics of commercial SiC platelets.

Supplier	C-Axis Technology SF grade	Millenium Materials 325 mesh	Carborundum Corporation
Maximum dimension (μm)	5–50	5–70	10–500
Thickness (μm)	2–5	0.5–5	1–5
Aspect ratio	8–10	8–15	10–15

opening reduces the stress intensity at the tip of the crack and stops or slows crack propagation.

The fiber or whisker pull out has the potential for the greatest toughening effect and consists in a debonding of fiber from the adjacent matrix consuming large energy. This mechanism depends strongly on the aspect ratio of the second phase and is favored by weak interface bonds.

SiC whisker also can be combined with metals (MMCs, metal matrix composites) to increase the high temperature strength of the material as well as provide a comparable substitution for heavier traditional materials, such as steel. MMCs can be fabricated by infiltrating a SiC whisker preform with aluminum or by addition of SiC whiskers to molten aluminum.

5.3.2.3 SiC Platelets

As a reinforcement material for all matrix materials, ceramics, metals and plastics, SiC platelets offer a similar potential as whiskers but at lower cost and without health hazard. Platelets are single-crystal, plate-like αSiC particles with an aspect ratio of about 8–15. SiC-platelets typically range in size from about 5 to 100 μm in diameter and 1 to 5 μm in thickness (see Table 3 and Fig. 10b). They are commercially produced from inexpensive raw materials (silica and carbon, micron-sized βSiC powders) at temperatures of 1900–2100°C under inert atmosphere [49]. Due to the presence of boron and aluminum dopants added, platelet-shaped crystals are formed. Aluminum enhances the growth in the [0001] direction and decelerates the growth perpendicular to the [0001] direction. Boron enhances the growth notoriously perpendicular to the [0001] direction [50]. As impurity elements in SiC platelets from Millenium Materials, aluminum of 0.04–0.45 weight-%, boron and nickel of 0.4–0.8 weight-% (each), and free silicon of 0.3–3.6 weight-% were analyzed [51]. With the aid of high resolution AES most impurity elements were found to be present on the outermost surfaces of the platelets, for example boron was found as about a 3 μm thick layer of B_2O_3, as about 1 μm thin layers consisting of boron or silicon borides and as 3–5 nm thick segregations near to growth steps [51].

Dense SiC platelet reinforced silicon carbide composites were recently fabricated [52]. The highest sintered densities of 97–98% TD were achieved with 20% platelet contents using hot molding as the thermoplastic forming technique [53].

In addition to reinforcing ceramics, SiC platelets are also used to increase the strength, wear resistance and thermal shock performance of aluminum matrices, and to enhance the properties of polymeric matrices.

Figure 10. (a) βSiC whisker, and (b) αSiC platelets (Courtesy S. Weaver, Millennium Materials, Inc., Knoxville, USA).

5.3.2.4 SiC Fibers

Together with boron filaments and carbon fibers, SiC fibers are important continuous inorganic reinforcement materials with high modulus. Owing to their high-temperature properties and the resistance to oxidation they are particularly well suited for structural parts where high stiffness at high temperatures is required. Three different types of SiC fibers are now available: substrate-based fibers (CVD-fibers), polymer-pyrolysis derived fibers (PP-fibers), and sintered powder derived fibers (SP-fibers). Defect size in amorphous or polycrystalline fibers is very small so that strength can be high. In Table 4 some commercially available SiC fibers are listed along with their mechanical characteristics.

5.3.2.4.1 Substrate-based Fibers (CVD-Fibers) βSiC filaments of 100–150 µm thickness can be prepared by chemical vapor deposition onto tungsten or carbon

Table 4. Mechanical characteristics of various SiC fibers

Supplier	Textron SCS-6	Nippon Carbon Nicalon	UBE Tyranno	Carborundum Sintered-Alpha
Crystal phase	βSiC	βSiC	amorphous	αSiC
Diameter (µm)	140	10–20	8–10	20–150
Tensile strength (MPa)	3920	>3000	>3000	up to 1380
Young modulus (GPa)	406	>420	>200	400

monofilaments of 40 μm thickness, which act as hot substrates during heterogeneous nucleation. Various carbon containing silanes have been used as reactants. Owing to high process costs and the large diameters of the resulting fibers, the method is disadvantageous. It has been found that faster SiC filament production is possible on a carbon filament substrate. The US company Textron produces βSiC fibers on a carbon-fiber-core with a surface layer of pyrolytic carbon, which is itself coated with silicon carbide. These fibers are given the designation SCS. The surface carbon layer provides a toughness-enhancing parting layer in composites having a brittle matrix.

5.3.2.4.2 Polymer Pyrolysis Derived SiC Fibers (PP-fibers) As shown 1976 by Yajima [54], βSiC fibers with a smaller diameter (8–30 μm) and without a central core can be manufactured by solid state pyrolysis of a polycarbosilane (PCS) precursor fiber. The melt-spun PCS fiber is first cured at 200°C in air to produce a thin layer to protect from melting later on, then heated up in inert atmosphere to 1500°C to convert the PCS in crystalline βSiC. The steps leading to the production of SiC can be summarized as follows:

$$n(CH_3)_2SiCl_2 \xrightarrow{Na} [-Si(CH_3)_2-]_n \xrightarrow[10\,MPa\,Ar]{470°C}$$

$$(-\underset{\underset{H}{|}}{Si(CH_3)} - CH_2-)_n \xrightarrow{1300°C} \beta SiC + CH_4 + H_2.$$

Dimethyldichlorsilane is dechlorinated with metallic sodium in the presence of xylene to polydimethylsilane $(R_2Si)_n$, which is converted to polycarbosilane (PCS) by polycondensation with introduction of silicon–carbon bonds into the chain. The pyrolysis of the PCS resin to βSiC is accomplished by a slow heating to 1300°C and is accompanied by an appreciable loss of mass ($\approx 60\%$) and considerable shrinkage. This melt spinning–curing–pyrolysis route has been adopted by Nippon Carbon to produce a fiber called Nicalon [55] which has a composition almost identical to Si_3C_4O thus, it is not pure SiC, but a Si–C–O composite, in which nanometer-sized βSiC, SiO_2 and carbon are uniformly dispersed. These impurities affect the thermal stability of the fiber and its strength, which has been reported to degrade above 1200°C. However, these fibers have successfully been used to reinforce aluminum, refractories and Li–Al–Silicate glass ceramics. The coating of fibers, for instance by carbon, allows a decrease in the interface bonding with the matrix, which increases the strength and the toughness by favoring pull-out mechanism.

UBE Chemicals synthesized amorphous Si–Ti–C–O fibers from the PCS-titanium alkoxide compound polymer. These so called Tyranno fibers show excellent properties and can be spun thinner than the Nicalon fibers (see Table 4). However heating above 1000°C results in a crystallization of the fiber. In the mid 1990s the German company Bayer AG synthesized an amorphous Si–B–N–C fiber, by pyrolysis of a polyborosilazane polymer [56]. This $SiBN_3C$ fiber (see Fig. 11) has a tensile strength of 3 GPa and maintains its amorphous character up to 1800°C. The advantage of the production route from liquid to solid to produce SiC has also attracted attention for

Figure 11. SEM of SiBN$_3$C fibers made by Bayer AG, Germany (courtesy H.P. Baldus, [56]).

the SiC film production in microelectronics or as protection layers. Amorphous, and polycrystalline films of high purity produced by dip-coating of substrates in PCS solutions and subsequent pyrolysis in inert gas atmosphere have been prepared [57].

5.3.2.4.3 Sintered Powder Derived αSiC Fibers (SP-fibers) Recently, the Carborundum Company developed a single phase polycrystalline αSiC fiber having a diameter of approximately 20–150 μm and density of over 96% of theoretical. It is made by first producing green filaments by melt extrusion or suspension spinning of plasticized mixtures of sinterable αSiC powder with organic additives, such as polyethylene or polyvinylbutyral plus novolac, respectively. The green filaments are subsequently debindered during free fall from the extruder in presintering furnaces and finally undergo pressureless sintered at 2100°C in argon atmosphere. These polycrystalline αSiC fibers are stable up to 1600°C in air and 2250°C in inert gas, far superior than most of the commerciallly available βSiC based fibers [49]. According to Prochazka [58], an increase in the strength of αSiC fibers is obtainable by using refined αSiC powders, having a surface area greater than 20 m^2 g^{-1}, a median SiC particle size ≤0.25 μm, and no particles larger than 1.5 μm.

5.3.3 Dense SiC Shapes

SiC can be converted into more or less dense parts either by using various bonding phases to bind the SiC particles or grains together at elevated temperatures without dimensional change of the initial compact, or by sintering (pressureless, sinter/HIP or hot pressing), that is, by mechanisms of mass transport between particles which lead to porosity elimination together with shrinkage of the overall body.

5.3.3.1 Ceramically Bonded Silicon Carbide

This product group is based on oxide- or nitride-bonded SiC grains and exhibits an apparent porosity of up to 20% by volume. A fracture surface of a silicon nitride bonded SiC is shown in Fig. 12a.

Coarse silicon carbide powder fractions are mixed with clay, or pure oxides like SiO_2 and/or Al_2O_3, formed to 'green' shapes by conventional ceramic methods, and fired at about 1400°C. The fired clay or oxidic raw materials bind the silicon carbides particles and, although these are only loosely bonded, this is sufficient for many types of applications such as refractory furnace bricks and abrasive disc. Firing mixtures of silicon carbide particles and elemental silicon in a nitrogen atmosphere, with or without the addition of air and possibly Al_2O_3 produces silicon nitride-, silicon oxynitride- or SiAlON (silicon aluminum oxynitride)-bonded SiC products with enhanced high-temperature strength [59–61].

5.3.3.2 Recrystallized Silicon Carbide

Silicon carbide shapes with a high green density made by slip-casting or press-molding silicon carbide powders of bimodal particle size distribution are fired in an electric furnace at temperatures of up to 2500°C, under the exclusion of air [62, 63]. Evaporation and condensation [64] take place at temperatures above 2100°C, resulting in a selfbonded structure without shrinkage. The initial (green) density and final density remain the same and solid SiC bonds are produced between the crystals. The final RSiC product, which shows transgranular fracture (see Fig. 12b), consists of 100% SiC and can have densities of up to $2.6\,\mathrm{g\,cm^{-1}}$, with a porosity of approximately 20%.

5.3.3.3 Reaction-Bonded Silicon Carbide

In reaction sintering or reaction bonding silicon carbide is produced to some extent by chemical synthesis in the molding. A preform consisting of silicon carbide, carbon, and a carbon-containing binder is made by one of the normal ceramic shaping techniques (dry press molding, slip casting, extrusion, or injection molding, see Table 5). The elemental carbon in the preform is then reacted with gaseous or liquid silicon. The process can be controlled in such a way that a porous but pure SiC part is produced, or one whose pores are filled with excess silicon [65–73]. Since the latter (SiSiC), which is porefree dense (high density, showing no residual porosity, see Fig. 12c), is much stronger it is this which is most frequently made, usually by infiltration with liquid silicon. Silicon infiltration can also be used to fill the pores of recrystallized SiC moldings at a later stage [74].

The amount of free silicon in infiltrated silicon carbide is usually between 10% and 15%. It is difficult to bring this below 5%. Infiltration is relatively easy, because of the good wettability of the material, although there are problems if the bodies are large or thick, because of the exothermic reaction and the expansion of silicon on freezing, which make perfect impregnation difficult.

Reaction sintering can, however, usually be controlled so that the change in volume is negligible, resulting in dimensionally accurate components. This is

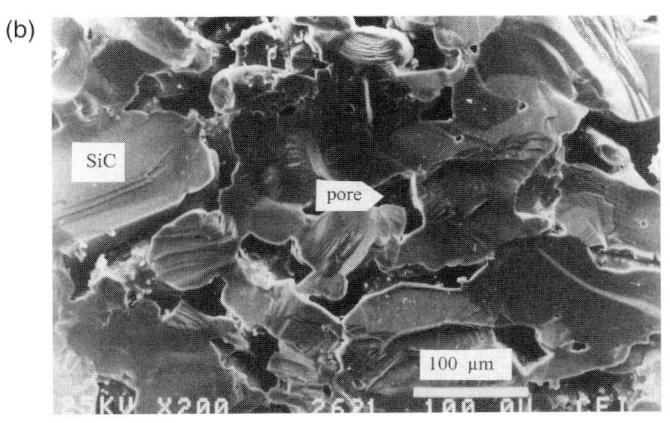

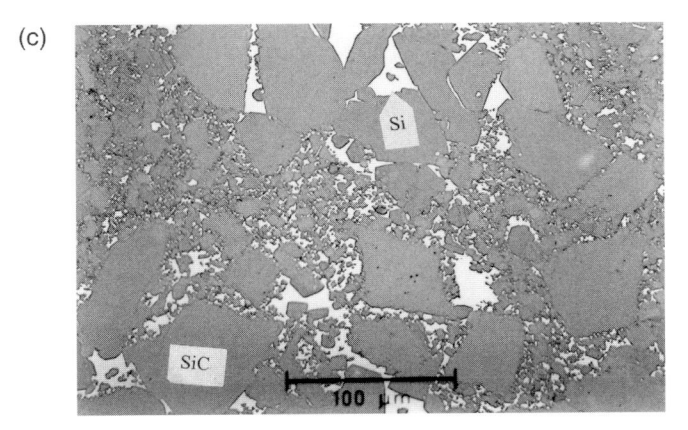

Figure 12. Microstructure of various SiC ceramics; (a) KSiC, nitride bonded SiC (AnnaSicon 25); (b) RSiC, recrystallized SiC (AnnaNox CK); (c) SiSiC silicon infiltrated SiC (AnnaNox CD), (d) SSiC, B/C doped (Ekasic F), (e) LPSSiC, equiaxed SiC grains (Ekasic T), (f) LPSSiC, platelet SiC grains (ESK), (g) HPSiC,Al-doped (ESK), (h) HIPSiC, undoped (ESK).

(d)

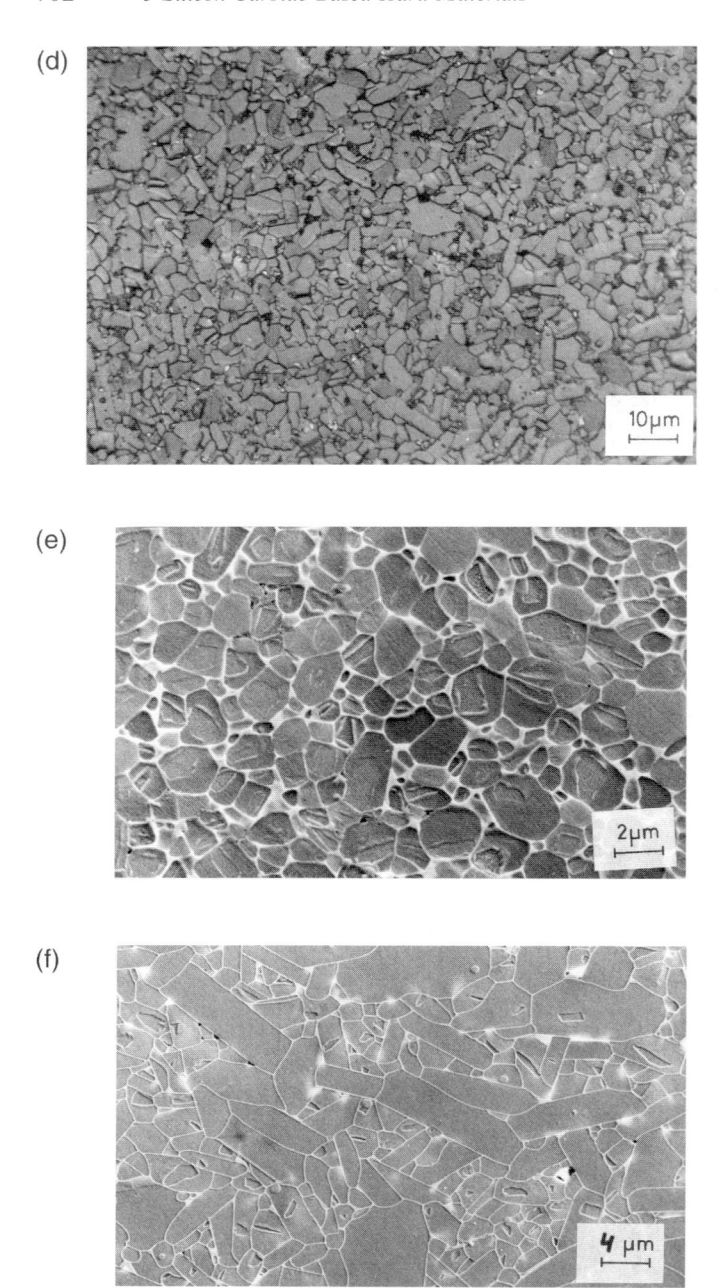

(e)

(f)

Figure 12. Continued

(g)

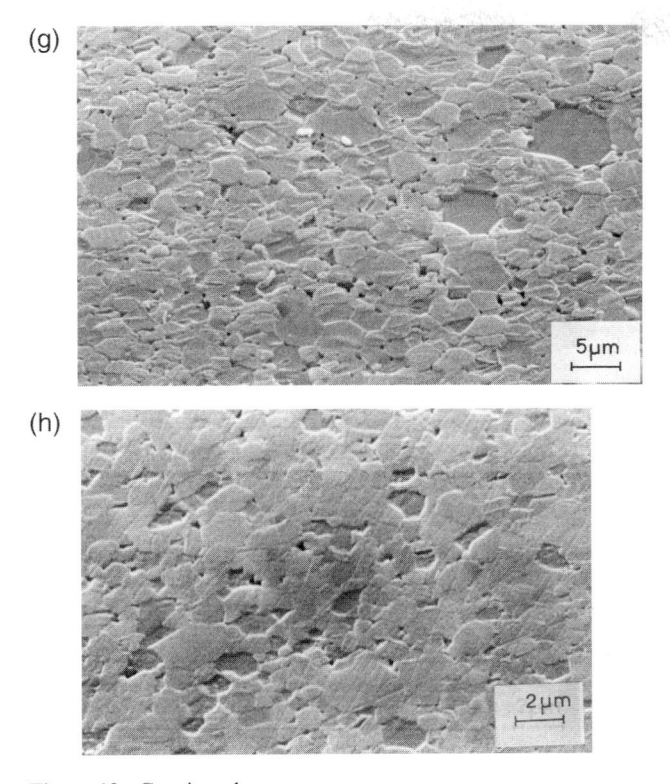

(h)

Figure 12. Continued

achieved by choosing the right SiC and carbon contents in the initial mix, and ensuring correct porosity in the preform and by proper temperature control.

5.3.3.4 Sintered Silicon Carbide

5.3.3.4.1 Solid-state Sintered SiC Once it was thought that silicon carbide was unsinterable. It was only in the early 1970s that it became possible to sinter silicon carbide without applied pressure and thus to achieve densities more than 95% of the theoretical density. This was largely due to the pioneering work of Prochazka [75] at the General Electric Corporation Research Center in Schenectady. The starting materials were submicrometer βSiC [76] or later αSiC [77] with the simultaneous addition of up to 2% carbon and boron. A minimum of 0.3 weight-% boron was required, although larger additions did not improve the density, which Prochazka attributed to the need to exceed the solubility limit of boron in SiC (0.2 weight-% according to Shaffer [78]). Carbon is added in excess to ensure complete deoxidation, that is removal of the SiO_2 film from the SiC grains, which in turn leads to the presence of carbon inclusions. Additions of excess carbon are observed to inhibit

Table 5. Shaping techniques for manufacture of SiC preforms (green bodies).

Moulding process	Diagram function scheme	Feasible molds, recommended dimensions	Type of tool, tool life	Yield for average parts in pieces/8 h
Automatic or dry pressing		Pellet-like parts in pressing directions not higher than three times the diameter; hole distances not less than 0.2 diameter; hole diameter >0.1 height of part	Cemented carbide molds more than 10^6 pieces	5000 pieces using single mold of average dimensions
Extrusion		Extrusions which can be cut to size, diameters 1.5–40 mm wall thickness > 1 mm, lengths up to 1000 mm	Extrusion die made of steel, minimal wear	Approx. 300 kg of compound
Slip casting		Complex shapes with unit weights of up to 5 kg and measuring up to 400 mm	Plaster molds from wooden patterns, up to 40 castings depending on mold	1–10 castings per mold, depending on wall thickness
Injection molding		Two-part molds, suitable for unit weights of up to 250 g	Usually multicavity molds made of steel, up to 50 000 shots	About 500 pieces per single-cavity mold
Isostatic pressing and dry grinding		Preferably axially symmetric bodies, maximum diameter 320 mm, maximum length 800 mm	Simple rubber molds	About 75 isostatically press molded parts

grain growth, and produce a more equiaxed grain structure [79], probably as a result of the energy required from the grain boundaries to overcome the carbon inclusions. Instead of boron and its compounds [80–83] one can also use aluminum and aluminum compounds [84–90], or beryllium and its compounds [91], the net result being the same. Sintering of Al/C doped [92] or Al_2O_3-doped [90] SiC powders is believed to occur in two stages: transient liquid phase sintering; then solid state sintering during the final stage.

This opened up an inexpensive method of producing dense and complex parts consisting of pure silicon carbide (SSiC). The powders (see Fig. 8) can be molded into a green body by any of the methods used in ceramic molding [93, 94], depending on the shape required and the number of pieces involved (see Table 5). Sintering is carried out in an inert gas atmosphere or in a vacuum at temperatures between 1900°C and 2200°C.

Linear shrinkage exceeds 15% and depends on the green density of the molding. During the sintering process, there is SiC-polytype transformation as well as grain growth, whose extent depends on the type and amount of sintering additives, as well as the sintering temperature. Boron-doped βSiC powders tend towards secondary recystallization (exaggerated grain growth [95] during 3C → 6H polytype transformation) whereas, if one starts with αSiC powders, a homogeneous, fine-grained microstructure can be obtained (see Fig. 12d). αSiC powder, which predominantly consists of the 6H-type, usually shows 6H → 4H polytype transformation during sintering with aluminum, as well as boron containing dopants. Aluminum promotes 6H → 4H transformation at lower sintering temperatures (2050°C), boron most easily at higher temperatures (2200°C). According to investigations of Schwetz et al. [96] there seems to be no fundamental correlation between SiC-polytype transformation and microstructure development during sintering of SiC (see Table 6).

Improved reliability in sliding bearing and rotating seal application have been achieved with microstructure-taylored solid-state sintered SiC materials, which contain pores [97–100], a combination of pores and graphite [101], and/or very large grained SiC crystallites [102].

5.3.3.4.2 Liquid-phase Sintered SiC Liquid-phase sintered SiC (LPSSiC) is a very new, dual-phase silicon carbide ceramic, with typically ≈ 95% SiC content and ultra-fine microstructure (grain size ≈ 1 μm, see Fig. 12e). Y_2O_3–Al_2O_3 [103–108] or Y_2O_3–AlN [109–111] based compositions were found to be the most effective densifying aids for even standard grade SiC powders and a pressureless- or gas-pressure sintered material can be produced with properties matching the best hot-pressed silicon carbide (see Table 8). Compared with SSiC, LPSSiC exhibits an increase of up to 100% in fracture toughness to 6.0 MPa m$^{1/2}$ and an increase of 70% in flexural strength to 730 MPa. Microcracking at the SiC–YAG interface ahead of a propagating crack is proposed as the dominant toughening mechanism.

The major part of the secondary phase is YAG (yttrium aluminate, $Y_3Al_5O_{12}$) which crystallizes together with reprecipitated SiC on solidification from the liquid [112]. The fracture toughness of LPS SiC has been recently further improved by anisotropic grain growth, that is growth of larger, platelike SiC grains (*in situ* toughening) in a matrix of finer equiaxial SiC grains [113, 114], see Fig. 12f.

Although LPSSiC may not be suitable as an universal tribology material like SSiC, the initial assessment of properties suggests that it will be suitable for high-stress engineering components and in many wear applications, where strength, reliability, and toughness are demanded.

5.3.3.5 Hot-pressed Silicon Carbide

Dense parts made by hot-pressing are generally considered to have a very fine microstructure (see Fig. 12g) and the best mechanical properties. Completely pure silicon carbide does not exhibit any noticeable plastic behavior up to its decomposition temperature and can therefore be densified only under diamond synthesis conditions

Table 6. Microstructure and SiC-polytype phase composition in sintered silicon carbide (B/C, Al/C and B/Al/C doped SSiC materials).

Sample no.	Powder SiC-modification (colour)	Powder Addition to SiC (weight-%)	T Sintering temperature (°C)[a]	Chemical analysis (weight-%) Al	B	Sintered body Polytype analysis (weight-%) 6H	15R	4H	3C	Micro-structure[b]
1	β	1 B, 2 C	2050	–	1.1	13.3	6.4	1.2	79.1	FG
2	β	2.5 AlN, 2 C	2050	1.0	–	3.9	8.9	74.6	12.6	FG
3	α-black	2 C	2050	0.44	–	5.6	1.5	92.9	–	FG
4	α-black	1 B, 2C	2050	0.65	0.86	–	6.5	93.5	–	FG
5	α-black	2 AlN, 2 C	2050	1.34	–	–	2.9	97.6	–	FG
6	α-green-I	1.2 B, 2 C	2050	–	1.12	100	–	–	–	FG
7	α-green-I	0.2 B, 2 AlN, 2 C	2050	1.03	0.22	73.2	–	26.8	–	FG
8	α-green-I	0.2 B, 1 Al, 2 C	2050	0.38	0.21	92.6	–	7.4	–	FG
9	α-green-I	0.2 B, 2 C	2050	–	0.25	100	–	–	–	FG
10	α-green-II	1 B, 2 C	2050	–	0.92	83.5	1.3	15.1	–	FG
11	α-green-II	1 B, 2 C	2200	–	0.93	40.9	14.3	44.8	–	AGG
12	α-green-II	1 AlN, 2 C	2050	0.49	–	85.9	2.1	12.0	–	FG
13	α-green-II	1 AlN, 2 C	2200	0.45	–	83.2	1.2	15.6	–	FG
14	α-green-II	2.5 AlN, 2 C	2050	1.20	–	63.7	5.3	31.0	–	FG
15	α-green-II	2.5 AlN, 2 C	2200	1.29	–	48.8	7.0	44.1	–	FG
16	α-green-II	1 Al, 2 C	2050	0.76	–	76.1	2.4	21.4	–	FG
17	α-green-II	1 Al, 2 C	2200	0.74	–	68.5	3.2	28.3	–	FG

(a) 1 bar argon, graphite-tube-resistance furnace, 15 min hold
(b) FG ... FINE GRAINED   AGG ... ABNORMAL grain growth
Characterization of Starting Powders (Specific Surface Area: 15–20 m^2 g^{-1}):
β   carbotherm. red. silica, 100 w/o 3 C, 20 p.p.m. Al-content.
α-black   Milled Acheson – SiC, 89.9 w/o 4 H – 2.5 w/o 15 R – 7.6 w/o 6 H, 0.71 w/o Al
α-green-I Milled Acheson – SiC, 100 w/o 6 H; 30 p.p.m. Al
α-green-II Milled Acheson – SiC, 6.5 w/o 4 H – 6.4 w/o 15 R – 87.9 w/o 6 H; 200 p.p.m. Al

(35 kbar, 2300°C) to 100% of the theoretical density [115]. In axial hot-pressing with graphite tools it is, however, necessary to use small amounts of sintering aids: 0.5–3% boron, aluminum, aluminum oxide, beryllium oxide, yttrium oxide, or tungsten carbide are added to fine αSiC powder before molding [26, 116–123]. The molding temperature is 1900–2000°C, depending on the particle size of the powder and the amount of sintering additives used. The pressure that can be used is limited by the strength of the graphite, the maximum being 50 MPa. The conventional method requires large amounts of energy and mold material and can normally be used only for parts with simple, uncomplicated shapes. Owing to this limitation, precision parts can only be produced by machining with diamond tools [124] these, of course, are expensive.

A more convenient method than conventional hot-pressing, also still more expensive, is hot-isostatic pressing (HIP). During the development of this technique [125–127] it proved possible to be prepare high-purity SiC products with silicon carbide contents of more than 99.5% by hot isostatic pressing of SiC powder or SiC preforms in vacuum-sealed casings to a final density exactly equivalent to the theoretical density and with an uniform, fine-grained microstructure (see Fig. 12h). Owing to the higher isostatic pressure of 2 kbar (argon gas), compared with normal hot-pressing, no sintering aids need to be added; that is, the thin film of silica on the silicon carbide particles (which represents residual oxygen impurity of the SiC powder) is sufficient to achieve complete consolidation.

The postdensification of SSiC or LPSSiC by hot isostatic pressing is less complicated because no gastight encapsulation is necessary. Not only is it possible to achieve more than 99% of the theoretical density, but the variation of density and strength can also be reduced. Pressureless sintering and hot isostatic pressing can now be performed in the same cycle (see Fig. 13). This new process, called sinter-HIP, offers economic benefits [128–130].

The importance of hot-pressed silicon carbide has decreased considerably since the introduction of pressureless sintering, although it is currently still the most suitable method of obtaining the best mechanical properties for pure, monolithic SiC.

5.3.3.6 Chemical-Physical-Vapor-Deposited SiC

CVD-SiC is obtained by the chemical reaction of volatile silicon- and carbon-containing compounds in the presence of hydrogen and in the temperature range 1000–1800°C, e.g.

$$CH_3SiCl_3 \xrightarrow{H_2} SiC + 3HCl.$$

Besides the formation of SiC powders, SiC fibers, and SiC thin or thick films, nowadays monolithic bodies in sections up to 25 mm thick can be prepared by this method.

CVD-SiC is of high purity (99.999%), cubic β-modification, very fined grained, and shows the lowest of oxidation rates. Bulk cubic phase CVD-SiC is produced for applications such as electronic packaging, components for wafer handling, kiln furniture for diode manufacture, laser optics for high temperatures, and substrates for computer storage media [131, 132]. It can be polished to variations of less than 0.3 nm. Coatings of CVD-SiC were initially developed in the 1960s

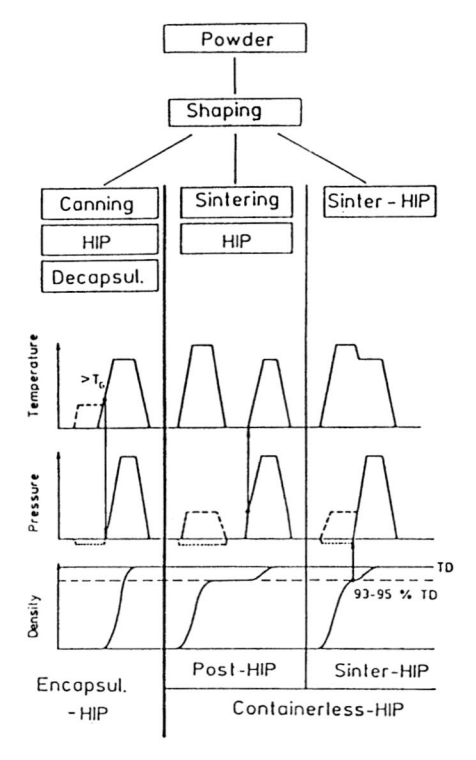

Figure 13. HIP processing routes (courtesy R. Oberacker, IKM-Karlsruhe/Germany, [130]).

for nuclear fuel particles employed to reduce the diffusive release of metallic fission products from the fuel kernel.

Today SiC thin films (<1 μm thickness) can also be produced by physical vapor deposition (PVD), for example by sputtering, which method allows lower substrate temperatures, but works more slowly. Electrically conductive B/N-doped sintered αSiC with up to 9 weight-% free carbon has been developed as target material, [133]. Novel applications for PVDSiC include films for computer storage media, protective coating for lenses, and microwaveable packaging for food.

5.3.3.7 SiC Wafers

Since 1990 high quality SiC wafers of 35 mm diameter are commercially available from single crystal 6H-SiC boules, produced via a seeded-sublimation growth technique [134]. In this process, nucleation occurs on a SiC seed crystal located at the top or bottom of a cylindrical growth cavity. As in the Lely process [135], SiC sublimes from a polycrystalline source at temperatures $>2200°C$ under vacuum to form Si, Si_2C, and SiC_2 vapors

$$4SiC_{(s)} = Si_{(g)} + Si_2C_{(g)} + SiC_{2(g)} + C_{(s)}.$$

The gases diffuse through a porous graphite retainer and along carefully programmed thermal and pressure gradients. The primary gaseous species of silicon reacts with the graphite walls of the growth cell to from additional Si_2C and SiC_2 which recombine on the growing crystal according to a 'double condensation' reaction

$$2SiC_2 + 2Si_2C = (SiC)_6 = 6H\ SiC.$$

Hence the silicon plays the role of a carbon-transporting agent. Water-clear boules of the pure 6H polytype having diameters > 50 mm and lengths > 60 mm have been grown by this technique. It is believed that by the year 2001 the size of the boules will have increased to 15 cm in diameter.

Based on the growth of epitaxial thin films of single crystal polytypes on boule-grown substrates, SiC is now becoming the material of choice for high-power, high temperature (> 500°C), and high-frequency devices [134].

As a spin-off, currently colorless SiC gemstones, 1/2 to 1 carat in size, cut from 6H-SiC wafers are entering the jewellery market at about 10–15% of the price of diamonds.

5.3.3.8 SiC Nanoceramics

Sintering nanosized powders with the aim of flaw avoidence from nanosized sintered bodies (grain size ≤ 100 nm) is recognized as a promising way of improving mechanical properties and reliability of SiC ceramics. However, nanosized SiC powders are not easy to process [42, 45, 136], and several difficulties must be overcome: the powder flows badly, exhibits low oxidation resistance [40], has low filling and compaction density [40] and is currently too expensive for large scale use.

5.3.3.8.1 Fabrication by Solid State Sintering In 1991 Vassen *et al.* [137] showed that polycrystalline βSiC bodies with a density of at least 95% of the theoretical density and a fine grain size of 150 nm can be prepared by encapsulated HIPing of B/C-doped laser-synthesized powders with particle sizes below 20 nm. Necessary HIP-temperatures were 1500°C, which is 250°C below the temperature needed to densify conventional submicron powders. This fabrication process was later optimized [138] by preheating the shaped SiC bodies in vacuum during an annealing step before encapsulation. Owing to this additional annealing step, residual oxygen contents were minimized and a mean grain size of only 60 nm could be obtained in the HIPed SiC shapes. It was found that the reduced final grain size has a strong effect on mechanical properties; as a consequence of grain size reduction from 1 μm to 150 nm the Vickers hardness increased from 2000 HV10 to 2500 HV10, whereas fracture toughness decreased from 4 to 3 $MPa\,m^{1/2}$, respectively [139]. However, by adoption of a bimodal grain-size distribution, with introduction of larger sized SiC grains into a nano-sized matrix, an incrase in fracture toughness to 6 $MPa\,m^{1/2}$ was achieved [140, 141]. For a fine grained HIPSiC (300 nm) even under a stress of 100 MPa at 1600°C very moderate creep rates of $1 \times 10^{-6}\,s^{-1}$ were measured [142].

5.3.3.8.2 Fabrication by Liquid Phase Sintering The preparation of βSiC nanoceramics with an average grain size of 110 nm by liquid phase sintering has been demonstrated by Mitomo et al. [143], who were subsequently awarded a patent called 'Superplastic SiC Sintered Body' [144].

Ultrafine βSiC powder with an average particle size of 90 nm was axially hot-pressed with additions of Al_2O_3, Y_2O_3, and CaO at 1750°C. The SiC nanoceramic showed large deformation with high strain rate: $5.0 \times 10^{-4} s^{-1}$ at 1700°C. On the basis of their results [143], the maximum temperature and the minimum deformation rate for nano-βSiC might be defined as 1800°C and $10^{-4} s^{-1}$, respectively. The superplastic deformation at temperatures as low as 1700°C is based on the fine-grained microstructure and the presence of a glassy phase at grain boundaries.

Thus a new technology has been developed whereby nano-SiC parts can be subjected to plastic deformation as in the case of metals, and can be made into complicated shapes with near net-shape quality, that is without the need for an expensive postprocessing stage such as diamond machining.

5.3.3.9 SiC-based Composites

Alloying of SiC is and has been done basically for two reasons: either to improve properties (toughness, wear, etc.) by the formation of tailored composites/solid solutions or to improve processing.

Improvements in processing can occur: (1) in solid state sintering, where the second phase acts simultanously as a sintering aid for SiC, accelerating material transport by grain boundary and/or lattice diffusion; (2) in reactive liquid sintering due to reduced sintering temperatures ('transient liquid phase sintering').

In the latter case SiC and/or additions are reacted to an intermediate liquid which not only provides densification at reduced temperatures but since it is consumed in the reaction yields a SiC-based material without glass at the grain boundaries.

Fabrication of SiC composites by second-phase dispersion is widely applied to improve material toughness. The various toughening mechanisms [93] that have high potential to reduce crack extension in SiC- composite materials are: crack deflection, microcrack formation, crack bridging by reinforcement with metallic ligaments (e.g. TiC, TiB_2), and crack bridging and pull out by platelet- or fiber-reinforcement.

Tensile fracture in SiC-based composites will only occur after a large enough load is applied to exceed the compressive stress in the process zone formed by cracking mechanisms along the crack path (see Fig. 14).

To achieve increased crack deflection and crack-wake interaction in SiC the microstructure can be modified in various ways:

– by addition of a second phase with an elongated or fibrous grain structure (SiC-based composites),
– by reinforcement with SiC platelets [52] or continuous SiC-fibers,
– by inducing growth of elongated SiC grains (*in situ* toughening of LPS-SiC [111, 113, 114]).

For optimum toughening by crack deflection Telle et al. [145] pointed out that geometric factors like grain size, volume fraction, orientation and morphology of the added or *in situ* grown phases as well as the grain boundary strength have to

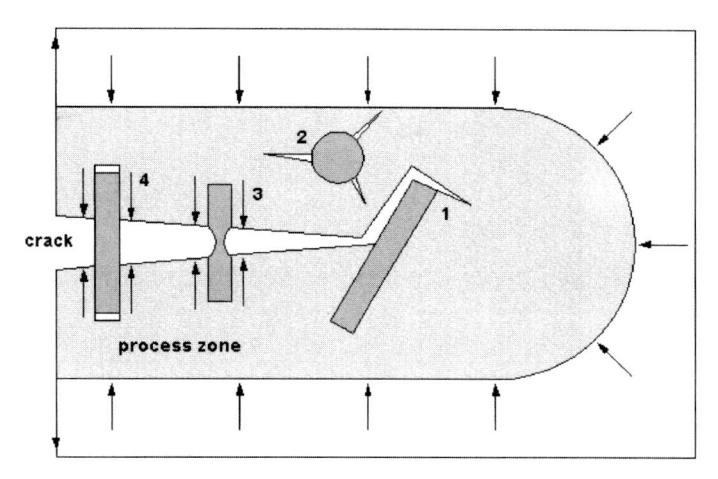

Figure 14. Potential toughening mechanisms for SiC based ceramics: 1 crack deflection, 2 microcrack formation, 3 crack bridging with metallic ligaments, e.g. TiB_2, 4 crack bridging and pullout by platelet or fiber – reinforcement.

be considered (see Fig. 15). In the following sections today's state of the art in some more or less important SiC-nonoxide composites is reviewed.

5.3.3.9.1 SiC–TiC Very promising composites have been developed in the SiC–TiC system with SiC as the matrix phase [146–148]. Dispersed TiC particles significantly

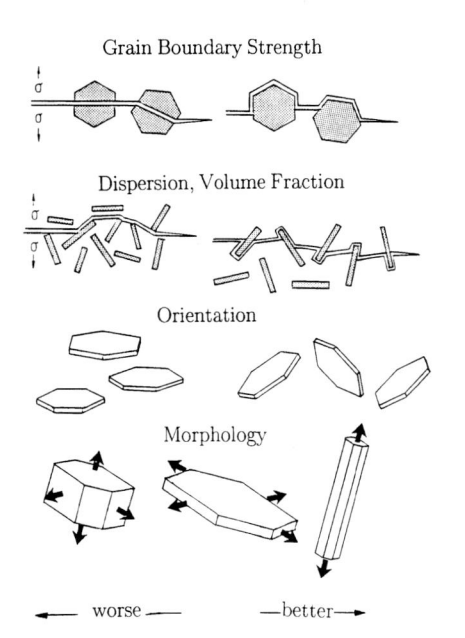

Figure 15. Microstructural design for optimized crack deflection in composites (Courtesy R. Telle, RWTH Aachen/Germany [145].

◄──── Crack propagation ────►

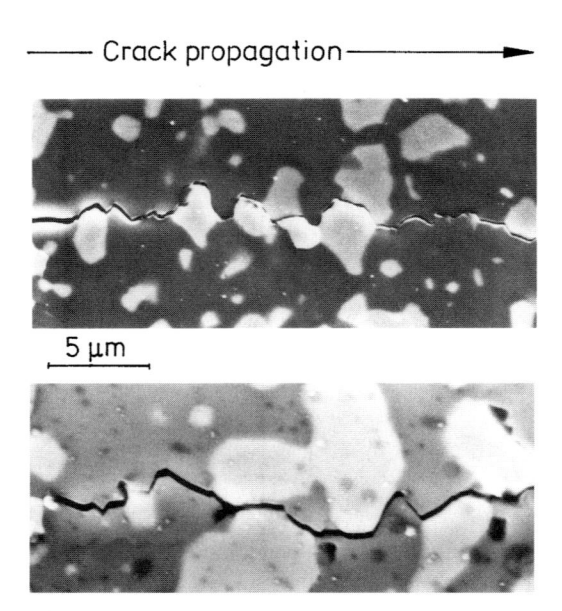

5 μm

Crack deflection in
SiC/TiC and SiC/TiB$_2$ composites

Figure 16. SEM micrograph showing crack deflection and crack bridging in SiC–TiC and SiC–TiB$_2$ composites: light areas = TiC or TiB$_2$, dark matrix = SiC
(Courtesy D.Ly Ngoc, MPI Stuttgart/Germany [149]).

improve both the strength and the toughness. Although an addition of TiC does not reduce the densification temperature significantly below 2100°C, the coarsening of SiC is completely retarded which raises the strength to 700–800 MPa [146–149]. The increase in K_{IC} to 6.5–7.5 MPa m$^{1/2}$ is attributed to the misfit of the thermal expansion coefficients of TiC and SiC, introducing considerable radial tensile stresses at the phase boundaries and hoop compressive stresses in the matrix. These stresses enable crack deflection (see Fig. 16), crack branching, and microcracking above a critical particle size of $\approx 3\,\mu$m. The optimum volume content of TiC ranges between 20 and 30 vol-%.

5.3.3.9.2 SiC–TiB$_2$ SiC-based composites with transition metal diboride (TiB$_2$, ZrB$_2$, etc.) particulates have been developed for electroconductive applications such as heating elements and ignitiors [150, 151], and also as wear resistant structural parts for high temperatures such as valve-train components and rocker arm pads in super-hot running engines [147]. These composites combine the high thermal and electrical conductivity of TiB$_2$ and ZrB$_2$ with the oxidation resistance of SiC. Additionally, due to thermal mismatch stresses of the order of 2 GPa toughening mechanisms such as crack deflection and stress-induced microcracking with a pronounced process zone as well as flank friction have been proven to occur. Cai et al. [152] and Faber et al. [153] have presented a detailed analysis of the contributions of the particular mechanisms to the total fracture toughness, stating that the stress induced microcracking is operational in a process zone of approximately 150 μm width.

Typical conditions for densification by axial hot-pressing are 2000–2100°C, at a pressure of 20–60 MPa for 30–60 min which results in 96–99.8% density. The particle sizes of the matrix and dispersed phases range between 1–5 and 4–8 μm, respectively. An optimum volume fraction of reinforcing particles of 25–30 vol-% has been reported, yielding a flexural strength of 710 MPa and a fracture toughness of 5.0–5.7 MPa m$^{1/2}$ [149]. Composites with a lower TiB_2 content of 15 vol-% exhibit a mean strength of 485 MPa combined with a K_{IC} of 4.5 MPa [147].

The strength of SiC based materials with 50 vol-% ZrB_2, HfB_2, NbB_2 or TaB_2 particles also ranges between 400 and 500 MPa [150]. Similar strength values (480 MPa) combined with a exceptionally higher fracture toughness of 7–9 MPa m$^{1/2}$ have been reported for large scale lots of pressureless sintered 16 vol-% TiB_2 composites [151]. Since the sintering was carried out with temperatures exceeding 2000°C yielding 98–99% of the theoretical density and an average particle size of 2.0 μm, it is obvious that the reinforcing phase also acts as a grain growth inhibitor for SiC. The high temperature strength of SiC–TiB_2 and SiC–ZrB_2 composites was found to remain nearly constant at 480 MPa up to 1200°C, and is hence superior to that of many sialons [150, 151].

Tani and Wada [154] fabricated optimized SiC–TiB_2 composites by reactive sintering starting from an intimate mixture of SiC, TiO_2, B_4C, and C powders. The mixture was either hot pressed or pressureless sintered and post-HIPed at temperatures of >1900°C. Titanium diboride was formed *in situ* according to the reaction

$$TiO_2 + 0.5B_4C + 1.5C \rightarrow TiB_2 + 2C$$

during an intermediate heating step at 1400–1500°C in vacuum or argon atmosphere. Overstoichiometric amounts of B_4C and C (1–2 weight-% each) can be adjusted to aid sintering. The primary advantages claimed for this reaction sintering process are the use of water in powder processing due to the disuse of highly reactive, preformed TiB_2 powders, and the very small size of reinforcing TiB_2 particles formed *in situ* due to the use of ultrafine TiO_2, B_4C, and C starting powders.

Effective reactive pressureless sintering of SiC–TiB_2 composites was recently reported by Blanc, Thevenot, and Treheux [155]. In addition they studied the tribological behavior using a pin on flat configuration (flat: SiC, pin: SiC–TiB_2). In dry conditions the composites showed less wear resistance than monolithic SiC, however, with water as lubricant the opposite was the case.

In the very recent study of Kuo and Kriven [156] indentation-strength tests were used to determine the retained strength, flaw tolerance, and toughness-curve characteristics of two kinds of SiC–TiB_2 composites. βSiC–TiB_2 composites which were hot-pressed with an Al_2O_3 sintering aid, were compared with the well-studied αSiC–TiB_2 composites, which were pressureless sintered with boron and carbon additives. TiB_2 (15 vol-%) in the B- and C-doped αSiC only increased the retained strength without a significant improvement in the toughening. On the other hand, TiB_2 (30 vol-%) along with the effect of Al_2O_3 sintering aid for the βSiC–TiB_2 composite greatly improved properties with a higher retained strength in long crack regions, better flaw-tolerance behavior and a sharply rising toughness vs. crack size curve. The

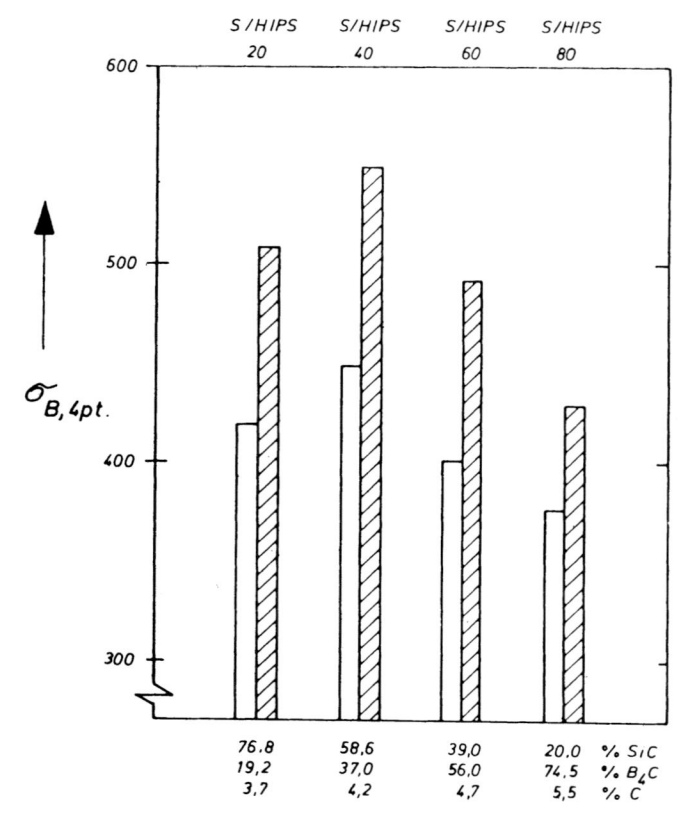

Figure 17. Flexural strength of pressureless sintered (=S, unhatched bars) and post-HIPed (=HIPS, hatched bars) SiC–B$_4$C particulate composites [157].

different toughening behaviors for α and βSiC–TiB$_2$ were related to the weak nature of the SiC–SiC and SiC–TiB$_2$ interfaces as well as the fraction and size of TiB$_2$.

5.3.3.9.3 SiC–B$_4$C In the mid 1980's pressureless sintering and posthipping were developed by Schwetz et al. [157] to produce 100% dense SiC–B$_4$C composite materials having SiC:B$_4$C weight ratios within the range of from 90:10 to 10:90 and a free carbon content of 4–5 weight-%.

These composites combine the good thermal shock resistance and oxidation resistance of silicon carbide, with the hardness, wear resistance, and low specific gravity of boron carbide. In this way a maximum strength of 550 MPa (four point bend) was achieved for a composite of 59 weight-% SiC–37 weight-% B$_4$C–4 weight-% C (see Fig. 17). The composite can be used in oxidizing atmospheres up to 1200 °C. Its microstructure is characterized by equiaxial B$_4$C and graphite grains of <1 μm diameter, which were embedded in a matrix of SiC grains with an average grain size of 1.5 μm (see Fig. 18). However, no improvement in fracture toughness was achieved, since the fracture mode was almost 100% transgranular. Similar results on sintered SiC–B$_4$C composites were obtained by Thevenot [158] and later by Tomlinson *et al.* [159], who observed a 20% increase in strength when 25 vol-% B_4C was added to SiC.

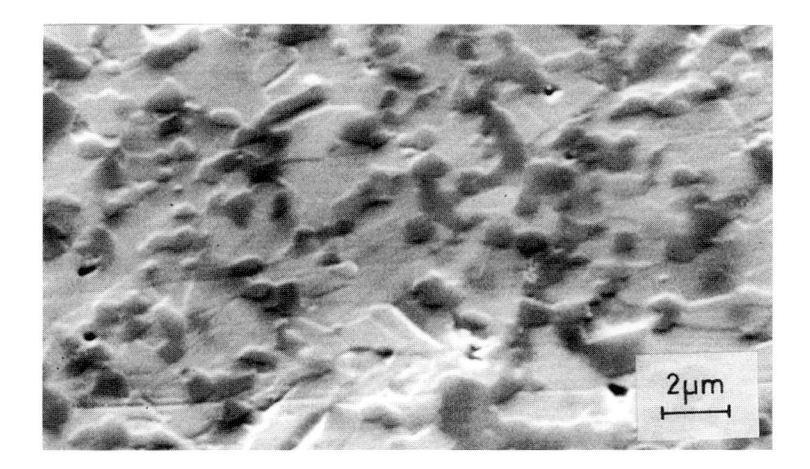

Excellent tribological properties for SiC–B_4C–C composite materials were recently encountered by Kevorkiijan *et al.* [160].

In this study SiC–B_4C–C seal rings for magnetic pumps were prepared by pressureless sintering and characterized by a pin-on-disc method (medium: water, pressure: 16–25 MPa, speed: 35–75 m s^{-1}). The introduction of a lower level of B_4C (5–20 weight-%) particles into the SiC matrix resulted in an almost linear decrease of wear rate. Moreover, further addition of B_4C (20–40 weight-%) led to an almost parabolic wear rate response. For example, with 40 weight-% B_4C a decrease of the relative wear volume of seal rings by more than 55% was achieved. These results recommend use of wear resistant SiC–B_4C composites for heavily loaded mechanical face seals in the pairing hard/hard against themselves. They may be likewise suitable for the production of shaft protection sleeves and components for sliding bearings whose wear resistance is to be improved.

5.3.3.9.4 SiC–AlN A series of solid solutions between SiC and AlN over the whole composition range was concurrently discovered at the Universities of Utah and Newcastle upon Tyne [161], and has since received considerable attention [162–165].

A 2H wurtzite-type structure is formed by the carbothermal reduction of fine SiO_2 and Al_2O_3 with a carbon source under nitrogen atmosphere at 1600 °C:

$$3SiO_2 + 0.5Al_2O_3 + 4.5C + N_2 \xrightarrow{1600°C} 3SiC.AlN + 1.5CO + 0.5N_2.$$

Kinetically favored is the carbothermal reduction of α' SiAlONs or α' SiAlON – precursor mixtures ($3Si_3N_4 + 3AlN + CaO$):

$$CaSi_9Al_3ON_{15} + 10C \xrightarrow{1800°C} 3(3SiC.AlN) + Ca + CO + 6N_2.$$

Because the diffusion coefficients in covalent solids are extremely small, solid solution was thought unlikely to be obtained by heating and annealing of the powdered solid components.

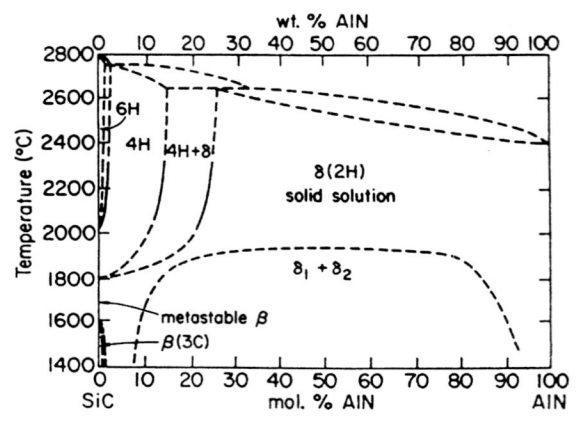

Figure 19. Tentative SiC–AlN phase diagram after Zangvil and Ruh [166].

Zangvil and Ruh [166] however, obtained SiC–AlN solid solutions by hot pressing powder mixtures

$$(1 - x)\text{SiC} + x\text{AlN} \rightarrow (\text{SiC})_{1-x}(\text{AlN})_x$$

According to the phase diagram proposed by Zangvil and Ruh (see Fig. 19) at temperatures above 2000°C, a 6H–4H–2H series of solid solutions appears with increasing amounts of AlN. AlN strongly stabilizes specific polytypes (4H and 2H) at certain composition ranges, enabling the engineering of single-phase SiC materials with discrete physical properties. At temperatures below about 1900°C, a miscibility gap was first proposed by Rafaniello et al. [163], a suggestion supported by several later studies. Xu et al. [165] obtained strong SiC–AlN materials with flexure strength up to 1 GPa. Several mechanisms of grain refinement resulting from SiC-polytype transformations into a wurtzite (2H) solid solution, were reported. Kuo et al. [167] discovered the formation of modulated structures within the miscibility gap and Lee and Wei [168] reported that pressureless sintering with 2 weight-% Y_2O_3 as sintering aid at 2050°C produced a duplex structure composed of large $(\text{SiC})_x(\text{AlN})_y$ grains and small SiC grains. Further solid solution treatment (>2225°C) followed by annealing within the miscibility gap (1860°C) resulted in spinoidal decomposition, giving various duplex/modulated structures with improved fractured toughness of the alloys.

In addition to their possible use as high temperature structural ceramics, materials in the system SiC-AlN have potential as wide band-gap semiconductors and for opto-electronic applications [166].

5.3.3.9.5 SiC–Al$_2$OC Extensive SiC–Al$_2$OC solid solutions have been found by Cutler et al. [161] for 1–100% Al$_2$OC. Moreover, they showed that wurtzite 2HSiC can incorporate substantial amounts of AlN and Al$_2$OC in solid solution and coined the acronym 'SiCAlON' to describe these materials by analogy with 'SiAlON' ceramics.

Jackson et al. [169] have sintered SiC at temperatures between 1850°C and 1950°C using a transient liquid phase produced by the carbothermal reduction of Al_2O_3 by

Al_4C_3. The resulting ceramic was fine grained (average grain size less than 5 µm) and consisted of SiC (starting polytypes) and Al_2OC as the two major phases. The properties of the hot pressed ceramics varied with the amount of Al_2OC, but at an optimum composition of about 5–10 weight-% Al_2OC, the strength (660 MPa) and fracture toughness ($K_{IC} = 3.1$ MPa m$^{1/2}$) obtained were comparable or superior to the corresponding properties of commercial grades of sintered SiC. Huang et al. [170] found encapsulation to be necessary for effective sintering with additions of Al_2O_3 and Al_4C_3, the densification occurring above $\approx 1860°C$. They attributed densification to a transient liquid phase in the system Al_2O_3–Al_4O_4C in the route to forming Al_2OC with an eutectic temperature of $\approx 1840°C$. Four point bend strength, hardness and fracture toughness for the SiCAlON materials have been reported [170]. The strength decreased with Al_2OC content in hot-pressed samples, from ≈ 600 MPa at 10 weight-% Al_2OC to around 250 MPa at $\approx 50\%$ Al_2OC. Most significantly, the fracture toughness of some SiCAlON compositions appeared to be higher than that of SiC (≈ 4.2 compared with 3.0 MPa m$^{1/2}$), using an indentation technique. Lihrmann and Tirlocq [171] proposed fabrication of sintered or hot-pressed SiC based composites containing SiC as well as 5–30 weight-% solid solutions composed of Al_2OC and AlN. Starting from SiC–AlN–Al_4C_3–Al_2O_3 powder mixtures, densification was greatly enhanced by occurrence of a transient liquid phase originating in the Al_2O_3–Al_4C_3 system at temperatures above 1800°C. The composites with 5–10% (Al_2OC–AlN) solid solution exhibit a mean grain size <2 µm and a mean strength of 620–670 MPa combined with a K_{IC} of 5.1–6.8 MPa m$^{1/2}$. Strength and fracture toughness both retained their values up to 1400°C, before weakening, thus demonstrating the highly refractory nature of the Al_2OC–AlN second phase.

5.3.3.9.6 SiC–SiC and SiC–C (Continuous Fiber Reinforced SiC Matrix Composites)

Fabrication by Chemical Vapor Infiltration In chemical vapor infiltration (CVI) silicon carbide is vapor deposited inside a porous preform (40–60% porosity) made of layers of woven cloth, from high strength C- or SiC fibers (CVI–SiC$_{fiber}$/SiC$_{matrix}$ or CVI–C$_{fiber}$/SiC$_{matrix}$). Isothermal CVI infiltration of the fibers fills the porosity with pure SiC and leads to a composite with up to 90% of the theoretical density (TD), which fractures in a noncatastrophic mode at a typical flexural stress of 300–400 MPa and a toughness of over 20 MPa m$^{1/2}$. Using the forced flow thermal-gradient CVI process developed at Oak Ridge National Laboratory, USA, the infiltration time is reduced from weeks (isothermal CVI at Societé Europeenne de Propulsion/France) to less than 24 h (ORNL) and final densities for composites of $>90\%$ TD [172] are reached.

Material data of CVI–SiC/SiC and CVI–C/SiC composites available from MAN – Technologie, Germany, are listed in Table 7 [173]. Figure 20 shows the fracture surface and fiber pullout of a CVI–SiC/SiC composite.

Prior to the matrix infiltration one ore more layers of pyrolytic carbon or boron nitride are usually applied to the fibers to provide a means of fiber debonding and toughening by pull-out and crack bridging.

Table 7. Mechanical and physical data for continuous fiber reinforced SiC matrix composites (MAN-Technology, Germany [173].

Property	Units	Gradient CVI SiC/SiC	Gradient CVI C/SiC	LPI C/SiC
Fiber fraction	Vol%	42–47	42–47	42–44
Density	g cm^{-3}	2.3–2.5	2.1–2.2	1.7–1.8
Porosity	%	10–15	10–15	15–20
Tensile strength	MPa	280–340	300–320	240–270
Strain	%	0.5–0.7	0.6–0.9	0.8–1.1
E-modulus	GPa	190–210	90–100	60–80
Flexual strength	MPa	450–550	450–500	330–370
Compressive strength	MPa	600–650	450–570	430–450
ILS[a]	MPa	45–48	44–48	35
CTE[b]	10^{-6} K^{-1} ∥	4	3	3
	⊥	4	5	4

(a) ILS = Interlaminar strength. (b) CTE = Coefficient thermal expansion.

At high temperatures (>600°C) the composites degrade in strength and toughness due to oxidation of C fibers and/or these interface layers and prevention of fiber pullout (brittle fracture mode). Studies are in progress to increase oxidation resistance by use of a CVD–SiC overlayer which seals the surface of the porous composites.

Fabrication by Liquid Polymer Infiltration (LPI) In the first step of the LPI process, a carbon-fiber preform is infiltrated with resin (e.g. polycarbosilane), to bind the fibers together. Then the polymer is pyrolyzed to form SiC. These process steps are repeated a number of times until the pores are narrow enough that further

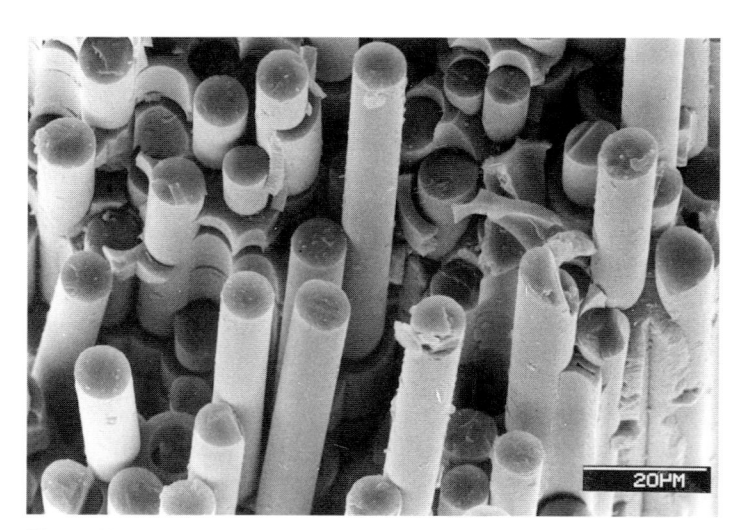

Figure 20. Fracture surface of a CVI–SiC/SiC composite showing fiber pullout (Courtesy M. Leuchs, MAN Technology/Germany).

infiltration ceases [174]. Finally the body is heated to temperatures between 1000 and 1500°C for crystallization of the SiC matrix (LPI–C/SiC).

Fabrication by Liquid Silicon Infiltration (reaction bonding) (LSI) A leading candidate for use in industrial gas turbine engine is a SiC matrix composite named 'toughened Silcomp' [175]. It is produced by melt infiltration of molten silicon into a porous preform containing carbon as well as BN-coated SiC fibers (e.g. Textron SCS – 6). The composites thus produced consist of a fully dense matrix of SiC + Si, reinforced with continuous SiC fibers. Moreover, the melt infiltration process is net shape and fast. Ultimate strength and strain at ultimate strength are 220 MPa and 0.8%, respectively at room temperature (LSI–SiC/SiC–Si).

5.4 Properties of Silicon Carbide

5.4.1 Physical Properties

5.4.1.1 Color

Pure αSiC is colorless while the cubic β modification is yellow. The only other elements that can be included in the SiC crystal lattice in amounts >1p.p.m. are N, Al, and B. Nitrogen gives a green color to 3C and 6H, and a yellow color to 4H and 15R. The presence of the trivalent elements boron and aluminum gives all the modifications and polytypes a blue-black color [176].

5.4.1.2 Optical properties

αSiC is birefringent due to its crystal structure: $n_O = 2.648$–2.649, and $n_E = 2.688$–2.893 (Na 589 nm, 20°C) [177]. For βSiC, a refractive index of ≈ 2.63 (Li 671 nm) has been reported [7].

5.4.1.3 Electrical properties

Silicon carbide is a semiconductor. The most important electronic properties of SiC are its wide energy band gap of 3.26 eV for 4H-SiC and 3.03 eV for 6HSiC, high breakdown electric field of 2.2×10^6 V cm^{-1} for 100 V operation, and high saturated electron drift velocity of 2×10^7 cm s^{-1}. Doping with the trivalent elements aluminum and boron gives the SiC *p character*, while the pentavalent element nitrogen produces *n character* when incorporated into the SiC crystal lattice. The resistivity can be varied between 0.1 Ω cm and 10^{12} Ω cm, depending on the concentration of the dopant [178].

Whereas compact, homogeneous SiC obeys Ohm's Law, aggregates of SiC grains show nonlinear current–voltage behavior. At low applied voltages they behave as insulators, but when the applied voltage is increased above a certain value the current increases exponentially. Thus, the points of contact between the grains cause the electrical resistance to be voltage dependent [178, 179].

5.4.1.4 Thermal and Calorific Properties

For a ceramic material, silicon carbide has an unusually high thermal conductivity: 150 W mK^{-1} at 20°C and 54 W mK^{-1} at 1400°C [180]. The high thermal conductivity and low thermal expansion (4.7 × 10^{-6} K^{-1} for 20–1400°C) explain why the material has such good resistance to thermal shock.

The specific heat capacity of SiC is 0.67 J g^{-1} K^{-1} at room temperature, and 1.27 J g^{-1} K^{-1} at 1000°C. The standard enthalpy of formation ΔH^0_{298K} is -71.6 ± 6.3 kJ mol^{-1}, and the entropy S^0_{298K} is 16.50 ± 0.13 J mol^{-1} K^{-1} [181].

5.4.1.5 Mechanical Properties

Silicon carbide is noted for its extreme hardness [182–184], its high abrasive power, high modulus of elasticity (450 GPa), high temperature resistance up to above 1500°C, as well as high resistance to abrasion. The industrial importance of silicon carbide is mainly due to its extreme hardness of 9.5–9.75 on the Mohs scale. Only diamond, cubic boron nitride, and boron carbide are harder. The Knoop microhardness number HK-0.1, that is the hardness measured with a load of 0.1 kp (≈ 0.98 N), is ≈ 2600 (2000 for αAl$_2$O$_3$, 3000 for B$_4$C, 4700 for cubic BN, and 7000–8000 for diamond). Silicon carbide is very brittle, and can therefore be crushed comparatively easily in spite of its great hardness. Table 8 summarizes some typical physical properties of the SiC ceramics.

Since the microstructural grain size (Fig. 12a–h), pore content, and chemical composition of the various ceramic products differ considerably, it follows that the properties are also different.

Recrystallized SiC is much stronger than the ceramically bonded material, but its high residual porosity imposes limits as far as mechanical strength is concerned [185]. Reaction-sintered SiSiC is still stronger, but only up to 1400°C (Fig. 21), the softening point of the accompanying silicon phase [186].

The best mechanical strength is exhibited by sintered SiC and hot-pressed materials [109, 187, 188].

Solid-state sintered, hot-pressed, and isostatically hot-pressed materials offer considerable advantages over all other ceramic materials in plastic deformation under a sustained load (creep), because of the low content or almost complete absence [189, 190] of sintering aids.

Compared to solid state sintered silicon carbide (SSiC) and dense alumina (Al$_2$O$_3$), liquid phase sintered SiC (LPS-SiC) features improved edge toughness (see Fig. 22) close to the edge flaking resistance of sintered silicon nitride (Si$_3$N$_4$). The latter property indicates the sensitivity of edges against mechanical chipping, a quantity which is most important for safer handling of ceramics in grinding, clamping, transportation etc.

5.4.2 Chemical Properties

One of the outstanding characteristics of silicon carbide is its chemical resistance, which is due to the high affinity of silicon for oxygen. The reaction of silicon with

Table 8. Physical properties of various SiC ceramics.

SiC material type	SiC content (weight-%)	Density (g cm^{-3})	Porosity (%)	Young modulus (GPa)	Thermal expansion 30–1500°C (10^{-6} K^{-1})	Thermal conductivity at 600°C (W mK^{-1})	Flexural strength 20°C (MPa)	Flexural strength 1400°C (MPa)
Ceramic bonded SiC KSiC	up to 95	2.55	20	100	5.8	16	30	20
Recrystallized SiC RSiC	100	2.60	20	250	5.0	28	100	100
Reaction bonded SiC RBSiC	100	2.60	20	250	5.0	25	250	250
Infiltrated SiC SiSiC	90	3.12	<1	400	4.3	60	350	200
Sintered SiC (solid state) SSiC	98	3.15	<2	410	4.9	50	430	450
Sintered SiC (liquid phase) LPSSiC	95	3.21	<1	420	4.5	50	730	400
Hot-pressed SiC HPSiC	98	3.20	0	450	4.5	55	640	650
Hot-isostatic-pressed SiC HIPSiC	>99.5	3.21	0	450	4.5	75	640	610
HIP-post densified SSiC HIPSSiC	98	3.19	0	430	4.8	50	450	450

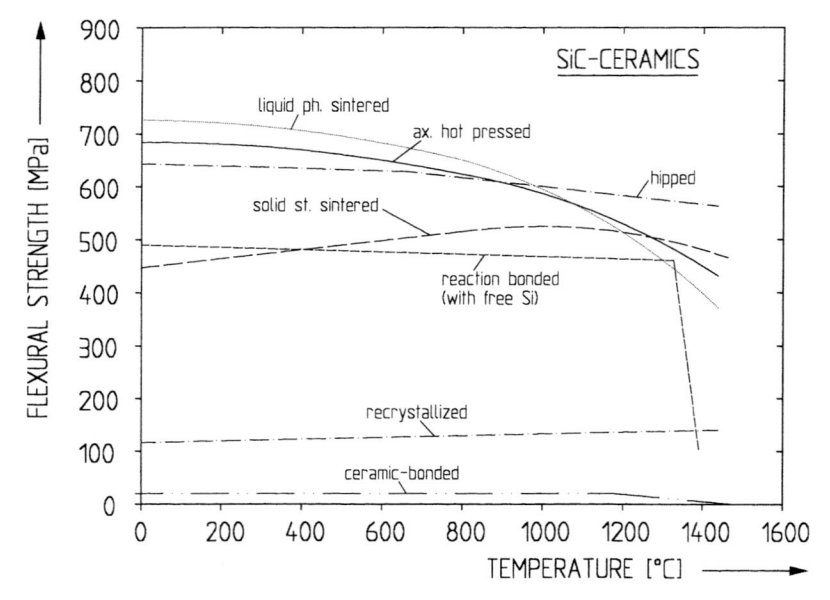

Figure 21. Flexural strength of SiC ceramics as a function of temperature.

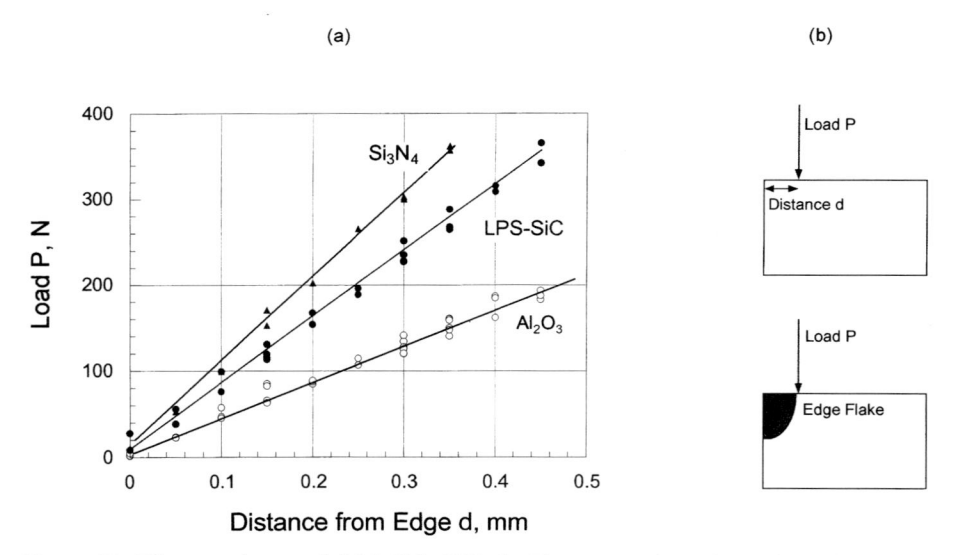

Figure 22. Edge toughness of LPS–SiC (EKasic T) compared to sintered Al_2O_3 and Si_3N_4: (a) flaking load against distance from edge, (b) schematic of experimental setup (Courtesy L.S.Sigl, ESK-Kempten/Germany).

oxygen in an aqueous medium causes passivity and, if exposed to thermal oxidation, for example, in air, leads to the formation of glassy silica films [191–193]. The oxidation of pure SiC begins at around 600°C, forming a coating of SiO_2 on the surface of the SiC that prevents further oxidation [194]:

$$SiC + 2O_2 \rightarrow SiO_2 + CO_2.$$

The reaction rate varies with time according to a parabolic law [191]. The kinetics are determined by the diffusion of oxygen through the SiO_2 layer. The temperature dependence of oxidation follows the Arrhenius equation.

The 'active' oxidation of SiC is distinguished from the 'passive' oxidation reaction described above:

$$SiC + O_2 \rightarrow SiO + CO.$$

Active oxidation takes place under conditions of oxygen deficiency above 1000°C and leads to decomposition of the SiC and formation of silicon monoxide [194, 192]. The two forms of corrosion (active, passive) and the conditions for the boundary have been recently discussed by Nickel *et al.* [195] using examples from SiC- and also Si_3N_4-based ceramics. A likely high temperature boundary for SiC is ≈ 1700–$1800°C$, where a secondary active-to-passive transition by bubble formation and spalling occurs.

Thus silicon carbide is attacked and decomposed by oxidizing agents (e.g., $Na_2CO_3 + Na_2O_2$ or $Na_2CO_3 + KNO_3$) if the protective layer of SiO_2 is removed, thereby enabling the reaction to proceed unhindered.

Pure silicon carbide is twice as oxidation resistant, even at 1500°C, than the best current superalloys at their maximum service temperature of 1200°C.

Silicon carbide is resistant to most chemicals, resisting acids and alkalis and even aqua regia and fuming nitric acid.

A mixture of hydrofluoric adic, nitric acid, and sulfuric acid slowly attacks silicon carbide βSiC being somewhat more reactive than αSiC [196]. However, complete dissolution only takes place if the SiC is very finely divided, under pressure, and at a elevated temperature (e.g., 250°C for 16 h). Alkali melts [197, 198] will attack it in the presence of oxidizing agents. Oxides, molten metals, and water vapor have a destructive effect on SiC at temperatures of 1000°C and over. Chlorine reacts exothermally with silicon carbide above 800°C with the formation of silicon tetrachloride ($SiCl_4$) and carbon.

Silicon carbide behaves in various ways towards molten metals. It is not attacked by molten zinc or zinc vapor [199]. Molten aluminum attacks SiC slowly, forming Al_4C_3 and silicon, but as the silicon content increases, the reaction eventually ceases because an equilibrium is established [200]. Molten iron desolves SiC, forming iron carbide and iron silicide.

5.4.3 Tribological Properties

5.4.3.1 Review on Tribology Work in Sintered αSiC Ceramics

Tribology is the study of adhesion, friction, wear and lubricated behavior of materials in solid state contact. It was in the early 1980s that the advantage of sintered SiC as seal face material was determined [201, 202]. Since then, the mechanical engineering industry has focused attention on this material and tribological studies have been simulated to understand better those physical and chemical properties of SSiC that will affect its behavior when in contact with itself, other ceramics, or metals. Model investigations involving SiC have shown the coefficients of frictions against various materials, even itself, to be a function of contact stress [203] and have documented the anisotropic wear behavior of monocrystalline SiC crystals [204]. Lashway et al. [205] found that a controlled amount of porosity improved the ability of sinteredSiC to retain a hydrodynamic film with lower friction. Seal tests also indicated lower power dissipation at varying pressure-velocity values for combinations with sinteredSiC seal face materials.

Consistently positive practical experience gathered by Knoch et al. on sinteredSiC seal rings [206, 207] has helped the material to gain popularity rapidly for use in situations involving wear problems.

Excellent results have been achieved in developing sliding bearings for hermetically sealed pumps [208]. All of them are absolutely leak-proof, whether in operation or shut down and they are therefore of great value environmentally. They all have the same design requirement that the sliding bearing of the pump shaft must be flushed and lubricated by the pumped medium. Traditional materials simply could not cope with the harsh conditions of the chemical industry and were quickly destroyed by corrosion and abrasion. Sintered silicon carbide solved the problems. Liquids containing abrasive particles do not restrict the use of sintered SiC.

Many sliding-wear problems occurring in the field are attributable to interruption of the ideal (that is properly lubricated) running conditions, in which case the sliding faces of the bearing or seal in question make contact with each other, giving rise to solid-state or dry friction marked by a pronounced increase in the coefficients of friction. Local frictional heat leads to peak thermal stresses that may be of such intensity as to cause a breakout of microstructural constituents. Then, when lubrication (and cooling) is restored, the material is in danger of cracking or fracturing due to thermal shock. Sintered silicon carbide, however, is better able to cope with such situations than other ceramic materials, because it is stronger and has a lower thermal expansion coefficient, and a higher thermal conductivity. Consequently, brief periods of dry running can be survived.

The nonlubricated wear behavior of sintered silicon carbide under unidirectional sliding at room temperature was studied by Derby et al. [209] as a function of load and sliding time. At low loads polishing and ploughing mechanisms were observed and the microstructure revealed an etched appearance. With increased loads and sliding times microcracking at the grain boundaries occurred leading to subsequent grain pull-out. In the study of Cranmer [210] it was shown that for sintered silicon carbide sliding against itself also surface plastic deformation, ploughing and cracking were operative as wear mechanisms. Miyoshi et al. [211] carefully studied friction and wear of the pair SiC/iron in vacuum environment at temperatures up to 1500°C. By using X-ray photoelectron spectroscopy it was found that the surface chemistry of sintered silicon carbide (graphite and SiO_2) as well as the alteration of this chemistry by the temperature are very influential in dramatically changing the friction coefficient. Breznak et al. [212] quantified the role of initial surface roughness on the friction behavior of sintered silicon carbide rubbing against itself in cyclic oscillatory motion. It was shown that the coefficient of friction decreased from an initial value of 0.40 to 0.25 as a result of surface polishing. Further work by Breval et al. [213] indicated that adding graphite to SiC reduces the wear of SiC/SiC couples somewhat, but improving their initial surface finish has the opposite effect: it leads to a greater loss of material during running in. Wear debris of SiC/SiC couples exhibited a bimodal particle size distribution: some particles were micrometer size and the others ranged in nanosize from 5 to 50 nm. Using pin on disk, as well as abrasion wheel test, Wu et al. [214] observed that the amount of wear increases with increasing grain size for sintered silicon carbide.

Smythe and Richerson [215] conducted experiments in which the dynamic sliding contact behavior at temperatures greater than 1000°C was studied. It was found that surface film formation at higher temperatures governs the friction behavior. It is believed that SiO_2 or SiO_2 modified by impurities from the SiC or the environment with flow properties contributed to the lowering of the friction coefficient [216].

Tomizawa et al. [217] performed pin-on-disk experiments on friction of SSiC against itself in water at room temperature. The authors found a friction coefficient of 0.26 and noted that wear of sintered silicon carbide occurs by a combination of tribochemical dissolution and the formation of pits by fracture of SiC grains. The amount of material removal varied from one SiC grain to the other, due to a strong dependence of tribochemical wear on crystallographic orientation of SiC grains.

Knoch and Kracker [218] observed that this anisotropic tribological behavior is most conspicuous with EKasic D sintered silicon carbide, which has a bimodal grain structure: about 30 vol-% of larger platelets of hexagonal α grains (about 100 μm long) and about 70 vol-% of smaller α grains (about 10 μm long) as shown in Fig. 23a. This material shows superior performance particularly if paired against a softer carbon material. Figure 23b shows the surface topography developed in a finely lapped surface under in-service conditions, as water lubricated EKasic/carbon seal. This relief structure, with the depressions in a textured surface, obviously acts as reservoirs for lubricant, thereby improving the emergency running properties of the components. Figure 23c shows a sketch of the process. Boch et al. [219] studied the dry friction behavior of SiC/SiC couples (fixed ball rotating disk) at temperatures from 20 to 1000°C and observed a rapid decrease in the coefficient of friction from 0.45 to 0.16 when temperatures increases over 400°C. From 400 to 700°C, the debris agglomerates in the form of SiO_2 rolls, which arrange themselves perpendicularly to the sliding direction. These rolls act as minute roller-bearings decreasing the coefficient of friction and preventing the formation of cracks in the wear track.

Habig and Woydt [220] studied unlubricated friction and wear of selfmated SSiC sliding couples (EKasic D) at temperatures between 22 and 1000°C using a stationary pin and a rotating disk assembly. As shown in Fig. 24, at room temperature the coefficient of friction decreases with increasing sliding velocity from $f = 0.8$ to 0.6. At temperatures $\geq 400°C$ a scatter of friction coefficients between 0.2 and 0.7 is observed. By using small-spot-ESCA (electron spectroscopy for chemical analysis) it was shown that the low friction coefficients of SiC/SiC were due to formation of thin oxide layers formed by tribo-oxidation, while the higher friction coefficients accompanied by higher wear coefficients were related to thicker oxide layers.

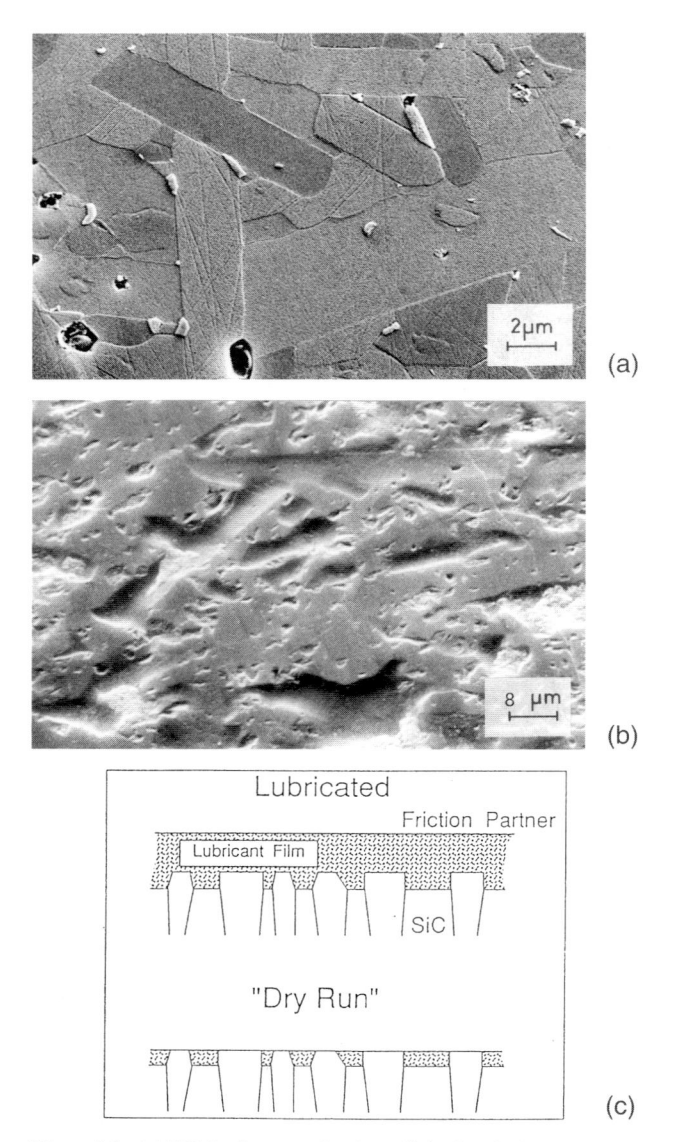

Figure 23. (a) SEM micrograph of a polished and plasma-etched section of sintered SiC, 'EKasic D', note: bimodal grain size distribution. (b) SEM micrograph of EKasic D sliding face after about 4000 h of successful service as a mechanical seal face (SiC/carbon pair, lubricated with water, maximum pressure 50 bar). (c) Schematic of the operational condition of a relief structured sliding surface in the lubricated state and after breakdown of the lubricant film.

These results are in excellent agreement with those of Yamamoto et al. [221], who observed a decrease in the coefficient of friction when SiC/SiC couples had been heated for 1 h at 1000°C with formation of a thin oxidation layer. However, at a higher oxidation level the coefficient of friction increased rapidly. A low coefficient

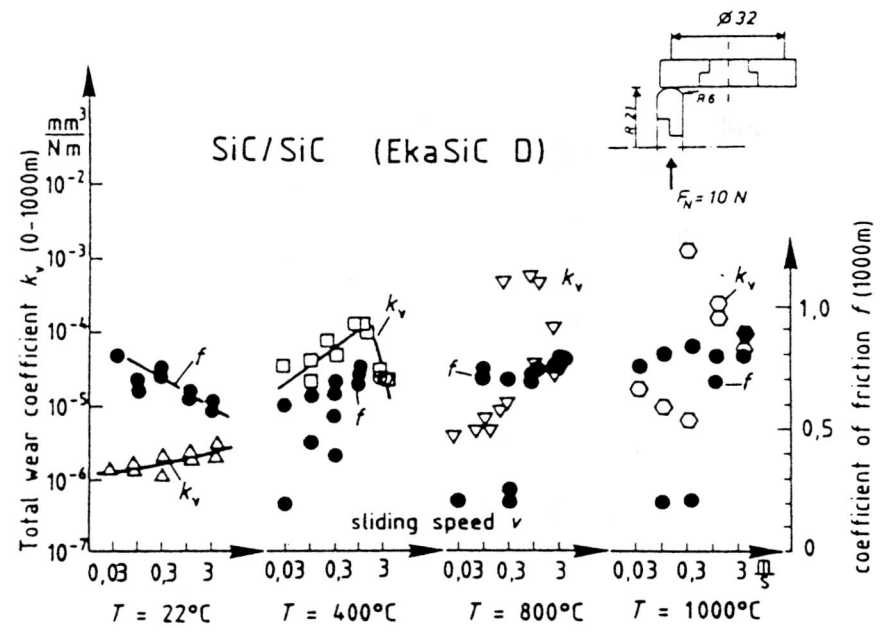

Figure 24. Friction coefficient, f, and wear coefficient, k, of SiC/SiC sliding pairs as a function of sliding velocity with 10 N for various temperatures (sintered SiC 'EKasic D', after Habig and Woydt, BAM Berlin/Germany [220]).

of friction for the SiC/SiC couple was also measured by Martin et al. [222] who studied friction under an oxygen partial pressure of 50 mPa, which should admit only a very thin oxidation layer.

Sasaki [223] studied the influence of humidity on friction and wear behavior of SiC/SiC couples. It was shown that humidity decreased the coefficient of friction from 0.5 (dry air) to 0.2 (wet air), simultaneously a decrease in the coefficient of wear was observed from 10^{-5} (dry air) to 10^{-6} mm^3 N^{-1} m^{-1}.

Denape and Lamon [224] considered the important contribution of wear particles to the wear mechanisms. The main findings were: polishing at low loads (5 N) is due to fine individual wear particles smaller than 1 µm circulating in the sliding interface, and abrasion and grain pull-out (at high loads of 20 N) are associated with large accumulations of particles adherent to the sliding phases. Wear tests performed under water showed again an anisotropic wear of individual SiC crystallites. It was concluded that the circulation of wear debris in the sliding interface is controlling both the wear rate and the friction response.

Woydt et al. [225] evaluated the tribological behavior of SiC–TiC and SiC–TiB$_2$ ceramic-matrix composites using a special high temperature tribometer. In an normal ambient environment these composites are able to form 'lubricious oxide' reaction layers on the hard substrate; the low friction coefficient of the couple SiC–TiC/SiC–TiC at room temperature with values of $f = 0.2$. and 0.3 was explained by the formation of self-lubricating layers of TiO$_2$ and SiC$_x$O$_y$. The friction coefficient of self-mated SiC–TiB$_2$ sliding couples was in the range 0.56–0.72 and compared to sintered αSiC materials no improvement was observed in the wear resistance of the SiC–TiB$_2$ composite [226].

In a study on the friction and wear behavior of lubricated ceramic journal bearings it was shown by Maurin-Perrier et al. [227] that both SiC/SiC and SiC/Si$_3$N$_4$ couples provide a significantly better behavior in terms of film stability compared with the classical materials used for water- or hydrocarbon-lubricated bearings. A self-improvement of surface roughness during the running-in period by tribochemical reaction significantly increases the range of stability of the film; consequently it is not necessary to produce a highly expensive surface finish to increase the performance of the SiC/SiC couple. The lifetime of the components is mainly determined by the behavior during the start and the stop phases where boundary lubrication occurs. Sliding wear on sintered silicon carbide leads to a very smooth surface between residual porosity. Further it was found [228] that in dry run situations, sintered SiC of bimodal grain structure (EKasic D) outperforms other SiC materials. The bimodal EKasic D shows the lowest development of heat, and therefore it has the lowest coefficient of friction. In a mayor US study on 'Tribological Fundamentals of Solid Lubricated Ceramics' [229] the bimodal grain size/shape distribution was judged best in terms of wear resistance, when compared with other sintered αSiC materials.

Löffelbein et al. [230] tested self-mated sliding couples of SSi$_3$N$_4$, HIP-RBSi$_3$N$_4$, SSiC, SiSiC, MgO–ZrO$_2$ and Al$_2$O$_3$ in different aqueous solutions (H$_2$O, NaOH, KOH, NH$_4$OH, HNO$_3$, H$_2$SO$_4$, H$_3$PO$_4$, CH$_3$COOH, HCl, HClO$_4$) under conditions of boundary or mixed lubrication. The best frictional behavior was observed with couples of SSiC, with steady state friction coefficients of 0.05. The lowest wear coefficients were measured for couples of the two types Si$_3$N$_4$, SSiC and Al$_2$O$_3$ with values of approximately 10^{-7} mm^3 N^{-1} m^{-1}. If low friction and low wear are required, couples of SSiC seem to be the best choice in most aqueous solutions. A recent study by Anderson and Blomberg [231], based on tests with sintered silicon carbide sliding unlubricated on itself in point (pin-on-disc), line (journal bearing) and plane (mechanical face seal) contacts. Tribo-oxidation and surface fracture were identified as the dominating deterioration mechanisms. The oxidation products formed silicon dioxide and, within narrow operational regimes, silicon monoxide. The highest wear rates occured in the pin-on-disc configuration, while the lowest rates were obtained in the journal bearing tests.

Kitaoka et al. [232] investigated the effects of temperature and sliding speed on the tribological behavior of sintered silicon carbide (0.1 weight-% B/1.0 weight-% C) by sliding on the same material in deoxygenated water from room temperature to 300 °C under high vapor pressures (120 °C/2 bar, 300 °C/85 bar). The wear mechanism appears to consist of hydrothermal oxidation of SiC according to the equations:

$$SiC + 2H_2O \rightarrow SiO_2 + CH_4,$$

$$SiC + 4H_2O \rightarrow SiO_2 + CO_2 + 4H_2$$

$$SiC + 2H_2O \rightarrow SiO_2 + 0.5C_2H_6 + 0.5H_2,$$

$$SiC + 2H_2O \rightarrow SiO_2 + C + 2H_2,$$

and dissolution of reaction products such as silica. Fine mirrorlike worn surfaces were observed without wear debris under all sliding conditions.

For description of the tribotechnically used machine-elements and more information about tribology, the book 'Tribology Handbuch – Reibung und Verschleiß' [233] is recommended to the reader.

5.4.3.2 Sintered SiC Material Development in Sliding Wear Applications

The coefficients of friction of hard and wear resistant ceramic materials and of most surface coatings are always greater than 0.1 under dry run conditions. This value is far too high for dry running bearings to be designed. Frictional heat develops, and if loads are high then very high temperatures are generated. These will not affect the sintered silicon carbide but the housing and the overall structure may well distort and become damaged leading to failure [234]. Since materials research shows that it is impossible to develop hard ceramic material with coefficients of friction below 0.01 (which is the typical value for a lubricated bearing) then the targets for product development must include:

- stabilizing the hydrodynamic lubrication film,
- preventing dry running,
- reducing friction and wear if the lubrication film breaks down.

There are obviously two routes towards stabilizing the hydrodynamic film. One is to optimize the design of the component including the most suitable bearing surface characteristics. The other is to optimize the silicon carbide material itself. In following this second route, it was assumed that the tribological performance can be improved by tailoring the microstructure of SSiC. This can be done by introducing isolated pores which act as lubricant pockets or incorporating graphite particles, which act as dry lubricants.

Fig. 25 shows typical sliding faces (left) as well as the microstructures (right) of the improved materials in comparison with standard EKasic D. The material containing pores (diameter about 40 µm) was named TRIBO 2000, the second material containing pores and graphite particles, in the range 40–60 µm, was named TRIBO 2000-1. Properties of these SiC materials are shown in Fig. 26 in comparison with EKasic D.

Both pores and graphite inclusions act as internal flaws, so it is no surprise that strength is reduced. On the other hand since these imperfections are of regular size it follows that the Weibull modulus is increased which is important for design. The Young modulus is slightly reduced. Friction under boundary lubrication conditions decreases as intended, while the corrosion resistance is unchanged.

TRIBO 2000 and TRIBO 2000-1 were tested in product lubricated bearings and mechanical seals in comparison with EKasic D as the basis sintered SiC material. The Stribeck testing method was used to evaluate the coefficient of friction as a function of sliding velocity (see Fig. 27). This is an axial bearing test in which a circular segment sits (lubricated) on a rotating ring of the same material. The torque between the segment and ring is measured. The load bearing capacity of the hydrodynamic lubricant film is the key factor. The better tribological material is the one which

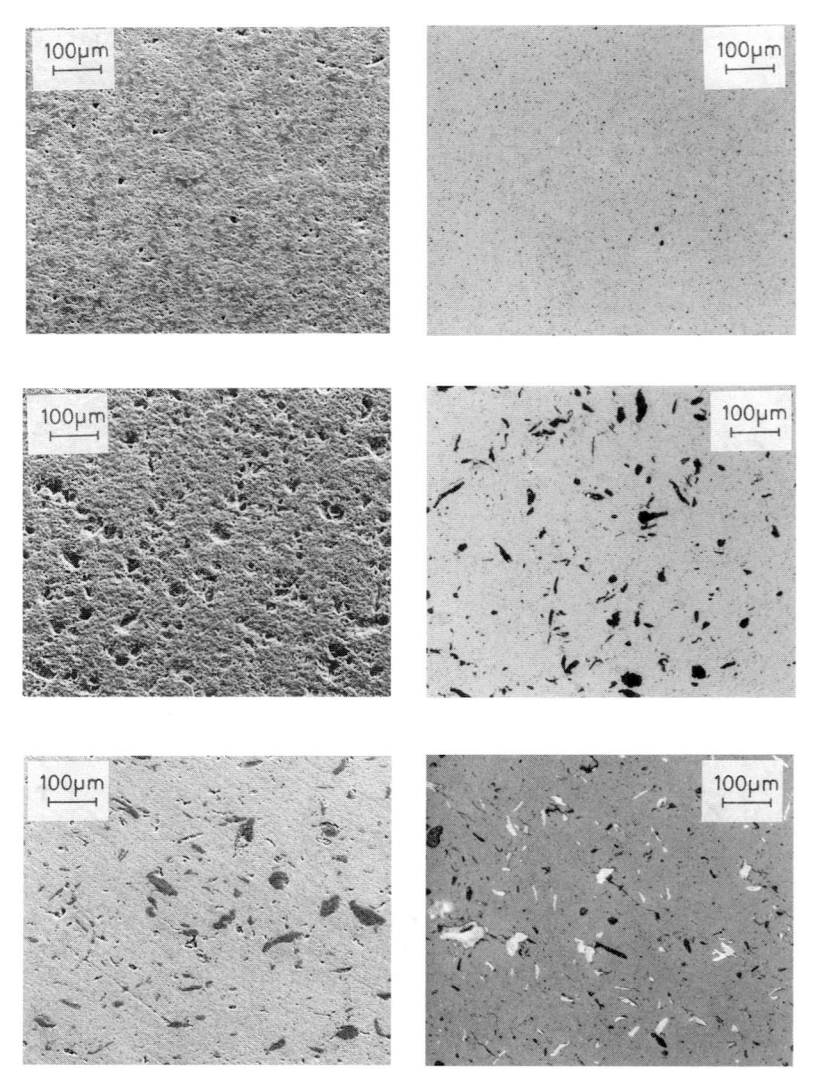

Figure 25. SEM micrographs of sliding faces (left) and polished sections (right) of sinteredSiC grades: EKasic D (top), TRIBO 2000 (middle), and TRIBO 2000-1 (bottom). The graphite in the TRIBO 2000-1 microstructure was revealed as white particles after coating the polished section with a carbon film.

shows stable hydrodynamics down to slower sliding velocities, before the lubricant film breaks down and the bearing comes into solid state contact. Figure 28 is an example.

In this case, perhaps unexpectedly, the TRIBO 2000-1 material shows both a higher hydrodynamic friction and a deviation with increasing velocity when compared with TRIBO 2000 and EKasic D.

5.4 Properties of Silicon Carbide

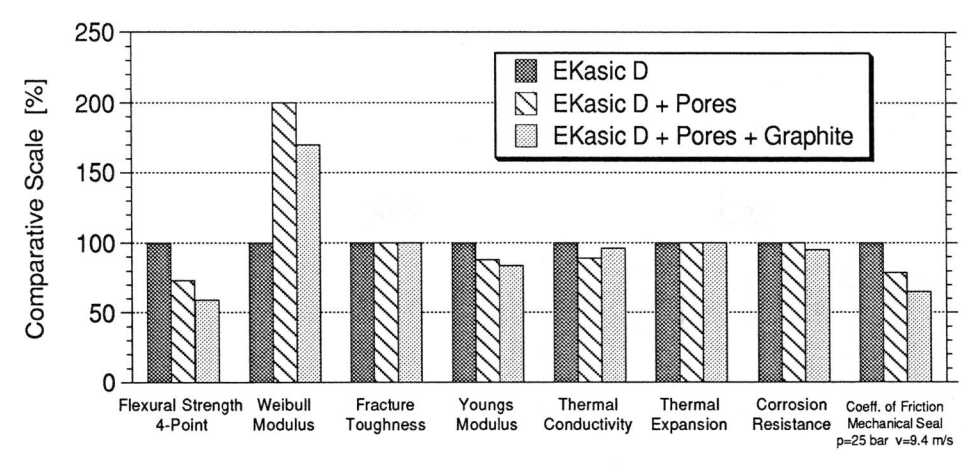

Figure 26. Comparison of properties of the pore-containing material (TRIBO 2000) and the pores-plus-graphite-particles-containing material (TRIBO 2000-1) with EKasic D.

Summarizing all Stribeck test data (load, media, and up to 10000 repeats) the TRIBO 2000 material shows the best performance in terms of stabilizing the hydrodynamic film down to the lowest velocities. This is important for sliding bearings and has been confirmed many times. The poorer performance of the TRIBO 2000-1 material in this typical sliding bearing test is not clearly understood, but it

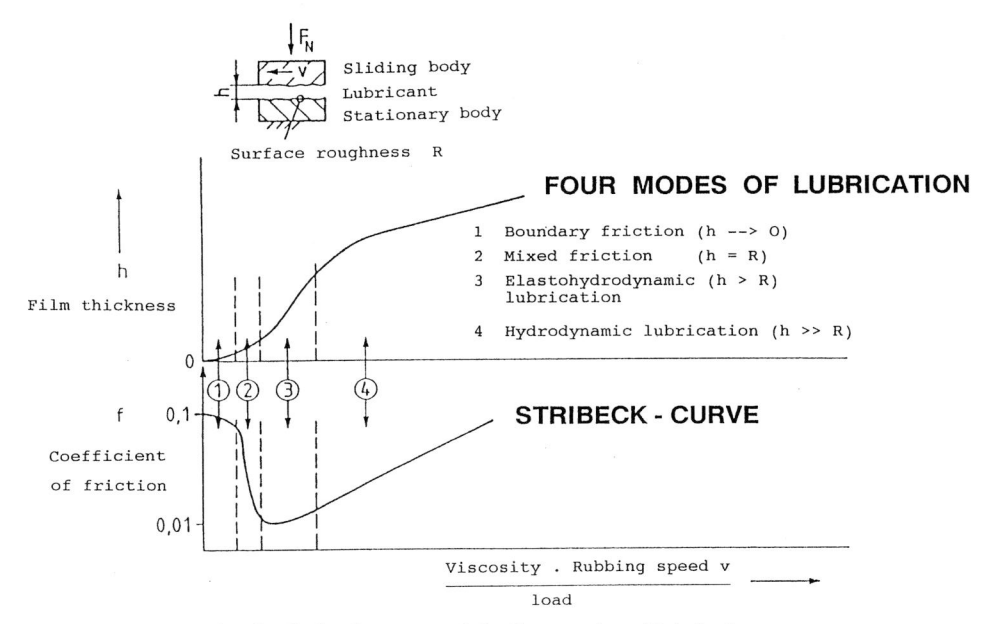

Figure 27. Schematic of a Stribeck curve and the four modes of lubrication.

FOUR MODES OF LUBRICATION

1. Boundary friction ($h \rightarrow 0$)
2. Mixed friction ($h = R$)
3. Elastohydrodynamic ($h > R$) lubrication
4. Hydrodynamic lubrication ($h \gg R$)

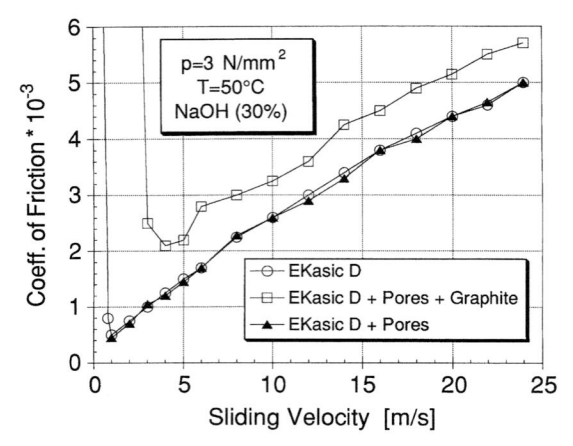

Figure 28. Coefficient of friction as a function of sliding velocity for self-mated sliding pairs of sintered SiC materials.

may have something to do with the poor wetting behavior of graphite in water under normal pressure.

In mechanical seal applications, under differential pressures of less than 15 bar the same ranking of materials is observed in terms of friction and wear. Under extreme situations, however, like very high differential pressure (producing continuous boundary lubrication) or pump cavitational run (causing breakdown of the hydrodynamic film under frequent violent loads), the TRIBO 2000-1 material showed the best performance (see Fig. 29).

To diminish the effects of hydrodynamic grain-boundary corrosion in applications with hot water, such as chipping out of fine grains in hot spots, tribochemical reaction with water

$$SiC + 2H_2O = SiO_2 + CH_4$$

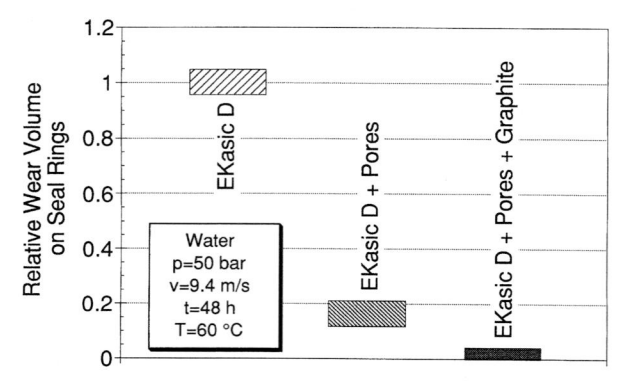

Figure 29. Relative wear volume on seal rings of sintered SiC material (self-mated couples).

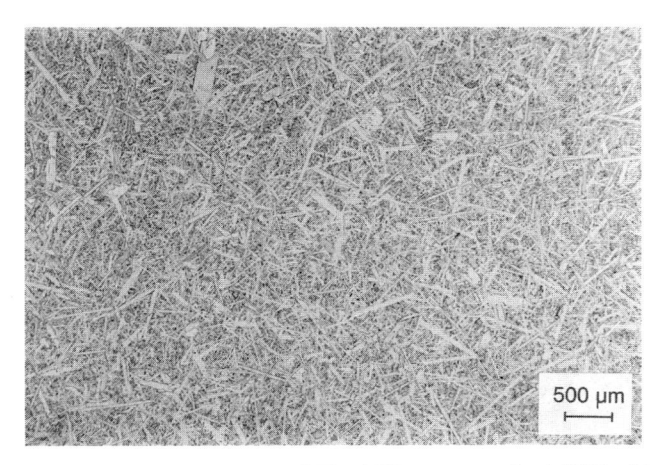

Figure 30. Microstructure of EkasicW: coarse-grained SSiC with platelet structure for mechanical seals and bearings (hard/hard combinations) in hot water applications.

and formation of damaging SiO_2 layers on the sliding surfaces, another two new modified SSiC materials with a predominantly coarse-grained, bimodal platelet structure have been developed [235]. Catastrophic failure of coponents is avoided since the large SiC platelets near the surface are anchored to a depth at which there is no grain-boundary corrosion.

EKasic W silicon carbide is a dense material with a predominantly coarse-grained bimodal platelet structure. The effectiveness of the coarser microstructure (see Fig. 30) in improving the corrosion resistance has been clearly demonstrated in practical tests on a mechanical seal test rig. Even after 500 h of testing (deionized water, 60 °C, 6 bar, hard/hard couples) no SiO_2 layer was formed on the functional surface.

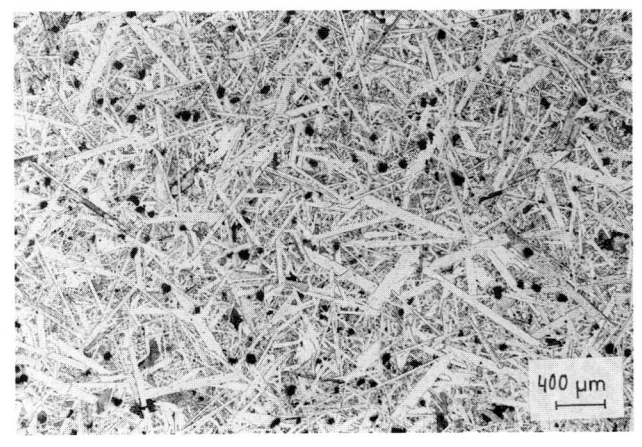

Figure 31. Microstructure of EKasic HW: coarse-grained SSiC with platelet structure and graphite particles (black phase 40–60 µm in diameter) for mechanical seals and bearings (hard/hard combinations) in hot water applications.

EKasic HW silicon carbide, on the other hand, is a dense material with a predominantly coarse-grained, bimodal platelet structure and graphite particles of 40–60 µm in size (see Fig. 31). This new tribological material, which is still in its test phase, should be particularly suitable for mechanical seals and sliding bearings in applications in contact with hot water containing solids.

5.5 Quality Control

The SiC content of silicon carbide products is now usually determined by measuring the carbon contents. The total carbon content is determined by combustion of the sample in a stream of oxygen at 1050°C in the presence of lead borate. The CO_2 produced is absorbed in $Ba(ClO_4)_2$ solution and determined by coulometry [236, 237]. An alternative technique is to oxidize the SiC with oxygen in a high-frequency induction furnace containing a flux metal, and to detect the CO_2 produced by IR absorption.

Free carbon can be directly determined coulometrically by combustion at 850°C without any additives, provided that SiC itself is not appreciably oxidized. If the SiC samples are very finely divided, that is, if they have large specific surfaces, and if long heating times are required due to high free carbon contents, this method can not be used. In such cases, the weight change on combustion is measured and the decrease in total carbon content is determined. This enables the effect of the unavoidable oxidation of the SiC to be eliminated, and the free carbon content can be calculated [236, 237]. For the determination of free carbon in SiC sintering powders or in more-or-less contaminated SiC, containing iron or silicon impurities or small amounts of sintering aids etc., the use of a wet chemical method, based on the wet chemical dissolution of free carbon in a hot chromic-sulfuric-acid mixture proved to be very useful [238].

Analyzers with IR detection are also suitable for the direct determination of free carbon provided that they allow precise temperature control during combustion of the free carbon.

The SiC content is calculated from the difference between total and free carbon:

$$\text{weight-\% SiC} = (\text{weight-\%C}_{\text{total}} - \text{weight-\%C}_{\text{free}}) \times 3.3383$$

The free silicon content is determined by measuring the volume of hydrogen produced on treatment with sodium hydroxide solution. Alternatively, metallic silver is precipitated from a silver fluoride solution, dissolved in nitric acid, and determined by a standard method [236, 237]. In the chemical analysis of SiC abrasive grits, emphasis is placed on the determination of accompanying materials such as carbon, silicon, silicon dioxide, and metallic oxides. The SiC content (plus the free carbon) is found by weighing the residue after treatment with a mixture of hydrofluoric, nitric, and sulfuric acids [237]. For special grades of SiC for refractory and metallurgical applications, volatile components can interfere with the SiC determination. In this case, the sample must be annealed under argon [239].

In order to characterize SiC powders and sintered ceramics the total oxygen and nitrogen content as well as the contents of metallic impurities are analyzed. Total oxygen and nitrogen contents are usually determined by an inert gas fusion method (Leco TC 436) using powdered samples, whereas metallic impurities (Na, K, Ca, Mg, V, Fe, Ti, Al, Cr and Ni) and boron content are determined in acidic solutions by inductive plasma emission (ICP) spectroscopy [240–242].

Besides chemical analysis, physical properties such as particle size, particle size distribution, and bulk density are also important.

The particle size distribution of coarse abrasive materials (5 mm–50 µm) is determined by sieving. For particle sizes less than 50 µm, sedimentation and laser diffraction methods are used.

Both the particle size distributions and the methods for their determination are standardized. The standard for bonded abrasive applications [243] differ from those for coated abrasives [244]. The bulk density depends both on the particle size distribution and on the particle shape. It is measured by weighing a known volume of SiC grains [245]. The packed density is often determined instead of the bulk density [246].

According to the analysis of the microstructure [247] and the physical properties measurements, the reader is referred to the literature [93, 248–251]. The major sequential steps in conducting analysis and properties evaluation for SiC ceramics are shown in Fig. 32.

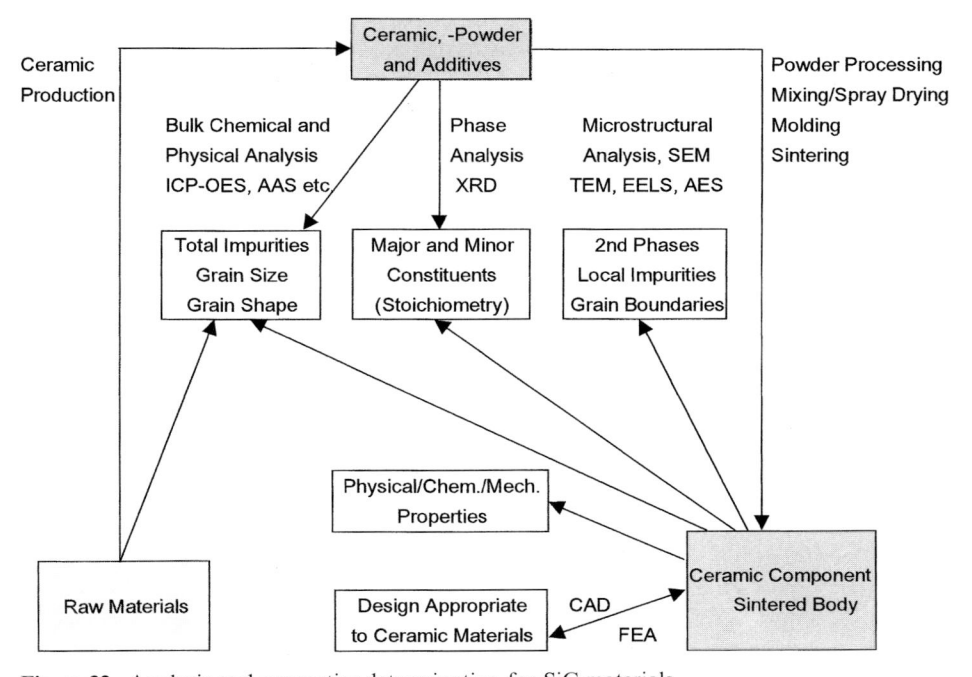

Figure 32. Analysis and properties determination for SiC materials.

5.6 Toxicology and Occupational Health

Silicon carbide is nontoxic, and is therefore a nonhazardous material as defined by GefStoffV [252].

For fiber-free finely divided SiC, the MAK value is $4\,\text{mg}\,\text{m}^{-3}$ [253]. If the MAK ('*M*aximalzulässige–*A*rbeitsplatz–*K*onzentration') is exceeded a dust mask with a P1 filter (for inert dusts) must be worn. Recent investigations, however, have shown that these SiC dusts are not fibrogenic [254], so that the need for a MAK value is questionable.

SiC whiskers have no known routes of entry into the body except as airborne particles.

Conclusive experimental studies on the effects of respirable fibrous SiC dusts (whiskers) are not yet available [255]. In the United States ASTM has published recommendations for safe handling procedures related to all ceramic fibers, including silicon carbide whiskers [256].

5.7 Uses of Silicon Carbide

Of the $\approx 700\,000$ tons SiC produced per year, about 33% is used in metallurgy as a deoxidizing plus alloying agent, and about 50% in the abrasive industry [257]. The remainder is used in the refractory and structural ceramics industries and to a small extent also in electric and electronic industries as heating elements, thermistors, varistors, light-emitting diodes, and attenuator material for microwave devices.

In its loose granular form silicon carbide is used for cutting and grinding precious and semiprecious stones and fine grinding and lapping of metals and optical glasses [258–260]. Bound with synthetic resins and ceramic binders SiC grits are used in grinding wheels, whetstones, hones, abrasive cutting-off wheels, and monofiles for machining of metals, ceramics, plastics, coal-based materials, and so on [261].

Coated abrasives include abrasive paper and cloth in sheet or band form. They are produced by strewing the SiC grains onto a substrate coated with glue or bonding resin and then covering with a second layer of bonding agent [257].

The addition of silicon carbide during the melting of cast iron aids carburization and siliconization, and improves the quality of the cast iron as a result of its seeding action [262]. In the production of steel in an arc furnace, silicon carbide acts as deoxidant and helps in slag melting.

The need to control thermal expansion and to increase the strength and Young modulus of aluminum alloys produced new Al–SiC composites alloys containing up to 50 vol-% SiC particles [263]. The market is poised for rapid expansion in this area over the next years as developmental products move into commercialization stages. For reinforcement applications SiC is also used in the form of whiskers [264], platelets and fibres.

The resistance of ceramically bonded and recrystallized silicon carbide to thermal shock, oxidation, and corrosion is utilized in its use as a refractory construction material, for example, in linings and skid rails for furnaces and hot cyclones, and as a kiln furniture, especially in saggers [61, 265–269]. The good electrical conductivity of the material at high temperatures, coupled with its outstanding oxidation resistance, led to its early use in the electric heating industry [270–272], which markets its products mainly in the form of rods and tubes that operate up to 1500°C. Recrystallized SiC igniters are used in home gas appliances, replacing pilot lights. High-purity SiC shapes are used in the electronic industry as furnace components for processing of silicon wafers. The thermoelectric properties of SiC suggest the use of sintered SiC rods as high-temperature thermoelements [273] and as Seebeck elements [274] for high-temperature thermoelectric energy conversion. Voltage-dependent resistors (varistors) consist of ceramic- or polymer-bonded SiC and are used in overvoltage protection equipment.

Silicon carbide is an outstanding material for the construction of electronic equipment. Blue light-emitting diodes having an improved 470 nm peak wavelength are being produced and marketed as the first commercial SiC semiconductor device.

The continual development of the deposition of SiC thin films and of large diameter single crystal SiC wafers, the associated technologies of doping, etching and electrical contacts have culminated in a host of new solid state devices including field effect transistors capable of operation up to 650°C [275].

High-density, high-strength SiSiC, SSiC, HP-, and HIPSiC materials, which have been on the market only since the 1970s, have opened up a new field of application [276], namely, in mechanical and high-temperature engineering. SiSiC and, especially SSiC, are displacing the chemically less resistant tungsten carbide (hard metal) and the thermal-shock-sensitive aluminum oxide in modern mechanical seals where they are used in the form of slide rings (see Fig. 33). The excellent

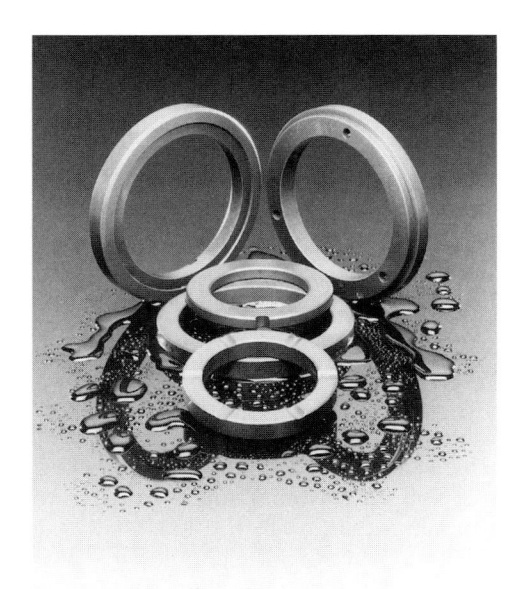

Figure 33. SSiC slide rings (ESK-Kempten, Germany).

Figure 34. SSiC sliding bearings (ESK-Kempten, Germany).

wear resistance of sintered silicon carbide, its excellent chemical resistance, and outstanding tribological characteristics ensure that mechanical seals made of this material last longer, resulting in much reduced maintenance and production costs for pump-dependent processes in the chemical industry [277].

SiSiC seal rings (in contrast with those made of pure SSiC) can only be used in acid media because of the accompanying silicon phase, which is attacked by alkalis [207].

For similar requirements involving radial loads, sliding bearings (see Fig. 34) are manufactured from SSiC. The erosion and chemical resistance of SiC enable the designer to position the bearings in the medium to be transported; that is, to eliminate lubrication and sealing problems. Other components include shaft protection sleeves for waste gas exhaust fans, and precision spheres for dosing and regulating valves.

Due to the tailored properties of liquid phase sintered silicon carbide (LPSSiC) it is used as dewatering elements in the paper machinery and as rings for highly stressed gas seals. It is a price competitive alternative to silicon nitride materials and outperforms alumina and tungsten carbide materials. In addition, LPSSiC is proposed as neutral matrix in ceramic matrix composites containing plutonium to burn the world's stockpiles of military plutonium in thermal or fast reactors [278].

Hot-pressed SiC is the preferred material to replace oxide ceramics for rods fixtures, and punches in high-temperature strength testing equipment. In view of the low level of plasma contamination, the low induced radioactivity, and excellent high-temperature resistance of silicon carbide, it is the ideal material (especially in its isostatically hot-pressed form) for use in fusion reactors [279].

Combustion tubes made of slip-cast SiSiC [280] have better resistance to corrosion, high-temperatures, and thermal shock, so that they will last far longer than, for example tubes made of heat resistant steel. The heat treatment industry has begun using SSiC radiant tubes in indirect gas-fired heat treating operations. In such systems, the tubes are internally heated by combustion burners and radiant heat to some external work load, such as an ingot of alloy, which is isolated from the combustion atmosphere [281]. SiSiC and SSiC are destined for use in heat exchanger systems because of their high-thermal conductivity and corrosion resistance [282–286].

Sintered silicon carbide bonded to plastic laminate substrates reinforced with glass or Kevlar ^{TM} fabrics can be used as ceramic armor to defeat armor piercing projectiles [287].

Sintered and isostatically hot-pressed SiC materials, as well as silicon nitride (Si_3N_4) are playing an important role in the development of ceramic components (see Fig. 35) for motor vehicle engines and gas turbines [288–293]. Real technical success with sintered SiC components in the field of high temperature gas turbines has not yet been achieved [294, 295]. However, SiC and Si_3N_4 gas turbine parts are in field tests and strong development efforts continue in several countries.

Developments for the application of continuous fiber reinforced SiC matrix composites (CMCs) have started with and are concentrating on hot components in military and space technology: hot gas ducts and thermal heat shields for space

Figure 35. Monolithic SSiC gas turbine rotor (courtesy A.Lipp, ESK-Kempten/Germany).

reentry vehicles [173]. Due to the high thermal shock capability of this class of materials and to the high fracture toughness of some of these composites some civil applications are gaining importance: highly loaded brake discs, for high speed trains, and highly loaded journal bearings; in both applications the conventional materials (metals for brakes, monolithic ceramics for bearings) are not applicable because of the thermal loads and the brittle failure mode respectively.

Acknowledgments

Some parts of the text appeared in the earlier review by K.A.S. (Silicon Carbide in: *Encyclopedia of Advanced Materials* pp. 2455–2461, **1994**) and Elsevier Science Ltd., Kidlington OX5 16B, UK is kindly thanked for permission to use them.

References

1. R. Kieffer and F. Benesovsky, *Hartstoffe*, Springer Verlag, New York, 1963.
2. J. J. Berzelius, *Am. Phys. Chem.* 1824, **1**, 169–230.
3. C. M. Despretz, *Compt. Rend.* 1849, **29**, 709–724.
4. P. Schützenberger and A. Colson, *Compt. Rend. Akad. Sci.* 1881, **92**, 1508.
5. H. Moissan, *Der elektrische Ofen*, Verlag M. Krayer, Berlin, 1900.
6. E. G. Acheson US Pat. 492 767 (28 Feb. 1893), DE 76629 (1894), DE 85195 (1896).
7. *Gmelin Handbook of Inorganic Chemistry*, 8th edition, Silicon Supplement Vo. B 3, System No. 15, H. Katscher et al. (eds.), Springer Verlag, Berlin, 1986.
8. H. Moissan, *Compt. Rend. Akad. Sci.* 1905, **140**, 405.
9. A. J. Regis, Land . B. Sand, *Bull. Geol. Soc. Am.* 1958, **69**, 1633.
10. A. P. Bobrievich, V. A. Kalyuzhnyi, and G. I. Smirnov, *Proc. Akad. Sci. USSR: Geol. Sci.* 1957, **115**, 757.
11. J. Bauer, J. Fiala, and R. Hrichová, *Am. Miner.* 1963, **48**, 620–34.
12. H. Kleykamp and G. Schumacher, *Ber. Bunsenges. Phys. Chem.* 1993, **97**, 799–805.
13. A. Dietzel, H. Jagodzinski, and H. Scholze, *Ber. Dtsch. Keram. Ges.* 1960, **37**, 524–537.
14. R. Kieffer, E. Gugel, P. Ettmayer, and A. Schmidt, *Ber. Dtsch. Keram. Ges.* 1966, **43**, 621–623.
15. N. W. Jeeps and T. F. Page, *J. Am. Ceram. Soc.* 1981, **64**, C177–C178.
16. R. F. Adamsky and K. M. Merz, *Z. Kristallogr.* 1959, **111**, 350–361.
17. H. Ott, *Z. Kristallogr.* 1925, **61**, 515–532; **62**, 210–218; **63**, 1–19.
18. L. S. Ramsdell, *Am. Miner.* 1947, **32**, 64–82.
19. J. Ruska, L. J. Gauckler, J. Lorenz, and H. U. Rexer, *J. Mater. Sci.* 1979, **14**, 2013–2017.
20. Y. Inomata, Z. Inoue, M. Mitomo, and H. Suzuki, Yogyo-Kyokai-Shi, 1968, **76**, 313–319.
21. D. Lundquist, *Acta. Chem. Scand.* 1948, **2**, 177–191.
22. P. T. B. Shaffer, *Mater. Res. Bull.* 1969, **4**, S13–24.
23. M. Mitomo, Y. Inomata, and M. Kumanomido, *Yogyo-Kyokai-Shi* 1970, **78**, 365–369.
24. M. Mitomo, Y. Inomata, and H. Tanaka, *Mater. Res. Bull.* 1971, **6**, 759–764.
25. M. M. Patience, Silicon Carbide Alloys, PhD. Thesis, University of Newcastle upon Tyne, 1983.
26. D. Foster, Densification of Silicon Carbide with Mixed Oxide Additives, University of Newcastle upon Tyne, 1996.
27. N. W. Jeeps and T. F. Page, *J. Microscopy* 1979, **116**, 159–171; 1980, **119**, 177.

28. Anon, *Ind. Heating*, 1954, 992–1004.
29. W. Poch and A. Dietzel, *Ber. Dtsch. Keram. Ges.* 1962, **39**, 413–426.
30. K. Liethschmidt, Siliciumcarbid in *Chemische Technologie, Band 2, Anorg. Techn. 1, 4. Aufl.* K. Winnacker and L. Küchler (Eds), Hanser-Verlag, München-Wien, 1982, pp. 626–629.
31. K. H. Mehrwald, *Ber. Dtsch. Keram. Ges.* 1992, **69**, 72–81.
32. A. Graf von Matuschka and G. Schönbrunn, Stand der Mikrokorngrößenbestimmung von Elektrokorund und SiC, *ZwF* 1981, **76**, 38–45.
33. W. Boecker, H. Landfermann, and H. Hausner, *Powder Met. Int.* 1981, **13**, 37–39.
34. S. Prochazka, Techn. Report 86-CRD-158, General Electric, Schenectady NY, 1986.
35. P. Matje and K. A. Schwetz, *Proc. 1st Internat. Conf. Ceramic Powder Processing Science*, Nov. 87, Orlando FL, Gary Messing et al. (Eds), *Ceram. Trans. 1988*, **1**, 460–468.
36. A. W. Weimer, Carbothermal reduction synthesis processes, in *Carbide, Nitride and Boride Materials Synthesis and Processing*, A. W. Weimer (Ed.), Chapman & Hall, London, 1997, pp. 75–180.
37. J. S. Prener, US Patent 3 085 863, 1960.
38. T. L. O'Connor and W. A. McRae, US Patent 3 236 673, 1963.
39. S. Prochazka, Final Report SRD 72-171, General Electric, Schenectady NY, 1972.
40. K. A. Schwetz and A. Lipp, *Radex-Rundschau*, 1978, **2**, 479–498.
41. W. M. Goldberger, A. K. Reed, and R. Morse, Synthesis and characterization of HSC silicon carbide, in *SiC '87*, J. D. Cawley (Ed.), *Ceram. Trans.* 1989, **2**, 93–104.
42. F. Hahn, G. Rudakoff, and H.J. Tiller, *Hermsdorfer Techn. Mitt.* 1990, **79**, 2546–2550.
43. H. R. Baumgartner and B. R. Rossing, Pressureless sintering and properties of plasma synthesized SiC powder, in *SiC '87*, J. D. Cawley (Ed.), *Ceram. Trans.* 1989, **2**, 3–16.
44. F. G. Stroke (PPG), US Patent 4 295 890, 1981.
45. W. R. Cannon, S. C. Danforth, J. H. Flint, J. S. Haggerty, and R. A. Marra, *J. Am. Ceram. Soc.*, 1982, **65**, 324–330.
46. W. Böcker and H. Hausner, *Ber. Dtsch. Keram. Ges.* 1978, **55**, 233–237.
47. P. T. B. Shaffer, SiC Whiskers, in *Handbook of Advanced Ceramic Materials*, Advanced Refractory Technologies, Buffalo, NY, 1994.
48. T. N. Tiegs and T. F. Becher, *Am. Ceram. Soc. Bull.* 1987, **66**, 339–342.
49. W. D. G. Böcker, S. Chwastiak, F. Frechette, and S. K. Lau, Single phase αSiC reinforcements for composites, in: *SiC '87*, J. D. Cawley (Ed.), *Ceram. Trans.* 1989, **2**, 407–420.
50. P. A. Kistler-De Coppi, and W. Richarz, *Int. J. High Technol. Ceram.* 1986, **2**, 99–113.
51. B. Meier, H. Ramminger, and E. Nold, *Microchim. Acta* 1990, **II**, 195–205.
52. R. Lenk, J. Adler, Influence of forming technique on the SiC platelet orientation in a liquid-phase sintered SiC-matrix, in *Fourth Euroceramics, Vol. 2, Part II, Basic Science*, C. Galassi (Ed.), Faenca Editrice Italy, pp. 407–414.
53. R. Lenk, A. F. Kriwoschepov, and K. Große, Heißgießen von drucklos gesintertem SiC, *Sprechsaal Ceram. Mater.* 1995, **128**, 17–20.
54. S. Yajima, J. Heyashi, M. Omori, and K. Okamura, *Nature*, 1976, **261**, 683.
55. T. Ishikawa, SiC continuous fiber, Nicalon, in *SiC Ceramics-2*, S. Somiya and Y. Inomata (Eds), Elsevier, Amsterdam, 1991, pp. 81–98.
56. H.-P. Baldus and M. Jansen, *Angew. Chem.* 1997, **109**, 338–354.
57. D. Heimann, T. Wagner, J. Bill, F. Aldinger, and F. F. Lange, *J. Mater. Res.* 1997, **12**, 3099–3101.
58. S. Prochazka, Sintered SiC and method of making, US Patent 566 8068, 1997.
59. K. K. Kappmeyer, D. H. Hubble, and H.W, Powers, *Am. Ceram. Soc. Bull.* 1966, **45**, 1060–1064.
60. M. E. Washburn and R. W. Love, *Am. Ceram. Soc. Bull.* 1962, **41**, 447–449.
61. A. Fickel, SiC Materials, in *Refractory Materials – Pocket Manual*, G. Routschka (Ed.), Vulkan Verlag, Essen, 1997, pp. 74–80.
62. R. Van der Beck and J. O'Connor, *Ceram. Ind.* 1957, **No. 3**, 96–98.
63. R. A. Alliegro, Processing and fabrication of non-hotpressed SiC, in *Ceramics for High Performance Applications*, J. J. Burke et al., (Eds), Metals and Ceramics Inf. Center, Columbus, OH, 1974, pp. 253–263.
64. J. Kriegesmann, *Interceram.* 1988, **No. 2**, 27–30 and *Powder Metall. Int.* 1986, **18**, 341–343.
65. P. Popper and D. G. S. Davies, *Powder Metall.* 1961, **8**, 113.

66. K. M. Taylor, Improved K TSiC for high temperature parts, *Mater. Meth.* 1956, Vol. 44, No. 4, 92–95; US Patent 3 189 472, 1965.
67. R. Kieffer, E. Gugel, and A. Schmidt, Verfahren zur Herstellung eines dichten Formkörpers auf Basis SiC, DE-OS 1 671 092, 1967.
68. P. Kennedy and J. V. Shennan, Engineering applications of REFEL silicon carbide, in *Silicon Carbide 1973*, R. C. Marhall et al. (Eds), Univ. of South Carolina Press, Columbia, 1974, pp. 359–366.
69. G. Q. Weaver, H. R. Baumgartner, and M. L. Torti, *Special Ceramics 6*, P. Popper (Ed.), Brit. Ceram. Res. Association, UK, pp. 261–281.
70. W. B. Hillig, R. L. Mehan, C. R. Morelock, V. J. De Carlo, and W. Laskow, *Am. Ceram. Soc. Bull.* 1975, **54**, 1054–1056.
71. T. Hase et al., *J. Nucl. Mater.* 1976, **59**, 42–48.
72. Anon. A super ceramic called REFEL, *Interceram.* 1982, **No. 1**, 50–51.
73. G. Willmann and W. Heider, *Werkstofftechnik* 1983, **14**, 158.
74. K. Taylor, Dense silicon carbide, US Patent 3 205 043, 1962.
75. S. Prochazka, The sintering process for SiC, a review, Techn. Report 81-CRD-314, General Electric, Schenectady NY, 1981, pp. 16.
76. S. Prochazka, Sintering of SiC, in *Ceramics for High Performance Applications*, J. J. Burke et al. (eds.), Metals and Ceramic Info. Center, Columbus, OH, 1974, pp. 239–252.
77. J. A. Coppola and G. H. McMurtry, Substitution of ceramics for ductile materials in design, *Nat. Symp. on Ceramics in the Service of Man*, Carnegie Institution, Washington DC, 1976.
78. P. T. B. Shaffer, The SiC phase in the system SiC–B_4C–C, *Mater. Res. Bull.* 1969, **4**, 213–220.
79. R. Hamminger, Carbon inclusions in sintered SiC, *J. Am. Ceram. Soc.* 1989, **72**, 1741–1744.
80. Y. Murata and R. H. Smoak, Densification of SiC by the addition of BN, BP and B_4C, and correlation to their solid solubilities, in *Proc. Int. Symp. Densification and Sintering, Oct. 1978*, Hakone/Japan, S. Somiya and S. Saito (Eds), Gakujutsu Bunken, Tukyu-Kai, Tokyo, 1979, pp. 382–399.
81. W. Boecker and H. Hausner, *Powder Metall. Int.* 1978, **10**, 87–89.
82. D. R. Stutz, S. Prochazka, and J. Lorentz, *J. Am. Ceram. Soc.* 1985, **68**, 479–482.
83. T. Fetahagic and D. Kolar, *Ceram. Acta* 1990, **2**, 31–37.
84. K. A. Schwetz, A. Lipp, The effect of boron and aluminum sintering additives on the properties of dense sintered αSiC, *Science of Ceramics 10*, Berchtesgaden Sept. 1979, H. Hausner (Ed.), Verlag Deutsche Keramiske Ges 1980, pp. 149–158.
85. W. Grellner, K. A. Schwetz, and A. Lipp, Fracture phenomena of sintered alphaSiC, in *Proc. 7th Symp. on Special Ceramics*, Bedford College, London, Dec. 1980, D. Taylor and P. Popper (Eds), Br. Ceram. Res. Assoc., Stoke on Trent 1981, pp. 27–36.
86. W. Boecker, H. Landfermann, and H. Hausner, *Powder Metall. Int.* 1979, **11**, 83–85.
87. H. Hausner, Pressureless sintering of non-oxide ceramics, in *4th CIMTEC, Energy and Ceramics*, St. Vincent, Italy 1979, P. Vincenzini (Ed.), Elsevier Scientific Publ. Comp., Amsterdam, 1980, pp. 582–595.
88. H. Tanaka, Y. Inomata, K. Hara, and H. Hasegawa, *J. Mater. Sci. Lett.* 1985, **4**, 315–317.
89. K. Suzuki, Pressureless sintering of SiC with addition of Al_2O_3, *Rep. Res. Lab. Asahi Glass Co.* 1986, **36**, 25–36.
90. K. Suzuki, Pressureless sintering of SiC with addition of Al_2O_3, in *SiC Ceramics – 2*, S. Somiya and Y. Inomata (Eds), Elsevier, London, 1991, pp. 163–182.
91. R. H. Smoak, Pressureless sintering beryllium containing SiC powder composition, DE-OS 27 51 851, 1977.
92. A. Mohr, Untersuchungen zur Minimierung der Additivgehalte für die drucklose Sinterung von αSiC, Diplomarbeit Institut für Keramik im Maschinenbau, Universität Karlsruhe, 1989.
93. D. W. Richerson, *Modern Ceramic Engineering*, Dekker, New York, 1992.
94. F. Thümmler and R. Oberacker, *Introduction to Powder Metallurgy*, The Institute of Materials Series, University Press, Cambridge, 1993.
95. C. A. Johnson and S. Prochazka, Microstructures of sintered SiC, in *Ceramic Microstructures 1976*, R. M. Fullrath and J. P. Pask (Eds), Westview Press, Boulder, CO, 1977, pp. 366–378.
96. K. A. Schwetz, F. Isemann, and A. Lipp, Injection molded sintered turbine components of Al-

doped alpha-SiC; in *Proc. 1st Internat. Symp. Ceramic Comp. For Engine*, S. Somiya et al. (Eds), KZK Scientific, Tokyo, 1984, pp. 583–594.
97. Showa Denko, DE 3 927 300, 1988.
98. Ceramiques et Composites, European Patent 486 336, 1991.
99. Carborundum Company, US Patent 5 589 428, 1996; US Patent 5 635 430, 1997.
100. ESK-GmbH, European Patent 685 437, 1995.
101. Carborundum Company, European Patent 145 496, 1984.
102. J. Greim, L. Sigl, and H. Thaler, Sintered SiC for high-performance bearings and seals, *Magazine of Wacker Chemie-GmbH Werk + Wirken: Int. Edn*, 1997, **1**, 23–25.
103. M. Omori and H. Takei, *J. Am. Ceram. Soc.* 1982, **65**, C-92; and US Patent 5 439 853, 1995.
104. E. Kostic, *Powder Metall. Int. 1988*, **20**, 28–29.
105. R. A. Cutler and T. B. Jackson, Liquid phase sintered SiC, in *Ceramic Materials and Components for Engines, Proc. 3rd. Int. Symp.*, V. J. Tennery (Ed.), Am. Ceram. Soc., 1988, pp. 309–318; and US Patent 4 829 027, 1988.
106. W. Boecker and R. Hamminger, *Interceram.* 1991, **40**, 520–525.
107. F. K. Van Dijen and E. Mayer, *J. Europ. Ceram. Soc.* 1996, **16**, 413–420.
108. C. Wolf, H. Hübner, and J. Adler, Mechanical behavior of pressureless sintered SiC at high temperature, in *3rd Euro-ceramics, Vol. 3*, P. Duran and J. F. Fernandez (Eds), Faenze Editrice Iberica, Castellon, Spain, 1993, pp. 465–470.
109. K. Y. Chia and S. K. Lau, *Ceram. Eng. Sci. Proc.* 1991, **12**, 1845–1861.
110. K. A. Schwetz, P. Matje, M. Mohr, and K. P. Martin, Development of a pressureless sintering process for ceramics in the system SiC–AlN (O), Final Report on German BMFT Project No. 03-zC-153, 1986 (unpublished).
111. I. Wiedmann, M. Nader, M. J. Hoffmann, and F. Aldinger, in Symposium 7, Materialwissenschaftliche Grundlagen, *Werkstoffwoche '96*, F. Aldinger and H. Mughrabi (Eds), DGM Informationsgesellschaft mbH, Oberursel, 1997, pp. 515–520.
112. L. S. Sigl and H. J. Kleebe, *J. Am. Ceram. Soc.* 1993, **76**, 773–776.
113. M. P. Padture, *J. Am. Ceram. Soc.* 1994, **77**, 519–523; 2518–2522.
114. S. K. Lee and C. H. Kim, *J. Am. Ceram. Soc.* 1994, **77**, 1655–1658.
115. J. S. Nadeau, *Ceram. Bull.* 1973, **52**, 170–174.
116. R. A. Alliegro, L. B. Coffin, and J. R. Tinklepaugh, *J. Am. Ceram. Soc.* 1956, **39**, 386–389.
117. S. Prochazka and R. J. Charles, *Am. Ceram. Soc. Bull.* 1973, **52**, 885–891.
118. J. Kriegesmann, *Ber. Dtsch. Keram. Ges.* 1978, **55**, 391–397.
119. F. F. Lange, *J. Mater. Sci.* 1975, **10**, 314–320.
120. J. M. Bind and J. V. Biggers, *J. Appl. Phys*, 1976, **47, 5171–5174.**
121. D. Broussaud, Independence of composition of hot-pressed SiC, in *Ceramic Microstructures '76*, R. M. Fullrath and J. A. Pask, (Eds), Westview Press, Boulder, CO, 1976, pp. 679–688.
122. T. Iseki, K. Arakawa, H. Matsuzaki, and H. Suzuki, *Yogyo-Kyokai-Shi* 1983, **91**, 349.
123. K. Nakamura and O. Asai, *Kagaku Kogyo* 1982, **33**, 977.
124. H. Kessel and E. Gugel, *Industrie Diamanten Rundschau* 1978, **12**, 180–185.
125. J. Kriegesmann, K. Hunold, A. Lipp, K. Reinmuth, and K. A. Schwetz, European Patent 71241, 1981.
126. T. J. Whalen, R. M. Williams, and B. N. Juterbock, HIP of SiC and Si_3N_4 structural ceramics, in *10th Plansee Seminar Proceedings*, Reutte, Tyrol, 1985, p. 783.
127. K. Hunold, *Powder Metall. Int.* 1984, **16**, 236–238; 1985, **17**, 91–93.
128. K. Hunold, Sinter/HIP of SiC, in *Proc. Adv. Mat. Technology Ceramic Workshop No. 4, Advances in Materials, Processing and Manufacturing Science*, Nagoya Japan, March 3–4, Japan Fine Ceramics Center,1988, pp. 49–62.
129. R. Oberacker, A. Kühne, and F. Thümmler, *Powder Metall. Int.* 1987, **19**, 43–50.
130. F. Fetahagic, R. Oberacker, and F. Thümmler, Process development for sinter-HIPing of SiC, in *Ber. KFA-Jühlich 1989, Juel. Conf. -77, Emerging Mat. Adv. Process.*, 9th German-Yugoslavian Meeting on Materials Science Development, April 16–19, 1989, pp. 313–325.
131. T. Hirai and M. Sasaki, SiC prepared by CVD, in *SiC Ceramics-1*, S. Somiya and Y. Inomata (Eds), Elsevier, London, 1991, pp. 77–98.
132. Anon, Material Innovations: CVD scaled up for commercial production of bulk SiC, *Am.*

Ceram. Soc. Bull. 1993, **72**, 74–78.
133. M. Tenhover, I. Ruppel, S. S. Lyle, and L. J. Pilione, DC-magnetron sputterable SiC, in *Proc. 36th Annual Techn. Conf. Society of Vacuum Coaters*, 1993, pp. 362–365.
134. Anon, SiC electronic materials and devices, in *MRS Bull.* 1997, **22**, No. 3 (Special Issue); Cree Research Inc., World Patent 97/28297, Growth of colorless SiC crystals, 1997.
135. J. A. Lely, *Ber. Dtsch. Keram. Ges.* 1955, **32**, 229–231.
136. R. Fantoni, E. Borsella, S. Piccirillo, R. Ceccato, and S. Enzo, *J. Mater. Res.* 1990, **5**, 143–150.
137. R. Vassen, D. Stöver, and J. Uhlenbusch, Sintering and grain growth of ultrafine amorphous SiC-Si-Powder mixtures, in *Euro-Ceramics II, Vol. 2, Structural Ceramics and Composites, Proc. 2nd European Ceram. Soc. Conf. Augsburg, Sept. 1991*, G. Ziegler and H. Hausner (Eds), Deutsche Keram. Ges., Köln, pp. 791–797, 1993.
138. R. Vassen, H. P. Buchkremer, and D. Stöver, Verfahren zum Herstellen feinkristalliner Siliciumkarbidkörper, DE 196 42 753 (1999).
139. R. Vassen and D. Stöver, *Philos. Mag. B* 1997, **76**, 585.
140. J. Förster, R. Vassen, and D. Stöver, *J. Mater. Sci. Lett.* 1995, **14**, 214–216.
141. R. Vassen, J. Förster, and D. Stöver, *NanoStructured Mater.* 1995, **6**, 889–892.
142. A. Kaiser, R. Vassen, T. Stöver, and H. Buchkremer, *NanoStructured Mater.* 1997, **8**, 489–497.
143. M. Mitomo, Y.-W. Kim, and H. Hirotsuru, *J. Mater. Res.* 1996, **11**, 1601–1604.
144. M. Mitomo, H. Hirotsuru, and Y.-W. Kim, US Patent 5 591 685, 7 Jan. 1997.
145. R. Telle, R. J. Brook, and G.Petzow, *J. Hard Mater.* 1991, **2**, 79–114.
146. G. C. Wei and P. F. Becher, *J. Am. Ceram. Soc.* 1984, **67**, 571.
147. M. A. Janney, *Am. Ceram. Soc. Bull.* 1987, **66**, 322.
148. D. J. Jiang, J. H. Wang, Y. L. Li, and L. T. Ma, *Mater. Sci. Eng.* 1989, **A109**, 401.
149. D. Ly Ngoc, Gefügeverstärkung von SiC Keramiken, doctoral thesis, University of Stuttgart, Germany, 1989.
150. R. Jimbou, K. Takahashi, and Y. Matsushita, *Adv. Ceram. Mater.* 1986, **1**, 341.
151. C. H. McMurtry, W. D. G. Böcker, S. G. Seshadri, J. S. Zanghi, and J. E. Garnier, *Am. Ceram. Soc. Bull.* 1987, **66**, 325.
152. H. Cai, W. H. Gu, and K. T. Faber, in *5th Techn. Conf. On Composite Materials, Proc. Am. Soc. Comp.*, 1990, pp. 892–901.
153. K. T. Faber, W. H. Gu, H. Cai, R. A. Winholtz, and D. J. Magleg, in *Toughening Mechanisms in Quasi-Brittle Materials*, S. P. Shah (Ed.), Kluwer, Dordrecht, 1991, pp. 3–17.
154. T. Tani, S. Wada, European Patent 303192, 1988.
155. C. Blanc, F. Thevenot, and D. Treheux, Elaboration and characterization of submicronic SiC-TiB$_2$ composites by reactive pressureless sintering, in *Key Eng. Mater.* 1997, **132–136**, 968–971.
156. D. H. Kuo and W. M. Kriven, *J. Europ. Ceram. Soc.* 1998, **18**, 51–57.
157. K. A. Schwetz, K. Reinmuth, and A. Lipp, *World Ceramics*, 1985, **2**, 70–84.
158. F. Thevenot, Sintering of B$_4$C and B$_4$C–SiC two-phase materials and their properties, in *Proc. 9th Int. Symp. on Boron, Borides and Related Compounds*, H. Werheit (Ed.), University of Duisburg Germany,1987, pp. 246–256.
159. W. J. Tomlinson and J. C. Whitney, *Ceram. Int.* 1992, **18**, 207–211.
160. V. Kevorkijan, A. Bizjak, J. Vizintin, F. Thevenot, G. Interdonato, and C. Reimondi, B$_4$C–SiC based material for wear applications, in *4th Euro Ceramics Vol. 4, Basic Science*, A. Bellosi (Ed.), Faenza Editrice, 1995, pp. 209–216.
161. I. B. Cutler, P. D. Miller, W. Rafaniello, H. K. Park, D. P. Thompson, and K. H. Jack, *Nature* 1978, **275**, 434–435.
162. W. Rafaniello, K. Cho, and A. V. Virkar, *J. Mater. Sci.* 1981, **16**, 3479–3488.
163. W. Rafaniello, M. R. Plichta, and A. V. Virkar, *J. Am. Cerm. Soc.* 1983, **66**, 772–76.
164. R. Ruh and A. Zangvil, *J. Am. Ceram. Soc.* 1982, **65**, 260–265.
165. Y. Xu, A. Zangvil, M. Landon, and F. Thevenot, *J. Am. Ceram. Soc.* 1992, **75**, 325–333.
166. A. Zangvil and R. Ruh, Alloying of SiC with other ceramic compounds: A Review, in *Silicon Carbide '87*, J. D. Cawley and L. E. Semler (Eds), Am. Ceram. Soc., Westerville, OH, 1989, pp. 63–82.
167. S. Y. Kuo, A. V. Virkar, and W. Rafaniello, *J. Am. Ceram. Soc.* 1987, **70**, C–125.
168. R. R. Lee and W. C. Wei, *Ceram. Eng. Sci. Proc.* 1990, **11**, 1094–1121.
169. T. B. Jackson, A. C. Hurford, S. L. Brunner, and R. A. Cutler, SiC-based-ceramics with

improved strength, in *Silicon Carbide '87*, J. W. Cawley and C. Semler (Eds), Am. Ceram. Soc., Columbus, OH, 1989, pp. 227–240.
170. J. L. Huang, A. C. Herford, R. A. Cutler, and A. V. Virkar, *J. Mater. Sci.* 1986, **21**, 1448–1456.
171. J. M. Lihrmann and J. Tirlocq, Process for producing dense products based on SiC and composite products thus obtained, World Patent, PCT, WO 97/06119, 1997.
172. D. P. Stinton, R. A. Lowdon, and R. H. Krabill, Mechanical property characterization of fiber-reinforced SiC matrix composites, ORNL/TM-11524, April 1990.
173. H. Wurtinger and A. Mühlratzer, Cost effective manufacturing methods for structural ceramic matrix composite (CMC) components, *ASME* Paper 96-GT-296, 1996.
174. E. Fitzer and R. Gadow, Fiber reinforced SiC, *Am. Ceram. Soc. Bull.* 1986, **65**, 326–335.
175. G. S. Corman, M. K. Brun, P. J. Meschter, K. L. Luthra, and R. Eldrid, Toughened silcomp ceramic composites for gas turbine applications, *Proc. 39th Internat. SAMPE Symposium*, April 11–13 1994, pp. 2300–2313.
176. Ullmanns Encyclopedia of Technical Chemistry (E. Bartholomé *et al.*, eds.), 4th edn., **21**, Verlag Chemie, Weinheim, 1982, 431–438.
177. P. Wecht, Feuerfest-Siliciumcarbid, *Applied Mineralogy, Vol. 11*, Springer Verlag, New York, 1977.
178. K. Zückler, in *Halbleiterprobleme III*, W. Schottky (Ed.), 1956, pp. 207–229.
179. S. H. Hagen, *Ber. Dtsch. Keram. Ges.* 1970, **47**, 630–634.
180. E. Gugel, P. Schuster, and G. Senftleben, *Stahl Eisen* 1972, **92**, 144–149.
181. Anon, *JANAF Thermochemical Tables, 2nd edn.*, NSRDS-NBS37, Washington DC, June 1971.
182. K. Niihara, *Am. Ceram. Soc. Bull.* 1984, **63**, 1160–1164.
183. T. F. Page, *Proc. Brit. Ceram. Soc.* 1978, **26**, 193–208.
184. W. Kollenberg, B. Mössner, and K. Schwetz, *VDI-Berichte* 1990, **Nr. 804**, 347–358.
185. M. E. Washburn and W. S. Coblenz, *Ceram. Bull.* 1988, **67**, 356–363.
186. F. Thümmler, Sintering and high temperature properties of Si_3N_4 and SiC, in *Sintering Processes*, G. C. Kuczynski (Ed.), Plenum, New York, 1980, pp. 247–277.
187. J. Kriegesmann, A. Lipp, K. Reinmuth, and K. A. Schwetz, Strength and fracture toughness of SiC, in *Ceramics for High Performance Appl. III*, E. M. Lenoe *et al.* (Eds), Plenum, New York, 1984, pp. 737–751.
188. R. Hamminger, R. G. Grathwohl, and F. Thümmler, Microchemistry and high temperature properties of sintered SiC, *Proc. 2nd Int. Conf. Science of Hard Materials, Rhodos Sept. 1984*, Inst. of Physics Conf. Series No. 75, Hilger, Bristol, 1986, pp. 279–292.
189. G. Grathwohl, Th. Reets, and F. Thümmler, Creep of hot-pressed and sintered SiC with different sintering additives, in *Sci. Ceram.* 1981, **11**, 425–431.
190. J. L. Chermant, R. Moussa, and F. Osterstock, *Rev. Int. Hautes Temp. Refract.* 1981, **18**, 5–55.
191. G. Wiebke, *Ber. Dtsch. Keram. Ges.* 1960, **37**, 219–226.
192. J. Schlichting, *Ber. Dtsch. Keram. Ges.* 1979, **56**, 196–199 and 256–261.
193. J. Schlichting and K. Schwetz, *High Temp. High Press.* 1982, **14**, 219–223.
194. B. Frisch, W. R. Thiele, R. Drumm, and B. Münnich, *Ber. Dtsch. Keram. Ges.* 1988, **65**, 277–284.
195. K. G. Nickel, Z. Fu, and P. Quirmbach, *Trans ASM: J. Eng. Gas Turb. Power* 1993, **115**, 76–82.
196. K. Konopicky, I. Patzak, and H. Dohr, *Glas Email Keramo. Tech.* 1972, **23**, 81–87.
197. D. W. McKee and D. Chatterji, *J. Am. Ceram. Soc.* 1976, **59**, 441–444.
198. N. W. Jepps and T. F. Page, *J. Microscopy* 1981, **127**, 227–237.
199. P. T. B. Shaffer, *Ceram. Age* 1966, **82**, 42–44.
200. T. Iseki, T. Kameda, and T. Maruyama, *J. Mater. Sci.* 1984, **19**, 1692–1698.
201. T. Labus, *Lubr. Eng.* 1981, **37**, 387–394.
202. J. H. Eisner, *Chemie-Anlagen + Verfahren*, 1982, p. 46, p. 51, p. 54.
203. D. W. Richerson, *Mater. Sci. Res.* 1981, **14**, 661–676.
204. K. Miyoshi, *Proc. Int. Conf. Wear of Mat.*, ASME, San Francisco 1981.
205. R. W. Lashway, S. G. Seshadri, and M. Srinivasan, Various forms of SiC and their effects on seal performance, *38th Ann. Mtg. of the ASLE, **Houston, Texas, 1983.***
206. H. Knoch, J. Kracker, and A. Schelken, *Chemie-Anlagen + Verfahren*, 1983, 28–30.
207. H. Knoch, J. Kracker, and A. Schelken, *Chemie-Anlagen + Verfahren*, 1985, 101–104.
208. H. Knoch, J. Kracker, and A. Schelken, *World Ceram*m 1985, Vol. 2, 96–98.

209. J. Derby, S. G. Seshadri, and M. Srinivasan, *Fracture Mechanics of Ceramics, Vol. 8*, R. C. Bradt *et al.* (Eds), Plenum, New York, 1986, pp. 113–125.
210. D. C. Cranmer, *J. Mater. Sci.* 1985, **20**, 2029–2037.
211. K. Miyoshi, D. H. Buckley, and M. Srinivasan, *Ceramic Bull.* 1983, **62**, 494–500.
212. J. Breznak, E. Breval, and N. H. Macmillan, *J. Mat. Sci.* 1985, **20**, 4657–4680.
213. E. Breval, J. Breznak, and N. H. Macmillan, *J. Mat. Sci.* 1986, **21**, 931–935.
214. C. C. Wu, R. W. Rice, B. A. Platt, and S. Carrir, *Ceram. Sci. Eng. Proc.* 1985, **6**, 1023.
215. J. R. Smythe and D. W. Richerson, *Ceram. Sci. Eng. Proc.* 1983, **4**, 663–673.
216. L. J. Lindberg and D. W. Richerson, *Ceram. Sci. Eng. Proc.* 1985, **6**, 1059–1066.
217. H. Tomizawa and T. E. Fisher, *ASLE Tans.* 1987, **30**, 41–46.
218. H. Knoch and J. Kracker, cfi *Dtsch. Keram. Ges.* 1987, **64**, 159–163.
219. P. Boch, F. Platon, and G. Kapelski, *J. Europ. Ceram. Soc.* 1989, **5**, 223–228.
220. K. H. Habig and M. Woydt, in *Proc. 5th Int. Congress Tribology, Vol. 3*, K. Holmberg and I. Nieminen (Eds), Lansi-Savo Oy, St Michel, Finland, 1989, 106.
221. Y. Yamamoto, K. Okamoto, and A. Ura, in *Proc. 5th Int. Congress Tribology, Vol. 3*, K. Holmberg and I. Nieminen (Eds), Lansi-Savo Oy, St Michel, Finland, 1989, 138.
222. J. M. Martin, T. LeMogne, H. Montes, and N. N. Gardos, in *Proc. 5th Int. Congress Tribology, Vol. 3*, K. Holmberg and I. Nieminen (Eds), Lansi-Savo Oy, St Michel, Finland, 1989, 132.
223. S. Sasaki, *Wear* 1989, **134**, 185–200.
224. J. Denape and J. Lamon. *J. Mater. Sci.* 1990, **25**, 3592–3604.
225. M. Woydt, A. Skopp, and R. Wäsche, *Proc. 4th Internat. Symp. Ceramic Materials and Components for Engines* R. Carlsson *et al.* (Eds), Elsevier, Amsterdam, 1992, pp. 1219–1239.
226. O. O. Ajayi, A. Erdemir, R. H. Lee, and F. A. Nichols, *J. Am. Ceram. Soc.* 1993, **76**, 511–517.
227. P. Maurin-Perrier, J. P. Farjandon, and M. Cartier, *Wear Mater.* 1991, **2**, 585–588.
228. P. Maurin-Perrier, BRITE-EURAM Project Proposal P-2231, Contract No. RJ-1b-295, 1992.
229. N. N. Gardos, Determination of the tribological fundamentals of solid/lubricated ceramics, WRDC-TR-90–4096, Hughes Aircraft, El Segundo, CA 90245, 1990.
230. B. Löffelbein, M.Woydt, and K. H. Habig, *Wear* 1993, 162–164, 220–228.
231. P. Andersson and A. Blomberg, *Wear* 1994, **174**, 1–7.
232. S. Kitaoka, T. Tsuji, T. Katoh, Y. Yamaguchi, and K. Kashiwagi, *J. Am. Ceram. Soc.* 1994, **77**, 1851–1856.
233. H. Czichos and K.-H. Habig, Tribologie Handbuch – Reibung und Verschleiβ, Vieweg Verlag, Braunschweig/Wiesbaden, 1992.
234. H. Knoch, L. Sigl, and W. D. Long, Product development with pressureless sintered SiC, 1 Oct. 1990, 37th Sagamore Army Mat. Res. Conf. AMTL-Watertown MA.
235. J. Greim, H. Thaler, and L. Sigl, Gesintertes Siliciumcarbid für hochbeanspruchte Komponenten in der Lager- und Dichtungstechnik, in *VDI – Bericht 1331*, Düsseldorf, 1997, pp. 153–159.
236. DIN 51 075 part 1–5, 1982, Beuth Verlag GmbH, Berlin/Germany.
237. ISO 9286, 1997, Beuth Verlag GmbH, Berlin/Germany.
238. K. A. Schwetz and J. Hassler, *J. Less Common Metals*, 1986, **117**, 7–15, and DIN 51079–3, 1997.
239. DIN 51 076, part 1, 1991, Beuth Verlag GmbH, Berlin/Germany.
240. F. F. van Dijen, *Interceram* 1993, **42**, 92–94.
241. J. A. C. Broekaert, R. Brandt, F. Leiβ, C. Pilger, D. Pollmann, P. Tschöpel, and G. Tölg, *J. Anal. Atom. Spectr.* 1994, **9**, 1063–1070.
242. G. Zaray, F. Leis, T. Kantor, J. Hassler, and G. Tölg, Analysis of silicon carbide powder by ETV-ICP-AES, *Fres. J. Anal. Chem.* 1993, **346**, 1042–1046.
243. DIN-ISO 8486, part 1–2, 1997, Beuth Verlag GmbH, Berlin/Germany.
244. DIN 69 176, part 1–3, 1985.
245. FEPA 44-D-1986, Fachverband Elektrokorund-und SiC-Hersteller, Fr.a.M.
246. ISO 787, part 11, 1991, Beuth Verlag GmbH, Berlin/Germany.
247. V. Carle, U. Schäfer, U. Täffner, F. Predel, R. Telle, and G. Petzow, Ceramography of high performance ceramics, Part II silicon carbide, *Pract. Met.* 1991, **28**, 420–434.
248. W. Kollenberg, Prüfverfahren keramischer Hochleistungswerkstoffe und deren Grundlagen, KfA-Bericht Nr. 470, Zentralbibiothek der Kernforschungsanlage Jülich GmbH, 1988.
249. L. L. Hench and R. W. Gould, *Characterization of Ceramics*, Dekker, New York, 1971.

250. O. Van der Biest, *Analysis of High Temperature Materials*, Applied Science Publishers, London and New York, 1983.
251. H. J. Hunger, *Werkstoffanalytische Verfahren*, Deutscher Verlag für Grundstoffindustrie, Leipzig/Stuttgart, 1995.
252. FEPA-ESK, SiC material safety data sheet, according to 91/155 EEC and ISO-Standard 11014, Edition 1, issued 22–05–97, 5 pages.
253. D. Henschler (Ed.), *Gesundheitsschädliche Arbeitsstoffe, Toxikologisch-arbeitsmedizinische Begründung von MAK-Werten*, VCH, Weinheim, 1987.
254. J. Bruch, Institut für Hygiene und Arbeitsmedizin, Universitätsklinikum Essen, 1993.
255. I. D. Birchall et al., *J. Mater. Sci. Lett.* 1988, **7**, 350–352.
256. ASTM, Standard Practice for Handling SiC whiskers, E 1437–91, Subcommittee E 34.70, American Society for Testing and Materials, Philadelphia, USA, 1991.
257. L. Coes, Jr., Abrasives, in *Applied Mineralogy, Vol. 1*, Springer Verlag, Wien, 1971.
258. K. Martin, Neue Erkenntnisse über den Werkstoffsabtragsvorgang beim Läppen, *Fachberichte Oberflächentechnik* 1972, **10**, 197–202.
259. G. Spur and I. Sabotka, *ZwF* 1987, **82**, 275–380.
260. G. Spur and D. Simpfendörfer, *ZwF* 1988, **83**, 207–212.
261. H. B. Britsch, *Ber. Dtsch. Keram. Ges.* 1976, **53**, 143–149.
262. T. Benecke, S. Venkateswaran, W. D. Schubert, and B. Lux, *Gießerei* 1993, **80**, 256–662.
263. S. V. Nair, J. K. Tien, and R. C. Bates, *Int. Met. Ref.* 1985, **30**, 275–290.
264. A. Lipp, *Feinwerktechnik* 1970, **74**, 150–154.
265. E. Gugel, *Ber. Dtsch. Keram. Ges.* 1966, **43**, 354–359.
266. A. F. Fickel, *Sprechsaal* 1980, **113**, 517–531; 1980, **113**, 737–747.
267. R. Rasch and H. Maatz, *Maschinenschaden* 1978, **51**, 145–147.
268. Z. Stavric and M. Hue, Crystar, *Keram. Z.* 1975, **27**, 125–128.
269. G. Bierbauer, *Keram. Z.* 1972, **24**, 142–145.
270. O. Rubisch and R. Schmitt, *Ber. Dtsch. Keram. Ges.* 1966, **43**, 173–179.
271. E. Buchner and O. Rubisch, Corrosion behavior of SiC heating elements, in *Silicon Carbide – 1973*, R. C. Marshall et al. (Eds), Univ. South Carolina Press, Columbia, SC, 1974, pp. 428–434.
272. Y. Nakamura and S. Yajima, *Am. Ceram. Soc. Bull.* 1982, **61**, 572–573.
273. W. Huether, Thermoelement zur Temperaturmessung und Verfahren zur Herstellung desselben, European Patent 72 430, 1982.
274. K. Kuomoto, M. Shimohigashi, S. Takeda, and H. Yanagida, *J. Mater. Sci. Lett.* 1987, **6**, 1453–1455.
275. R. Davis, Recent advances regarding the definition of the atomic environment, film growth and microelectronic device development in SiC, in *The Physics and Chemistry of Carbides, Nitrides and Borides*, R. Freer (Ed.), Kluwer, Dordrecht, 1990, pp. 589–623.
276. M. Srinivasan, The silicon carbide family of structural ceramics, in *Structural Ceramics – Treatise on Materials Science and Technology, Vol. 29*, J. B. Wachtman (Ed.), Academic, San Diego, CA, 1989, pp. 100–159.
277. D. Zeus, How the use of advanced ceramics as tribomaterial has effected the evolution of mechanical seals, cfi *Ber. Dtsch. Keram. Ges.* 1991, **68**, 36–45.
278. V. D. Krstic, M. D. Vlajic, and R. A. Verrall, SiC ceramics for nuclear applications, in *Advanced Ceramic Materials*, H. Mostaghaci (Ed.), *Key Eng. Mater.* 1996, **122–124**, 387–396.
279. F. Porz, G. Grathwohl, and R. Hamminger, *J. Nucl. Mater.* 1984, **124**, 195–214.
280. G. Heuschmann and G. Willmann, *Interceram.* 1986, **1**, 24–29.
281. D. P. Butt, R. E. Tressler, and K. E. Spear, Discontinuous phase formation and selective attack of SiC materials exposed to low oxygen partial pressure environments, in *Corrosion of Advanced Ceramics*, K. G. Nickel (Ed.), Kluwer, Dordrecht, 1994, pp. 153–164.
282. R. A. Penty and J. W. Bjerklie, SiC for high-temperature heat exchangers, in *Ceram. Eng. Sci. Proc.* 1982, **3**, 120–127.
283. J. Heinrich, J. Huber, S. Foster, and P. Quell, Advanced ceramics as heat exchangers in domestic and industrial appliances, in *High Tech Ceramics*, P. Vincenzini (Ed.), Elsevier, Amsterdam, 1987, pp. 2427–2440.
284. M. C. Kerr, Advanced ceramic heat exchangers utilizing Hexology-SA, single phase SiC tubes, in *High Tech Ceramics*, P. Vincenzini (Ed.), Elsevier, Amsterdam, 1987, pp. 2441–2449.

285. W. Heider, Verwendung von SiSiC im Apparatebau, *Proc. Techkeram. '87, Wiesbaden/FRG*, S. Schnabel and J. Kriegesmann (Eds), Demat Exposition Managing, 1987, pp. 18.01–18.19.
286. W. Hof, Hexoloy SiC SA – ein neuer Werkstoff für Rohrbündel-Wäremetauscher, *Chemi. Techn.* 1991, **20**, 18–22.
287. B. Matchen, Applications of ceramics in armor products, in *Advanced Ceramic Materials*, H. Mostaghaci (Ed.), Key Eng. Mat. 1996, **122–124**, 333–342.
288. R. S. Storm, R. W. Ohnsorg, and F. J. Frechette, Fabrication of injection molded sintered alpha-SiC turbine components, *Trans. ASME: J. Eng. Power.* 1982, **104**, 76.
289. K. A. Schwetz, W. Grellner, K. Hunold, A. Lipp, and M. Langer, HIP-treated sintered SiC turbocharger rotors, in *Proc. 2nd Intern. Symposium Ceramic Materials and Components for Engines*, W. Bunk, H. Hausner (Eds), DKG Verlag, 1986, pp. 1051–1062.
290. R. W. Ohnsorg and M. O. Ten Eyck, Fabrication of sintered α-SiC Turbine Engine Comp., in *Silicon Carbide '87*, J. D. Cawley and C. E. Semler (Eds), *Ceram. Trans.*, Vol. 2, Am. Ceram. Soc., Westerville, OH, pp. 367–386.
291. R. Westerheide, T. Hollstein, and K. A. Schwetz, Tension-compression testing of HIP-treated sintered SiC for gas turbine applications at temperatures between 1400 and 1600°C, *Proc. 6th Intern. Symp. on Ceramic Materials and Components for Engines*, October 1997, Arita, Japan, (K. Niihara *et al.*, eds.), 253–258, 1998.
292. Ch. Gutmann, A. Schulz, and S. Wittig, A new approach for a low-cooled ceramic nozzle vane, ASME-Paper 96-GT-232, 1996.
293. M. Dilzer, Ch. Gutmann, A. Schulz, S. Wittig, Testing of a low-cooled ceramic nozzle vane under transient conditions, paper presented at 43th ASME Gas Turbine and Aeroengine Technical Congress, June 2–5 1998, Stockholm/Schweden, Paper No. 98-GT-116, ASME Atlanta/Georgia.
294. G. Andrees, Entwicklung eines keram. Werkstoffes zur Auskleidung thermisch hochbeanspruchter Brennräume und Heißgasführungen, Final Report, Project BMFT 03M–2028, MTU Motoren- und Turbinen-Union München GmbH, p. 114.
295. L. S. Sigl and K. A. Schwetz, Fabrication and properties of HIP-treated sintered SiC for combustor liners of stationary gas turbines, in *Symposium 3, Werkstoffe für die Energietechnik, Werkstoffwoche '98/München*, A. Kranzmann and V. Gramberg (eds.), Wiley-VCH, Weinheim, NY, 1999, pp. 15–24.

6 Silicon Nitride Based Hard Materials

M. Herrmann, H. Klemm, Chr. Schubert

6.1 Introduction

Silicon nitride ceramic materials have been intensively studied for many years because of their great potential for use in structural applications at room and high temperatures. This is due to their excellent mechanical properties in combination with good corrosion and thermal shock resistance.

Over the past decade a continuous increase in the number of application fields has been observed. Besides parts for the automotive industry (cam rollers, valve plates for common rail systems), which have not been produced in the volume that was predicted 10 years ago, applications such as cutting tools, ball bearings, wear parts, applications at high temperatures, and application in the electronic industry are becoming increasingly important [1–4]. A new cooking system based on a silicon nitride plate is penetrating into the market at the moment [5].

This wide range of applications leads to an increase in the variety of materials with different microstructures (see Fig. 1), making precise control of the microstructure and properties of the materials necessary (Table 1).

The materials differ in the production technology used to make them, their phase content and properties. The different production technologies used are summarised in Table 2. A more detailed overview is given in [1, 6].

The reaction-bonded materials (RBSN) produced by the nitridation of compacts of silicon powder (see Table 2) are not dense, whereas the other kinds of materials in the table are dense.

The range of applications for RBSN and sintered reaction-bonded silicon nitride has shrunk over the last few years. This is connected with the decreasing cost of silicon nitride powders, reducing the production costs of SSN and the lower strength and lifetime of RBSN in comparison to SSN in many applications. For this reason, these materials will not be explained in detail in this work; reference is made to special reviews [6, 7]. The following paragraphs concentrate on dense silicon nitride materials made from silicon nitride powders.

The most common sintering technology used for silicon nitride materials is gas-pressure sintering. This technology leads to improved reliability and strength in comparison to normal gas-pressure sintering and is accompanied by only a moderate increase in cost. The HIP processes are only used for special applications (e.g. balls for ball bearings) due to the high cost of these technologies. Hot pressing is used for evaluation of materials or for some special applications requiring simple geometries and only a very limited number of parts.

Besides the different production technologies, the materials differ in their composition and microstructure (see Fig. 1). The wide variation of the microstructure and

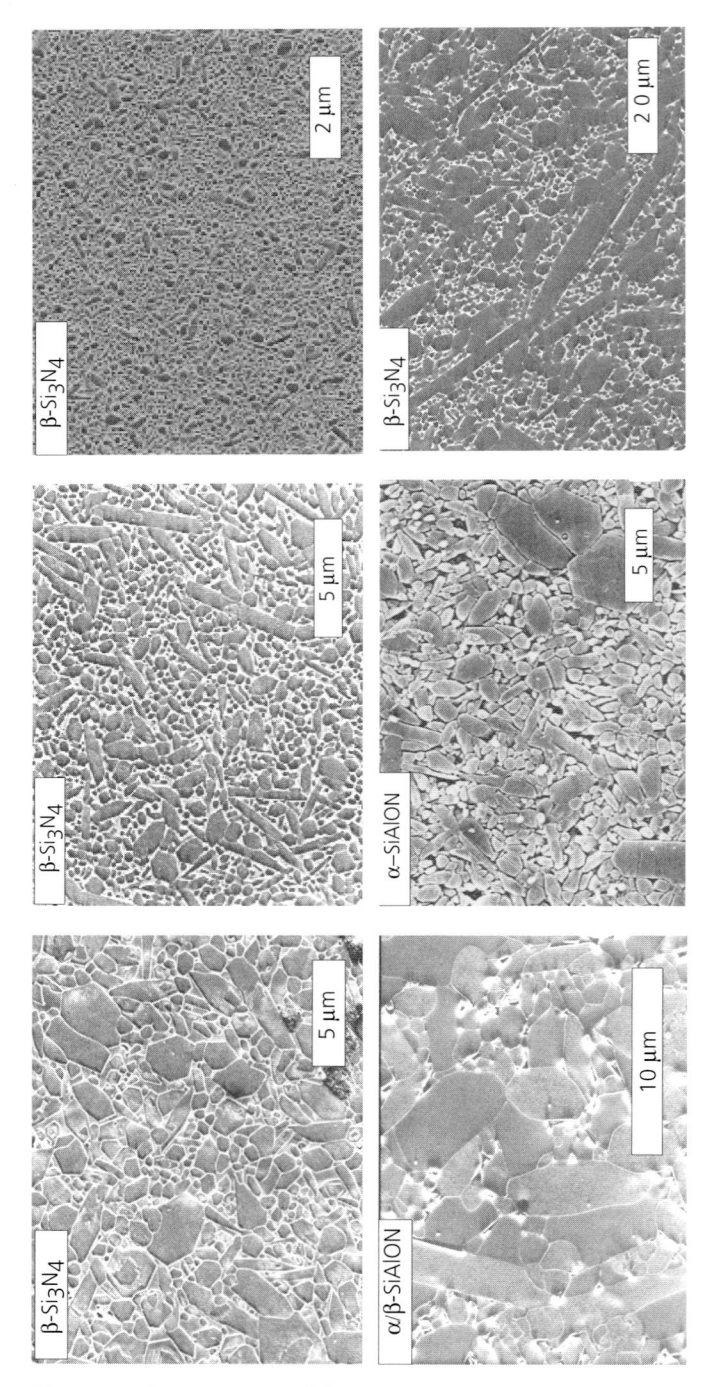

Figure 1. Microstructures of different gas pressure sintered dense Si_3N_4 materials.

Table 1. Overview of the microstructure-property relationships for dense Si_3N_4 materials.

Property	Microstructure	
	grain size/shape	grain-boundary phase
high strength up to 1000°C	fine-grained/needle-like grains	median content
high strength at T >1200°C	fine-grained/needle-like grains	Al_2O_3-free with special compositions
high fracture toughness	large, needle-like grains, or large, needle-like grains in a fine matrix	low Al_2O_3 and SiO_2 content
high hardness	fine-grained or α'-SiAlON	low additive content
high fatigue strength (cyclic mechanical load)	fine-grained/needle-like grains	
high heat conductivity	large grains	no components which can be incorporated into the silicon nitride crystal lattice, nitrides (e.g. Al, Be), low impurity content
high creep resistance	large, needle-like grains, composites with SiC or refractory silicides	Al_2O_3-free with special compositions or no sintering additives
high oxidation resistance at T >1200°C	large, needle-like grains, composites with SiC or refractory silicides	Al_2O_3-free with special compositions or no sintering additives
high corrosion resistance		special compositions depending on the corrosive media
good wear behaviour	fine-grained microstructure	homogeneous distribution

phase content, which is possible in all of the above-mentioned production technologies, is the reason for the wide variation of material properties produced by each of these production technologies.

Dense silicon nitride materials are not monophase materials because, for densification, additives, which remain as an amorphous or partially crystallised grain-boundary phase in the material, are necessary (see Section 6.3).

This grain-boundary phase has both positive and negative consequences for the material properties (see Section 6.5). The softening of the grain boundary has a negative influence on the high-temperature properties, but on the other hand the grain-boundary phase is a main factor causing the high strength and fracture toughness of Si_3N_4. The liquid-phase-sintering process and the wide range of possible additives make the tailoring of the microstructure and the adaptation of the microstructure to a given application possible.

The aim of this paper is to show how the different microstructures and compositions influence the properties at room and elevated temperatures.

Table 2. Different production technologies and resulting materials.

Material type	Reaction bonded silicon nitride (RBSN)	Sintered RBSN (SRBSN)	Sintered silicon nitride (SSN)	Gas pressure sintered silicon nitride (GPSN)	Hot-pressed silicon nitride (HPSN)	Sinter HIP silicon nitride (Sinter-HIPSN)	HIP-ed silicon nitride HIP-SN
Starting components	metallic Si	Si + additives	Si_3N_4 + additives	Si_3N_4 + additives	Si_3N_4 + additives	Si_3N_4 + additives	Si_3N_4 + additives
Heat treatment	1250–1450 °C up to 100 h	nitridation as for RBSN, Sintering as for GPSN	1700–1800 °C	1750–2000 °C nitrogen pressure up to 10 MPa	1500–1800 °C uniaxial pressure in a graphite die	1750–2000 °C gas pressure up to 200 MPa	1750–2000 °C gas pressure up to 200 MPa
Relative density	70–88%	95–99%	95–99%	98–100%	100%	100%	>99%
Main advantages	no shrinkage	low shrinkage, low cost raw material	low sintering cost, different shapes possible	different shapes possible, better reliability than SSN; lower additive content can be used	good densification, high reliability	different shapes possible, sintering of materials with low additive content (low sinterability), high reliability	Materials without additives can be produced
Main disadvantages	low strength (150–350 MPa)	expensive nitridation	Lower strenght than GPSN, good sinterabillity necessary		only simple shapes, low productivity	high sintering costs	high costs for encapsulation
Applications	refractories	the same as SSN/GPSN	wear parts	cutting tools, wear parts, ball bearings, seals, valves, turbo charger rotor	mainly used for evaluation of materials, prototypes with simple geometries	balls for ball bearings, wear parts	used for materials with very low additive content or without additives

6.2 Crystal Structure and Properties of the Si$_3$N$_4$ Modifications

Silicon nitride has three modifications. The α and β modifications can be produced under normal nitrogen pressure. Recently a high-pressure modification was produced under 15 GPa pressure by the technique of laser heating in a diamond cell [8] (see Table 3). The α-phase is metastable under sintering conditions (e.g. at 1400–2000°C and 0.1–100 MPaN$_2$), but is the main phase in the starting silicon nitride powders. The α and β modifications are based on SiN$_4$ tetrahedra connected at the corners. Every N belongs to 3 tetrahedra, i.e. the structure is made up of a three-dimensional network of tetrahedra. Only one layer of SiN$_4$ tetrahedra exists in β-Si$_3$N$_4$ (see Fig. 2), whereas two layers shifted with respect to each other exist in α-Si$_3$N$_4$. This leads to a doubling of the c-axes in the α-Si$_3$N$_4$ crystal lattice in comparison to the β-Si$_3$N$_4$ lattice.

The high-pressure phase has a completely different structure. On the basis of powder diffraction patterns, it is found that this phase has a spinel-type structure, in which one silicon atom is coordinated by four nitrogen and two silicon atoms by six nitrogen atoms (octahedra; see Fig. 2d).

The Si and N atoms in β-Si$_3$N$_4$ can be replaced by Al and O atoms to form the so-called β'-SiAlONs with the formula: Si$_{6-z}$Al$_z$N$_{8-z}$O$_z$. The range of the β' solid solution extends from $z = 0$ to 4.2 at 1750°C [1, 3], i.e. it includes pure Si$_3$N$_4$ (see Fig. 3). That is why a clear distinction between SSN and β-SiAlON materials cannot be made. Every Si$_3$N$_4$ material containing Al$_2$O$_3$ as a sintering additive (i.e. nearly every commercial material) is some kind of β-SiAlON. Most commonly the terminology β-SiAlON is used for z values >0.5. A similar solid solution can be

Table 3. Crystal structures of the modification of the Si$_3$N$_4$ modification and the resulting materials names.

Property	Modifications		
	β-Si$_3$N$_4$	α-Si$_3$N$_4$	high pressure modification [8]
Space group	P6$_3$/m No. 176 [9, 10]	P31c No. 159 [9]	Fd$\bar{3}$, or Fd$\bar{3}$m
Lattice parameter			
a, nm	0.7586	0.766	0.7738
c, nm	0.2902	0.5615	
Hardness	(100) plane: 2100	(110) plane: 2250	20–40 GPa
	(001) plane: 1326	(001) plane: 2200	
	HV 0.025 [12]	HV 0.300 [11]	
Coefficient of thermal expansion 0–1000°C, 10^{-6}/K	3.39 [13]	3.64 [13]	
materials based on the modification	nearly all SSN-, GPSN-materials, β'-SiAlON	α'-SiAlON α'/β'-SiAlON	?

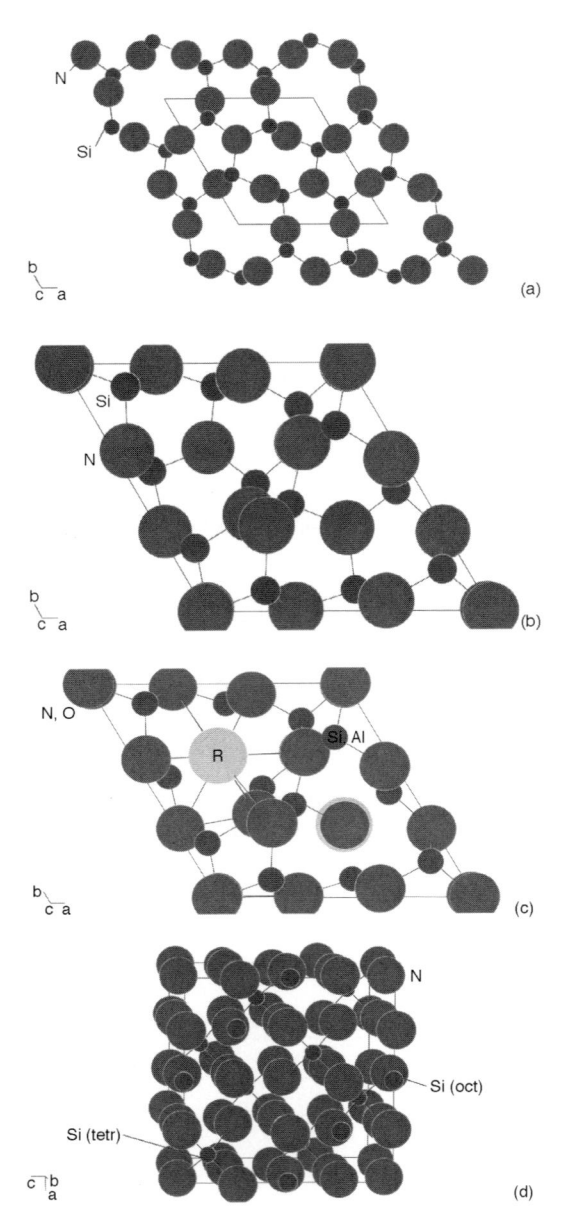

Figure 2. Crystal structures of the Si_3N_4 modifications.

formed with the addition of BeO/BeN [13]; however, materials made from these solid solutions have no practical use due to the toxicity of Be.

Analysis of the α-Si_3N_4 crystal structure shows that there is an empty position with a coordination number of eight (7 + 1). This position can be partially occupied

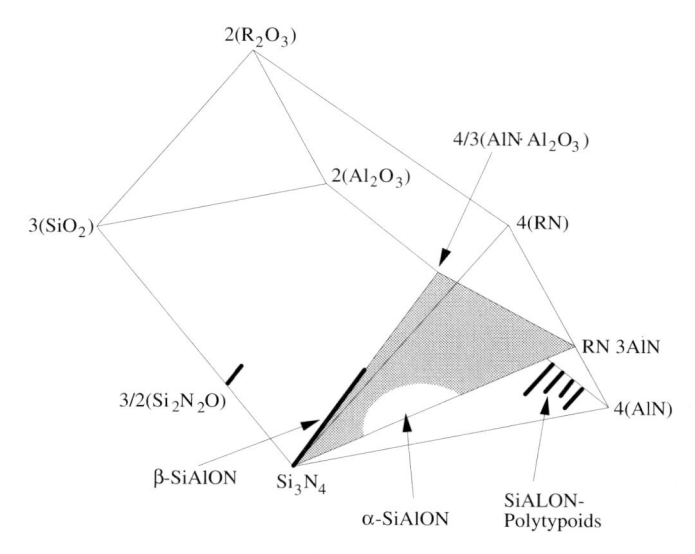

Figure 3. Stability range of α'- and β-SiAlONs in the system Y–Si–Al–O–N (see also Section 6.4.2 [1, 14, 15]).

by ions with an atomic radius of about 0.1 nm. Additionally, Si and N must be replaced by Al and O to obtain electroneutrality. The resulting phase is the so-called α-SiAlON with the formula: $R_x Si_{12-n-m} Al_{m+n} N_{16-n} O_n$. Occupation of the (7 + 1)-coordinated position leads to stabilisation of the metastable α-Si_3N_4 phase. The lowest x value for trivalent cations is 0.33, for Ca 0.3 [14–16]. This means that the α'-SiAlON solid solution does not include the composition of pure α-Si_3N_4.

The different kinds of the materials based on the different modifications are given in Table 3.

The hardness values of the crystals are given in Table 3. These data show that the β modifications have much lower hardnesses than the α modifications. No data for the chemical stability of the high-pressure modification have been published up to now. First measurements of the microhardness show that this structure must have a high hardness which is similar to that of diamond and c-BN.

The microhardness was determined for the different crystal planes of β-Si_3N_4. The data differ widely in different orientations.

6.3 Densification

The high energy of the covalent chemical bonds in the Si_3N_4 crystals, which is the basis for the excellent properties, presents a disadvantage in the fabrication of the materials. The self-diffusuion coefficient is very low in comparison to other ceramic materials. The mass transport necessary for sintering can only be achieved at temperatures where the decomposition of silicon nitride occurs to a great extent even at high pressure (i.e. >2000°C). Dense materials without sintering additives can only be produced by encapsulation-HIP processes. The green bodies which are thus formed are sealed in a dense glass container and densified at 1800–1900°C at 2000 bar [1]. The resulting material consists of Si_3N_4, with some SiO_2 at the grain boundary. These materials have good oxidation and creep resistance (see Section 6.5.2) but low fracture toughness and strength (see Section 6.5.1).

Usually the densification of Si_3N_4 is achieved through the acceleration of mass transport by the formation of a liquid phase during sintering. In the liquid-phase-sintering process, silicon nitride can be sintered in a temperature range of 1550–1900°C. The liquid phase used for sintering is an oxynitride liquid formed by the reaction of the sintering additives with the SiO_2 existing on the surface of the Si_3N_4 powder particles. The most commonly used additives are Al_2O_3 in combination with Y_2O_3, La_2O_3, lanthanoids or MgO. For high-temperature applications, pure rare-earth additives are used. AlN additives are necessary for the production of α'- or β'-SiAlON.

Under sintering conditions stable oxides or nitrides, which form a liquid phase with Si_3N_4 and SiO_2 during sintering, can be used as sintering additives. For example, alkalis and alkaline earths beside MgO can be used as sintering additives. The problem in using these elements is their high vapour pressure under sintering conditions.

Besides Y_2O_3, Sc_2O_3 and La_2O_3, the oxides of the d-elements, ZrO_2 or HfO_2, can be used as sintering aids. Under sintering conditions, TiO_2 reduces to TiN; hence TiO_2 can only accelerate the sintering during the initial stage of sintering. The d-elements in the 5th to 8th group form silicides under sintering conditions, forming separate inclusions.

The amount of liquid phase formed and the resulting grain-boundary phases can be predicted by the phase diagrams (see Fig. 3). Detailed information about the phase diagrams are given in [1, 3, 13, 17].

Densification is improved when the liquid is formed at low temperatures and has a low viscosity. For this reason, the densification behaviour worsens in the order $MgO/Al_2O_3 < MgO/R_2O_3$; MgO; $R_2O_3/Al_2O_3 \ll R_2O_3$ (R = Y, Sc, La and lanthanoids).

The mechanisms of liquid-phase sintering of ceramic materials are explained in detail in [19]. Figure 4 shows typical shrinkage curves of Si_3N_4 materials with different amounts of liquid phase of the same composition. Densification starts at 1200–1300°C. At these temperatures, the chemical reaction involving the additives starts and the formation of the liquid is initiated. The first maximum in the densification rate is connected with the rearrangement of the Si_3N_4 particles. At 1500–1800°C, densification by the solution/diffusion/precipitation mechanism takes places. The phase transition of α-Si_3N_4 to β-Si_3N_4 occurs simultaneously. In [18] it was shown that the α/β-Si_3N_4 ratio has practically no influence on the densification when the grain sizes of the α and β phases are the same. This is an indication that the solubilities of the α- and β-Si_3N_4 modifications in the liquid do not differ significantly [20].

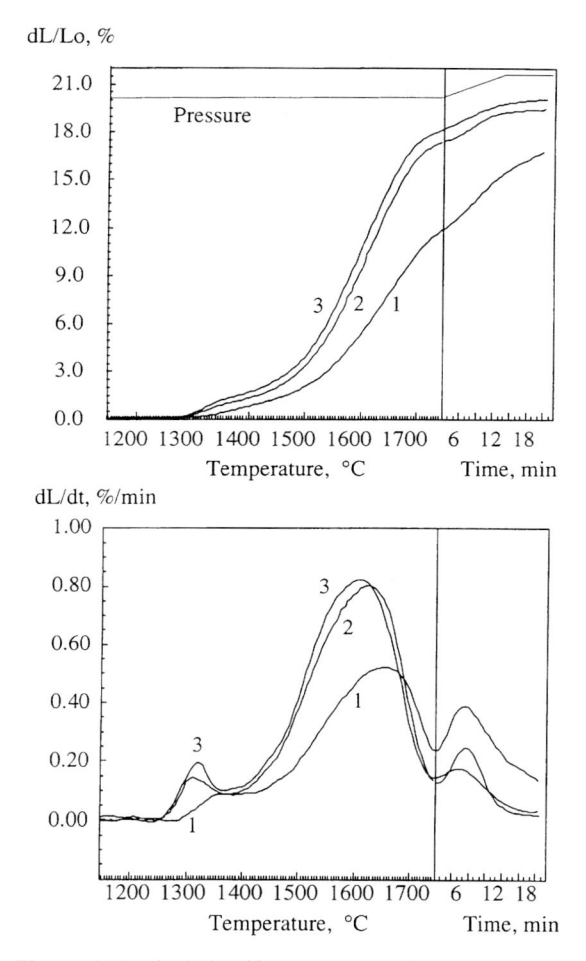

Figure 4. Typical densification curves for Si_3N_4 materials (1–3.5 vol.% additives, 2–6 vol.% additives; 3–8.5 vol.% additives; Al_2O_3/Y_2O_3 weight ratio 1:2).

The sintering behaviour is improved with increasing additive amount (see Fig. 4). However, for additive contents higher than 15–20 vol.%, formation of gas bubbles hinders densification. The bubbling is connected with the low solubility rate of Si_3N_4 and the increased formation of gaseous SiO [21, 22].

After sintering, the additives are situated at the grain boundary (3–20 vol.%). The majority of the grain-boundary phase is concentrated in the triple junctions. The thickness of the grain boundary films between two grains is in the range of 0.8–1.5 nm [41].

The grain-boundary phase significantly influences the material properties. Therefore, one usually attempts to reduce the amount of additives. However, a lower additive content also leads to a lower sinterability, a lower reliability and a broader distribution of the properties. Thus for materials for room-temperature applications

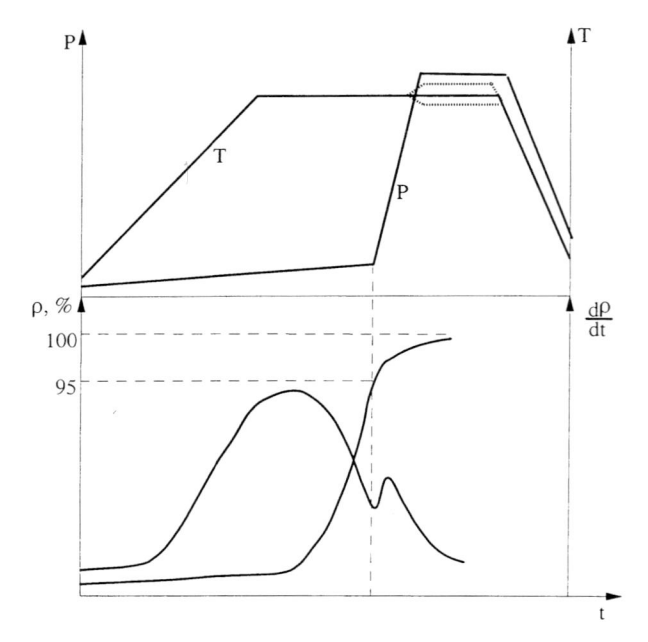

Figure 5. Time pressure sintering regime for gas-pressure sintering.

(e. g. wear parts), the additive content usually lies between 6 and 15 vol.%, representing a compromise between reliability and properties. For high-temperature applications, special refractory grain boundary compositions are used (see Section 6.5.2).

The most common method used for sintering silicon nitride materials is gas-pressure sintering. The principle of gas-pressure sintering is the same as that of the sinter-HIP process (see Fig. 5). The material is sintered to closed porosity (>95% theoretical density) at a low gas pressure. Then an outer gas pressure of up to 10 MPa (GPS: gas-pressure sintering) or 100–200 MPa (sinter-HIP) is applied. The difference between the outer pressure and the pressure in the closed pores is an additional driving force for densification, leading to a better healing of defects. The increase in cost for the GPS process in comparison to sintering under 1 atm (SSN) is relatively low. In many cases these costs can be offset by the better reliability and properties of the materials.

6.4 Microstructural Development

6.4.1 Microstructural development of β-Si_3N_4 materials

The adaptation of silicon nitride materials to various applications leads to an increase in the variety of microstructures (see Fig. 1). This was the impetus for the large number of investigations into the mechanisms of microstructural formation and controlling of the same in recent years.

The most important commercial silicon nitride powders have α-Si_3N_4, i.e. the phase which is metastable at high temperatures, as the main constituent [1]. The consequence of this is that during densification, a phase transformation takes place, having a great influence on the subsequent formation of the microstructure.

It was shown that high fracture toughness and strength can be achieved when the grains have a needle-like shape. Characterisation of the shape is usually done using two parameters: the thickness of the grains and the aspect ratio, i.e. the ratio of needle length to thickness [23–26].

For this reason the influence of different parameters such as additive content, grain size of the powder, β- and oxygen content on the sintering and microstructural formation was investigated [24–39]. In basic works from the 1980s [26–29], it was formulated that a needle-like microstructure, which is necessary for high fracture toughness, can only be achieved using starting powders which contain a high amount of α-Si_3N_4. The reason for this was suggested to be the needle-like grain growth of a limited number of β nuclei [28] or the homogeneous nucleation of β-Si_3N_4 [26, 27]. Later it was shown that needle-like microstructures can also be produced from fine β-Si_3N_4 powders [30, 32–34]. Despite intensive research, no direct evidence of homogeneous nucleation in the oxynitride liquid or heterogeneous nucleation of β-Si_3N_4 on α-Si_3N_4 could be found [30, 36–38]. The only process of epitaxial growth of one modification on the other was found in SiAlON systems with a high degree of substitution [40]. The lattice constants differ less in these materials than in pure α/β-Si_3N_4 due to the incorporation of other ions.

The reason for the absence of formation of β-Si_3N_4 nuclei during sintering seems to be connected with the minimal differences in the free energies of formation of β- and α-Si_3N_4, leading to a quite high critical size of nuclei which can grow (60–600 nm [38]). Thus the critical nucleation size is the same or even larger than the grain size of the β-Si_3N_4 crystallites in the powder. In recent works, it could be shown experimentally that the addition of small β-Si_3N_4 nuclei to a commercial α-Si_3N_4 powder does not result in a change in the microstructure, whereas the addition of β-Si_3N_4 nuclei of the same size as the nuclei existing in the starting α-Si_3N_4 powder alters the microstructure significantly (see Fig. 6). The result is a fine structure which is very similar to the structure made from the β-Si_3N_4 powder. In [20], it was shown that at low sintering times and temperatures, the microstructure correlates with the crystallite size in the starting powder. This provides indirect evidence of the absence of nucleation during sintering. The measurement of the β crystallite size in materials sintered at different temperatures shows a continuous increase in the β crystallite size, indicating that no nucleation takes place (see Fig. 7).

A detailed analysis of different microstructures [29–31, 33] shows that they can be explained by the number of β nuclei in the starting powder, even at very low β-Si_3N_4 contents. In Fig. 8 the calculated volume fractions of growing nuclei which are necessary in the starting composition to produce two different fine microstructures in the sintered materials are given. This calculated content is the lowest β-Si_3N_4 content in the starting powder required to explain the microstructure. For crystallite sizes between 50 and 100 nm [20, 35], the necessary amount of the β fraction in

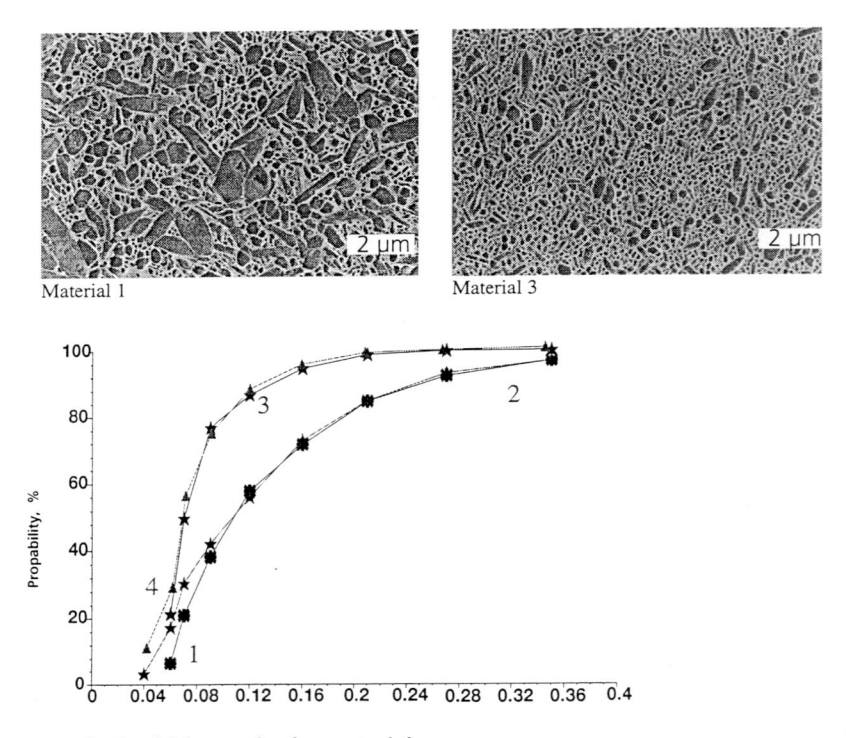

Figure 6. Change in the grain size distribution of materials with different β-Si_3N_4 amounts and crystallite sizes (material 1 produced from α-Si_3N_4 powder with 4% β-Si_3N_4; crystallite size β-Si_3N_4 46 nm; material 2 produced from a mixture of 60% α-Si_3N_4 powder and 40% β-Si_3N_4 powder (the β-powder contain 33% β-Si_3N_4; crystallite size 18 nm); material 3 made from 60% α-Si_3N_4 powder 40% β-Si_3N_4 powder (the β-powder contain 80% β-Si_3N_4; crystallite size 55 nm); material 4 produced from β-Si_3N_4 powder (crystallite size 18 nm) (according to [62]; corrected device function of the XRD-equipment).

the starting powder must be in the range >0.5–1%. This is the case for commercially available α-Si_3N_4 powders. In [39], it was found that at normal heating rates the nitrogen content in the oxynitride liquid during heating is lower than at low heating rates and low additive contents. This means that for normal heating rates, the liquid phase is undersaturated with respect to β-Si_3N_4 and no nucleation can take place.

These experimental data suggest that β-Si_3N_4 nucleation has no influence on the microstructure of silicon nitride materials under common sintering conditions and using common powders.

The consequence of this is that the different microstructures are a result of the different amounts and sizes of pre-existing β nuclei as well as the growing conditions.

The amount of growing β-Si_3N_4 nuclei depends on the β content in the starting powder, the crystallite size and the sintering conditions. In Fig. 9 the dependence of the equilibrium concentration in the oxynitride liquid during sintering on the

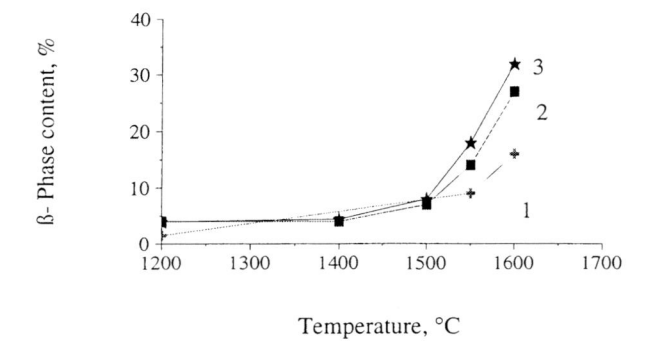

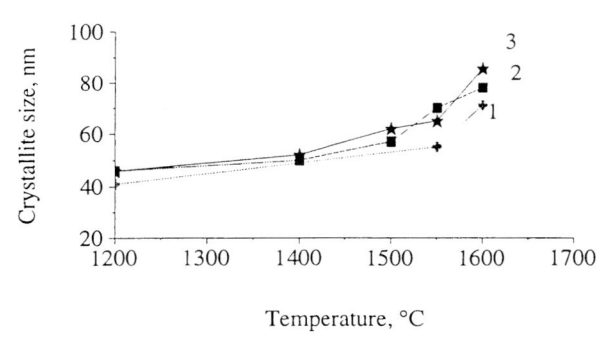

Figure 7. Change in crystallite size and phase content as a function of sintering temperature, amount of sintering additives and starting powder 1. diimid powder 5% Y_2O_3/Al_2O_3; **2.** gas phase synthesised powder 5% Y_2O_3/Al_2O_3; **3.** gas phase synthesised powder 11.6% Y_2O_3/Al_2O_3.

crystallite size of α- and β-Si_3N_4 is shown. Nuclei with a crystallite size lower than a critical value have a higher equilibrium concentration than the Si_3N_4 concentration in the liquid. These crystals dissolve in the liquid and cannot growth. During the α/β transformation under isothermal conditions, the critical size of β-Si_3N_4 is determined by the crystallite size of α-Si_3N_4 (r(crt1), see Fig. 9). During heating, especially for high heating rates, the oxynitride liquid is undersaturated (see Fig. 9, lines 2 and 3).

As a result of the undersaturation, the critical crystallite size becomes infinitely large and only the largest nuclei survive because they require more time to dissolve than smaller nuclei. The resulting microstructure of materials heated at high heating rates in the temperature range of 1200–1600°C is coarser than for materials heated at low heating rates [20, 29, 39].

After the α/β transformation, the critical crystallite size is determined by the β crystallite size distribution. The critical crystallite size increases with increasing sintering time and temperature, i.e. only the largest crystallites which were present in the starting powder determine the microstructure.

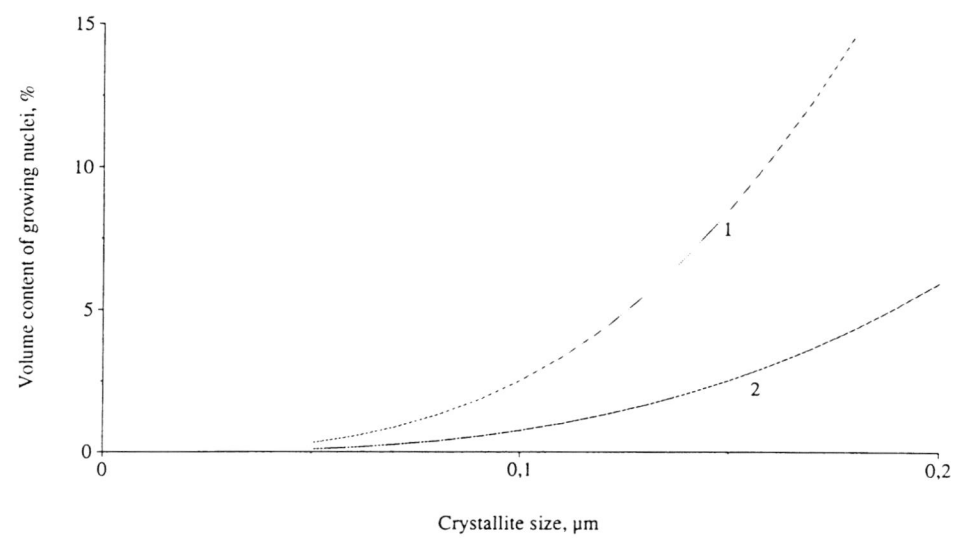

Figure 8. Calculated volume content of growing nuclei as a function of the size of the nuclei, if the resulting microstructure has a mean grain thickness of 0.2 μm (1) and 0.3 (2) and an aspect ratio of 5.

The needle-like grain shape seems to be kinetically determined and therefore not the equilibrium crystal shape [30, 38]. The estimated equilibrium aspect ratio (ratio of grain length to thickness) is 1.3, not as high as 20, as was found to be the case for the silicon nitride materials. The microstructural formation must therefore be explained on the basis of growth mechanisms.

The explanations for the microstructure given in the '80s were based on the assumption that needle-like grain growth can take place only during the α/β transformation. Later it could be shown that anisotropic grain growth could also take place after the α/β conversion [31, 32, 42] (see Fig. 10).

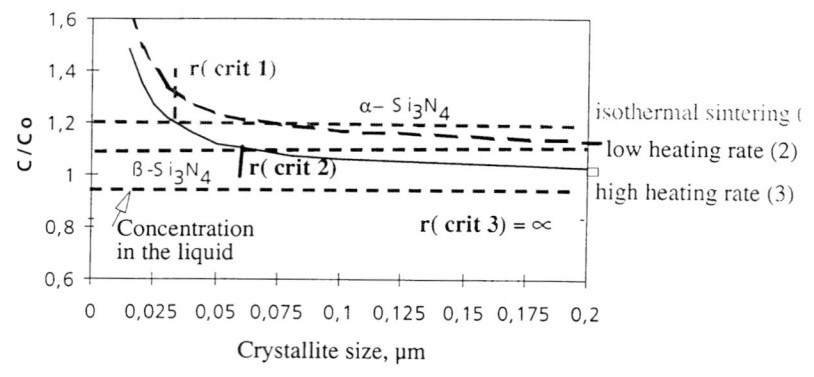

Figure 9. Dependence of the equilibrium concentration in the oxynitride liquid during sintering on the crystallite size of α- and β-Si_3N_4, according to [63].

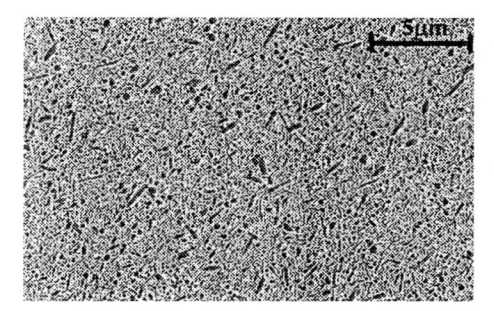

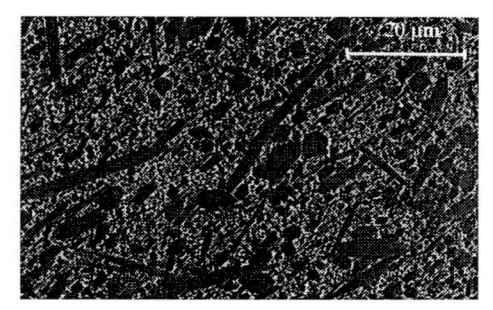

Figure 10. Microstructure of a material hot pressed at 1800°C (starting powder: fine β-Si$_3$N$_4$) and the same material after additionally heat treatment at 1900°C. It is observed that anisotropic grain growth takes place after α/β conversion.

Anisotropic grain growth occurs due to the different growth mechanisms in the different directions of the β-Si$_3$N$_4$ crystals. The (001) basal planes of the needle-like β-Si$_3$N$_4$ crystals are atomically rough; therefore, their growth is diffusion-controlled [42–44]. Transport occurs by two mechanisms: diffusion through the oxynitride liquid and surface diffusion in the (100) planes [44]. The (100) prismatic planes are atomically flat and therefore the growth is interface-controlled. The growth rate of the (001) plane has been assigned values of 30 [38] and greater than 27 [35]; thus, it is at least ten [42] times higher than the growth rate of the (100) plane. The first value seems to be the more exact one because it was determined by growth observation on crystals in oxynitride glasses. The different growth mechanisms are supported by TEM observations [43, 44], theoretical calculations [38] and by the observed crystal habits [45–47]. The relationship between growth rate and oversaturation is shown schematically in Fig. 11. For a high degree of oversaturation, the (100) planes become atomically rougher, causing the growth rates to approach the rate of the other planes, the growth to be isotropic and the grain morphology to be equiaxed. At the lower saturations found in common sintering conditions, extended anisotropic grain growth occurs. The boundary between the two regimes shifts to lower oversaturations with increasing temperature.

For a given crystal, the oversaturation depends on the size and is different for the different crystal planes. Using the differences in surface energy calculated by the broken bond method [38], the oversaturation as a function of grain thickness is given in Fig. 12. The equilibrium concentration above the basal planes depends only on the thickness of the grains, whereas the equilibrium concentration above the (100) planes depends on both the thickness and the length of the grains. Using these data and the dependencies of the growth rate on the oversaturation, it can be shown that the largest crystals with the highest aspect ratios grow anisotropically, whereas the small crystals with a low aspect ratio dissolve. Thin crystals with a high aspect ratio grow stably whereas crystals of the same thickness but a low aspect ratio dissolve. This behaviour was also determined experimentally [30, 48].

At first these results and the growth mechanisms discussed seem to be in contradiction to the results of the works from the 1980s showing a reduction in the aspect

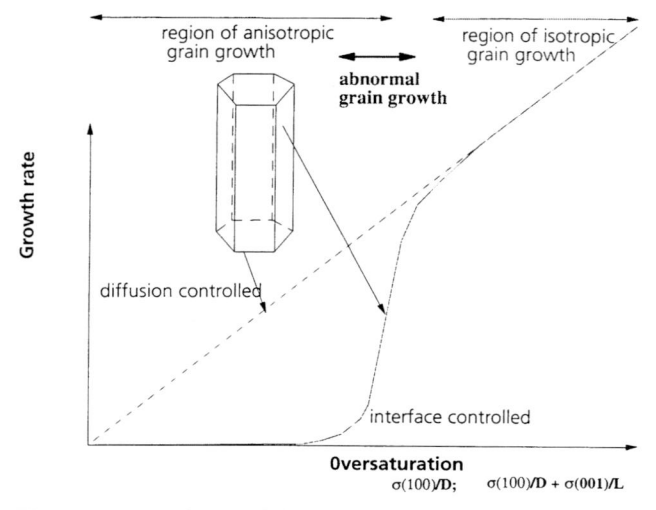

Figure 11. Growth rate of the basal plane (diffusion-controlled) and the prismatic planes (interface-controlled) as a function of oversaturation in the oxinitride liquid. The oversaturation after the α/β-transformation depends on the surface energy (σ), the grain thickness (2D) and the length of the needle like grains (2L).

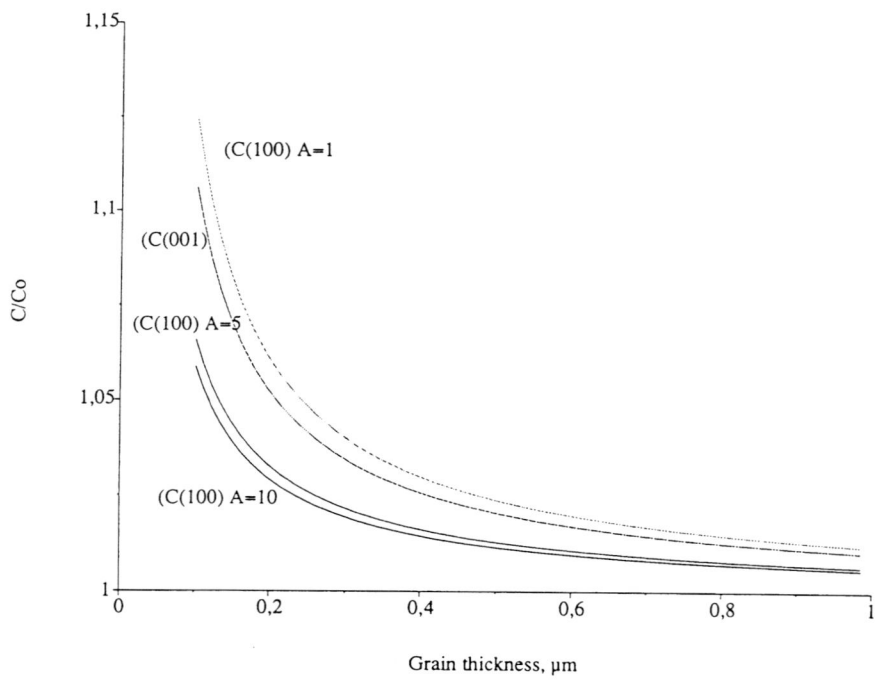

Figure 12. Calculated dependence of the oversaturation of the basal (001) plane and the prismatic (100) planes on the thickness of the grains, according to [30] using data [38, 63] (A = aspect ratio).

ratio with increasing β-Si$_3$N$_4$ content and at longer soaking times. These differences can be explained by taking the crystallite size of the β-Si$_3$N$_4$ in the starting powder into consideration. The powders used for these investigations were produced by direct nitridation of silicon. The higher β content of these powders was a result of the faster nitridation and larger crystallite size. The larger crystallite size has two consequences: 1. the oversaturation, which is inversely proportional to the grain size, decreases and consequently the grain growth rate also decreases; and 2. the thicker crystallites must undergo more growth to become needle-like. Both processes lead to a reduced aspect ratio. Additionally, the powders had a wide grain size distribution. With increasing sintering time, the critical crystallite size (see Fig. 9) increases and only the largest crystals survive. However, the largest crystallites had lower aspect ratios. The aspect ratio is the result of two competing processes: anisotropic grain growth (preferred growth in the direction of the needle), and the dissolution of small elongated grains. When the second process is the dominating process, the overall aspect ratio decreases.

An additional effect influencing the different microstructures which are formed is connected with the different grain sizes of the materials. The oversaturation depends on the mean grain size. With increasing grain size, the curve characterising the dependence of the oversaturation on grain size is flatter (see Fig. 9); for a given mean grain size, the oversaturation cannot compensate for the difference in surface energy between the (001) and (100) planes. As a result, anisotropic grain growth cannot take place and the grain shape approaches the more equiaxed equilibrium grain shape. Based on the investigations of microstructural formation, it can be concluded that these processes are only significant for mean grain thicknesses of 0.7–1.5 μm or larger. The materials investigated in the 1970s and early 1980s were characterised by mean grain thicknesses of 0.7–2 μm, whereas today's materials are based on finer powders having normally finer grain sizes.

Recent investigations [49–52] show that needle-like microstructures can be achieved from β-Si$_3$N$_4$ powders when the crystallite size of the starting powder is small enough or the sintering temperature and soaking time are large enough. Materials with high fracture toughnesses of up to 10–12 MPam$^{1/2}$ [48] are observed. Low-cost powders with higher β-Si$_3$N$_4$ contents, produced by special nitridation processes which allow the exothermic nitridation to be controlled [53], have relatively small crystallite sizes. Materials with strengths higher than 800 MPa and fracture toughnesses similar to materials based on α-Si$_3$N$_4$ powders can be produced from these powders [49].

At high sintering temperatures, abnormal grain growth can take place [35, 54, 55]. The reason for this is the presence of large β-Si$_3$N$_4$ nuclei (seeds) which can exist in the starting powder, formed during densification [35, 56, 62] or added as seeds [57–59]. The formation of large seeds during sintering can be intensified by low heating rates or dwell times at temperatures higher than 1500°C but lower than the temperature at which rapid densification occurs [56]. The amount of additional seeds needed to produce a reinforced material ranges from 1–5 or 10 vol.%. The seeds are usually between 10 and 100 times larger than the nuclei in the starting powder [56–58, 60, 61]. The addition of larger seeds is intended to increase the fracture toughness without drastic decreasing the strength (see Section 6.5.1).

The mechanism of accelerated grain growth can be explained by adapting the ideas developed for metals and hard materials [45–46]. According to [45], abnormal grain growth can take place under sintering conditions when the grain growth is interface-controlled (two-dimensional nuclei on an atomically flat surface) and the degree of supersaturation is high, sufficient to realise accelerated growth. The region of necessary oversaturation is given in Fig. 11. This grain growth is anisotropic and leads to the formation of large β-Si_3N_4 needles. With increasing temperature, the probability of formation of two-dimensional nuclei on the (100) surface (rate determine step of the growth) at a given oversaturation increases. As a result, the accelerated grain growth is more pronounced at higher sintering temperatures than at lower temperatures.

The oversaturation above a large nucleus depends on the mean grain size of the matrix (see Fig. 12); therefore, abnormal grain growth occurs to a greater extent in materials with a fine matrix.

When the density of large nuclei is too high, the mean oversaturation is reduced and no abnormal grain growth takes place. Thus the addition of more than 10–15 vol.% of large β-Si_3N_4 seeds to a starting α-Si_3N_4 powder does not result in a bimodal microstructure but in a coarse microstructure consisting of large grains with low aspect ratios.

If the seeds added have the same size as the crystallites in the starting powder, the resulting microstructure consists of fine β-Si_3N_4 grains. If the size is smaller than in the starting powder, there is no change in the microstructure.

The microstructural changes resulting from the addition of β-Si_3N_4 seeds are shown schematically in Fig. 13. The microstructural formation depends on the

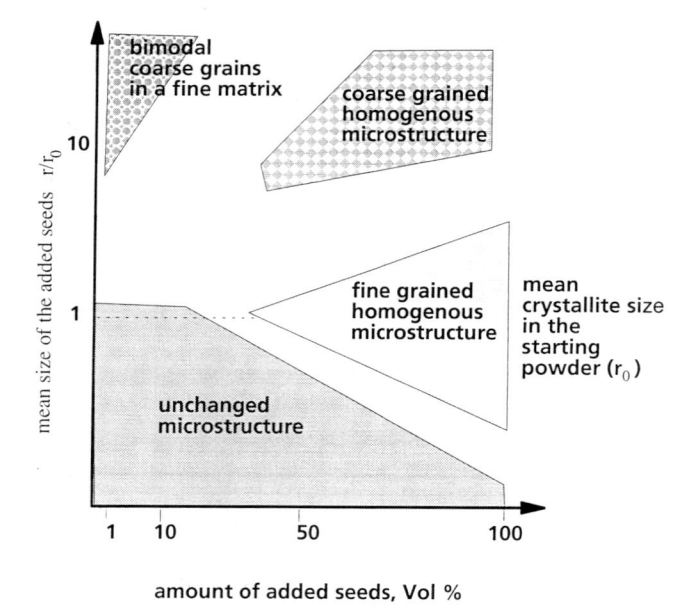

Figure 13. Microstructural changes resulting from the addition of β-Si_3N_4 seeds (schematically).

density of β nuclei. The density is proportional to the ratio of amount of β-Si_3N_4 to the mean crystallite size. Fig. 13 is based on the crystallite size of the common α-Si_3N_4 powders, which lies in the range of 50 to 100 nm. A change from one type of microstructure to an other can be achieved by changing the sintering conditions. To example a fine grained homogeneous microstructure can be changed in a bimodal microstructure by high temperature sintering (Fig. 10).

The scheme shown in Fig. 13 can be used for the design of the microstructure in β-Si_3N_4. The use of β powders with small grain size distributions allows the microstructure of the β-Si_3N_4 materials to be controlled more precisely. Presently such fine β-Si_3N_4 powders are only available in limited quantities and are not widely used [63]. By hot pressing or using special shaping technologies (tape casting), an anisotropic orientation of the needle-like grains can be achieved. This results in anisotropy of the fracture toughness, strength and thermal conductivity.

The microstructure, i.e. grain thickness and aspect ratio, depends on the type and amount of sintering additives used. The influence of additive type on the morphology was determined from a large number of investigations; only some important tendencies can be given as a result of these investigations. With increasing Al_2O_3 content and constant overall additive content, the Si_3N_4 grains become finer and more equiaxed. [24, 26–27, 64]. At the same time, partial incorporation of Al in the Si_3N_4 grains occurs. The incorporation of Al in the grains can be expressed by the reaction:

$$0.5z\, Al_2O_3 + (2 - 0.25z)Si_3N_4 \Leftrightarrow Si_{6-z}Al_zN_{8-z}O_z + 0.25z\, SiO_2$$

This process leads to a reduction of the Al_2O_3 content in the grain-boundary phase and to an increase in the SiO_2 content; i.e. Al_2O_3 functions as a buffer of the grain-boundary phase composition. The more SiO_2 evaporates during sintering, the more Al dissolves in the silicon nitride grains and reproduces SiO_2 [64]. The incorporation takes place according to the local equilibrium in the liquid. The diffusion in the silicon nitride grains is slow. Therefore Al gradients in the silicon nitride grains can be observed [64, 67].

The incorporation of Al in the silicon nitride crystal lattice leads to a reduction of the thermal conductivity of the materials [65, 66]. This means that the Al_2O_3/AlN content in the sintering additives can be used for the tailoring of the thermal conductivity; values between 10 and 150 W/mK are possible.

Investigations of the growth of the β-Si_3N_4 grains in oxynitride liquids with different rare-earth additives show that with decreasing melt viscosity (Yb to La), the aspect ratio of the grains increases and the thickness of the grains decreases. This was found in materials with rare earths and alumina as sintering additives and in materials with only rare-earth additives [68–69]. Very fine, needle-like microstructures were obtained for Sc_2O_3-containing additives. [70]. CaO and MgO were found to accelerate the densification and the anisotropic grain growth [71, 72], but were found to be detrimental to the high-temperature properties. The use of rare-earth additives or mixtures with MgO results in the formation of microstructures consisting of grains with high aspect ratios [1].

6.4.2 Microstructural development of α'-SiALON materials

α'-SiALONs are formed during liquid-phase sintering by the reaction of Si_3N_4, AlN, Al_2O_3 and an appropriate cation which can enter the structure of α'-SiALON ($R_xSi_{12-n-m}Al_{m+n}N_{16-n}O_n$). The amount of liquid available for densification is quickly reduced due to the formation of α-SiAlON solid solutions (Fig. 14). This reaction starts during heating at temperatures above 1450°C

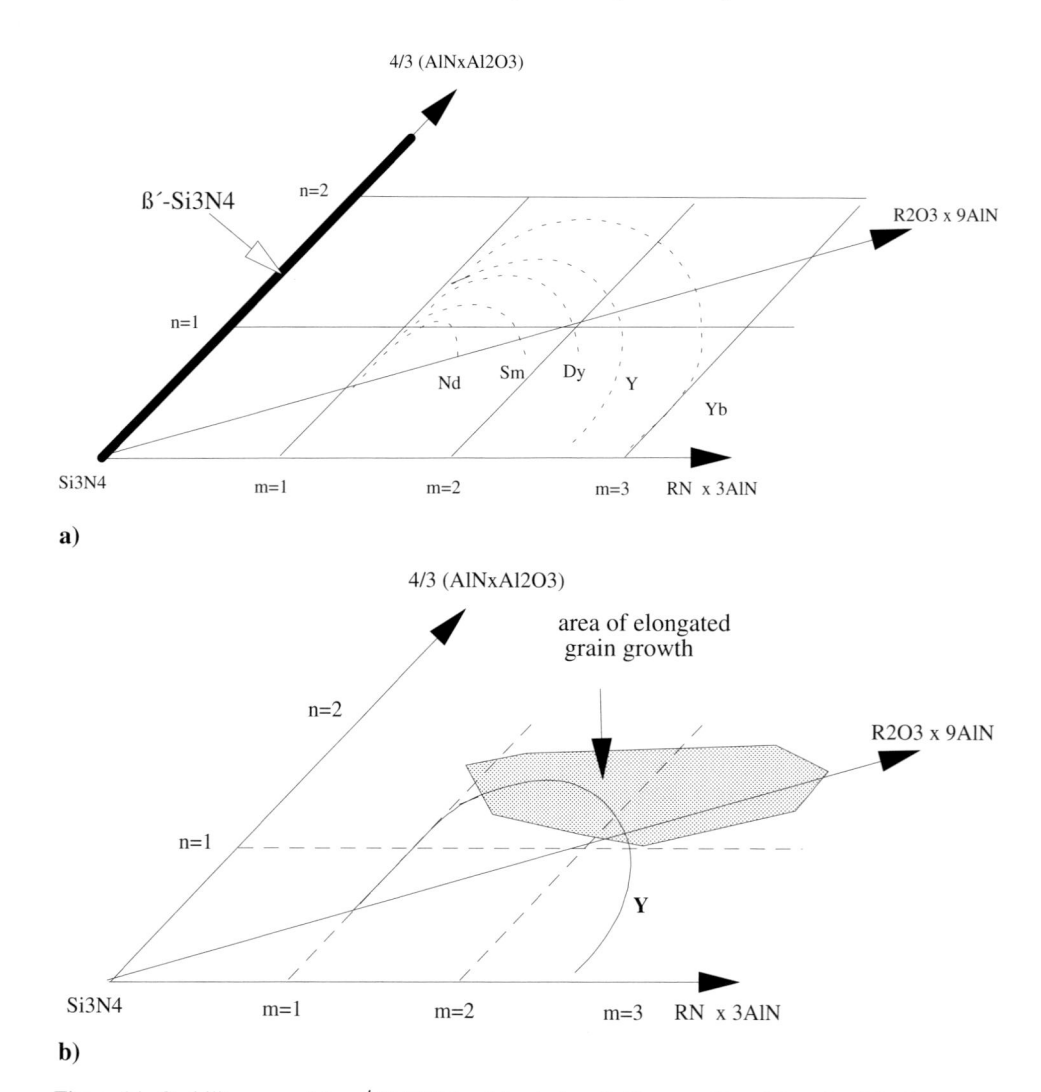

Figure 14. Stability area of the α'-SiAlON solid solution $R_xSi_{12-n-m}Al_{n+m}N_{16-n}O_n$ a) in the systems with R = Nd, Sm, Dy, Y and Yb at 1800–1900°C (after [14]) and b) in the area where elongated grain growth takes place in the α'-SiAlON/β'-SiAlON-plane (after [14, 15, 77, 79]).

[73, 74]. When the α'-SiAlON formation is completed, no more liquid exists for densification. This is the reason why pure α'-SiAlON materials with low degrees of substitution n and m are difficult to densify to full density. Better densification can be realised at higher n and m values (Fig. 14) [75–77] or in composite α'/β'-SiAlON materials [78].

The resultant materials had microstructures consisting of equiaxed grains, leading to reduced fracture toughness and strength. This was a reason for the limited use of these materials [1]. In the past few years materials with elongated grains have been produced by different research groups. The materials with elongated α-SiAlON-grains have relatively high n and m values (greater than 1.2) (Fig. 14) or mixed cations (rare earth's and Sr or Ca [14, 75]). In early studies it was suggested that elongated grain growth occurs only when β-Si_3N_4 powder is used [76]. However, later investigations have shown that the use of α-Si_3N_4 powder also leads to elongated grain growth [75, 14, 15, 79]. The reason for the anisotropic grain growth of α'-SiAlON is not completely clear because it was not investigated in the same manner as the β-Si_3N_4 grain growth. The α'-SiAlON materials with low n and m values and equiaxed grains possess a low grain size. This is the consequence of the fast disappearance of the liquid. In contrary, the microstructures of the α'-SiAlON materials with elongated grains show a coarser microstructure; i.e. more intensive grain growth takes place. This must be connected with the amount of liquid existing during sintering, which is higher when higher substitution levels of the α'-SiAlONs are used. Recently it could be shown that materials with elongated α'-SiAlON and low n and m values can be produced by gas pressure sintering using an appropriate liquid phase (see Fig. 15) [80]. This suggests, that the needle like grain shape is governed by the grain growth kinetics.

The stability of α'-SiAlON at elevated temperatures (1100–1500°C) depends on the nature of the cations R entering the α'-SiAlON structure. For the trivalent rare-earth cations (i.e. La–Yb; Y), the stability increases with decreasing atomic radius [14, 81]; i.e. the temperature below which a phase transformation $\alpha' \rightarrow \beta'$ can thermodynamically take place decreases. The stability also increases with increasing substitution level. Thermodynamic stabilisation of the α'-SiAlON with respect to the β'-SiAlON can be achieved by the use of mixed cations [14].

Besides the nature of the cation which is incorporated in the α'-SiAlON, the rate of α'-β'-transformation depends on the existence of β'-SiAlON nuclei and the amount and viscosity of the amorphous grain-boundary phase. The transformation between α'-SiAlON and β'-SiAlON is reconstructive and must occur via a liquid in which one phase can dissolve and the product precipitate. Therefore the transformation can be retarded by the absence of a glassy phase or by a very high viscosity [14, 81].

Better understanding of the microstructural formation of the α'-SiAlON materials in the future will be necessary for the improvement of the mechanical properties, especially fracture toughness and strength, and the wider application of this group of materials, having higher hardnesses than the β-Si_3N_4 and β-SiAlON materials (see Fig. 16).

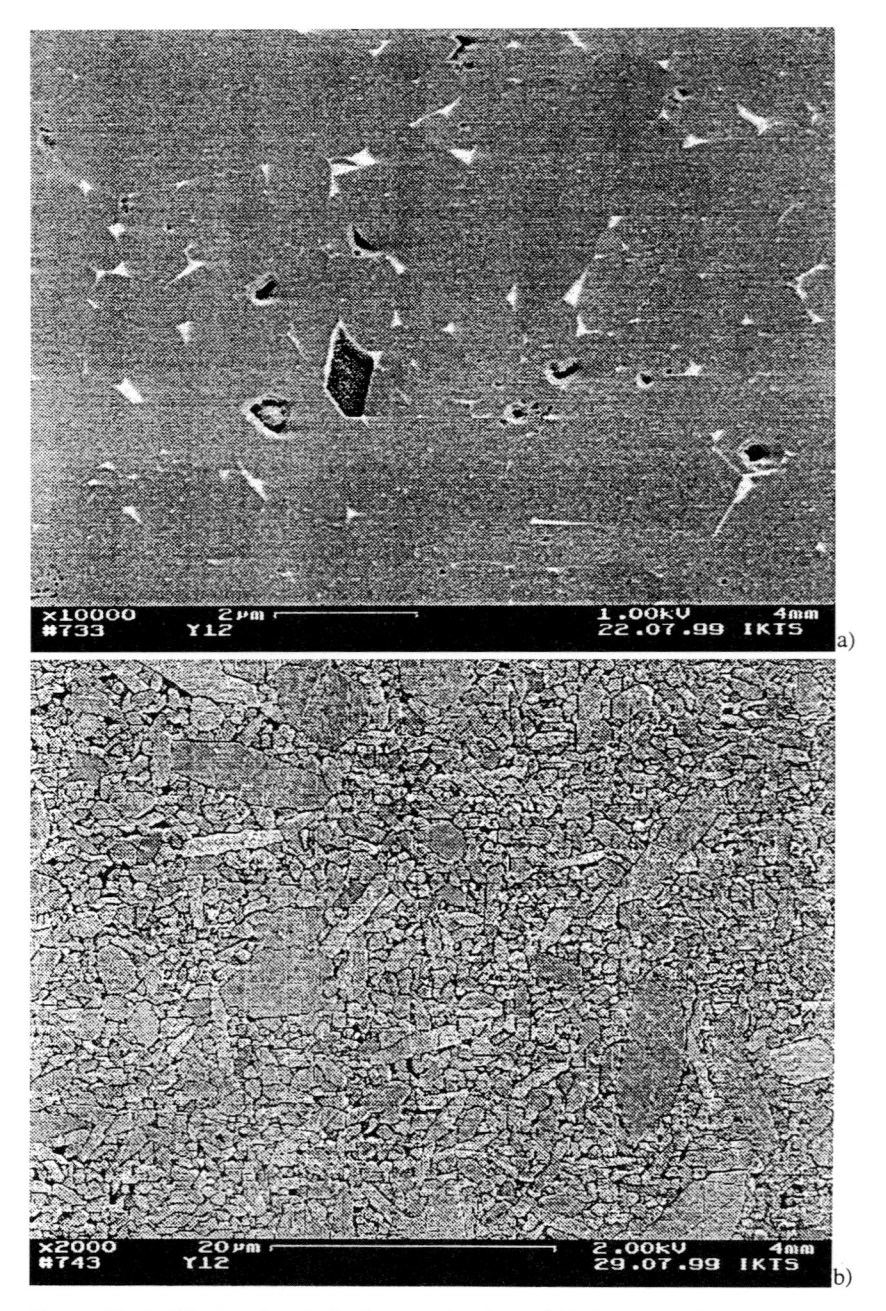

Figure 15. Polished and etched micrograph of an α'-SiAlON material with $n, m < 1$; **a)** plasma-chemically etched material (dark area: β-Si_3N_4; grey area: α'-SiAlON; bright area: grain-boundary phase); **b)** chemically etched material.

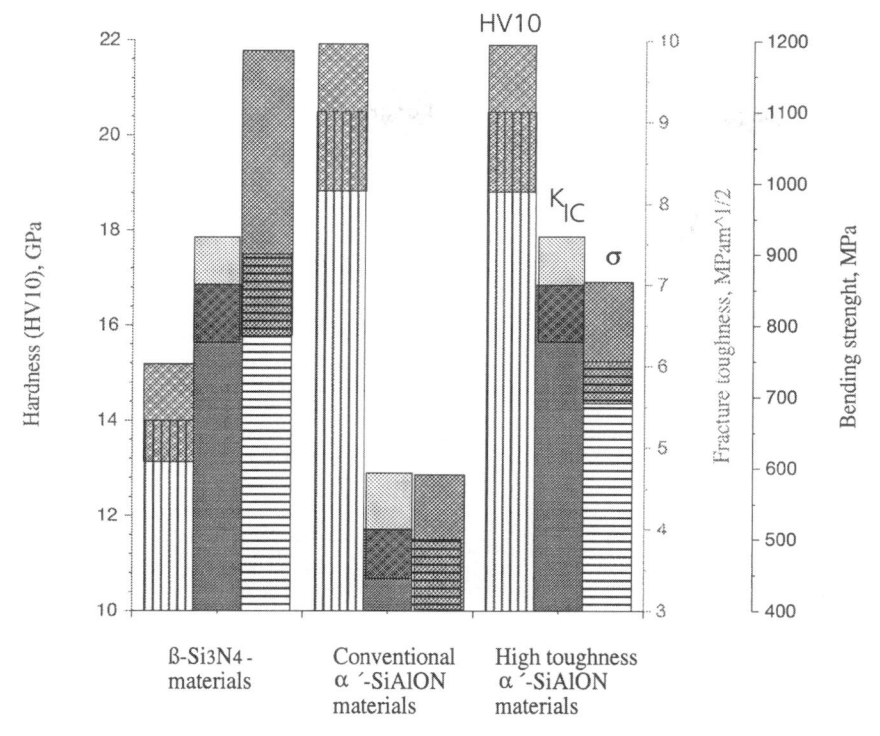

Figure 16. Comparison of the properties of the conventional α'-SiAlON materials with β-Si$_3$N$_4$ materials and with high toughness α'-SiAlON materials with elongated grains [80].

6.5 Properties of Si$_3$N$_4$ Materials

6.5.1 Mechanical properties at room temperatures

The strength of commercial silicon nitride materials lies in the range of 800 to 1400 MPa. The strength-determining defects are usually pores, inclusions such as iron silicides or agglomerates of glasses especially for low medium strengths. Besides these defects, large, elongated grains can be strength-determining defects for higher strengths or in specially developed, high-fracture-toughness materials. For strength levels higher than 1000 MPa, special surface finishing of the samples is necessary because surface defects play a decisive rule in determining the strength.

The strength of a brittle material is proportional to the fracture toughness and indirectly proportional to the square root of the defect size. The defect size of the materials can be reduced by optimised processing. [55, 82]. The highest measured mean three-point-bending strength for silicon nitride was 2000 MPa [85]. This corresponds to a defect size of about 5 μm. Materials with a strength level of 1400 to 1500 MPa usually have a defect size of 10 μm [55, 83], i.e. these materials have

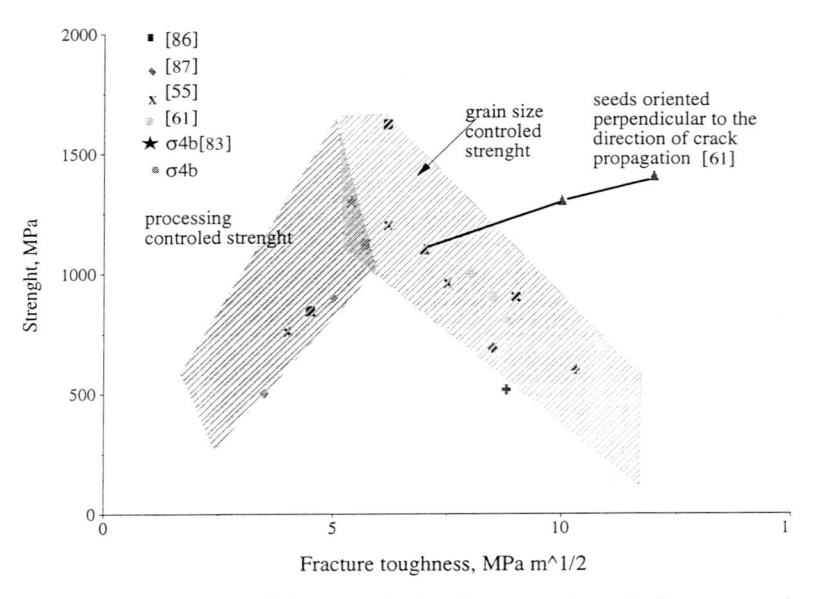

Figure 17. Dependence of the tree point bending strength on the fracture toughness.

to have a grain size lower than that of the defect, because larger grains can act as strength-determining defects. This means that high-strength materials must be fine-grained. On the other hand, the fracture toughness increases with the square root of the grain size [36, 69, 87, 88]. These two adverse dependencies produce the relationship between strength and fracture toughness shown in Fig. 17. For low fracture toughness values, flaws other than the largest grains are the strength-limiting defects. In this region of K_{IC}, further improvement of the strength is possible. For high K_{IC} values, the strength is determined by large elongated grains. Thus the structure itself determines the strength, not processing-dependent defects. Materials with large, needle-like grains in the matrix are usually called insitu-reinforced materials. Attempts to improve the strength of such materials by seeding a matrix with large, needle-like grains with a narrow size distribution are underway [57, 59, 60]. Results for tape-cast silicon nitride materials with large, oriented β-Si_3N_4 needles represent a possible approach [60] (Fig. 17) to increase the strenght and fracture toughness. However, such materials exhibit anisotropic properties, e.g. low strength in the direction of orientation of the needles.

Additionally, the insitu-reinforced silicon nitride materials exhibit an increased Weibull modulus [86]. This has two reasons: firstly, the concentration of the large, strength-limiting grains is so high that the defect size has a narrow size distribution; and secondly, these materials exhibit pronounced R-curve behaviour, which makes the material tolerant of larger cracks. The high thermal shock resistance of these materials also seems to be connected with this fact [30].

The subcritical crack growth rate in silicon nitride materials is relatively low in comparison to other ceramic materials. The growth exponent is in the range of

$n = 30–300$ for static loads and $n = 20–50$ for cyclic loading [90, 91, 97–100]. Up to now, no clear correlation between the microstructure and the growth exponent could be made. For cyclic loading it can be shown that materials with the lower grain size have the higher growth exponent, i.e. have a lower degradation of strength during loading and a higher lifetime at a given strength level [91, 88].

The fracture toughness depends strongly on the microstructure. Two main factors influence the fracture toughness: 1. the grain shape and size, and 2. the composition of the grain-boundary phase.

The high fracture toughness of silicon nitride materials in comparison to other ceramic materials is connected with toughening mechanisms which are similar to those in whisker-reinforced composite materials: grain bridging and pull-out, crack deflection and grain branching around large, elongated grains [88, 89, 94].

Due to these mechanisms, the fracture toughness increases with increasing volume fraction and square root of the mean grain thickness of the elongated grains (i.e. grains with aspect ratio >4) [37, 70, 88, 89].

Recent investigations show that fine-grained materials have a higher toughness than coarse-grained materials in the small crack region but lower toughness in the large crack region [82, 84, 90]. This can be important for the use of such materials under conditions in which high local stresses exist (e.g. for ball bearings). The dominant toughening mechanism depends on the grain thickness of the elongated grains. Elastic bridging and pull-out were observed for thin, needle-like grains (with thickness <1 µm). Crack deflection was mainly observed for thick, elongated grains (with thickness >1 µm), whereas frictional bridging was detected independently of the grain size [89].

The toughening mechanisms described above can only operate when the dominant fracture mode is intergranular. The ratio of transgranular to intergranular fracture depends on the relative strengths of the grain boundary and the grains. For a material with a high toughness, the grain boundary must be weak in comparison to the grains (see Fig. 18).

The strength of the grain boundary is connected with two different mechanisms: local residual stresses [92, 101] and special chemical interactions between the grain-boundary phase and the silicon nitride grains [93, 94].

The amorphous or partially crystallised grain-boundary phases generally have different thermal expansion coefficients than silicon nitride. When the thermal expansion coefficient of the grain boundary is higher than that of the silicon nitride, the grain-boundary phase is under tensile stress and the fraction of intergranular fracture increases. As a consequence the fracture toughness increases (see Fig. 19). Similarly, a material with a grain-boundary phase under compression (e.g. HIPSN without sintering additives) shows a low fracture toughness [102] due to a high percentage of transgranular fracture (see Fig. 18a).

The residual stresses can be influenced by the sintering cycle. This can be brought on by a change in the composition of the grain boundary (e.g. by evaporation of SiO_2) during sintering, by crystallisation of the grain boundary or by partial relaxation of the stresses (e.g. through slow cooling). These changes are usually outweighed by the grain size and shape or the starting composition of the material, which have a greater influence on the stress state and K_{IC}.

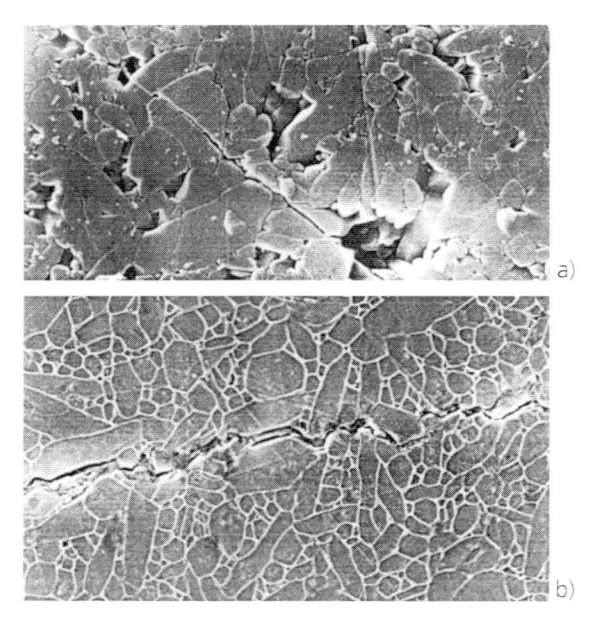

Figure 18a and b. Crack path in silicon nitride materials **a)** HIPed material (only SiO_2 sintering additive, mainly transgranular fracture; chemical etched), **b)** material with Y_2O_3/Al_2O_3 additives (mainly intergranular fracture, plasma etched).

Figure 19 shows the additional influence of the grain size (see curves 1–3). The composition of the grains also influences the fracture toughness by determining the special chemical interactions that occur between the grains and the grain boundaries. The formation of β-SiAlON layers on the Si_3N_4 grains was found to result in an increase in the amount of transgranular fracture and a decrease in the fracture toughness [93, 100]. This can be a reason for the lower fracture toughness of the β-SiAlONs with a high degree of substitution, in addition to the lower aspect ratio (see Fig. 18d).

The recently developed α-SiAlON materials have a microstructure consisting of relatively coarse, elongated grains (see Fig. 15) and show a high percentage of intergranular fracture; thus, these materials exhibit a high fracture toughness (see Fig. 16).

Summarising the fracture toughness data, it can be concluded that the fracture toughness of the silicon nitride materials depends strongly on the microstructure. For high fracture toughness values, elongated grains and a grain-boundary phase which promotes intergranular fracture are necessary.

The hardness of different silicon nitride materials is given in Fig. 20. Materials with a high $α-Si_3N_4$ content, i.e. materials with a high amount of metastable $α$-Si_3N_4 which did not transform during sintering (normally produced by hot pressing or HIP'ing [1, 59]), show a hardness which is as high as that of the $α'$-SiAlON materials.

Analysis of the hardness values of different $β-Si_3N_4$ materials reveals that the hardness increases with decreasing amount of grain-boundary phase and decreasing

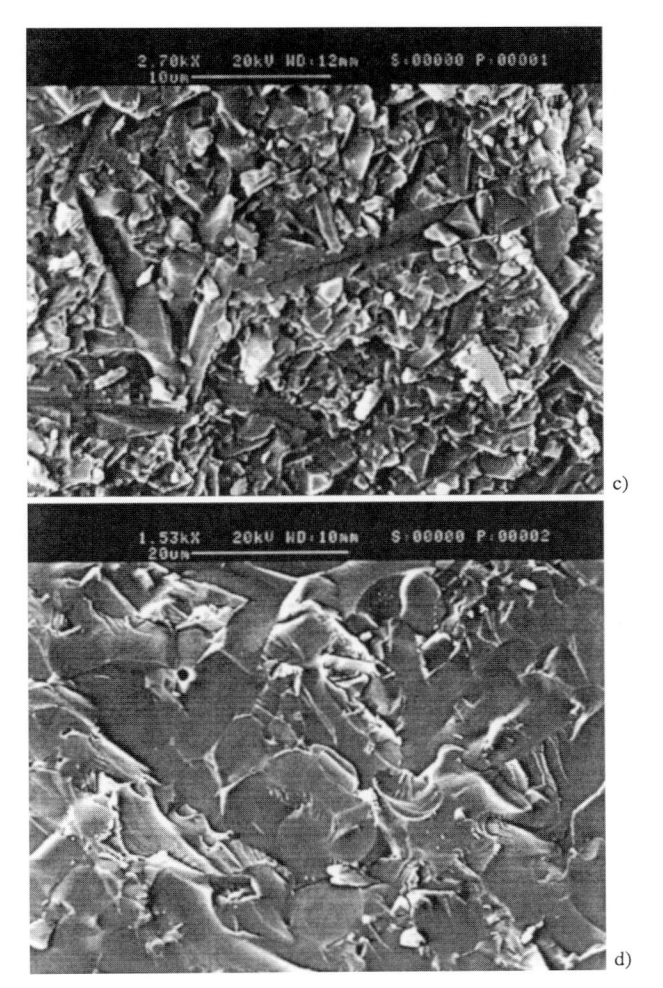

Figure 18c and d. Fracture surfaces of silicon nitride materials **c)** fracture surface of material with Y_2O_3/Al_2O_3 additives (mainly intergranular fracture), **d)** fracture surface of a β'-SiAlON material with a high degree of substitution z ($z = 4$) (mainly transgranular fracture).

grain size; i.e. the dependence of hardness on the microstructure is opposite to that of the fracture toughness.

Usually the hardness values of β-Si_3N_4 materials (HV10) are in the range of 12 GPa (coarse grain size, high additive content, residual porosity) to 16 GPa (fine grain size, low additive content). The hardness of the mixed α/β-SiAlONs change lineary with the phase content [131].

The hardness at elevated temperatures depends additionally on the softening of the glassy grain-boundary phase. Materials with MgO/Al_2O_3 as sintering additives show a faster degradation of the hardness than materials with more refractory grain-boundary phases (Fig. 20).

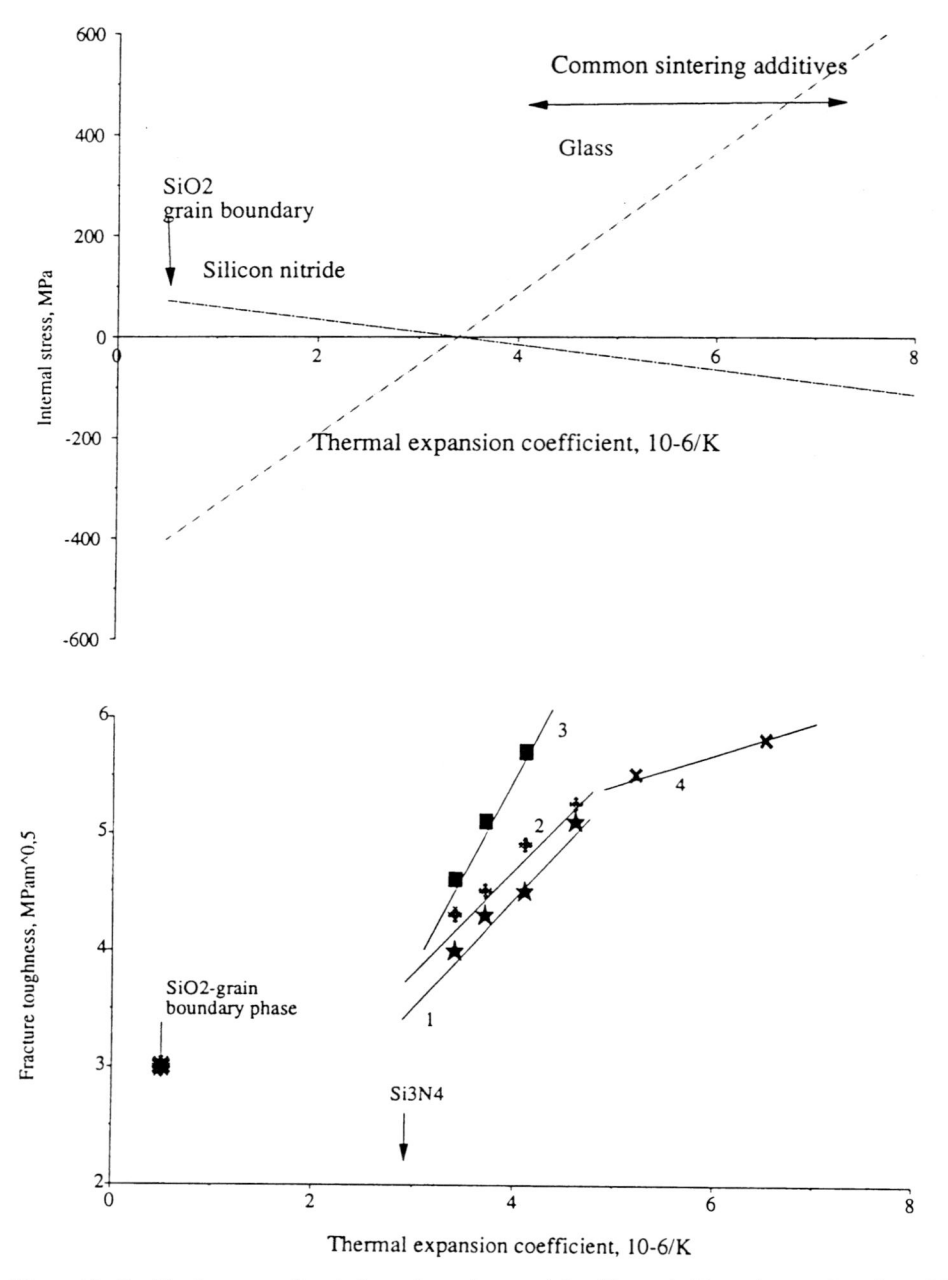

Figure 19. Residual stresses of grain-boundary phase and the silicon nitride grains as a function of the thermal expansion coefficient of the grain-boundary phase (calculated after [92]); used constants for Si_3N_4: $E = 320$, $\nu = 0.27$, $\alpha = 3.39 \times 10^{-6}$ K; and for the glass: $E = 140$, $\nu = 0.29$, volume fraction of glass $= 0.15$; a) and dependence of the fracture toughness on the thermal expansion coefficients (data 1–3 after [92]) **1.** mean thickness of the grains 0.55; **2.** 1–1.3; **3.** 1.5–1.7 (data 4 after [64]).

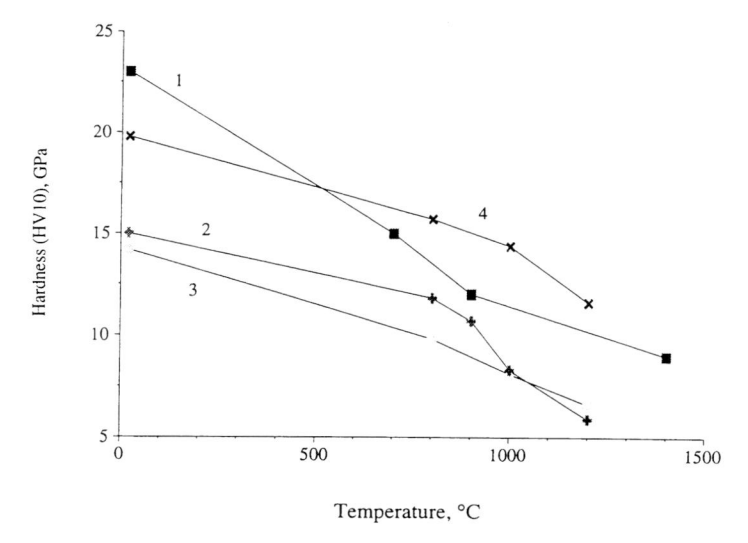

Figure 20. Hardness of different Si_3N_4 materials; **1.** additive free material with remaining α-Si_3N_4 [59], **2.** hot pressed with 10% CeO_2 and 6.6% AlN contain 20% α-SiAlON [84]; **3.** sintered β-silicon nitride material with 3% Y_2O_3 and 2% Al_2O_3; **4.** α'-SiAlON.

6.5.2 High-temperature properties of silicon nitride materials

Silicon nitride is among the most promising materials for high-temperature applications because of its combination of excellent mechanical properties at room and elevated temperatures, oxidation resistance, low coefficient of thermal expansion and low density in comparison to refractory metals. However, for the majority of Si_3N_4 materials these properties can only be obtained by a purposeful tailoring of chemical composition and microstructure. Special emphasis needs to be placed on the intergranular grain-boundary phase which is formed during the sintering process.

As already mentioned, Si_3N_4 materials are usually densified by the addition of sintering aids. They form a glassy silicate phase with the silica always present as an impurity in the Si_3N_4 raw powder; this silicate phase remains as an amorphous or crystalline grain-boundary phase between the Si_3N_4 grains after sintering.

While the properties of the Si_3N_4 grains do not change up to temperatures of 1600°C, as a consequence of the high degree of strong covalent bonding in Si_3N_4, the grain-boundary phase starts to become weak at elevated temperatures. Depending on the amount, composition and condition of the grain-boundary phase between the silicon nitride grains, various processes (diffusion, creep, slow crack growth, oxidation, corrosion) occur at elevated temperatures. The consequence of these high-temperature processes is the generation of a new defect population which determines the failure behaviour and ultimately limits the lifetime of the silicon nitride material. The extent to which these processes occurs is mainly influenced by the softening point and viscosity of the amorphous grain-boundary phase.

Table 4. High-temperature properties of various Si_3N_4 materials ($\sigma_{1400°C}$ – bending strength at 1400°C; $\dot{\varepsilon}$ – creep rate at 1400°C, 200 MPa; Δm_{ox} – weighted gain during oxidation at 1500°C; σ_{ox} – residual strength after oxidation).

	material	$\sigma_{1400°C}$/MPa	$\dot{\varepsilon}/h^{-1}$	$\Delta m_{ox}/mg/cm^{-2}$	σ_{ox}/MPa	σ_{ox}/σ_o
1	HIP-Si_3N_4	500	9×10^{-6}	0.7 (2500 h)	490	1
2	SN 5% Y_2O_3/3% Al_2O_3	450	–	4 (100 h)	< 200	< 0.2
3	SN 8% Y_2O_3/0.6% Al_2O_3	720	6×10^{-5}	3.7 (1000 h)	210	0.2
4	α'/β'-Sialon (Y)	380	5×10^{-5}	10 (100 h)	< 200	< 0.2
5	SN 8% Y_2O_3	730	2×10^{-5}	2.2 (2500 h)	340	0.4
6	Si_3N_4–SiC (8% Y_2O_3)	750	2×10^{-5}	3.3 (2500 h)	640	0.8
7	Si_3N_4–$MoSi_2$ (8% Y_2O_3)	730	2×10^{-5}	3.1 (2500 h)	630	0.8

In the following chapter fundamentals of the high-temperature behaviour of silicon nitrides and silicon nitride-based composites will be presented. Depending on the sintering additive used, silicon nitride materials with different high-temperature properties can be obtained. Some typical materials are chosen from the broad variety of Si_3N_4 ceramics that exist and summarized in Table 4.

The first Si_3N_4 ceramic (Table 4) is a material without additional sintering aids. The only liquid phase appearing during sintering is the silica from the high-purity Si_3N_4 powder. Because of the small amount and the high softening point of the grain-boundary phase, these materials can only be densified by encapsulation hot isostatic pressing (HIP). The materials exhibit excellent behaviour at elevated temperatures as the consequence of a clean grain-boundary phase consisting only of silica [102]. The oxidation resistance is used to demonstrate the excellent high-temperature properties of this material. Figure 21 shows the microstructure of a HIP'ed Si_3N_4 without additives after 2500 h oxidation at 1500°C. The oxidation behaviour of silicon-based nonoxide ceramics is mainly influenced by the protective layer which is formed at the surface of the material as a result of the oxidation

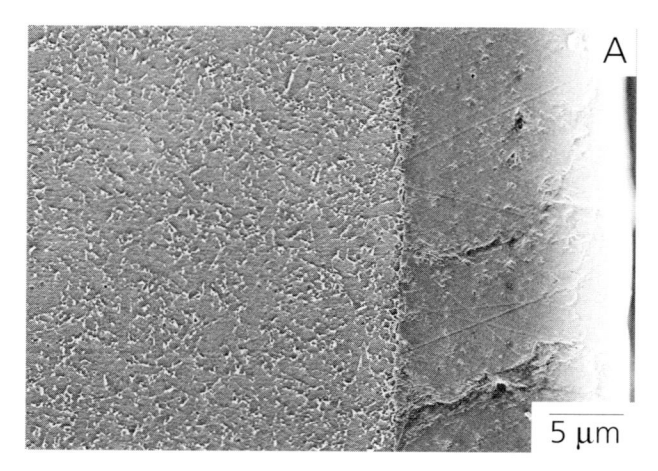

Figure 21. SEM images of polished cross sections of the Si_3N_4 material 1 (Table 1) after 2500 h of oxidation treatment at 1500°C.

process and the ability of this layer to prevent oxygen from diffusing into the material. In this case the formation of a surface layer of pure silica causes the rate of oxygen diffusion into the material to be very low [103, 104].

A more refractory grain boundary can only be obtained by removing the silica from the grain boundaries and triple junctions. Each addition of only a small amount of impurities or sintering additives will weaken the material by producing a change in the chemistry of the grain-boundary phase [105–107]. The result is degradation of the high-temperature behaviour of the materials. Besides the costly fabrication, the main disadvantage of these material is their inadequate mechanical properties (especially toughness) at room temperature. The microstructure, which is mainly globular, and the strong bonding of the intergranular phase do not allow toughening mechanisms like crack deflection or bridging to be active.

With the addition of sintering aids these disadvantageous properties can be improved; however, the improved densification, the microstructure with elongated grains, and the improved toughness are the consequence of a changed grain-boundary phase. Depending on the kind of sintering additive used, a silicate phase will be formed during sintering and will remain as an amorphous or partially crystallized phase in the grain boundaries and triple junctions between the Si_3N_4 grains. Besides the chemical composition and the amount, special emphasis is placed on the crystallinity of the intergranular silicate phase. With a high amount of crystallized silicates in the grain boundary, the amorphous phase is minimized; this is found to be especially beneficial to the creep and slow crack growth behaviour of the Si_3N_4 materials.

Glass-forming and stabilizing sintering additives which, with the silica of the Si_3N_4 powder, form a silicate phase with a low softening point and viscosity such as MgO, Al_2O_3 or AlN are principally not useful in the fabrication of high-temperature silicon nitride materials. More refractory silicates that form an intergranular phase with a high crystalline content were obtained by using yttria or the rare-earth oxides as sintering additives. Examples for these materials are given in Table 4. When Al^{3+} ions are present, the softening point and viscosity of the grain boundary are lowered [1, 3], with the consequence of degradation of the mechanical properties at elevated temperatures. Figure 22 summarizes the creep curves of the alumina-containing silicon nitride, material 3 (Table 4), in comparison with the additive-free material and the material with yttria as a sintering aid. The HIP'ed material exhibits the highest creep resistance. Superior creep behaviour is also found for Si_3N_4 with Y_2O_3 as a sintering additive. This is the consequence of the high degree of crystallized grain-boundary phase having the apatite structure. The creep resistance of the Si_3N_4 materials is found to be lowered by an amount which depends on the amount of Al^{3+} ions present. With the addition of a small amount, the crystalline grain-boundary phase does not change significantly; however, a degradation of the creep resistance is observed, caused by the concentration of Al^{3+} ions in the residual amorphous grain-boundary phase. In the material with 5% Y_2O_3 and 3% Al_2O_3, the grain-boundary phase is found to be mainly amorphous, with a lowered softening point and viscosity. The result is a very poor high-temperature resistance: the material failed after only a few minutes during creep testing at 1400°C and 200 MPa.

A similar tendency is found by comparison of the oxidation behaviour of the materials. With the addition of the sintering aid the composition of the protective

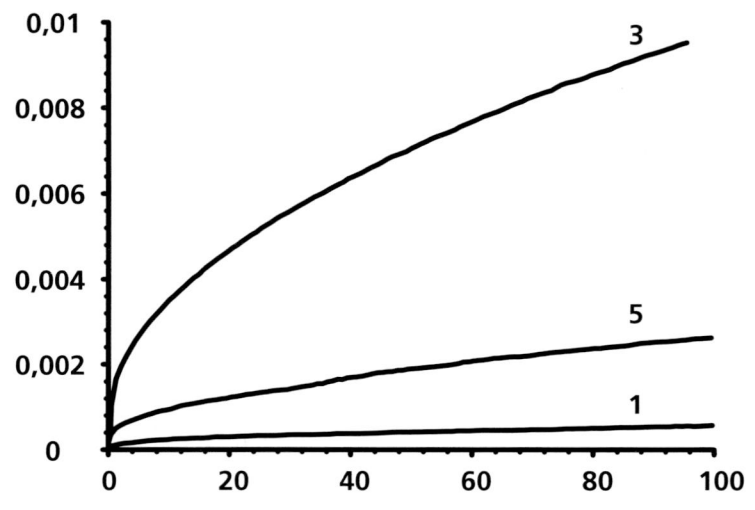

Figure 22. Comparison of the creep behaviour of the materials (1 in Table 4) HIP'ed Si_3N_4 without additives, (3 in Table 4) Si_3N_4 with 8% Y_2O_3/0.6% Al_2O_3 and (5 in Table 4) Si_3N_4 with 8% Y_2O_3 as sintering additives. The tests were conducted at 1400°C in air at 200 MPa.

oxidation layer formed during oxidation changes, with the consequence of a higher oxygen diffusion into the materials. The weight gain of the Si_3N_4 with Y_2O_3 is about three times higher than that of the HIP'ed Si_3N_4 without sintering aids, and that of the material with the small amount Al_2O_3 about one order of magnitude higher than that of the HIP'ed Si_3N_4 without sintering aids. Figure 23 shows the microstructure of the materials 3 (8% Y_2O_3 + 0.6% Al_2O_3) and 5 (8% Y_2O_3) after oxidation at 1500°C. Oxidation processes are observed in the bulk of both materials and lead finally to alteration of the microstructure. However, due to the addition of Al_2O_3, the oxidation of material 3 (in Table 4) occurs faster. Already after 1000 h oxidation the microstructure is found to be strongly degraded. The same damaging processes occur with the Y_2O_3-containing Si_3N_4 material; however, this material exhibits, due to its more refractory grain-boundary phase, a higher oxidation resistance (see

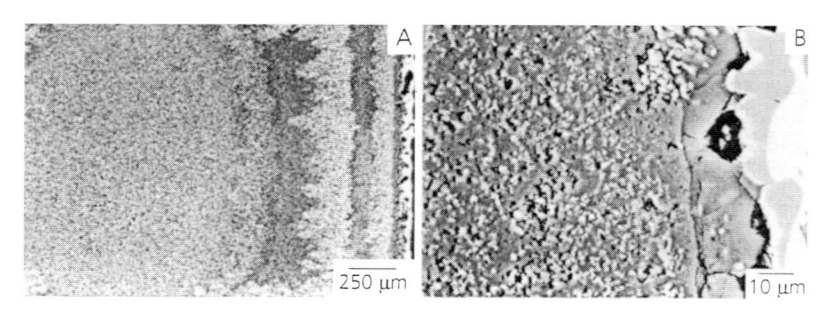

Figure 23. Microstructural damage in the surface region of the Si_3N_4 materials with Y_2O_3/Al_2O_3 (3 in Table 4) after 1000 h (A) and Y_2O_3 (5 in Table 4) after 2500 h (B) oxidation at 1500°C.

microstructure in Fig. 23 after 2500 h). As a result of the oxidation processes, degradation of the mechanical properties is observed; this is found to be dependent on the extent of oxidation as represented by the residual bending strength after oxidation (see Table 4).

Interesting materials with respect to high-temperature behaviour are mixed α'/β' SiAlON [3, 78]. Due to the ability of the α'-SiAlON grains to incorporate cations from the sintering aids into their lattice, it is possible to modify the grain boundary of the silicon nitride material. In a material with a high amount of α'-SiAlON, a skeleton from strong grain boundaries between the α'-SiAlON is formed due to the incorporation of the ions of the sintering aids into these grains, with the consequence of an improved creep resistance. The oxidation resistance cannot be improved by the increase in the α'-SiAlON content. The high amount of Al^{3+} ions in the material is the reason for the formation of a weak surface layer of alumosilicates during oxidation, allowing a high rate of diffusion of oxygen into the material [78].

One goal in recent studies in the field of high-temperature Si_3N_4 materials was the improvement of the behaviour at elevated temperatures by the formation of composites. Typical examples which have been reported in the literature are Si_3N_4–SiC micro- or nanocomposites or Si_3N_4–$MoSi_2$ composite materials. In the case of Si_3N_4–SiC nanocomposites, a structural synergism between the matrix Si_3N_4 and the nano-SiC was supposed to be responsible for the high level of the properties of these composites at elevated temperatures [109]. Recent studies, however, showed that the improvement of the mechanical properties at high temperatures was the consequence of a chemical modification of the composition of the grain-boundary phase and a significant improvement of the oxidation resistance [104, 108, 110, 111]. Similar behaviour was found for Si_3N_4–$MoSi_2$ composite materials [112]. After long-term oxidation tests at 1500°C, a considerably less damaged microstructure was observed. Figure 24 shows the comparison of the material 5 (Si_3N_4 with 8% Y_2O_3) with a Si_3N_4–$MoSi_2$ composite (material 7). Apart from the 10% $MoSi_2$ in the composite, both materials had the same composition. The microstructural development of the materials was found to be different. A significantly lower amount of damage was observed in the Si_3N_4–$MoSi_2$ composite, although the oxidation processes also occurred in the bulk of the composite.

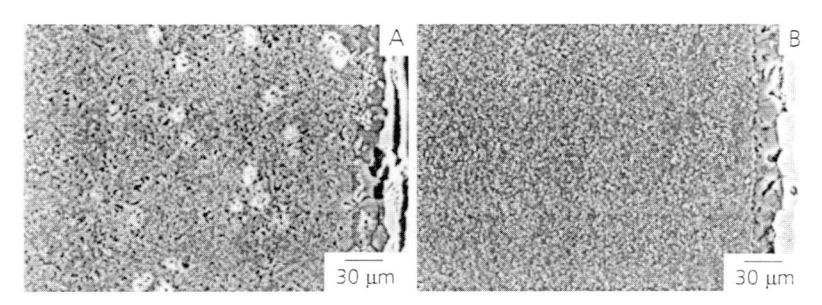

Figure 24. Comparison of the microstructure of (**A**) the Si_3N_4 material (5 in Table 4) and (**B**) the Si_3N_4–$MoSi_2$ composite (7 in Table 4) after 5000 h oxidation at 1500°C in air.

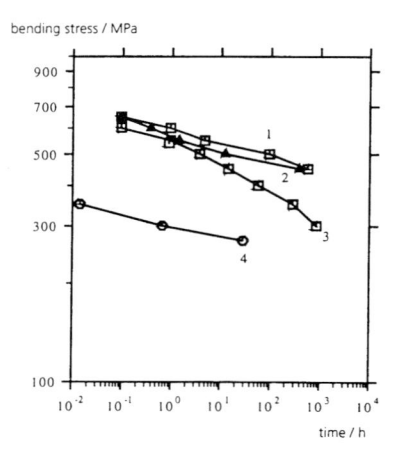

Figure 25. Comparison of time to failure behaviour of silicon based nonoxiode materials at 1400°C in bending (curve 1: material 7; curve 2: material 6; curve 3: material 5; curve 4: SSiC).

This behaviour was the consequence of different processes in the surface region of the materials, resulting in a changed oxidation mechanism in the composites. The important feature of the Si_3N_4–$MoSi_2$ and Si_3N_4–SiC composite was the increased formation of Si_2ON_2 instead of SiO_2, producing an additional layer between the oxidation surface and the bulk of the silicon nitride composite materials.

The significantly reduced defect size gives an idea of the potential of these composite materials in terms of stability and time-to-failure behaviour at elevated temperatures (Fig. 25). Principally, the amount, composition and condition of the grain-boundary phase are the key factors which must be considered for the successful development of Si_3N_4 materials for applications at elevated temperatures. Materials without sintering additives exhibit the highest stability at elevated temperatures due to the high refractory grain-boundary phase of pure SiO_2. The addition of sintering aids ultimately leads to a degradation of the properties at elevated temperatures. However, by a purposeful design of the grain-boundary phase and the formation of high temperature stable composites which are stable at high temperatures, it is possible to obtain silicon nitride materials which can be used up to temperatures of 1500°C.

6.5.3 Wear resistance of Si_3N_4 materials

The excellent wear properties of silicon nitride materials are the basis for many applications of these materials, such as ceramic or hybrid rolling bearings, cutting tools, wear parts in automotive engines (valves, cam rollers, valve plates for common rail systems), and parts for metal shaping.

Wear is a property of the system, not only of the material. Therefore a few special applications will be presented to show the influence of the microstructure on the wear properties.

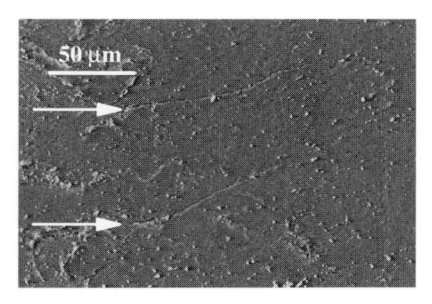

Figure 26. Typical cracks on the track of ball bearings (dry running under 1.75 GPa Hertzian pressure, 18 10^{-6} turns).

Silicon nitride is the preferred ceramic material for use in ceramic and hybrid ball bearings due to its high strength, fracture toughness and high resistance to subcritical crack growth. The advantages of ceramic ball bearings in comparison to metal ball bearings are: reduced wear, higher hardness, better temperature and corrosion resistance, lower specific weight and higher electrical resistance. This combination of properties makes lubrication of such bearings either almost or completely unnecessary [2, 113, 115, 117]. The field of application of such bearings is fast growing. The bearings show a good performance and a high reliability. To improve the behaviour a development of silicon nitride materials for ball bearing was started in the last few years [113–116, 118]. The main failure mechanism of ceramic ball bearings are connected with local and temporal changing high tensile and compressive stresses during operation. These local and temporally changing stresses result in the formation of cracks perpendicular to the rolling direction of the balls in the ball bearing (see Fig. 26) [114].

The lifetime and load capacity of a ball bearing can be increased by optimisation of the geometry of the track [113–115, 118], by reduction of the material damage due to cyclic Hertzian contact, and by reduction of the friction coefficient [114, 116].

The influence of the microstructure on the Hertzian contact damage was investigated in detail [119, 120]. Coarse-grained silicon nitride materials are subjected to an accelerated fatigue under cyclic Hertzian pressure, unlike fine-grained materials. Therefore materials for ball bearings have to be fine-grained. Apart from that, the materials must exhibit high strength (i.e. >800 MPa) and a high Weibull modulus because every defect on the track can be a starting point for failure.

The influence of the friction coefficient on the load capacity is connected with the fact that, besides rolling, sliding always exists in the ball bearing [114, 115]. This friction of the ball on the track is the reason for an increase in the local tensile stress in the track during operating. Therefore reduced friction will increase the load capacity and the lifetime. The load capacity of ball bearings operating in wet conditions is more than 2 times higher than in dry conditions. This is due to the lubricating effect of the water [115]. The development of Si_3N_4 composites with small amounts of submicro-TiN or BN inclusions leads to an increase in the load capacity of up to 60% under dry run (see Fig. 27) [115]. Further improvements can be expected through surface modification by ion implantation [117], the use

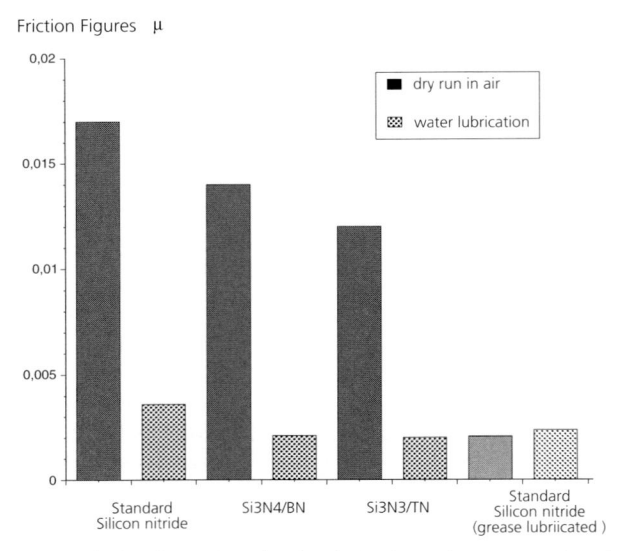

Figure 27. Friction figure (maximal value) of angulare contact bearings 7006 E under dry running in air and in water at 500 N preload [115].

of special thin coatings or by the development of Si_3N_4 materials with high strength and low grain sizes < 100 nm with improved wear resistance [63].

The friction of silicon nitride materials depends strongly on the wear conditions. At low loads the grain boundary is mainly affected (see Fig. 28a), whereas at higher loads silicon nitride grains can break (see Fig. 28b). The wear products can form lubricating layers or rolls that can reduce the friction (see Fig. 28c) [121–124]. Additions of TiN and BN can further reduce the friction and wear under special conditions. An overview of the wear properties under sliding conditions of different Si_3N_4 materials is given in [121–124, 63]. Up to now, the silicon nitride materials have not been optimised for use under different wear conditions. Optimisation of the grain size and composition and amount of the grain boundary can lead to improved wear behaviour of silicon nitride materials in the future.

Silicon nitride materials are used as cutting tools for cutting cast iron and superalloys [125–131]. Ceramic tools and particularly silicon nitride cutting tools exhibit higher hardness at high temperatures than cemented carbides. That allows the use of ceramic tools for machining at higher cutting speeds than for cemented carbides. The typical increase in metal removal rate due to the use of Si_3N_4 in comparison to cemented carbide tools is more than 200% [129].

The high temperatures and mechanical load that are generated at the cutting edge cause wear by a variety of mechanisms:

– abrasive wear due to hard inclusions in the workpiece or chemical interaction with the workpiece or the ambient atmosphere,

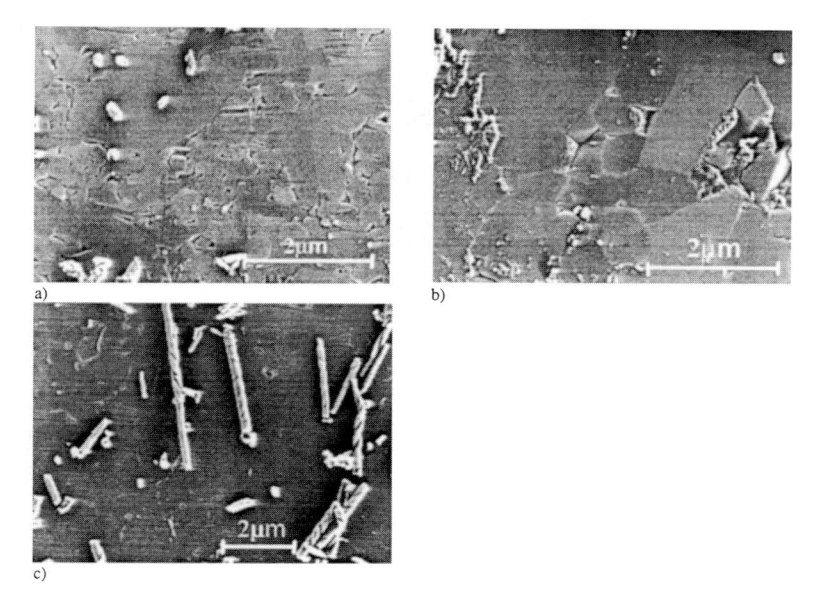

Figure 28. Surface of silicon nitride materials after oscillating sliding tests.

- thermal deformation or degradation by thermal shock, subcritical crack growth and, consequently, fracture of the tool. Tool failure due to fracture must be avoided because it leads to an unpredictable behaviour of the tool. Therefore the cutting materials must have high strength, fracture toughness and thermal shock resistance.

The first two mechanisms lead to more or less continuous wear of the cutting tool over its lifetime.

The temperature of the cutting tool edge can reach about 800°C for turning cast iron [130, 134]; temperatures of up to 1350°C were measured for the cutting of steel [134]. The enhanced chemical interaction and softening of the grain-boundary phase due to the high temperatures during steel cutting are the reasons why silicon nitride cutting tools cannot be used effectively for the cutting of steel.

Chemical interaction is also a dominant wear mechanism during turning and milling of cast iron [127, 130]. On the surface of the cutting tool, different silicates, which have a low hardness and which can therefore be easily removed during machining, can be formed. Chemical interaction seems to be the reason for the relatively rapid rounding of the cutting edge at the beginning of the cutting process. This produces an increase in the cutting forces and is the reason for using silicon nitride tools only for rough machining.

The combination of the different wear mechanisms during cutting makes understanding of the cutting behaviour difficult. Up to now, many open questions exist and the cutting behaviour cannot be predicted using knowledge of the mechanical properties and microstructure. However, some overall relationships between cutting behaviour and material properties can be given. The cutting materials must exhibit high strength, hardness and fracture toughness. These properties are necessary but not sufficient for a good cutting behaviour: there are several Si_3N_4 materials which fulfil these conditions, but which have a bad cutting behaviour due to enhanced chemical interaction with the workpiece.

A wide range of substances in different ratios are used as sintering additives for cutting tools: CaO, MgO, Al_2O_3, Y_2O_3, ZrO_2 and HfO_2 [2, 171, 125–131]. The use of sintering additives that reduce the high-temperature strength is surprising, but is less so when one considers that the softening of the glassy grain-boundary phase starts at temperatures above 800°C, whereas the temperature at the cutting edge in the machining of cast iron is lower than 800°C. An increased SiO_2 content in the grain-boundary phase reduces the lifetime of the cutting tools [133]. This seems to be connected with the acceleration of the chemical interaction.

A decrease in the amount of sintering additives improves the cutting behaviour [125]. The problem of optimising the cutting tool material is that on the one hand it is necessary to minimise the amount of the grain-boundary phase and on the other hand it is necessary to attain a reliable densification to achieve full density, high strength and high reliability. One possibility to improve the densification of the materials with a low additive content is to use MgO-containing additives [71, 129]. Another way to improve the sintering behaviour without increasing the amount of the residual grain-boundary phase is to form β'- or β'/α'-SiAlONs [130]. The substitution grade z in the β-SiALON must be small because high z values cause the fracture toughness and strength to be reduced. In [132] an optimum for $Si_{6-z}Al_zO_zN_{8-z}$ is found with $z = 1$–1.5, but most of the commercial cutting tools have an lower substitution degree. β'/α'-SiAlON composite materials are used especially for cutting superalloys. They have a slightly reduced strength and fracture toughness [128, 131].

A low grain size is favoured for cutting tools because the hardness increases with decreasing grain size. Nevertheless excellent cutting performance is found for self-reinforced Si_3N_4 materials with large, elongated grains [71].

The examples given above show that it is necessary to find a compromise between the different material properties. The compromise strongly depends on technological possibilities (e. g. improvement of the sintering behaviour of the powder) and on changes in the metal machining. The chemical stability of the cutting tools can be increased through the use of coatings, e. g. Al_2O_3/TiN [129].

6.5.4 Corrosion resistance of Si_3N_4

Silicon nitride-based ceramics are promising engineering materials for application under corrosive and wear conditions [135–142]. However, the use of silicon nitride materials in hot acids under hydrothermal conditions and in bases is often limited by degradation of the materials. This is the impetus for current efforts for better understanding of the influence of the microstructure on the corrosion behaviour of silicon nitride materials and to improve the materials' behaviour on its basis. Up to now, only a small amount of data about the mechanisms of corrosion and pitting formation exists.

Table 5. Classification of corrosion conditions and main features of the corrosion behaviour.

Conditions	Corrosion	Literature
Organic components (oil, hydrocarbons)	Work as lubricants, reducing the wear	113, 115
Acids (HCl, H_2SO_4, HNO_3 ...)	Main attack at the grain boundary phase. Corrosion resistance can be improved significantly by tailoring the composition	135–146, 148
Media-intensive solving SiO_2 protective layers, HF alkaline melts; concentrated alkaline solutions at high temperatures >100 to 150°C; hydrothermal conditions at ≥250°C	Dissolution of the grain boundary and the Si_3N_4 grains. Intensive corrosion	135, 144–146, 149
Bases at medium temperatures <100 to 150°C	Main attack at the grain boundary phase. Corrosion resistance can be improved significantly by tailoring the composition	135, 138, 140, 144, 149, 150
Hydrothermal conditions at $T \leq 200°C$	Main attack at the grain boundary phase. Corrosion resistance can be improved significantly by tailoring the composition	136, 141, 151–155

Nevertheless, some correlation's between the microstructure and the corrosion behaviour are known. The corrosion behaviour of silicon nitride materials in liquids is mainly controlled by the stability of the grain-boundary phase. Therefore the corrosion resistance of the silicon nitride materials in different media can be altered by more than two orders of magnitude by changing the composition of the material. The corrosion behaviour can be organised into a few main classes (see Table 5).

Standard silicon nitride materials with Y_2O_3/Al_2O_3 sintering additives degrade quite strongly in hot acids of medium concentration (see Figures 29–31). With increasing temperature the corrosion resistance decreases, whereas with decreasing additive content the corrosion resistance increases. The best corrosion resistance in acids was obtained for a HIP'ed Si_3N_4 material without additives. Si_3N_4 materials are more stable in concentrated acids (>5 N) than in solute acids [135, 136, 138, 148].

The MgO-containing materials possess better corrosion resistance than the Y_2O_3/Al_2O_3-containing materials. The reasons for this are not clear [136, 141]. The corrosion behaviour is different in H_3PO_4 than in other acids due to formation of a protective phosphate layer [138, 144].

Corrosion in HF-containing solutions is much more intensive than in other acids. This is connected with the ability of HF to dissolve the SiO_2 protective layers. Therefore HF solutions will not only attack the grain-boundary phase but also dissolve the Si_3N_4 grains.

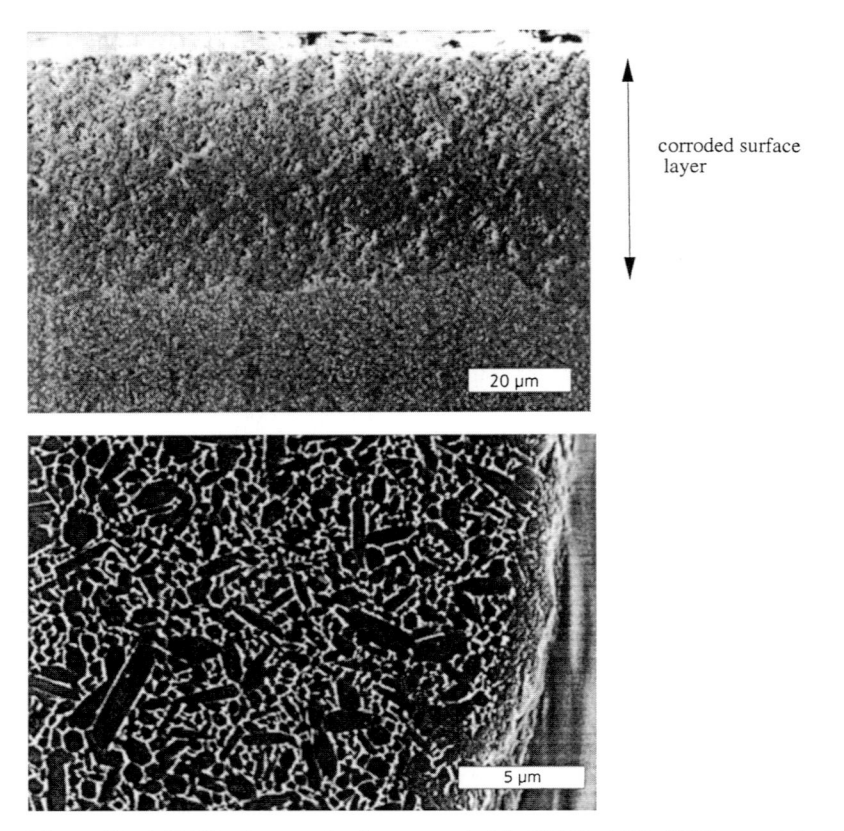

Figure 29. a) SEM micrograph of an corroded in H_2SO_4 100 h 90°C material with Y_2O_3/Al_2O_3 additives and b) SEM micrograph of an material with optimized grain boundary corroded in H_2SO_4 500 h 90°C material (plasmaetched).

Recent investigations show that, besides the amount of the additives, the main parameter governing the corrosion resistance of the silicon nitride materials is the SiO_2 content in the grain-boundary phase (see Figures 29–31). A material with a high SiO_2 content in the grain-boundary phase is more stable in acids than one with a low SiO_2 content. This correlates with the stability of the grain-boundary phase. The acid corrosion of glasses shows a similar behaviour [147, 155]. The network modifier is leached from the glass network by the acids and the glass network hydrated. Therefore a network with a low amount of network modifiers has a higher stability [147]. Additional to the factors influencing the corrosion of glass, the corrosion of silicon nitride is affected by the fact that corrosion takes place only in small channels between the Si_3N_4 grains. During corrosion the Si_3N_4 grains also produce hydrated SiO_2, which can reduce the corrosion rate and change the corrosion mechanism. This seems to be one of the reasons for the low corrosion rate of Si_3N_4 materials with a low amount of sintering additives [141].

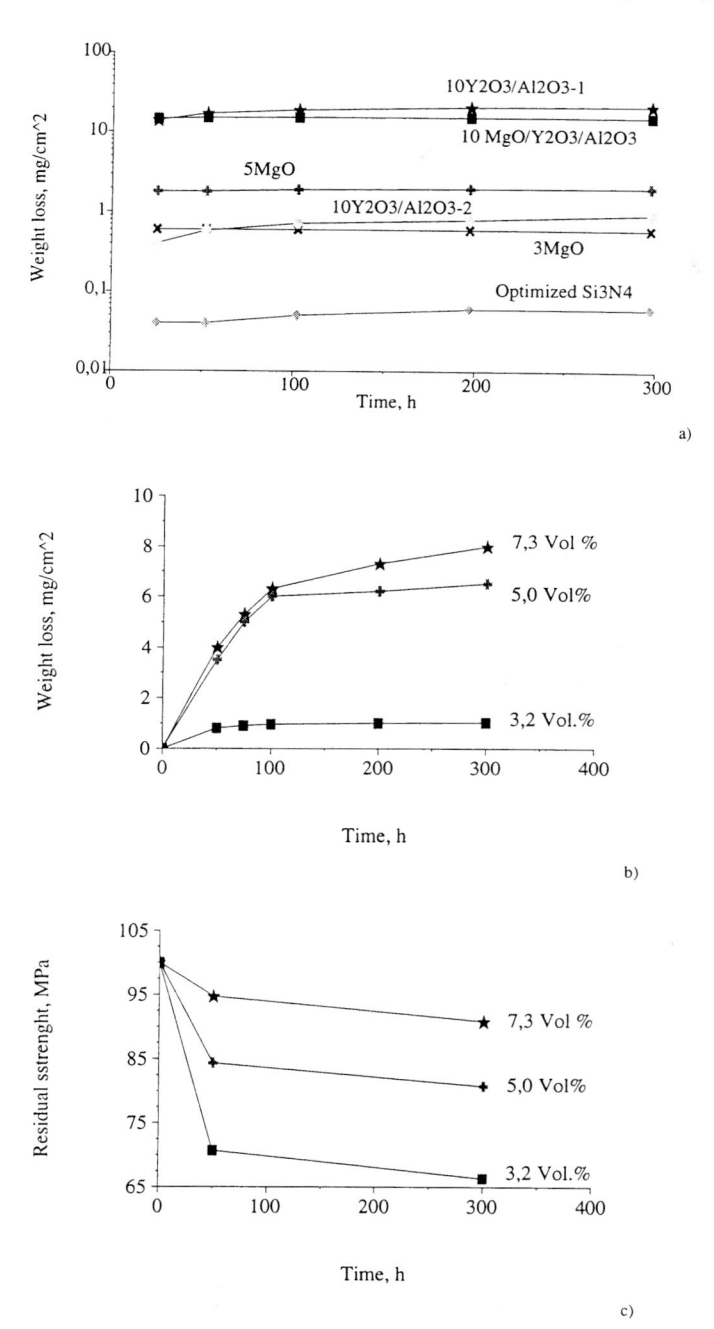

Figure 30. Weight loss of different Si_3N_4-materials in H_2SO_4 at 60°C (additive composition is given at the curves) **a)** and weight loss **b)** and residual strength **c)** of different materials with different amount of Y_2O_3/Al_2O_3 sintering additives in HCl at 60°C.

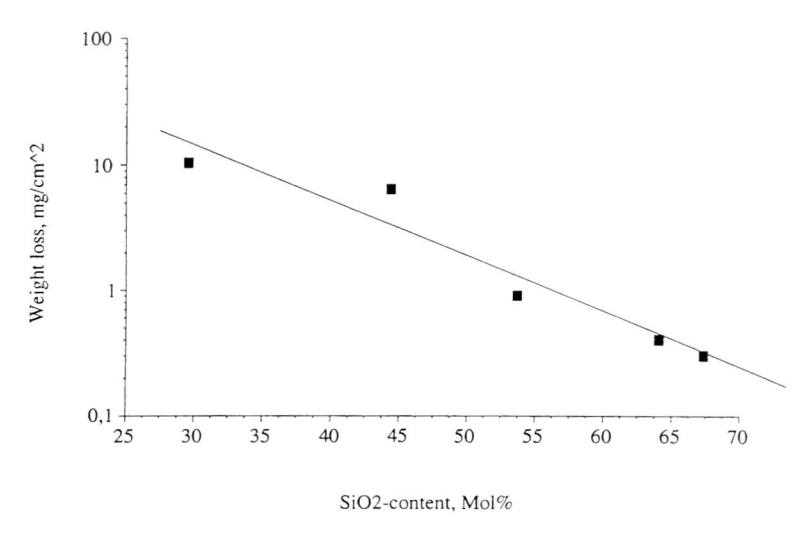

Figure 31. Weight loss of silicon nitride materials with different SiO_2 content in the grain boundary (200 h corrosion in H_2SO_4 at 60°C).

Additionally, the concentration of the other constituents in the glassy grain-boundary phase has an influence on the corrosion behaviour more detailed investigations are necessary to confirm this.

The kinetics of the corrosion of Si_3N_4 materials in acids and the corrosion mechanisms are not clear. Data about diffusion and reaction-controlled mechanisms exist. It is possible that the mechanism changes, depending on the corrosion depth, the acid concentration, temperature and microstructural parameters, but further investigations are necessary to confirm this [141]. The reduction of strength after corrosion in acids for a short time (pitting formation) cannot be correlated with the weight loss. Acid corrosion have an influence on the subcritical crack growth of the materials [138].

The corrosion resistance of selected Si_3N_4 materials in NaOH is given in Fig. 32. The silicon nitride materials are less attacked by bases than by acids. The extent of corrosion increases with increasing temperature and concentration of the bases [150]. Materials which are less stable in acids are more stable in bases and under hydrothermal conditions.

The most stable materials under hydrothermal conditions are materials with a low SiO_2 content in the grain-boundary phase (see Fig. 33). The HIP'ed materials without additives are, besides the MgO-containing materials, less stable than the Y_2O_3/Al_2O_3-containing materials. Under hydrothermal conditions the grain-boundary phase of these materials is dissolved completely, leading to removal of the silicon nitride grains from the surface. This behaviour is different to the corrosion in acids, where the grain boundary is not completely dissolved and the corroded structure is quite strong and stable. This is connected with the stability of the glass network. Under corrosion conditions in acids mainly the network modifiers are dissolved, whereas under hydrothermal corrosion conditions and in strong basic solutions the SiO_x network dissolves. Therefore materials with a high SiO_2 content

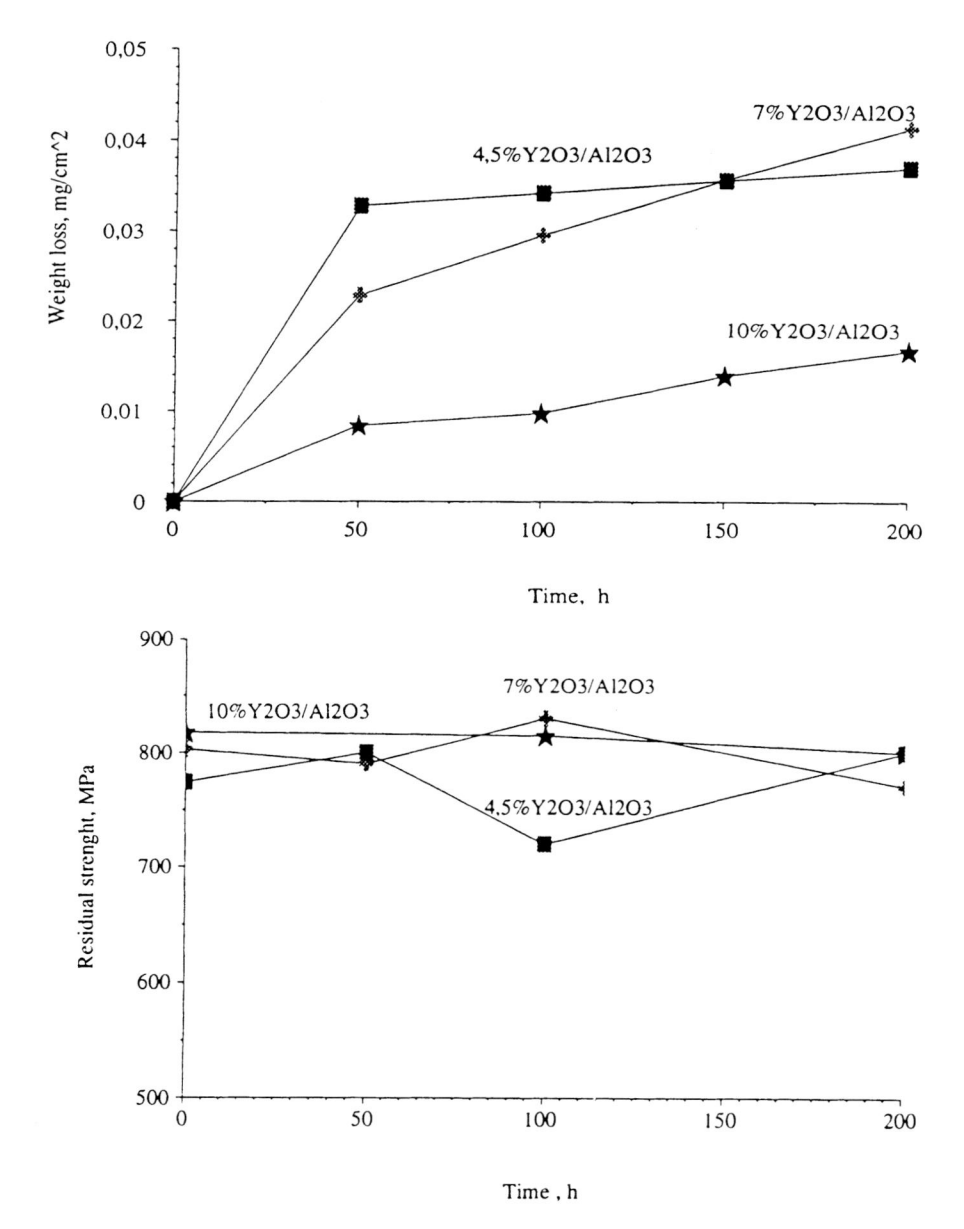

Figure 32. Weight loss and residual strength of silicon nitride materials in 1 N NaOH at 60°C.

in the grain boundary are less stable under hydrothermal conditions than materials with a low SiO_2 content in the grain-boundary phase.

Under hydrothermal corrosion at 270°C, a significant dissolution of the Si_3N_4 grains takes place. In materials with Y_2O_3/Al_2O_3 as sintering additives, the dissolu-

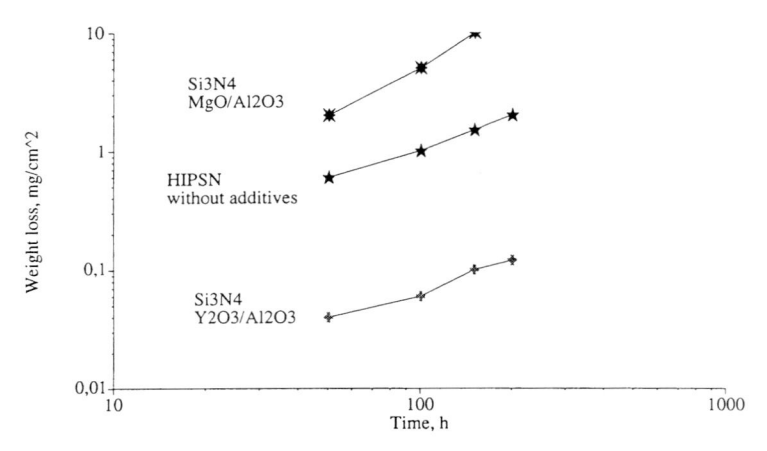

Figure 33. Weight loss of silicon nitride materials under hydrothermal conditions at 210°C.

tion rate of the grains is higher than that of the grain-boundary phase (see Fig. 34).

The weight loss due to hydrothermal corrosion depends linearly on time. Only materials with high corrosion rates (MgO-containing and SiO_2-rich materials) show an increase of the corrosion rate with time.

Investigations of the corrosion behaviour of silicon nitride materials in acids, bases and under hydrothermal conditions show that a main parameter governing the corrosion resistance is the composition and amount of the grain-boundary phase. Materials with a high SiO_2 content in the grain-boundary phase are stable in acids and less stable in bases and under hydrothermal conditions. Materials with a low SiO_2 content in the grain-boundary phase are stable in bases and under hydrothermal condition but less stable in acids.

Up to now no materials which are simultaneously stable in acids, bases and under hydrothermal conditions exist.

A more detailed understanding of the corrosion mechanisms and the influence of the medium on the lifetime of the materials is necessary before silicon nitride materials can be used in a broad range of applications under corrosive conditions.

The results of the investigations indicate that tailoring of the microstructure and the grain boundary composition is necessary for the application of silicon nitride ceramics in corrosive environments.

6.6 Conclusions/Further potential of silicon nitride materials

Silicon nitride ceramic materials exhibit a high potential for structural applications at room and elevated temperatures. This is due to their excellent mechanical properties in combination with good corrosion and thermal shock resistance.

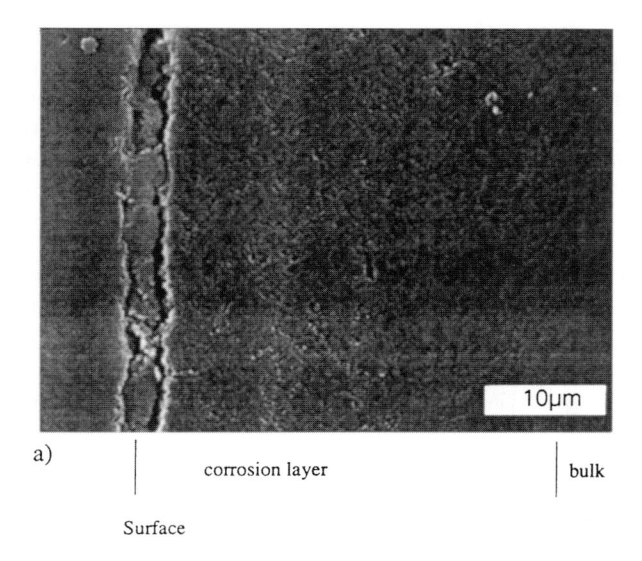

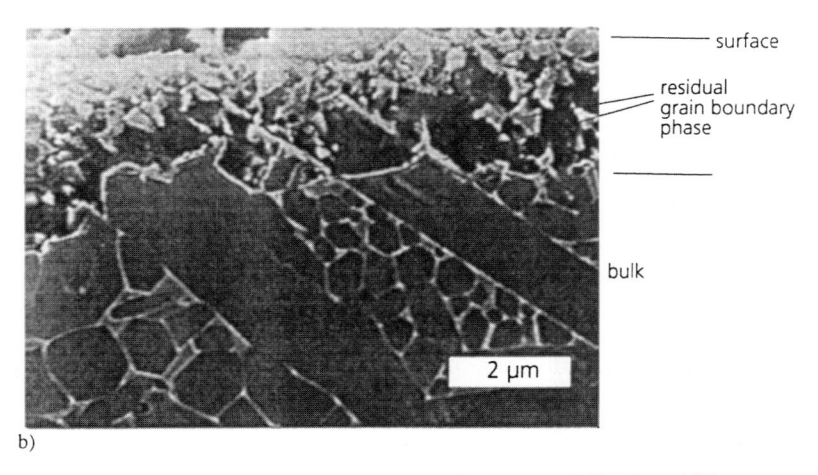

Figure 34. Corrosion layer of Si_3N_4 materials **a)** MgO/Al_2O_3-additive system, hydrothermal corrosion conditions: 210°C, 200 h under at 250°C; **b)** Y_2O_3/Al_2O_3-additive system, hydrothermal corrosion conditions: 270°C, 200 h (the grain-boundary phase solves less than the Si_3N_4 grains and are visible as bright areas).

The development of the silicon nitride materials in the future will be divided in two main directions; the increased penetration in the market of the state of the art silicon nitride materials by improving the technology and reliability and decreasing the production costs and the development of new materials with improved wear, corrosion, high temperature properties.

In the last decade a continuously increase in the number of application fields was observed. The technical feasibility and the reliability of the materials and components were shown in many applications, such as valves and other parts for the automotive industry, bearings, household applications. For the wider penetration into the market there is a engineering work necessary to optimise the production, stabilise the reliability and reduce the production cost. First steps in this direction were made by using low cost powders and optimising processing, sintering and finishing.

The majority of applications can be realised with strength levels lower 1000–1200 MPa.

Higher strength level need a special more expensive finishing procedure increasing the costs of parts. Additionally a strength level of 1200 MPa corresponds to a defect size of 10–15 µm (at the existing K_{IC} levels) which is in the range of defects created during operating of the materials.

Whereas the direction of further material development mentioned above is connected with optimising the technology, the second direction is connected with adapting of the microstructure and properties of silicon nitride materials on different applications. As like as exists different kinds of steel, which are well designed for the given applications, as well in the future will exist different classes of silicon nitride materials aimed for applications under different conditions e.g. wear, corrosion, operation at high temperature and so on (see Table 1).

These differentiation processes stay only at the beginning and need as a bases a controlled formation of the microstructures and the knowledge of the relation between the microstructure and properties.

The main factors and processes influencing the microstructure of Si_3N_4 materials are known and the microstructure can be tailored in a wide range. In the future the use of different β-silicon nitride powders as seeds for a precise tailoring of the microstructure will be used more intensively. New aspects can be expected by the development of fine grained materials. The correlation between microstructure, strength and toughness are well understood, whereas the understanding of the correlation between microstructure on the one side and wear, corrosion properties and long term behaviour on the other side need further intensive interdisciplinary investigations.

An intensive development of high fracture toughness α'-SiAlON-materials can be expected. The better understanding of the microstructure formation of these materials and a further improvement of strength and toughness by a reduced grain size are main topics of the investigations.

Frequently, there is a need to make a compromise between the different properties due to the opposite dependencies on the microstructure. How such compromise looks like depends on the technological possibilities and on the powder properties, which developments offer new possibilities.

An improvement of the materials behaviour can be expected by the development of homogeneous composite materials based on silicon nitride, as it was shown for $Si_3N_4/MoSi_2$ or Si_3N_4/SiC composites for the high temperature long term applications. Especially in this area the use of metallorganic precursors can lead to an improvement of the materials. The reduction of sintering temperature and in some modifications the shrinkage (concept of active filler) can allow the production of composites of thermodynamic unstable components. The development in this direction is only at the beginning.

Acknowledgements

The authors are grateful to the colleges of the IKTS which take part in the technical completion of the paper especially A. Bales and D. Hermannutz. They are indebted to the coworkers in the IKTS especially to Dr. I. Schulz, Dr. T. Reich, A. Bales and G. Michael involved in research projects in the field of silicon nitride materials which results were included in this paper. The work is based on different projects supported by the BMBF, AiF and the DFG in the field of silicon nitride materials.

References

1. S. Hampshire, "Nitride Ceramics in Structure and Properties of Ceramics", ed. R. W. Cahn, P. Haasen and E. J. Kramer, *Materials Science and Technology*, **11**, 121–168.
2. R. N. Katz, "Application of Silicon Nitride based Ceramics in the U.S.", *Mat. Res. Soc. Symp. Proc.*, **287**, 1993, 197.
3. K. H. Jack, "SiALON Ceramics: Retrospect and Prospect", *Mat. Res. Soc. Symp. Proc.*, **287**, 1993, 15–18.
4. J. G. Heinrich and H. Krüner, "Silicon Nitride Materials for Engine Applications", M. J. Hoffmann and G. Petzow (eds.), *Tailoring of Mechanical Properties of Si_3N_4 Ceramics*, Kluwer Academic Publishers, 1994, 19–41.
5. "Der Kochherd der Zukunft arbeitet mit Si_3N_4-Platten", *cfi/Ber. DKG*, **76**, 1999, D 18–20.
6. J. Heinrich, "Siliciumnitridkeramik", editor: J. Kriegsmann, *DKG Technische Keramische Werkstoffe*, 1999, 4.3.1.0.
7. H. M. Jennings, "Review on Reactions between Silicon and Nitrogen, Part 1 Mechanisms", *Journal of Material Science*, **18**, 1983, 951–967.
8. A. Zerr, G. Miehe, G. Serghiou a. o., "Synthesis of Cubic Silicon Nitride", *Nature*, **400**, 1999, 340–344.
9. W. Ching, Y. Xu, J. Gale and M. Rühle, "Ab-Inito Total Energy Calculation of α- and β-silicon nitride and the Derivation of Effective Pair Potentials with Application to Lattice Dynamics", *J. Am. Ceram. Soc*, **81**, 1998, 3189–3196.
10. I. Kohatsu and J. W. McCauley, "Re-Examination of the Crystal structure of α-Si_3N_4", *Mat. Res. Bull*, **9**, 1974, 917–920.
11. H. Suematsu, J. J. Petrovic and T. E. Mitchell, "Plastic deformation of silicon nitride single crystals" *Materials Science and Engineering*, A **209**, 1996, 97–102.
12. J. Dusza, T. Eschner and K. Rundgren, "Hardness anisotropy in bimodal grained gas pressure sintered Si_3N_4", *Journal of Materials Science Letters*, **16**, 1997, 1664–1667.
13. Andrievskii, Spivak I. I. , "Nitrid kremnija i materiali na evo osnove", *Moskau, Metallurgija*, 1984, 19–20.
14. H. Mandal, D. P. Thompson and K. H. Jack, "α ↔ β Phase Transformations in Silicon Nitride and Sialons", preprint.
15. L.-O. Nordberg, "α-Sialon Ceramics and Y-α-Sialon Composites; Composition, Microstructures and Properties", *Doctorial Dissertation, 1997*, 1–124.

16. K. H. Jack, "α-SiAlON Ceramics", *Nature*, **274**, 1978, 880–882.
17. T. Y. Tien, G. Petzow, L. J. Gauckler and J. Weiss, "Phase Equilibrium Studies in Si_3N_4-Metal Oxides Systems", *Progress in Nitrogen Ceramics*, 1983, 89–99.
18. D. D. Lee, S. J. Kang, G. Petzow and D. N. Yoon, "Effect of α to β Phase Transition on the Sintering of Silicon Nitride Ceramics", *J. Am. Ceram. Soc.*, **73**, 1990, 767–9.
19. M. N. Rahaman, "Ceramic Processing and Sintering", *Marcel Dekker Inc., Department of Ceramic Engineering University of Missouri*, 1995.
20. M. Herrmann und Chr. Schubert, "Grundlagen der Gefügeausbildung in β-Siliciumnitridwerkstoffen", editor: J. Kriegsmann, *DKG Technische Keramische Werkstoffe*, Sep. 1999, 5.1.3.1.
21. P. Greil, "Einfluß der intergranularen Glasphase auf die Hochtemperatureigenschaften von β-SiALONen" *Dissertation Stuttgart* 1982.
22. M. Herrmann, G. Putzky, S. Siegel and W. Hermel, "Einfluß von Zersetzungsreaktionen auf die Sinterung", *cfi/Ber DKG*, **69**, 1992, 375–382.
23. P. Obenaus and M. Herrmann, "Methode zur quantitativen Charakterisierung von Stengelkristalliten in Siliziumnitridkeramik", *Prakt. Met.*, **27**, 1990, 503–513.
24. H. Björklund, L. K. Falk, K. Rundgren and J. Wasen, "α-Si_3N_4 Grain Growth, Part I: Effect of Metal Oxide Sintering Additives" *Journal of the European Ceramic Society*, **17**, 1997, 1285–1299.
25. H. Björklund and L. K. L. Falk, "α-Si_3N_4 Grain Growth, Part II: Intergranular Glass Chemistry", *Journal of the European Ceramic Society*, **17**, 1997, 1301–1308.
26. G. Wötting, B. Kanka and G. Ziegler, "Microstructural Development, Micccrostructural Characterization and Relation to Mechanical Properties of Dense Silicon Nitride", *Nonoxide Ceramic*, 1986, 83–95.
27. G. Ziegler, J. Heinrich and G. Wötting, "Relationship between Processing, Microstructure and Properties of dense and Reaction-Bonded Silicon Nitride", *J. of Mat. Sci.*, **22**, 1987, 3041–3086.
28. F. F. Lange, "Fracture Toughness of Si_3N_4 as a Function of the Initial a-Phase Content", *J. Americ. Ceram. Soc.*, **62**, 1979, Nr. 7, 428–429.
29. G. Woetting, H. Feuer and E. Gugel, "The Influence of Powder and Processing Methods on Microstructure and Properties of Dense Silicon Nitride", *Mat. Res. Soc. Symp. Proc.*, **287**, 1993, 133–146.
30. S. Kessler, M. Herrmann and Chr. Schubert, "The α-β-Transformation and anisotropic growth of β-grains during sintering of Si_3N_4", *Mat. Sci. Forum*, **94–96**, Trans. tech. Publ., 1991, 821–827.
31. G. Petzow and M. J. Hoffmann, "Grain Growth Studies in Si_3N_4-Ceramics", *Mat. Sci. Forum*, **113–115**, 1993, 91–102.
32. M. Herrmann, S. Kessler and Chr. Schubert, "Microstructural Design of Dense Si_3N_4", *Euro-Ceramics II, Vol.2*, Edited by G. Ziegler and H. Hausner, 1991, 847–851.
33. W. Hermel, M. Herrmann and Ch. Schubert, "Have the limits of the development of β-Si_3N_4-Materials Already been reached", In: *Third Euro-Ceramics, V3*, edited by P. Duranand, J. F. Fernandez, Faenca Edititrice Iberica S. L., 1993, 391–396.
34. W. Hermel, M. Herrmann and I. Schulz, "Sintering and Microstructure of Si_3N_4-Materials", *Silicon Nitride 93*, Edited by M. J. Hoffmann, P. F. Becher, G. Petzow, Trans tech Publications, 1993, 181–186.
35. W. Dressler, H.-J. Kleebe, M. J. Hoffmann, M. Rühle and G. Petzow, "Model Experiments Concerning Abnormal Grain Growth in Silicon Nitride", *J. of the Europ. Ceram. Soc.*, **16**, 1996, 3–14.
36. M. Krämer, M. J. Hoffmann and G. Petzow, "Grain Growth Studies of Silicon nitride Dispersed in an Oxynitride glass", *J. Am. Ceram. Soc.*, **76**, 1993, 2778–84.
37. W. Dressler and R. Riedel, "Progress in Silicon Based Non-Oxide Structural Ceramics", *Int. J. of Refractory hard Materials*, **15**, 1997, 13–47.
38. M. Krämer, "Untersuchungen zur Wachstumskinetik von β-Si_3N_4 in Keramiken und Oxinitirdgläsern", *Dissertation*, Stuttgart, 1992.
39. G. Ziegler, W. Lehner and H.-J. Kleebe, "Nucleation affecting microstructure development of Si_3N_4 ceramics", *Brit. Ceram. Proc.*, **60**, 1999, 5–7.
40. S.-L. Hwang and I.-W. Chen, "Nucleation and Growth of α'-SiAlON on α-Si_3N_4", *J. Am. Ceram. Soc.*, **77**, 1994, 1711–18.

41. H.-J. Kleebe, "Structure and Chemistry of Interfaces in Si_3N_4 Ceramics Studied by Transmission Electron Microscopy", *Journal of the Ceramics Society of Japan*, **105**, 1997, 453–475.
42. T.-Y. Tien, "Silicon Nitride Ceramics-Alloy Design", *Mat. Res. Soc. Symp. Proc.*, **287**, 1993.
43. T. Y. Tien and C. J. Hwang, "Microstructural Development in Silicon Nitride Ceramics", *Mater. Sci. Forum*, **47**, 1989, 84–109.
44. L. L. Wang and T. Y. Tien, "Morphology of Silicon Nitride Grown from a Liquid Phase", *J. Am. Ceram. Soc*, **81**, 1998, 10, 2677–2686.
45. D. Y. Yoon, Korean Advanced Institute of Science and Engineering, Report 16. 6. 1999 in the IKTS.
46. Y.-J. Park, N.-M. Hwang and D. Y. Yoon, "Abnormal Growth of Faceted (WC) Grains in a (Co) Liquid Matrix", *Metall. Mater. Trans.*, **27A**, 1996, 1–11.
47. D.-D. Lee, S.-J. L. Kang and D. N. Yoon, "Mechanism of Grain Growth and α-β' Transformation During Liquid-Phase Sintering of β'-Sialon", *J. Am. Ceram. Soc.*, **71**, 1988, 803–806.
48. N. Hirosaki, Y. Akimune and M. Mitomo, "Microstructural Design by Selective Grain Growth of α-Si_3N_4", *Mat. Res. Soc. Symp. Proc.*, **287**, 1993, 405–410.
49. M. Herrmann, I. Schulz and J. Hintermayer, "Materials From Low Cost Silicon Nitride Powders", *Proc. 4th EcerS Conf., Riccione, Ed. by C. Galassi, Gruppo Editoriale Faenza Editrice*, 1995, 211–216.
50. N. Hirosaki, M. Ando, Y. Akimune and M. Mitomo, "Gas-Pressure Sintering of β-Silicon nitride Containing Y_2O_3 and Nd_2O_3", *J. of Ceram. Soc. of Japan*, **100**, 1992, 817–820.
51. M. Mitomo, M. Tsutsumi, H. Tanaka, S. Uenosono and F. Saito, "Grain Growth During Gas-Pressure Sintering of β-Silicon Nitride", *J. Am. Ceram. Soc.*, **73**, 1990, 2441–45.
52. M. Mitomo, H. Hirotsuru, H. Suematsu and T. Nishimura, "Fine-Grained Silcon Nitride Ceramics Prepared from β-Powder", *J. Am. Ceram. Soc.*, **78**, 1995, 211–14.
53. J. Hintermayer, G. Ernst, W. Gmöhling, G.Schroll and W. Kobler, "Verfahren zur Herstellung von Siliziumnitrid" EP 0377132B1, 1994.
54. E. Hidseyuki and M. Mitomo, "Control and Characterization of Abnormally Grown Grains in Silicon Nitride Ceramics", *Journal of the European Ceramic Society*, **17**, 1997, 797–804.
55. M. Mitomo, "In-situ Microstructural Control in Engineering Ceramics", *Key Engineering Materials*, **161–163**, 1999, 53–58.
56. P. Bränvall and K. Rundgren, "Grain growth mechanisms in *in situ* reinforced Si_3N_4", *Brit. Ceram. Proc.*, **60**, 1999, 7–9.
57. K. Hirao, T. Nagaoka, M. E. Brito and S. Kanzaki, "Microstructure Control of Silicon Nitride by Seeding with Rodlike β-Silicon Nitride Particles", *J. Am. Ceram. Soc.*, **77**, 1994, 1857–62.
58. K. Hirao, T. Nagaoka, M. E. Brito and S. Kanzaki, "Mechanical Properties of Silicon Nitrides with Tailored Microstructure By Seeding", *Journal of the Ceramic Society of Japan*, **104**, 1996, 55–59.
59. R. A. Andrievski, "Some high-temperature properties of silicon nitride", *High Temperature – High Pressure*, **26**, 1994, 451–455.
60. H. Emoto and H. Hirotsuri, "Microstructure Control of Silicon Nitride Ceramics Fabricated from α-Powder Containing Fine β-Nuclei", *Key Engineering Materials*, **161–163**, 1999, 209–212.
61. K. Hirao, H. Imamura, K. Watari, M. E. Brito, M. Toriyama and S. Kanzaki, "Seeded Silicon Nitride: Microstructure and Performance", *Key Engineering Materials*, **161–163**, 1999, 469–474.
62. M. Herrmann, I. Schulz, Chr. Schubert and W. Hermel, "Silicon Nitride Materials with Low Friction Coefficients", *Key Engineering Materials*, **161–163**, 1999, 599–602.
63. M. Herrmann, I. Schulz, Chr. Schubert, I. Zalite and G. Ziegler, "Ultrafine Si_3N_4-Material with Low Coefficient's of Friction and Wear Rates", *cfi/Ber. DKG*, **75**, No 4, 1998, 38–45.
64. G. Riedel, H. Bestgen and M. Herrmann, "Influence of Sintering Additives with Differing Proportions of Y_2O_3/Al_2O_3 on the Sintering and Material Properties of Si_3N_4 Ceramics", *cfi, Ber. DKG*, **75**, No. 1–2, 1998, 30–34.
65. K. Watari, K. Ishizaki, S. Cao and K. Mori, "The Relationship between Thermal Conductivity and Microstructure in Si_3N_4 Ceramics", *Key Engineering Materials*, **161–163**, 1999, 213–216.

66. N. Hirosaki, Y. Okamoto, M. Ando, F. Munakata and Y. Akimune, "Effect of Grain Growth on the Thermal Conductivity of Silicon Nitride", *Journal of the Ceramic Society of Japan*, **104**, 1996, 50–54.
67. D. A. Bonnell, M. Rühle and T.-Y. Tien, "Redistribution of Aluminum Ions During Processing of Sialon Ceramics", *J. Am. Ceram. Soc.*, **69**, 1986, 623–27.
68. M. Kitayama, K. Hirano, M. Toriyama and S. Kanzaki, "Anisotropic Ostwald Ripening in β-Si_3N_4 With Different Lantanide Additives", *G. L. Messing, F. F. Lange, S.-I. Hirano: Ceramic Processing Science*, 517–24.
69. M. J. Hoffmann, "Analysis of Microstructural Development and Mechanical Properties of Si_3N_4 Ceramics", *M. J. Hoffmann, G. Petzow, Tailoring of Mechanical Properties of Si_3N_4 Ceramics*, Kluwer Academic Publishers, 1994, 59–72.
70. H. J. Kleebe, G. Pezzotti and G. Ziegler, "Microstructure and Fructure Toughness of Si_3N_4 Ceramics", *J. Am. Ceram. Soc.*, **82**, 1999, 1857–1860.
71. A. R. Prunier and A. J. Pyzik, "Self reinforced Silicon Nitride for Cutting Tool Application", *Key Eng. Mat.*, **89–91**, 1994, 129–134.
72. A. J. Pyzik and D. R. Beaman, "Microstructure and Properties of Self-Reinforced Silicon Nitride", *J. Am. Ceram. Soc.*, **76**, 1993, 2737–44.
73. M. Menon and I.-W. Chen, "Reaction Densification of α'- Sialon: I, Wetting Behavior and Acid-Base Reactions", *Journal of the American Ceramic Society*, **78**, 1995, 545–552.
74. M. Menon and I.-W. Chen, "Reaction Densification of α'-Sialon: II, Densification Behavior", *Journal of the American Ceramic Society*, **78**, 1995, 553–559.
75. H. Mandel and M. J. Hoffmann, "Novel developments in α-SiALON ceramics", *Brit. Ceram. Proc.*, **60**, 1999, 11–13.
76. Schutzrecht USA WO 98/23554 (1998-06-04).
77. I-W. Chen and A. Rosenflanz, "A tough SiAlON ceramic based on α-Si_3N_4 with a whisker-like microstructure", *Nature*, **389**, 1997, 701–704.
78. H. Klemm, M. Herrmann, T. Reich, C. Schubert, L. Frassek, G. Wötting, E. Gugel and G. Nietfeld, "High-Temperature Properties of mixed α'/β'-Sialon Materials", *Journal of the American Ceramic Society*, **81**, 1998, 1141–1148.
79. T.-S. Sheu, "Microstructure and Mechanical Properties of In Situ β-Si_3N_4/α'-Sialon Composite", *Journal of the American Ceramic Society*, **77**, 1994, 2345–2353.
80. M. Herrmann, annual report 1998, *IKTS*, 1998.
81. N. Camascu, D. P. Thompson and H. Mandal, "Effect of Starting Composition, Type of Rare Earth Sintering Additive and Amount of Liquid Phase on α'/β-Sialon Transformation", *Journal of the European Ceramic Society*, **17**, 1997, 599–613.
82. K. Urashima, Y. Ikai and S. Iwase, "Features of a Superior Strength Si_3N_4 Ceramic and its Applications", *6th International Symposium on Ceramic Materials and Components for Engines*, 1997, 167–172.
83. G. Riedel and H. Krüner, "Si_3N_4-material with high strength (1400 MPa)", *Third Euro Ceramics*, Madrid, 1993, 453–458.
84. R. F. Silva and J. M. Vieira, "Hot hardness of Si_3N_4-based materials", *Journal of Material Science*, **30**, 1995, 5531–5536.
85. M. Yoshimura, T. Nishioka, A. Yamakawa and M. Miyake, "Grain Size Controlled High-Strength Silicon Nitride Ceramics", *J. Ceram. Soc. Japan*, 1995, 407.
86. T. Nishioka, K. Matsunuma, T. Yamamoto, A. Yamakawa and M. Miyake, "Development of High Strength Si_3N_4 Material for Automobile Parts", *Sumitomo Electric Technical Review*, **36**, 1998, 77.
87. N. Hirosaki, Y. Akimune and M. Mitomo, "Effect of Grain Growth of β-Silicon Nitride on Strength, Weibull Modulus and Fracture Toughness", *J. Am. Ceram. Soc.*, **76**, 1993, 1892–94.
88. P. F. Becher, H. T. Lin, S. L. Hwang, M. J. Hoffmann and I. W. Chen, "The Influence of Microstructure an the Mechanical Behaviour of Silicon Nitride Ceramics", *Mat. Res. Soc. Symp. Proc.*, **287**, 1993, 147–157.
89. P. Sajgalik, J. Dusza and M. J. Hoffmann, "Relationship between Microstructure, Toughening Mechanism, and Fracture Toughness of Reinforeced Silicon Nitride Ceramics", *J. Am. Ceram. Soc.*, **78**, 1995, 2619–24.

90. C. W. Li, C. J. Gasdaska, J. Goldacker and S. C. Lui, "Damage Resistance of in Situ Reinforced Silicon Nitride", *Mat. Res. Soc. Symp. Proc.*, **287**, 1993, S. 473–480.
91. B. Speicher, G. A. Schneider, W. Dreβler, G. Lindemann, H. Böder and V. Knoblauch, "Reliability of Ceramic Valve plates for Common-Rail Injection Pumps", *preprint Materialica 1999*, 1999.
92. I. Peterson and T. Tien, "Effect of the Grain Boundary Thermal Expansion Coefficient on the Fracture Toughness in Silicon Nitride", *J. Am. Ceram. Soc.*, **78**, [9], 1995, 2345–52.
93. E. Y. Sun, P. F. Becher and K. P. Plucknett, "Microstructural Design of Silicon Nitride with Improved Fracture Toughness: II, Effects of Yttria and Alumina Additives", *J. Am. Ceram. Soc.*, **81**, 1998, 2831–2840.
94. A. G. Evans, "Perspective on the Development of High-Toughness Ceramics", *J. Am. Ceram. Soc.*, **73**, 1990, 187–206.
95. H. Kessler, H. Kleebe, R. Cannon and W. Pompe, "Influence of Internal stresses on Crystallization of Intergranular Phases in Ceramics", *Acta metall mater.*, **40**, 1992, 2233–2245.
96. I. Tanaka, G. Pezotti, Y. Miyamoto and T. Okamoto, "Fracture Toughness of Si_3N_4 and its Si_3N_4 Whisker Composite Without Sintering Aids", *J. of Mat. Sci.*, **26**, 1991, 208–210.
97. J. Zhihao and Q. Guanjun, "Fatigue Behavior of some Advanced Ceramic Materials", *6th International Symposium on Ceramic Materials and Components for Engines*, 1997, 760–765.
98. T. Ogasawara and Y. Mabuchi, "Effect of Humidity and Cyclic Loading Condition on Fatigue behavior of Silicon Nitride", *J. Ceram. Soc. Jap., Int. Edition*, **101**, 1993, 1122–1127.
99. D. Jacobs and I. Chen, "Mechanical and Environmental Factors in the Cyclic and Static Fatigue of Silicon Nitride", *J. Am. Ceram. Soc.*, **77**, 1994, 1153–61.
100. P. F. Becher, E. Y. Sun, K. P. Plucknett, K. B. Alexander, C. H. Hsueh, H. T. Lin, S. B. Waters and C. G. Westmoreland, "Microstructural Design of Silicon Nitride with Improved Fracture Toughness: I, Effects of Grain Shape and Size", *J. Am. Ceram. Soc.*, **81**, 1998, 2821–30.
101. R. Schober, M. Herrmann, W. Krehler, H.-J. Richter, W. Hermel and G. Naumann, "Verdichtungsverhalten und Eigenschaften von nachgesintertem reaktionsgebundenem Siliziumnitrid (SRBSN)", editor: J. Kriegsmann, *DKG Technische Keramische Werkstoffe*, 3. Erg.-Lfg. 1999.
102. I. Tanaka, G. Pezzotti, T. Okamoto, Y. Miyamoto and M. Koizumi, "Hot Isostatic Press Sintering and Properties of Silicon Nitride at Elevated Temperatures", *J. Am. Ceram. Soc.*, **74**, 1992, 752–59.
103. H. Klemm, M. Herrmann and Chr. Schubert, "High Temperature Oxidation and Corrosion of Silicon-Based Nonoxide Ceramics", *ASME Turbo Expo '98*, Stockholm, Sweden 1998, ASME Paper 98-GT-480.
104. H. Klemm, M. Herrmann and Chr. Schubert, "High Temperature Oxidation of Silicon Nitride Based Ceramic Materials", *Proc. 6th Internat. Symp. on Ceram. Mater. & Comp. for Engines*, Arita, Japan, 1997, 576–81.
105. I. Tanaka, G. Pezzotti, K. Matsushita, Y. Miyamoto and T. Okamoto, "Impurity Enhanced Cavity Formation in Si_3N_4 at Elevated Temperatures", *J. Am. Ceram. Soc.*, **74**, 1992, 752–59.
106. H. Klemm and G. Pezzotti, "Fracture Toughness and Time-Dependent Strength Behavior of Low- Doped Silicon Nitrides for Applications at 1400°C", *J. Am. Ceram. Soc.*, **77**, 1994, 553–61.
107. G. Pezzotti, K. Ota and H.-J. Kleebe, "Viscous Slip along Grain Boundaries in Chlorine-Doped Silicon Nitride," *J. Am. Ceram. Soc.*, **80**, 1997, 2341–48.
108. M. Herrmann, Chr. Schubert, A. Rendtel and H. Hübner, "Silicon Nitride/Silicon Carbide Nanocomposite Materials: I, Fabrication and Mechanical Properties at Room Temperature", *J. Am. Ceram. Soc.*, **81**, 1998, 1095–108.
109. K. Niihara, "New Design Concept of Structural Ceramics – Ceramic Nanocomposites", *The Centennial Memorial Issue of the Ceramic Society of Japan*, **99**, 1991, 974.
110. M. Herrmann, H. Klemm, Chr. Schubert and W. Hermel, "Long-Term Behavior of SiC/Si_3N_4-Nanocomposites at 1400–1500°C", *Key Engineering Mater.*, **132–136**, 1977–80, 1997.
111. G. Pezzotti and M. Sakai, "Effect of a Silicon Carbide 'Nano-Dispersion' on the Mechanical Properties of Silicon Nitride", *J. Am. Ceram. Soc.*, **77**, 1994, 3039–41.
112. H. Klemm, K. Tangermann, Chr. Schubert and W. Hermel, "Influence of Molybdenum Silicide Additions on High-Temperature Oxidation Resistance of Silicon Nitride Materials", *J. Am. Ceram. Soc.*, **79**, 1996, 2429–35.

113. R. Sternagel and M. Popp, "Innovationen für Gleitlager, Wälzlager, Dichtungen und Führungen", *VDI-Ber.*, **1331**, 1997, 131–138.
114. M. Rombach, R. Schöfer and W. Pfeiffer, "Tragfähigkeit und Lebensdauer für Wälzlager aus Hochleistungskeramik", *Report of a BMBF- project 03 M2107B*, 1996.
115. M. Popp, R. Sternagel and G.Wötting, "Hybride- and Ceramic Rolling Bearings with modified Surface and low Friction Rolling Contact", Euromat 1999, 1999.
116. Ch. Schubert and M. Herrmann, "Materials development of silicon nitride for ceramic ball bearings", *Report of the BMBF- project Nr. 03 M2107B4*, 1996.
117. M. Herrmann, I. Schulz, Chr. Schubert and W. Hermel, "Siliziumnitridwerkstoffe mit niedrigem Reibkoeffizienten und hohem Verschleißwiderstand", *Werkstoffwoche 98, Vol. V, Symposium 6*, 1998, 241–247.
118. M. Popp and R. Sternnagel, "Neue Entwicklungen bei Wälzlagern aus Hochleistungskeramik", *Werkstoffwoche 98, Vol. V, Symposium 6*, 1998, 247–253.
119. B. R. Lawn, S. K. Lee and K. S. Lee, "Fracture and Deformation Damage Accumulation in Tough Ceramics", *Key Engineering Materials*, **161–163**, 1999, 3–8.
120. S. K. Lee and B. R. Lawn, "Role of Microstructure in Hertzian Contact Damage in Silicon Ntride: II, Strength Degradation", *Journal of the American Ceramic Society*, **81**, 1998, Nr. 4, 997–1003.
121. S. W. Lee, "Tribological Characterization in Ceramic Based Composite", *Key Engineering Materials*, **161–163**, 1999, 593–598.
122. K. H. Habig, "Tribologisches Verhalten von Ingenieur-Keramik", *Ingenieur-Werkstoffe*, **1**, 1989, 78–83.
123. A. Skopp, "Tribologisches Verhalten von Siliziumnitridwerkstoffen bei Festkörpergleitreibung zwischen 22°C und 1000°C", *Dissertation* 1993.
124. S. K. Lee and B. R. Lawn: Role of Microstructure in Hertzian Contact Damage in Silicon Ntride: II, Strength Degradation. In: *Journal of the American Ceramic Society*, **81**, 1998, 997–1003.
125. H. Tanaka, "A Recent Tendency of Si_3N_4 Cutting Tools", *Advanced Materials: Ceramics, Powders, corrosion and Advanced Processing*, **14A**, 1994, 541–545.
126. H. K. Tönshoff and B. Denkena, "Wear of Ceramic Tools in Milling", *Journal of the Society of Tribologists and Lubrication Engineers*, **47**, 1991, 772–778.
127. H. K. Tönshoff, Chr. Blawit, M. Rodewald and H.-G. Wobker, "Entwicklung und Verschleißverhalten von Si_3N_4-Schneidkeramik", *Mat.-wiss. u. Werkstofftech.*, **26**, 1995, 255–262.
128. G. Brandt, "Development of ceramic cutting tools", *Materiaux & Techniques*, **85**, 1997, 3–12.
129. R. D. Nixon and P. K. Mehrotra, "Optimization of Sintering Aids in Silicon Nitrides for Cutting Tool Application", *Ceramic Engineering & Science Proceedings*, **18**, 1997, 457–464.
130. H. K. Tönshoff and C. Blawit, "Wear mechanism of silicon nitride in dependence of workpiece material", BMFT- Project, Final Report No. 03T0028A, 1993, 1–96.
131. H. Miao, L. Qi and G. Cui: Silicon Nitride Ceramic Cutting – Tools and their Applications", *Key Engineering Materials*, **114**, 1995, S. 135–172.
132. G. Brandt, "Wear- and thermal shock-resistant SIALON cutting tool material", EP 97-850038 970311.
133. US- Patent 432 325.
134. Y. R. Liu, J. J. Liu, B. L. Zhu, Z. R. Zhu, L. Vincent and P. Kapsa, "Wear Maps of Si_3N_4 ceramic cutting tools", *J. Mater. Eng. and Perform.*, **6**, 1997, 671–675.
135. Gmelin Handbook, *Si-Suppl*, **b5d2.**, 110–117; 1995, 184–210.
136. K. Komeya, T. Meguro, S. Atago, C.-H. Lin, Y. Abe and M. Komatsu, "Corrosion Resistance of Silicon Nitride Ceramics", *Key Engineering Materials*, **161–163**, 1999, 235–238.
137. M. Herrmann, A. Krell, J. Adler, G. Wötting, T. Hollstein, W. Pfeifer and M. Rombach, "Innovationen für Gleitlager, Wälzlager, Dichtungen und Führungen", *VDI-Ber.*, **1331**, 1997, 251–258.
138. T. Hollstein, T. Graas, K. Bundschuh and M. Schütze, *Keramische Zeitschrift*, **50**, 1998, 416–421.
139. T. Sato, Y. Tokunaga, T. Endo, M. Shimada, K. Komeya, *et al.*, "Corrosion of Silicon Nitride Ceramics in Aqueous Hydrogen Cloride Solutions", *J. Mat. Science*, **23**, 1988, 3440–3446.

140. S. Iio, A. Okada, A. Akira, A. Tetsuo and Y. Masahiro, "Corrosion of silicon nitride ceramics in solutions. Part 3. Corrosion behavior in hot sulfuricacid and microstructure of the corroded layer", *Journal of the Ceramic Society of Japan, Int. Edition*, **100**, 1992, July, 954–956.
141. M. Herrmann, Chr. Schubert and G. Michael, "Korrosionsstabile keramische Werkstoffe für Anwendungen in Wälzlagern und im Anlagenbau", *Fortschrittsberichte der DKG*, **14**, 1999, 130–151.
142. M. Herrmann and G. Michael, "Corrosion Behaviour of Engineering Ceramics in Acids and Basic Solutions", *British Ceramic Proceedings*, **60**, extended Abstracts, Vol **1**, 1999, 455–457.
143. A. Okada and M. Yoshimura, "Mechanical Degradation of Silicon Nitride Ceramics in Corrosive Solutions of Boiling Sulphuric Acid", *Key Engineering Materials*, **113**, 1996, 227–236.
144. Y. G. Gogozi and V. A. Lavrenko, "Corrosion of High Performance Ceramics", *Springer Verlag, Berlin*, 1992, 76–78.
145. M. Shimada and T. Sato, "Corrosion of Silicon Nitride Ceramics in HF and HCl Solutions" *Ceram. Trans.*, **10**, 1989, 355–365.
146. Q. Fang, P. S. Sidky and M. G. Hocking, "The Effect of Corrosion and Erosion on Ceramic Materials", *Corrosion Science*, **39**, 1997, 511–527.
147. A. Paul, "Chemical Durability of Glasses: a Thermodynamic Aproach", *J. Mat. Sci.*, **12**, 1977, 2246–2268.
148. K. Kanbara, N. Uchida, K. Uematsu, T. Kurita, K. Yoshimoto and Y. Suzuki, "Corrosion of Silicon Nitride Ceramics by Nitric Acid", *Mat. Res. Soc. Symp. Proc.*, **287**, 1993, 533–538.
149. T. Sato, Y. Tokunaga, T. Endo, M. Shimada, K Komeya, K. Nishida, M. Komatsu and T. Kameda, "Corrosion of Silicon Ceramics in Aqueous HF Solutions", *J. of Mat. Sci.*, **23**, 1988, 3440–3446.
150. T. Sato, S. Sato, K. Tamura and A. Okuwaki, "Corrrosion Behaviour of SiliconNitride Ceramics in Caustic Alkaline Solutions at High Temperatures", *Br. Ceram. Trans. J.*, **91**, 1992, 117–120.
151. T. Sato, T. Murakami, E. Shimada, K. Komeya, T. Komeda and M. Komatsu, "Corrosion of Silicon nitride ceramics under hydrothermal conditions" *J. Mat. Science*, **26**, 1991, 1749–1754.
152. M. Yoshimura and S. Yamamoto, "Corrosion and alteration of ceramic materials in high temperature-high pressure water", *Seramikkusu*, **30**, 1995, 995–998.
153. K. Oda, T. Yoshio, Y. Miyamoto and M. Koizumi, "Hydrothermal corrosion of pure, hot isostatically pressed silicon nitride", *J. Am. Ceram. Soc.*, **76**, 1993, 1365–1368.
154. K. Oda and T. Yoshio, "Properties of Y_2O_3-Al_2O_3-SiO_2 Glasses as a Model System of Grain Boundary Phase of Si_3N_4 Ceramics (Part 2)", *J. Ceramic Soc. of Japan*, **99**, 1991, 1110–1112.
155. S. Iio, A. Okada, T. Asano and M. Yoshimura, "Corrosion Behavior of Silicon Nitride Ceramics in Aqueous Solutions (Part 3)", *J. of the Ceramic Soc. of Japan*, **100**, 1992, 965–967.

7 Boride-Based Hard Materials

R. Telle, L. S. Sigl, and K. Takagi

7.1 Introduction

Materials based on boron compounds have been explored for many decades because of their exceptional properties in respect to chemical bonding, crystal structure, and phonon and electron conduction. Especially in the field of energy conversion, electron emission, and neutron absorption, borides occupy many niches of application for which no other material can be employed. Until approximately 1980, the main interest in borides always came, however, from basic research aimed at the understanding of their electronic structure, being either responsible for the unique transport properties or the peculiarities in chemical bonding. It is, therefore, no wonder that the most information about borides was at that time created from the viewpoint of physicists and chemists.

Although even some of the interesting thermal and mechanical properties of borides, e.g., the generally high melting point and high hardness, have already been exploited and led to application as wear-resistant parts or grinding grits a long time ago, the interest in borides increased dramatically together with the fundamental understanding of technical ceramics. The exploitation of materials with exceptional mechanical, chemical, electrical and thermal properties yielded boron compounds as potential candidates for "high-tech" applications besides the well-developed oxide ceramics, silicon nitride, silicon carbide, and hard metals or cermets.

For a successful economic application, these advanced ceramics should be available without any limitations; prepared by relatively simple methods, they should exhibit a low specific gravity, high reliability, long lifetime and of course they should be available at low costs. Unfortunately, most boron-based materials do not satisfy all of these requirements and, hence, disadvantages and disappointments have to be tolerated in development and application. In particular, densification of powder-derived parts is extremely difficult due to their high amount of covalent bonding resulting in low diffusion coefficients and, therefore, costly. Furthermore, oxide impurities on surfaces and grain boundaries create undesired and unexpected effects on the properties whereas the affinity of boron to oxygen limits its use at higher temperature in an oxidizing atmosphere. Additionally, phase diagrams and crystal structures of the particular compounds are the subject of permanent revision since improvement of the purity of raw materials, more sophisticated synthesis procedures and characterization methods, such as precursor-based processing, advance chemical analysis, atomic-resolution transmission electron microscopy, and high-temperature X-ray diffraction, yielded more precise but also more confusing information about the true characteristics of the boride structures and phases.

Besides the relatively high costs a lack in understanding is the main reason why boride-based materials are not yet in as widespread use as the other ceramics materials, although some properties such as extreme hardness make them superior to oxides and nitrides. Thus, some "eras" of boride exploration can be distinguished, one supported by the needs of the nuclear energy and weapon industry around 1960, another between 1970 and 1980 for thermo-electric power generation and coming up again at the present, the next driven by the need for exceptional wear properties between 1980 and 1990, and finally between 1985 and 1992 triggered by aerospace and military research. The interest in borides rising from a political and strategic point of view has contributed to the fact that information about borides is still difficult to obtain and even more difficult to interpret since most of the data have been restricted for a long time and their sources and quality cannot be ascertained in all cases. Thus, sometimes contradictory data are available that cannot be easily discerned as being right or wrong.

Besides the "high-tech" applications of novel materials, it should not be forgotten that there was and still is a tremendous market for borides in the metallurgy of steel and iron, e.g., for antioxidizing additives in refractory linings or as alloying ingredients for the metals.

Thus, this contribution is aimed at the state of the art in boride ceramics with their problems in densification, microstructural peculiarities and exceptional mechanical properties. Starting with the unique interaction of metallic, covalent and ionic types of bonding and the crystal structures of technically important compounds, phase diagrams will be presented as far as they are of technical interest. The major part consists of the description of the synthesis and properties of ceramics and cermets, reflecting the development of suitable sintering procedures and the consequent improvement of the thermal and mechanical properties.

7.2 Chemical Bonding and Crystal Chemistry of Borides

The nature of the chemical bonding is the key to the physical and chemical behavior of matter. Borides possess exceptional properties due to a high amount of covalent bonding in combination with small band gaps or even metal-like transport properties. The unique interaction of metallic, covalent and ionic types of bonding results in a high melting point combined with semiconductivity or metallic electric and thermal conductivity and excellent wetting by metallic melts. The formation of many unique crystal structures found only in borides reflects again the outstanding nature of these materials.

7.2.1 Chemical Bonding of Borides

The nature of the chemical bonding in boron compounds is governed by the well-known two-electron-three-center bond, i.e., three boron atoms share two common electrons. These electrons are thus more or less delocalized. The resulting sp² hybridization leads to the plane B_3X_3 hexagon as the main structural element in BN, B_2O_3, H_3BO_3 and related compounds, and to the B_3 triangle as a fraction of the typical five-fold symmetric icosahedron of elemental boron, the group of boranes and their derivatives. Depending on the saturation of the electron deficiency, soft and non-conducting, salt-like compounds or semimetallic to metallic materials of exceptionally high melting point and hardness and excellent electrical conductivity exist. As pointed out in the following section, the latter boron compounds may contain ionic, metallic and covalent fractions of bonding forming very stable compounds due to the well-balanced electron transfer between metal and boron sublattice.

7.2.2 The Crystal Structure of Borides

Similar to silicates, the crystal structures of borides can easily be classified according to the arrangement of the boron atoms. Boron may occur as an isolated atom or form B-B bonds with an increasing degree of interconnection in the chains, double chains, layers and frameworks and combinations thereof (Fig. 1). Due to the strong covalent bonding between the boron atoms and the electron deficiency of the three-center bond a number of complex and unique structures result which

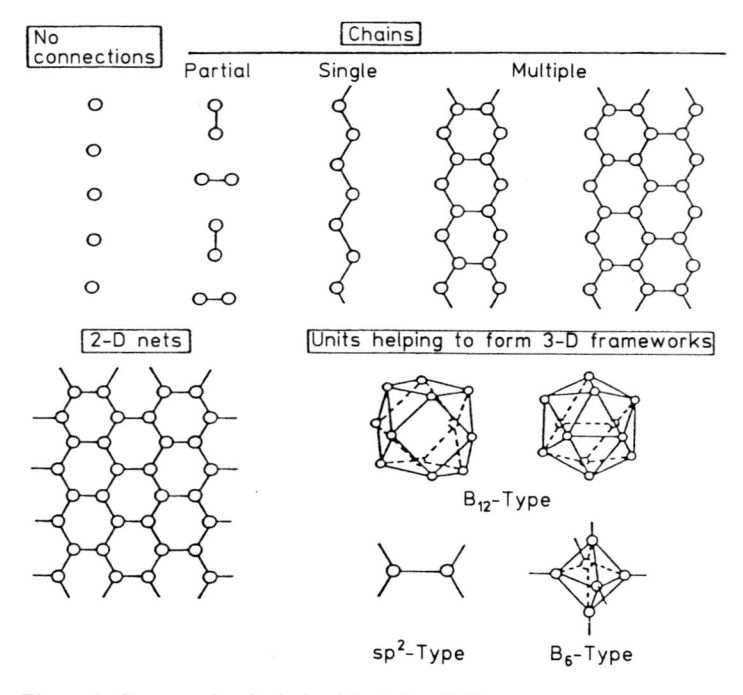

Figure 1. Structural units in borides (after [64]).

have been the subjects of investigation for many years [1, 2]. In general, compounds with a boron-to-metal ratio of less than 1.0 are built up of isolated boron atoms or pairs with a low B-B interaction (e.g., Ni_3B, Ru_7B_3, Fe_2B, Cr_5B_3), in zigzag chains with additional, isolated B (e.g., o-Ni_4B_3). At a ratio of 1.0 to 1.3, infinite chains are formed which may be parallel to one or even two crystallographic axes (e.g., m-Ni_4B_3, FeB, CrB, MoB), whereas in M_3B_4 borides double chains are predominant (e.g., Cr_3B_4). With increasing boron content, two-dimensional nets are stable, yielding preferential stoichiometries between M_2B_3 and MB_4. The most important structure type group thereof is the AlB_2 type, which is covered in more detail later on. Three-dimensional frameworks exist in so-called higher borides with typical stoichiometries of MB_4, MB_6, MB_{12}, and MB_{66}. Channels with rectangular cross-sections were found in, e.g., CrB_4 and MnB_4, which is unique for the three-center bond of boron [3]. A rigid boron skeleton consisting of B_6 octahedra is a characteristic of the CaB_6 structure group (important member: LaB_6), whereas the UB_{12} structure contains B_{12} cubo-octahedra. Other borides of MB_6 and MB_{12} stoichiometry or a higher boron-to-metal ratio, especially the main group element borides, can be derived from the trigonal rhombohedral α-boron or β-boron structure with the B_{12} icosahedron as the most important structural unit. The SiB_6 structure consists of a special type of boron arrangement built up by 18 boron icosahedra (B_{12} units), 4 icosihexaedra (B_{15} units) and 8 single atoms where some of the boron positions are occupied by Si [4]. A structural curiosity are the yttrium borides with $B/Y > 25$. The member being richest in boron, YB_{66} crystallizes in cubic symmetry and contains 1584 boron atoms and approximately 24 yttrium atoms (Fig. 2). The boron is arranged in 27 $(B_{12})_{13}$ units made up by 13 interconnected icosahedra and in 8 B_{80} clusters being occupied by approximately 42 B atoms. The yttrium is arranged in Y-Y pairs of most probably Y^{3+} state, providing all the electrons necessary to stabilize this framework [5, 6]. Even this compound is of high technical importance for monochromators of synchrotron radiation.

Some general systematics on chemical bonding and crystal chemistry have been published by Matkovich and Economy [7], and Aronsson et al. [8, 9], who also refer to the structural similarities in silicides and phosphides.

7.2.2.1 AlB$_2$-Type Structures

The transition metal borides of the AlB_2 structure type group are of great technical interest for ceramics, as are the ternary ω, φ and τ type borides as compounds for cermets and coatings. The AlB_2 structure is conveniently described as a sequence of alternating metal and boron layers of hexagonal symmetry. The metal layers are close-packed and stacked in an A-A-A sequence, resulting in a basal-centered unit cell. The boron atoms are six-fold coordinated and situated in the center of trigonal prisms of metal atoms (H position). Hence they generate a planar primitive hexagonal, two-dimensional, graphite-like network (Fig. 3). The total stacking sequence is then AHA-HAH... and belongs to the space group $P6/mmm$. The unit cell contains one formula unit, MB_2. Since this structure is very versatile at accommodating metal atoms of various sizes and electron configurations, M could be Mg, Al, group IVa, Va, VIa, actinide or lanthanide elements.

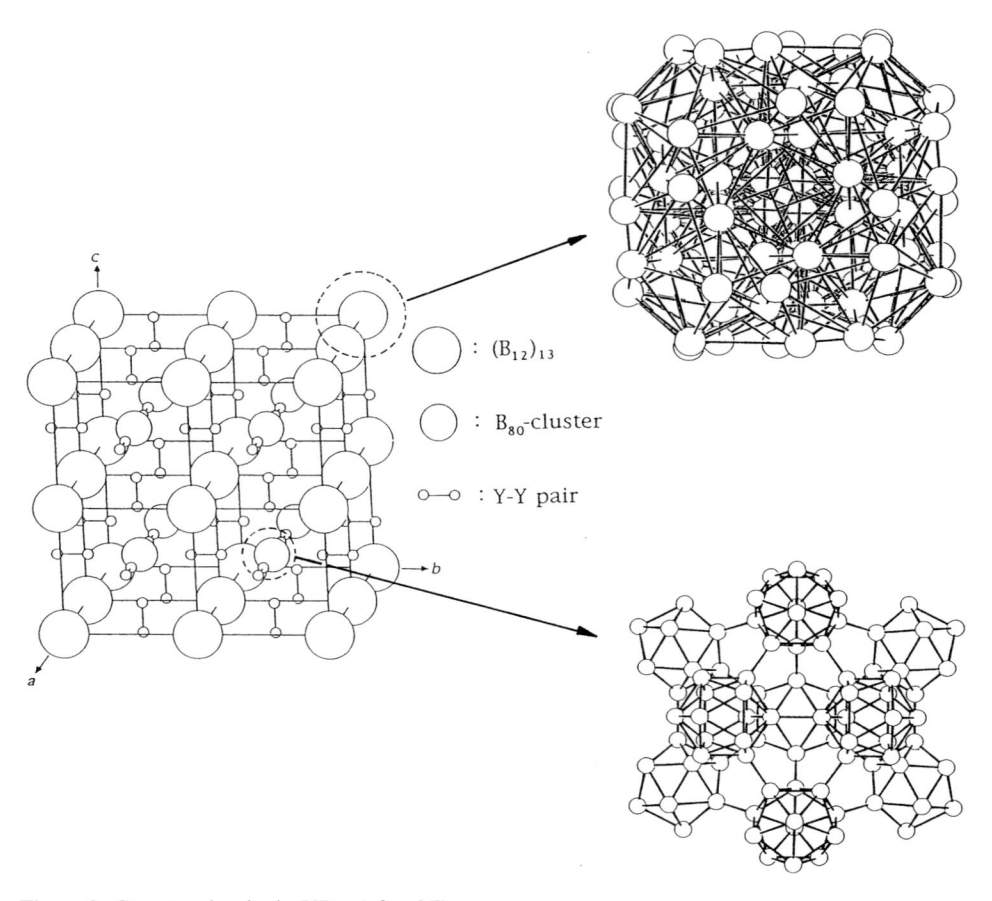

Figure 2. Structural units in YB$_{66}$ (after [6]).

Furthermore, other transition metal borides of various stoichiometries can be derived from the AlB$_2$ structure type by introducing the metal layer positions B and C in analogy to close packings and the boron layer types K and K', which are slightly puckered. By allowing stacking sequences such as $AHAK$-$BHBK$-$CHCK$... ("Mo$_2$B$_5$"-type), $AHAK'$-$BHBK'$... ("W$_2$B$_5$"-type), or $AH'AK'$-$BH'BK'$... (Ru$_2$B$_3$-type) and $AK'BK'$... (ReB$_2$-type), and vacancies in the boron K layers, other structures and symmetries can be generated (Fig. 4) [9, 10]. The particular molybdenum and tungsten borides have formerly often been denoted Mo$_2$B$_5$ and W$_2$B$_5$, respectively, but there is evidence that the homogeneity range is narrow and close to the 1:2 stoichiometry [9–12, 141]. High-resolution TEM micrographs of WB$_2$-containing ceramics show, however, that the puckered B layer really exists regularly alternating with the plain boron layer of the AlB$_2$ structure (Fig. 5), yielding the space group $P6/mmc$. Thus, the stoichiometry range from WB$_{2.0}$ [13] to WB$_{2.27}$ [14] arises from a boron deficiency in both kinds of layers. As

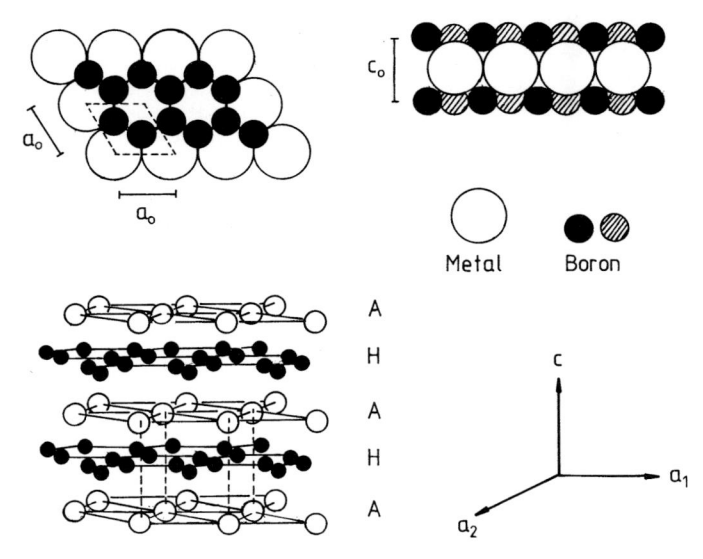

Figure 3. The AlB$_2$ structure type.

an exception, AlB$_2$-type WB$_2$ has been produced under non-equilibrium conditions by chemical vapor reaction of WCl$_6$ with a boron wire at 800°C [15] with entirely different unit cell dimensions.

Calculation of the band structures of AlB$_2$-type compounds shows that no band gaps are present, and all the compounds are predicted to be electron conductors, which is in agreement with experimental results. For the main-group element

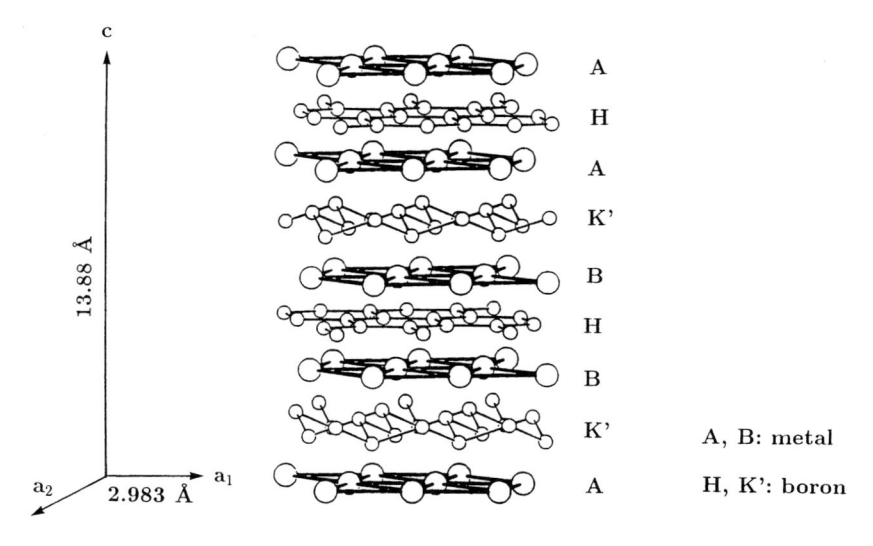

A, B: metal
H, K': boron

Figure 4. The Mo$_2$B$_5$ structure type (after [11]).

Figure 5. High-resolution TEM micrograph of W_2B_5 stacks, view of [110] zone axis, distance of stacks: 13.9 Å (courtesy, B. Freitag [368]).

diborides the boron $2p\sigma$ and $2p\pi$ orbitals are the main constituents of the states at the Fermi edge, while for the transition metal diborides it is the localized metal 3d orbitals which are the predominant component of the valence and conduction bands. Since the boron sublattice is electron deficient all diborides exhibit an electron transfer from the metal atom to the boron, which gives rise to a strong ionic contribution to the bonding. In the transition metal diborides, the charge transfer decreases from 2.28 electrons in ScB_2 to 1.09 electrons in MnB_2 [16]; lower values have been presented by Samsonov and Kovenskaya [17, 18]. The additional electrons occupy the $2p\pi$ orbitals of the boron where the electrons are involved in both the B-B bonding as well as the metal–boron interactions. Cluster calculations of main group element diborides show that the metal–metal bonds are weak, the metal–boron interaction is significant and the boron–boron interactions are very strong. In the transition metal diborides the metal–metal bonds within the layers are considerably stronger than in main-group diborides and reaches a maximum for VB_2. This internal bonding within the layers is clearly of a metallic type and is thus responsible for the metallic transport properties. The metal–metal interlayer bonding, as well as the metal–boron interactions also increase from ScB_2 to MnB_2, whereas the contribution of the boron–boron bonding decreases in this order. Due to the existence of vacancies in the boron layer and the possible occupation of interstitial sites by additional boron atoms, the boron sheets may also exhibit some metallic or semimetallic conductivity. The considered metallic fraction does not, however, account very much for the transport properties. In contrast, the interaction between metal and boron layers contains a more efficient metallic portion, which explains the electric conductivity along the c-axis [19]. In ideal boron layers, the donor capability of the metal governs the extent of electron localization in the sp states of the boron atoms. Thus the covalent character of the B-B bonds decreases from group IV to group VI metal diborides [20].

As it has been established that the boron network is rather rigid and governs the lattice expansion in the a direction whereas the lattice dilatation perpendicular to the metallic layers strongly depends on the metal species, it seems likely that the metal atoms are distorted in some cases [21, 22]. The c/a ratio is thus a function of the r_{metal}/r_{boron} ratio and depends furthermore on the valency electron concentration [19].

Since the layers of the AlB_2 type structure consist of very distinct kinds of material, namely metals and boron, exhibiting entirely different binding characteristics, more attention was recently paid to the (0001) terminating surface layers of single crystals. Hayami et al. [23] and Souda et al. [24] have studied HfB_2, TaB_2, and WB_2 single crystals by impact-collision ion scattering spectroscopy at 5×10^{-8} Pa inert gas pressure, which gives information on the ultimate surface structure by atomic shadowing effects of scattered He^+ and Li^+ ion beams at low incident angles. As a result, the HfB_2 basal plane is shown terminated by metal atoms whereas B is entirely absent even in some layers down further. In contrast, TaB_2 and WB_2 basal planes exhibit boron surface layers emerging from boron diffusion to the surface at above approximately 500°C, and compensating for the boron defects due to ion bombardment. This effect is attributed to the comparatively stable graphite-like boron layers and the higher affinity of Hf to oxygen compared to Ta and W.

7.2.2.2 Crystal Structure of Boron Carbide and Isotopic Compounds

Boron carbide, referred to as (B_4C) in brackets indicating the solid solution in contrast to $B_{13}C_2$, for example, meaning the stoichiometry, crystallizes in the trigonal-rhombohedral space group $R3m$. The unit cell is shown in Fig. 6. The structure may be described as a cubic primitive lattice elongated in the direction of the space diagonal with almost regular icosahedra at the corners. Parallel to the space diagonal, which becomes the c-axis in hexagonal notation, a linear chain consisting of three atoms interconnects the adjacent icosahedra. Thus the unit cell contains twelve icosahedral sites and three sites on the linear chain. If B atoms are attributed to the icosahedral positions and C atoms are considered to be situated in the linear chain, a stoichiometry of $B_{12}C_3$, i.e., B_4C, results. The icosahedra exhibit two topologically different positions, first the $B1$ position (also known as $6h_1$) which consists of a planar array of three atoms perpendicular to the linear chain around the outer atoms. Thus, this position occurs six times in the unit cell. The second distinguishable icosahedral site is the $B2$ (or $6h_2$) position, which is situated in the middle of the edges of the rhombohedral unit cell and accounts for a further six atoms. A special position is the centrosymmetric $B3$ (or $1b$) site in the linear chain, which is considered to be preferentially occupied by larger atoms such as Al and Si that form solid solutions [25–29]. For the binary solid solution series the question is whether the linear chain is formed by a C-C-C, a C-B-C, a C-C-B, a B-C-B, or a B-B-C array throughout the homogeneity range and how to explain the variations in the

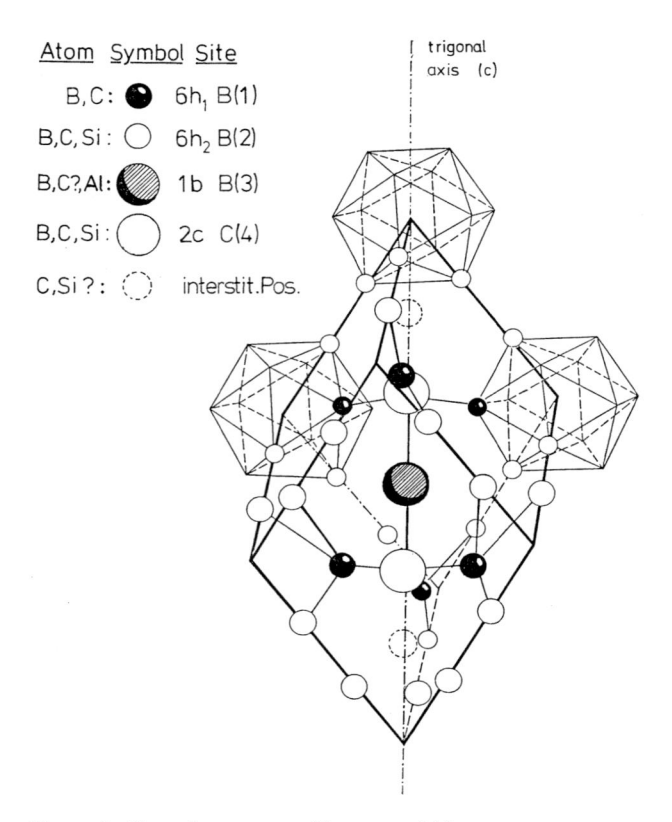

Atom	Symbol	Site
B,C:	●	$6h_1$ B(1)
B,C,Si:	○	$6h_2$ B(2)
B,C?,Al:	◐	1b B(3)
B,C,Si:	○	2c C(4)
C,Si?:	○	interstit.Pos.

Figure 6. Crystal structure of boron carbide.

stoichiometry [30–42]. At that time, the majority of the authors agreed that the linear chain is of the C-B-C type and does not change with the C content, which fits the stoichiometry of the most stable compound in the system, $B_{13}C_2$. Accordingly, the linear chain thus contains a closed shell of ten valence electrons, which is achieved by a charge transfer to the B_{12} icosahedra to which 38 valence electrons are formally assigned. The charge difference between B and C within the chain results in even stronger electrostatic binding forces [35], but the energetic differences estimated by density-of-state calculations are very small [43]. Since there is a deficiency of one electron in the icosahedron, additional C as an electron donor preferentially replaces B in one of the icosahedral sites [42, 44, 45]. The $B2$ site was established as the most favorable position for such a replacement [41, 43, 46–49]. Thus the total structure can be written as $(B_{11}C)^-(CBC)^+$.

To re-investigate the carbon-rich limit of solid solubility, boron carbide being in eutectic equilibrium with graphite was prepared by arc fusion or diffusion couples and subsequently analyzed with a microprobe [50] yielding a composition of $B_{4.3}C$ (i.e., a maximum carbon concentration of 18.87 at.-%). Optical absorption spectroscopy of this material indicated that 81.4% of the linear chains had a C-B-C structure

Table 1

Phase	Sublattice structure	Reference
B_4C	$(B)_{12}(C_3, C_2B, B_4, C_2Si)$	[52]
$B_{4+\delta}C$	$(B, C)_{12}(C_2B, CB_2, B_2V)$	[53]

and 18.6% consisted of a C-B-B array; the chains were statistically distributed [51]. Concluding from Fourier transform infrared (FTIR) spectroscopic data, Kuhlmann et al. [51] argue that in contrast to the generally accepted continuous substitution of B_{12} for $B_{11}C$ icosahedra and C-B-B for C-B-C chains with decreasing carbon content, the structure consists of statistical mixtures of these units. With decreasing C content, a growing portion of unit cells without any central linear chain is formed. Earlier, the boron-rich side of the homogeneity range was established at 8.7 mol.-% C (i.e., B:C = 10.4) by Bouchacourt and Thevenot (1981) [49] who assigned the structural composition $B_{12-x}C_x(C-B_xC_{1-x}-C)$ with 15.33 atoms per unit cell to the resulting formula $B_{10.4}C$. The density of boron carbide at the boron-rich end is with 2.465 g cm^{-3} lower than that at the carbon-rich corner with 2.51 g cm^{-3}.

Together with the assessment of the binary B-C phase diagram the homogeneity range of boron carbide and the other compounds was modeled by means of the Compound Energy Formalism using the following sublattice models (Table 1) where vacancies are denoted as V:

In comparison to Lim and Lukas [52] who still consider the existence of a linear C-C-C chain, Kasper [53] favors the model taking vacancies in the chain into account. His calculation reveal a prediction of the occupation state of the icosahedral and chain position (Fig. 7) being in accordance with experimental results on IR absorption bands and heat capacity.

As already mentioned, the linear chain can accommodate other main group elements such as Al, Si, P, As, and O without a change in the structure type. Solid solutions formed with (B_4C), however, are only known for Al and Si, which can partially occupy one of the positions within the C-B-C chain, e.g., Al is placed in the centrosymmetric *B*3 site which causes a slight kinking of this linear array, whereas Si replaces up to one third of the carbon sites at the ends of the chain. The total solid solubility of both species is comparatively small (max. 2.5 at.-%) [28, 54], although a complete solid solution series between $B_{12}(C, B, Si)_3$ and the silicon boride "SiB_4" was considered [55]. Due to the large size of the Si atom, however, the formation of a two-atom chain is favored, as in the cases of P, As, and O. The isotopic binary borides of these elements thus fit into the general stoichiometry $B_{12}(XVX)$. This arrangement favors a charge transfer to the B_{12} icosahedron. The ideal structure of $B_{12}O_2$ may be written as $(B_{12})^{2-}(OO)^{2+}$, which exactly yields a filled-band configuration. In reality, for all these compounds a considerable disordering has to be taken into account, which results in the compositions $B_{12}P_{1.8}$, $B_{12}As_{2.0}$, and $B_{12}O_{1.82}$ (corresponding to $B_{6.6}O$) [25]. In the case of "SiB_4", the real stoichiometry is $(B_{10.4}Si_{1.6})(SiSi) = SiB_{2.89}$, which is attributed to a Si-Si chain and two sites in the icosahedron being partially occupied by Si [7, 21, 56]. Sublattice structure modeling carried our by Kasper [53] to establish

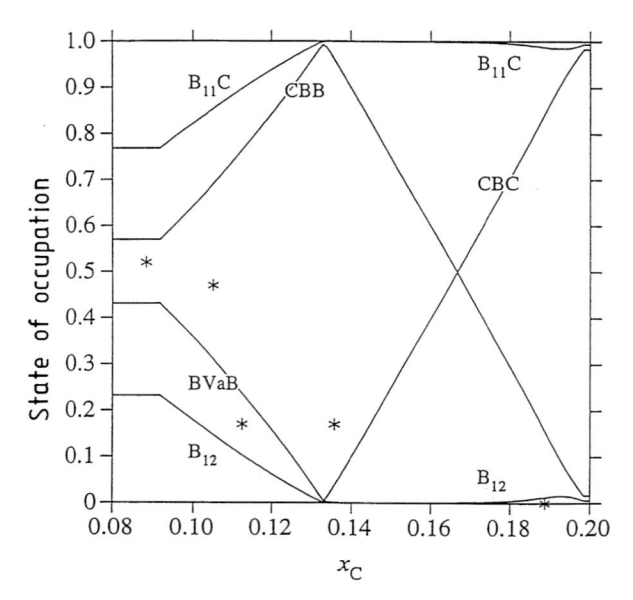

Figure 7. Calculated occupation of sites in the icosahedra and the chain of B_4C solid solution [53].

the ternary B-C-Si system considered that the Si-Si chain may also at least partially be the structural unit for the Si-containing boron carbide solid solution [Eq. (1)]:

$$B_{4+\delta}C(Si) = (B,C)_{12}(C_2B, CB_2, B_2V, Si_2) \qquad (1)$$

The solid solubility limit was then simulated by a statistical combination of all the kinds of chain occupancies known.

7.3 Phase Systems

Knowledge of the phase diagrams for compounds of technical interest and of the environmental phases in contact with these compounds is the key for materials development and for the understanding of materials behavior in application. Not only can the thermal stability of particular phases be calculated by means of thermodynamic data, but suitable sintering procedures can also easily be considered, and decomposition in aggressive media can be predicted. Generally recommended data books on binary and ternary systems are, e.g., those by Hansen [57], Elliott [58], Shunk [59], Mofatt [60, 61], Massalski [62], and Petzow and Effenberg [63]. Nowadays, the thermodynamic data of most of the important phases are available in publications or databases and can be readily used for thermochemical calculations.

In the following paragraphs, phase diagrams of the most important boron-containing phases will be presented and discussed, starting with the binary systems, then selected ternary systems which are of technical interest. In the subsequent sections particular phase systems will be treated in respect to sintering of B_4C and TiB_2 or in the context of microstructural design and mechanical strengthening.

7.3.1 Binary Phase Diagrams of Technically Important Systems

Many attempts have been made to correlate the binary metal boride phase equilibria, the boride crystal chemistry, and the ranking of the elements in the Periodic Table [64]; larger metals and those with unfilled d-shells favor the formation of boron-rich phases with two- or three-dimensional boron frameworks. The smaller metals having a high number of d-electrons prefer the formation of metal-rich phases with only a few boron-rich bonds, whereas more noble metals such as Ru, Rh, and Pd tend to generate defect structures. Systematic work on crystal structures and stabilities has been performed by Kiessling [1], Aronsson *et al.* [8], and Lundström [2, 65]. Spear [64, 66] presented a systematic compilation of binary phase diagrams resulting in predictions of phase relations. Guillermet and Grimvall [67, 68] systematized thermodynamic data of transition metal diborides with emphasis on the enthalpy of formation, the vibrational entropy, and the melting temperature, in order to account for the transition from stable to metastable phases with increasing atomic number. As a result, an entropy-related free energy term was introduced which correlates linearly with cohesive energies and melting points. By this means, stabilities of boride, carbide, and some nitride phases have been successfully predicted. In the following sections some binary systems of technical interest are presented.

7.3.1.1 The B-C System

In contrast to early publications by Samsonov and Schuravlev [69] and Schuravlev and Makarenko [70] considering several boron carbide phases it is generally accepted today that only one binary phase $B_{13}C_{2\pm x}$ exists with a wide homogeneity range of 8.8 to 20.0 at.-% C, depending on temperature. This phase melts congruently at 2450°C [58] at the composition $B_{13}C_2$ (18.5 at.-% C, 20.2 weight-% C). For the B-rich corner of the phase diagram, Bouchacourt and Thevenot [71] proposed a degenerated peritectic with elemental boron at 2075°C, according to measured element distribution coefficients. In this diagram the melting point of boron is placed at 2020°C. Since the melting point of B accepted today is 2092°C the resulting reaction with boron carbide should be an eutectic one, assuming that the non-variant equilibrium at 2075°C is correct. The maximum carbon content is usually given as 20.0 at.-%, corresponding to the stoichiometry of B_4C. Beauvy [72] suggested a carbon content steadily increasing with temperature from 21.4 at.-% (20°C) to 23.1 at.-% (2375°C). Recent microprobe analyses by Schwetz and Karduck [50] indicated, however, that the maximum carbon content of fused boron carbide being in equilibrium with graphite is only 19.2 at.-% at the eutectic with carbon, and 18.5 at.-% at 1000°C, corresponding to the formula $B_{4.3}C$. The eutectic with

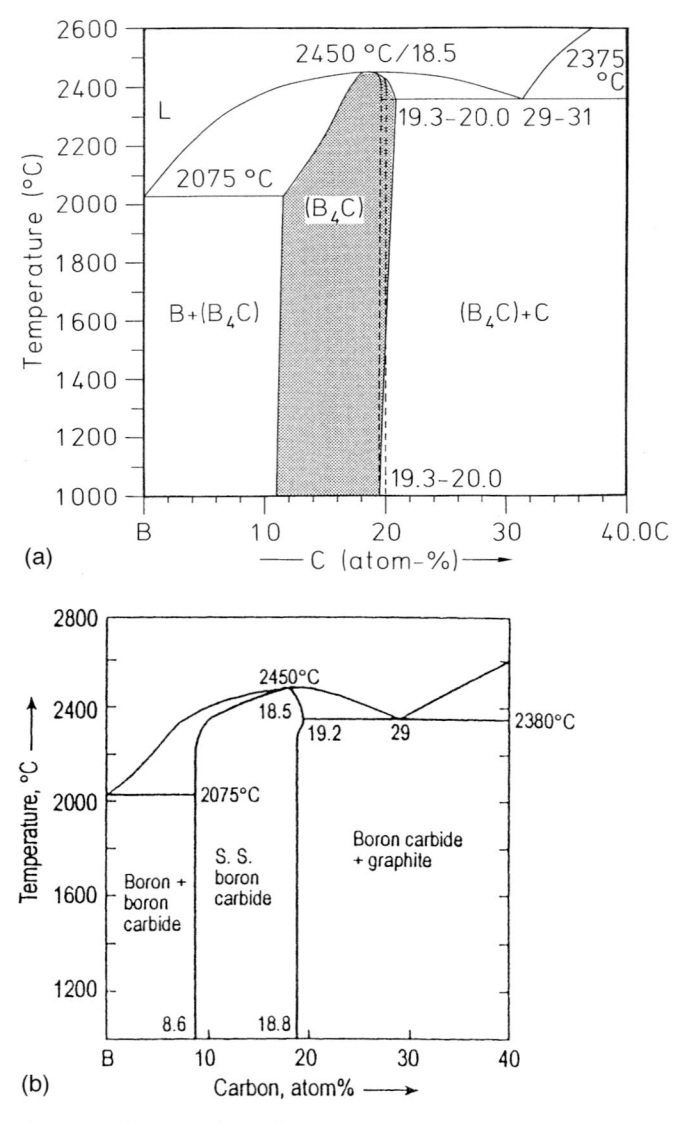

Figure 8. The B–C phase diagram (a) calculated, (b) according to Schwetz and Karduck [50].

carbon is given at 2375 ± 5°C and 29–31 at.-% C, which is in a good agreement with thermodynamic calculations [53, 73] stating 29.1 at.-% as the eutectic composition. The phase diagram is presented in Fig. 8 comparing the calculated version (Fig. 8a) with the one (Fig. 8b) suggested by Schwetz and Karduck [50]. The reaction of boron carbide with elemental boron was modeled as a eutectic at 0.2 at.-% C and 2073°C, i.e., 2 K below the melting point of boron.

Table 2. Experimentally observed and calculated reactions in the B-Si system.

Reaction type	Temperature °C					
	Calculated			Experimental		
	Dörner [78]	Lim and Lukas [52]	Kasper [53]	Armas [82]	Olesinski and Abbaschian [80]	Telle [83]
$SiB_n \leftrightarrow \beta\text{-B} + \text{liquid}$	2020	2020	2054	1929	2020	2060
$SiB_6 \leftrightarrow SiB_n + \text{liquid}$	1898	1850	1850	1989	1850	1900
$SiB_6 + (Si)_{solid} \leftrightarrow \text{liquid}$	–	1385	1384	–	1385	1385
$SiB_3 \leftrightarrow SiB_6 + \text{liquid}$	1377	–	–	1377	–	–
$SiB_3 + (Si) \leftrightarrow \text{liquid}$	1340	–	–	1340	–	–
$SiB_3 \leftrightarrow (Si)_{solid} + SiB_6$	–	1270	1198	–	1270	1358

7.3.1.2 The B-Si System

The B-Si system is of particular importance for the understanding of Si as a sintering aid for boron carbide because of its chemical relationship to B and C since hot pressing of Si bearing boron carbide results in a significant reduction of the grain size and, therefore, in an improved fracture toughness and strength of 500–600 MPa [74, 75]. Moreover, the silicon borides have been periodically studied for their thermoelectric and thermomechanical properties.

Uncertainties in the interpretation of the system arise from the three silicon borides, $SiB_{2.89-3.65}$, SiB_6, and SiB_{12-14}. Both the homogeneity range and the decomposition temperatures have not yet been completely established (e.g., Elliott [58]; Ettmayer et al. [76]; Lugscheider [77]; Dörner [78]; Olesinski and Abbashian [80]). Experimental problems in both binary and ternary phase studies are related to the comparatively high vapor pressure of Si at temperatures exceeding 1400°C. The Si-richest $SiB_{2.89-3.65}$, often also denoted SiB_3 or SiB_4, can be derived from the B_4C structure by an arrangement of $B_{12-x}Si_x$ icosahedra and a Si-Si chain. It is not known for certain whether "SiB_3" decomposes peritectoidally at 1270°C to form SiB_6 and solid Si or shows a eutectic reaction with Si at 1340–1385°C and decomposes peritectically to SiB_6 and a boron-rich Si melt at 1377°C.

SiB_6 crystallizes in an own structure type of space group *Pnnm*, which is built up by 18 boron icosahedra (B_{12} units), 4 icosihexaedra (B_{15} units) and 8 single atoms where some of the boron positions are occupied by Si [4]. In spite of this versatility in B-Si exchange, its solid solubility for Si and B is generally considered less than 1 at.-%. It is agreed upon that SiB_6 reacts peritectically to the next higher boride SiB_{12-14} and a Si liquid containing 65.3 at.-% B.

SiB_n with $n = 12–14\ldots 23$ comprises a homogeneity range between 93.3 and 97 at.-% B and can structurally be derived from the β-B structure. The close similarity to elemental boron led formerly to the conclusion that SiB_n is in fact a Si-rich solid solution of boron. SiB_n undergoes a peritectic reaction with B containing 3 at.-% Si and liquid containing 9.3 at.-% Si at $2020 \pm 15°C$ as compiled by Olesinski and Abbashian [80] and calculated by Lim and Lukas [52]. Table 2 presents the experimental and calculated data of the various reactions, Figs. 9a,b show the corresponding types of phase diagrams.

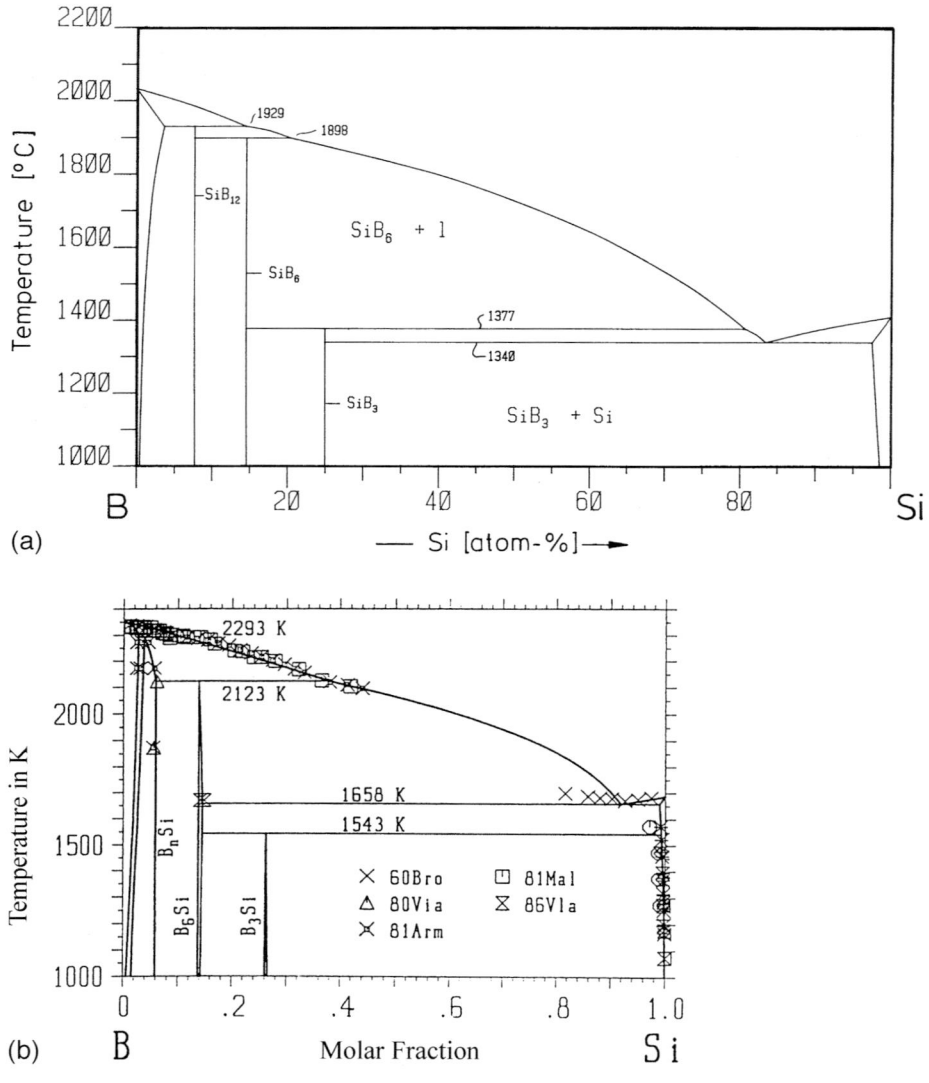

Figure 9. (a) The binary B-Si system according to experimental results with powders of technical purity. Homogeneity fields of the silicon borides according to Ettmayer et al. [76], Lugscheider et al. [77], Armas et al. 1981 [82]. (b) The binary B-Si system according to recent calculations by Lim and Lukas [52].

7.3.1.3 The Ti-B System

The most recent compilation on the Ti-B system was published by Murray et al. [84]. The assessed phase diagram (Fig. 10), being in good agreement with thermodynamic calculations, consists of three intermediate phases, orthorhombic TiB (FeB type structure), orthorhombic Ti_3B_4 (Ta_3B_4 structure), and hexagonal TiB_2

(AlB$_2$ structure). While TiB and Ti$_3$B$_4$ decompose peritectically at 2180 and 2200°C, respectively, TiB$_2$ melts congruently at 3225 ± 25°C [85]. TiB has a narrow homogeneity range of about 49–50 at.-% B [86] and reacts eutectically with Ti solid solution at 1540°C and 7 at.-% B. The existence of the Ti$_3$B$_4$ phase was proven by Fenish [86] and its peritectic reaction with TiB$_2$ and liquid was placed at 2020°C. Rudy and Windisch [85], however, omitted this phase from their binary diagram, probably because it could not be observed in melt-derived samples close to the decomposition point of TiB. In 1981, Ti$_3$B$_4$ was re-investigated by Neronov et al. [87] in reaction layers between Ti and B; later, in 1986, it was confirmed by Spear et al. [88] by arc melting and annealing studies. TiB$_2$ reacts eutectically with elemental boron at 2080 ± 20°C and approximately 98 at.-% B. Bätzner [89] re-calculated the system with slightly different liquidus lines and omitting the little solid solubilities of TiB and TiB$_2$, respectively, reported by Murray et al. [84]. The congruent melting point of TiB$_2$ was set to 3216°C, the eutectic between TiB$_2$ and B to 2059°C instead of 2080 ± 20°C after Murray et al.

7.3.1.4 The Zr-B System and Other Transition Metal–Boron Systems

Similar to the Ti-B system, ZrB$_2$ is an important phase having an AlB$_2$-type structure and a melting point of 3250°C (Fig. 11) [62]. It reacts eutectically with elemental Zr at approximately 1680°C and 86 at.-% Zr. In contrast to the Ti-B system, no ZrB or Zr$_3$B$_4$ phases exist, but there is a ZrB$_{12}$ phase with a UB$_{12}$ structure which melts incongruently at 2030°C forming ZrB$_2$ and liquid. ZrB$_{12}$ forms a eutectic with elemental boron at approximately 1990°C which is not yet firmly established.

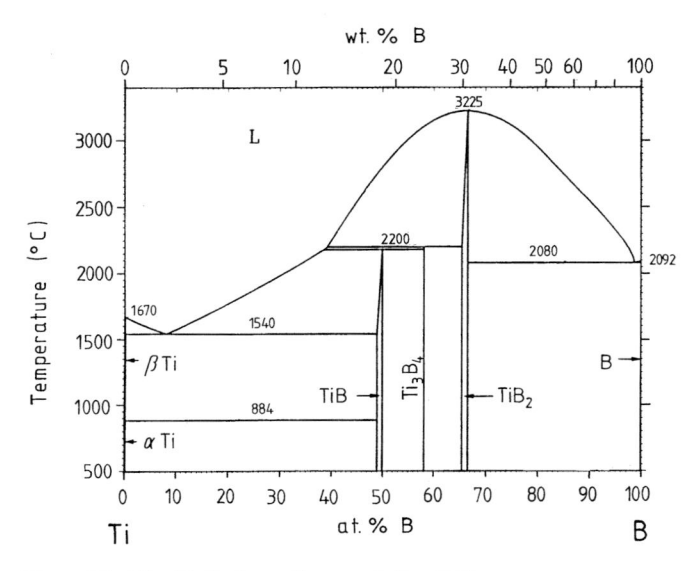

Figure 10. The Ti-B phase diagram (after [84]).

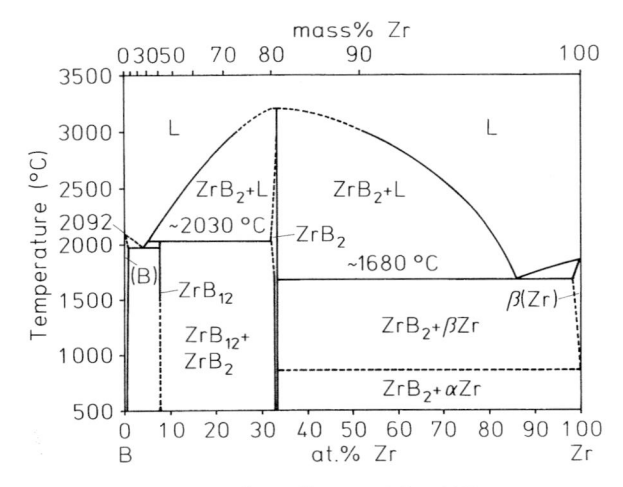

Figure 11. The Zr-B phase diagram (after [62]).

In the other transition metal boron phase systems of groups III, IV, and V, the MB_2 phase is also the dominating compound with respect to the high melting point. Exceptions are the Y-B, lanthanide metal–B and actinide metal–B systems, which possess very stable MB_4- and MB_6-type compounds. In the Y-B system additional higher borides of the stoichiometry YB_{12}, YB_{25}, and YB_{66} exist. In group V the number of known phases with a B/M ratio < 2 increases, and the tendency for the MB phase to be more stable than MB_2 is obvious on advancing from V to Ta. Destabilization of the AlB_2 structure to the benefit of the MB structure is also evident in group VI where the MB_2 phase forms an individual structure type that can be derived from the AlB_2 structure by the introduction of variations in the stacking sequences. In the VIIth and VIIIth group metal–boron systems the melting points of the borides decrease becoming significantly lower than the melting points of the elements. A summary of the known boron-containing binary systems is given by Spear [64].

7.3.2 Ternary and Higher Order Systems

Many ternary, quaternary and higher systems containing borides have been intensively investigated basically for three reasons: firstly, to elaborate suitable sintering systems for these high-melting and thus difficult to densify compounds; secondly, to avoid probable chemical complications such as phase changes and decomposition during application; and thirdly, to investigate ways of optimizing materials properties by the fabrication of tailored composites or solid solutions. Most of these investigations have been concerned with military-, nuclear- or aerospace-related research for new high-temperature materials, fabrication of cutting tools (e.g., transition metal borides for hard metals and cermets), or wear resistant parts (high-strength and high-toughness structural ceramics based on composites). For the last decade, more systematic studies related to the edition of alphabetic volumes on ternary systems, or to the investigation of peculiarities of chemical bonding in multicomponent phases, have also been started. Hence the data available can be divided into three groups: boron–carbon–metal/semi-metal systems (basically ceramics: sintering systems, composites), transition metal boride systems with low melting metals (for densification of hard metals, cermets and other cutting tools), and systems with two transition metals and boron (development of solid solutions of exceptional properties or with emphasis on the substitution of tungsten and other strategic metals). In the following sections some ternary systems are selected as technically important examples of these three categories.

7.3.2.1 Boron–Carbon–Metal Systems

Aluminum is an effective sintering aid for B_4C and SiC ceramics if combined with elemental boron and carbon. Phase relations in the B-C-Al-Si system may hence indicate suitable procedures to initiate transient liquid phase or enhanced solid-state sintering. Furthermore, liquid Al may be used to infiltrate porous B_4C bodies acting as a reinforcing phase.

Although the binary boundary phase diagrams of the Al-B-C system are rather well established there is only limited experimental information on the ternary equilibria [73, 90]. Six ternary phases have been discovered, $B_{40}AlC_4$ and $B_{48}Al_2C_8$, which have a B_4C structure and thus are probably $B_{12}(B,C,Al)_3$ solid solutions, orthorhombic $B_{51}Al_2C_8$, hexagonal and orthorhombic $B_{48}Al_3C_8$, and hexagonal B_4Al_8C, denoted T. The temperature stabilities of these phases are not known. The ternary solid solubility of $B_{12}(B,C,Al)_3$ was discovered by Lipp and Röder [27] and described in more detail by Neidhard et al. [32]. An isothermal section calculated by Lukas [73] neglecting all boron-rich ternary phases except T is presented in Fig. 12, stating that $B_{12}(B,C,Al)_3$ is in equilibrium with a boron-rich Al melt at above 1000°C.

The *B-C-Si system* was first treated by Kieffer et al. [91]. The experimental data indicated a ternary equilibrium between B_4C, SiB_6 and SiC up to temperatures exceeding 1900°C.

Thermodynamic calculations by Dörner [78], Lukas [93] and Lim and Lukas [52], however, clearly demonstrated the existence of a binary phase equilibrium of boron carbide and a Si- and B-containing melt above 1560°C. The theoretical results were confirmed by hot pressing, liquid phase sintering and infiltration experiments by Lange and Holleck [75], Telle [83], Telle and Petzow [94], and Telle [54], which also yielded more details on the extension of the homogeneity field of boron carbide towards the Si-rich corner of the system B-C-Si.

Discrepancies between previous and recent experiments as well as calculations arise from the binary B-Si system, in particular from the various plausible equilibria at the Si-rich corner. Furthermore, experimental problems in the ternary phase studies are related to the high Si vapor pressure at $> 1400°C$.

The characteristics of the B-C-Si system as assessed today, comprise the stability of a $B_{12}(B,C,Si)_3$ solid solution with a maximum of 2.5 at.-% Si [28, 82, 95] being

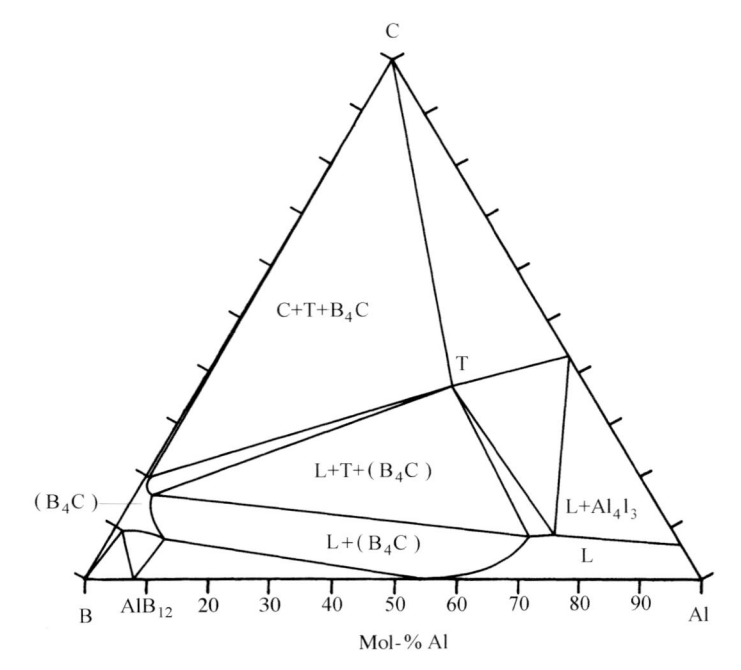

Figure 12. The B–C–Al system at 1400°C (after [73]).

in equilibrium with a boron-saturated silicon melt [54, 75]. No ternary phase exists. The unlimited solid solubility between B_4C and $SiB_{2.89}$, as postulated by Meerson et al. [55], could not be verified. According to Secrist [96] and Shaffer [97], the B_4C-SiC section is of a quasi-binary type with a eutectic equilibrium between 2250 and 2420°C and 30–35 mol.-% SiC. This section is, however, of a real ternary type since the eutectic melt postulated by Secrist and Shaffer is in fact a binary equilibrium between solid carbon and Si-B-C liquid. The formation of a solid solution of $B_{12}(B,C,Si)_3$ is accompanied by the precipitation of β-SiC which melts eutectically with SiB_6 and residual solid Si above 1380°C. As a Si incorporation into the boron carbide lattice always results in the release of C or the simultaneous formation of SiC if Si is present in excess, it may be concluded that Si substitutes for C assuming that there is no carbon in boron carbide on interstitial sites. This assumption is supported by X-ray analysis and electron energy loss spectroscopy in TEM. Lim and Lukas [52], and Kasper [53] have refined again the B-C-Si system by thermodynamic calculations. A consistent data set is now available which reproduces the maximum solid solubility of boron carbide for silicon and the peritectic decomposition of the particular silicon borides. The calculated phase diagram shows some differences to that in [54], arising from the new information about the B-Si system where now the eutectic reaction $SiB_6 + Si \leftrightarrow$ liquid is preferred rather than the $SiB_3 + Si \leftrightarrow$ liquid (Fig. 9b and 13). This means that SiB_3 is formed peritectoidally by

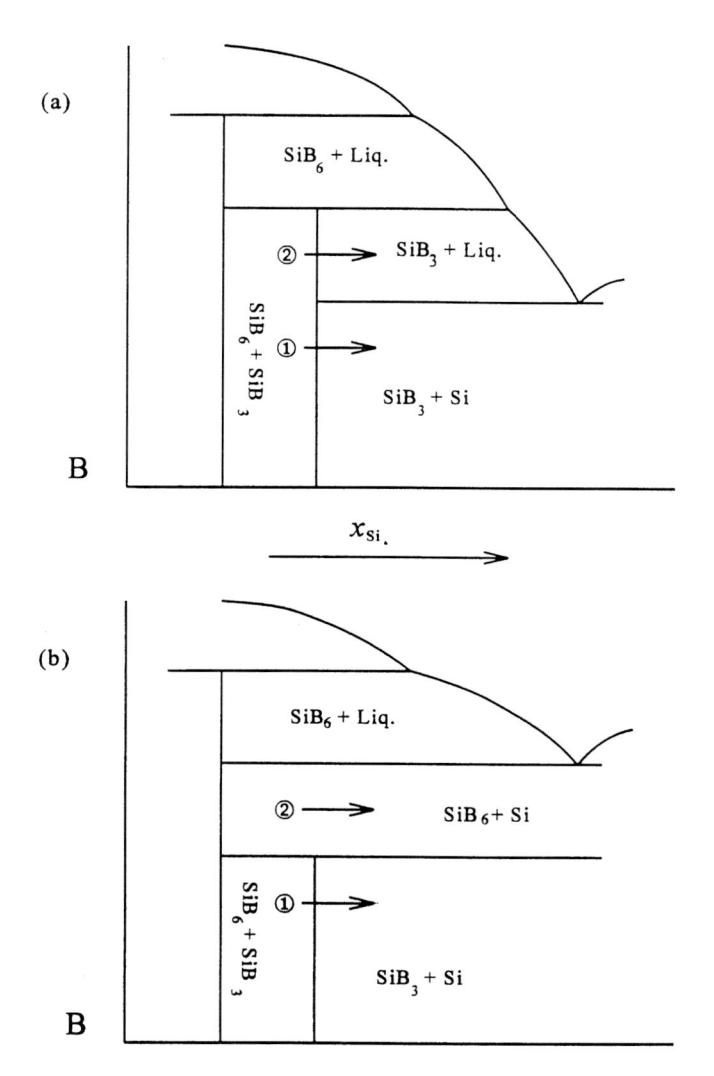

Figure 13. Two possible view of the B-Si system (after [53]).

$SiB_3 \leftrightarrow SiB_6 + Si_{solid}$. Fig. 14 shows a calculated isothermal section of the system at 2000 K according to Lim and Lukas [52].

The ternary B-C-Ti system was intensively studied by Rudy et al. [99] and has been considered for the fabrication of ceramic cutting tools [100–103]. The most recent critical assessment was published by Duschanek et al. [104, 105] who, in contrast to the previous authors, take the re-established Ti_3B_4 phase into account. No ternary phases have been discovered. The ternary solid solubility of the particular binary compounds except $TiC_{0.81}B_{0.17}$ is generally less than 1 at.-%. TiB_2 coexists with

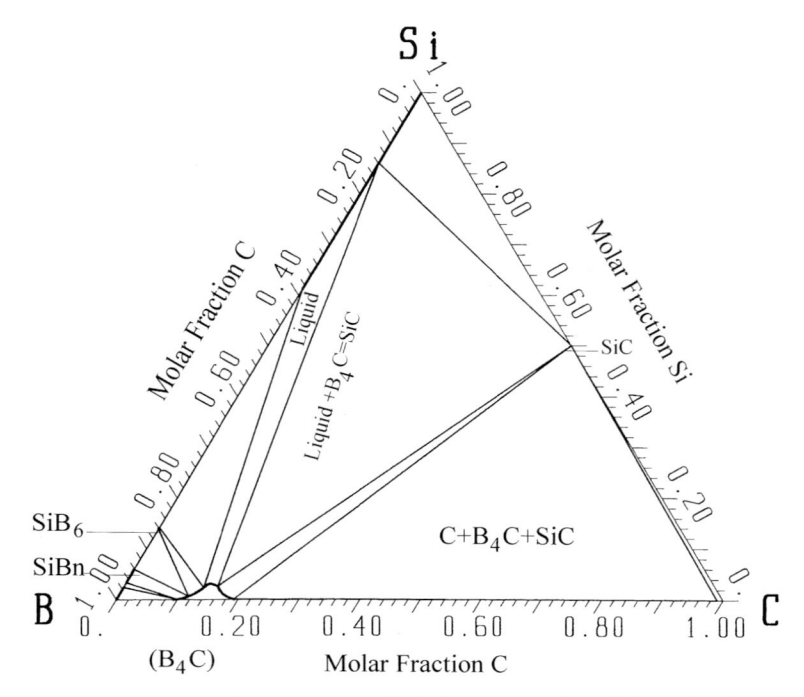

Figure 14. Isothermal section of the B-C-Si system at 2000 K (calculated by Lim and Lukas [52]).

TiC_{1-x} in a quasi-binary eutectic equilibrium at 2620 ± 15°C (calc.: 2643°C) and 57 ± 2 mol.-% TiC_{1-x} (Fig. 15, [99]; calculated by Duschanek et al. [104, 105]). TiB_2 also forms a quasi-binary eutectic with B_4C at 2310°C (calc.: 2381°C) and 88 ± 3 mol.-% B_4C, as well as with C at 2507°C (calc.: 2456°C) and 33 ± 3 at.-% C. This means, that TiC is not stable in the presence of B_4C but reacts to form TiB_2 + C. On the other hand, TiB_2-B_4C composites can be fabricated from TiC

Table 3. Experimentally observed and calculated reactions in the B-C-Si system.

Reaction type	Temperature °C			
	Calculated		Experimental	
	Dörner [78]	Kasper [53]	Secrist [96]	Telle [83]
$B_4C + SiC \leftrightarrow C +$ liquid	2166	2286	>2250	n.d.
$B_4C + SiB_n \leftrightarrow B +$ liquid	2058	2038		n.d.
$B_4C + SiB_6 \leftrightarrow SiB_n +$ liquid	1896	1840		1900
$SiC + SiB_6 \leftrightarrow B_4C +$ liquid	1560	–		1560
$B_4C + Si_{solid} \leftrightarrow SiC +$ liquid	–	1396		–
$Si + SiB_6 + SiC \leftrightarrow$ liquid	1380	–		1380
$Si + SiB_6 \leftrightarrow SiB_3 + SiC$	1345	–		1345

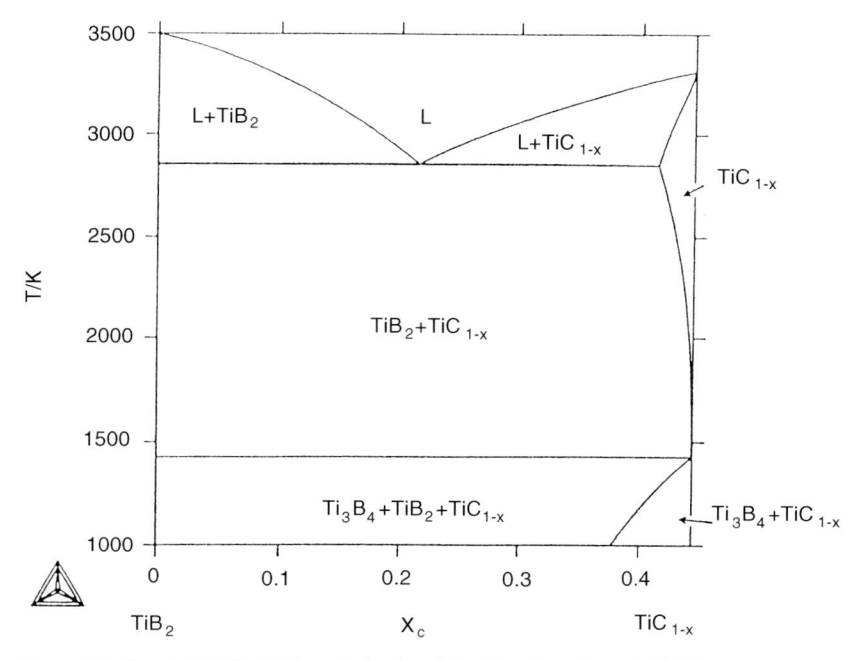

Figure 15. Isopleth TiB$_2$-TiC$_{1-x}$ (calculated by Duschanek *et al.* [105]).

and B by reaction sintering. Composites of TiC$_{1-x}$ and TiB$_2$ have been investigated for coherent grain boundaries [103]. Furthermore, TiC$_{1-x}$ is stable together with TiB up to approximately 2160°C (calc.: 2113°C) and then decomposes to Ti$_3$B$_4$ and liquid. The previously established ternary peritectic at 2160°C (TiB$_2$ + TiC$_{1-x}$ → TiB + liquid, [99]) has also to be replaced by three transition equilibria due to the existence of Ti$_3$B$_4$. Table 4 presents the ternary reactions accordingly [105].

Table 4. Ternary Reactions in the B-C-Ti system [104, 105].

Reaction Type	Reaction	Experimental temperature [°C]	Calculated temperature [°C]
quasibinary eutectic	TiB$_2$ + TiC$_{1-x}$ ↔ liquid	2620	2643
quasibinary eutectic	TiB$_2$ + C ↔ liquid	2507	2456
quasibinary eutectic	liquid ↔ TiB$_2$ + B$_4$C	2310	2381
eutectic	liquid ↔ TiB$_2$ + TiC$_{1-x}$ + C	2400	2394
eutectic	liquid ↔ TiB$_2$ + B$_4$C + C	2240	2246
transition	liquid + TiB$_2$ ↔ Ti$_3$B$_4$ + TiC$_{1-x}$	2180	2121
transition	liquid + Ti$_3$B$_4$ ↔ TiB + TiC$_{1-x}$	2160	2113
eutectic	TiB$_2$ + B$_4$C$_{ss}$ + B ↔ liquid	2016	2058
eutectic	TiB + TiC$_{1-x}$ + βTi ↔ liquid	1510	1535
transition	βTi + TiC$_{1-x}$ ↔ TiB + αTi	890	927

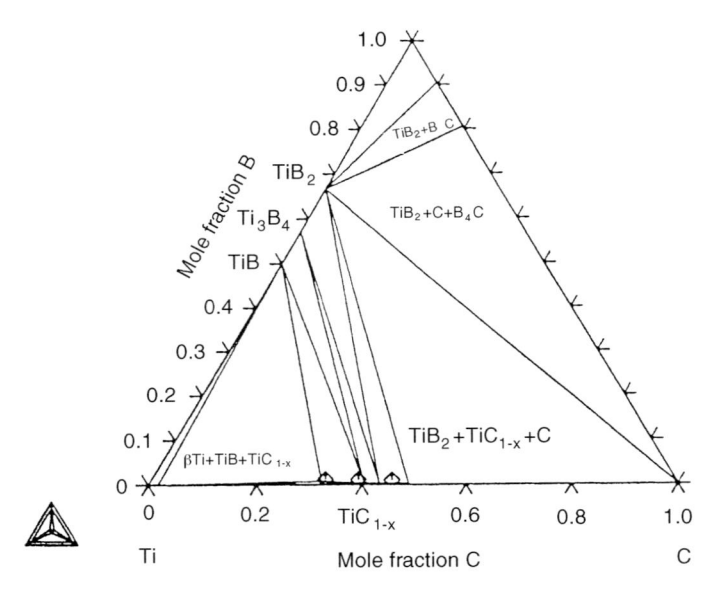

Figure 16. Isothermal section of the B–C-Ti system at 1400°C [105].

Figures 16 and 17 show isothermal sections of the Ti-B-C system at 1400°C and 2100°C, respectively [104, 105]. The diagrams based upon the work of Rudy et al. [99] are wrong below 2100°C as they neglect the Ti_3B_4 phase. Technically interesting isopleths are shown in Fig. 18 (TiC-B), Fig. 19 (TiB_2-B_4C), Fig. 20 (B_4C-Ti), and Fig. 21 (B_4C-TiC). The equilibria of the various titanium borides with TiC_{1-x} have been studied in more detail by Brodkin and Barsoum [106].

The quaternary B-C-Si-Ti system has not yet been established completely. Evidently TiB_2 and SiC form a eutectic and can be cast after arc-melting [107]. TiB_2-$B_{12}(B, C, Si)_3$-SiC composites are chemically stable and may form a quasiternary eutectic. Composites of these three hard materials can be prepared either by direct hot pressing or by reactive infiltration of porous B_4C bodies with an Si-$TiSi_2$ eutectic melt [108, 109]. Of technical interest is the ternary boundary system C-Si-Ti studied by Brukl [110], Borisova et al. [111], and Holleck [112], calculated by Touanen et al. [113] and experimentally re-examined by Wakelkamp et al. [114].

7.3.2.2 Ternary Systems with Boron and Metals

High-temperature equilibria of the extraordinarily hard borides with metallic melts bring about the opportunity for a pressureless liquid phase sintering and the fabrication of hard and simultaneously tough composites similar to hard metals but are also of interest in ceramic systems or coatings. In this section emphasis is put on binary and ternary borides which are in equilibrium with transition metals.

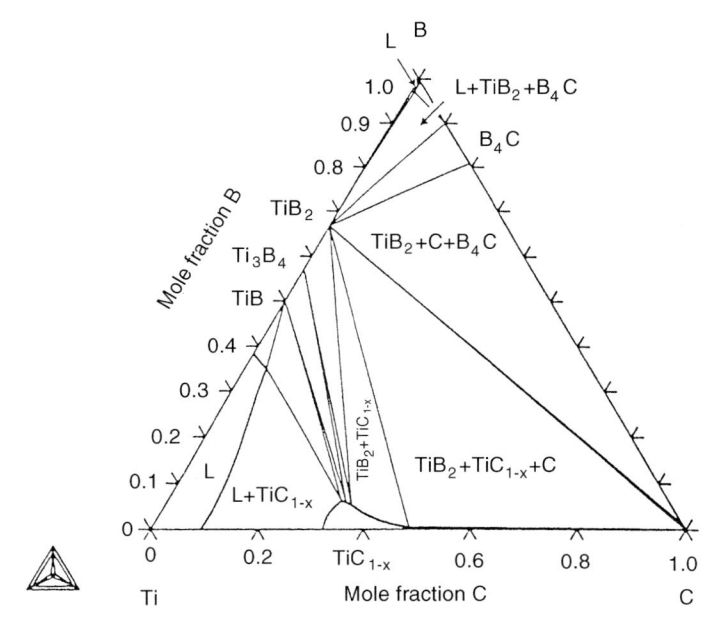

Figure 17. Isothermal section of the B–C–Ti system at 2100°C [105].

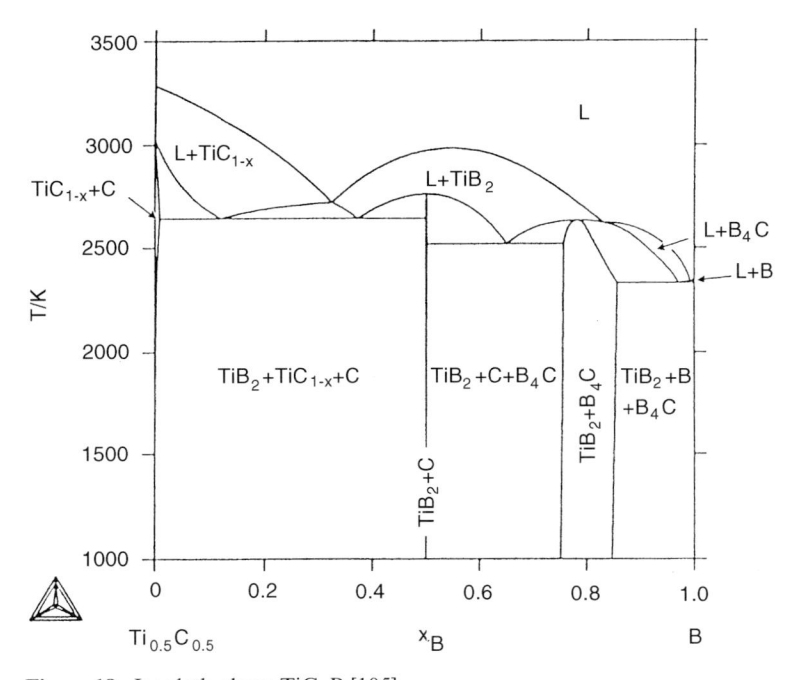

Figure 18. Isopleth along TiC–B [105].

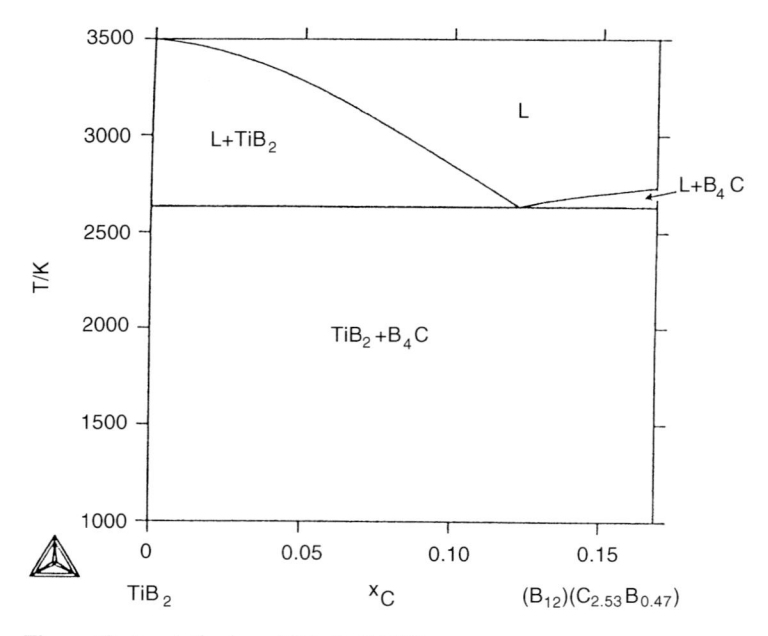

Figure 19. Isopleth along TiB$_2$-B$_4$C [105].

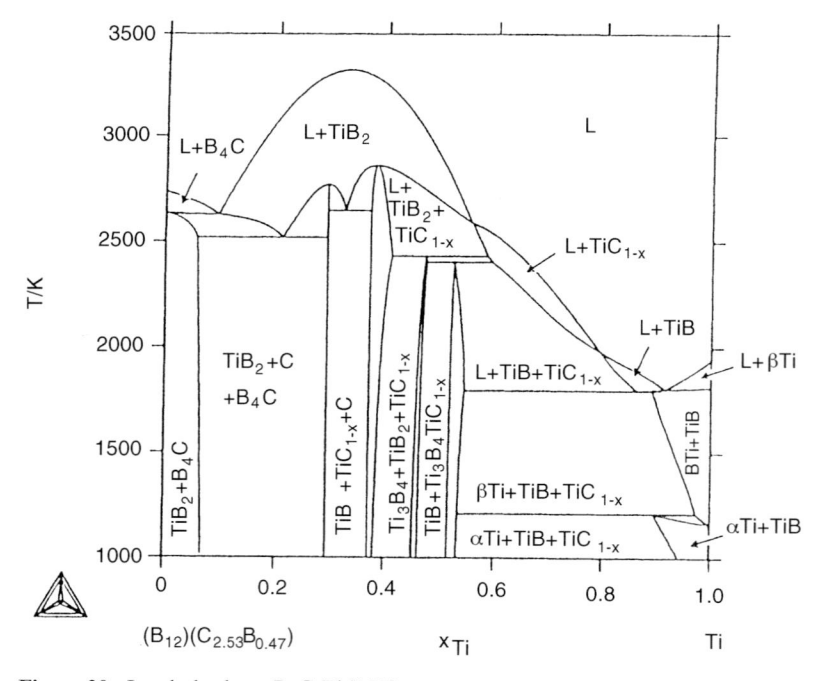

Figure 20. Isopleth along B$_4$C-Ti [105].

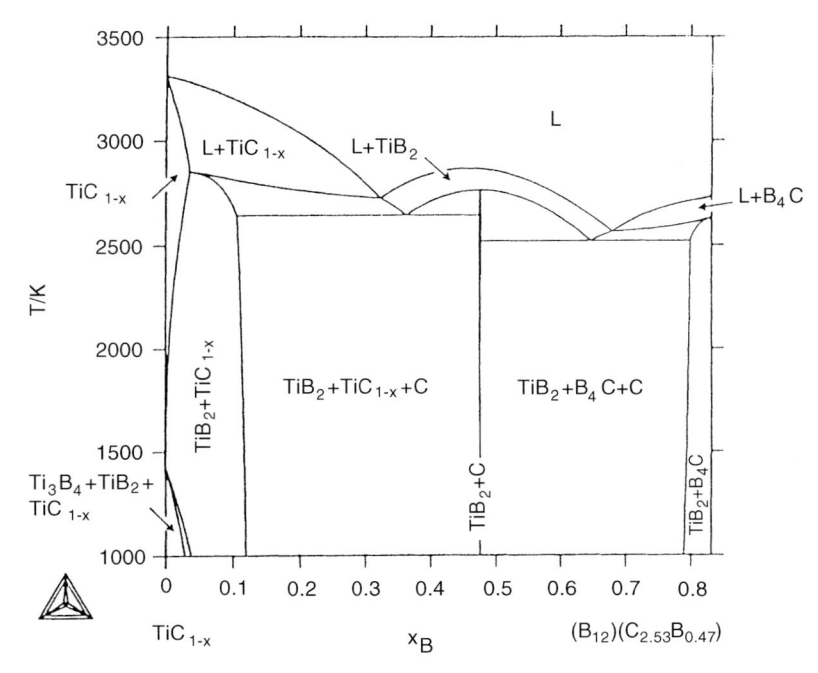

Figure 21. Isopleth along B_4C-TiC [105].

The Ti-Fe-B System

The fabrication of TiB_2-based cermets was recently achieved by using Fe as a binder phase [115–119]. The ternary Ti-Fe-B system was studied first by Federov and Kuzma [120] who established that no ternary phases exist. Below 1100°C, TiB_2 is in equilibrium with FeB, Fe_2B, α- or γ-Fe, Fe_2Ti, and FeTi, whereas TiB is stable together with FeTi and α-Ti (Fig. 22). Although there are still some controversies concerning the phase diagram at higher temperatures [121–124], TiB_2 is in an eutectic equilibrium with liquid Fe at 1340°C (eutectic concentration 6.3 mol.-% TiB_2), which enables liquid phase sintering. Discrepancies exist for the phase equilibria at temperatures between 1100 and 1300°C because of the problem of whether the observed, undesired brittle Fe_2B is an equilibrium phase or results from impurities in the starting materials used (Figs. 23 and 24). It is, however, obvious that oxygen and carbon contaminants introduced by the manufacturing processes of the starting powders significantly affect the composition of the liquid phase by the precipitation of Ti-rich oxides and TiC, respectively. Since the solid solubility of B in δ-Fe is less than 0.5 at.-%, any slight Ti deficiency of the liquid phase composition will move the overall composition from the binary TiB_2-Fe equilibrium into the ternary TiB_2-Fe-Fe_2B field. Both constituents, oxygen and carbon, do, therefore, indeed cause dramatic changes of the sintering kinetics [119].

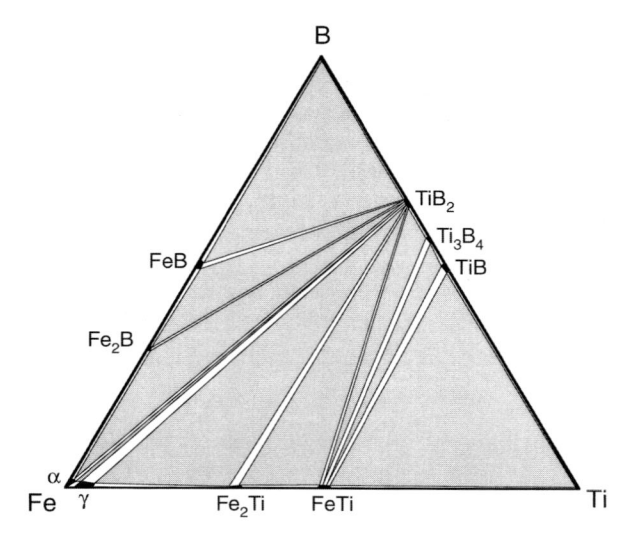

Figure 22. Isothermal section of the Fe-Ti-B system at 1050°C (calculated by Golczewski and Aldinger [320]).

The Ti-Ni-B System

In order to achieve lower sintering temperatures, also many other metallic additives such as Ni, Co, and Cr, or borides of these elements have been used at higher concentrations to allow liquid phase sintering of TiB_2. These transition metals

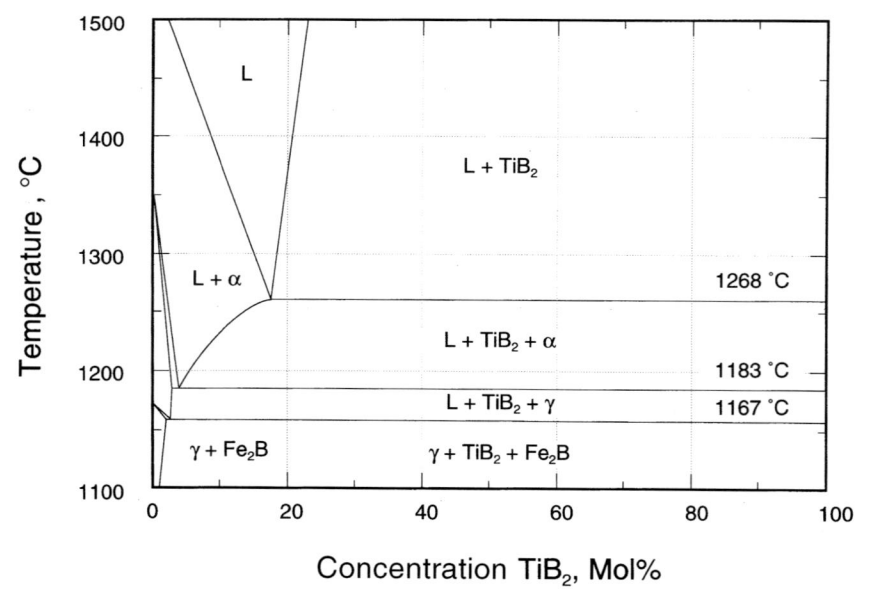

Figure 23. Isopleth of the Fe-TiB_2 section [121].

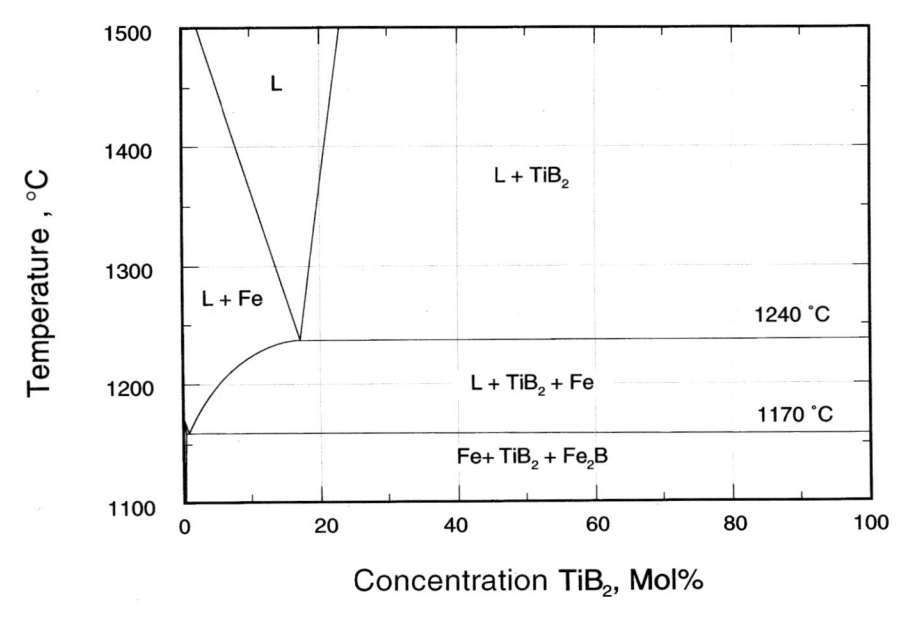

Figure 24. Isopleth of the Fe-TiB$_2$ section [124]. both 1992 refs?

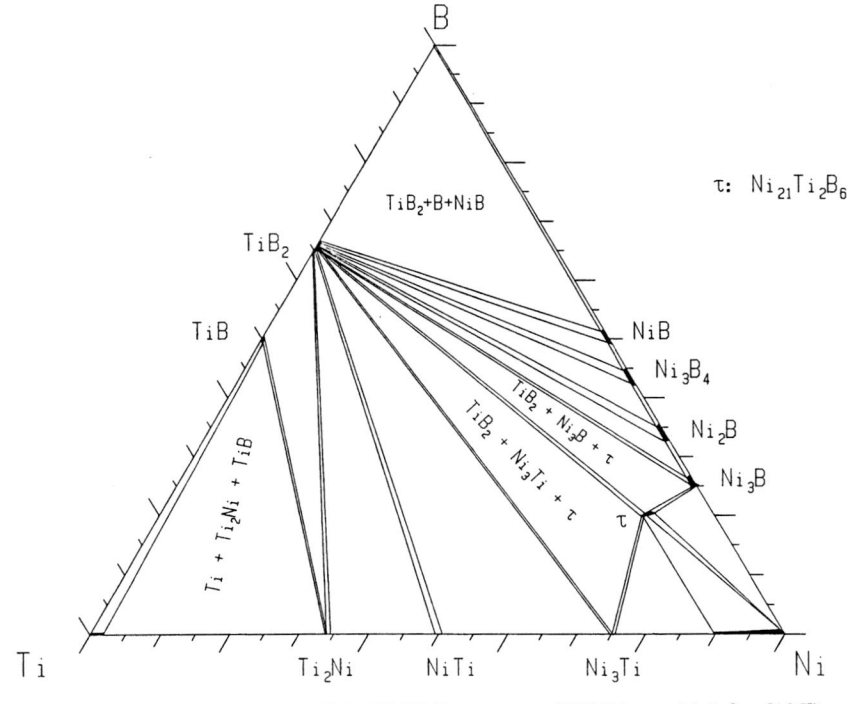

Figure 25. Isothermal section of the Ti-Ni-B system at 800°C in at.-% (after [125]).

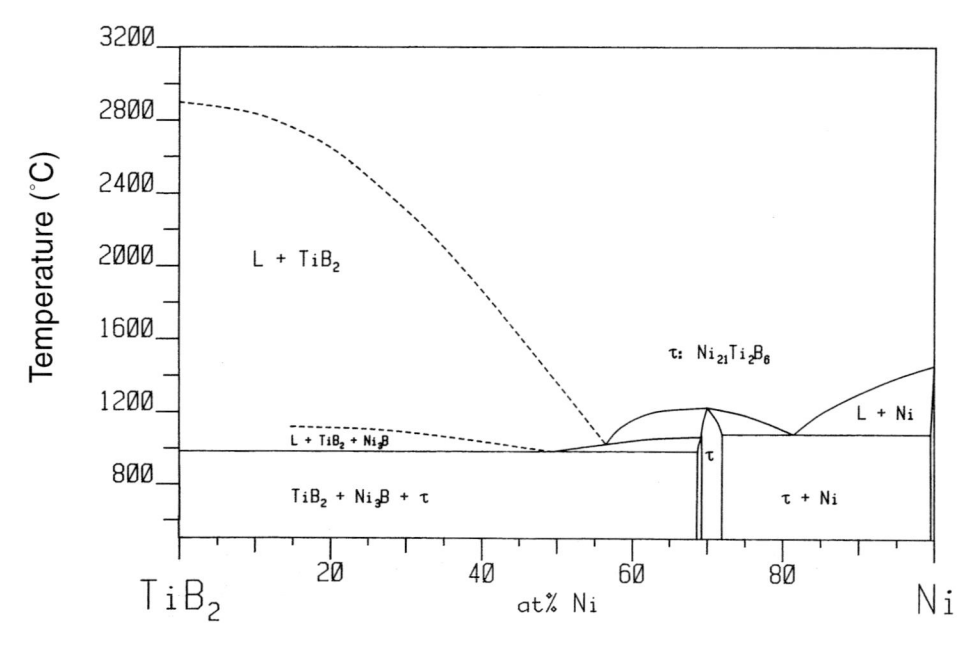

Figure 26. Isopleth of the Ni-TiB$_2$ section (after [127]).

react forming various metal borides with a low melting point (approx. 900–1100°C) and a suitable wetting behavior.

In contrast to the very convenient Ti-Fe-B system, the Ti-Ni-B phase diagram contains a congruently melting ternary phase, denoted τ, with a composition of Ni$_{21}$Ti$_2$B$_6$ and a Cr$_{23}$C$_6$ structure (Fig. 25) [125–127]. At 800°C, the τ-phase is in equilibrium with Ni, Ni$_3$B, Ni$_3$Ti, and TiB$_2$. A pseudo-binary eutectic with Ni exists at 1077 ± 5°C, whereas the relationships in the TiB$_2$-rich corner are more complicated. An isopleth (Fig. 26) across the line Ni-TiB$_2$ reveals a solid-state equilibrium below 980°C involving TiB$_2$, τ, and Ni$_3$B [127]. A liquid phase forms above that temperature because of the decomposition of τ. Above 1100°C, Ni$_3$B also decomposes completely in the presence of TiB$_2$, which increases the amount of liquid phase and thus accelerates the densification.

The Boron–Transition Metal Systems of Group (IV–VI) and Group VIII Elements

Cemented borides with a metallic matrix have also been fabricated successfully from the ternary transition metal borides of so-called τ-, φ- and ω-types since these composites can easily be liquid phase sintered with metallic melts.

The τ-phase with a general composition of $M^I_{21}M^{II}_2B_6$ or $(M^I, M^{II})_{23}B_6$ has been observed in ternary systems with M^I = group VIIIB elements such as Fe, Co, or Ni and M^{II} = group IVB–VIB elements such as Ti, Zr, Hf, Nb, Ta, Mo or W, and with M^I as the liquid phase [125–128]. It forms in coatings or particulate reinforced metals as a consequence of the reaction between diborides such as TiB$_2$, ZrB$_2$,

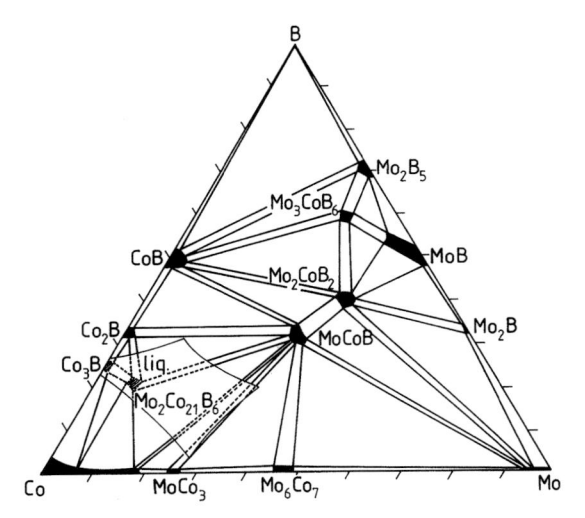

Figure 27. Isothermal section of the B–Co-Mo system at 1000°C with indicated field of liquid phase formation [128].

HfB$_2$, NbB$_2$, and TaB$_2$ and Ni-Co-Cr based alloys. The refractory and extraordinarily hard MB$_2$ phases are hence in equilibrium with a comparatively soft and brittle ternary compounds.

The ternary systems of these kinds of metals with boron reveal a more complex structure because of the presence of many other ternary phases denoted to as φ- and ω-phases. The stoichiometries of these φ- and ω-phases are M^{I}M^{II}B and M$_2^I$M^{II}B$_2$, respectively, where M^I represents Cr, Mo, Ta, or W and M^{II} holds for Fe, Co, or Ni and solid solutions thereof. Other ternary phases have the composition and M$_x^I$M^{II}B$_{2x}$, e.g., TaNiB$_2$, Mo$_2$FeB$_4$, and Mo$_3$CoB$_6$. As an example, an isothermal section of the B-Co-Mo system is shown in Fig. 27 in which both the τ- and the φ-phases are linked with Co as the binder [128]. However, in systems with Fe replacing Co, a φ-phase does not exist. Hence ω is in equilibrium with liquid metal and is thus likely to form a cermet material with Fe like the τ-phase Mo$_2$Fe$_{13}$B$_5$ (Fig. 28). Phase compositions located in the pseudo-binary equilibria with a metal can easily be pressureless liquid phase sintered at temperatures between 1500°C and 1700°C. Wear-resistant parts have been developed from Mo$_2$FeB$_2$-Fe cermets with Ni or Cr additives [129–131, 307]. Figure 29 presents an isothermal section of the Ni-Ta-B system at 950°C [126] with three ternary phases where only τ is in equilibrium with metallic Ni.

Quasibinary Systems of Ti, Cr, and W Diborides

Boron-based ternary phase diagrams with two transition metals have been investigated for exploitation of the diborides with an AlB$_2$ structure, which have high hardness and a high melting point. Due to the identical crystal structure, most of the transition metal diborides have been considered of high mutual solubility

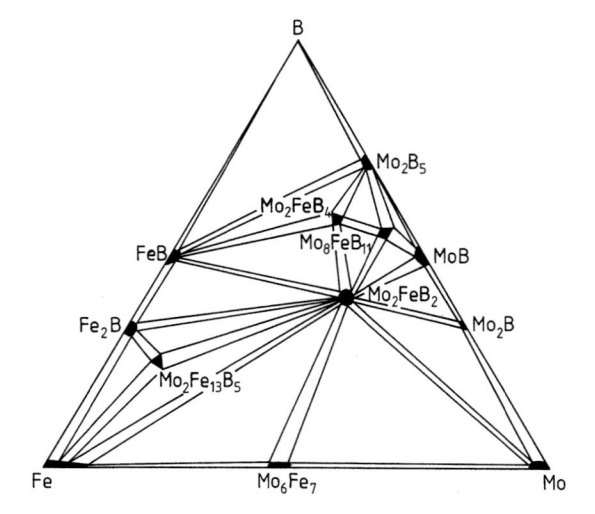

Figure 28. Isothermal section of the Mo-Fe-B system at 1000°C [128].

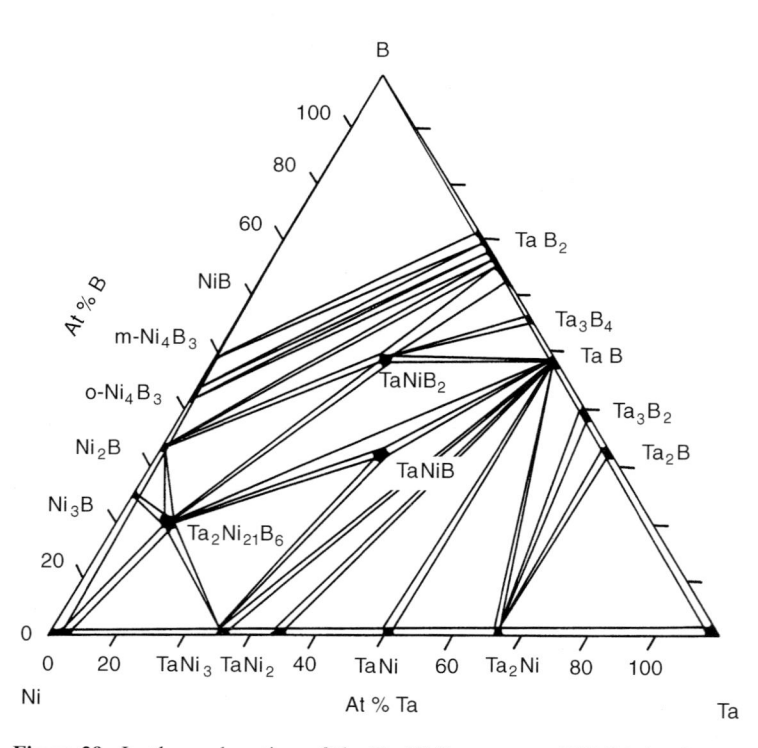

Figure 29. Isothermal section of the Ta-Ni-B system at 950°C (after [126]).

[132–136]. Precise experimental data are, however, rare or are not readily available. Makarenko [137] simply mentions that most of the transition metal diborides of group IV–VI are fully soluble between 2000 and 3000°C and probably also at room temperature. As the only limiting factor the difference in atomic radii was considered which should not exceed 15% like in case of V and Cr, and Hf and Zr [132]. This general statement should be taken with care since low-temperature miscibility gaps have been observed for $(Cr,Nb)B_2$ and $(Cr,Ta)B_2$ [133, 138] and other quasi-binary systems were proved of eutectic character. Complete solid solubility with an miscibility gap at lower temperatures was also proven for the TiB_2-CrB_2 system, limited low-temperature boundary solubility and large homogeneity ranges at high temperatures have been observed for the TiB_2-WB_2, and CrB_2-WB_2 systems which will be discussed in more detail.

Studies in the TiB_2-CrB_2 system are difficult due to the comparatively high vapor pressure of Cr and the little densification of powder blends at temperatures below 2000°C. Hot-pressing at 2000°C for 30–120 min yields a homogeneous $(Ti,Cr)B_2$ solid solution [136]. Koval'chenko et al. [139] have also studied the sintering behavior of pre-reacted powders and the physical properties of a $(Ti,Cr)B_2$ solid solution containing 20 mol.-% CrB_2. Klimenko and Shunkowski [140] have investigated the activated sintering of a "titanium–chromium mixed boride" at 2000 and 2200°C being obviously a completely single phase. Unfortunately, no X-ray nor chemical analyses have been reported for the particular annealing steps. In spite of the problems upon obtaining equilibrium conditions in the overall material chemical and X-ray analyses by Telle et al. [136] yielded evidence for the existence of an immiscibility gap below 2000°C in the TiB_2-rich corner. In the temperature–concentration range between 1500°C and 2000°C and correspondingly 0 to 55 mol.-% CrB_2 no solid solutions possessing an intermediate CrB_2 concentration were observed whereas homogeneous $(Cr,Ti)B_2$ particles with CrB_2 contents exceeding 55 mol.-% at 1500°C could be detected in contact with $(Ti,Cr)B_2$ phase of a CrB_2 content less than 2 mol.-%. Samples with overall concentrations of 35, 75, and 90 mol.-% CrB_2 have been treated by dilatometry and hot-pressing up to 2200°C to prove the formation of a liquid phase but no melting could be monitored at temperatures lower than the melting point of CrB_2. This indicates that the quasi-binary system does not contain a liquidus temperature minimum nor is of eutectic character. From the equilibrium limits a room temperature solubility of TiB_2 in CrB_2 to 15–20 mol.-% is estimated. The melting points of TiB_2 and CrB_2 in Fig. 30 have been adopted from Rudy and Windisch [85] and Liao and Spear [142]. These results confirm the existence of a continuous solid solubility above 2100°C as stated by Post et al. [132]. Zdaniewski [135] also reported on properties of a continuous series of $(Cr,Ti)B_2$ solid solutions, however, without giving details on the temperature range.

Post et al. [132] have investigated the CrB_2-WB_2 system at $2100 \pm 100°C$ but note that no exact limits of the homogeneity range could be determined. Telegus and Kuz'ma [143] studied 15 compositions in the quasi-binary CrB_2-WB_2 section represented as a part of an isothermal section of the B-Cr-W system at 1500°C. They report a maximum solid solubility of 6 at.-% Cr in W_2B_5 corresponding to 3 mol.-% CrB_2 in WB_2 whereas the published section shows 16.7 mol.-% CrB_2 being dissolved in the W_2B_5 structure. This work allows an extrapolation of

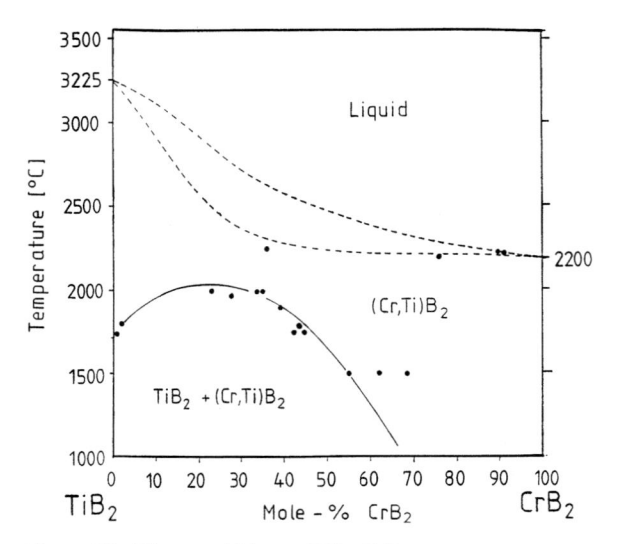

Figure 30. The quasi-binary TiB_2-CrB_2 system.

the homogeneity range of $(W, Cr)B_2$ to about 89 mol.-% CrB_2 or less at 1500°C. Moreover, in contrast to Rudy [104, 144] the binary homogeneity range of the "W_2B_5" phase is more extended. In the CrB_2 corner Telegus and Kuz'ma report less than 2 at.-% W in solution which could not be confirmed in the materials investigated by Telle *et al.* [136] who synthesized the system up to 2070°C. Unfortunately, the liquidus equilibria were not established, so the accurate composition of the eutectic is not determined yet. At temperatures between 1750°C and 2000°C samples with overall compositions between 40 and 86 mol.-% WB_2 consist of two phases, namely $(Cr, W)B_2$ solid solutions with WB_2 concentrations up to 38–40 mol.-% and $(W, Cr)B_2$ phases containing always less than 9–14 mol.-% CrB_2. These results allow the formulation of the boundaries of the homogeneity fields

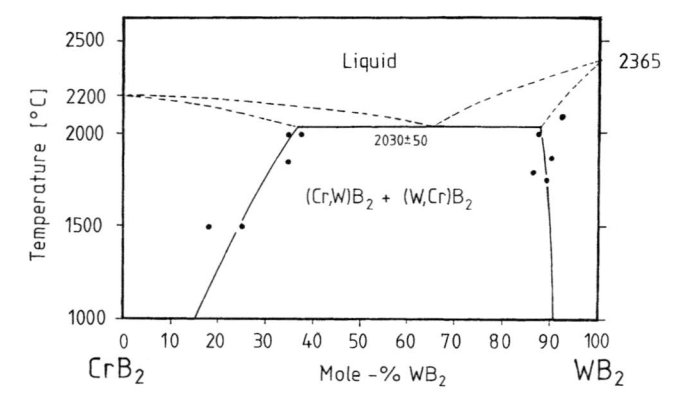

Figure 31. The quasi-binary CrB_2-WB_2 system.

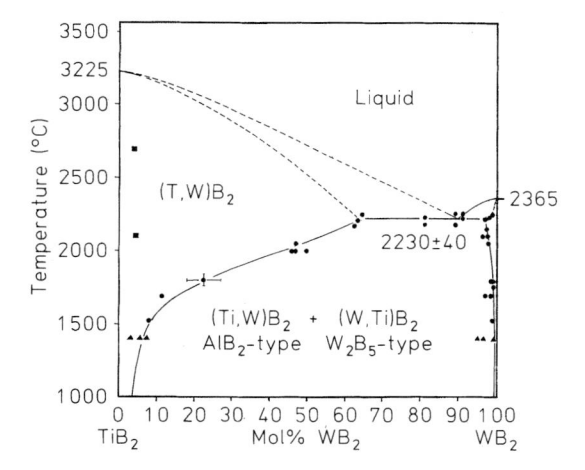

Figure 32. The quasi-binary TiB_2-WB_2 system.

of $(Cr, W)B_2$ and $(W, Cr)B_2$ phases with AlB_2 and W_2B_5 structure, respectively. As shown in Fig. 31, the CrB_2-WB_2 system is of a simple eutectic type ($T_e = 2030 \pm 50\,°C$) with a solid solubility of CrB_2 in WB_2 of 10 mol.-% and WB_2 in CrB_2 of 37 mol.-% at eutectic temperature. The tungsten content in CrB_2 decreases strongly with decreasing temperature whereas the chromium content in the W_2B_5 structure is almost temperature independent.

The quasibinary TiB_2-WB_2 system as a part of the Ti-W-B system has been treated by Telle et al. [136] up to 2250°C. The borders of the $(Ti, W)B_2$ homogeneity range have been intensively studied between 1500°C and 1700°C, at 2000°C and around the quasi-binary eutectic temperature. Thus a huge solid solubility up to approximately 63 mol.-% WB_2 could be established. In the W-rich corner of the phase system, the TiB_2 concentration in $(W, Ti)B_2$ is with 1–2 mol.-% almost constant up to 1700°C. At higher temperatures the solid solubility of the W_2B_5-type structure for TiB_2 increases slightly to approximately 3 mol.-% at the eutectic temperature. A further decrease in TiB_2 along the supersolidus equilibrium could be established up to 2250°C as expected according the phase rules. Taking inaccuracies in the temperature measurement of the dilatometric experiments into account, TiB_2 and WB_2 react eutectically at $T_e = 2230 \pm 40\,°C$ and 90 ± 3 mol.-% TiB_2 (Fig. 32). The solid solubility of WB_2 in TiB_2 at that temperature is approximately 63 mol.-% whereas the solid solution of $(W, Ti)_2B_5$ type contains only 3 mol.-% TiB_2 at the eutectic equilibrium. The homogeneity range of the $(Ti, W)B_2$ solid solution narrows significantly with decreasing temperature and is 46–49 mol.-% at 2000°C and 8–10 mol.-% WB_2 at 1500°C.

Post et al. [132] have treated this system before and found a homogeneity range of the AlB_2 structure type to at least 50 mol.-% WB_2. The preparation of a pure single phase WB_2 of AlB_2 type structure has failed. Pastor [134] also states that the WB_2 concentration in $(Ti, W)B_2$ may exceed 50 mol.-%. Yasinskaya and Groisberg [145] have examined the interaction of TiB_2 with metallic W up to 2700°C. Although no data are given in the text a maximum solubility of 5 at.-% metallic W can be taken

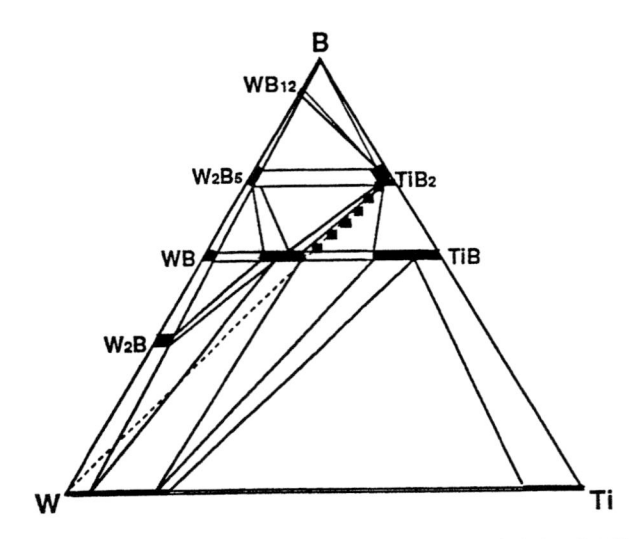

Figure 33. The ternary Ti-W-B system at 1400°C (after [146]).

from the given pseudo-binary section TiB_2-W. The corresponding points are shown as a quadrangles in Figs. 32 and 33. The isothermal section of the B-Ti-W system presented by Kuz'ma et al. [146] reveals a maximum of 6–8 mol.-% WB_2 dissolved in $(Ti, W)B_2$ and 3–4 mol.-% TiB_2 in $(W, Ti)B_2$ at 1400°C (Fig. 33). These data are more or less consistent with the phase diagram presented by Telle et al. [136] and are thus shown as triangles in Fig. 32. Kosterova and Ordan'yan [147] have retreated the ternary B-Ti-W at 1400°C in order to specify more accurately the phase equilibria in the metal-rich region but have basically found the same situation in the TiB_2-WB_2 section as Kuz'ma et al. According to their isothermal section the TiB_2 content in $(W, Ti)B_2$ is considered in the range of 10 mol.-%, which is certainly overestimated. Contrary to Kuz'ma et al. [146], Kosterova and Ordan'yan [147] take the existence of a WB_4 phase into account being in equilibrium with a W-rich TiB_2 and WB_2. The presentation of a metastable B-Ti-W phase diagram at room temperature by Ariel et al. [148] notes a maximum W content in $(Ti, W)B_2$ of 2 at.-%, which was obviously measured on samples of the TiB_2-W section and thus corresponds to the data of Kosterova and Ordanyan [147]. Unfortunately, the more recent publication on the formation of AlB_2 structure type solid solutions [135] does not give any data on temperature–concentration relations in this system. A partial re-treatment of the Ti-W-B system by Ahn et al. [149] for the development of TiB_2-W cermets was based on the isothermal section of Kuz'ma et al. [146] but proved the existence of a distinct $WTiB_2$ phase in the WB-TiB monoboride section. Hot isostatic pressing of TiB_2 with 1–30 vol.-% W at 1900°C resulted in TiB_2-$WTiB_2$ composites whereas a higher amount of W yielded $WTiB_2$-W cermets with little residual TiB_2. The $WTiB_2$ possesses β-WB structure and contains 25.7 at.-% W, 24.8 at.-% Ti, and 49.3 at.-% B which is close to the stoichiometry 1:1:2. In spite of this, the isothermal section given by Ahn et al. [149] show an extension of the homogeneity range to WB.

Current EXAFS- and HR-TEM studies in the TiB_2-WB_2 by Pohl *et al.* [150, 151], Mitra and Telle [152, 153], and Schmalzried and Telle [154] indicate, however, that annealing of $(Ti, W)B_2$ solid solutions in the TiB_2-"W_2B_5" two phase region at lower temperatures results not only in the precipitation of "W_2B_5" particles but, besides segregation of metallic glasses and elemental boron, also of metastable β-WB-type phases and spinodal ex-solutions of W-rich AlB_2-type phases which have formerly been erroneously interpreted as "W_2B_5" [136, 155]. This is attributed to the much faster diffusion of boron compared to Ti and W [150] and shows that equilibrium conditions are difficult to obtain.

The quasi-ternary CrB_2-TiB_2-WB_2 system is of special interest in respect to anisotropy effects of the thermophysical properties of the diboride solid solutions and has been studied between 1900°C and 2100°C by Telle *et al.* [136] up to now. It is obvious that homogeneous ternary $(Ti, Cr, W)B_2$ solid solutions of AlB_2 type structure are formed in a wide range of composition. The miscibility gap emerging from the TiB_2-CrB_2 edge extends to towards the TiB_2-WB_2 edge but disappears at < 80 mol.-% CrB_2, in the WB_2 rich corner a W-rich ternary liquid was observed. It is concluded that the quasi-ternary system exhibits a ternary eutectic at 1800 ± 50°C. Close to the TiB_2-WB_2 edge, a relatively small amount of chromium stabilizes a $(Ti, Cr, W)B_2$ solid solution with a high tungsten content exceeding by far that of the quasi-binary compound. Also the extend of TiB_2 solid solubility in "W_2B_5" is increased by the presence of CrB_2. But also in this system, the previously mentioned metastable phases have been observed which make the interpretation of the entire equilibria difficult.

7.4 Boron Carbide Ceramics

Boron carbide was first prepared by Joly [156] and labeled as B_6C by Moisson [157]. In 1934, Ridgeway [158] suggested the composition B_4C, which is still under controversial discussion. As discussed in Section 7.2.2.2 in connection with the crystal structure, the composition as assessed today ranges from $B_{4.3}C$ to $B_{10.4}C$. Due to the fabrication process where carbon is used to reduce boron oxide, the composition of the commercially available boron carbide is close to B_4C.

7.4.1 Preparation of Boron Carbide

Boron carbide is prepared for most purposes according to a technical-scale process providing comparatively cheap powder which is used for grinding and lapping grits or for the boronization of steel, superalloys and other non-ferrous metals. For the production of ceramics this material must be processed further on by intensive milling, favorably ball-milling, and subsequent cleaning from impurities. Other more sophisticated methods, especially vapor phase reactions and synthesis starting with metallo–organic precursors are very costly and, therefore, restricted to scientific applications or special purposes of high profit gain.

7.4.1.1 Technical Scale Production

Boron carbide powder is produced on a technical scale by the carbothermic reduction of boron oxide with graphite or petroleum coke

$$2\,B_2O_3 + 7\,C \rightarrow B_4C + 6\,CO \uparrow \tag{2}$$

The process is carried out in huge electric arc or resistance furnaces and is comparable to the Acheson process. The reaction takes place between 1500 and 2500°C, is strongly endothermic and requires 1812 kJ/mol, i.e., 9.1 kWh/kg [159, 160]. Since large quantities of carbon monoxide (approximately 2.3 m^3/kg) are formed, the reaction of Eq. (2) is accelerated to the benefit of B_4C. Both volatilized boron oxides and carbon monoxide generate an internal Boudouard equilibrium within the raw material mixture and thus contribute to a self-propagating purification process, which can be expressed by

$$B_2O_3 + 3\,CO \rightarrow 2\,B + 3\,CO_2 \tag{3}$$

$$2\,CO \leftrightarrow CO_2 + C \tag{4}$$

$$4\,B + C \rightarrow B_4C \tag{5}$$

The furnace is usually cooled externally to limit the loss of volatile materials and hence the outer mantle stays unreacted. The core contains blocky boron carbide of relatively high purity (total metallic impurities <0.5 mass-%), reproducible stoichiometry (B/C ratio $= 4.3$) [50], and several percent of residual graphite. The chunks are crushed and milled to the final grain size.

A similar process with lower productivity is used for the synthesis of high-purity B_4C of controlled stoichiometry. At temperatures of 1600–1800°C, hydroboric acid reacts with acetylene black, high purity sugar, or ethylene glycol in a vented tube furnace:

$$4\,H_3BO_3 + 7\,C \rightarrow B_4C + 6\,H_2O + 6\,CO \tag{6}$$

Powders of 0.5–5 µm particle size are obtained [161].

Boron oxide can also be converted to boron carbide by exothermic magnesiothermic reduction in the presence of carbon black at 1000–1800°C [162].

$$2\,B_2O_3 + 6\,Mg + C \rightarrow B_4C + 6\,MgO \tag{7}$$

The process is performed by single point ignition (*thermite process*) or in a carbon tube furnace in a hydrogen atmosphere. The problem is the removal of magnesia, magnesium borides, and unreacted magnesium metal which are usually extracted by hydrochloric or sulfuric acid. Since MgO acts as a grain growth inhibitor, submicron powders with Mg compounds as the only impurities are produced [163, 164]. Further chemical refinement by high-temperature vacuum treatment, however, induces an undesirable coarsening of the particles. The 1990 total annual production of boron carbide in the western world is estimated at approximately 500–600 t.

7.4.1.2 High-Purity Material

In laboratory-scale production, boron carbide can also be synthesized in the form of high-purity powders or coatings (e.g., [165]):

(i) from the elements by arc melting at 2500°C, or self-propagating synthesis above 1100°C:

$$4\,B + C \rightarrow B_4C \tag{8}$$

(ii) by chemical vapor deposition reducing boron trichloride in the presence of carbon in a hydrogen atmosphere:

$$4\,BCl_3 + 6\,H_2 + C \rightarrow B_4C + 12\,HCl \tag{9}$$

(iii) by pyrolysis of boron trihalides with methane or carbon tetraiodide as carbon carriers, in high-frequency furnaces:

$$4\,B(Cl,\,Br)_3 + CH_4 + 4\,H_2 \xrightarrow{900-1800°C} B_4C + 12\,HCl(HBr) \tag{10}$$

$$4\,BI_3 + CI_4 \xrightarrow{900-1100°C} B_4C + 8\,I_2 \tag{11}$$

The latter methods yield boron-enriched solid solutions with a maximum of 20.4 mass-% carbon.

Very fine boron carbide powders of spherical shape and 20–30 nm in size have been prepared by chemical vapor deposition according to (iii). In an Ar-H_2-CH_2-BCl_3 atmosphere a radio frequency plasma produces stoichiometries between $B_{15.8}C$ and $B_{3.9}C$ [33, 166]. Also laser-induced pyrolysis of similar gas mixtures with or without acetylene has been employed for the preparation of nano-sized particles [167]. With similar success, composites of B_4C and SiC have been produced by the pyrolysis of boron-containing polysilanes [168].

The general problem associated with the production of submicron powders by pyrolysis is the comparatively low yield of these highly expensive procedures and the excess of free carbon which cannot usually be avoided. The advantages of high purity and well-defined composition are limited due to the pick-up of oxygen by the large and hence extremely reactive surface area of the particles when exposed to air.

7.4.2 Sintering of Boron Carbide

Sintering of covalently bonded materials is generally much more difficult than densification of oxide ceramics or metals. This is not only due to the low self-diffusion (poor tendency towards grain boundary and volume diffusion), high ratio of grain boundary-to-surface energies and high vapor pressure of particular constituents (strong tendency towards surface diffusion and evaporation recondensation), but also due to their extreme sensitivity to environmental factors such as sintering atmosphere, traces of contaminants, particle size and shape distribution, temperature gradients, etc. The phenomenon of a "terminal" density, i.e., the density obtained after sintering which is far below the theoretical density for pore closure, above which neither an increase in the temperature nor a prolonged sintering time would assist further densification, was frequently observed for B_4C, SiC and Si_3N_4. The reason for this, as proposed by DeHoff et al. [169], Greskovich and Rosolowski [170], and Prochazka [171], is that upon sintering the decrease in

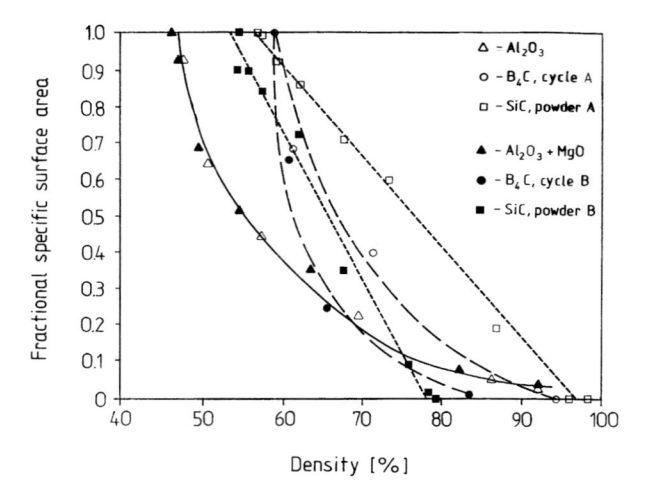

Figure 34. De-Hoff diagram showing the loss of specific surface area during densification of various ceramics (after [171]).

the specific surface area (driving force for densification) is consumed to a much greater extent for pore and particle coarsening (Ostwald ripening) than for grain boundary movement and pore removal. Figure 34 shows a so-called DeHoff diagram correlating the specific surface area and the fractional density on which the path of an ideally densifying material is illustrated by the diagonal line. The lines plotted for Al_2O_3, SiC and B_4C powders make it obvious that in the first step of sintering, the surface energy is dissipated very fast due to coarsening [171]. Since a doubling in particle size corresponds to a decrease in the densification rate by a factor of ten, it is no wonder that densification comes to an end before pore closure is achieved. As pores are favorably removed by grain boundary movement, it is essential to generate a pore size distribution below a critical size above which pores are stable or even tend to grow (i.e., the driving force for pore shrinkage is ≤ 0) and to induce grain growth at moderate rates so that vacancies may be suitably removed from the surface of the pores.

Another reason for the poor sinterability is the extraordinarily high vapor pressure of boron oxides and suboxides. Since boron carbide powders are generally coated by a B_2O_3 layer [172] which quickly reacts to form boric acid, H_3BO_3, in humid atmosphere, vapor phase reactions are active at higher temperatures, in particular above 1500°C, providing a fast transport of boron compounds. Redox reactions such as

$$5\,B_2O_{3\,vap.} + B_4C_{solid} \leftrightarrow 14\,BO_{vap.} + CO_{vap.} \qquad (12)$$

may be shifted to the benefit of the left or right side of Eq. (12) depending on the local chemical potential of the particle surface being defined by the local curvature. The Gibbs–Thomson equation correlates not only the particular chemical potential μ with the local surface radii but also the local vapor pressure. As a result, B_2O_3 may pick up boron from B_4C particle areas with convex surface curvatures and move it as BO to areas of concave curvature where B_4C is deposited again due to the lower vapor pressure compared to the starting point. This reaction releases B_2O_3 again, which starts the process cycle another time. Depending on the oxygen vapor pressure also other boron suboxides such as B_2O_2 and B_2O may be involved. This mechanism, denoted as evaporation/reprecipitation in the science of sintering, is known to contribute significantly to grain growth without any shrinkage (e.g., [173]).

In conclusion, the sintering of boron carbide requires (i) oxygen-removing additives such as graphite, carbon black or organic deoxidation agents such as formaldehyde resin, (ii) very fine powders of high surface area and therefore high driving force, (iii) high temperatures to enable grain boundary and volume diffusion, and, if everything fails, (iv) high pressures.

7.4.2.1 Pressureless Sintering

Densification of pure stoichiometric boron carbide is extremely difficult. Due to the high fraction of covalent bonding (>90%), pore eliminating mass transport mechanisms such as grain boundary and volume diffusion become effective at temperatures above 2000°C, i.e., at temperatures close to the melting point. At lower temperatures, surface diffusion and the already discussed evaporation–recondensation reactions are the favored mechanisms, resulting in neck formation (increase of contact area), pore coalescence and particle rounding (decrease of specific surface area), or euhedral growth of particles by vapor phase reactions, respectively. Grabchuk and Kislyi [174] proved that the regime of predominant surface diffusion extends from 1500°C to 1800°C, whereas sublimation occurs above 1800°C with boron being the more volatile species. Only the latter sintering mechanism causes an enhanced shrinkage of the ceramic body. However, a poor tendency towards plastic deformation, a high resistance to grain boundary sliding, and low surface energies hinder considerable particle rearrangement or shape accommodation before grain boundary or volume diffusion is effective. Even submicron powders cannot thus readily be densified completely by pressureless sintering if they are not mechanically or chemically activated. The general preconditions for the densification of pure stoichiometric boron carbide are to start with very fine powders (preferably $\ll 3\,\mu m$) of low oxygen content and to use temperatures in the range of 2250–2350°C. Above 2000°C, a rapid coarsening occurs, which usually results in unremovable, entrapped residual porosity. Sintering parameters and obtained densities have been presented for many compositions in the homogeneity range of boron carbide by, e.g., Adlassnig [175] (2250–2300°C; 80–87% density; 2450°C: >90%) and Grabchuk and Kislyi [176] (2300°C: 99–99.5% density) who used finer powders. A considerable reduction in temperatures can be achieved by microwave sintering using 2.45 GHz radiation. After a 12 min treatment at 2000°C, 95% of the theoretical density was obtained [177]. The energy conservation compared to hot-pressing is, however, rather low.

Starting from powders with a smaller particle size, e.g., <1 μm, results in lower sintering temperatures and higher final densities. Boron carbide, however, becomes pyrophoric with increasing specific surface area and is hence strongly oxygen-loaded or even dangerous in handling.

Activation of grain boundary and volume diffusion and thus densification at lower temperatures is possible by increasing the density of point defects or dislocations: (i) mechanically by high-energy milling (attrition milling); (ii) by doping with trivalent ions, which substitute for carbon and thus introduce electron deficiencies and vacancies, e.g., by adding boron or aluminum; (iii) by introducing sintering additives, which remove oxide layers on the surface of the boron carbide particles and thus increase the surface energy, e.g., by adding carbon, aluminum carbide, silicon carbide or related compounds, which also inhibit exaggerated grain growth [178]. Other methods make use of additives which possess a comparatively low melting point and have a suitable wetting behavior on boron carbide to provide a rapid path for mass transport via the melt and thus to initiate liquid phase sintering. Dense bodies of boron carbide have also been obtained by liquid phase infiltration of highly porous powder compacts or presintered ceramic bodies.

Kislyi and Grabchuk [179] reported that volume diffusion is enhanced in the boron-rich area of the homogeneity range of boron carbide due to the generation of point defects. In fact, pressureless sintering with boron additives results in an onset of shrinkage at temperatures which are 300 K lower than those required for stoichiometric B_4C. Since aluminum also substitutes for carbon, a similar mechanism may be activated. Accordingly, 95–99.2% of the theoretical density is obtained at 2100–2200°C with 3–15%, preferentially <1% Al additive (e.g., [180]). Other Al-providing sintering additives are Al_4C_3, Al_2O_3 and AlF_3 [180–182], which also use carbon or fluorine as deoxidizing agents. The use of metallic additives is limited due to the low thermodynamic stability of boron carbide reacting with metals to form metal borides and free carbon, except in the case of Cu, Zn, Sn, Ag, and Pb. Nevertheless, Mg, Cr, Co, and Ni have been used [181, 183, 184], with minor success. Stibbs et al. [185] have proposed additions of 5–10 mass-% Al, Mg or TiB_2 to obtain >99% density between 2150 and 2250°C. TiB_2, CrB_2, and W_2B_5 additives inhibit grain growth by grain boundary pinning or, as in the case of W_2B_5 at >2220°C, initiate liquid phase sintering if a eutectic reaction occurs [186]. Sintering of submicron powder with an addition of 1 mass-% Be_2C resulted in 94% density when sintered between 2200 and 2280°C [187]. Kriegesmann [188] made use of the previously discussed vapor phase reactions and considered a "recrystallization" process to produce intentionally porous boron carbide resembling "recrystallized" SiC. Since the vapor pressure of the boron species increases with higher surface area an smaller radii of curvature the starting material consists of powders with two maxima of grain size distribution. The smaller portion of the particles serves as a feeding material for the volatile boron species, which are deposited at the necks between the coarser particles. The process is carried out at >2000°C and a high-purity ceramic of defined pore size and grain size is produced.

The only technically important sintering additive for boron carbide is carbon, as discovered almost simultaneously by Schwetz and Vogt [189], Henney and Jones [190], and Suzuki et al. [191], which today allows the routine production of dense parts. An amount ranging from 1 to 6 mass-% is sufficient to obtain almost theoretical density. Schwetz and Grellner [192] added *phenolic resin* (corresponding to 1–3 mass-% C) to a submicron B_4C powder and obtained >98% density at 2150°C. The resins are pyrolyzed at temperatures up to ≈1000°C and they leave a homogeneous layer of amorphous carbon on the B_4C surface, which promotes sintering. The sinter activation was attributed to an increase in surface energy due to the removal of oxide layers. Moreover, residual graphite particles which have been observed at the grain boundaries may inhibit surface diffusion and evaporation and may also control the grain boundary movement [178, 193]. Firing of B_4C with 6 mass-% C additive at 2220°C results in a 97% dense microstructure of 1–5 μm equiaxed particles, i.e., almost no coarsening has occurred. Abnormal growth of individual grains to 10–30 μm starts above 2235°C; at 2250°C, extensive Ostwald ripening and twinning is observed. The local growth of faceted grains exceeding 500 μm in size was attributed to liquid phase sintering processes due to the presence of low-melting impurities [178].

The method of in-situ pyrolysis of organic additives such as *novolac-type resins* to amorphous carbon was also studied by Bougoin *et al.* [194]. The advantage of the precursor method is the improved homogeneity of the carbon distribution and the extraordinarily fine resulting average grain size of 2 μm and less. Furthermore, the resin may act as a molding aid upon cold isostatic pressing, or may even be the plasticizer for injection molding. Thus complicated parts can be fabricated easily and subsequently pyrolyzed and pressureless sintered.

The drawbacks of the resin route, i.e., the fairly complex handling of the powders and environmental problems during pyrolysis, have triggered research towards suitable resin substitutes. The direct blending of amorphous carbon with B_4C has been suggested as an alternative, but until recently rather porous microstructures were observed when B_4C is sintered with *carbon black*, the suspected reason being the insufficient distribution of the sintering aid. An improved route of dispersing amorphous carbon in B_4C powders has been developed recently by Matje [195] and studied further by Sigl and Schwetz [116, 197]. Boron carbide powder was doped with carbon black resulting in 5, 6, and 7 mass-% of free carbon in the as sintered bodies. Sintering was carried out in a 10 mbar Ar atmosphere for 2 h at temperatures between 2050 and 2175°C. All compositions sinter at surprisingly low temperatures and approach nearly full density at around 2150°C. These sintering temperatures are significantly lower ($\approx$100 K) than those reported for the phenolic resin as a dopant. As shown in Fig. 35 densification is hardly affected by the amount of free carbon. This observation agrees well with the results of Dole and Prochaska [178] who found that the density of B_4C becomes independent of the carbon addition above $\approx$3 mass-% C.

A typical microstructure of B_4C doped with 5 mass-% free carbon and sintered at 2150°C is shown in Fig. 36. In general the microstructure is very uniform and fine grained, but some grains have already entered the regime of discontinuous growth. This process is more clearly reflected in Fig. 37: While the grain size increases rather slowly below 2150°C, a rapid rise of the mean intercept length is observed beyond that temperature.

Combined additives consisting of *carbon and a metal carbide or boride* make use of both the deoxidizing effect of carbon and the diffusion enhancement by the metal, or the grain growth inhibiting and reinforcing effect of nonreacting phases, e.g., B + C, SiC + C, SiC + Al, or TiB_2 + C (e.g., [194, 198, 199]). Weaver [200, 201] sintered relatively coarse (average size 9 μm) boron carbide powders with 2–40 mass-%

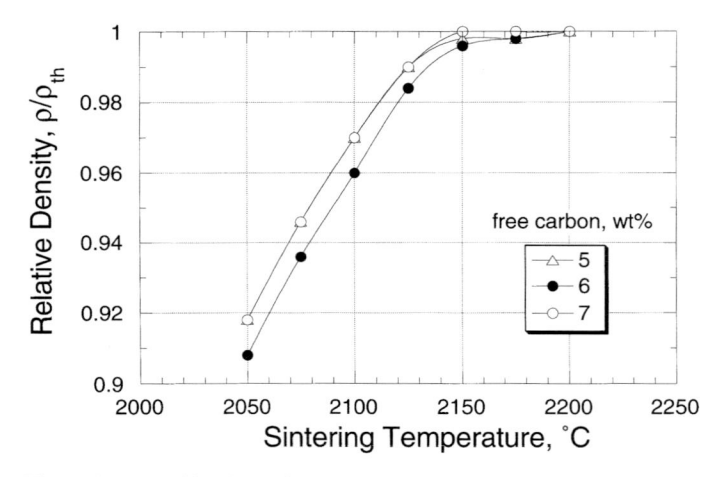

Figure 35. Densification of boron carbide with 5–7 mass-% C (after [197]).

SiC and 0–10 mass-% Al additives to >85% density. In similarity to the decomposition of an Al_4C_3 addition, metallic Al is dispersed very homogeneously by evaporation and condensation in the still porous ceramic body [180, 202]. Starting from submicron powders, Schwetz et al. [203] prepared composite materials consisting preferentially of 9–10 mass-% SiC and 1–3 mass-% C with 97–99.7% density at 2000–2100°C. Residual porosity was removed completely by a post-HIP (hot isostatic pressing) treatment at 1950–2050°C. Both C and SiC may also be introduced in the form of organometallic precursors, e.g., by infiltration of a porous B_4C body with polycarbosilane and phenolic resin, or by coating of boron carbide powders followed by pressing and subsequent pyrolysis. [107, 204]. Bougoin and Thevenot

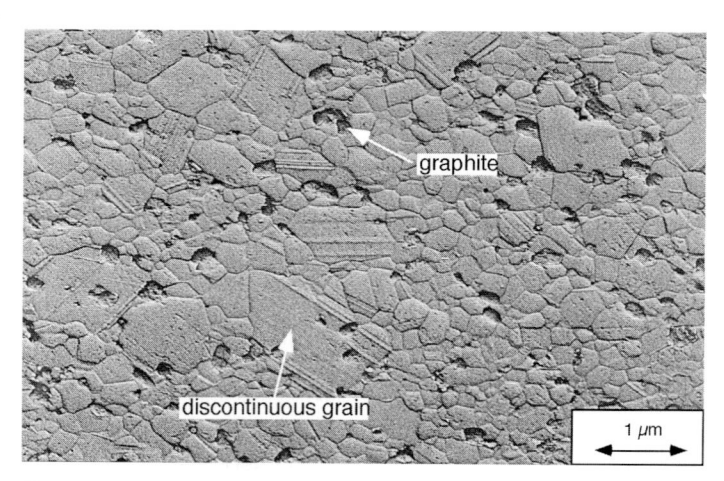

Figure 36. Microstructure of boron carbide pressurelessly sintered with carbon black (after [197]).

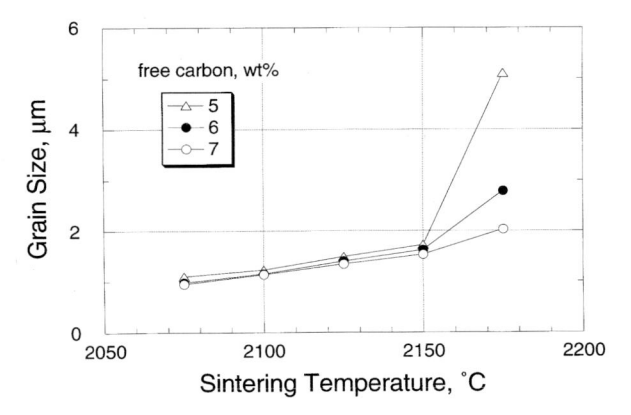

Figure 37. Grain growth in boron carbide pressurelessly sintered with carbon (after [197]).

[205] reported on the fabrication of composite bodies containing 5 mass-% SiC residue but no free graphite. Sintering for 15 min at 2175°C results in a density of >92%. The microstructure of 7.5 mass-% polycarbosilane material exhibits relatively large, faceted B_4C particles (20–50 μm) with entrapped pores and local enrichments of SiC implying that liquid phase sintering may be active. Increasing the amount of polycarbosilane to 17.5 mass-% results in a more uniform microstructure that is characterized by β-to-α transformed SiC platelets of 50 μm size.

Pressureless sintering with liquid phases was studied in the B_4C-Al and B_4C-Si systems. Since Al melts below 600°C and exhibits a significant vapor pressure at only slightly higher temperatures, the equilibrium between 1000 and 1880°C, at which liquid Al is stable with an Al-saturated $B_{12}(B, C, Al)_3$ solid solution (Fig. 12) [73], cannot readily be utilized for liquid phase sintering. Moreover, problems in wetting due to oxide layers on the surface of both Al and B_4C powder particles have to be overcome. As shown by Halverson et al. [206], it is more effective to infiltrate compacted or presintered porous B_4C bodies with liquid Al. Since the resulting material is a metal-reinforced B_4C cermet rather than a liquid-phase sintered B_4C ceramic, it will be treated in detail in one of the following sections.

According to the B-C-Si phase diagram, liquid phase sintering of B_4C should generally be possible above 1560°C with a B-rich Si liquid [54, 94]. Starting from powder mixtures of B_4C, B, and Si, the first unit is generated at 1380°C, which is in equilibrium with SiB_6 and SiC and thus may cause the partial decomposition of B_4C. Above 1560°C, however, a $B_{12}(B, C, Si)_3$ solid solution is in equilibrium with the liquid. Besides the complications due to iterative changes of the wetting behavior due to dissolution and precipitation reactions upon heating, a strong limitation on the final densification arises from the continuous evaporation of Si, which may cause degassing channels and thus even open porosity.

Recently, boron carbide was successfully densified with TiC by pressureless sintering [207]. Since TiC reacts with B_4C by the formation of TiB_2 and free carbon,

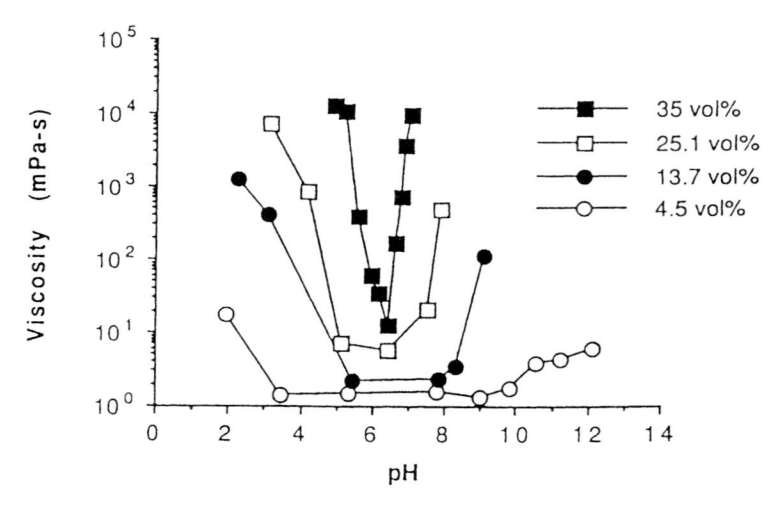

Figure 38. Viscosity as a function of pH and volume fraction of B_4C in aqueous suspension (after [208]).

densification is enhanced by the same mechanisms described before. Details about the resulting B_4C-TiB_2 composites will be discussed later on.

Since pressureless sintering to relatively high density allows the fabrication of complex parts compared to hot-pressing research on appropriate molding procedures for boron carbide has increased since 1990. Slip casting was studied by, e.g., Williams and Hawn [208] using aqueous solutions. Surface contamination by hydroboric acid was removed by washing the powders with water or alcohol. Electrophoresis indicated that B_4C is negatively charged in water above pH 1. Stable dispersions with > 30 vol.-% solid fraction and low viscosity were obtained and successfully slip-cast at pH > 6 but boric acid destabilized suspensions above pH 7. A slight reoxidation after storing the powders in air for 6 months caused an increase of viscosity at pH > 9 but yielded > 58% green density compared to 45% of as-received powders. The highest casting density of 63% was obtained after washing of the powder. Fig. 38 shows the viscosity as a function of pH and volume fraction of solids whereas Fig. 39 presents the influence of surface conditions to the viscosity for a 35 vol.-% B_4C suspension.

The zeta potential of B_4C is generally negative above pH 1, indicating the isoelectric point is at pH $\ll$ 1 (Fig. 40). The results by Williams and Hawn [208] correspond to similar findings by Pyzik *et al.* [209] who have prepared slips at pH values between 9 and 10. This behavior is attributed to the strong shielding of surface charges by dissolved boric acid or, at higher pH, by the formation of polyborates.

Injection molding of boron carbide with 2–5 mass.-% carbon black was developed by Schwetz *et al.* [210]. Like in conventional processes known for oxide and nitride ceramics, the spray-dried powder blend was mixed with 18 mass-% organic binder and molded at 120°C and 45 MPa. Dewaxing was accomplished by heating in an atmosphere-controlled furnace at 100 mbar. The binder components decomposed thermally by cracking and evaporated within four days and temperatures up to

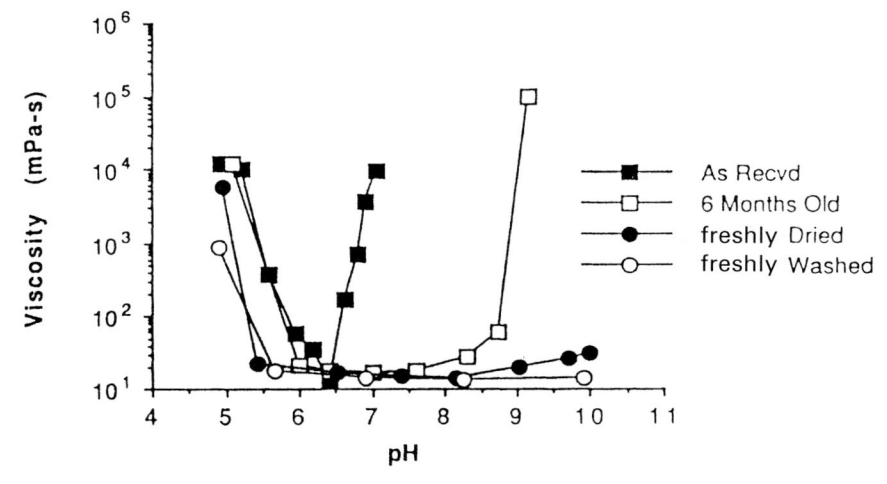

Figure 39. Influence of surface conditions to the viscosity for a 35 vol.-% B_4C suspension (after [208]).

450°C. Vacuum sintering between 2125 and 2225°C for 2h yielded >96% of theoretical density which was improved to >99.5% by subsequent hot isostatic pressing at 2050°C at 200 MPa.

7.4.2.2 Hot Pressing and Hot Isostatic Pressing

Since pressureless sintering allows the fabrication of complex shapes but results in coarse microstructures and some residual porosity, this process is only applicable

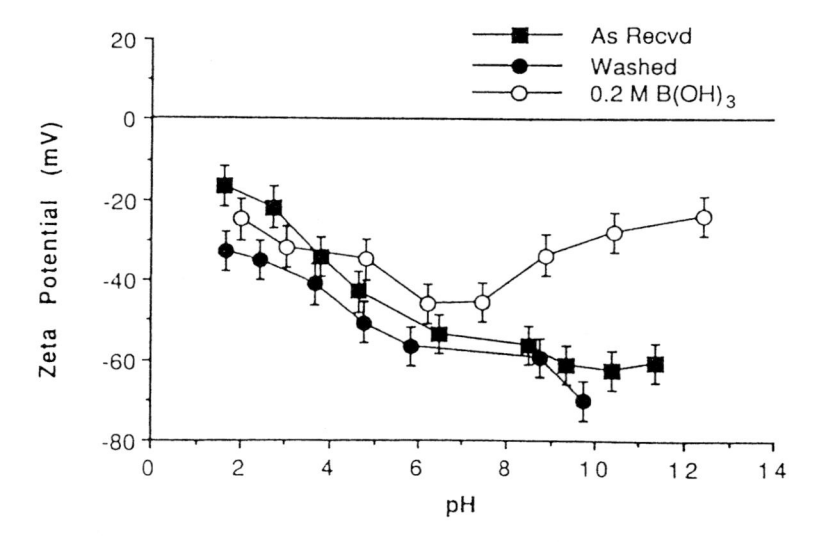

Figure 40. Zeta potential of B_4C (0.01 mass-% suspension) as a function of pH [208].

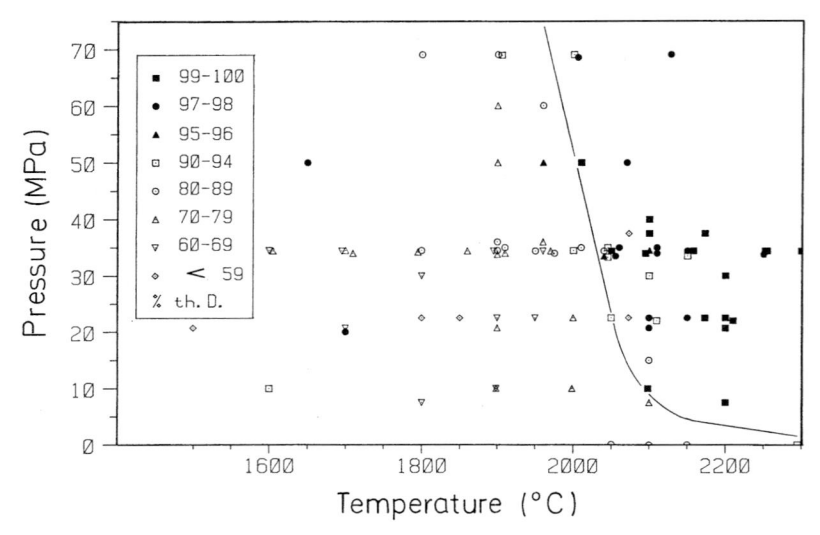

Figure 41. Hot-pressing parameters and densities obtained from undoped boron carbide (literature survey).

for wear parts or shieldings which are not subjected to high stresses because these materials exhibit a low strength ($\sigma_b < 300$ MPa) and a low fracture toughness ($K_{Ic} < 3$ MPa m$^{1/2}$). Hence, for high densification at reasonable temperatures a hot-pressing treatment that causes particle rearrangement and plastic flow is required. Grain boundary sliding, strain-induced twinning, creep and, at a later stage, bulk diffusion combined with recrystallization were identified as the mechanisms of mass transport [211–213]. Densification maps and diffusion diagrams of B-rich boron carbide and C have been established by Beauvy and Angers [214], and Bouchacourt et al. [215]. Suitable preconditions are (i) the use of submicron powders, (ii) temperatures in the range 2100–2200°C, (iii) pressures of 25–40 MPa, (iv) 15–20 min hold, and (v) a vacuum or an argon atmosphere. To resist the high pressures at these temperatures and to provide carbon as a sintering aid, the use of boron nitride-coated, graphite crucibles is favored. In Fig. 41, literature data on the obtained fractional densities of undoped B$_4$C are related to the particular hot-pressing conditions. It is obvious that both high temperatures and high pressures are required to achieve a density of >95%. Only formation of sinter necks is obtained at 20 MPa pressure and 2000°C (Fig. 42). However, a strong coarsening has to be taken into account at higher temperatures and average particle sizes of >100 μm in commercial ceramics are not rare.

Similar to pressureless sintering, additives may be used for hot pressing of boron carbide to reduce temperatures required for grain boundary and bulk diffusion and to retard grain growth. Figure 43 shows a typical microstructure of a hot-pressed B-doped material with strain-induced polysynthetic twinning. Suitable dopants are B [216,217], almost doubling the strength, C [192], Mg, Al, Si, Ti, V, Cr, Fe, Ni, and Cu [94,183–185,216]. As demonstrated by Telle and Petzow [218], combined B-Si or B-Si-Ti additions lubricate the grain boundary sliding and prevent coarsening by

Figure 42. Formation of sinter necks in undoped boron carbide after hot-pressing at 20 MPa pressure and 2000°C.

forming a thin SiC or TiB_2 grain boundary phase (Fig. 44), which pins the grain boundary movement and controls the surface diffusion. Compounds used as additives are various glasses, alumina, sodium silicate with $Mg(NO_3)_2$, and Fe_2O_3, which may reduce the hot-pressing temperature down to 1750°C [219]. MgF_2, AlF_3 [181] and ethyl silicate [220] are other additives which are active at particle surfaces and grain boundaries: Hot pressing with 1–5 mass-% of the above-mentioned additives usually requires a temperature of 1750–1900°C to obtain >95% density. In most cases, grain size refinement and distributed second phases result in improved mechanical properties such as strength and fracture toughness.

Figure 43. Hot-pressed B-doped boron carbide with strain-induced polysynthetic twinning.

Figure 44. SiC intergranular phase in hot-pressed B_4C-Si-B powder blends.

Hot isostatic pressing (HIP) of boron-containing ceramic powders creates special difficulties due to the choice of the canning material. In general, containers made from metals or usual glasses cannot be used because of reactions with the sample material. In the presence of metals, boron carbide decomposes forming metal borides plus graphite, which embrittles the capsule. In the case of silica glass, boron diffusion from the outer layers of the specimen into the glass strongly changes the viscosity and the glass transformation temperature. Hence the softening of the container and the pressure transfer to the specimen cannot be controlled reliably. Moreover, boron oxide gas may be released from both the capsule and the sample and hence result in blowing of the container.

Promising techniques have been developed by Asea Cerama AB, Sweden, and Elektroschmelzwerk Kempten, Germany, using diffusion barriers and a special type of boron oxide glasses [221]. These methods are also applicable to silicon nitride and silicon carbide ceramics and make the fabrication of complex parts, e.g., injection-molded sand blasting nozzles, in large-scale production feasible. In the case of boron carbide, this treatment was applied to additive-free submicron powder obtained by sedimentation of commercial powder in an aqueous suspension by changing the pH value from 10 for dispersion to 3 for flocculation. The sedimented powder exhibited a particle diameter $\ll 3\,\mu m$ and yielded a final density of 100% after HIP above 1700°C for 60 min at a hydrostatic pressure of 200 MPa. In Fig. 45, a fracture surface is shown which illustrates that no grain growth has occurred during the heat treatment. A three-point bending strength of 714 MPa was reported with a Weibull modulus m of 8.3 [221]. The increase in strength compared to normal hot-pressed material is almost three-fold. The fracture toughness was, however, not influenced at all since values of $K_{Ic} = 2.5$–$3.2\,MPa\,m^{1/2}$ have been measured in all cases by the indentation method.

Generally, all presintered or hot-pressed materials with closed porosity (i.e., >95% density) can be fully densified by a post-sintering HIP-treatment [210, 222]. Best results for C-SiC-doped B_4C are obtained at 2000°C and 200 MPa isostatic pressure. Post-HIP of injection molded B_4C-C powder mixtures yielded >99.6% density. The four-point bending strength of 470–485 MPa for 2% C samples sintered between 2150 and 2175°C was improved to 560–580 MPa. At higher sintering or post-HIP temperatures the strength decreases due to the exaggerated grain

Figure 45. Fracture surface of pure B_4C hot isostatically pressed at 1700°C.

growth to 350 MPa and less. The fracture toughness as studied by Schwetz et al. [210] by means of the bridge method (indentation and subsequent quasi-three-point loading) was found to be grain size dependent and ranges between 2.5 and 3 MPa m$^{1/2}$ at particles sizes <2 μm and between 3.7 and 4 MPa m$^{1/2}$ at 10 m mean grain size, which can be attributed to grain size dependent residual stresses in the microstructure causing crack deflection.

7.4.3 Properties of Boron Carbide

The physical and chemical properties of boron carbide have been reviewed by Lipp [159], Thevenot and Bouchacourt [256], Thevenot [164,165], and Schwetz (1999) [223]. Special problems while presenting the physical properties arise from the large homogeneity range of boron carbide. Furthermore, its poor sinterability requests additives that are usually unspecified and results in residual porosity and various grain sizes which are often also not considered in the publications. Most variation and discrepancies in the properties reported come from the undefined composition of the materials studied.

The microstructure and dopant-dependent mechanical properties have been addressed in the previous section. The strong variability of Young's modulus, strength, hardness and toughness become more comprehensive if compared with the sintering temperature and, e.g., free carbon content and grain size (Figs. 46–48). The intrinsic mechanical properties given in Table 5 are based on the most reliable measurements essentially by Schwetz et al. [210]. An excellent overview of hardness measurements as a function of sample preparation and testing load was presented by Bouchacourt and Thevenot [71] stating that the hardness values rise exponentially with decreasing load. A more or less constant hardness is measured at loads >20 N, which results in a large indentation diameter integrating across the average microstructure. Depending on porosity and free carbon content, however, parasitic cracks may be generated by this and higher loads which make the evaluation of the indents again impossible. At very low loads, e.g., at 0.25 N hardness values in the range of 58 GPa have been measured with, however, a high standard deviation. Figure 49 compares the data by Bouchacourt and

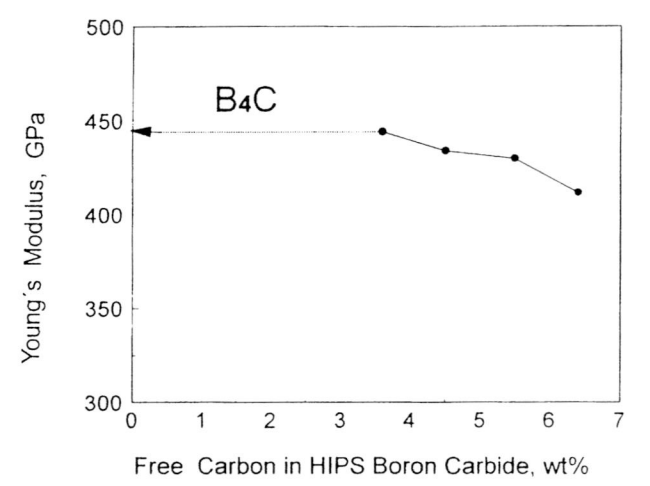

Figure 46. Young's modulus vs. free carbon in post-HIP B_4C sintered at 2175°C (after [210]).

Thevenot [71] with those of Si-doped both single crystalline and polycrystalline B_4C [83].

At room temperature the hardness of B_4C is only inferior to diamond and cubic boron nitride, which tend to weaken above 500–600°C due to the beginning of the transformation from the diamond structure into the graphite structure. Above

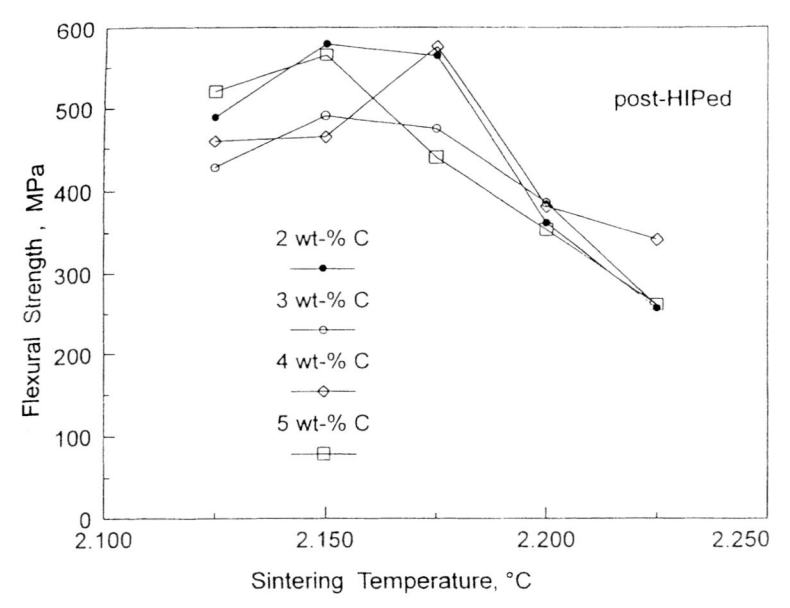

Figure 47. Four-point bending strength vs. sintering temperature and free carbon content (after [210]).

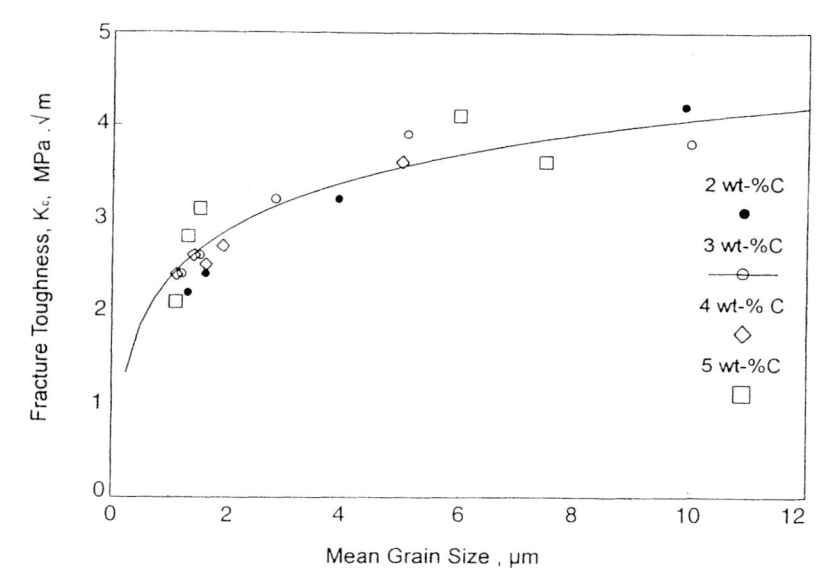

Figure 48. Fracture toughness (Bridge method) vs. mean grain size and free carbon content (after [210]).

1100°C, and in a non-oxidizing atmosphere, B_4C is the hardest compound known up to now. B_4C is thus used for wear-resistant parts and inserts for mortars and ball mills, wear plates, sand blasting nozzles, dressing tools for grinding wheels, lightweight armor plates for helicopters, tanks, and in composites of glass fiber-reinforced plastics as bullet-proof protection for personnel.

Depending on the B:C ratio boron carbide possesses remarkable conduction properties. It is a high-temperature p-type semiconductor with a forbidden band width of 0.8 eV. The electric resistivity is with 0.1–100 Ωcm in the range of SiC. The temperature coefficient of electric resistivity is negative but it shows a high increase of thermoelectric power (e.g., [224]). The extraordinarily high thermoelectric power of boron carbide was subject of research for many years for direct power conversion plants. The application of this unique property is today restricted to high-temperature thermocouples in connection with a graphite electrode. Temperatures up to 2300°C can reliably measured by this device. Thermal, optical, and electrical properties have been intensively studied in order to understand the nature of atomic bonding and the homogeneity range by, e.g., Bouchacourt and Thevenot [225], Werheit et al. [226], Wood et al. [227], Emin [79], and Aselage et al. [39,228] in hundreds of papers since 1980 so that no general view can be given in this book. The data listed in Table 5 have been taken from these few publications listed above.

The large cross section for thermal neutrons makes boron carbide an interesting candidate for absorption or retardation of neutron radiation in power plants and as first-wall coating in fusion reactors. The cross section for ^{10}B is approximately 4000 barn, which is naturally present in boron carbide at 19.9%.

Table 5. Physical properties of boron carbide.

Property	Unit	High carbon	High boron
Formula	–	$B_{4.3}C$	$B_{10.4}C$
Density	g/cm^3	2.52	2.465
Composition B, C	at.%	81.1, 18.9	91.3, 8.7
Crystal structure	Space group	$R3m$	$R3m$
Lattice constants a, c	nm	0.5607, 1.2095	0.5651, 1.2196
Cell volume	nm^3	0.32894	0.33938
Melting point	°C	2380 (eutectic with C)	2250 (liquid + C-rich)
Congruent melting point	°C	2450	
Thermal expansion coefficient α	10^{-6}/K	$2.6 + 4.5 \cdot T$ (25–800°C) 4.5 (600°C) 4.6 (25–800°C) $3.016 + 4.3 \cdot 10^{-3} T - 9.18 \cdot 10^{-7} T^2$	
Thermal conductivity λ	W/m·K	35 (25°C) 28 (200°C) 23 (400°C) 19 (600°C) 16 (800°C)	4.0 (20°C) B_9C 4.7 (200°C) 5–7 (400°C) 5–6.9 (600°C) 5–6.7 (800°C)
Thermal diffusivity α	cm^2 s^{-1}	1.0×10^{-1} (400 K) 3.0×10^{-2} (600 K) 2.1×10^{-2} (800 K)	1×10^{-2} (600 K) 0.8×10^{-2} (800 K)
Electric resistivity ρ	Ωcm	0.1–100	
Thermal coeff. of electric conductivity	K^{-1}	3.2×10^{-5}	
Seebeck coefficient of thermoelectric power	μV/K	100 (400 K) 140 (750 K) 196 (1000 K) 225 (1300 K)	220 (400 K) 233 (750 K) 256 (1000 K)
Self diffusion coefficient: boron	cm^{-2}	5.18×10^{-10} (1000°C)	
neutron absorption cross section	barn	400–750 at 0.025 eV	
Fracture energy γ_s	mJ	5.45 ± 0.44	
Hardness $HK_{0.1}$	kp/mm^2	2950	
Strength (4-pt. bending)	MPa	450–550	
Fracture Toughness (bridge method)	MPam$^{1/2}$	2.5–3 for < 2 µm grain size 3.5–4 for > 2 µm grain size	
Young's modulus E	GPa	440–460 (25°C) 430 (250°C) 420 (500°C) 415 (750°C) 400 (1000°C)	
Porosity dependence of E	GPa	$E_p = E_o [(1-P)/(1+2.999P)]$, $E_o = 460$ GPa	
Poisson ratio	–	0.15	
Sound velocity	ms^{-1}	14.000	

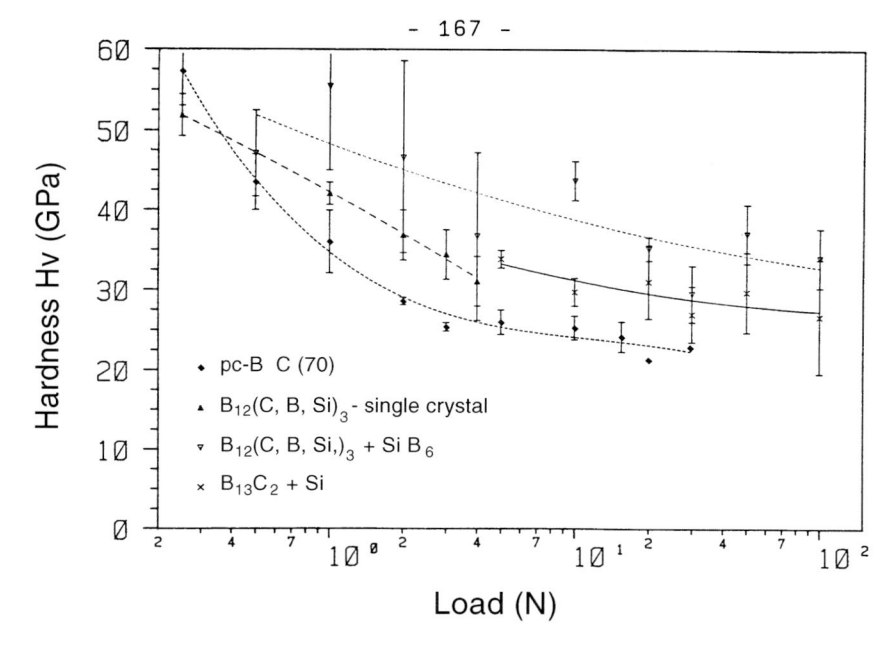

Figure 49. Hardness vs. test load (after [71, 83]).

7.4.4 Chemical Properties and Oxidation of Boron Carbide

The chemical properties are characterized by a pronounced stability in acids or alkali liquids. Boron carbide reacts slowly with $HF-H_2SO_4$ or $HF-HNO_3$ mixes, i.e., in strongly oxidizing environment. Due to this stability, impurities of metal or other boron compounds can readily be removed by chemical leaching. In contact with molten alkalis boron carbide reacts to form borates. A disadvantage of the application of boron carbide ceramics is their instability against metals at high temperatures, in particular with metallic melts. Depending on the affinity of the metal to boron or carbon, the particular metal borides and carbides are formed. If no stable metal carbide exists, free carbon is released by the reaction. Boron carbide is also capable of reducing many oxides to form metal borides and carbon monoxide under reducing conditions. In air, the particular metal borates are generated. In hydrogen it reacts slowly above 1200°C to form borane and methane, which prohibits sintering in hydrogen atmosphere. In nitrogen a decomposition to BN occurs above 1800°C. Boron halides are evolved by the reaction with gaseous chlorine and bromine above 600°C and 800°C, respectively.

Boron carbide in air is immediately, i.e., within seconds, coated by a B_2O_3 layer, in presence of water by a hydroboric acid layer [172, 229]. Oxidation starts at 500–600°C and accelerates significantly above 800–1000°C, depending on the humidity. The weight gain of boron carbide powder being surface-cleaned in glove boxes of 1 ppm O_2 and 1 ppm H_2O and subsequently Si-sputtered, Ar^+ ion implanted, or

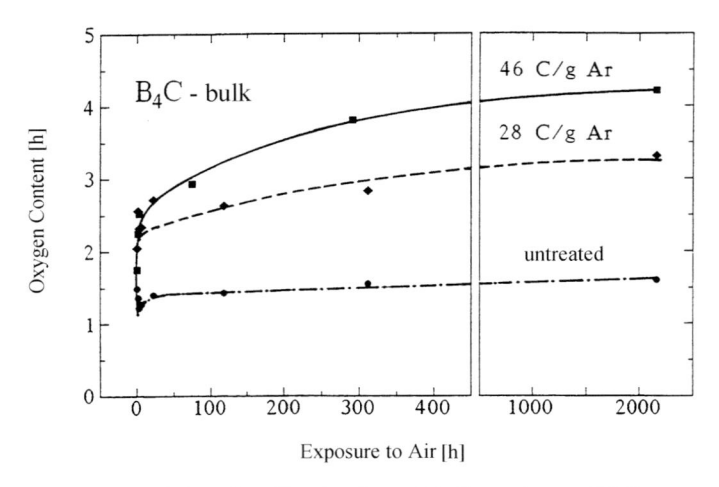

Figure 50. Weight gain of ion-implanted B_4C powders at 20°C.

left untreated at 20°C in air was monitored by Heuberger [230] and reported by Telle [231, 232]. Similar experiments were carried out by Matje and Schwetz [172]. Controlled exposure of Ar-sputtered and implanted powders resulted in a strong, immediate weight gain which was considerably higher for the treated powder compared to the untreated (Fig. 50). After approximately 20 h, the untreated and the Si-sputtered material was stable whereas the implanted powder exhibited a continuous weight gain with a final oxygen content of almost 4 mass-% compared to the contamination of 1.6 mass-% in the as-received powder. The studies by Matje and Schwetz [172] proved a similar parabolic weight-gain in atmosphere of 92% humidity, whereas the increase in weight was linearly dependent on time in air of 52% humidity (Fig. 51). The tremendous oxygen pick-up by Si-sputtered and

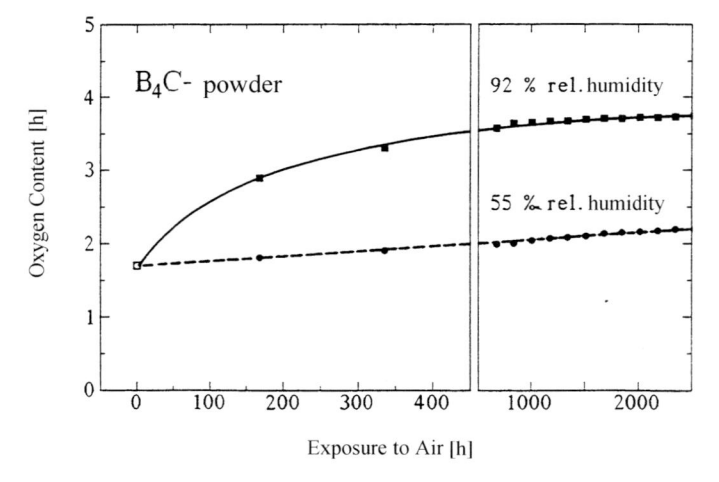

Figure 51. Weight gain of B_4C powders in dry and humid air [172].

Ar-implanted materials at room temperature is attributed to the enlarged surface area as well as due to the formation of structural defects which aid oxygen diffusion.

7.4.5 Boron Carbide–Based Composites

Since boron carbide is brittle and susceptible to oxidation in air a combination with other materials such as SiC, TiB_2, ZrB_2 was considered beneficial. As discussed earlier, boron carbide tends to grow exaggeratedly at the temperatures required for high densities. According to the theory of sintering isolated inert particles dispersed in the boron carbide matrix would also inhibit coarsening by pinning the grain boundaries.

7.4.5.1 Boron Carbide–Silicon Carbide Ceramics

In boron carbide–based composites, silicon carbide can be dispersed as isolated particles, e.g., by simple powder mixing [203], mechanical alloying or as a grain boundary phase which is formed in situ by liquid phase reactions [233]. Another method of coating B_4C with SiC which was mentioned in the previous section is the deposition of a polysilane precursor on powder particles prior to sintering which can be converted to SiC by a pyrolytic heat treatment [204]. In all the examples, the presence of SiC retards the strong coarsening of the matrix at temperatures above 1900°C (Figs. 44 and 52). In general, B_4C and SiC matrix ceramics can be toughened by the incorporation of SiC whiskers, but polytype changes are encountered because of the high temperatures required for complete densification, and the decomposition temperature of 2160°C may easily be reached in the B_4C-SiC system. Pressureless sintering of B_4C-SiC_{fibre} composites is difficult because of back-stresses which cause porosity in the vicinity of the fibers. Moreover, the toughening and strengthening effect is not very large since the thermal expansion coefficients of matrix and inclusions are about the same. Thus, misfit stresses are small, and only load transfer mechanisms due to differences in the Young's modulus may be operational rather than crack deflection. Thus the combination of B_4C and SiC

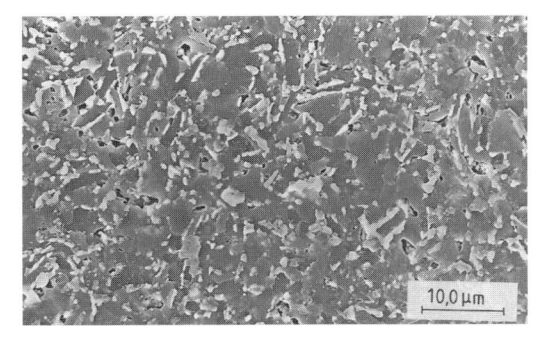

Figure 52. Microstructure of a B_4C-SiC composite derived from polysilane-coated B_4C powders.

are, for the time being, only useful for the fabrication of corrosion and oxidation resistant parts with, however, comparatively small fracture toughness.

The oxidation behavior of single-phase B_4C and B_4C-SiC composites in dry and in humid air was studied by Telle [231] in more detail depending on the microstructure. B_4C-SiC composites were prepared for this purpose (i) by the conventional powder route yielding isolated SiC particles in B_4C matrix and (ii) by the precursor route [204] giving SiC layers between the B_4C particles with the same volume fraction. Figure 53 illustrates the weight change for the heating period and the

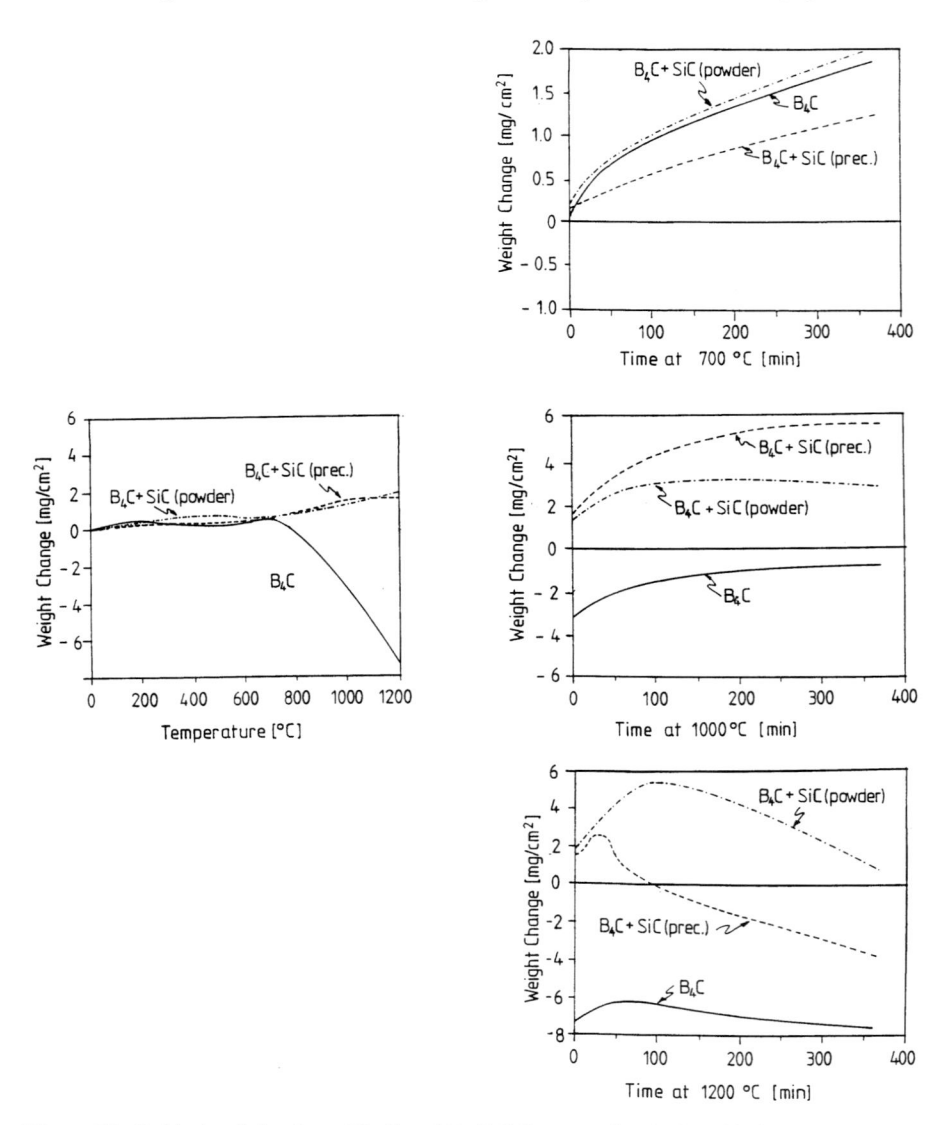

Figure 53. Oxidation behavior of B_4C and B_4C-SiC composites in humid air.

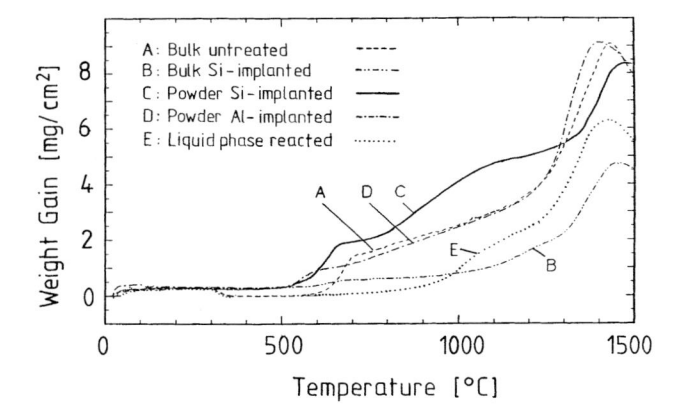

Figure 54. Weight gain of Si-sputtered, ion-implanted, and liquid-phase sintered B_4C in air.

subsequent isothermal annealing procedures between 700 °C and 1200 °C in humid air (dew point 0 °C). Upon heating to 700 °C, B_4C and powder-derived B_4C-SiC composites react with a slight sinusoidal weight gain and weight loss followed by a strong weight loss at higher temperatures. Precursor-derived B_4C-SiC composites exhibit a slight increase in weight starting from the very beginning but a decrease in rate is observed above 1000 °C. The isothermal treatment at 700 °C shows clearly that B_4C and B_4C-SiC_{powder} ceramics behave similarly, undergoing a parabolic weight gain whereas the B_4C-$SiC_{precursor}$ material exhibits less increase in weight. At 1000 °C, the initial weight loss of single phase B_4C during heating is compensated again by a slight weight gain approaching steady-state conditions. B_4C-SiC_{powder} samples again show almost the same reaction of stabilizing weight change whereas precursor-derived composites undergo a continuous weight gain. At 1200 °C, all materials suffer from a slight weight loss after 50–100 min of exposure.

Monitoring the weight change of continuously heated single-phase and SiC-containing boron carbide ceramics in dry air (dew point −20 °C) yielded again a step of weight gain at 650 °C for all materials except one with continuous SiC grain boundary phase prepared by liquid phase sintering, followed by a strong increase in mass at above 1200–1250 °C (Fig. 54).

A XPS analysis of the energies of Si_{2p} and B_{1s} bonds in oxidation layers from humid air compared to untreated reference materials revealed that in B_4C-SiC_{powder} composites the binding energy of Si stays constant between 700 °C and 1200 °C and is somewhat lower than that of crystalline silica, but a peak splitting is observed at 1200 °C indicating that a boro-silica glass is additionally formed. The B binding energies are close to that of glassy and crystalline boron oxide. Precursor-derived B_4C-SiC composites show a decreasing binding energy of Si with increase of oxidation temperature, indicating that a boro-silica glass forms at low temperatures becoming rich in Si with increasing temperature and time. Finally at 1200 °C, the binding energy is close to that of silica, and, indeed, isolated cristobalite and tridymite crystals can be found on the surface.

The B bonds decrease only slightly in strength confirming that a boro-silica glass of changing composition is present. X-ray analysis of oxidized single-phase B_4C and B_4C-SiC_{powder} composites reveals also the existence of H_3BO_3 and HBO_2 after cooling if treated at 700°C and 1000°C, respectively, whereas no derivatives of boric acid are found at room temperature after heating to 1200°C except of crystals formed newly during storage. In contrast, B_4C-$SiC_{precursor}$ did not exhibit any indication of the presence of boric acid.

It may be concluded that in the initial stage the oxidation of all materials is governed by the build up of a boron oxide layer, which in the case of single-phase B_4C and B_4C-SiC_{powder} immediately undergoes hydrolysis in humid air yielding boric acid, which is removed by evaporation and hence is responsible for the subsequent weight loss according to the reactions. Only in dry systems a stable liquid B_2O_3 layer may be formed before volatile suboxides are generated at locally low oxygen partial pressures and temperatures exceeding 1000°C:

$$B_4C + 4O_2 \rightarrow B_2O_{3(glass/cryst)} + CO_2 \quad (13)$$

dry air   *humid air*

$$B_2O_{3(glass)} \rightarrow B_2O_{2(vap)} + 1/2 O_2 \qquad B_2O_3 + 3H_2O \rightarrow 2H_3BO_{3(liq)} \quad (14a,b)$$

$$B_2O_{3(glass)} \rightarrow B_2O_{(vap)} + O_2 \qquad B_2O_3 + H_2O \rightarrow 2HBO_{2(liq)} \quad (15a,b)$$

$$B_2O_{3(glass)} \rightarrow B_2O_{3(liq)} \rightarrow B_2O_{3(vap)} \qquad B_2O_3 + H_3BO_3 \rightarrow liquid + H_2O \quad (16a,b)$$

In the case of B_4C-SiC_{powder} material, the B_4C matrix is exposed to air like pure B_4C, thus oxidizes according to Eqs. (13)–(16) and hence governs the weight change of the composites whereas a $SiO_{2(glass)}$ passivation layer is deposited on the isolated SiC particles. Thus, the oxygen attack is not retarded at all (active oxidation) unless a boro-silica glass is formed at 1200°C which tends to release B_2O_3 as a vapor phase. In contrary, B_4C particles which have been coated with precursor-derived SiC form a continuous boro-silica glass layer at low temperatures (400°C), which results in a progressive weight gain (passive oxidation) although B_2O_3 may also evaporate from the melt and cause a relative enrichment of SiO_2. At 1200°C, B_2O_3 and SiO_2 eventually segregate in all materials and form distinguishable SiO_2 and boro-silica glass areas. The oxidation then is accompanied by increasing loss of CO ($pCO = 10^{-3}$ at 1200°C), generating small fumaroles in the protecting cover. Surprisingly, thermodynamic calculations predict the occurrence of elemental boron as a stable reaction product. At temperatures above approximately 1400°C, SiO evaporates at the expense of the boro-silica layer whereas SiC is newly formed due to active carbothermic reactions close to the oxidation front involving SiO_2 and CO.

The rate-controlling mechanism after the formation of boro-silica glass layers is molecular oxygen diffusion through that layer. The activation energy for that diffusion in silica is reported to range between 83 and 125 kJ/mol [234] whereas that in B_2O_3-SiO_2 glass is 139 kJ/mol. The activation energy of B_4C oxidation in dry air was determined by Telle [231] and Litz [235] to be 134 and 108 kJ/mol, respectively, which is considerably higher than that determined for humid air in that work (51 kJ/mol).

As expected, B_4C-SiC composites exhibit an intermediate activation energy of 77 kJ/mol, whereas precursor-derived B_4C-SiC composites with 112 kJ/mol are close to oxygen diffusion in SiC or boro-silica glass, which confirms this mechanism to be the rate-controlling step in dry systems. For comparison, the formation of Si-O-H...-O-Si bonds (i.e., H_2 diffusion in $SiO_{2(glass)}$) requires 71–75 kJ/mol for activation [236] and may become important for humid environments.

The microstructural appearance of SiC in B_4C-SiC composites clearly influences the oxidation kinetics. The formation of a continuous boro-silica layer at rather low temperatures is helpful in preventing active oxidation during which always new material is exposed to oxygen attack by the evaporation of volatiles. Humid atmosphere is generally more detrimental to the oxidation behavior of B_4C-containing materials than a dry one. A uniform boro-silica glass layer is only formed if SiC is homogeneously distributed as an intergranular phase, which can be accomplished by (i) liquid phase reaction sintering, (ii) coating of B_4C powder with polysilane-derived SiC, and (iii) ion beam assisted sputtering of Si or SiC on B_4C powders if the powders are processed under inert gas atmosphere.

7.4.4.2 Boron Carbide–Transition Metal Diboride Ceramics

Boron carbide–based composites with transition metal diborides – in particular with TiB_2 – have been extensively studied for cutting tools and wear parts [71, 98, 100, 103, 218, 237–240]. Since both phases are thermodynamically stable up to 2300°C composites can be prepared either by pressureless sintering with an Fe additive at 2175°C [241], or by hot-pressing and HIP without additives. Nishiyama and Umekawa [240] have obtained full density by pressureless sintering blends of 20–60 vol.-% TiB_2 at 2100°C for 1 h in a vacuum. A maximum three-point bending strength of 620 MPa was measured on composites with 35 vol.-% TiB_2 combined with hardness HR_A of 93.8. During cutting of a Al25Si alloy, the composites exhibited the same performance as cBN and K15 hard metal tools and were clearly superior to Ti(C,N)- and alumina-based materials.

Another method of densification makes use of reaction hot-pressing or self-propagating combustion sintering of MC-B powder mixtures under pressure:

$$MC + 6B \rightarrow MB_2 + B_4C \qquad (17)$$

Since sintering of MB_2-B_4C powder mixtures yields similar complications to the sintering of the pure compounds due to favored surface diffusion and evaporation–recondensation reactions, the combustion route is more likely because heat is generated inside the sample due to the exothermic conversion, the bulk diffusion is significantly enhanced and a grain size refinement occurs as the carbide phase decomposes. A certain risk is the evaporation of volatiles such as CO, CO_2, B_2O_3 as deoxidation products, or even of B_{gas}, which may be formed because of the high heat release. Temperatures exceeding 2300°C have been reported during fabrication of TiC and TiB_2 from the elements [242–246]. The reaction velocity can be retarded by the addition of the final conversion product to the starting powder which then behaves as if inert. Thus in the case of TiC/B mixtures, TiB_2 is added

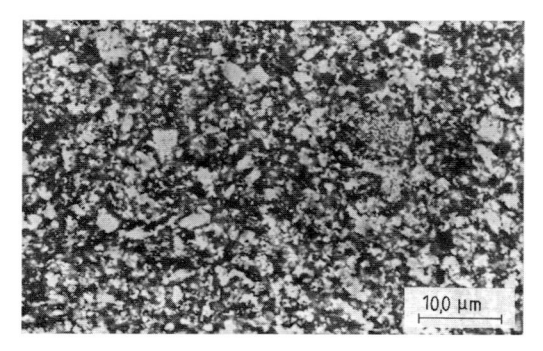

Figure 55. SEM micrograph of a reaction hot-pressed TiC–B powder blend. Light: TiB_2, dark: B_4C.

or B_4C, which also takes part in the reaction:

$$2\,MC + B_4C \rightarrow 2MB_2 + 3\,C \tag{18}$$

$$3\,C + 12\,B \rightarrow 3\,B_4C \tag{19}$$

In this case, B_4C also undergoes a grain size refinement which is very beneficial for the mechanical properties. In Fig. 55, a micrograph of a reacted TiC-B powder mixture is presented, which still exhibits TiB_2-B_2C agglomerates of the size of the initial TiC particles. Note that the average particle size of both reaction products is approximately 1 μm. Generally, this reaction can be employed for most of the transition metal boride–boron carbide composites since the borides are usually more stable than the particular carbides [92].

Dense composites of MB_2 and B_4C, in particular of TiB_2 and B_4C, regardless of their fabrication technique exhibit improved mechanical properties compared to the particular single-phase materials. The increase in strength of hot-pressed or HIPed materials to $\sigma_b = 600$–800 MPa is mostly attributed to a retardation of the grain growth, whereas the improved toughness is due to crack deflection around TiB_2 particles. K_{Ic} values of 5–7.3 MPa m$^{1/2}$ have been reported for B_4C-based composites with LaB_6, TiB_2, ZrB_2, NbB_2 and W_2B_5 [75]. At 2150°C, pressureless sintered B_4C/TiB_2 composites with 1 mass-% Fe additive exhibit a maximum bending strength of $\sigma_b = 420$ MPa at an optimum volume fraction of 20% TiB_2. The lower strength compared to the hot-pressed material is mainly attributed to the embrittling FeB intergranular phase. With increasing sintering temperature and amount of additive, the strength even drops to 100–250 MPa due to the exaggerated coarsening of the B_4C matrix by one order of magnitude.

Another example of successful materials development is B_4C-TiB_2-W_2B_5 composite ceramics prepared by reaction hot pressing [218, 232, 237]. The initial powders consisted of B_4C, B, and Si and were mechanically alloyed with WC, TiC, and Co. During sintering or hot pressing, the carbides react with B_4C and free boron to the desired transition metal borides. This reaction is accompanied by a strong bloating. Above the eutectic a boron- and metal-containing Si melt is generated which promotes liquid phase sintering depending on the volume fraction. Generally, hot pressing is, however, required to overcome the swelling. The reaction was

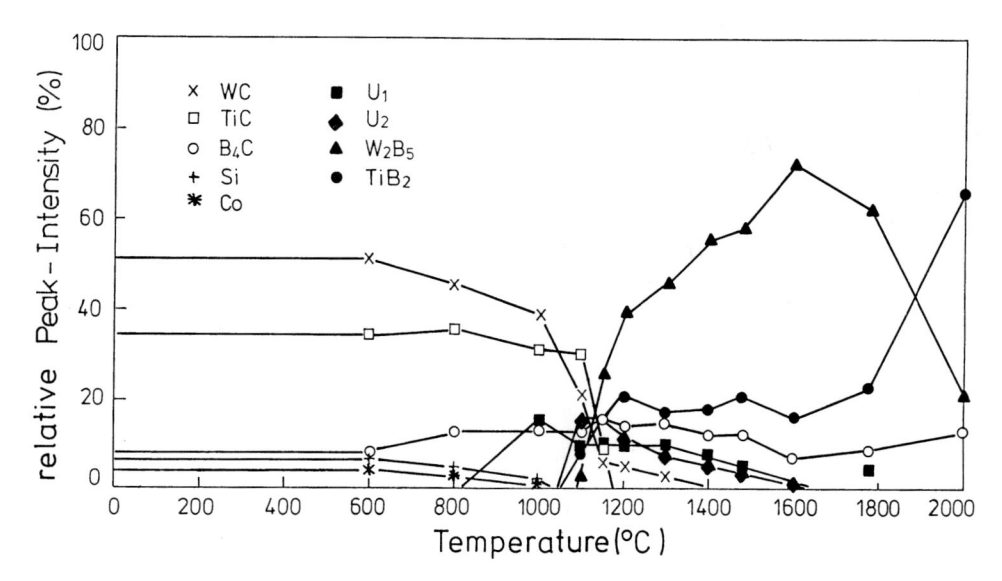

Figure 56. X-ray analysis of sintering reactions during hot pressing of B_4C-Si-B-TiC-WC-Co blends. Holding time at the particular temperature step 1 min.

monitored by heat treatments at temperatures ranging from 600°C up to 2100°C with pressure of 47 and 65 MPa and isothermal sintering times ranging from 1 min to 8 h. As expected from the phase diagrams, WC and TiC react with elemental boron and B_4C-forming borides such as WB_4, W_2B_5, and TiB_2. As shown by the relative X-ray intensities of the phases in Fig. 56, this reaction takes place in the temperature range from 1000°C to 1200°C. Si coming from the additive and Co resulting from the binder material of the hard metal milling balls also react forming a B-rich Si-Co melt at approximately 1100°C. At the same temperature, unknown phases denoted as U_1 and U_2 appear which are possibly silicides or ternary borides and decompose at 1600°C. Above 1600°C, the dominating phases are B_4C, W_2B_5, and TiB_2. Individual Si or Co phases are not detectable any more. Si and Co segregate at or close to the grain boundaries between the transition metal borides and the boron carbide matrix where they dissolve into TiB_2 and W_2B_5 particles. A solid solubility of TiB_2 for Si of at least 8 mass-% was documented in model experiments. Free carbon can be observed only in powders which were milled for 3 h if the initial boron content was not increased. The strong weight loss observed during heating is due to the evaporation of Si, Co, and B species as well as due to deoxidation products such as H_3BO_3, B_2O_3, CO, and CO_2.

The following chemical reactions may occur:

$$2\,WC + 13\,B \xrightarrow{900-1100°C} W_2B_5 + 2\,B_4C \tag{20}$$

$$TiC + 6\,B \xrightarrow{900-1100°C} TiB_2 + B_4C \tag{21}$$

$$8\,WC + 7\,B_4C \xrightarrow{1000-1100°C} 4\,W_2B_5 + 2\,B_4C + 13\,C \tag{22}$$

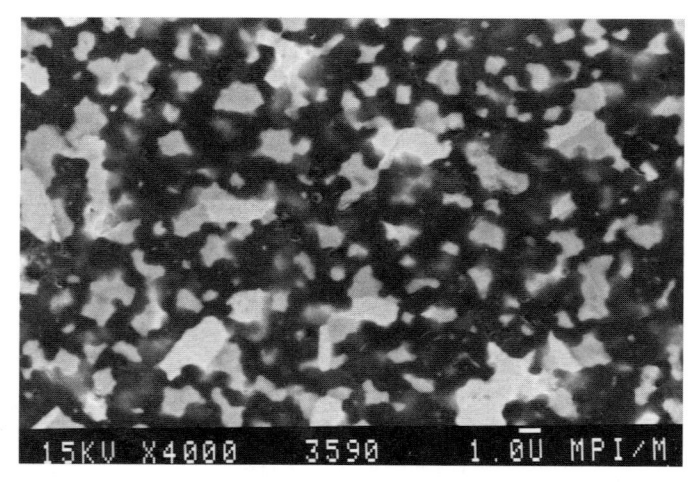

Figure 57. Microstructure of hot-pressed material. Dark matrix: boron carbide, gray: TiB$_2$-solid solution, white: W$_2$B$_5$.

$$2\,\text{TiC} + \text{B}_4\text{C} \xrightarrow{1000-1100°C} 2\,\text{TiB}_2 + 3\,\text{C} \tag{23}$$

$$4\,\text{B} + \text{C} \xrightarrow{>600°C} \text{B}_4\text{C} \tag{24}$$

These reactions are strongly exothermic and belong to the type of materials synthesis denoted as "self-propagating high-temperature combustion synthesis". The standard Gibbs Free Energy of the reaction (21) is with $\Delta G^0 = -200$ kJ/mol B$_4$C insensitive to the MC compound involved and fairly independent of temperature. Furthermore, W$_2$B$_5$ tends to decompose above 1600°C in the presence of TiB$_2$ forming a solid solution (Ti,W)B$_2$ with increasing solubility of W by increasing temperature:

$$n\text{TiB}_2 + m\text{W}_2\text{B}_2 \xrightarrow{>1600°C} (\text{Ti}_n\text{W}_m)\text{B}_2 \quad \text{with } n = 1 - m \text{ and } m_{\max} = 0.63 \text{ at } 2230°C \tag{25}$$

The microstructure of a hot-pressed sample is characterized by a homogeneous distribution of B$_4$C, W$_2$B$_5$, and (Ti,W)B$_2$ (Fig. 57). The average grain size is 0.8–1.0 µm. The final phase composition consists of 72 vol.-% B$_4$C, 20 vol.-% (Ti,W)B$_2$, and 8 vol.-% W$_2$B$_5$. The porosity is less than 3 vol.-%. Boron carbide and titanium diboride particles are of more or less equiaxed morphology whereas it is a characteristic of the W$_2$B$_5$ phase to grow in an elongated shape. Hot pressing at temperatures above 1900°C results in an exaggerated grain growth of boron carbide and the transition metal borides. The average grain size may reach more than 10 µm if fired at 2000°C. Boron carbide grows most probably by transient liquid film sintering, consuming smaller particles, whereas the transition metal borides grow by the dissolution of W$_2$B$_5$ in the liquid film and precipitation as (Ti,W)B$_2$ solid solution. The liquid phase consists possibly of a Si-Co-Ti-W-B alloy which is consumed by the formation of B$_{12}$(B,C,Si)$_3$ and (Ti,W,Co,Si)B$_2$ solid solution in the areas close to the grain boundaries.

Table 6. Mechanical properties of B_4C-TiB_2-W_2B_5 composites hot pressed at standard conditions.

Property	Value	Unit
Hardness HV_{10}	32.1 ± 3.2	GPa
4-pt. bending strength	830 ± 113	MPa
Fracture toughness		
ISB	4.2 ± 0.4	$MPa\,m^{1/2}$
ICL, 10 N	5.0 ± 0.8	
Young's modulus	425 ± 10	GPa
Shear modulus	185 ± 5	GPa
Poisson's ratio	0.15 ± 0.01	–

The hardness (HV_{10}) reaches 32 GPa as a maximum. The maximum bend strength of 830 MPa is obtained by hot pressing at 1820 °C. Compared to single-phase boron carbide, the strength is increased more than two times. In comparison to other TiB_2-reinforced boron carbide ceramics, the increment in strength is still 150%. Higher hot-pressing temperatures result, however, in a decrease to 600 MPa due to coarsening. Fracture toughness was also improved from 2.5 to 4.2–5.0 $MPam^{1/2}$ by crack deflection. The properties are summarized in Table 6. Figure 58 shows the microstructure of the composite in Fig. 57 failed by dynamic fracture toughness measurement applying the multiple impact test (courtesy of B. Ilschner and R. Zohner).

The microstructure of hot isostatically pressed samples does not differ so much from that of hot-pressed material. The average grain size is slightly smaller than

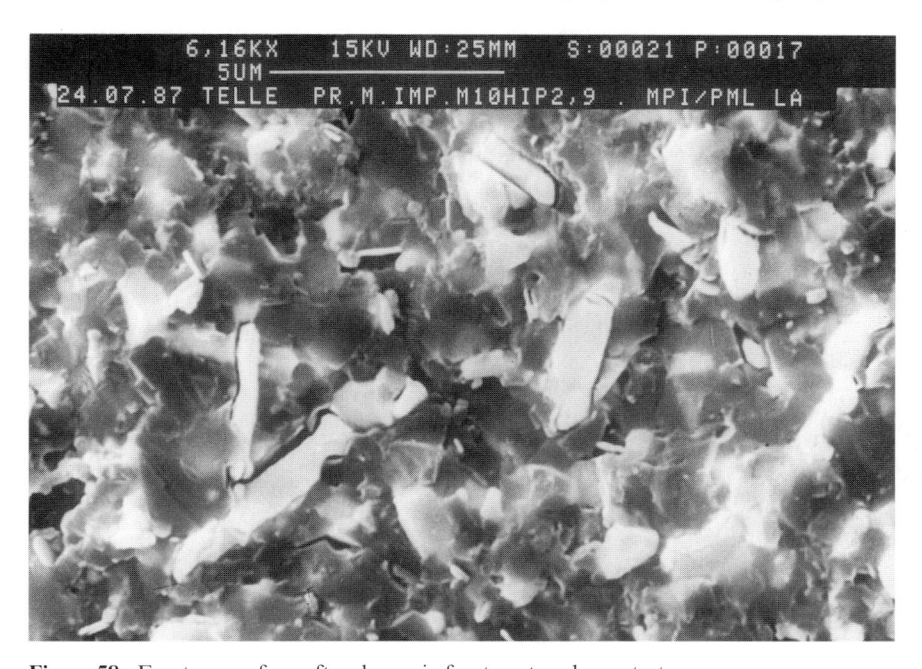

Figure 58. Fracture surface after dynamic fracture toughness test.

Table 7. Mechanical properties of B_4C-TiB_2-W_2B_5 composites HIPed at 1700°C, 200 MPa, 60 min.

Silicon content [mass-%]	Porosity [vol.-%]	Hardness HV_1 [GPa]	4-pt. Bending strength [MPa]	Fracture toughness [MPa m$^{1/2}$]	Young's modulus [GPa]	Poisson ratio
0.0	3.4	17.8 ± 2.4	102 ± 965	4.5 ± 0.2	384	0.17
3.0	1.7	37.3 ± 0.2	894 ± 73	3.5 ± 0.7	412	0.17
7.5	0.2	29.8 ± 2.4	1129 ± 85	5.2 ± 0.8	431	0.17

that of hot-pressed material, i.e., in the range of 0.8 μm if treated at 1600°C and 1.2 μm if HIPed at 1900°C. Porosities of < 0.55% at 1600°C have been achieved. The bending strength is further enhanced to 1129 ± 85 MPa. The fracture toughness exhibits a remarkable maximum of 5.2 MPam$^{1/2}$ at 1700°C HIP temperature. The entire data set is listed in Table 7.

A further improvement of both hot-pressing cycles and additives, in particular concerning the transition metal diborides resulted in a microstructural optimization with in-situ grown W_2B_5 particles. After annealing, the average of the diameter of the W_2B_5 particles was about 4.5–7 μm (Fig. 59).

According to the proceeding growth of W_2B_5 platelets an increase of toughness with increasing sintering temperature and time was observed. As predicted by the theory of particle reinforcement of brittle matrices introduced by Faber and Evans [247] the increasing aspect ratio makes crack deflection a more efficient toughening mechanism (Fig. 60).

In Figs. 61 and 62 the data of high-temperature strength and K_{Ic} measurements are shown. At 1000°C the strength is only half of the room temperature strength. The toughness at this temperature is still more than 60% of the room temperature toughness and stays constant for higher temperatures. Since crack deflection responsible for the relatively high fracture toughness is mainly caused by internal stresses emerging from the misfit of thermal expansion between matrix phase and

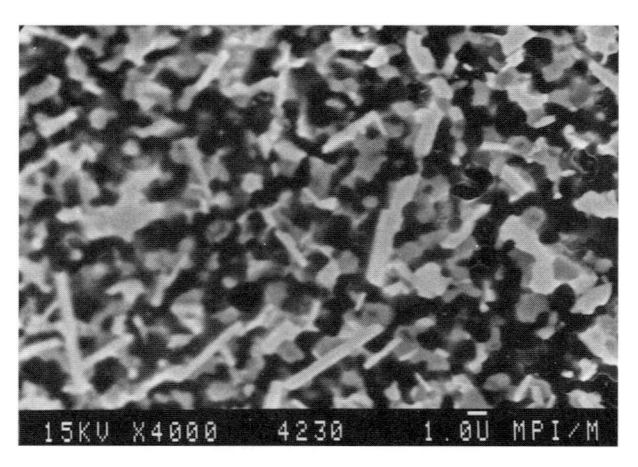

Figure 59. SEM micrograph of annealed B_4C-TiB_2-W_2B_5 sample. Note the elongated W_2B_5 particles.

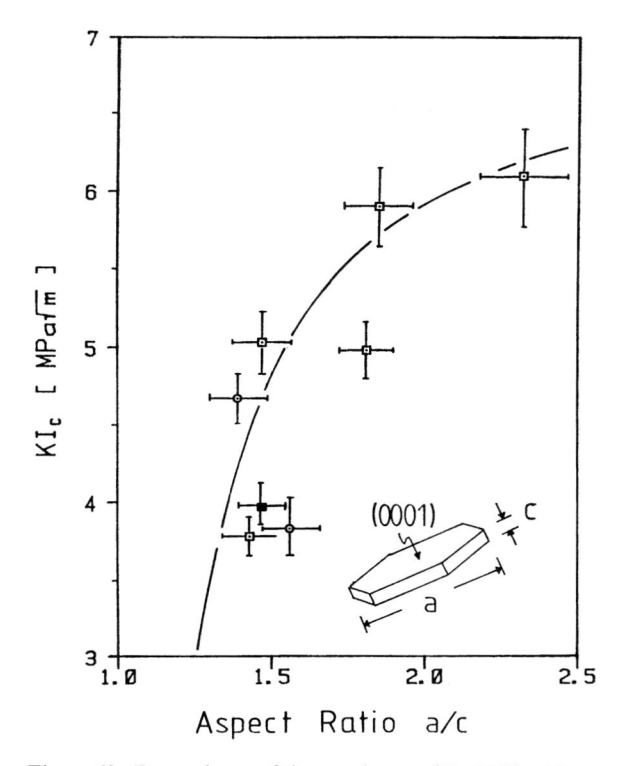

Figure 60. Dependence of the toughness of B_4C/diboride composites on the aspect ratio of platelets.

dispersed particles, it is evident that during an increase in temperature the stresses are reduced to zero at 1000°C. At that temperature no stresses contribute to the deviation of the crack path, and the composite possesses the same toughness as the unreinforced boron carbide matrix.

Simple stress calculation according to Selsing [248] in the B_4C-TiB_2-WB_2-CrB_2 system have shown possible stress variations from -10 to 2500 MPa if the thermal

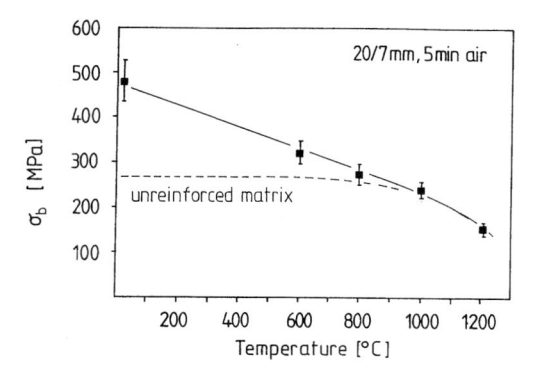

Figure 61. High-temperature strength of hot isostatically pressed $B_4C/TiB_2/W_2B_5$ composite.

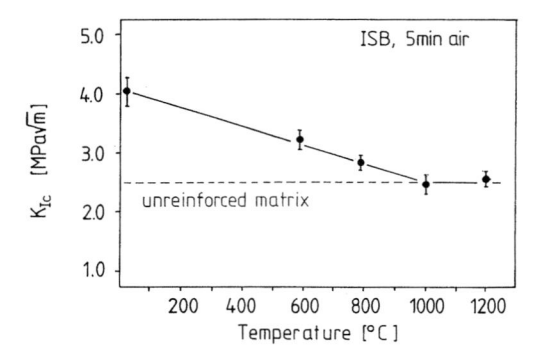

Figure 62. High-temperature fracture toughness of hot isostatically pressed $B_4C/TiB_2/W_2B_5$ composite measured by the ISB method.

expansion of the diborides is changed by the formation of solid solutions [232, 249, 250]. Since measurements show that the thermal expansion of $(Ti, Cr)B_2$ solid solutions do not obey the linear rule of mixture and that the anisotropy of α may even change in sign [251], the radial matrix stresses around these kinds of particles may be tensile or compressive in nature. Calculations of the stress levels indicate, however, that microcracking is not likely to occur due to the comparatively small grain size of the hot-pressed and HIPed materials [250].

As introduced above, the reaction of boron carbide with metal carbides can be used to fabricate metal borides or metal boride/boron carbide composites in a controlled way during densification if boron carbide or free boron is used in excess, or if carbon is bonded by another additive. Although the incompatibility of B_4C and metal carbides is well known, many attempts have been undertaken to produce composites or coatings thereof but failed as soon as equilibrium conditions were approached. Physical or chemical vapor deposition of B_4C on hard metal substrates, or WC coatings on boron carbides are typical problems (e.g., [252]). In both cases, interlayers of graphite form and hence result in an unsatisfactory adhesion of the deposited coating to the substrate.

Recently, Sigl [207] made use of this conversion and reacted B_4C with TiC to fabricate a B_4C-TiB_2 composite while using the emerging free carbon for the reduction of oxide layers and thus for activated pressureless sintering of boron carbide powders. The most striking advantage associated with the processing of MC sintering aids is their chemical similarity to boron carbide. Thus, unlike resins or amorphous carbon, metal carbides exhibit significantly fewer problems with long-term stability, dispersability of the sintering agent, or flow behavior of spray-dried granules into die cavities. Furthermore, the in-situ synthesized TiB_2 is anticipated to support sintering similar to the beneficial effects such as grain growth inhibition that have been observed in SiC-TiB_2 or B_4C-TiB_2 composites. Submicron boron carbide powder was doped with various amounts of fine TiC to yield samples with nominal contents of 1.5, 3, 4.5, and 6 mass-% free carbon after the reaction. The B_4C-TiC reaction took place in a 60 min holding step at 1250°C. The specimens were finally sintered in a 10 mbar Ar atmosphere for 2 h

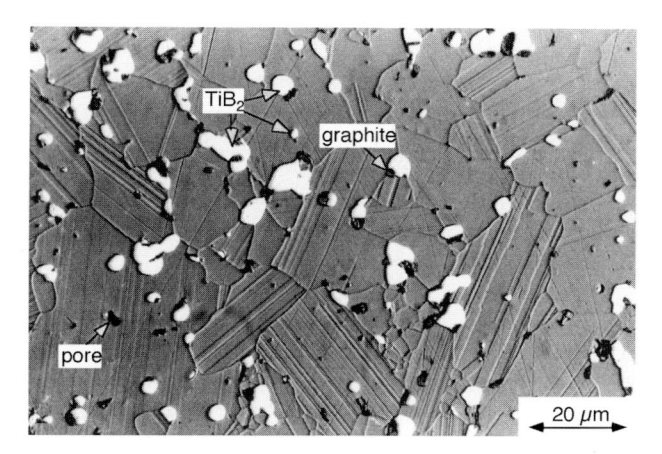

Figure 63. Microstructure of B_4C-TiC sintered with 3 mass-% C. White: TiB_2, black: free C (after [207]).

at various temperatures (2125, 2150, 2175, and 2200°C). Samples with closed porosity were further densified in a post-HIP cycle at 2050°C (30 min., 200 MPa Ar).

Figure 63 shows representative microstructures of the composite sintered at 2150°C with 3 mass-% C. It is notable that the 1.5 mass-% C composites consist of a small fraction of pores, two major solid phases, i.e., TiB_2 grains in a B_4C matrix, and traces of the dark graphite. Significantly more graphite becomes visible in the high carbon materials. Figure 63 shows, that TiB_2 particles and most graphite are in direct contact, indicating the simultaneous formation of both compounds consistent with the reaction considered. The mean grain size of B_4C in all compositions is plotted in Fig. 64. As expected, there is a pronounced tendency for grain

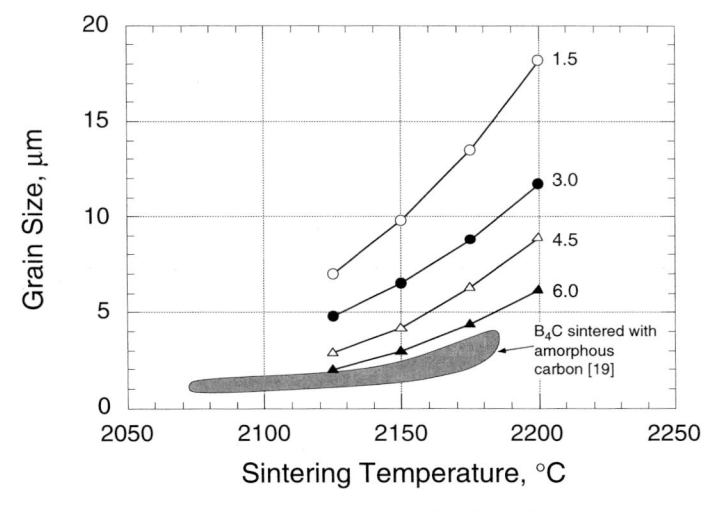

Figure 64. Dependence of the grain size of B_4C on free carbon content and sintering temperature (after [207]).

Table 8. Mean strength and toughness of sintered and post-HIPed B_4C-TiB_2 composites as a function of sintering temperature; n.d. = not determined.

Residual C content [mass-%]	Flexural strength [MPa] Sintering temperature [°C]			Fracture toughness [MPa$\sqrt{m}$]		
	2150°C	2175°C	2200°C	2150°C	2175°C	2200°C
1.5	292	286	266	4.2	4.0	3.7
3.0	368	n.d.	n.d.	3.6	n.d.	n.d.
4.5	454	n.d.	n.d.	3.2	n.d.	n.d.
6.0	502	n.d.	n.d.	2.9	n.d.	n.d.

growth in B_4C with increasing sintering temperature. However, the coarsening is suppressed by an increasing amount of sintering aid as documented in Fig. 64. This plot, together with the observation that the number of second phase particles at grain boundaries increases with initial TiC content, provides evidence that both TiB_2 and graphite precipitates are effective in pinning moving grain boundaries. The TiB_2 particles coarsen with increasing sintering temperature as well. This coarsening happens most likely by coalescence [253].

All compositions, except the 1.5 mass-% carbon material, exceed a relative density of 95% at temperatures above 2150°C. In high carbon materials, density increases further as the sintering temperature is raised to 2200°C, whereas the density of materials with 1.5 and 3 mass-% carbon declines at 2200°C. Composites exceeding 95% of the theoretical density were post-HIPed to full density, virtually independent of the initial sintering temperature.

The range of strength and toughness in B_4C-TiC-derived composites is generally comparable to B_4C materials sintered on the resin route. Data for fracture toughness, K_c, and flexural strength, σ_f, are disclosed in Table 8 as a function of the nominal carbon content. An inverse trend becomes, however, evident from Fig. 65

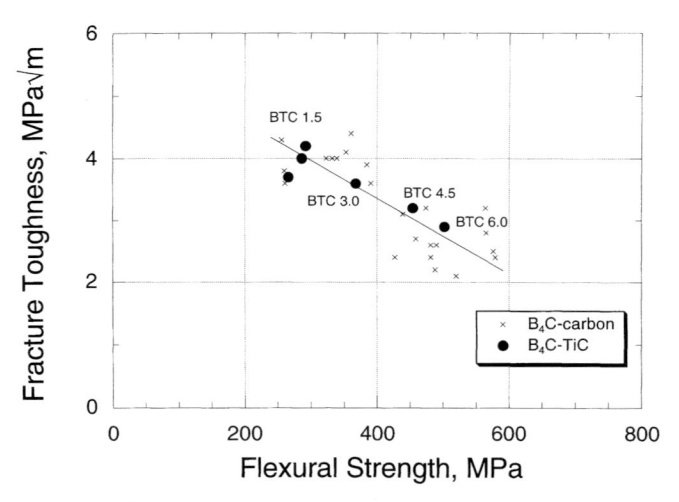

Figure 65. Fracture toughness and strength of B_4C-TiC materials (after [207]).

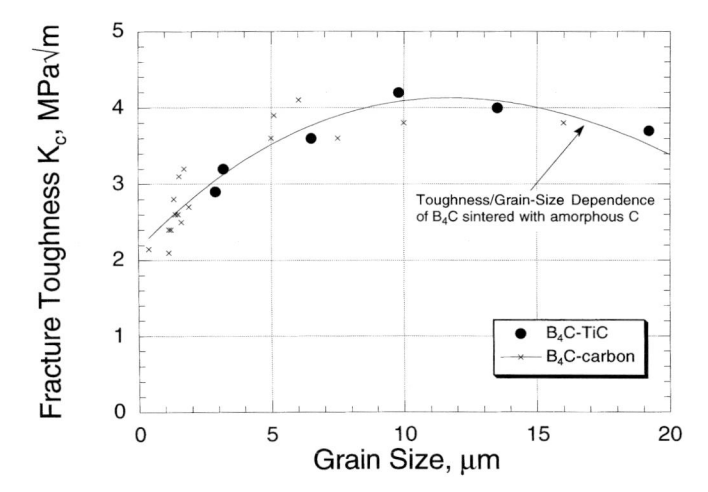

Figure 66. Fracture toughness–grain size relation (after [207]).

where the flexural strength is plotted versus toughness: This plot reveals a general trend for the strength of boron carbide materials to drop with increasing toughness.

Figure 66 shows a plot of the asymptotic fracture toughness of B_4C materials versus grain size. The fracture resistance data of B_4C-TiC materials closely follow a scatter band, which is also characteristic for B_4C doped with amorphous carbon. K_{Ic} starts at a toughness of 2 MPa$\sqrt{m}$, reflecting the inherent fracture resistance of B_4C, and increases up to a shallow maximum of ≈ 4 MPa$\sqrt{m}$ at a grain size of about 10–15 µm, until K_{Ic} gradually drops again. From a microstructural point of view, this behavior is accompanied by a continuous switch from inter- to transgranular fracture. This trend is outlined in Fig. 67, which summarizes the dependence of flexural strength on grain size for B_4C materials doped both with TiC and amorphous carbon. As in many ceramic materials, the strength data could be fitted to a grain size dependence, where strength scales with the inverse square root of grain size, $d^{-1/2}$.

Since toughening mechanisms by changing the crack path from transgranular to intergranular fracture depend on both internal stresses and interfacial strength it is worth while studying the grain boundary structure in B_4C-TiB_2 materials in more detail as well as to explore the possibility for microcrack formation. In contrast to Telle and Petzow [218] who have used transient liquid film hot pressing providing a relatively strong B_4C-MB_2 interface, Sigl and Schwetz [254] and Sigl and Kleebe [255] have exploited the opportunity for crack deflection and microcracking by weakening the B_4C-TiB_2 phase boundary by the incorporation of free carbon. It was suggested that microcracking accounts most for the toughness increment.

Sintering and post-HIP of B_4C blends 20 and 40 vol.-% TiB_2 and 0, 1.2, and 4.8 mass-% free carbon derived from phenolic resins yielded full density samples which have been examined by SEM and TEM. A SEM micrograph of a 20 vol.-% TiB_2 composite is shown in Fig. 68. Despite the fact that 1.2 mass-% free

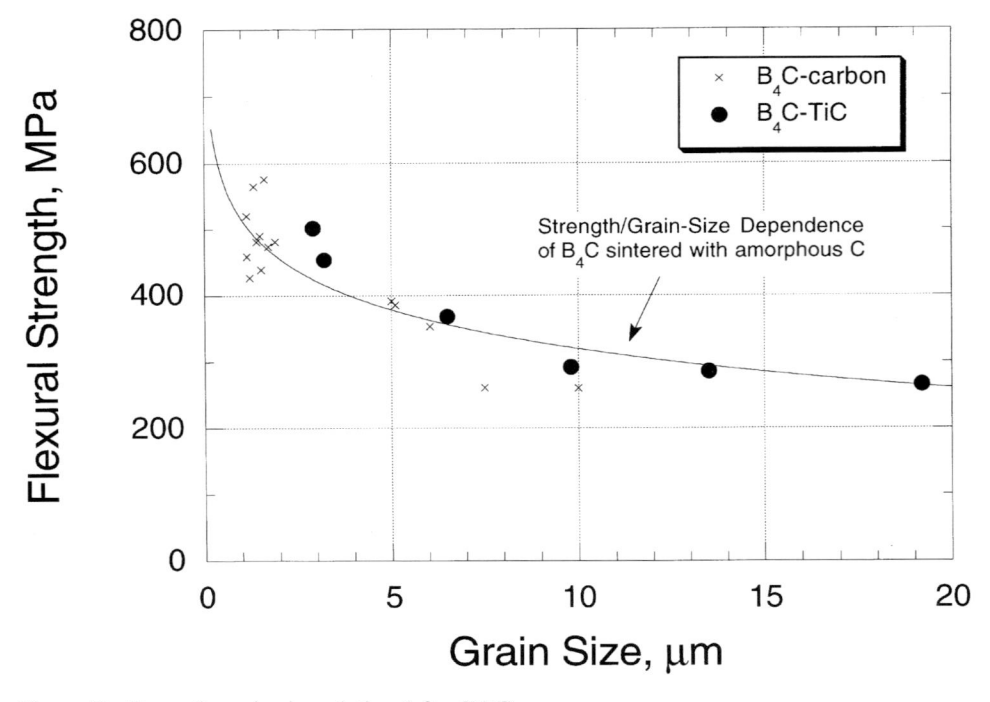

Figure 67. Strength–grain size relation (after [207]).

carbon should be present according to the chemical analysis, only two phases, i.e., TiB_2 (white) and boron carbide (gray), are visible. Particulate carbon precipitates are obviously absent in B_4C-TiB_2 composites with <2 mass-% free carbon. Discrete graphite particles appear in 4.8 mass-%-free carbon composites

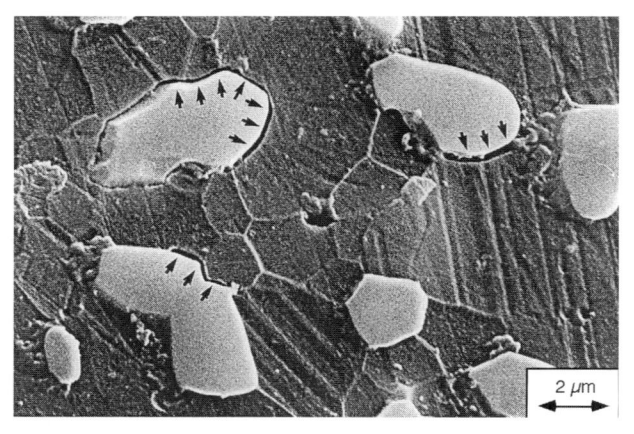

Figure 68. SEM micrograph of post-HIPed 1.2% free carbon containing B_4C-TiB_2 composite. Arrows indicate circumferential microcracks after cooling from sintering temperature (after [255]).

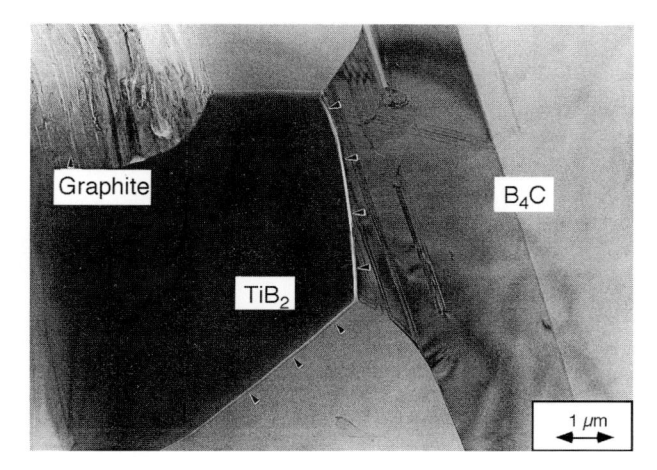

Figure 69. TEM micrograph of microcracked B_4C-TiB_2 grain boundary (after [255]).

(Fig. 69). Additionally, the analysis of hetero-phase boundaries by TEM yields evidence of thin carbon interlayers (Fig. 70) with a thickness between ≈ 5 and 10 nm.

Composites with free carbon contain microcracks which have formed spontaneously upon cooling from sintering temperature. Such cracks are predominantly detected at B_4C/TiB_2 phase boundaries (Fig. 68), and in rare cases also at TiB_2/TiB_2 grain boundaries. Notably, microcracks are not present at all phase boundaries but restricted to a few interfaces feeling residual tension above a critical threshold stress. Remarkably, microcracking coincides with the position of carbon interlayers which is in accordance with the fact that microcracks were not observed in composites without free carbon. Another origin of microcracking is occasionally observed inside graphite particles, which are usually located along B_4C-TiB_2

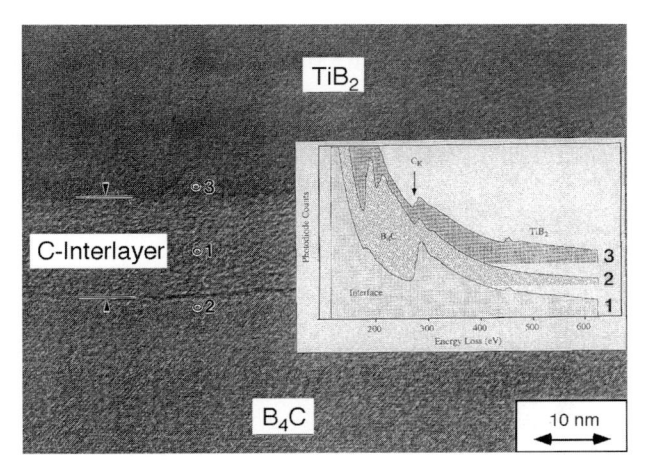

Figure 70. TEM micrograph of C interlayer between B_4C and TiB_2 particles (after [255]). The inserted diagram presents EELS-analyses of the corresponding points 1–3 of the micrograph.

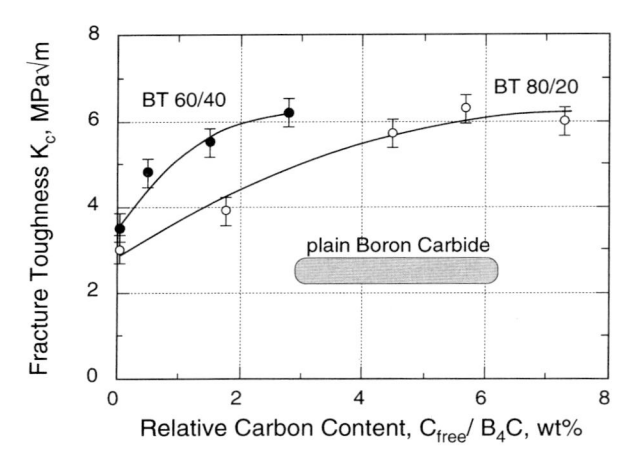

Figure 71. Fracture toughness–free carbon relation (after [255]).

phase boundaries. Figure 71 compares the fracture toughness of B_4C-TiB_2 composites with plain boron carbide as a function of free carbon content. Two effects obviously control the toughness of these composites: (i) Particle effect: The toughness of plain boron carbide varies around 2.2–2.5 MPam$^{1/2}$. Upon the addition of 20 and 40 vol.-% TiB_2, the fracture resistance increases to 3.0 and 3.6 MPam$^{1/2}$, respectively, with the trend to increase with rising volume fraction of particles. (ii) Free carbon effect: A further increase in toughness is observed upon the addition of free carbon such that the fracture resistance exceeds 6.0 MPam$^{1/2}$ at elevated carbon contents. At still higher carbon quantities, toughness tends to remain constant. In conclusion, strong B_4C-B_4C interfaces with comparatively low tensile stresses coexist together with weak but highly stressed B_4C-TiB_2 phase boundaries with carbon interlayer or graphite precipitates which allow not only crack deflection but also microcracking. Cracking and the relatively large graphite particles obviously contribute to a decrease in strength since no data are provided by Sigl and Kleebe [255].

7.5 Transition Metal Boride Ceramics

Transition metal boride ceramics are mainly based on TiB_2 or ZrB_2 due to their high hardness and relatively high toughness. Because of their poor sinterability, additives have been employed consisting of other transition metal diborides, carbides or iron-group metals and their borides. Since the introduction of a second phase as usually beneficial effects to the mechanical properties, a large variety of composites has been created to reinforce TiB_2-based ceramics but, vice versa, TiB_2 was also used to strengthen other matrix phase materials. The fabrication of cemented borides is presented separately in Section 1.8.

7.5.1 Preparation of Transition Metal Borides

Large-scale production of metal borides occurs preferentially in electric furnaces by the following high-temperature reactions:

(i) *Carbothermic reduction* of the metal oxide, graphite or carbon black

$$MO_2 + B_2O_3 + 5C \rightarrow MB_2 + 5CO \qquad (26)$$

The carbothermic method yields carbon contaminated powders and is suitable for materials in which a C content of up to 3 mass-% can be tolerated. For instance, TiB_2, ZrB_2, and the technically important hexaboride CaB_6 are synthesized by this method.

(ii) Reduction of metal oxides with carbon and/or boron carbide, known as the *boron carbide process*

$$2MO_2 + B_4C + 3C \rightarrow 2MB_2 + 4CO \qquad (27)$$

$$M_2O_3 + 3B_4C \rightarrow MB_6 + 3CO \qquad (28)$$

where M = rare earth elements.

The boron carbide process can also start from blends of metal carbides, metal hydrides, boron oxide, boron carbide and carbon black:

$$3MO_2 + B_4C + B_2O_3 + 8C \rightarrow 3MB_2 + 9CO \uparrow \qquad (29)$$

$$MC + MO_2 + B_4C \rightarrow 2MB_2 + 2CO \uparrow \qquad (30)$$

This material usually contains only small amounts of residual carbon or boron carbide but no metals, and is thus the favored process for the technical synthesis of less contaminated borides. The process is carried out in tunnel furnaces under hydrogen or in a vacuum at 1600–2000°C, i.e., below the melting point of the boride. It is thus a reaction sintering procedure yielding a high-porosity product which can easily be crushed and milled. Additional refinement is obtained by multiple vacuum treatments with metallic or B_4C additives to compensate non-stoichiometries. The final product is then called "vacuum quality".

(iii) *Aluminothermic, silicothermic, magnesiothermic reduction* of mixtures of metal oxides and boric acid

$$MO_2 + B_2O_3 + Al(Si, Mg) \rightarrow MB_2 + Al_2O_3(SiO_2, MgO) \qquad (31)$$

The yield is usually contaminated by residual metals or oxides and thus has to be purified by subsequent leaching, or a high-temperature vacuum treatment.

High-quality borides of the transition metals with-defined stoichiometry and crystal structure are synthesized by the following laboratory-scale methods:

(i) From the elements or metal hydrides by fusion in an arc or resistance furnace, or by diffusion during sintering or hot pressing

$$M + 2B \rightarrow MB_2 \qquad (32)$$

$$MH_2 + 2B \rightarrow MB_2 + H_2 \qquad (33)$$

(ii) Borothermic reduction of metal oxides

$$MO_2 + 4\,B \rightarrow MB_2 + B_2O_2 \uparrow \tag{34}$$

(iii) Conversion of metal carbides with boron and/or boron carbide yielding powder mixtures or, carried out during powder metallurgical densification, i.e., sintering or hot pressing, composites

$$MC + 2\,B \rightarrow MB_2 + C \tag{35}$$

$$MC + 6\,B \rightarrow MB_2 + B_4C \tag{36}$$

$$2\,MC + B_4C \rightarrow 2\,MB_2 + 3\,C \tag{37}$$

(iv) Electrolysis of fused salts containing metal oxides, boron oxide or hydroboric acid plus alkaline borates and fluorides.

(v) Molten metal/boron dissolved in Al, Cu, Sn, or Pb melts (auxiliary-metal bath method). This procedure is based upon the growth of large particles at the expenses of small particles of high surface area which are dissolved in the melt. Furthermore, solid solutions may be precipitated starting from a mixtures of the pure materials. The grain size can readily be controlled by the ultimate temperature of soaking and the cooling rate. Upon cooling down the solubility of the feedstock material gets smaller, which is hence preferentially precipitated. The auxiliary bath method is well-suited for the fabrication of single crystals of borides.

(vi) Chemical vapor reaction of metal halides and boron halides in a hydrogen atmosphere under plasma conditions. This method, however, yields material of varying stoichiometry and crystallinity.

Limiting factors for commercial fabrication are the relatively high costs of elemental boron and the low production rate in the reactors.

7.5.2 Densification of Transition Metal Borides

The densification of single-phase and pure ceramics of transition metal diborides is complicated by two characteristics of these compounds, the high melting point and the comparatively high vapor pressure of the constituents. As a rule, sintering temperatures exceeding 70% of the absolute melting temperature have to be applied.

7.5.2.1 Pressureless Sintering

Titanium diboride, TiB_2 ($T_m = 3250°C$), requires firing temperatures of the order of 1800–2300°C to initiate grain boundary and volume diffusion, and thus to obtain more than 95% of the theoretical density. One disadvantage is that the borides undergo a similar abnormal grain growth at high temperatures to B_4C or SiC. Furthermore, at lower temperatures evaporation of B and boron suboxides enhance the grain growth without shrinkage by evaporation and recondensation mechanisms. Since TiB_2 is strongly anisotropic due to its layered structure well-faceted particles are formed and, together with pore trapping inside large grains, the final densities of pure TiB$_2$ do not exceed 90% even if heated between 1900 and 2500°C (e.g., [257, 258]). Thus it is nearly impossible to achieve completely dense bodies by pressureless sintering, as no shrinkage between 1900°C and 2100°C occurs, and further densification by volume diffusion and plastic flow, is accompanied by exaggerated grain growth. This behavior was attributed by Coble and Hobbs [259] and Kislyi and Zaverukha [260] to the competing mechanisms of evaporation–recondensation and volume diffusion exhibiting the same rate of mass transport, whereas gas-transport reactions are favored due to the lower activation energy. The powder size–dependent sintering behavior of TiB$_2$ was studied by Kislyi *et al.* [261]. Starting with high-purity submicron-size powder, synthesized from TiCl$_4$ and BCl$_3$ in hydrogen in a plasma-arc heater, Baumgartner and Steiger [262] achieved densities of 98.4–99.4% at 2000–2100°C combined with a comparatively fine microstructure (average grain size 1–18 µm) due to TiC and TiO inclusions. Further heating or a prolonged holding time generate exaggerated grain growth to 80 µm, whereas the density does not improve. This is attributed to entrapped porosity, which can only be removed by volume diffusion. In contradiction to reports of the other authors, no significant weight loss was observed even after several hours hold at 2250°C, which could be related to active evaporation reactions. Since the small grain size of 1 µm could be retained up to 2000°C, the porosity was preferentially removed by grain boundary diffusion. Thus, contrary to carbothermically produced TiB$_2$ powder, high densities have been obtained below the critical temperature promoting rapid grain growth. This implies that both the initial particle size and the presence of impurities significantly influence the densification kinetics. Baik and Becher [263] have studied the effect of oxygen contamination of submicron TiB$_2$ powders and concluded that in the case of hot-pressing between 1400 and 1700°C, oxygen promotes grain coarsening by enhanced evaporation–recondensation of B$_2$O$_3$. Upon pressureless sintering between 1700 and 2050°C oxygen remains primarily as titanium oxides and suboxides, which increase the surface diffusivity and thus the pore and particle coarsening rather than the densification. A maximum total amount of oxygen of less than 0.5 mass-% in the powder or reducing additives such as carbon is recommended. Sintering is usually carried out under vacuum in a resistance furnace with a graphite, tantalum or tungsten resistor or in a high-frequency furnace with a graphite susceptor. Ar or H$_2$ gas atmospheres can also be used. If carbon crucibles are used, boron nitride diffusion barriers have to be inserted to prevent eutectic melting of the borides and carbon in the temperature range of 2000–2500°C. The considerable losses of volatile boron or boride species may be reduced by powder-bed sintering [134].

7.5.2.2 Hot Pressing and Activated Sintering

Densities above 95% have been achieved by axial hot-pressing at pressures exceeding 20 MPa and temperatures above 1800°C. The microstructures consist typically of particles of >20 µm in size. Another problem is related to the hexagonal layered structure of the AlB$_2$-type borides. Because of the strong anisotropic behavior of the physical properties, especially of the coefficients of thermal expansion, the coarsening can be very detrimental to the mechanical properties, by producing spontaneous microcracking and residual strains.

Pressureless sintering and hot-pressing of transition metal borides can be generally activated either physically by starting from submicron powders or by extensive milling, i.e., by increasing the specific surface area and introducing defects, or chemically by doping with small additions (0.3–3 mass-%) of transition metals such as Fe, Ni, Co, Cr, Pt, or their halides. The mechanisms of sinter activation by doping are not yet readily understood but an increase of driving forces and volume diffusion and a retardation of evaporation seems to be likely. Crystallographic studies on Co-, Nb-, Cr-, and Re-doped ZrB_2 by Czech *et al.* [264] indicate that the metals substitute for Zr in the metal sublattice, which is also confirmed for Mo and W [265]. Other borides such as VB_2, NbB_2, TaB_2 or W_2B_5 and Mo_2B_5 could not be satisfactorily densified by pressureless sintering. For an extensive survey on powder molding, compaction and sintering of various transition metal borides, containing detailed descriptions of additives, the reader is referred to [134].

Reaction sintering starting from chemically incompatible compounds may also lead to high densities, especially if combined with hot-pressing, since that synthesis is strongly exothermic and provides high internal temperatures [242]. The so-called self-propagating high-temperature combustion synthesis was used for TiB_2, ZrB_2, NbB_2, and TaB_2 starting from the elements, B_4C mixtures with metallic Ti, Zr, Cr, or Nb [266], and Ti-B-TiB_2 blends [244, 267]. In the case of other reactants, e.g., blends of metal oxides and boron carbide or Al-TiO_2-B_2O_3 mixtures [243], the formation of gaseous by-products often prevents complete densification but may result in bodies of well-defined porosity. The kinetics of the combustion synthesis of TiB_2 from the elements have been studied by Holt *et al.* [245]. Ouabdesselam and Munir [268] investigated the sinterability of directly synthesized TiB_2 powder but could not find any significant difference to carbothermically produced powders.

7.5.3 Properties of Transition Metal Borides Ceramics

Transition metal borides are mainly explored for their mechanical properties. Since they exhibit metallic transport properties such as high electric and thermal conductivity with a negative temperature coefficient they are also of interest as electrode materials, for heating elements and sensors.

7.5.3.1 Single Phase Ceramics

Single phase TiB_2 ceramics with high density have been prepared almost exclusively by hot-pressing. Applying temperatures between 1800 and 2300°C densities of >95% may be achieved [134]. Small amounts of additives may, however, significantly improve the sintering behavior. While between 1950 and 1965 the fabrication of single phase TiB_2 was aimed at applications in the nuclear industry, multiphase ceramics produced since then have basically been aimed at wear applications. But also the electric properties have been studied extensively and led to highly developed electrode materials ready for use in, e.g., aluminum production.

Table 9. Physical properties of TiB_2 and ZrB_2

Property	Unit	TiB_2	ZrB_2
Density	g/cm^3	4.52	6.09–6.17
Crystal structure	Space group	C_6/mmm	C_6/mmm
Lattice constants a, c	nm	0.3028, 0.3228	0.3167, 0.3529
Standard enthalpy of formation	kJ/mol	279.49	308.78
Melting point	°C	322 ± 520	324 ± 518
Thermal expansion coefficient α_a,	10^{-6}/K	$5.107 + 1.997 \times 1^{-3}\ T$	
Thermal expansion coefficient α_c,	10^{-6}/K 25–1600°C	$7.443 + 2.26110^{-3}\ T$	
Thermal conductivity λ	W/m·K	24–59	23–24
Electric resistivity ρ	$\mu\Omega$ cm	20.4 (25°C)	9.2 (25°C)
		26 (200°C)	10 (200°C)
		36 (400°C)	11 (400°C)
		46 (700°C)	13.5 (700°C)
		56 (1000°C)	17 (1000°C)
		68 (1300°C)	20 (1300°C)
Thermal coeff. of electric conductivity	K^{-1}	4.76×10^{-3}	6.32×10^{-3}
Debye temperature ϑ	K	807–820	585
Hardness $HK_{0.1}$ ceramics, >95% dense	kp/mm^2	2600 (25°C)	2100 (25°C)
		2400 (200°C)	1850 (200°C)
		1800 (400°C)	1000 (400°C)
		1050 (600°C)	900 (600°C)
		700 (1000°C)	800 (1000°C)
Young's modulus E	GPa	560	490
Shear modulus G	GPa	490	220
Poisson ratio	–	0.327	0.3

The physical properties of diborides have been reviewed in many papers, e.g., by Clougherty and Pober [269], Samsonov et al. [270], and Castaign and Costa [271]. The reported value depend, however, on purity, final porosity, grain size, and other factors that are usually not well-documented. Therefore, a comparison of the data is difficult. Especially the mechanical properties differ very much as the anisotropy of thermal expansion eventually results in microcracking during cooling from sintering temperature, or at least to high stress concentrations at grain boundaries. Hardness and fracture toughness, for instance, vary therefore with grain size and testing conditions. Table 9 presents a data set for TiB_2 and ZrB_2. Only data from polycrystalline materials with densities >95% and preferentially small grain size, if available, have been used. Thermal expansion of some diborides have been measured by high-temperature X-ray analysis by Lönnberg [272] and by Fendler et al. [250] and Telle [232].

The electric resistivity in the solid solution system TiB_2-ZrB_2 was studied by Rahman et al. [273]. Billehaug and Øye [274] present a study of several transition metal diborides for cathode materials in Hall–Héroult cells and come to the conclusion that TiB_2 should be an excellent candidate because of its stability against the aluminum–kryolithe melt and high thermal and electric conductivity. The sensitivity to thermal shock, the infiltration of liquid phase along grain boundaries and, finally, the high price are, however, the limiting factors. In combination with other, cheaper or properties-adjusting materials, there are excellent chances for this kind of application. Sørlie and Øye [275] favor today a dense TiB_2 coating on graphite to reduce costs and to gain full advantage of the corrosion resistance and energy saving of TiB_2. In connection with BN and AlN the excellent electric conductivity of TiB_2 and the perfect wetting by Al has been used for the manufacturing of resistance-heated evaporator boats for the metal deposition on plastic sheets. Data about the corrosion behavior have been presented by Bannister and Swain [276].

7.5.3.2 TiB_2-TiC Composites

As shown in the phase diagram in Fig. 19, the combination of TiC and TiB_2 is thermodynamically stable up to 2500°C undergoing a quasi-binary eutectic reaction [101] In that system, excellent wear-resistant materials have been produced by hot-pressing or even by pressureless sintering of eutectic compositions at 1600–1700°C [103]. A Vickers hardness of $HV_2 = 23$ GPa was measured at room temperature, which is lower than that of the pure materials with values of 27.5 GPa for TiC and 28.5 GPa for TiB_2. At 600°C, however, the hardness of the composite, 8.3 GPa, far exceeds the hardness of monolithic TiC and TiB_2, which decrease to 6.8 GPa and 7.8 GPa, respectively. The fracture toughness is notably improved to 7.1 MPa m$^{1/2}$. The significant decrease in wear during turning or milling of steel compared to the monolithic materials was mainly attributed to "phase boundary toughening" due to the favored occurrence of common coherent $(111)_{TiC}/(0001)_{TiB2}$ particle interfaces [103]. Besides this very sophisticated toughening effect, mechanisms which influence the crack propagation such as crack deflection or crack impediment due to thermal misfit effects between the boride and carbide phases certainly contribute to the increase in toughness, whereas grain growth retardation due to the pinning of grain boundaries by incorporated particles affects the strength positively. As another example, WC is used for grain size refinement of TiB_2, and, vice versa, TiB_2 is used as an additive for WC-based materials [257, 277].

High-temperature reinforcement by in situ precipitation of TiC and TiB_2 from supersaturated solid solutions has already been used with interesting results. In the TiC-TiB_2 system, either TiC or TiB_2 can be the host crystal for the corresponding minority phase or the precipitate [278–280]. The addition of a small fraction of boron to TiC can increase the critical resolved shear stress at 1600°C by a factor of six if TiB_2 precipitates are formed at the (111) slip plane of TiC.

7.5.3.3 TiB_2-B_4C Composites

Kang and Kim [281] have investigated the improvement of TiB_2 with a dispersion of B_4C particles. Using 1 mass-% Fe as a reactive additive, hot-pressing at 1700°C for 60 min at 35 MPa resulted in 99% dense composites with a clear maximum in strength of 700 MPa at 10 vol.-% B_4C and in K_{Ic} of 7.6 MPa m$^{1/2}$ at 20 vol.-% B_4C. This optimizing effect was attributed to both grain growth inhibition and change in fracture mode from transgranular to intergranular by the B_4C addition. Since studies on the B_4C-rich side of this system also indicate optimum properties at approximately 60–70 vol.-% B_4C a change in strengthening and toughening mechanisms most occur at a composition between 40 and 50 vol.-% B_4C. The total system was investigated by Nishiyama and Umekawa [240] by pressureless sintering of ultrafine B_4C and TiB_2 powders. Besides other properties such as oxidation and wear resistance, they report a maximum in strength of 650 MPa at 35 vol.-% TiB_2 and an optimum hardness of $HR_A = 94$ at 20 vol.-% TiB_2. Pressureless sintering of B_4C with additions up to 16.7 mass-% TiC to initiate the reaction to TiB_2 and the simultaneous release of C for deoxidation was studied by Sigl [207] and discussed already in Section 7.4.4.2 like all other B_4C-TiB_2 composites with B_4C as the majority compound.

7.5.3.4 Transition Metal Diboride-SiC Composites

Reinforcement of TiB_2 by dispersed SiC particles is generally possible since both materials are chemically compatible. As a result, crack impediment is obtained but the increase in strength and toughness is small [282]. This composite material has, however, not yet been studied extensively.

Silicon carbide-based composites with transition metal boride particulates have been developed for electroconductive applications such as heating elements and igniters [283–285] but also as wear-resistant structural parts for high temperatures such as valve-train components and rocker arm pads in super-hot running engines [286, 287]. These composites combine the high thermal and electric conductivity of, e.g., TiB_2 and ZrB_2 with the oxidation resistance of SiC. Additionally, due to thermal mismatch stresses of the order of 2 GPa, toughening mechanisms such as crack deflection and stress-induced microcracking with a pronounced process zone, as well as crack flank friction have been proven to occur. Cai et al. [288] and Faber et al. [289], have presented a detailed analysis of the contributions of the particular mechanisms to the total fracture toughness, stating that stress-induced microcracking is operational in a process zone of approximately 150 µm width.

Typical conditions for densification by axial hot-pressing are 2000–2100°C, at a pressure of 20–60 MPa for 30–60 min which results in 96–99.8% density. The particle sizes of the matrix and dispersed phases range from 1–5 and 4–8 µm, respectively. An optimum volume fraction of reinforcing particulates of 25–30 vol.-% has been reported, yielding a flexural strength of 710 MPa and a fracture toughness of 5.0–5.7 MPa $m^{1/2}$, as shown in Fig. 72 [282]. Composites with a lower TiB_2 content of 15 vol.-% exhibit a mean strength of 485 MPa combined with a K_{Ic} of 4.5 MPa $m^{1/2}$ [287]. The strength of SiC-based materials with 50 vol.-% ZrB_2, HfB_2, NbB_2 or TaB_2 particles also ranges between 400 and 500 MPa [283]. Similar strength values (480 MPa) combined with an exceptionally higher fracture toughness of 7–9 MPa $m^{1/2}$ have been reported for large-scale lots of pressureless sintered 16 vol.-% TiB_2 composites [284]. Since the sintering was carried out with temperatures exceeding 2000°C (no details given) yielding 98–99% of the theoretical density and an average TiB_2 particle size of 2.0 µm, it is obvious that the reinforcing phase also acts as a grain growth inhibitor for SiC. The high-temperature

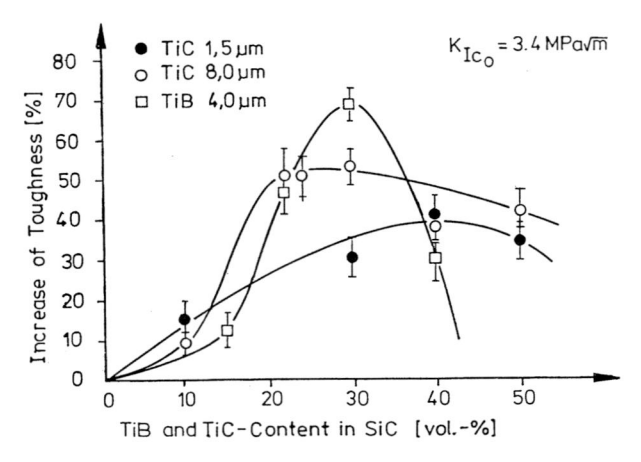

Figure 72. Volume-dependent mechanical properties of SiC-TiB$_2$composites [282].

strength of SiC/TiB$_2$ and SiC/ZrB$_2$ composites was found to remain nearly constant at 480 MPa up to 1200°C, and is hence superior to that of many sialons [283, 284]. SiC-TiB$_2$ composites have been prepared a in situ synthesis by Ohya et al. [253] by adding TiC and B powder to SiC. If the dopants are well-balanced, these incompatible compounds react between 1000 and 1600°C to form isolated and homogeneously distributed TiB$_2$particles. The reaction is, however, accompanied by a strong expansion, which can be overcome by isothermal soaking at 1500°C and final pressureless sintering at 2200°C for 30 min. The fracture toughness of the >98% dense samples is around 3.7 MPa m$^{1/2}$ at 10 vol.-% TiB$_2$.

A systematic study on the rule of mixtures for the mechanical and electrical properties of TiB$_2$-TiC-SiC composites was carried out by de Mestral and Thevenot [290]. They modeled "iso-property" curves in the quasi-ternary phase diagram for mechanical parameters by fitting 20 independent coefficients of a third-order polynomial developed by Phan-Tan-Luu et al. [291] to the results of experimental test points. Calculated iso-bend strength curves as well as tests on hot-pressed materials indicate a maximum of 1100 MPa close to the TiB$_2$-TiC binary edge of the system (Fig. 73). The best fracture toughness value of 6.4 MPa m$^{1/2}$ was obtained on the binary SiC-TiC edge (measured and calculated), on the binary TiB$_2$-SiC edge and in the ternary region close to the TiB$_2$ phase (67 mol.-% TiB$_2$, and 16.5 mol.-% SiC and TiC, respectively) (Fig. 74). The calculated rule of mixtures could also be confirmed in the case of hardness measurements.

SiC-based composites with W$_2$B$_5$ have been discussed by Telle [232]. The fabrication method was similar by starting with mechanically alloyed WC-B-containing SiC powders. Due to the strong tendency of W$_2$B$_5$ to grow anisotropically in a platelet shape crack deflection and crack bridging was efficiently applied. Several kinds of diboride additives have been studied by Tanaka and Iyi [292] to reinforce SiC. Pressureless sintering of β-SiC composites with 15–17 vol.-% NbB$_2$, TaB$_2$, TiB$_2$, and ZrB$_2$ at >2000°C resulted in >99% density. Surprisingly, the additives strongly influenced the β-α transformation of SiC during sintering. TaB$_2$ addition

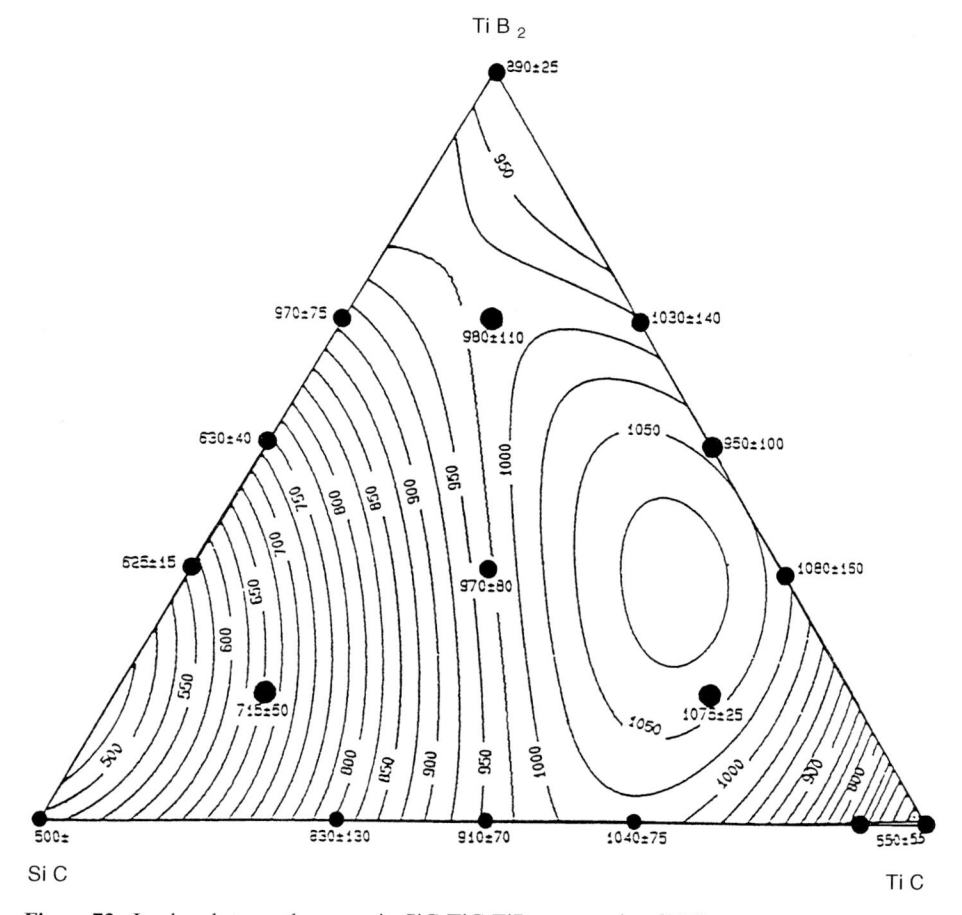

Figure 73. Iso-bend strength curves in SiC-TiC-TiB$_2$ composites [290].

clearly stabilized the 3C polytype up to 2200°C whereas ZrB$_2$ addition extended the stability of the 15R structure. NbB$_2$ doping results in a very smooth transition from 3C to 6H between 2000 and 2200°C. The 4H polytype was only found in TaB$_2$ containing materials at a larger content. Although the grain size and shape of the SiC matrix is heavily influenced the increment in toughness reaches only 20%. The highest value was obtained for 15 vol.-% TaB$_2$ with $K_{Ic} = 4.75$ MPam$^{1/2}$.

7.5.3.5 TiB$_2$-Transition Metal Diboride Composites

Combinations of diborides of different transition metal borides have been studied, especially in the TiB$_2$/CrB$_2$ and TiB$_2$/W$_2$B$_5$ systems, for wear applications and to a minor extent for electrodes in Hall–Heroult cells [135, 139, 140, 293–295]. Since the transition metal diborides crystallize in the same structure type, namely the AlB$_2$

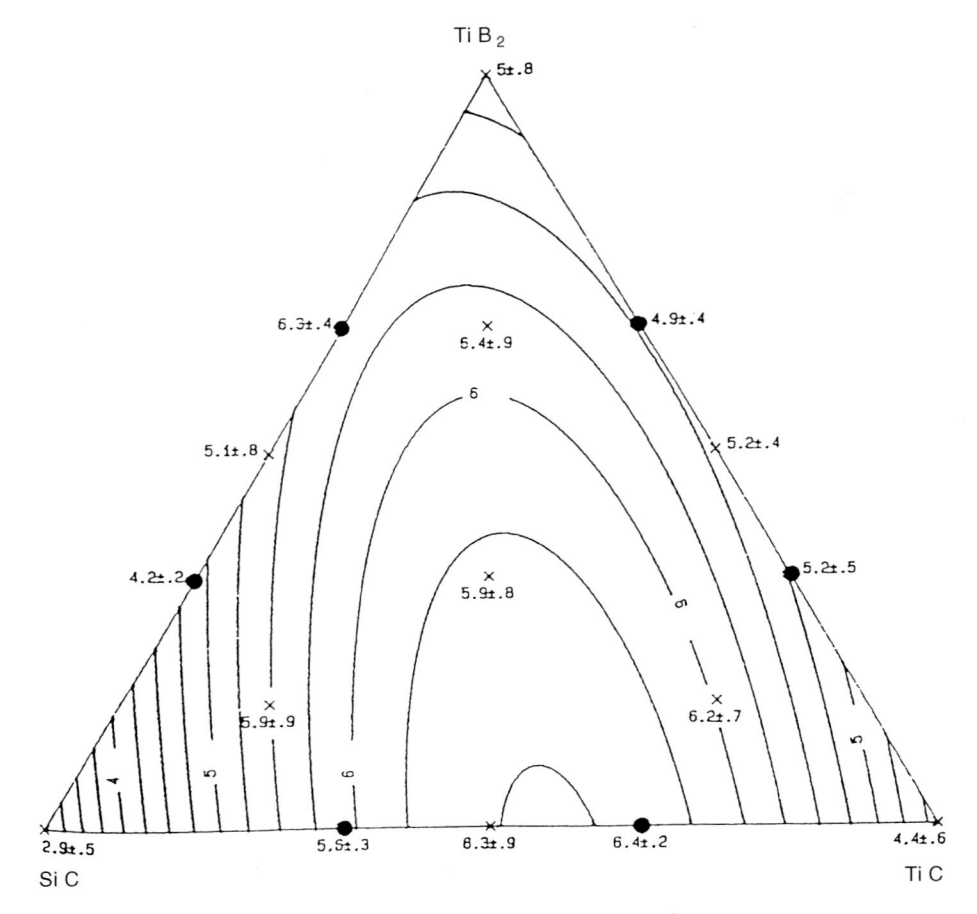

Figure 74. Iso-toughness curves in SiC-TiC-TiB$_2$ composites [290].

layered structure, the formation of solid solutions has been extensively investigated and used for hardening effects. As an example, the quasi-binary system CrB$_2$-TiB$_2$ exhibits a continuous mutual solid solubility approximately between 2000°C and 2100 ± 50°C (Fig. 30) [132, 136], but there is evidence of a solubility gap below 2000°C where the solubility of TiB$_2$ in CrB$_2$ is about 40 mol.-% at 1500°C and the solubility of CrB$_2$ in TiB$_2$ is less than 1 mol.-% below approximately 1800°C. The presence of CrB$_2$ aids the densification of TiB$_2$ due to its higher diffusion coefficient. Above 2100°C, CrB$_2$ containing materials partially melts, which is due to an almost horizontal solidus line between approximately 40 mol.-% CrB$_2$ and pure CrB$_2$. This fact enables liquid phase sintering of TiB$_2$ but with the risk of exaggerated grain growth and evaporation of chromium and chromium borides, since the vapor pressure of Cr is four orders of magnitude higher than that of Ti. Pre-reacted and hot-pressed materials of that system exhibit a flexural strength of 350–500 MPa [139, 140].

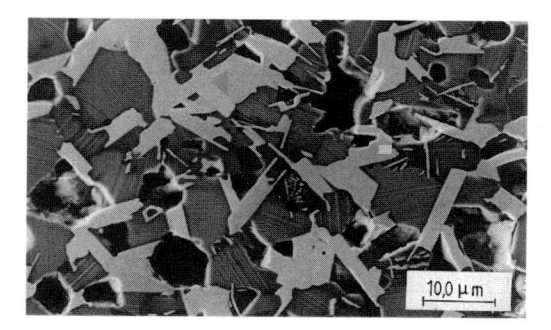

Figure 75. SEM micrograph of large W_2B_5 precipitates white and thin $TiWB_2$ layers in $(Ti,W)B_2$ solid solution matrix.

In the TiB_2-W_2B_5 system, the borders of the $(Ti,W)B_2$ homogeneity range have been intensively studied between 1500°C and 1700°C, at 2000°C and around the quasi-binary eutectic temperature. TiB_2 and WB_2 react eutectically at $T_e = 2230 \pm 40$°C and 90 ± 3 mol.-% WB_2 (Fig. 32). The solid solubility of WB_2 in TiB_2 at this temperature is approximately 63 mol.-%, whereas the solid solution of the $(W,Ti)_2B_5$-type contains only 3 mol-% TiB_2 at the eutectic equilibrium. The homogeneity range of the $(Ti,W)B_2$ solid solution narrows significantly with decreasing temperature and is 46–49 mol-% at 2000°C and 8–10 mol-% WB_2 at 1500°C.

A high-temperature treatment of TiB_2-W_2B_5 powder mixtures inside the solid solubility range of $(Ti,W)B_2$ at above, e.g., 2000°C, for 30–720 min results theoretically in a uniform, single phase microstructure. Subsequent annealing at, e.g., 1500–1700°C, causes the epitaxial precipitation of very fine platelets of the metastable $WTiB_2$ monoboride phase with β-WB structure onto the prism plains of the host crystal (Fig. 75). After 30–240 min annealing, these precipitates measure 0.5–5 µm in length and 0.01–0.2 µm in thickness and can be aged by prolonged heating or by the choice of a higher temperature (Fig. 76). The growth of W_2B_5 platelets can also be initiated by heterogeneous nucleation close to a grain boundaries of the host crystals [81, 152, 153]. The precipitate is then able to grow across the grain boundary into a neighboring W-rich $(Ti,W)B_2$ grain and thus create an

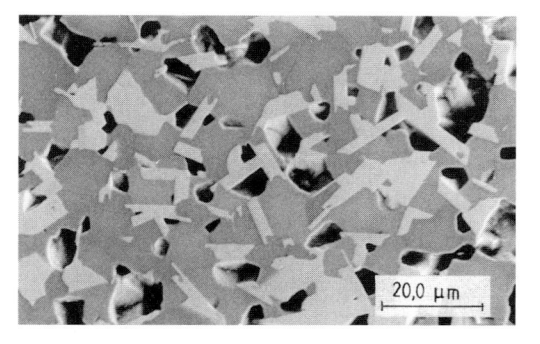

Figure 76. Aged in-situ reacted $(Ti,W)B_2$-WB_2 composite with grown WB_2 particles.

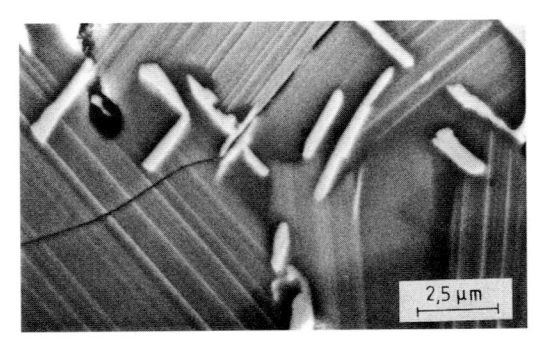

Figure 77. Interlocking grain boundaries bridged by WB_2 particles. Note the thin $TiWB_2$ stacks in the host crystals.

interlocking microstructure, as shown in Fig. 77. Crack propagation studies confirm that crack deflection is operational around the W_2B_5-type phases. This process is assisted by differences between the Young's moduli of the particular phases, the differences in the thermal expansion coefficients and their anisotropic behavior generating residual misfit stresses. Of similar importance for crack interactions are the grain boundaries of the W-depleted host crystals and neighboring W-rich solid solutions. Here, an active crack deflection was observed, which indicates that both the elastic constants and thermal misfit stresses of TiB_2-type solid solutions vary significantly with composition. High-temperature X-ray diffraction measurements of the lattice constants of $(Ti,W)B_2$ solid solutions confirm this observation.

Hot-pressed composite materials developed from the more complex systems of the type TiB_2-$M^I B_2$-M^{II} with M^I being Hf, V, Nb, Ta, Mo, or Mn and M^{II} being sintering additives such as Co and Ni, exhibit bending strengths between 850 and 1000 MPa which are due to the grain growth inhibiting influence of the 1–5 mass-% of $M^{II}B_2$ particulates (Fig. 78) [293–296]. During liquid phase sintering in a Co- or Ni-boride melt, both TiB_2 and M^I are partially dissolved and reprecipitated as a solid solution. The effect of grain growth retardation as well as of strength and hardness increments is attributed to stresses at the $TiB_2/(Ti,M^I)B_2$ phase boundaries generated by the mismatch of the lattice parameters between the unreacted TiB_2 acting as a nucleus and the epitaxially precipitated $(Ti,M^I)B_2$ solid solution. In the case of a TiB_2-5 mass-% W_2B_5/TaB_2 material with 1 mass-% CoB binder the lattice strain ranges between 9×10^{-4} and 14×10^{-4} depending on the hot-pressing temperature [293]. The addition of 1.7% TiC to the above-mentioned base composition reduced the porosity from 0.3–0.7 to 0.1–0.2 vol.-% after hot-pressing at 1500°C and a pressure of 20 MPa for 1 h. The improved sintering behavior was achieved by intensive ball milling resulting in an average particle size of 1 μm, but increased oxygen contamination. Watanabe and Shoubu [297] reported the formation of a $(Ti,Ta)(C,O)$ solid solution which is considered to initiate the improved densification resulting in a flexural strength of 1000 MPa. In a similar multiphase system, transition metal carbides were used as additives for pressureless sintering of TiB_2 yielding composites of binary and ternary borides [218, 296]. Attrition milled powder mixtures of TiB_2 with 3–10 mass-% Co or Ni and 20–35 mass-% WC have been sintered in a vacuum at temperatures between

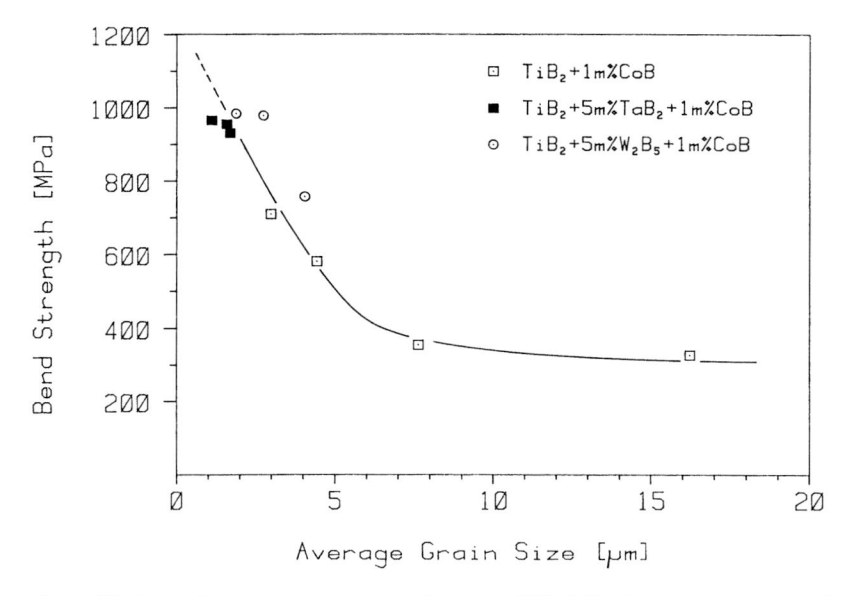

Figure 78. Strength–grain size relation of various TiB_2-MB_2-M composites (data from [293]).

1500 and 1700°C for 60 to 120 min yielding 98–99% relative density. Densification starts above 980°C due to the formation of a liquid phase in the Ti-B-Co/Ni system (Fig. 25). At this early stage, a rigid skeleton of TiB_2 and WC develops. Due to dissolution and reprecipitation, a $(Ti,W)B_2$ solid solution grows on the residual TiB_2 particles. Subsequently, crystals of ω-phase (see Section 7.8.3) form with compositions of WCoB or W_2NiB_2, respectively. Upon cooling, the residual liquid phase crystallizes as C-and Ti-enriched Co_3B or Ni_3B solid solution. A typical microstructure is shown in Fig. 79. Sintering at 1700°C for 2 h yields an average particle size of

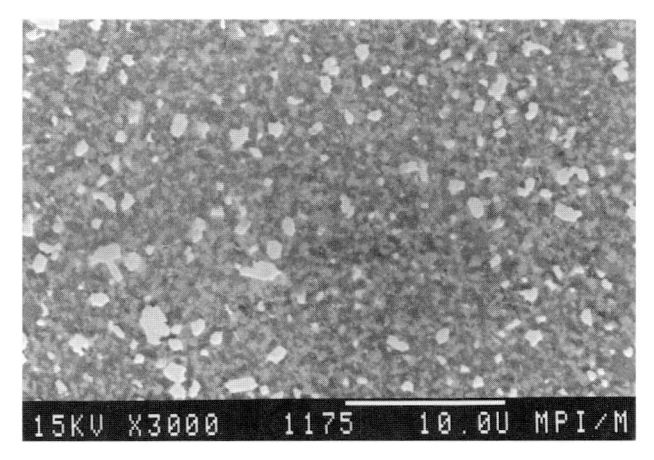

Figure 79. SEM micrograph of a $(Ti,W)B_2$-W_2NiB_2-Ni_3B composite. Light: W_2NiB_2, intermediate: Ni_3B, dark: TiB_2.

0.8 μm resulting in a flexural strength $\sigma_b = 600$–680 MPa. The K_{Ic} of 6.5–7.5 MPa m$^{1/2}$ is mainly attributed to crack deflection because of the weak Co- or Ni-boride intergranular phases.

7.6 Multiphase Hard Materials Based on Carbide–Nitride–Boride–Silicide Composites

Ternary composites of Ti(C,N)-TiB$_2$-MoSi$_2$ were studied by Shobu and Watanabe [298] in order to improve the oxidation resistance of Ti(C,N) TiB$_2$ materials. Full density was obtained after sintering composites with less than 80 mass-% TiB$_2$ and less than 60 mass-% MoSi$_2$ at 1750°C. The oxidation resistance above 1000°C was good for small Ti(C,N) concentrations, i.e., when all the carbonitride particles were surrounded by a phase of either TiB$_2$ or MoSi$_2$. The formation of rutile (TiO$_2$) and silicate glass was observed and considered to prevent further oxygen diffusion. TiB$_2$-20 mass-% MoSi$_2$ composites sintered at 1800°C in a vacuum exhibited a flexural strength of 600 MPa and a hardness of HV 2100, whereas the fracture toughness was only 3.7 MPa m$^{1/2}$. Composites of 70 mass-% Ti(C,N)-30 mass-% TiB$_2$ showed a three-point bending strength of 800 MPa and a K_{Ic} of 5 MPa m$^{1/2}$. With a hardness of HV > 2500, the material was tested as a cutting tool and exhibited a longer lifetime upon machining plain carbon steel at 300 m/min than conventional hard metals or cermets [298].

7.7 Boride–Zirconia Composites

The most important toughening strategy for oxide ceramics, namely the so-called transformation toughening by dispersed tetragonal zirconia particles, is not applicable to borides and carbides to a similar extent. Chemical interactions between ZrO$_2$, with its pronounced tendency for oxygen loss, and, in presence of oxygen, the thermodynamically less stable borides lead to the formation of boron oxides or carboxides, respectively, which in some cases result in the total degradation of the composite. This is particularly the case under reducing conditions and at high temperatures, which are both required for a successful densification of hard materials. For example, boron carbide decomposes in the presence of zirconia according to Lange and Holleck [75]:

$$B_4C + 2\,ZrO_2 \rightarrow 2\,ZrB_2 + CO + B_2O_3 \tag{38}$$

A pronounced bloating of the samples is observed due to the release of gaseous compounds. Eq. (38) resembles the so-called boron carbide route for the production of the particular transition metal borides. Another limiting factor is the chemical, geometrical and mechanical destabilization of tetragonal zirconia if combined

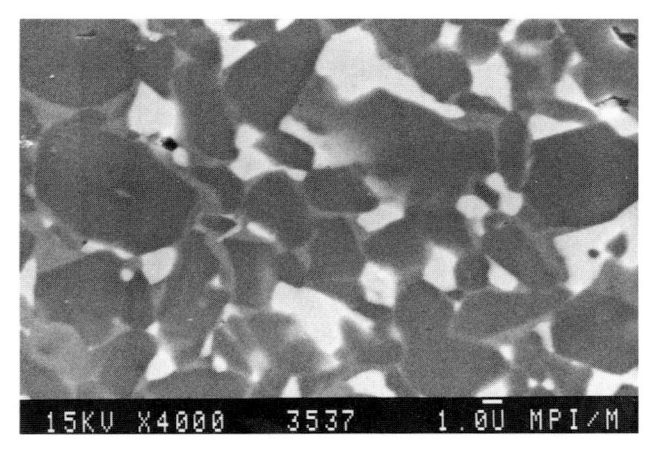

Figure 80. SEM micrograph of a $(Ti,Zr)B_2$-$(Zr,Ti)O_2$ material. Note the core-rim structure of the dark TiB_2 particles.

with transition metal diborides. Stabilizing additives such as MgO or Y_2O_3 tend to migrate into grain boundaries since these bivalent or trivalent cations in the zirconia lattice are substituted by the more favored, such as Ti^{4+}. A geometrical destabilization results from the strong coalescence of ZrO_2 causing a particle coarsening due to the high sintering temperatures. Large crystallites exceeding a critical size cannot be retained in the tetragonal modification upon cooling to room temperature and hence transform spontaneously to the monoclinic modification. Moreover, if associated in clusters, a transforming zirconia particle may trigger the transformation of all the other crystals by an autocatalytic reaction. This mechanical destabilization results from the anisotropy of the thermal expansion of the diborides, which introduces radial tensile stresses in the vicinity of the zirconia inclusions. This initiates the spontaneous tetragonal-to-monoclinic transformation or at least reduces the contribution of the ZrO_2 volume expansion during stress-induced transformation to toughening [299].

TiB_2-ZrO_2 ceramics have been studied intensively as possible candidates for active transformation toughening [239, 300–305]. Composites with ZrO_2 additives show an improved densification behavior and a grain growth inhibiting effect for the TiB_2 (Fig. 80). Hot-pressing of composites with 22–60 mass-% ZrO_2 between 1700°C and 1900°C at 20 MPa yields densities exceeding 99.8% [300, 305], whereas 98% of the theoretical density is obtained by pressureless sintering at 2100°C [239]. Volume fractions of *unstabilized* ZrO_2 between 15 and 30% result in a significant increase in both the strength and the toughness. Depending upon the microstructure and the density, a maximum σ_b of 700–800 MPa is measured at 22 or 35 vol.-%, respectively, and the maximum K_{Ic} varies between 5 and 9.5 MPa m$^{1/2}$ (Figs. 81 and 82). The hardness decreases linearly with the amount of ZrO_2 additive and is thus of the order of 16–18 GPa at 20–30 mass-% ZrO_2 being optimum for both strength and toughness (Fig. 83). This improvement in the mechanical properties is attributed to enhanced sintering and grain size refinement of TiB_2, active transfor-

890   7 Boride-Based Hard Materials

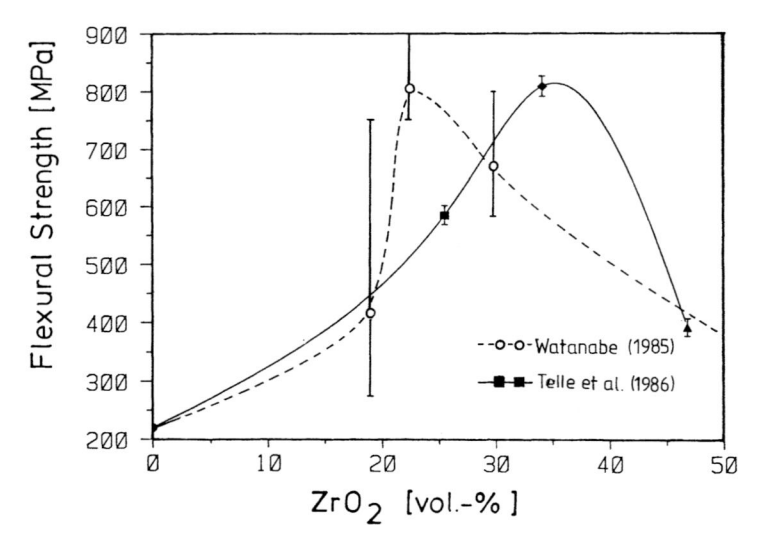

Figure 81. Strength of TiB_2-ZrO_2 composites (data from [218, 300]).

mation toughening [300], crack deflection and microcracking [239, 301]. Müller [305] quantified the transformable amount of tetragonal ZrO_2 by X-raying of as-sintered and as-ground samples. Depending on the processing and sintering conditions, the remaining tetragonal fraction ranges between 0 and 80%, which could almost be entirely transformed to monoclinic during machining. Between 20 and 30 mass-% ZrO_2, there is, however, about 15 mass-% untransformable tetragonal ZrO_2. The presence of these untransformable tetragonal ZrO_2 particles without an yttria addi-

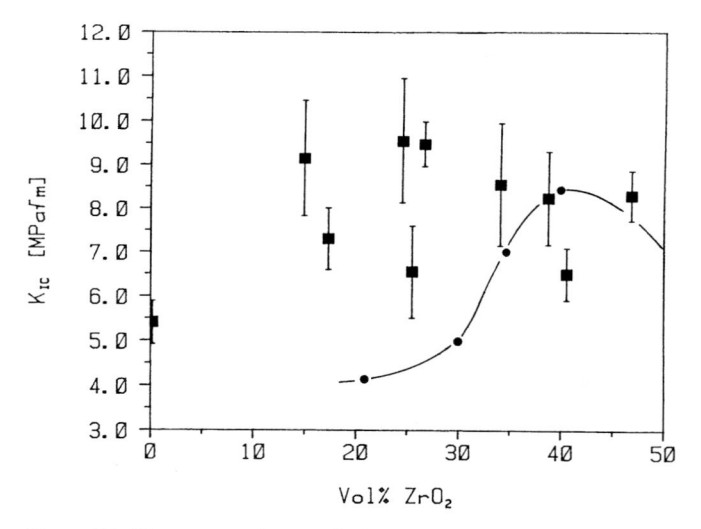

Figure 82. Facture toughness of TiB_2-ZrO_2 composites (dots: data from [300], squares: data from [218]).

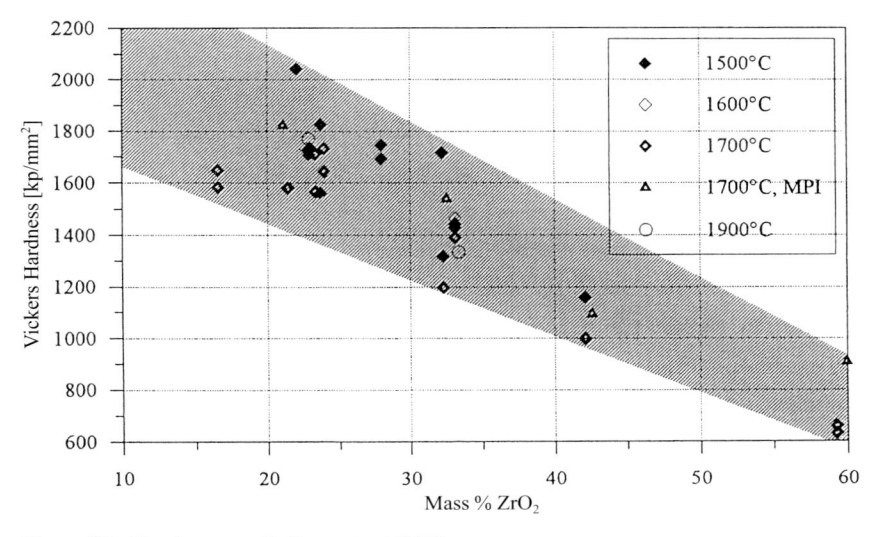

Figure 83. Hardness vs. ZrO$_2$ content [305].

tion can be explained by a pronounced mutual Ti and Zr interdiffusion, resulting in (Ti,Zr)B$_2$ and (Zr,Ti)O$_2$ solid solutions. The (Zr,Ti)O$_2$ may contain 14–16 mol-% TiO$_2$ at 1700°C in outer layers of the ZrO$_2$ particles and hence result in a stabilization of the tetragonal modification. The maximum ZrB$_2$ content in TiB$_2$ at 1700°C was found to be 3.2 mol-%. Moreover, after sintering between 2000°C and 2100°C there is evidence of an intergranular phase between adjacent TiB$_2$ and ZrO$_2$ particles consisting of zirconium titanate (Zr,Ti)$_2$O$_4$ which forms peritectically from a TiO$_2$-ZrO$_2$ melt [239, 306]. This embrittling phase can be avoided by the substitution of TiB$_2$ by ZrB$_2$ since zirconium zirconates do not exist. These composites are, however, also very sensitive to a spontaneous tetragonal-to-monoclinic transformation resulting in extensive microcracking (Fig. 84).

If *yttria-stabilized* ZrO$_2$ is applied, the contribution of stress-induced transforma-

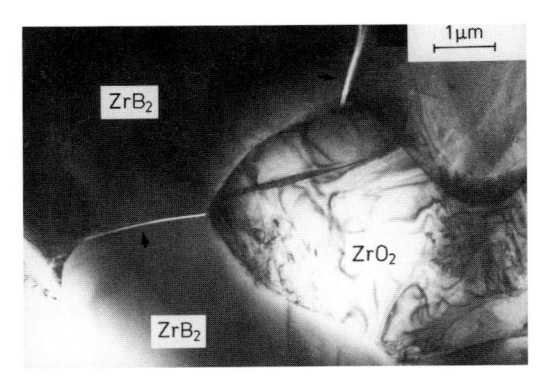

Figure 84. TEM micrograph of spontaneously cracked ZrB$_2$-ZrO$_2$ composites.

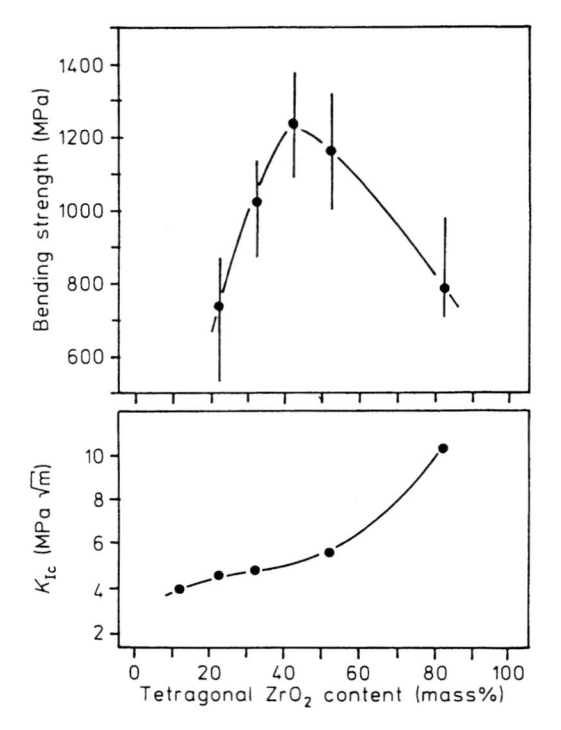

Figure 85. Mechanical Properties of TiB$_2$-YTZP composites sintered at 1500°C [301].

tion to the toughening is more likely. An average bending strength of 1250 MPa (maximum σ_b = 1500 MPa) is obtained a 40 mass-% of 1.94 mol-% Y$_2$O$_3$-doped ZrO$_2$, whereas the fracture toughness is, however, only 4–6 MPa m$^{1/2}$ but increases with higher Y-ZrO$_2$ contents (Fig. 85) [301]. The reason for this comparatively small increase in toughness was attributed to the high stiffness of the matrix phase, which reduces the dilatational strain associated with the tetragonal-to-monoclinic transformation of ZrO$_2$ [302, 308]. In contrast to these results Müller [305] observes a parabolic increment of toughness as a function of the 3Y-stabilized tetragonal zirconia content (Fig. 86). The obtained maximum fracture toughness was 7.5–8 MPa m$^{1/2}$ at 25–40 mass-% ZrO$_2$ if the sample was hot-pressed at least 1700°C. Müller reports a maximum strength of 750 MPa at 20–25 mass-% ZrO$_2$. At higher 3Y-TZP fractions the strength decreases drastically, at >40 mass-% the samples ruptured spontaneously during cooling from hot-pressing temperature. X-ray analysis of as-sintered and as-ground samples proved that during cooling 20–90% of the Y-TZP transforms to the monoclinic state which explains the severe cracking of the material. The residual tetragonal phase can almost be fully transformed to monoclinic by grinding [305]. The reasons for the spontaneous transformation are the diffusion of Y$_2$O$_3$ from the ZrO$_2$ solid solutions into the grain boundaries where, especially at temperatures >1800°C, a Y$_2$O$_3$-B$_2$O$_3$ eutectic is found. On the other hand, the formation of ZrO$_2$ clusters with increasing sintering temperature and

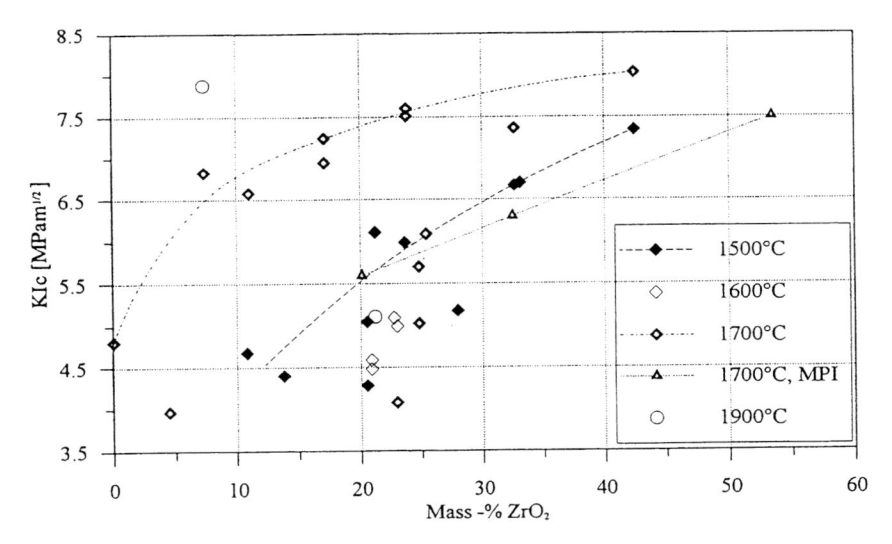

Figure 86. Parabolic increase in toughness in TiB$_2$-3YTZP-composites [305].

time is obvious [239, 305]. Composites of comparably fine grain size and homogeneous phase distribution, i.e., isolated ZrO$_2$ particles of ≈ 1–2 μ size, undergo a continuous phase redistribution and coarsening if hot-pressed up to 1900°C. Zirconia is then arranged in clusters with favored ZrO$_2$-ZrO$_2$ interfaces and has grown to >5 μm size. As in the case of unstabilized ZrO$_2$, Ti is found in the surface-near areas of the ZrO$_2$ particles and, vice versa, Zr is also present in the outer rims of the TiB$_2$ particles. Thus, a grain boundary diffusion of both Zr and Ti, and presumably also B and O is most likely. Plasma etching makes an intergranular phase clearly visible. This kind of coalescence resulting in ZrO$_2$ clusters incorporated by a contiguous TiB$_2$ matrix with an optimized number of TiB$_2$-TiB$_2$ grain boundaries is considered the reason for geometrical and mechanical destabilization of the tetragonal phase. The spontaneous t-m transformation is assisted by the large ZrO$_2$ grain size and the radial tensile stresses at the contact to TiB$_2$. Since these interfaces are the weakest anyway, microcracking occurs here preferentially, and the residual tetragonal ZrO$_2$ particles in the cluster transform by autocatalysis.

At higher temperatures, coarsening continues up to 2100°C where the material decomposes by internal oxidation [239]. Both boron and oxygen are then very mobile and exhibit a high vapor pressure so that the material is blown up drastically. Volume changes of more than 200% can be observed. The ceramic then consists of TiO$_2$, titanium borates, and oxygen-deficient ZrO$_2$-TiO$_2$ solid solutions whereas B$_2$O$_3$ vapor creates round entrapped pores causing the strong bloating. Rupture was not observed, most probably because the grain boundary phase is already liquid or at least viscous at this temperature.

The high enthalpy of formation of both ZrO$_2$ and transition metal borides can be used to enhance densification by a chemical driving force starting from, e.g., TiO$_2$ and ZrB$_2$ [303, 304]:

$$TiO_2 + ZrB_2 \rightarrow TiB_2 + ZrO_2 \tag{39}$$

This reciprocal salt-couple reaction yields a tremendous grain size refinement since all powder particles are involved in the conversion. After reaction the microstructure is characterized by an average grain size of 1–2 μm, approximately, and the preformation of solid solutions. Thus Eq. (39) has to written more precisely as follows:

$$TiO_2 + ZrB_2 \rightarrow (Ti, Zr)B_2 + (Zr, Ti)O_2 \tag{40}$$

The mechanical properties of the reaction product have not been exploited as yet.

Besides TiB_2, also WC, ZrC, TiC and mixtures thereof have been transformation-toughened with Y-stabilized ZrO_2 [309]. Hot-pressing of WC-TZP blends with up to

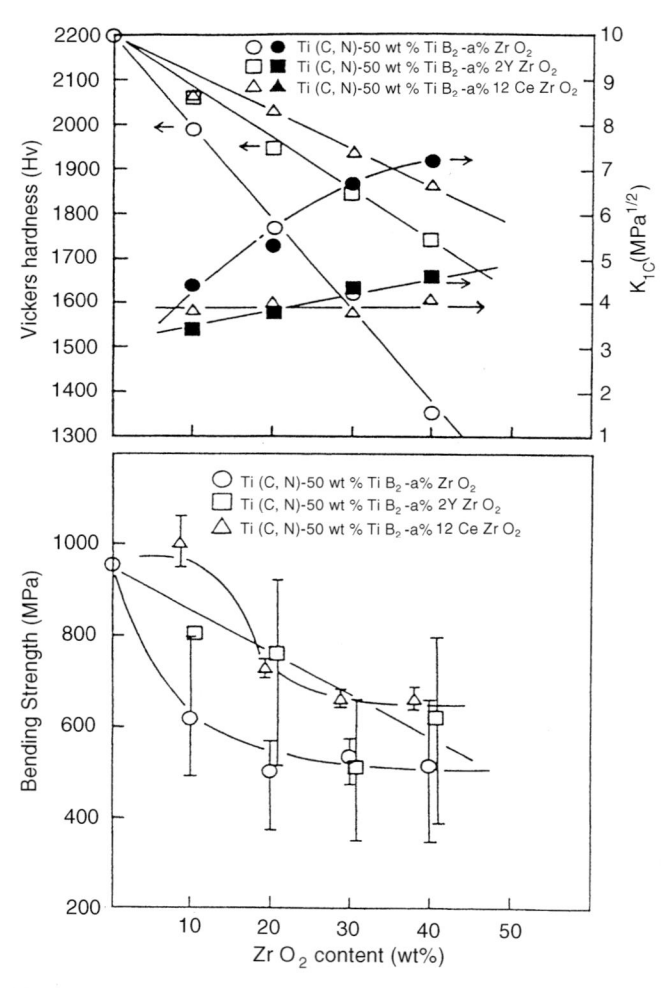

Figure 87. Mechanical properties of Ti(C,N)–50 mass-% TiB_2-ZrO_2 composites (after [310]).

50 vol.-% ZrO_2 at 1600°C yields almost dense composites with a flexural strength of 2300 MPa and a hardness of HV 1800 ($HR_A = 93.4$). A reaction between zirconia and tungsten carbide has not been observed but the appearance of a liquid phase is reported. A similar behavior of the mechanical properties has been observed for composites of Ti(C,N) containing Y-doped tetragonal zirconia composites where a significant increase in the strength and hardness with Y-ZrO_2 content is, however, accompanied by a modest increment in the toughness. In Ti(C,N)-50 mass-% TiB_2-ZrO_2 a decrease in hardness and strength combined with a parabolic increase in toughness was measured with increasing ZrO_2 content [310]. The best performance was achieved by 12% ceria-stabilized ZrO_2 (Fig. 87). A characteristic of all the ZrO_2-reinforced borides is the comparatively high scatter of the mechanical properties measured.

7.8 Cemented Borides

Due to their remarkable hardness, borides are interesting candidates for the development of metal-matrix composites resembling the hard metals based on cemented carbides or cermets such as WC-Co or Ti(C,N)-Ni composites. However, since metal-boron systems usually contain many binary and ternary borides a suitable choice of materials is difficult and requires full knowledge of the particular phase diagrams and compound properties. Moreover, to gain full advantage of the chosen metal-boride system, problems in densification such as wetting have to be overcome by appropriate dopants, and, finally, the comparatively little corrosion and oxidation resistance of borides and matrix phase at high temperatures have to be taken into consideration for an application.

7.8.1 Boron Carbide-Based Cermets

The fabrication of metal matrix cermets with boron carbide as a dispersed phase is very limited under equilibrium conditions since B_4C reacts with all metals, except Ag, Cu, Sn, and Zn, forming metal borides and graphite or metal carbides (e.g., [311]). In systems with slow reaction kinetics, however, complex low-temperature materials with interesting mechanical properties have been investigated. A development from Kiew makes use of a Ti-containing bronze as a binder phase in which the reaction of Ti with B_4C to give TiB_2 is employed for active brazing and improvement of the wetting behavior. The use of pure Cu, Sn, or Zn, or alloys thereof for the infiltration of B_4C power compacts usually fails since the wetting behavior is rather poor (wetting angle $>90°$), but this can be improved by adding Cr or other metals, which may react with the B_4C when approaching equilibrium conditions.

Other metal matrix composites with B_4C particulates have been obtained using aluminum because of slow reaction kinetics. The process is based on an infiltration of liquid Al into a porous body of B_4C at temperatures between 700°C and 1200°C. Since Al melts at 600°C and exhibits a significant vapor pressure at only slightly higher temperatures, the equilibrium between 1000 and 1880°C at which liquid Al is stable with an Al-saturated $B_{12}(B,C,Al)_3$-solid solution [73] cannot readily be utilized for liquid-phase sintering with small volume fractions of liquid. As shown by Halverson et al. [206], it is more effective to infiltrate compacted, or presintered, porous B_4C bodies with liquid Al. The resulting material is a metal-reinforced B_4C cermet rather than a liquid-phase sintered B_4C ceramic. The wetting behavior is strongly influenced by oxidation layers formed on the surface of the B_4C particles [312], but can be improved by superheating the melt. Between the melting point of Al and approximately 1000°C, wetting angles of 100–150° are observed, which decrease to reasonable values with prolonged soaking for thousands of hours [206]. Hence, in that temperature range, only hot-pressing or hot isostatic pressing result in high-density cermets. Above 1000–1200°C, a suitable wetting behavior is obtained within minutes of annealing. Due to capillary forces and phase reactions both densification and adhesion of the metal–ceramic interface are excellent. During infiltration, reactions of Al with B_4C occur. Below 1200°C, Al_4BC, AlB_2, AlB_{12}, and $AlB_{12}C_2$ are formed within tens of hours whereas above 1200°C the generation of Al_4C_3, AlB_{12}, and $AlB_{24}C_4$ is more favored [206]. If the composite is prepared by fast heating, infiltration, and rapid cooling, most of the aluminum matrix is retained unreacted. The matrix can then be hardened by a subsequent heat treatment at 800°C for 20 hours due to the precipitation of aluminum carbides and borides. Since the mechanical properties are determined by the Al matrix, a K_{Ic} of 5–16 MPa m$^{1/2}$ and a flexural strength of 200–680 MPa can be obtained, depending on the quality and volume fraction of the metallic binder. The Vickers microhardness of 15.7 GPa for a 31 vol.-% Al composite is improved by annealing to 19.4 GPa. A re-investigation of the Al-B-C system with special emphasis on B_4C-Al cermets was published by Pyzik and Beaman [90]. The composites consisted of both isolated and interconnected B_4C particles and Al matrix and only a small amount of ternary aluminum carbo-borides by fast firing and a post-sintering heat treatment between 600 and 1050°C. Cermets with a high B_4C contiguity had the highest strength of 550 MPa if treated at <600°C. Then the strengths drops to 400–420 MPa and remains constant at higher heat treatment temperatures while structures with isolated B_4C particles show a maximum strength of 420 MPa after a 600°C treatment with a similar decrease at higher temperatures. As expected, the latter composite with the continuous Al-matrix phase gives the higher fracture toughness of 11.7 MPa m$^{1/2}$ after a 600°C annealing for 20 hours compared to 8.5 MPa m$^{1/2}$ for the contiguous B_4C microstructures, being best without any heat treatment (Fig. 88).

Reactions of B_4C with metals have also been employed for the bonding of boron carbide-based ceramics to metallic substrates. Nishiyama and Umekawa [313] have studied the intermediate phases in B_4C-Fe couples with Al and Al/Mo interlayers. Depending on the treatment temperature and thickness of the foils, $B_{12}C_2Al$ and intermetallic phases are formed that allow reactive brazing with a tensile strength of 55–60 MPa. In a similar way, porous boron carbide of several B:C ratios was infiltrated by Si-TiSi$_2$ alloys with a 1330°C melting point [94,109,233]. After wetting and

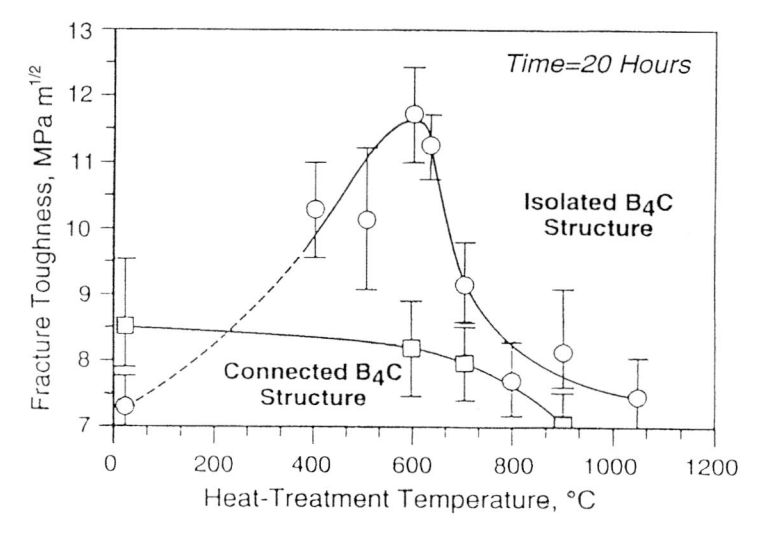

Figure 88. Effect of heat-treatment temperature on fracture toughness of B_4C-Al composites (after [90]).

infiltration, the liquid phase reacts with boron carbide to form TiB_2 and SiC. Therefore, these composites were discussed in Section 7.4.4.2.

7.8.2 Titanium Diboride-Based Cermets

TiB_2 is a candidate material for wear-resistant parts and cutting tools because of its high hardness, very high Young's modulus, considerable high-temperature strength and remarkable chemical stability. The liquid-phase sintering of TiB_2 with suitable metallic binder phases is intended to produce cermets which combine the desired properties with sufficient toughness. Although TiB_2 is stable together with liquid Al and Cu, these systems have not been studied seriously for cermets but extensively for particle hardening of aluminum and copper alloys by TiB_2. In the Al-Ti-B system it is, for instance, still uncertain whether a continuous $(Ti,Al)B_2$ solid solution exists which is in equilibrium with Al [89,314] or whether AlB_2 and TiB_2 coexist as separate phases [315]. Pastor [134] summarizes the results of other systems based on the iron group and other transition metals such as Cr, Zr, Mo, W, and Re. Fe, Co, and Ni have, however, the highest potential for this purpose.

7.8.2.1 TiB$_2$-Fe Composites

In various studies Fe was investigated as a binder phase for TiB_2 [316, 317], but Samsonov [318] and Kieffer and Benesovsky [319] mentioned, that Fe_2B is generated during the liquid phase sintering of TiB_2-Fe powder blends. Therefore, the fabrication of suitable TiB_2-Fe cermets was considered impossible. Other studies by

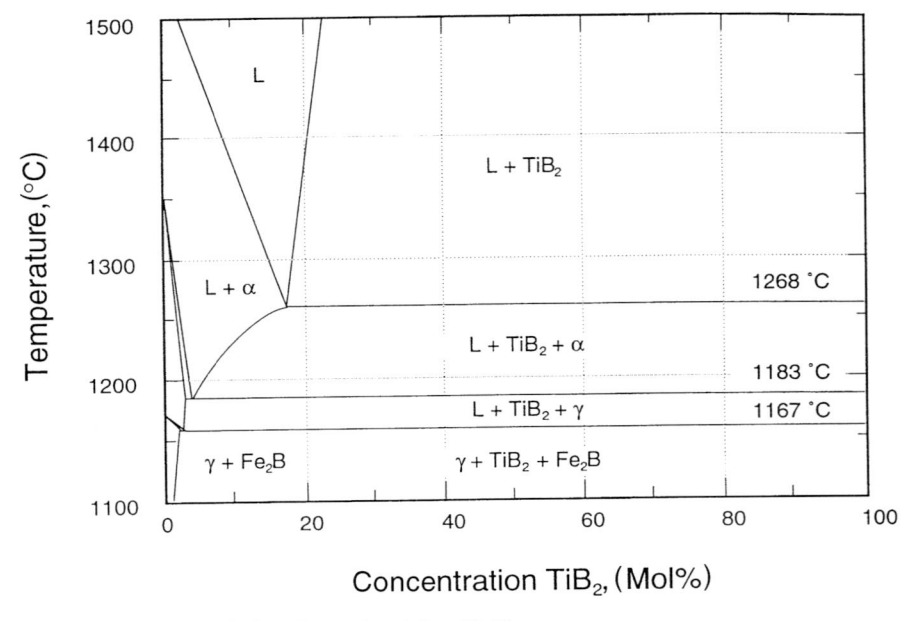

Figure 89. Isopleth of TiB$_2$-Fe section (after [320]).

Federov and Kuzma [120] and Shurin and Panarin [121], however, showed evidence for the existence of a pseudo-binary Fe-TiB$_2$ equilibrium, which is characterized by a eutectic point at a temperature of 1340°C and 6.3 mol.-% TiB$_2$ (Fig. 23). After Smid and Kny [122] the two phase equilibrium is limited at 800°C by the tie-lines TiB$_2$-Fe$_2$B and TiB$_2$-Fe (5 at.-% Ti). Ottavi et al. [123, 124] finally proved that not pure Fe is in equilibrium with TiB$_2$ but a Fe solid solution with Ti. The boundary tie-line interconnects TiB$_2$ with Fe + 0.5 at.-% Ti. Accordingly, a small excess of boron by result in the formation of FeB$_2$. The first liquid phase in the pseudo-binary section appears at 1170°C coming from the Fe-FeB$_2$ eutectic, the ternary eutectic is set to 1240°C (Fig. 24). The contradictions in these experiments have been attributed by Sigl and Schwetz [116, 119, 196] to the instability of the system TiB$_2$-Fe in the presence of C or B$_4$C impurities, which originate from the TiB$_2$ synthesis by the carbothermic reduction of TiO$_2$-B$_2$O$_3$ or TiO$_2$-B$_4$C mixtures. These impurities react with Fe and TiB$_2$ and form Fe$_2$B and TiC. Thermodynamic calculations by Golczewski and Aldinger [320], however, take the γ–α transition of iron more precisely into account and indicate that TiB$_2$ is compatible with Ti-bearing ferrite only below 900°C (Figs. 89 and 90). The pseudo-binary section TiB$_2$–pure Fe therefore contains FeB$_2$ below 1167°C, and between 1167 and 1268°C solid γ- or α-Fe instead plus liquid. At above 1268°C, the two phase equilibrium TiB$_2$–liquid is entered (Fig. 89) whereas in the TiB$_2$-Fe (2 at.-% Ti) section FeB$_2$ and γ-Fe are absent (Fig. 90). Although there are still some uncertainties on the phase diagram, which the synthesis of pure two-phase cermets is based upon, the authors agree that the presence of oxygen and carbon impurities is detrimental to the wetting behavior

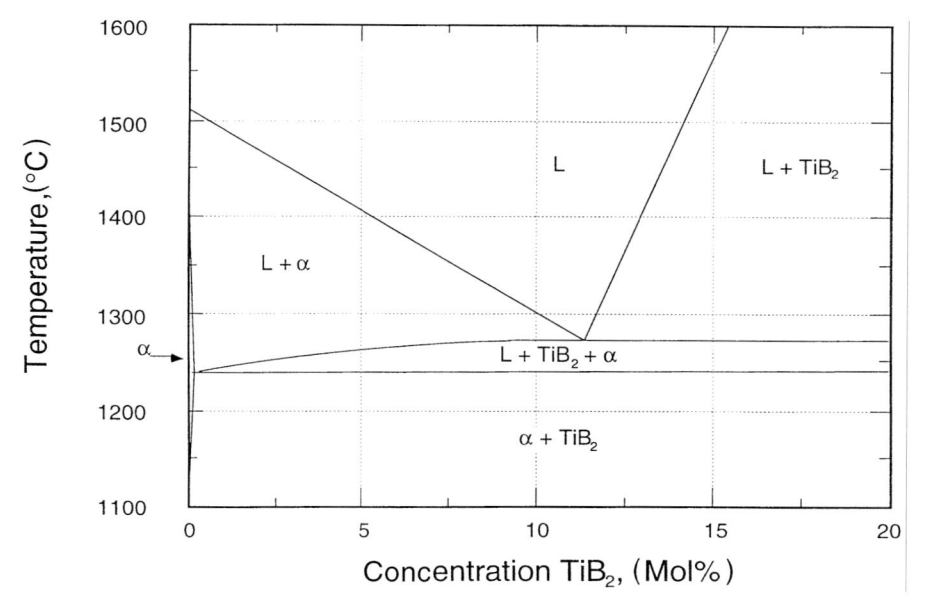

Figure 90. Isopleth of TiB_2-(Fe + 2 at.-% Ti) section (after [320]). Note the difference to Figure 89 concerning the stability of $\alpha \rightarrow \gamma$ Fe and the appearance of the liquid phase.

and responsible for the presence of the embrittling but hard Fe_2B phase, which also controls the sintering behavior and thus the properties.

After the development of cemented borides with a Fe matrix by Funke et al. [316] and Funke and Yudkovskii [321], this cermet system was recently intensively studied again by Yuriditsky [115], Sigl and Schwetz [116, 196], and Jüngling et al. [322, 323], and, in connection with an enhanced European research activity, independently by Ottavi et al. [123, 124], Ghetta et al. [117], Pastor et al. [324, 325], and Sánchez et al. [326].

Liquid Phase Sintering of TiB_2-Fe Composites

A disadvantage of TiB_2-Fe composites is their delicate densification behavior, which has been attributed to the oxygen impurities present in commercial TiB_2 powders. At solidus temperature, oxide impurities may give rise to evaporation and recondensation reactions involving volatile boron suboxides, which result in an exaggerated grain growth of TiB_2 without densification. Coarsening by means of vapor phase or surface reactions consumes surface energy and, therefore, sinter activity. Consequently, much effort was put into the removal of oxide contamination from TiB_2 powders, e.g. by reduction with BCl_3 [263, 327].

As already mentioned before, oxides and carbon impurities are also responsible for the wetting behavior of the liquid. Figure 91 illustrates both complete and imperfect wetting of TiB_2 by liquid iron. Thin TiO_2 layers covering parts of the TiB_2 grains are assumed to account for the poor wetting [116, 196]. Experiments

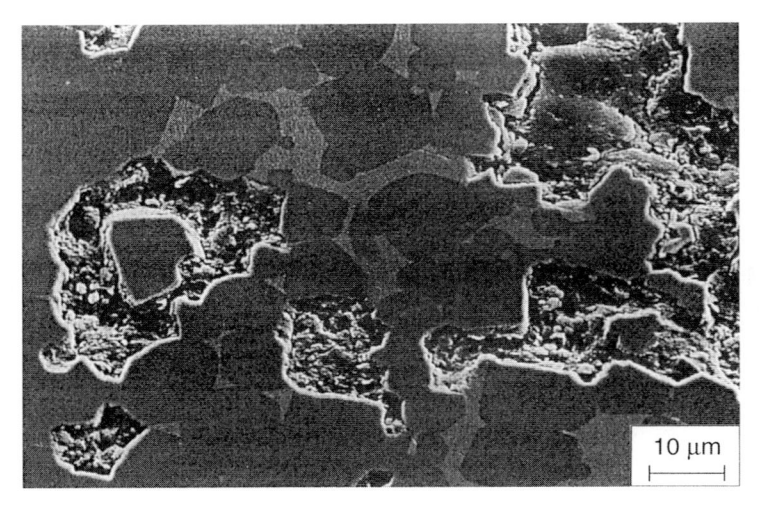

Figure 91. SEM micrograph of an 80 vol.-% TiB_2 + 20 vol.-% Fe cermet sintered at 1600°C. Note both the incomplete and complete wetting of TiB_2 by liquid Fe.

by Ghetta et al. [328] have demonstrated that the spreading of liquid Fe in fact worsens with increasing oxygen content of TiB_2 substrates (Fig. 92) whereas powders with <0.26 mass-% oxygen show a wetting angle of <5° after 10 min soaking time.

Generally, transition-metal diborides should be wetted well by liquid metals due to their missing gap between the valency band and the conduction band whereas oxides exhibit bad wetting behavior due to their partially ionic and partially covalent character in bonding [329–331]. Although an improvement of wetting by liquid Fe may be obtained by addition of Cr or Ni [332], breaking up the TiO_2-layer is thus considered the key to well-processible TiB_2 powders.

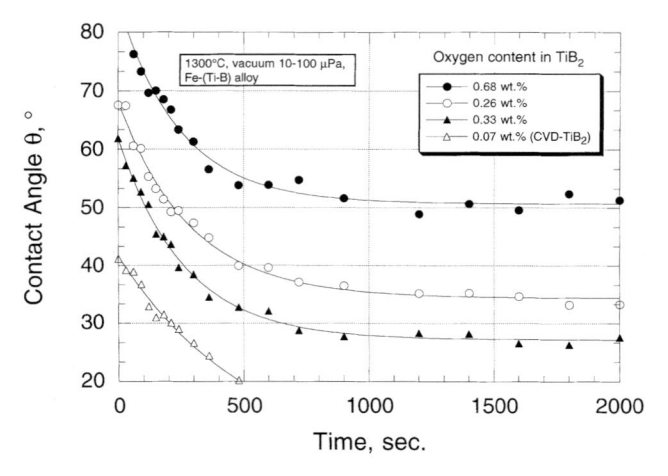

Figure 92. Oxygen contamination affecting the wetting behavior of Fe liquid on sintered TiB_2 at 1300°C (redrawn after [117]).

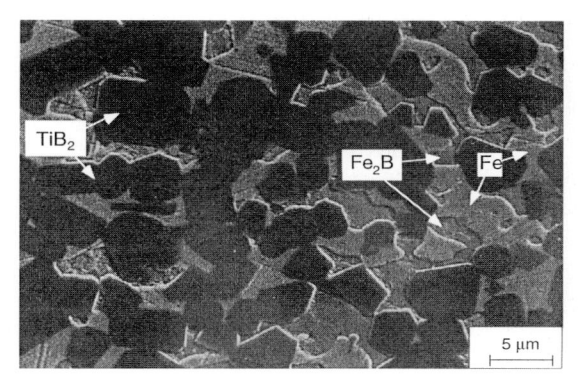

Figure 93. SEM micrograph of a TiB_2-Fe cermet with a nominal composition of 80 vol.-% TiB_2 + 20 vol.-% Fe. Note the existence of Fe_2B in the binder phase (after [119]).

It also appears that the impurities in commercial TiB_2 promote the formation of iron boride, Fe_2B. A TiB_2-Fe alloy with a typical two-phase microstructure of the binder, i.e. with both Fe and Fe_2B making up the matrix between the TiB_2 grains, is shown in Fig. 93. Fe_2B is a hard but fairly brittle compound [333] and its formation consumes a substantial fraction of the ductile Fe phase such that Fe_2B is believed to embrittle the composite seriously. Avoiding that phase has consequently been a major goal for developing tough TiB_2 cermets.

Commercial TiB_2 powders are typically produced by the carbothermic reduction of TiO_2 with B_2O_3 or B_4C. Both reactions

$$TiO_2 + B_2O_3 + 5C \rightarrow TiB_2 + 5CO \qquad (41)$$

and

$$2TiO_2 + B_4C + 3C \rightarrow 2TiB_2 + 4CO \qquad (42)$$

are being industrially utilized. A typical analysis of the as-received TiB_2 is listed in Table 10.

Since this powder originates from the B_4C-processing route, i.e. Reaction (41), it is not surprising that about 90% of the total carbon is present as boron carbide and only 10% is free carbon. Assigning the oxygen content to particular compounds is not unambiguous, but it is estimated that approximately 1/2 of the oxygen is present as TiO_2 and B_2O_3, respectively. Nitrogen is completely bonded as TiN, as BN is usually not found. It is therefore concluded that B_4C, B_2O_3, free carbon, and

Table 10. Chemical composition of the as-received TiB_2 and Fe powders.

Powder	Spec. surf. [m^2/g]	Total O [weight-%]	Total C [weight-%]	Total N [weight-%]	B_4C [weight-%]	Free C [weight-%]	TiO_2 [weight-%]	B_2O_3 [weight-%]
TiB	3.2	2.1	0.20	0.05	0.85	0.02	2.6	1.5
Fe	0.5	0.2	0.05	–	–	–	–	–

Table 11. Chemical compatibility of the compounds in TiB_2-Fe powder mixtures

	Fe	C	B_4C	TiO_2	B_2O_3
TiB_2	Compatible	Compatible	Compatible	Compatible	Compatible
Fe	–	Solution reaction	Reaction	Compatible	?
C	–	–	Compatible	Reaction	Reaction
B_4C	–	–	–	Reaction	Compatible
TiO_2	–	–	–	–	Compatible

TiO_2, generally make up the critical impurities in the TiB_2 powder. More carbon is introduced to the powder blend when organic binders are added and it should be noted that the oxygen content increases significantly upon milling [118].

The key questions for a successful liquid phase sintering of TiB_2 with iron are, therefore, (i) how to break up the TiO_2-films on the TiB_2 grains and (ii) how to avoid the formation of Fe_2B.

Sigl and Jüngling [119] have shown that powder compacts of TiB_2 + 20 vol.-% Fe heated to 1000°C contained Fe_2B and Ti_2O_3. Notably, their concentration did not increase significantly at temperatures up to 1600°C, suggesting that the major part of the reaction had been completed at that time already. Since Fe_2B and Ti_2O_3 already appear at low temperature, solid-state reactions are considered responsible for the formation of this compound. Of particular interest are the reactions of Fe with the boron-containing impurities, i.e. with B_4C and B_2O_3. According to Table 11 the following reactions are candidates to develop Fe_2B:

$$B_4C + 8\,Fe \rightarrow 4Fe_2B + C \qquad (43)$$

$$2\,B_2O_3 + 7\,Fe \rightarrow 2\,Fe_2B + 3\,FeO \qquad (44)$$

The change in Gibbs Free Energy, ΔG, has been calculated for Eqs. (43) and (44) with the data from Janaf [334]. The results are plotted versus temperature in Fig. 94. Obviously a reaction between Fe and B_2O_3 is unlikely in the whole temperature range of interest, which has been confirmed by corresponding model reactions. It is, however, well known that Fe and B_4C react heavily even at low temperatures forming Fe_2B and free carbon [335]. Due to the favorable properties of Fe_2B, this fact is commercially utilized for depositing thin layers on steel surfaces [336].

As mentioned above, borides can be synthesized by the carbothermic reduction of B_2O_3 with carbon and a metal or a metal oxide [337], e.g. according to Eq. (40). Since free carbon is present in the as-milled powder mixture and also develops during Reaction (43), the following process must also be considered a potential source of Fe_2B:

$$B_2O_3 + 4\,Fe + 3\,C \rightarrow 2\,Fe_2B + 3\,CO. \qquad (45)$$

Although the ΔG, which is plotted as a dashed line in Fig. 94 appears to favor Reaction (45) above 800°C, it is still considered unlikely. According to Sigl and Jüngling [119], the following steps rather than the gross Reaction (45) are proposed to operate instead

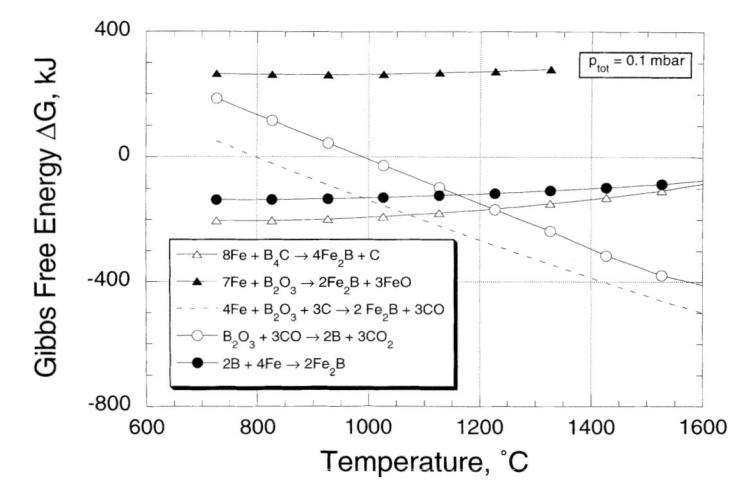

Figure 94. The change in Gibb's free energy for reactions yielding Fe_2B at 0.1 mbar total pressure.

$$B_2O + 3\,CO \rightarrow 2\,B + 3\,CO_2 \quad (46)$$

$$2\,B + 4\,Fe \rightarrow 2Fe_2B \quad (47)$$

$$3\,CO_2 + 3\,C \rightleftharpoons 6\,CO \quad (48)$$

Given the CO/CO_2 equilibrium according to Eq. (48), the Gibbs Free Energy for Reactions (45) and (47) can be calculated and is plotted in Fig. 94 at a total pressure of 0.1 mbar. Though Fe and B would favorably react, CO is unable to reduce B_2O_3 below 1000°C and therefore the carbothermal formation of Fe_2B is operating only beyond 1000°C. This hypothesis has been confirmed experimentally: While a B_2O_3-Fe-C powder blend with a composition according to Eq. (45) does not change its phase composition after annealing the mixture at 1000°C for 1 h, the X-ray diffraction of identical samples heated up to 1600°C shows ample evidence for Fe_2B. This suggests that processes involving liquid iron can play an important role during the formation of Fe_2B as elaborated below.

In the regime of liquid-phase sintering TiB_2 is dissolved in the Fe melt according to

$$TiB_2 \rightarrow \{Ti\} + 2\,\{B\} \quad (49)$$

where the brackets {} denote the dissolved state of an element in liquid Fe. Obviously impurity compounds such as TiO_2, Ti_2O_3, C, B_2O_3, TiN, and Fe_2B will also dissolve in the liquid. During cooling some of these compounds reprecipitate from the melt in their original composition, e.g. TiN or Ti_2O_3 and hence do not influence the equilibrium concentration of Ti and B in the liquid. Species such as carbon, TiO_2, and B_2O_3 which introduce a lot of {C}, {O}, and {B} but only little {Ti} into the melt do, however, precipitate as compounds that are richer in Ti than the previously dissolved species, e.g. as

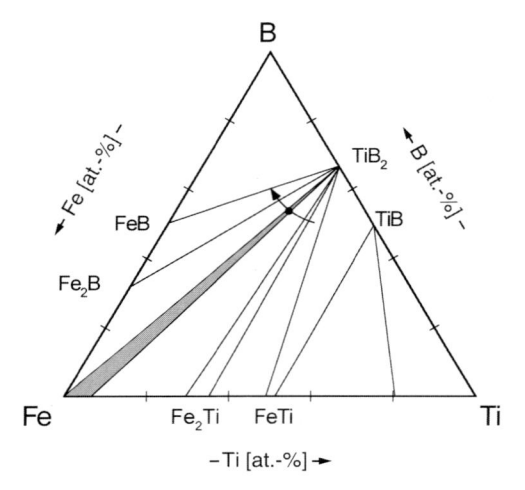

Figure 95. Isothermal section of the Fe-Ti-B system at 1000°C. The point indicates the cermet composition and the arrow the change in composition if the titanium content in the liquid phase is decreasing.

$$\{Ti\} + 2\{B\} \rightarrow TiB_2 \tag{50}$$

$$\{Ti\} + \{C\} \rightarrow TiC \tag{51}$$

$$2\{Ti\} + 3\{O\} \rightarrow Ti_2O_3 \tag{52}$$

Thus, these processes extract a lot more {Ti} from the melt than has been introduced during the dissolution. Consequently, the solubility product ($a_{Ti} \cdot a_B^2$) being constant at a given temperature forces the composition of the liquid to shift towards low {Ti} and high {B} concentrations upon cooling. Finally, a {B}-rich Fe melt is left behind. This conclusion is supported by the observation that a liquid phase is stable down to 1150°C [335], i.e. the eutectic temperature of Fe-Fe$_2$B, although the TiB$_2$-Fe eutectic solidifies at 1350°C. Eventually, the remaining melt crystallizes as a mixture of Fe and Fe$_2$B (Fig. 93). An experimental confirmation of this conclusion was obtained by Sigl and Jüngling [119]. These processes shift the composition of the cermets from their intended place in the (TiB$_2$ + Fe) two-phase field into the (TiB$_2$ + Fe + Fe$_2$B) three-phase equilibrium of the Fe-B-Ti ternary system, either by introducing B or by extracting Ti (Fig. 95). Thus the Ti concentration in TiB$_2$ cermets is a variable to be strictly controlled during sintering, very similar to the C content of cemented carbides, which must be adjusted well to avoid graphite or η-phase formation.

Figure 91 suggests that liquid Fe cannot penetrate in between the TiB$_2$ grains, unless the TiO$_2$ layer is broken up. Total wetting is nevertheless feasible. Figure 96 indicates that this process is closely related to the precipitation of particulate Ti$_2$O$_3$. It is generally accepted that the reduction of TiO$_2$ involves various titanium suboxides, i.e. Ti$_3$O$_5$, Ti$_2$O$_3$, and TiO, until Ti is finally obtained [334]. The compound that actually evolves from this process depends on how well the reducing agent can overcome the affinity of oxygen in the Ti oxides. Since the TiB$_2$ powder of this study contains only free carbon as a suitable reducing agent, reactions such as

7.8 Cemented Borides

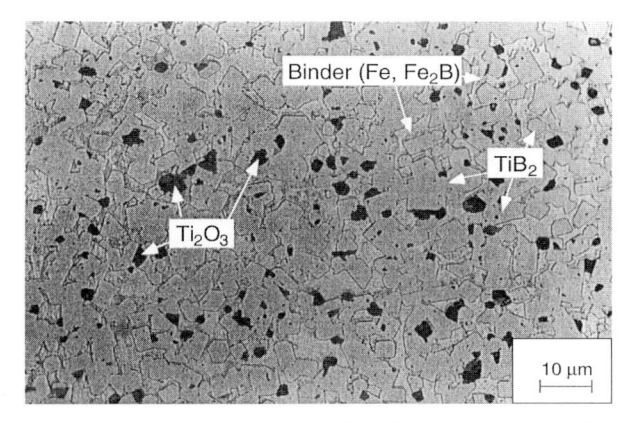

Figure 96. Optical micrograph of a TiB$_2$-Fe cermet containing Ti$_2$O$_3$ emerging from powder contaminations.

$$2\,TiO_2 + C \rightarrow Ti_2O_3 + CO \tag{53}$$

need to be studied. The Gibbs free energy for Reaction (52) is plotted in Fig. 97 as a function of temperature at a total pressure of 0.1 mbar. Though carbon should be able to reduce TiO$_2$ at temperatures above 850 °C according to the thermodynamic calculations, a reaction between a thin oxide film and solid carbon is considered unlikely. Instead, carbon monoxide rather than free carbon is believed to deoxidize TiO$_2$ according to

$$2\,TiO_2 + CO \rightarrow Ti_2O_3 + CO_2. \tag{54}$$

since the driving forces for Reactions (52) and (53) are identical. A further reduction

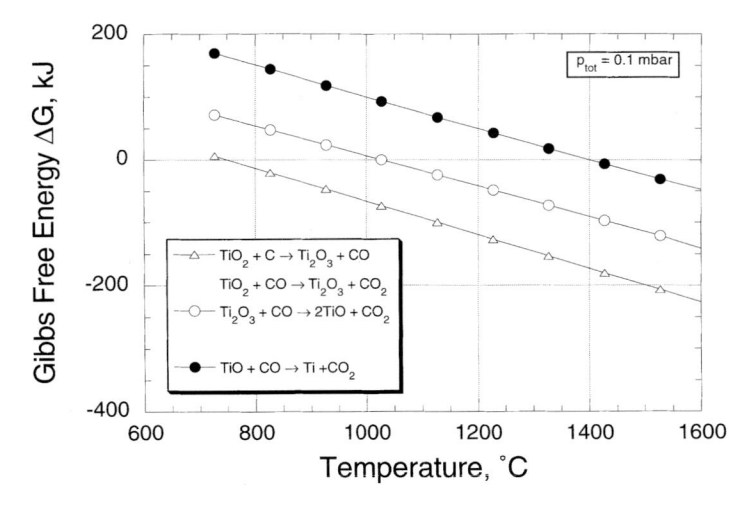

Figure 97. The change in Gibb's free energy for the reduction of titanium oxides by carbon monoxide at 0.1 mbar total pressure.

of Ti_2O_3 according to

$$Ti_2O_3 + CO \rightarrow 2\,TiO + CO_2 \tag{55}$$

is also supported by thermodynamic arguments (Fig. 97), but CO seems unable to reduce TiO to Ti at a pressure of 0.1 mbar.

Since experiments have shown that TiB_2 powders with more than 2 weight-% oxygen can be sintered to full density [116, 196, 322] and that Ti_2O_3 particles precipitate during sintering, it is considered that the free carbon in the as-received TiB_2 powder and the carbon being set free from the reaction of Fe with B_4C generates CO, which eventually breaks up the oxide layers on the TiB_2 grains. The result that Ti_2O_3 appears instead of TiO, as may be inferred from Fig. 97, is attributed to the small quantity of carbon normally available. Principally, a carbon quantity sufficient to form either Ti_2O_3 or TiO should promote the wetting and consequently the sintering behavior of TiB_2 powders. This conclusion is supported by Fig. 96, which suggests that Ti_2O_3 particles hardly disturb the wetting by liquid Fe. Excess carbon can, however, lead to the formation of Fe_2B via the liquid-phase processes described above. Since small quantities as in the previous case will be used up during the reduction of TiO_2, the carbothermic formation of Fe_2B from Fe and B_2O_3 according to Eq. (45) cannot occur. Hence it is concluded that Fe_2B originates mainly from the reaction of Fe with B_4C and that the gross reaction

$$B_4C + 2\,TiO_2 + 8\,Fe \rightarrow 4Fe_2B + Ti_2O_3 + CO \uparrow \tag{56}$$

operates at temperatures below 1000°C. Though a good part of Fe_2B is due to solid-phase reactions, it is anticipated that B_2O_3 will also give rise to both Fe_2B and Ti_2O_3. However, the role of B_2O_3 is as yet not fully understood.

In conclusion, deoxidizing the thin TiO_2 films appears to be an absolute prerequisite for the sintering of TiB_2 with liquid iron. Though carbon monoxide cannot fully reduce TiO_2 at the sintering conditions (0.1 mbar), the formation of particulate lower oxides such as Ti_2O_3 seems to be sufficient for supplying a good wetting behavior.

Kinetics of Liquid-Phase Sintering of TiB_2 with Fe

The sintering kinetics and phase development during sintering were investigated by Jüngling et al. [322] who demonstrated that Fe_2B already forms at low temperatures, i.e. before shrinkage starts (Eq. (55)). The deoxidation of TiB_2 grains and the simultaneous formation of Fe_2B contribute significantly to the densification behavior of TiB_2-Fe-materials, i.e. nearly full densities (>98.5% T.D.) were achieved by pressureless sintering. However, the formation of Fe_2B consumes a considerably high amount of the Fe binder, which deteriorates the fracture toughness of the composite seriously. For these reasons Jüngling et al. [322] and Sigl and Jüngling [338] suggested the addition of Ti to balance the excess boron in order to obtain materials with a ductile binder:

$$2\,TiO_2 + 2\,Ti + B_4C \rightarrow 2\,TiB_2 + Ti_2O_3 + CO \tag{57}$$

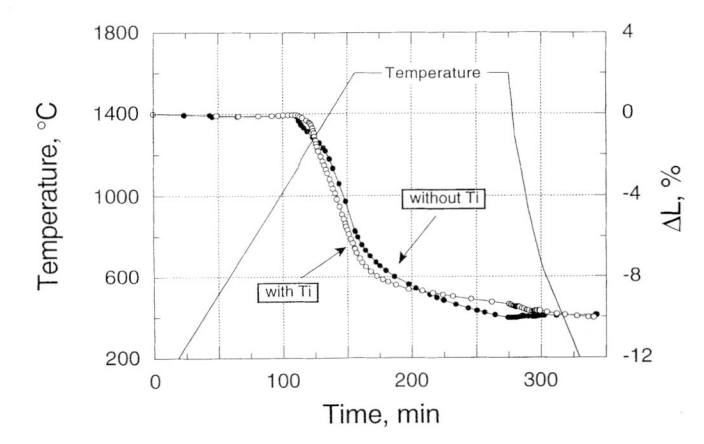

Figure 98. Densification behavior of model alloys with 30 vol.-% binder with and without Ti addition (after [340]).

This reaction as well as Eq. (56) were confirmed by thermodynamic considerations. Ghetta et al. [328] and Ottavi et al. [339] concluded from wettability studies that the addition of Ti deoxidizes the TiB_2 grain surface by the formation of a parasitic Ti(O,C,N) phase. Nevertheless, the sintered density of a material with Fe_2Ti addition was rather poor (<89% of theoretical density). The addition of $NdNi_5$ enhanced the density to 96.7% of theoretical density and resulted in a parasitic Nd_2O_3 phase.

Jüngling et al. [323] report on model experiments to study the sintering kinetics of TiB_2 with 20 vol.-% carbonylic Fe and 20 vol.-% carbonylic Fe + 7 mass-% Ti up to 1700°C in 0.1 mbar Ar. Sintering parameters were optimized for materials with 15 and 20 vol.-% Fe and 20 vol.-% Fe-Cr-Ni binder and for materials with 5–20 vol.-% Fe-Cr-Ni binder phase with Ti addition to prevent the formation of the Fe_2B phase. Materials with Ti addition were pressurelessly sintered at 1650°C to a density of about 95% T.D. followed by a hot isostatic pressing step at 1460°C with a pressure of 100 bar Ar.

Figures 98 and 99 illustrate the shrinkage behavior and densification rates of the alloys with and without Ti additions. Soon after the spontaneous start of densification, the sintering rate decreases drastically, but eventually begins to increase again. Without Ti addition shrinkage starts at 1140°C but in the alloy containing Ti, sintering begins at 1250°C. After this temporary decrease, the densification rate increases in both materials above 1320°C. Densification comes to an end after 2 h isothermal sintering at 1700°C.

Unlike the material with pure Fe, the alloy with Fe + 7 mass-% Ti does not develop any Fe_2B. Again Ti_2O_3 is formed at 1000°C but the intensity ratio Ti_2O_3/TiB_2 is smaller than without Ti addition and increases with increasing temperature. Additionally, X-ray diffraction presents evidence for the formation of very small amounts of TiC in the sintered sample.

Figure 100 displays the density as a function of the sintering temperature without holding time. Samples with 15 or 20 vol.-% Fe show similar densities in

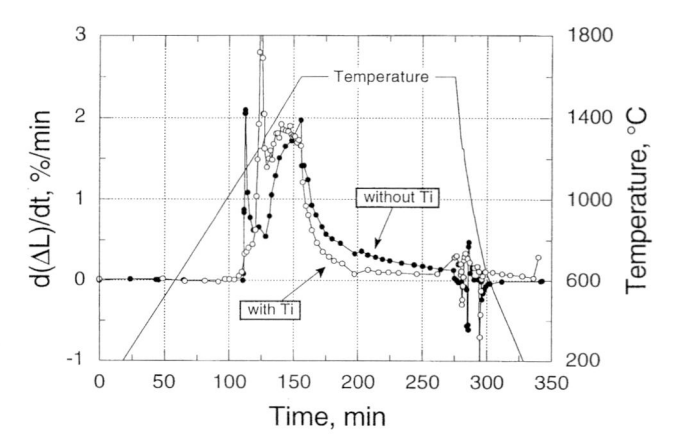

Figure 99. Densification rate of model alloys with 30 vol.-% binder with and without Ti addition (after [340]).

the temperature range of 1500–1700°C and reach nearly 99% T.D. The addition of 7 mass-% Ti results in a strong decrease in the density of as-sintered materials, but at >1650°C a density of >95% T.D. and closed porosity are obtained. Further densification can be achieved by subsequent hot isostatic pressing. Materials with 5, 10, and 20 vol.-% Fe-Cr-Ni-binder with 7 mass-% Ti addition nearly approach their theoretical densities by a post-HIP treatment at 1460°C with a pressure of 100 bar Ar (Table 13).

The spontaneous shrinkage for materials with Fe_2B at 1140°C is explained by the formation of a liquid phase and a subsequent rearrangement of particles. This conclusion is compatible with the eutectic temperature of 1177°C in the system $Fe-Fe_2B$

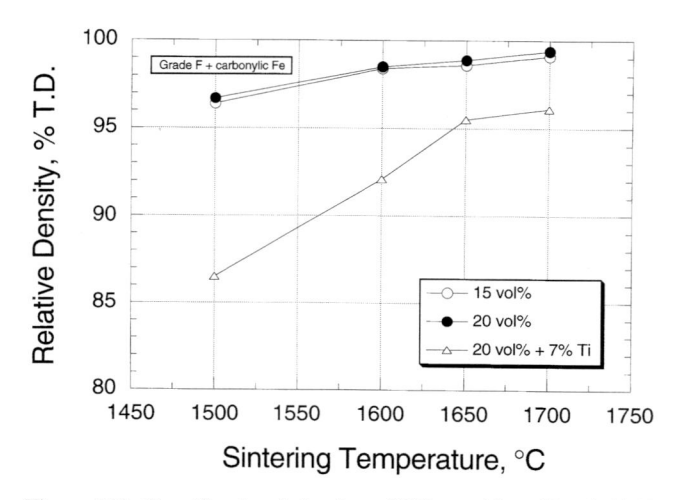

Figure 100. Densification behavior of TiB_2 + 15 or 20 vol.-% Fe cermets (after [340]).

Table 12. Solid-state and vapor-phase reactions during sintering.

Without Ti additive	With 7 mass-% Ti	Eq.
$B_4C + 8\,Fe \rightarrow 4\,Fe_2B + C$	–	(57)
	$B_4C + 2\,Ti \rightarrow 2\,TiB_2 + C$	(58)
$C + CO_2 \rightarrow 2\,CO$	$C + CO_2 \rightarrow 2\,CO$	(59)
$2\,TiO_2 + CO \rightarrow Ti_2O_3 + CO_2$	$2\,TiO_2 + CO \rightarrow Ti_2O_3 + CO_2$	(60)

[341]. In this way Fe_2B contributes a lot to the excellent densification behavior of these materials but it also consumes a major part of the ductile Fe-binder phase.

In materials without Fe_2B the beginning of shrinkage is delayed. A first maximum in densification rate appears at 1250°C, about 90 K below the eutectic temperature of $Fe-TiB_2$. Again, this densification peak is attributed to the formation of a liquid phase, which coincides with the eutectic $Fe-Fe_2Ti$ at 1289°C rather than with the $Fe-TiB_2$ eutectic [342]. This can be explained by the solution of some Ti in the binder during heating. As proved by X-ray analysis, the addition of 7 mass-% Ti fully prevents the formation of Fe_2B. Smaller amounts of Ti yield corresponding quantities of Fe_2B. Table 12 compares the particular reactions of materials with and without Ti additions.

After initial particle rearrangement the densification rate decreases in both materials. It is accelerating again when a sufficient amount of liquid phase is formed due to the eutectic $Fe-TiB_2$ at about 1320°C. During cooling the liquid phase finally solidifies at 1140°C for samples with and at 1250°C for samples without Fe_2B as proved by exothermic reactions producing the small expansion peak in Fig. 99. The last expansion peak at 800°C is due to the $\alpha \rightarrow \gamma$ transformation of Fe. As expected, this peak is much higher for the composite that does not contain Fe_2B.

In conclusion, the formation of FeB_2 can be avoided by the control of powder impurities, sintering atmosphere, and compensation of excess boron and carbon by titanium addition. The wetting behavior depends strongly on oxide surface layers and is excellent if surface contamination of TiB_2 is entirely removed. The sintering temperature and Fe addition must ensure that a suitable amount of liquid phase is available at the lowest possible vapor pressure. Since the reducibility of TiO_2 and the vapor pressure of liquid Fe determine the atmospheric pressure, the interval $(0.01-0.1)$ mbar $< p < 1$ mbar is considered best suited. The optimum amount of liquid phase is generated at $1450 < T < 1650°C$ if grain growth is taken into account as another limiting factor. Accordingly, a feasibility diagram of suitable sintering conditions can be constructed (Fig. 101) showing the optimum environmental parameters.

Microstructure and Properties of TiB$_2$-Fe Composites

Since the eutectic concentration in the quasi-binary TiB_2-Fe system with 14 vol.-% TiB_2 is considerably closer to the metal corner than in the similar WC-Co system (32 vol.-% WC), a much smaller amount of liquid phase is generated upon sintering,

Table 13. Properties of optimized TiB$_2$–hard metals. ICL: Indendation-crack-length method, NB: notch-beam method.

Starting powders	TiB$_2$ + Fe			TiB$_2$ + Fe-Cr-Ni	TiB$_2$ + Fe-Cr-Ni + 7 mass-% Ti		
Binder composition	α-Fe + Fe$_2$B			γ-Fe-Cr-Ni + Fe$_2$B	γ-Fe-Cr-Ni		
Binder content [vol.-%]	5	15	20	20	5	10	20
Density of pressureless sintered material [% T.D.]	97.8	98.8	99.1	99.2	≈95	≈95	≈95
Density of post-HIP [% T.D.]	–	–	–	–	100	99.1	99.2
3-point bend strength [MPa]	610	760	720	790	900	1010	950
Young's Modulus [GPa]	510	465	465	465	515	510	465
Hardness [HV$_{10}$] 25°C	2260	1830	1510	1610	2040	1800	1450
Hot hardness [HV$_2$] 200°C	–	–	–	1130	1560	1270	1150
Hot hardness [HV$_2$] 400°C	–	–	–	870	1290	990	850
Hot hardness [HV$_2$] 600°C	–	–	–	730	1030	810	740
Hot hardness [HV$_2$] 800°C	–	–	–	480	660	580	500
K$_{Ic}$/ICL [MPa$\sqrt{m}$]	5.9	6.9	7.9	8.2–8.7	7.3–8.1	8.1–9.5	12.4
K$_{Ic}$/NB [MPa$\sqrt{m}$]	5.5	9.3	10.0	11.4–11.9	7.6	9.0	14.0

7.8 Cemented Borides

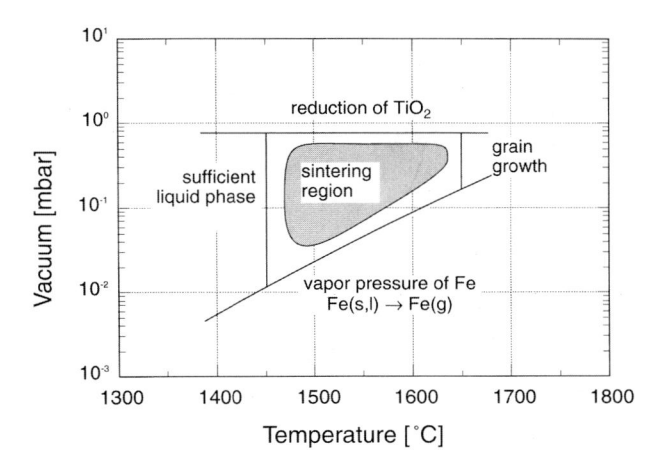

Figure 101. Optimum parameters for the pressureless sintering of TiB$_2$-Fe cermets limited by grain growth, evaporation, suboxide formation, and suitable liquid phase content.

which makes densification more difficult. A simple increase in temperature cannot satisfactorily balance the lack of liquid because it is accompanied by accelerated coarsening of TiB$_2$ due to Ostwald ripening [116]. The volume fraction of binder phase thus ranges between 10 and 30%. A typical microstructure is very similar to that of WC-Co hard metals. Euhedral TiB$_2$ particles are embedded in a continuous Fe matrix. The densification mechanisms are typically dissolution and reprecipitation as well as coalescence, i.e., rearrangement and intergrowth of particles with common faces of the same orientation. The latter mechanism is active if the volume fraction of liquid exceeds 30% but may result in the growth of elongated platelets.

The residual porosity after pressureless sintering between 1500°C and 1800°C depends upon the initial liquid phase composition. At 1500°C, 88% of the theoretical density has been obtained for the TiB$_2$-Fe system (99% at 1800°C), whereas at 1450°C Ti addition results in 98% and combined Ti-Nb additives result in 96.7%. Hot-pressing and hot isostatic pressing yield densities >98% at a lower binder content.

Figure 102 shows the microstructure of a plasma etched sample with 20 vol.-% binder. Besides TiB$_2$ with its hexagonal grain shape and Ti$_2$O$_3$ a binder consisting of Fe and Fe$_2$B is present. A Fe-Cr-Ni binder phase diminishes the grain sizes notably. In materials with 7 mass-% Ti addition to the Fe-Cr-Ni binder the FeB$_2$ phase is absent. The total carbon content of samples without Ti addition strongly decreases during sintering from 0.11% (starting powder) to 0.01% (sintered compact). The oxygen content slightly decreases from 2.8% to 2.4%.

Table 13 summarizes some properties of TiB$_2$ with 15 and 20 vol.-% Fe-Fe$_2$B binder, 20 vol.-% Fe-Cr-Ni-Fe$_2$B binder, and 5–20 vol.-% Fe-Cr-Ni binder. Samples with a two-phase binder reach densities of about 99% T.D. by pressureless sintering. However, alloys with a single-phase binder approach no more than 95% T.D. by pressureless sintering, yet can be densified to near theoretical density by a

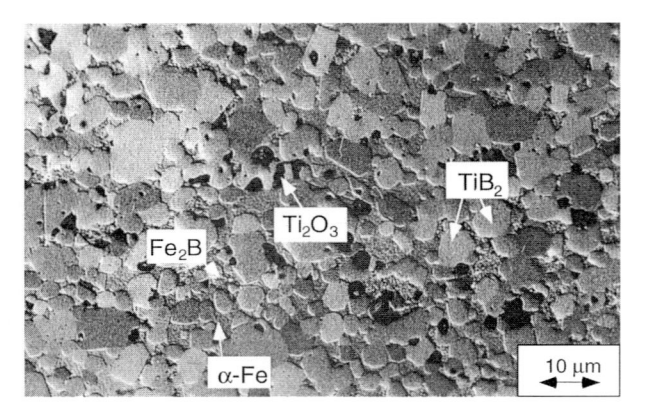

Figure 102. Plasma-etched microstructure of a cermet with 20 vol.-% α-Fe-Fe$_2$B-binder.

post-HIP treatment. As expected, samples with a low binder content are harder than those with high binder fraction, provided the binder composition remains unchanged. The observation that materials with Fe-Cr-Ni-Fe$_2$B binders are harder than materials with Fe-Fe$_2$B binders is attributed to the smaller grain size of the former materials (Fig. 103). Simultaneously, the fracture toughness increases. Thus avoiding the formation of Fe$_2$B strongly increases the fracture toughness but decreases the hardness only moderately, such that the combination of hardness and fracture toughness is clearly improved [343] compared to WC-Co-based hard metals.

The bending strength does not vary significantly with increasing binder content (Fig. 104) but the presence of FeB$_2$ results in considerably lower values. This not only attributed to the brittleness of this phase but is also due to the residual

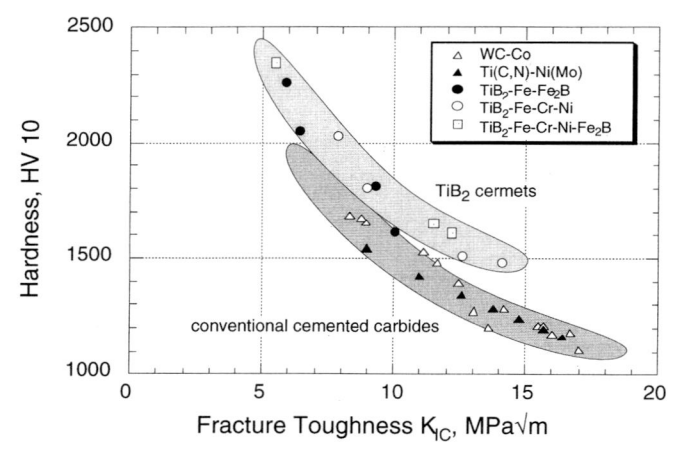

Figure 103. Hardness and fracture toughness of TiB$_2$-Fe cermets in comparison to conventional hard metals (after [340]).

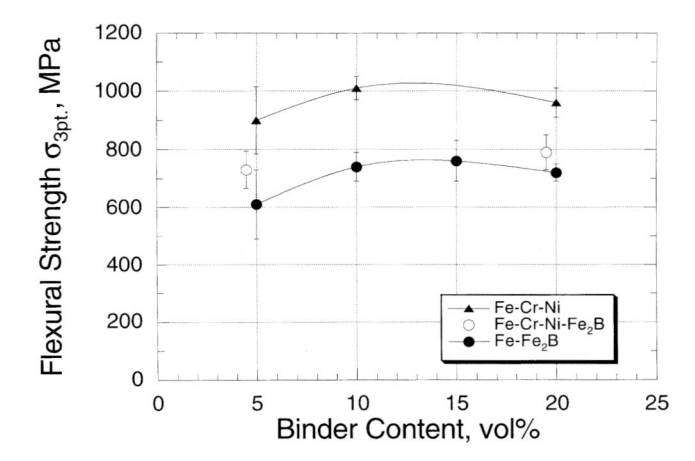

Figure 104. Flexural strength dependence on binder volume of TiB$_2$-Fe cermets with and without Fe$_2$B (after [343]).

2% porosity. A post-HIP treatment may overcome this problem and also yields a lower scattering of data. Additions of metals such as Mo, Cr, Ni, and Co to the Fe matrix may be used to fabricate composites with improved mechanical and corrosion properties. Figure 105 shows significant variations in the bending strength with increasing amounts of Mo in the binder phases of different volume fractions [115].

Figure 106 illustrates the hot-hardness of composites with various binder content, again in comparison to WC-Co and WC-Fe-Co-Ni hard metals [343]. It is evident that TiB$_2$-based materials with a small amount of binder are superior in the entire temperature range measured.

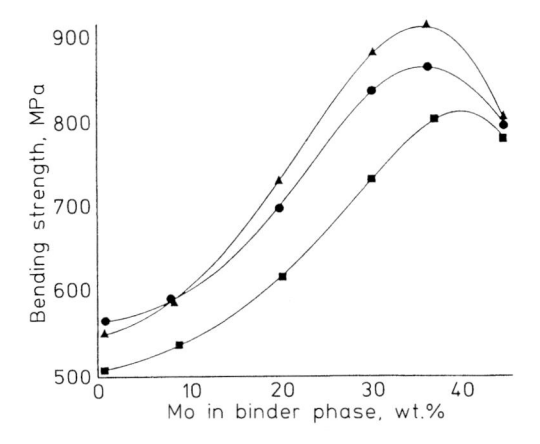

Figure 105. Variation of strength with Mo content in the binder phase of TiB$_2$-Fe cermets (after [115]). Binder volume: ▲, 12.5 vol.-%; ■, 15 vol.-%; ●, 17.5 vol.-%.

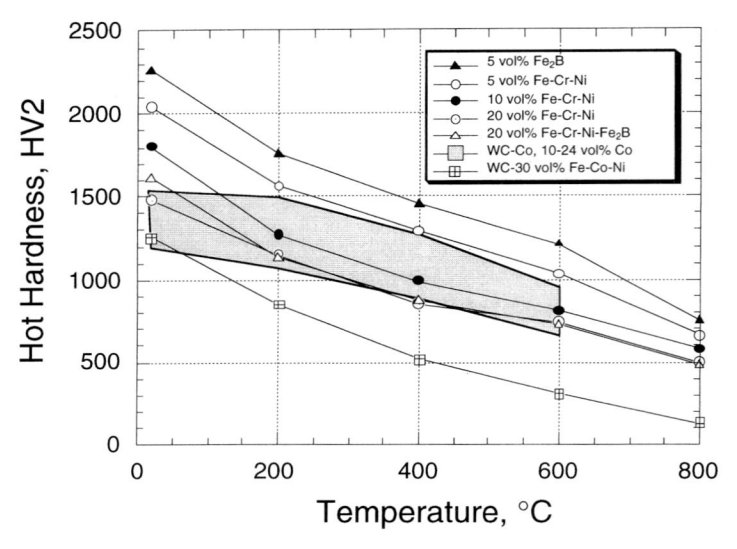

Figure 106. Room-temperature hardness HV_{10} and high-temperature hardness HV_2 of TiB_2-Fe cermets and conventional hard metals (after [340]).

7.8.2.2 TiB$_2$-Ni, Co Composites

Transition metals such as Co and Ni are useful for liquid-phase sintering of TiB_2-type borides causing rapid densification and grain growth by Ostwald ripening, but they react chemically to form M_xB_y-type phases or even more complex ternary phases that are very brittle and possess a comparatively low melting point $< 1500°C$. No metal-reinforced composites containing solely TiB_2 as the hard phase can thus be produced except where reactions can be at least partially avoided by fast heating during hot-pressing. In contrast, Co- and Ni-based alloys can be successfully improved for wear resistance by the incorporation of TiB_2 and CrB_2 particles if reaction layers of lower hardness can be tolerated.

In the case of Ni-bonded TiB_2, a ternary τ-phase with the composition $Ni_{21}Ti_2B_6$ forms by the dissolution of TiB_2. At 800°C, the τ-phase is in equilibrium with Ni, Ni_3B, Ni_3Ti, and TiB_2, as shown in the isothermal section in Fig. 25 [125]. During sintering or hot-pressing, the τ-phase is generally not obtained. The formation of τ may be suppressed either by the presence of TiO or TiO_2 [344], or for kinetic reasons by fast firing and quenching. Hence the residual matrix phase consists mainly of Ti-containing Ni_3B solid solution [218, 344, 345]. Since other Ni-borides such as Ni_2B and Ni_3B_4 and even metallic Ni are found after hot-pressing at 1600°C [94], equilibrium conditions are obviously not easy to obtain.

Typical metal contents required for a successful liquid phase hot-pressing of TiB_2 are 5–25 mass-% (i.e. 2–12 at.-%) Ni or Co. In order to avoid reactions consuming TiB_2, the borides of Ni or Co have also been used. By this method, the sintering temperatures have been decreased from 2100°C to 1400°C [134, 277, 294, 295, 346].

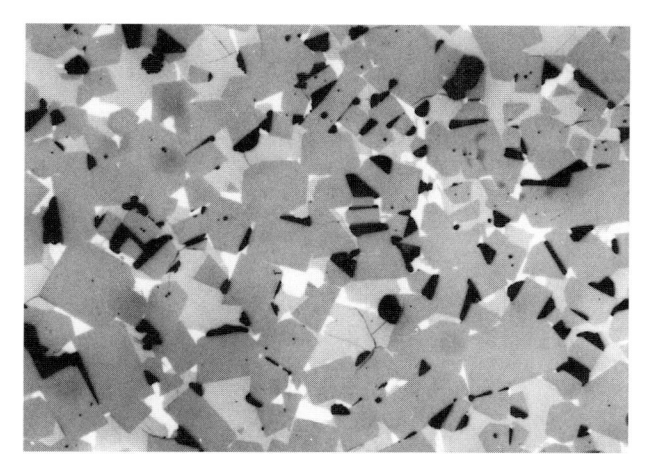

Figure 107. Optical micrograph of a TiB$_2$-Ni composite, pressurelessly sintered at 1600C. Note the incomplete wetting of TiB$_2$ by liquid Ni and the formation of Ni borides in the matrix phase.

The liquid phase intensifies the mass transport but causes an accelerated grain growth. The microstructures of composites prepared by liquid-phase sintering are similar to those of hard metals. The TiB$_2$ particles form a rigid skeleton of faceted crystals whereas the binder, e.g., Ni$_3$B, Ni$_2$B, Ni$_3$B$_4$, or comparable compounds of Fe, Cr, or Co, is the matrix phase. The TiB$_2$ grain size usually exceeds 20 µm (Fig. 107). Depending upon the wetting behavior, which is influenced by the surface oxidation of the hard material phase, round pores may accumulate at particle/matrix interfaces or close to triple junctions that have not been completely infiltrated by the liquid phase. Moreover, the evaporation of Fe-, Co-, or Ni-borides may cause entrapped gas pores. Hence, hot-pressing is still required for a homogeneous distribution of the liquid phase, particle rearrangement, and complete removal of the residual porosity. In contrast to hard metals, the matrix phase is very brittle, e.g., the K_{Ic} of Ni$_3$B equals 1.4–1.9 MPa m$^{1/2}$ [347], and hence does not improve the mechanical properties.

Nishiyama and Umekawa [348] report on sintering and wear application of a TiB$_2$ cermet with Ni$_7$Zr$_2$ binder. Pressureless sintering between 1450 and 2100°C in hydrogen results in an incomplete melting of the alloy added with wetting angles of around 90°. The phases detected are Ni$_4$B$_3$, Ni$_7$Zr$_2$, and solid solutions thereof with Ti. Since TiB$_2$ forms a rigid skeleton the hardness HV$_1$ reaches 30–33 GPa at binder contents of <8 mass-%, dropping almost linearly to 17 GPa with 30 mass-% Ni$_7$Zr$_2$ addition. Due to the toughness of the matrix phase, the 3-point bending strength improves from 500–600 MPa at <5 mass-% binder to 850–900 MPa with >15 mass-% of additives.

7.8.2.3 Transition-Metal Diboride Cermets with Co, Ni, Cr, Mo, and W

The research for high-strength/high-toughness composites in the TiB$_2$-transition metal systems resulted in many attempts to avoid brittle matrix phases and to introduce hardening particles while making use of the opportunity of liquid-phase

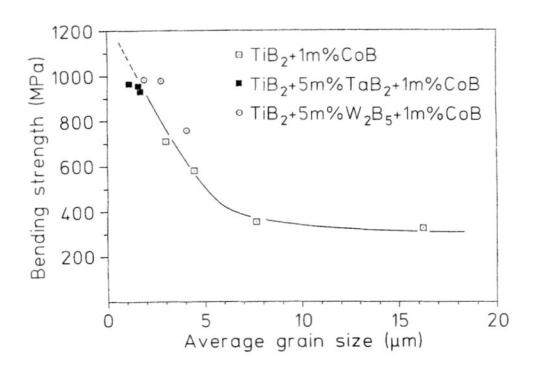

Figure 108. Strength–grain size relation in TiB_2-CoB-based composites (drawn according to data in [293]).

sintering. Pastor [134] reports on the manifold of diborides that have been sintered with Fe, Co, Ni, Cr, Mo, W, and even Re additives, yielding closed porosity or almost total density. Unfortunately no phase compositions were presented in most cases. In order to avoid brittle ternary phases, many liquid-phase sintering systems started with combinations of binder metals, sometimes even with their low melting boride phases, which will be reviewed in the following paragraph.

Hot-pressed composite materials developed from the more complex systems of the type TiB_2-M^IB_2-M^{II} with M^I being Hf, V, Nb, Ta, Mo, or Mn and M^{II} being sintering additives such as Co and Ni, exhibit bending strengths between 850 and 1000 MPa, which are due to the grain growth inhibiting influence of the 1–5 mass-% of $M^{II}B_2$ particulates (Fig. 108) [293–296]. During liquid-phase sintering in a Co- or Ni-boride melt, both TiB_2 and M^I are partially dissolved and reprecipitated as a solid solution. The effect of grain growth retardation as well as of strength and hardness increments is attributed to stresses at the $TiB_2/(Ti,M^I)B_2$ phase boundaries generated by the mismatch of the lattice parameters between the unreacted TiB_2 acting as a nucleus and the epitaxially precipitated $(Ti,M^I)B_2$ solid solution. In the case of a TiB_2-5 mass-% W_2B_5/TaB_2 material with 1 mass-% CoB binder the lattice strain ranges between 9×10^{-4} and 14×10^{-4}, depending on the hot-pressing temperature [293]. An improved sintering behavior of this material was obtained by the addition of 1.7% TiC, which reduced the porosity from 0.3–0.7 to 0.1–0.2 vol.-% after hot-pressing at 1500°C and by intensive ball milling resulting in an average particle size of 1 μm, but increased oxygen contamination. Watanabe and Shoubu [297] reported the formation of a (Ti,Ta)(C,O) solid solution, which is considered to initiate the improved densification resulting in a flexural strength of 1000 MPa. The strength–grain size correlation in Fig. 108 can be interpreted in the way that the fracture is flaw-controlled at average grain sizes < 7–8 μm and microcrack-controlled at > 8 μm.

In a similar multiphase system, *transition metal carbides* were used as additives for pressureless sintering of TiB_2, yielding composites of binary and ternary borides [218, 296]. Attrition milled powder mixtures of TiB_2 with 3–10 mass-% Co or Ni and 20–35 mass-% WC have been sintered in a vacuum at temperatures between

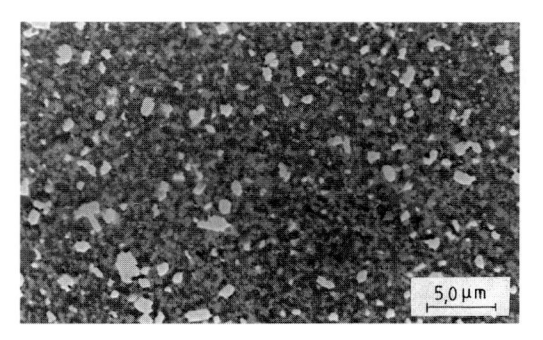

Figure 109. SEM micrograph of a $(Ti,W)B_2$-W_2NiB_2 (ω)-Ni_3B composite. Light areas: W_2NiB_2; intermediate: Ni_3B; dark: $(Ti,W)B_2$.

1500 and 1700°C for 60 to 120 min yielding 98–99% density. Densification starts above 980°C due to the formation of a liquid phase in the Ti-B-Co/Ni system (Fig. 26). At this early stage, a rigid skeleton of TiB_2 and WC develops. Due to dissolution and reprecipitation, a $(Ti,W)B_2$ solid solution grows on the residual TiB_2 particles. Subsequently, crystals of ω-phase form with compositions of WCoB or W_2NiB_2. Upon cooling, the residual liquid phase crystallizes as C- and Ti-enriched Co_3B or Ni_3B solid solution. A typical microstructure is shown in Fig. 109. Sintering at 1700°C for 2 h yields an average particle size of 0.8 µm, resulting in a flexural strength $\sigma_b = 600$–680 MPa. The K_{Ic} of 6.5–7.5 MPa m$^{1/2}$ is mainly attributed to crack deflection because of the weak Co- or Ni-boride intergranular phases.

As already shown, an interesting characteristic of the mixed diborides is the so-called core-rim structure, which is well known from complex cermets based on $(Ti,Ta\ldots)(C,N)$ compounds. During sintering or hot-pressing of the particular carbonitride powder blends solid solutions are formed at the phase contacts but the annealing time is insufficiently long to obtain equilibrium conditions. Consequently, interdiffusion is stopped, an unreacted core of, e.g., $Ti(C,N)$ remains whereas an outer layer, the rim, consists of, e.g. $(Ti,Ta)(C,N)$ solid solution with a very distinct boundary to the residual host crystal. Exactly the same effect is observed in borides of all kinds of composition if sintered for a not appropriately long time (Fig. 110). In case of liquid-phase sintered borides, this layer is formed by epitaxial precipitation of the thermodynamically more stable solid solution onto a nucleus of undissolved starting material. Recently, Telle et al. [349] observed dislocations at the interface of TiB_2-$(Ti,W)B_2$ solid solutions in hot-pressed TiB_2-WB_2 composites (Fig. 111), indicating that stresses are generated at the epitaxial interface as considered by Watanabe and Kouno [293].

Another approach to fabricate metal-matrix-based boride and carbide composites according to the Lanxide process starts with reactive blends of B_4C and Ti or Zr metal. Upon conversion to TiB_2 or ZrB_2, respectively, a large release of heat is observed, which can easily lead to partial melting of the composites. Depending upon the starting composition, residual metallic Ti or Zr, or B_4C may be found after reaction. Interesting microstructures can also be obtained if TiC or ZrC are added as fillers [350].

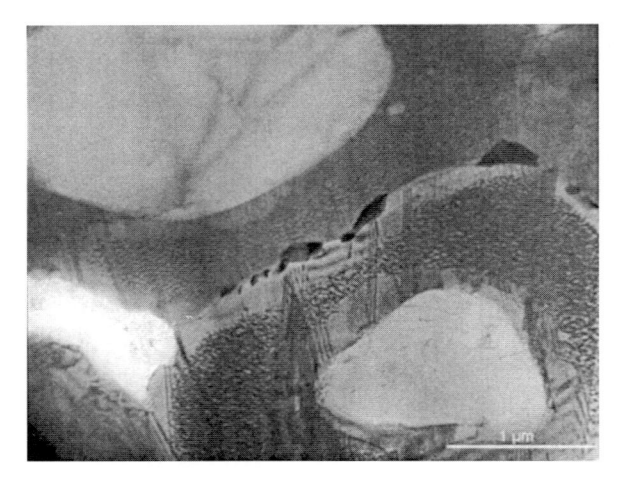

Figure 110. Core-rim structure of (Ti,W)B$_2$ solid solution on a TiB$_2$ host crystal.

Similar metal-matrix composites with B$_4$C, SiC, and TiB$_2$ as fillers have been fabricated by the so-called Lanxide- or Dimox-process (*di*rect *m*etal *ox*idation) where an Al- or Ti-based liquid mixed with ceramic particles – preferably of whisker or platelet shape – is slowly converted in air, oxygen, or nitrogen to alumina or titanium nitride, respectively [351–353]. This self-propagating reaction yields columnar crystals of the oxide or nitride phase, with B$_4$C, SiC, or TiB$_2$ inclusions and residual metal-filled channels, which contribute significantly to the strength and toughness. SiC-Al$_2$O$_3$ composites have an excellent fracture toughness of 8–15 MPa m$^{1/2}$ and a flexural strength of 500–800 MPa. They have also been demonstrated to be highly resistant against erosive wear [354].

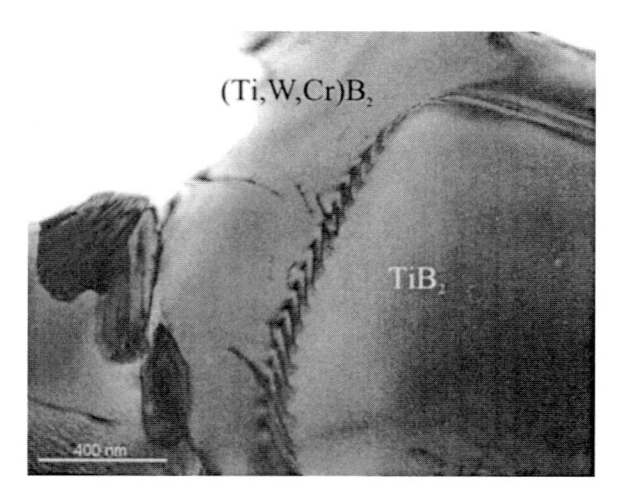

Figure 111. Dislocations at the boundary (Ti,W)B$_2$/TiB$_2$.

7.8.3 Cemented Ternary Borides

As discussed before, hard metal-like composites can be prepared by pressureless sintering of ternary borides with Fe, Ni, or Co melts. Materials with τ-phase ($M^I_{21}M^{II}_2B_6$, where M^I = Fe, Ni, or Co, and M^{II} = Zr, Hf, Nb, Ta, or W with M^I as the matrix phase) have not been developed for technical use but Ni-based alloys with τ are in applications as wear- and corrosion-resistant coatings on steels [355]. The τ phase is also used for the improvement of the creep resistance of Ni-based superalloys.

The other kind of ternary phases, φ and ω, have $M^IM^{II}B$ and $M^I_2M^{II}B_2$ stoichiometry, respectively, where M^I = Cr, Mo, Ta, or W and M^{II} = Fe, Ni, Co, or Cr and solid solutions thereof. In particular, the ω-type borides have been developed extensively by Takagi et al. [129–131] and Komai et al. [356], focusing on molybdenum rather than W or Ta due to its corrosion resistance.

7.8.3.1 Technically Important Systems and Structures

The crystal structure of the ω-type ternary borides $M^I_2M^{II}B_2$ was described first by Rieger et al. [357] as a distinct kind of structure similar to the U_3Si_2 structure. The lattice is orthorhombic, space group *Immm*, and contains two formula units, i.e. ten atomic sites. The cell is built up by two metal prisms containing four atoms of M^I and two atoms of M^{II}. These double prisms are piled-up along the *b*-axis. The boron atoms are arranged in pairs at the edges and in the center of the cell with their axes parallel to the *b*-axis. If Cr is inserted into the structure it is considered to preferentially replace Fe, Ni, or Co and thus occupies the M^{II}-position. Rietveld analysis of powder data by Ozaki et al. [358] proves that this kind of solid solutions creates a distortion of the orthorhombic structure, which then can be indexed according to a tetragonal lattice where the former *c*-axis of the orthorhombic cell becomes the tetragonal *c*-axis. The M^I sites are now arranged in a planar quadrangle surrounded by a tetragonal-face centered M^{II} coordination shell. Since the edges of the M^I quadrangle are not parallel to the axes of the cell, the boron pairs are also out of the *b*-alignment. Thus the space group *P4/mbm* is of comparatively low symmetry (Fig. 112).

Three kinds of ternary borides with metallic binders have been exploited for future applications in detail, the WCoB-Co cermet for its high heat resistance, the Mo_2NiB_2-Ni cermet because of its excellent corrosion resistance, and the Mo_2FeB_2-Fe composite for its wear resistance. In all cases, these borides are compatible with a melt consisting of M^{II} and alloys thereof. According to the particular phase diagrams, which have been treated in Section 7.3.2.2 (third part), the corresponding liquid phase must, however, contain some portion of M^I as well. The two phase equilibria φ- or ω-liquid, respectively, are limited by a tie line with the brittle τ-phase or, in systems in which (does not exist, with the particular binary M^{II} borides, e.g., with Fe_2B, Co_3B, etc., which have already been treated as a problem in TiB_2-Fe cermets. The other tie line interconnects the ternary borides of $M^I_xM^{II}_y$ type intermetallic phases, which may even be beneficial for the hardening of the consolidated binder phase.

(a) Orthorhombic-M3B2 : a=0.70945 b=0.45746 c=0.31733nm $\alpha=\beta=\gamma=90°$

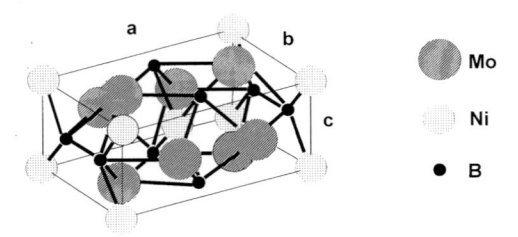

Mo
Ni
B

(b) Tetragonal-M3B2 : a=b=0.58042 c=0.31367nm $\alpha=\beta=\gamma=90°$

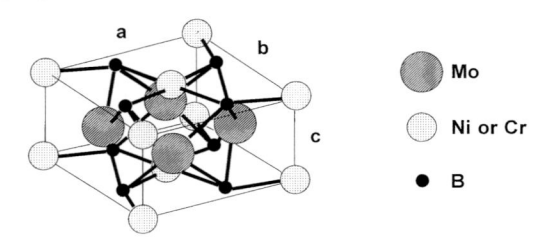

Mo
Ni or Cr
B

Figure 112. Crystal structure of ω-type ternary borides $M_2^I M^{II} B_2$ with M^I = Mo and M^{II} = Ni or Cr. Depending on the Cr content, the orthorhombic structure (a) or the tetragonal structure (b) are stable (after Ozaki et al., 1994 [358]).

7.8.3.2 Liquid-Phase Sintering and Phase Reactions

Since ternary borides are commercially not readily available the preferred sintering route is not a simple liquid-phase sintering, making use of powder blends of the equilibrium compounds. Takagi [359] introduced a reaction sintering process denoted as *reaction boronizing sintering*. This method has several benefits, (i) the starting powders are comparatively cheap, available in fine grain size with suitable purity and can easily be handled; (ii) the process takes advantage of chemical activation, i.e. the enthalpy of formation being released for sintering; (iii) the liquid phase forms after a solid-state reaction and is therefore homogeneously distributed throughout the microstructure; and (iv) the previous reaction brings about in-situ grain size refinement and removal of oxide impurities, which enables an almost perfect wetting by the liquid phase.

The process starts with a pressed powder mixture of one or more binary borides of M^I-type metals and the metals required for the M^{II} constituent of the ternary phase and for the binder phase. This method enables the fabrication of tailored M^I-M^{II}-solid solutions in the ternary boride as well as allows the introduction of high-performance Fe- or Ni-based alloys with exceptional properties. The metal powders are usually atomized alloys but may also consist of powder blends of the particular elements of interest. In Fig. 113 [360] the reaction process is schematically summarized showing the M^{II} alloy denoted as "metal". In principle, the first step is the boronization of the M^{II} powder particles by boron diffusion

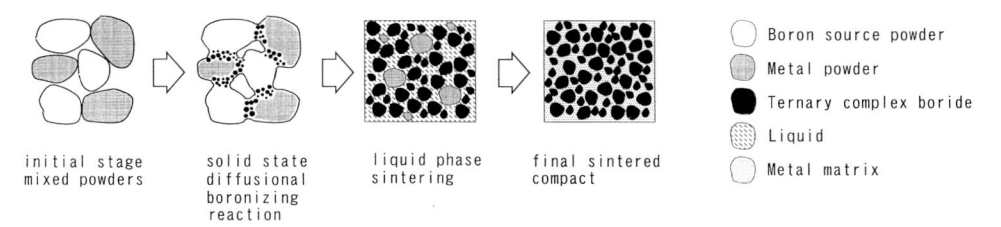

Figure 113. Schematic illustration of the "reaction boronizing sintering process" (Takagi, 1993) [359].

from the M^I boride across the contact points. This solid-state diffusion is enhanced by the concentration gradient in boron and both M^I and M^{II} metals as well as by the chemical driving force for the formation of the ternary $M_2^I M^{II} B_2$ phase. Both diffusion and heat release result in the growth of sinter necks between the starting particles and the precipitation of very small ternary boride particles being homogeneously arranged around the M^{II} grains. The source powder for boron is used up completely by this reaction, and, upon further heating, the liquid phase is created first on sites where the eutectic composition exists. Since the ternary borides have been precipitated in situ, no oxide layer is present, preventing the wetting. During spreading of the melt particle rearrangement by capillary forces is active and causes shrinkage and a more homogeneous distribution of all phases. Due to the higher temperature and the enhanced diffusion via the liquid phase, the residual metal powder dissolves in the melt, filling the residual pores and allowing grain growth of the stable ternary borides. Finally, after cooling, the cermet consists of the hard boride phase dispersed in a metallic matrix phase or with a binder phase at the triple points, depending on the volume fraction of binder. The microstructure resembles that of ordinary hard metals based on WC-Co with some peculiarities, however, which will be discussed later.

To enable the solid-state reaction at the beginning and to control wetting at higher temperatures, oxide impurities on the starting metal powders have to be removed. This can be performed by reducing agents such as carbon or by evaporation of boron oxides and suboxides during vacuum sintering. A hydrogen atmosphere can also be used for sintering. Pressureless sintering is carried out slightly above the quasi-binary eutectic temperature, usually between 1200 and 1350°C. The sintering time is comparatively short, e.g. 20–30 min to prevent grain growth by dissolution and reprecipitation.

A detailed study of the sintering phenomena of a Mo_2FeB_2-(Fe-alloy) cermet is presented by Sivaraman et al. [361] who focused on the commercially available KH-C50 material by Toyo Kohan Co., Ltd. consisting of 42.71 mass-% Mo, 39.28% Fe, 10.20% Cr, 4.96% B (as FeB), and 2.85% Ni. Sintering was followed by DTA, TG, dilatometry, and X-ray diffraction; the microstructure was observed after quenching from particular heating steps. As a result, the sintering reactions could easily be monitored by both shrinkage effects and exo- and endothermic effects. The first solid-state reaction is the formation of Fe_2B at 486°C followed by the precipitation of tetragonal Mo_2FeB_2, the reduction of metal oxides, the

Table 14. Sintering reactions of ternary boride cermets.

Temp. °C	Reaction	Remarks, Microstructure
480–500	$Fe + FeB \rightarrow Fe_2B$	Solid-state reaction
700–1000		Sinter neck formation
1000–1100	$2\,Mo + 2\,Fe_2B \rightarrow Mo_2FeB_2 + 3\,Fe$	Solid state reaction, enhanced growth of sinter necks, formation of spherical particles and clusters
1132	$y\,X + M_xO_y \rightarrow x\,M + y\,XO \uparrow$	Reduction of oxides, X = reducing agent, e.g., C
1150	$Mo_2FeB_2 + Cr \rightarrow Mo_2(Fe,Cr)B_2$	Solid-state reaction, later also in liquid state; high-density clusters with rounded surfaces, initial growth of faceted grains by surface and volume diffusion
1150–1180	$Fe + Fe_2B + Mo_2(Fe,Cr)B_2 \rightarrow$ Liquid	Ternary eutectic, wetting, spreading, cluster fragmentation and particle rearrangement
1200		Redistribution of phases, filling of pores
1225		Grain growth of faceted particles
1245		Continuous homogenization, formation of boride framework with high contiguity

formation of the liquid phase, and the incorporation of Cr into the Mo_2FeB_2. Table 14 summarizes the reaction steps and the resulting intermediate microstructures.

The sintering as monitored by dilatometry is illustrated in Fig. 114 [361]. Evidently, the solid-state reaction to form Fe_2B as well as the initial state sintering by neck growth and particle center-approach yield a small contribution to shrinkage. The most dramatic effect is, however, the generation of a liquid phase at above 1150°C which yields a shrinkage rate of 6%/min, i.e., after 5–10 min the final density of 99.6% of theoretical density is achieved. This fast sintering can be attributed to the sudden rearrangement of the boride particles of less than 1 µm in size. At temperatures exceeding 1200–1220°C only coarsening is observed by dissolution and reprecipitation mechanisms but the particle size remains around 1 µm. The change in particle size in Mo_2NiB_2 cermets with Cr-Ni binder was investigated in more detail by Matsuo et al. [362]. They started from MoB as a boron source to compare the influence of Cr addition on structural changes. Table 15 gives the composition of the boride phase aimed at whereas the binder content was kept constant at 25 mass-% (Ni ± 10% Cr).

Sintering at 1260–1300°C for 20 min revealed that MoB was consumed totally by the formation of Mo_2NiB_2, which exhibited an orthorhombic crystal structure, while the addition of Cr yielded $Mo_2(Ni,Cr)B_2$ with a tetragonal structure. The microstructures consist of blocky faceted crystals of somewhat elongated habit in all cases. It is, however, evident that Mo_2NiB_2 tends to coarsen strongly (mean particle size ($\approx$8–10 µm) compared to $Mo_2(Ni,Cr)B_2$ (mean particle size ($\approx$3–5 µm), Fig. 115. Another peculiarity is that the first phase forms euhedral crystals with sharp edges and corners whereas the latter phase shows rounded edges and corners under exactly the same conditions. This observation is attributed to the higher surface energy ani-

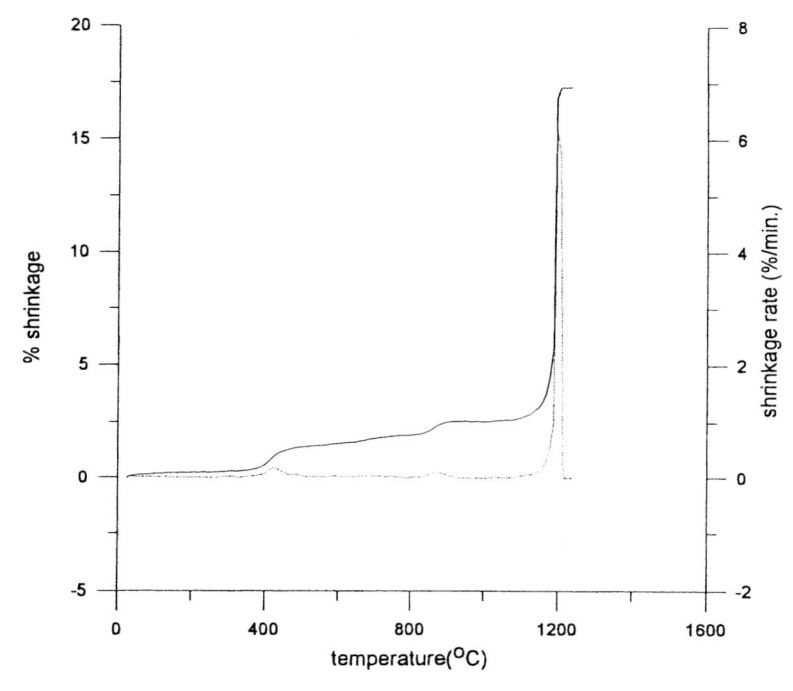

Figure 114. Sintering of Mo_2FeB_2 with Fe,Cr,Ni-binder (after [361]).

sotropy of Cr-free orthorhombic boride by which the formation of equilibrium low-energy facets are favored, as predicted by Warren [363]. The lower anisotropy in case of tetragonal $Mo_2(Ni,Cr)B_2$ is therefore also considered the reason for the observed retardation of grain growth by Ostwald ripening [358, 362].

7.8.3.3 Microstructures and Properties

A further study of Mo_2NiB_2-based cermets was dedicated to the influence of other transition-metal additives on the structure and habit of the ternary boride phase [365]. Taking the previously developed 35.4 mass-% Ni + 6 mass-% B + 58.6 mass-% Mo cermet as a starting point, 10 mass-% of Cr, V, Fe, Co, Ti, Mn, Zr, Nb, and W were added at the expenses of Ni. In another series of experiments, V was added in steps of 2.5 mass-% up to 12.5 mass-%. The mechanical properties

Table 15. Composition of Mo_2NiB_2 with Cr-Ni binder.

Boride composition		B	Mo	Ni	Cr
Mo_2NiB_2	mass-%	7.9	70.5	21.6	0
	at.-%	40	40	20	0
$Mo_2(Ni,Cr)B_2$	mass-%	8.0	71.4	10.6	10.0
	at.-%	40	40	9.7	10.3

Mo_2NiB_2 $(Mo, Cr, Ni)_3B_2$

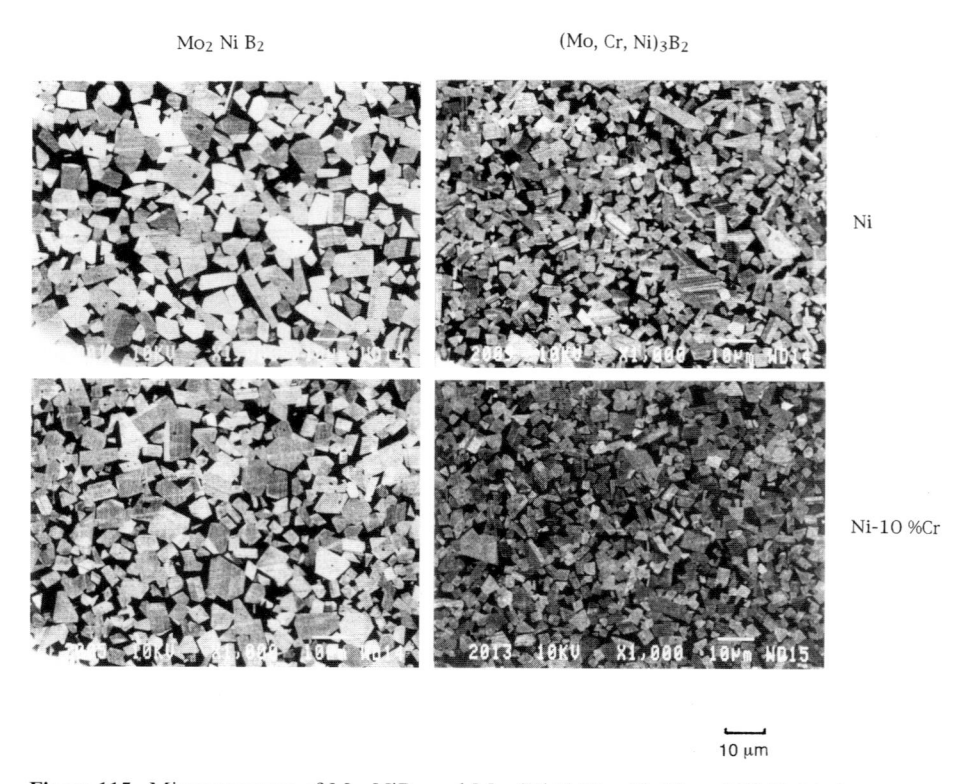

Ni

Ni-10 %Cr

10 µm

Figure 115. Microstructure of Mo_2NiB_2 and $Mo_2(Ni,Cr)B_2$ with Ni and Ni-Cr binder.

obtained clearly indicate that some of the elements have no effect on strength and hardness at all, namely, Fe and Co, whereas Ti, Mn, Zr, Nb, and W yielded even lower strength than the cermet without additives. Cr and V, however, caused more than 40% increase in strength with a slightly higher hardness. The addition of V gives rise to a steep increase in both strength and hardness if only 2.5 mass-% are added (Figs. 116–118).

It was proved that all boride solid solutions except the Cr- and V-containing one crystallize in the orthorhombic structure type. As mentioned before, the lower anisotropy of the surface energy of the tetragonal phase is the reason for the formation of more spherical particles. The inhibition of grain growth was accordingly found in Cr- and V-containing cermets but also in Nb- and W-bearing materials (Fig. 119). In the latter cases, however, the microstructure is very inhomogeneous, which may be the reason for the comparatively low strength. In case of Co and Mn addition the particles have grown considerably, which again explains the low strength since no other brittle phase was detected. If Ti and Zr are added, third phases have been observed but could not be identified.

A variation of the Cr content similar to the studies of the influence of V on the properties was carried out by Komai et al. [356]. Also in this case, an improvement of both strength and hardness was achieved as a function of the additive amount.

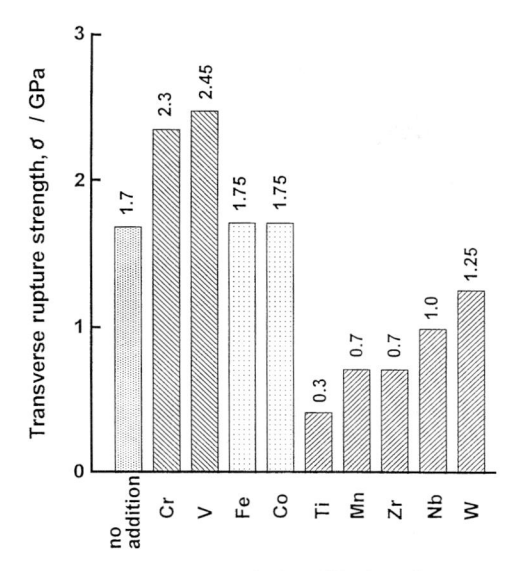

Figure 116. Strength of Mo_2NiB_2-based cermets with various additives (after [365]).

However, the strength shows a pronounced maximum at 2300 MPa at 10 mass-% and a decrease beyond the level of the cermet without Cr at 20 mass-%, while the hardness steadily increases on the order of 5% (Fig. 120).

Although the ternary phases are hard but brittle, the improved commercially available cermets exhibit excellent toughness ranging from 18 to 30 MPa m$^{1/2}$ and strength ($\sigma_b = 1.6$–2.2 GPa). Young's modulus reaches 290–350 GPa, and the hardness 83–89 R_A, depending on additives and binder content [359]. The coefficient of thermal expansion ranges between 8.5 and 11 $\times 10^{-6}$ K^{-1}. The cermets

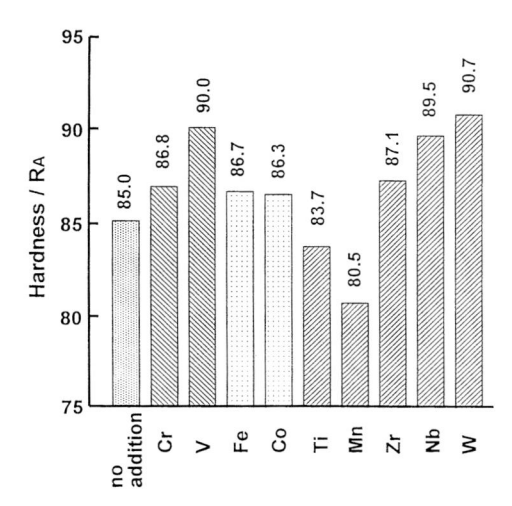

Figure 117. Hardness of Mo_2NiB_2-based cermets with various additives (after [365]).

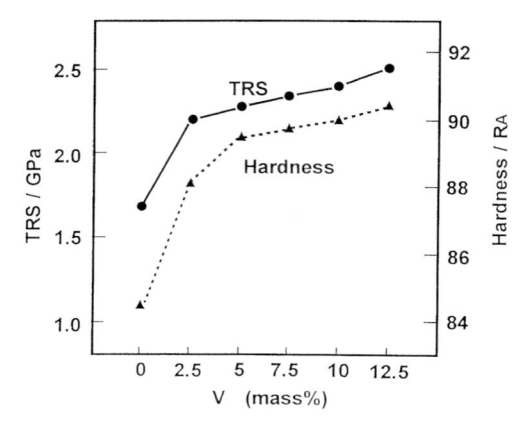

Figure 118. Transverse rupture strength (TRS) and hardness of V-containing Mo_2NiB_2 (after [365]).

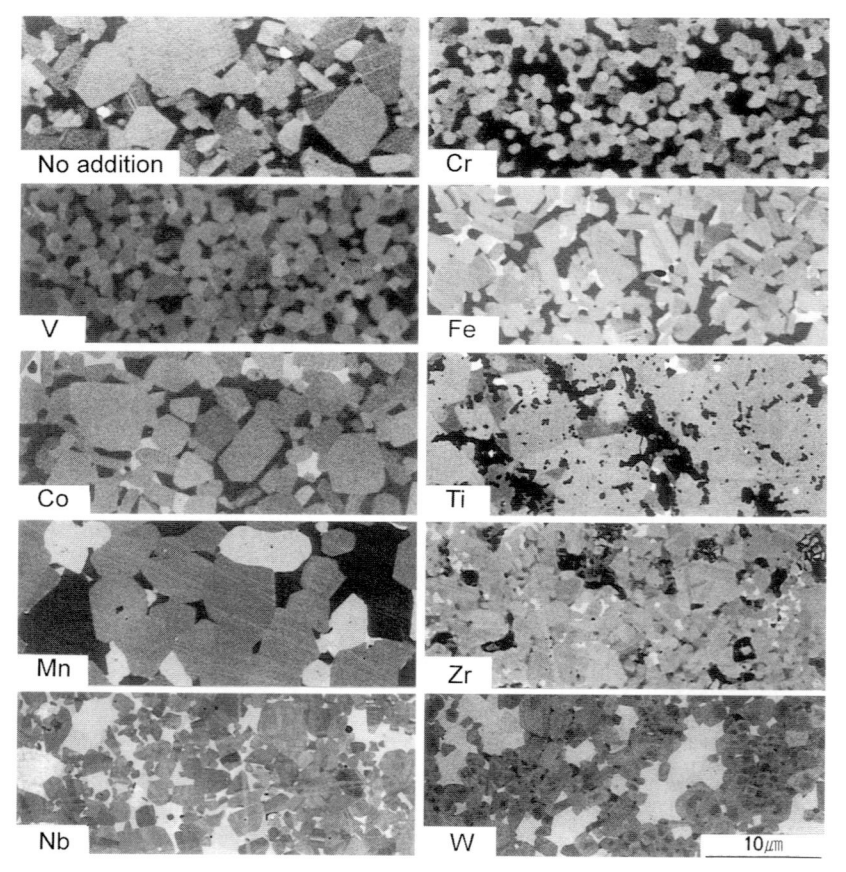

Figure 119. Back-scattered SEM micrographs of a Ni–6 mass-%, B–58.6 mass-%, Mo–10 mass-% metal cermet. Note the change in grain size and shape due to the metal additive (after [365]).

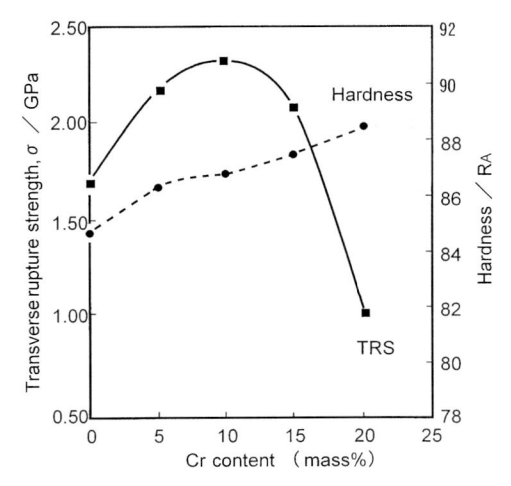

Figure 120. Transverse rupture strength and hardness as a function of Cr addition to Mo_2NiB_2 (after [356]).

of the Mo_2FeB_2 group exhibit an excellent corrosion resistance against organic and inorganic acids and alkalis [131] being comparable to or better than those of cemented carbides and stainless steels. Furthermore, the cemented borides are resistant in various molten metals such as Zn and Al. Consequently, these kinds of materials can also be used as anti-corrosion layers on other metals [361].

7.8.4 Potentials and Applications

Most of the developments in cemented borides aim at the high wear resistance due to the superior hardness compared to carbide-based hard metals. One of the most important applications are, therefore, the cutting tools where cemented boride have to compete with tungsten carbide, titanium carbonitride, and ceramic materials as well.

7.8.4.1 Cutting Tools

Bearing the potential application as a wear-resistant material in mind, the mechanical properties of the new cermets were assessed by measuring key mechanical properties such as fracture toughness and hardness both at ambient and elevated temperatures. Obviously, boride-based cermets integrate both excellent hardness and a considerable fracture toughness such that their hardness–toughness profile is superior to that of commercial hard metals. Of particular importance is the fact that TiB_2-Fe alloys are able to cover the high hardness regime beyond 2000 HV_{10}.

A test program to evaluate the potential of TiB_2 cermets as a cutting tool material was thus initiated by the European Community and many companies between 1990

Table 16. Mechanical properties of tool materials.

Tool material	Hardness HV 10	Fracture toughness [MPa$\sqrt{m}$]	Young's modulus [GPa]
TiB_2-α(Fe)-Fe_2B	1830	9.3	490
TiB_2-γ(Fe-Cr-Ni)	1800	9.0	510
Al_2O_3-ZrO_2	1700	4.5	380
Ti(C,N)-Ni	1550	8.1	440

and 1993. The tests were aimed at (i) screening the cutting properties of TiB_2-Fe and (ii) providing useful information for further development of these materials.

The properties of TiB_2 cermets suggest that workpiece materials that demand tools of the lower P, M, and K qualities would be most adequate for cutting tests. Thus a medium carbon steel Ck 45, a globular cast iron GGG 50, and a hypo-eutectic cast G-AlSi12 CuMgNi alloy were selected by Sigl et al. [366] for continuous turning experiments having a Brinell hardness of <210, <220, and <115, respectively.

Two series of TiB_2 cutting inserts with ferritic iron binders and austenitic steel binders (nominal composition 18 weight-% Cr, 10 weight-% Ni, bal. Fe) were tested. Due to a Ti deficiency, the ferritic TiB_2 grades with a nominal binder fraction of 15 vol.-% contained approximately 10 vol.-% α-Fe and 5 vol.-% Fe_2B. Ti was added to prevent the formation of Fe_2B in the steel cutting grades, such that the second series of TiB_2 cermets could be prepared with 10 vol.-% of fully austenitic binder. Uncoated ferrite-binder cermets were used to machine the Al and cast iron alloys, whereas the Ck 45 steel was treated with both uncoated and TiN-coated (3 μm CVD-TiN layer) inserts with austenite binder. Both composites represent "generation-I" TiB_2 cermets, in the sense that they feature a much coarser grain size (on the order of 5 μm) than the composites available today. The tests were carried out in comparison to conventional cutting tool materials. Thus the cast iron grade was machined with a commercial alumina–zirconia material (10 weight-% ZrO_2), while the carbon steel and the hypo-eutectic Al-Si alloy were treated with a commercial Ti(C,N)-Ni cermet. Important mechanical properties of the cutting tool materials are compared in Table 16. The actually tested combinations of workpiece and cutting tool materials are compiled in Table 17.

Continuous turning tests were performed. The SNGN 120408 tool geometry and parameters summarized in Table 18. Cutting was interrupted after a cutting length of 500 m for the GGG 50 and after 1000 m for the Ck 45 and G-AlSi12 alloys,

Table 17. Tested combinations of workpiece and cutting tool materials.

Tool materials	G-AlSi12	GGG 50	Ck 45
TiB_2-α(Fe)-Fe_2B	x	x	
TiB_2-γ(Fe-Cr-Ni)		x	x
Al_2O_3-ZrO_2		x	
Ti(C,N)-Ni	x		x

Table 18. Cutting parameters.

Cutting parameters	G-AlSi12 finishing	GGG 50 light roughing	GGG50 finishing	Ck 45 finishing
Feed f [mm/rev.]	0.2	0.3	0.1	0.1
Cutting depth a [mm]	1.0	1.5	0.5	2.0

respectively. The flank wear, v_B, the depth of cut notch depth, δ, the surface roughness of the workpiece material, the cutting forces, and the power consumption of the machine were monitored at these intervals. The test was performed until the failure criterion or a total cutting length of 8000 m was reached.

In all cutting materials the flank wear and the cutting forces increased continuously up to the ultimate lifetime. With very few exceptions of cutting edge fracture, all inserts faded due to flank wear. Failure due to excess depth-of-cut notch depth was not observed. Cratering at the rake face did occur, but was in no instance the reason for insert breakdown. The surface roughness of the workpiece materials was rated acceptable in all cases. Thus a flank wear, v_B of 0.4 mm was defined as lifetime criterion. The cutting results using this criterion are summarized in Figs. 121–124 for the various combinations of tool and workpiece materials.

Ranking the cutting tool materials, TiB_2 with a ferritic binder is superior to the Ti(C,N) cermet and the alumina–zirconia ceramic for light cutting operations such as finishing of GGG 50 and G-AlSi12. However, increased thermal loads favor cutting materials with enhanced high-temperature endurance, i.e., the Ti(C,N) cermets and the alumina–zirconia ceramics. Figures 121–124 illustrate the details of these general observations: TiB_2 cermets outperform commercial Ti(C,N) tool materials

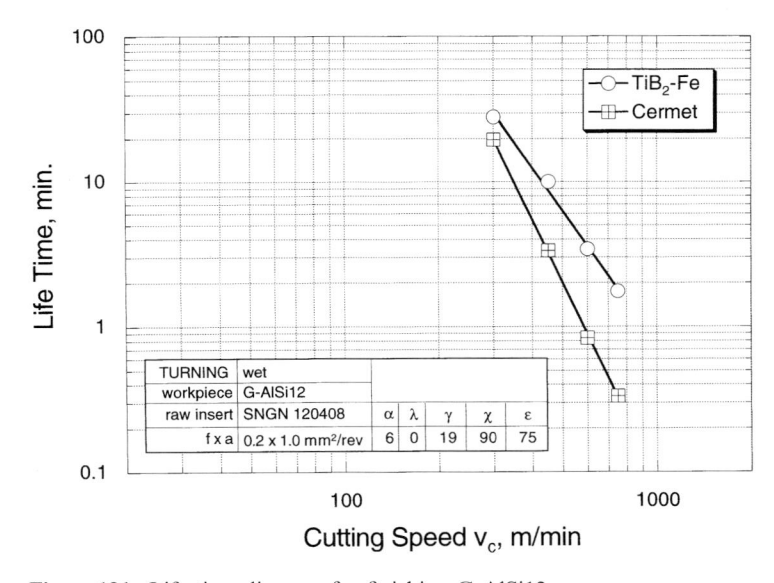

Figure 121. Life-time diagram for finishing G-AlSi12.

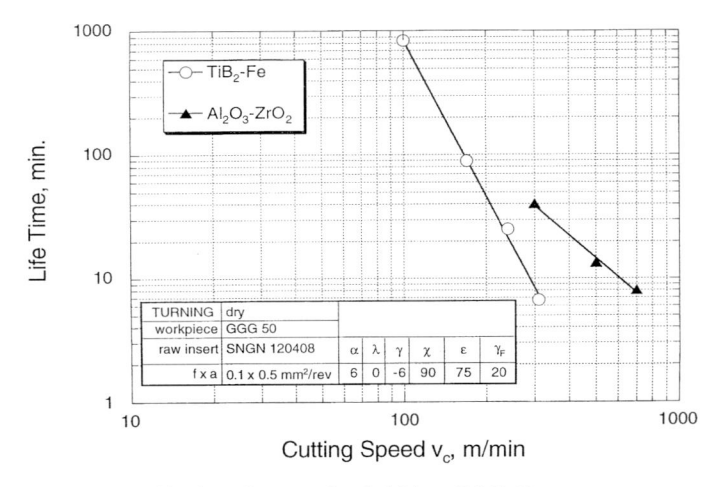

Figure 122. Life-time diagram for finishing GGG 50.

in Al cutting over the whole range of cutting speeds up to 700 m/min, and they work well at medium cutting speeds up to 200 m/min during finishing operations of gray cast iron. Tool life decreases dramatically, however, upon increased cutting loads, i.e., for light roughing of GGG 50 and for steel finishing operations (Figs. 123, 124). In particular, the Taylor slope of TiB_2-Fe in cast iron cutting is much steeper than for alumina–zirconia tools (Figs. 122, 123). A similar, yet less pronounced difference in Taylor slopes is observed between Ti(C,N) and TiB_2 steel cutting grades (Fig. 124). The TiN coating significantly improves the life of TiB_2 tools, but it cannot fill the gap of the exceptional quality of Ti(C,N) cermets.

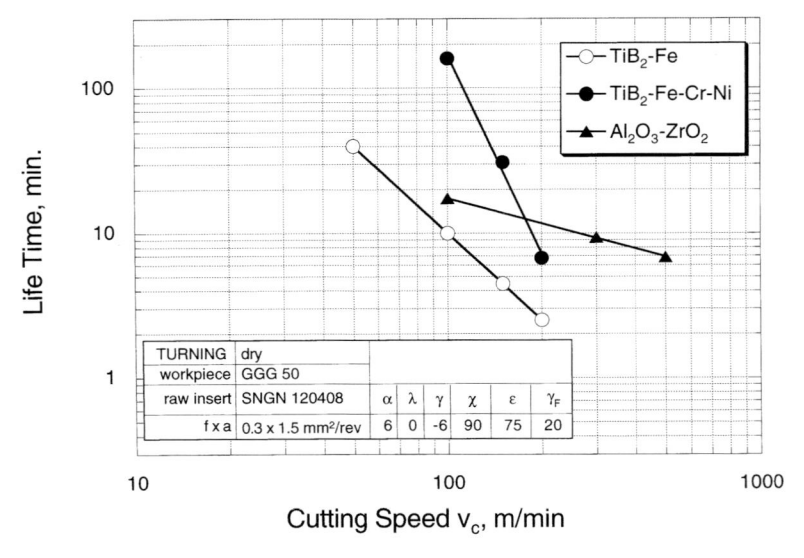

Figure 123. Life-time diagram for light roughing GGG 50.

7.8 Cemented Borides

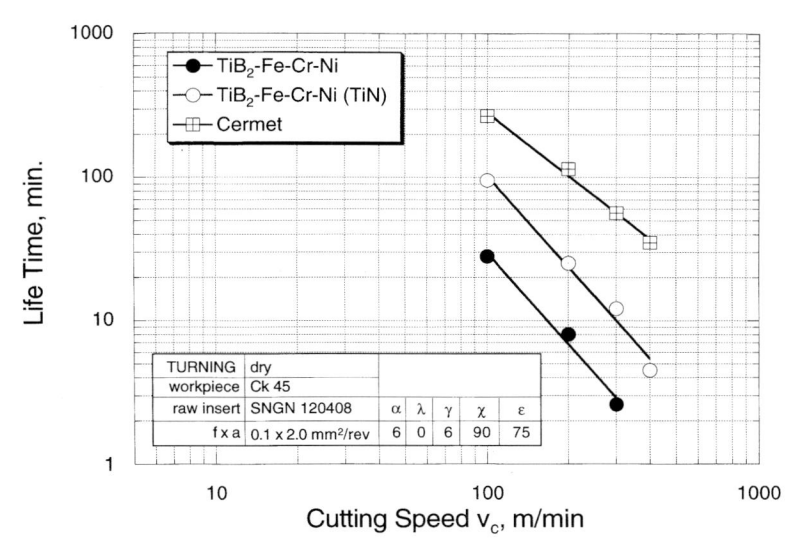

Figure 124. Life-time diagram for finishing Ck 45.

The following issues must be considered in a discussion of the cutting results:

(i) TiB_2 is a very creep and plasticity resistant compound, particularly at elevated temperatures [367].
(ii) The ferrite binder in the Al and cast iron grades contains only Ti and B in solid solution and its high-temperature strength is expected to be rather limited. The austenite binder with 18% Cr and 10% Ni should perform significantly better at high temperatures, but is probably less creep resistant than Ni(Mo) binders.
(iii) Coarse microstructures are known to favor erosive wear, even in very hard and tough materials.

According to (iii), a generally improved cutting performance of TiB_2 cermets, both at high and low temperatures, is expected from a reduction of the boride grain size. A further enhancement, specifically at high cutting loads, is anticipated from binders with improved creep resistance since both high cutting speed and increasing feed and depth of cut tend to raise the cutting tool temperatures. Thus the reduced cutting performance at rising loads is primarily attributed to the limited high-temperature strength of the ferrous binders. This conclusion is supported by the improved tool life of TiN-coated inserts because the coating reduces the friction between tool and chip and consequently keeps the temperature level of the cutting insert low.

In conclusion, TiB_2 is an excellent hard compound for continuous cutting operations already matching or surpassing conventional tool materials in applications where limited temperature loads evolve. Specifically, ferritic binders perform well in the cutting of cast Al alloys and cast iron. TiB_2 austenite cermets do show potential for the cutting of steel, but as yet they cannot match the exceptional performance of Ti(C,N) cermets.

An optimization of the microstructure towards fine grain size of TiB_2 and binders with sufficient high-temperature strength will further improve the cutting capacity of TiB_2 cermets specifically for steel machining operations. The major disadvantage that has prevented the production of TiB_2-Fe composites on a commercial scale is their delicate processing behavior. Thus the cost–performance relation is still the critical point for the current market conditions.

7.8.4.2 Wear Parts and Molding Tools

The applications of cemented ternary borides aim at other industrial branches but also make use of the unique combination of hardness, toughness, and corrosion resistance. Sliding wear tests carried out by Takagi [359] in a rotating-ring-on-block arrangement have shown that the wear rate of a Mo_2FeB_2 (Fe-alloy) cermet is one order of magnitude less than that of WC-10Co. The characterization of the wear debris yielded a small amount of MoO_2, B_2O_3, and Fe_3O_4 at sliding velocities >0.6 m/s, whereas chips produced by the WC-Co hard metal did not give evidence for any oxide formation. The higher wear rate of WC-Co is attributed to WC-particle pull-out, providing abrasive grains at the interface between the wear couples.

Wear-resistant parts such as injection-molding machine parts for fabrication of fiber-reinforced polymers, metals, and ceramics, drawing dies for metallic wires, bearings, liners, and other parts for protection against sliding wear have been developed from Mo_2FeB_2-Fe or Mo_2NiB_2-Ni cermets (Fig. 125). They are commercially available and already widely used in Japan. Moreover, ternary borides have been successfully applied as coatings on steel for surface hardening and improvement of the corrosion resistance [360]. Thus this new class of materials has to

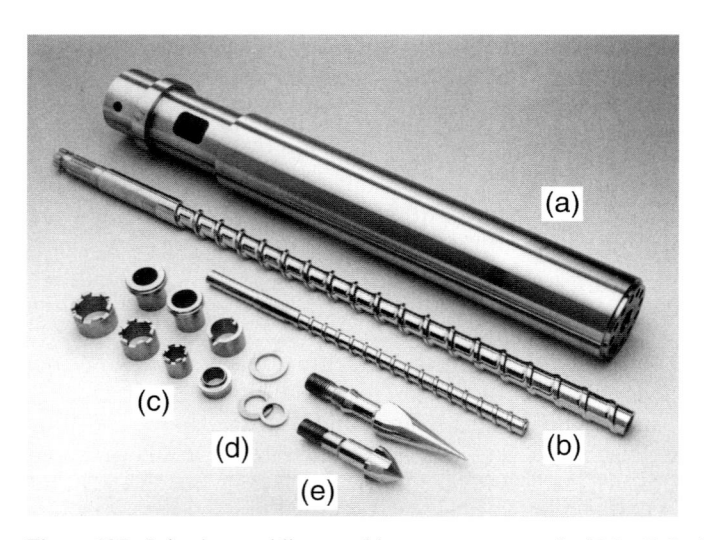

Figure 125. Injection molding machine parts composed of Mo_2FeB_2-based alloys: (a) cylinder; (b) auger; (c) check valve; (d) seal ring; (e) auger head.

compete with all kinds of hard metals and cermets with respect to performance and costs.

7.9 Future Prospects and Fields of Application

At the time this article was written, optimism concerning the application of high-performance advanced ceramics was still considerably low after the boom between 1980 and 1990, especially in respect to structural parts for automotive engines or substitutes for other high-strength, high-toughness materials. This tendency was and is not restricted to Europe or the USA., but is also valid for Japan and can easily be attributed to the following problems: (i) difficulties related to reliable large-scale production at reasonable cost and quality (problems of "zero-flaw" processing and part testing), (ii) difficulties in a material-appropriate design of parts (problems of substitution and new construction taking the relative brittleness of ceramics into account), (iii) the still comparatively high price of high-quality starting materials and manufacturing processes (problems of affordable high-purity and fine grain size), and (iv) the current worldwide economic recession in production stopping any activities in materials research and development. Considering these facts, the following paragraph will give some basic ideas on the actual and most probable applications, which have already been demonstrated up to now to be feasible. Besides the well-known application of grits (e.g., as grinding, lapping, and polishing particles, as oxidation inhibitors in refractories, and for alloying of special metals), boride ceramics are already widely used in mechanical, chemical, thermal, and electrical areas.

Most of the applications of borides related to mechanical aspects make use of their extraordinarily high hardness and wear resistance. Also, their high Young's modulus (stiffness) and excellent high-temperature behavior are very attractive properties. The high-temperature hardness of the most important ceramics and hard materials is shown in Fig. 126. It is obvious that, besides carbides, borides are superior to most oxide ceramics. At high temperatures, however, the metallic transition-metal borides suffer from a strong decrease in hardness due to enhanced plastic deformability (thermal initiation of slip systems), which is basically due to their comparatively primitive crystal structure and their high metallic portion in bonding. Additionally, oxidation has to be taken into account at temperatures above 600°C.

The poor corrosion resistance of borides may be partially overcome in composites with SiC if the microstructure allows passive oxidation kinetics. In combination with SiC the borides retain their high electrical and thermal conductivity and thus suitable thermal shock resistance. SiC-TiB$_2$ composites have been extensively developed for wear parts in machinery such as sliding rings, valves, valve seats, roller and ball bearings, plungers, and rocker arm pads.

Besides the well-known cemented carbides and cermet cutting tools, boron carbide- and titanium diboride-containing cermets have been proven suitable for

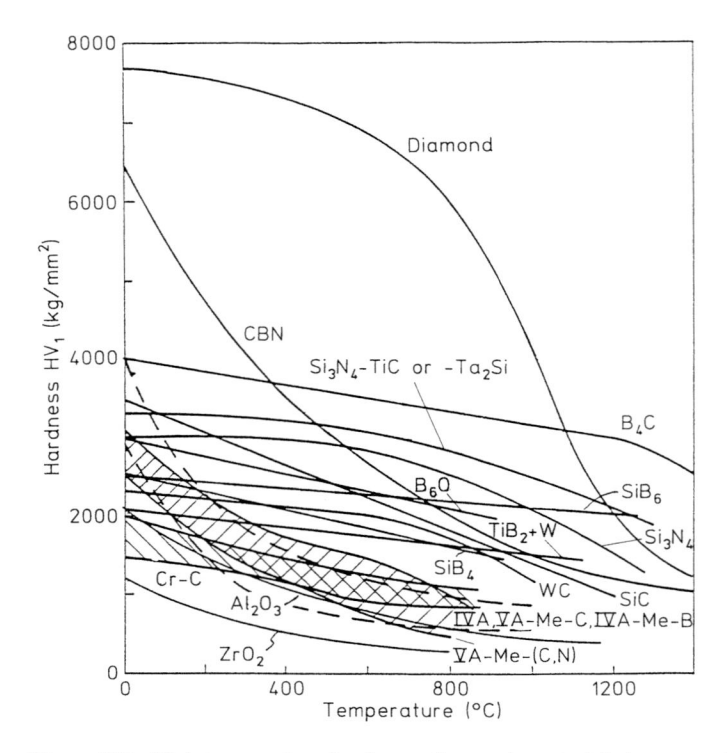

Figure 126. High-temperature hardness of ceramic materials in comparison to diamond and cubic boron nitride.

machining aluminum alloys, whereas ternary boride cermets have been successfully developed for injection molding nozzles, bearings, wire drawing cones, and other wear resistant parts that are not exposed to high temperatures. Composites of TiB_2/TiC, TiB_2/TiN, $TiB_2/ZrB_2/TaN$, as well as B_4C/TiB_2, B_4C/SiC, and $B_4C/SiC/TiB_2$ have been exploited as cutting tools for brass, bronze, and Al alloys, drilling tools for rocks, and concrete due to their comparatively high toughness.

An obvious conclusion from the various efforts in strengthening and toughening of boride-based materials is that the maximum strength obtained by crack-propagation influencing mechanisms operational in composites is about 1–1.2 GPa and the fracture toughness usually obtained in optimized microstructures is about 6–7 MPa m$^{1/2}$. It is also clear that simple two-phase composites are generally superior to materials of more complex composition. Metal-matrix composites may exhibit a much higher toughness than ceramics but are limited in high-temperature use due to their sensitivity towards creep and corrosion. With the possible exception of the very promising TiB_2-Fe-based and ternary boride–based cermets, borides are thus currently of more interest for strengthening metals similar to oxide-dispersed superalloys. A certain challenge is still the transformation toughening of borides by Y-stabilized ZrO_2, increasing the toughness by a factor of two compared to simple particulate reinforcement. In competition with oxide and nitride ceramics the borides have thus been developed to a similar excellent mechanical performance but still suffer from their considerably high price and poor oxidation resistance at high temperatures.

Generally speaking, the high reactivity of the boron compounds is the most important, and sometime the most limiting factor. Although the wear behavior is very promising due to the high hardness and the high thermal conductivity, boron compounds, except for some transition-metal diborides, cannot be used in contact with iron-based alloys but for wear contacts with aluminum and, in some cases, nickel-based materials. On the other hand, reaction sintering can be widely employed for the fabrication of unique microstructures, which has not yet been exploited entirely.

Although mechanical applications are currently by far the most important aims for the development of materials and parts based on carbides and borides, this may easily change to the benefit of thermal and electrical applications if there is an improvement in the understanding and control of the semiconductivity of B_4C and a breakthrough in the manufacture of suitable parts. Moreover, thermal applications may increase considerably in importance as more efficient energy conversion or recovery becomes a significant requirement.

Excellent opportunities for applications of transition-metal borides such as TiB_2 are thus more a result of their high electrical conductivity and resistance against metallic melts combined with an excellent wetting behavior. Well-known examples are TiB_2-AlN-BN evaporator dies (e.g., for the evaporation of Al, Cd, Ge, Pb, Bi, Cr, Cu, Ag, Au, Mg, Fe, Zn, and Sn for the coating of plastics, paper, and parts) and electrodes for the aluminum electrolysis in Hall–Hérould cells. In the latter case, there is a very high demand for such a material having a high electric conductivity and extraordinary corrosion resistance against kryolithe melts, but some undesired grain boundary reactions (infiltration of the melt along impurity-loaded grain boundaries) and the poor thermal shock resistance (strongly anisotropic thermal expansion) are the limiting factors for a successful replacement of graphite electrodes. Other diborides, such as ZrB_2, have been studied for use as structural parts, such as bearings, nozzles, and molds for injection molding, valves, and sealings, but further development is still necessary for an industrial breakthrough. For example, ZrB_2-$MoSi_2$ composites exhibit remarkable corrosion resistance up to 2000°C.

Functional applications of borides are also related to the high cross-section of boron for the absorption or retardation of thermal neutrons. B_4C sintered products have been used as moderating elements and radiation-protecting components in nuclear reactors of several designs. Of course, this market is declining with the decreasing political support for nuclear power and nuclear research.

In conclusion, boride ceramics are materials of high potential for use in mechanical and electric fields. Due to the stagnating materials markets, however, most research on borides is focused on the basic questions of atomic bonding and physical transport properties rather than on application-related problems. Only a few highly specialized research groups are active worldwide and promote the science rather than the technology of borides. Depending upon the economic situation and the demand for future key technologies, there is a chance that these materials may increase in importance.

References

1. Kiessling, R. *Acta Chem. Scand.* 1950, **4**, 209.
2. Lundström, T. *Arkiv Kemi* 1969, **31(19)**, 227.
3. Andersson, S. and Lundström, T. *Acta Chem. Scand.* 1968, **12**, 3103.
4. Vlasse, M., Slack, G. H., Garbauskas, M., Kasper, J. S., and Viala, J. C. *J. Solid State Chem.* 1986, **63**, 31.
5. Richards, S. M. and Kasper, J. S. *Acta Crystallogr.* 1969, **B25**, 237.
6. Higashi, I., Kobayashi, K., Tanaka, T., and Ishizawa, Y. *J. Solid State Chem.* 1997, **133**, 16.
7. Matkovich, V. I. and Economy, J. in *Boron and Refractory Borides*, Matkovich, V. I. (Ed.), Springer, Berlin, 1977, pp. 77–94.
8. Aronsson, B., Lundström, T., and Rundqvist, S. *Borides, Silicides and Phosphides: A Critical Review of Their Preparation, Properties and Crystal Chemistry*, John Wiley & Sons Inc., New York, 1965.
9. Aronsson, B., Lundström, T., and Engström, I. in *Anisotropy in Single Crystal Refractory Compounds*, Plenum Press, New York, 1968, Vol. 1, pp. 3–22.
10. Lundström, T. *Arkiv Kemi* 1969, **31(19)**, 227.
11. Higashi, I. and Takahashi, Y. *J. Less-Common Met.* 1986, **123**, 277.
12. Okada, S., Atoda, T., Higashi, I., and Takahashi, Y., *J. Mater. Sci.* 1987, **22**, 2993.
13. Lundström, T. *Arkiv Kemi* 1968, **30(11)**, 115.
14. Romans, P. A. and Krug, M. P. *Acta Crystallogr.* 1966, **20**, 313.
15. Woods, H. P., Wawner, F. E., and Fox, B. G. *Science* 1966, **151**, 75.
16. Armstrong, D. R. in *Proc. 9th Int. Symp. on Boron, Borides, and Related Compounds*, Werheit, H. (Ed.), University Press, Duisburg, 1987, pp. 125–131.
17. Samsonov, G. V. and Kovenskaya, B. A. in *Boron and Refractory Borides*, Matkovich, V. I. (Ed.), Springer, Berlin, 1977, pp. 5–18.
18. Samsonov, G. V. and Kovenskaya, B. A. in *Boron and Refractory Borides*, Matkovich, V. I. (Ed.), Springer, Berlin, 1977, pp. 19–30.
19. Aravamudham, R. *Z. Metallkd.* 1967, **58**, 179.
20. Samsonov, G. V., Goryachev, Yu. M., and Kovenskaya, B. A. *Izv. Vuz Fiz.* 1972, **6**, 37.
21. Hoard, J. L. and Hughes, R. E. in *The Chemistry of Boron and Its Compounds*, Muetterties, E. L. (Ed.), Wiley, New York, 1967, pp. 25–154.
22. Spear, K. E. *J. Less-Common Met.* 1976, **47**, 195.
23. Hayami, W., Souda, R., Aizawa, T., and Tanaka, T. *Surf. Sci.* 1998, **415**, 433.
24. Souda, R., Asari, E., Kawanowa, H., Otani, S., and Tanaka, T. *Surf. Sci.* 1998, **414(1–2)**, 77.
25. La Placa, S. and Post, B., *Planseeberichte für Pulvermetallurgie* 1961, **9**, 109.
26. Lipp, A. *Ber. Deutsche. Keram. Ges.* 1966, **43(1)**, 60.
27. Lipp, A. and Röder, M. *Z. Anorg. Allg. Chem.* 1966, **343**, 1.
28. Lipp, A. and Röder, M. *Z. Anorg. Allg. Chem.* 1966, **344**, 225.
29. Matkovich, V. I. and Economy, J. in *Boron and Refractory Borides*, Matkovich, V. I. (Ed.), Springer, Berlin, 1977, pp. 98–106.
30. Clark, H. K. and Hoard, J. L. *J. Am. Ceram. Soc.* 1943, **65**, 2115.
31. Allen, R. D. *J. Am. Chem. Soc.* 1953, **75**, 3582.
32. Neidhard, H., Mattes, R., and Becher, H. J. *Acta Crystallogr.* 1970, **26B**, 315.
33. Ploog, K. *J. Less-Common Met.* 1974, **25**, 115.
34. Will, G. and Kossobutski, K. H. *J. Less-Common Met.* 1976, **47**, 43.
35. Kirfel, A., Gupta, A., and Will, G. *Acta Crystallogr.* 1979, **B35**, 1052.
36. Werheit, H. and de Groot, K. *Phys. Status Solidi* 1980, **97**, 229.
37. Conard, J., Bouchacourt, M., and Thevenot, F. *J. Less-Common Met.* 1986, **117**, 51.

38. Tallant, D. R., Aselage, T. L., and Campbell, A. N. *Phys. Rev. B* 1989, **40**, 5649.
39. Aselage, T. L., Emin, D., and Wood, C. in *Trans. 6th Symp. Space Nuclear Power*, Albuquerque, NM, January 8–12, 1989, pp. 430–433.
40. Aselage, T. L., Tallant, D. R., Gieske, J. H., Van Deusen, S. B., and Tissot, R. G. in *The Physics and Chemistry of Carbides, Nitrides, and Borides*, Freer, R. (Ed.), NATO ASI Ser. E, Vol. 185, Kluwer, Dordrecht, The Netherlands, 1990, pp. 97–112.
41. Aselage, T. L. and Emin, D. in *Boron-Rich Solids*, Proc. 10th Int. Symp. on Boron, Borides, and Related Compounds, Emin, D., Aselage, T. L., Switendick, A. C., Morosin, B., and Beckel, C. L. (Eds), AIP Conf. Proc. 231, American Institute of Physics, New York, 1991, pp. 177–185.
42. Morosin, B., Aselage, T. L., and Emin, D. in *Boron-Rich Solids*, Proc. 10th Int. Symp. on Boron, Borides, and Related Compounds, Emin, D., Aselage, T. L., Switendick, A. C., Morosin, B., and Beckel, C. L (Eds), AIP Conf. Proc. 231, American Institute of Physics, New York, 1991, pp. 193–196.
43. Kleinman, L. in *Boron-Rich Solids*, Proc. 10th Int. Symp. on Boron, Borides, and Related Compounds, Emin, D., Aselage, T. L., Switendick, A. C., Morosin, B., and Beckel, C. L. (Eds), AIP Conf. Proc. 231, American Institute of Physics, New York, 1991, pp. 13–20.
44. Armstrong, D. R., Bolland, J. and Perkins, P. G. *Acta Crystallogr.* 1983, **39**, 324.
45. Emin, D. in *Boron-Rich Solids*, Emin, D., Aselage, T, Beckel, C. L., Howard, I. A., and Wood, C. (Eds), AIP Conf. Proc, American Institute of Physics, New York, 1986, pp. 189–205.
46. Silver, A. H., Bray, P. J. *J. Chem. Phys.* 1959, **31**, 247.
47. Hynes, T. V., Alexander, M. N. *J. Chem. Phys.* 1974, **54**, 5296.
48. Becher, H. J., Thevenot, F. *Z. Anorg. Allg. Chem.* 1974, **410**, 274.
49. (a) Bouchacourt, M. and Thevenot, F. *J. Less-Common Met.* 1981, **82**, 219. (b) Bouchacourt, M. and Thevenot, F. *J. Less-Common Met.* 1981, **82**, 227.
50. Schwetz, K. A. and Karduck, P., *J. Less-Common Metals* 1991, **175**, 1.
51. Kuhlmann, U., Werheit, H., and Schwetz, K. A. *J. Alloys Compd.* 1992, **189**, 249.
52. Lim, S.-K. and Lukas, H.-L., in *Hochleistungskeramiken – Herstellung, Aufbau, Eigenschaften*, Petzow, G., Tobolski, J., and Telle, R. (Eds), VCH, Weinheim, and Deutsche Forschungsgemeinschaft, 1996, pp. 605–616.
53. Kasper, B. *Dr. rer. nat. Thesis*, University of Stuttgart, 1996.
54. Telle, R. in *The Physics and Chemistry of Carbides, Nitrides, and Borides*, Proc. Freer, R. (Ed.), NATO ASI Ser. E, Vol. 185, Kluwer, Dordrecht, The Netherlands, 1990, pp. 249–268.
55. Meerson, G. A., Kiparisov, S. S., and Gurevich, M. A. *Sov. Powder Metall. Met. Ceram.* 1966, **5**, 223.
56. Magnusson, B. and Brosset, C. *Acta Chem. Scand.* 1962, **16**, 449.
57. Hansen, M. *Constitution of Binary Alloys*, McGraw-Hill, New York, 1958.
58. Elliott, R. P. *Constitution of Binary Alloys*, 1st suppl., McGraw-Hill, New York, 1965.
59. Shunk, F. A. *Constitution of Binary Alloys*, 2nd suppl., Mc-Graw-Hill, New York, 1969.
60. Mofatt, W. G. *The Handbook of Binary Phase Diagrams*, General Electric Co., Schenectady, NY, 1976.
61. Mofatt, W. G. *The Index to the Binary Phase Collections*, General Electric Co., Schenectady, NY, 1979.
62. *Binary Alloy Phase Diagrams*, Massalski, T. B. (Ed.), ASM Int., Materials Park, OH, 1990, Vol. 1: Alloys, Vol. 2: Phase Diagrams.
63. *Ternary Alloys*, Petzow, G., Effenberg, G. (Eds), VCH, Weinheim, 1988ff, Vols. 1 ff.
64. Spear, K. E. in *Application of Phase Diagrams in Metallurgy and Ceramics*, National Bureau of Standards, Gaithersburgh, 1977, Special Publication SP-496, pp. 744–762.
65. Lundström, T. in *Boron and Refractory Borides*, Matkovich, V. I. (Ed.), Springer, Berlin, 1977, pp. 351–376.
66. Spear, K. E. in *Phase Diagrams*, Materials Science and Technology, Vol. IV, Alper, A. M. (Ed.), Academic Press, New York, 1976, Ch. 11, pp. 91–159.
67. Guillermet, A. F. and Grimvall, G. *Phys. Rev. B* 1989, **40**, 1521.
68. Grimvall, G. and Guillermet, A. F. *Boron-Rich Solids*, Proc. 10th Int. Symp. on Boron, Borides, and Related Compounds, Emin, D., Aselage, T. L., Switendick, A. C., Morosin, B., and Beckel, C. L. (Eds), AIP Conf. Proc. 231, American Institute of Physics, New York, 1991, pp. 423–430.

69. Samsonov, G. V. and Schuravlev, N. N. *Fiz. Met. Metallovad. Akad. Nauk, SSSR Ural Filial 3* 1956, 109.
70. Schuravlev, N. N. and Makarenko, G. N. *Ed. Acad. Sci. USSR, OTN, Metallurgia i Toplivo 1*, 1961, 133.
71. Bouchacourt, M. and Thevenot, F. *J. Less-Common Met.* 1979, **67**, 327.
72. Beauvy, M. in *Conf. Abstract, 8th Int. Symp. on Boron, Borides, Carbides and Related Compounds*, Tsagareishvili, G. V. (Ed.), Acad. Sci. Georg. S.S.R., Tbilisi, 1984, p. 25.
73. Lukas, H. L. in *Constitution of Ternary Alloys 3*, Petzow, G. and Effenberg, G. (Eds), VCH, Weinheim, 1990, pp. 140–146.
74. Ekbom, L. B. *Sci. Ceram.* 1977, **9**, 183.
75. Lange, D. and Holleck, H., *Proc. 11th Int. Plansee-Conf.*, Planseewerke, Reutte, 1985, **HM 50**, 747.
76. Ettmayer, P., Horn, H. C., and Schwetz, K. A. *Microchim. Acta*, 1970, Suppl. IV, 87.
77. Lugscheider, E., Reimann, H., and Quaddakers, W.-J., *Ber. Deutsche Keram. Ges.* 1979, **65(10)**, 301.
78. Dörner, P. *Dr. rer. nat. Thesis*, University of Stuttgart, 1982.
79. Emin, D. in *The Physics and Chemistry of Carbides, Nitrides and Borides*, Freer, R. (Ed.), Kluwer, Dordrecht, The Netherlands, 1990, pp. 691–704.
80. Olesinski, R. W. and Abbaschian, G. J. *Bull. Alloy Phase Diagrams* 1984, **5**, 478.
81. Mitra, I., Büttner, D., Telle, R., Babushkin, O., and Lindbäck, T. in *Proc. 8th CIMTEC, Ceramics Charting the Future, Advances in Science and Technology*, Vol. 3B, Vincenzini, P. (Ed.), Techna Srl., Milano, 1995, pp. 787–792.
82. Armas, B., Male, G., and Salanoubat, D. *J. Less-Common Metals* 1981, **82**, 245.
83. Telle, R. *Dr. rer. nat. Thesis*, University of Stuttgart, 1985.
84. Murray, J. L., Liao, P. K., and Spear, K. E. *Bull. Alloy Phase Diagrams* 1986, **7**, 550.
85. Rudy, E. and Windisch, S. in *Ternary Phase Equilibria in Transition Metal–Boron–Carbon–Silicon Systems*, Part I, Vol. Vll, Technical Report No. AFML-TR-65–2, Wright Patterson Air Force Base, OH, 1966.
86. Fenish, R. G. *Trans. AIME* 1966, **236**, 804.
87. Neronov, V. A., Korchagin, M. A., and Aleksandrov, V. V. *J. Less-Common Met.* 1981, **82**, 125.
88. Spear, K. E., McDowell, P., and McMahon, F. *J. Am. Ceram. Soc.* 1986, **69**, C4.
89. Bätzner, C. *Dr.rer.nat. Thesis*, University of Stuttgart, 1994.
90. Pyzik, A. J. and Beaman, D. R. *J. Am. Ceram. Soc.* 1995, **78**, 305.
91. Kieffer, R., Gugel, E., Leimer, G., and Ettmayer, P. *Ber. Deutsche. Keram. Ges.* 1972, **49(2)**, 44.
92. Mogilevsky, P., Gutmanas, E. Y., Gotman, I., and Telle, R. *J. Eur. Ceram. Soc.* 1995, **15**, 527.
93. Lukas, H. L. Final Report, Deutsche Forschungsgemeinschaft Project, Contract No. Lu 283/2–2, 1992.
94. Telle, R. and Petzow, G. in *High Tech Ceramics*, Vinzencini, P. (Ed.), Material Science Monographs, Elsevier, Amsterdam, 1987, pp. 961–973.
95. Werheit, H., Kuhlmann, U., Laux, M., and Telle, R. *J. Alloys and Compounds* 1994, **209**, 181.
96. Secrist, D. R. *J. Am. Ceram. Soc.* 1964, **47**, 127.
97. Shaffer, P. T. B. *Mater. Res. Soc. Bull.* 1969, **4**, 213.
98. Nishiyama, K., Mitra, I., Momozawa, N., Watanabe, T., Abe, M., and Telle, R. *J. Jpn. Res. Inst. Mater. Technol.* 1997, **15**, 292.
99. Rudy, E. and Windisch, S. *Ternary Phase Equilibria in Transition Metal-Boron-Carbon-Silicon Systems,* Part II. Related Tenary System Vol. VIII, TI-B-C System, Technical Report No. AFML-TR-65–2, Wright Patterson Air Force Base, OH 1966.
100. Nowotny, H., Benesovsky, F., and Brukl, C. *Monatsh. Chem.* 1961, **92**, 403.
101. Ordan'yan, S. S. and Unrod, V. V. *Sov. Powder Metall. Met. Ceram.* 1975, **14**, 729.
102. Holleck, H. *Planseeber. Pulvermetall.* 1982, **2**, 849.
103. Holleck, H., Leiste, H., and Schneider, W. in *High Tech Ceramics*, Proc. 6th CIMTEC, Vincenzini, P. (Ed.), Elsevier, Amsterdam, 1987, pp. 2609–2622.
104. Rudy, E. *Ternary Phase Equilibria in Transition Metal-Boron-Carbon-Silicon Systems*, Vol. V, Compendium of Diagram Data, US Atomic Energy Comm. Publ. AFML-TR-65-2, Wright Patterson Air Force Base, OH, 1969.

105. Duschanek, H. and Rogl, P., Lukas H. L. *J. Phase Equilibria* 1995, **16**, 46.
106. Brodkin, D. and Barsoum, M.W. *J. Am. Ceram. Soc.* 1996, **79**, 785.
107. Telle, R., Brook, R. J., and Petzow, G. *J. Hard Mater.* 1991, **2**, 79.
108. Täffner, U., Telle, R., and Schäfer, U. *Z. Prakt. Metallogr.* 1988, **21**, 17.
109. Klingler, H., Franz, E.-D., and Telle, R. *Sprechsaal* 1995, **128**, 11.
110. Brukl, C. E. *Ternary Phase Equilibria in Transition Metal-Boron-Carbon-Silicon Systems Part II: Ternary Systems*, Wright, Patterson Air Force Base, OH, 1965, Vol. VII, Tech. Rep. No. AFML-TR-65–2.
111. Borisova, A. L., Borisov, Yu. S., Polyanin, B. A., Shvedova, L. K., Kalinovskii, V. R., and Gorbatov, I. N. *Sov. Powder Metall. Met. Ceram.* 1986, **25**, 769.
112. Holleck, H. in *Proc. 12th Planseeseminar of Powder Metallurgy 3*, Bildstein, H., Ortner, H. M. (Eds), Planseewerke, Reutte, 1989, pp. C3, 1–12.
113. Touanen, M., Teyssandier, F., and Ducarroir, M. in *Proc. 7th Eur. Conf. on Chemical Vapour Deposition*, Ducarroir, M., Bernard, C., and Vandenbulcke, L. (Eds), *J. Phys.* 1989, **50**, 105.
114. Wakelkamp, W. J. J., van Loo, F. J. J., and Metselaar, R. *J. Eur. Ceram. Soc.* 1991, **8**, 135.
115. Yuriditsky, B. Y. *Refractory Mater. Hard Mater.* 1990, **3**, 32.
116. Sigl, L. S. and Schwetz, K. A. in *Boron-Rich Solids*, Proc. 10th Int. Symp. Boron, Borides, and Related Compounds, Emin, D., Aselage, T L., Switendick, A. C, Morosin, B., and Beekel, C. L. (Eds), AIP Conf. Proc. 231, American Institute of Physics, New York, 1991, pp. 468–472.
117. Ghetta, V., Gayraud, N., and Eustathopoulos, N. *Solid State Phenomena* 1992, **25/26**, 105.
118. Jüngling, T., Oberacker, R., Thümmler, F., Sigl, L. S., and Schwetz, K. A. *Powder Metall. Int.* 1991, **23(5)**, 296.
119. Sigl, L. S. and Jüngling, T. *J. Hard Mater.* 1992, **3**, 39.
120. Federov, T. F. and Kuzma, Y. B. *Izvest. Akad. Nauk SSSR – Neorg. Mater.* 1967, **3**, 1489.
121. Shurin, A. K. and Panarin, V. E. *Izvest. Akad. Nauk. SSSR – Metally* 1974, **4**, 235.
122. Smid, I. and Kny, E. *Int. J. Refractory Met. Hard Mater.* 1988, **8**, 135.
123. Ottavi, L., Chaix, J. M., Allibert, C., and Pastor, H. *Solid State Phenomena* 1992, **25/26**, 543.
124. Ottavi, L., Saint-Jours, C., Valignant, N., and Allibert, C. *Z. Metallkd.* 1992, **83**, 80.
125. Schöbel, J. D. and Stadelmaier, H. H. *Metallwissenschaft Tech.* 1965, **19**, 715.
126. Lugscheider, E., Reimann, H., and Pankert, R. *Z. Metallkd* 1980, **71**, 654.
127. Lugscheider, E., Reimann, H., and Pankert, E. *Metall.* 1982, **36**, 247.
128. Haschke, H., Nowotny, H., and Benesovsky, F. *Monatsh. Chem.* 1966, **97**, 1459.
129. Takagi, K., Ohira, S., and Ide, T. in *Modern Developments in Powder Metallurgy*, Vol. 16: Ferrous and Non-ferrous Materials, Aqua, E. N., and Whitman, C.I. (Eds), MPI & APMI Publ., Princeton, NJ, pp. 153–166.
130. Takagi, K. *et al. Int. J. Powder Met.* 1987, **23**, 157.
131. Takagi, K., Ohira, S., Ide, T., Watanabe, T., and Kondo, Y. *Met. Powder Rep.* 1987, **42**, 483.
132. Post, B., Glaser, F. W, and Moskowitz, D. *Acta Metall.* 1954, **2**, 20.
133. Kuz'ma, Y. B., Telegus, V. S., and Kovalyk, D. A. *Poroshkovaya Metallurgiya* 1969, **5(77)**, 79; *Sov. Powder Metall. Met. Ceram.* 1969, **4**, 403.
134. Pastor, H. in *Boron and Refractory Borides*, Matkovich, V.I. (Ed.), Springer, Berlin, 1977, pp. 457–493.
135. Zdaniewski, W. A. *J. Am. Ceram. Soc.* 1987, **70**, 793.
136. Telle, R., Fendler, E., and Petzow, G. *J. Hard Mater.* 1992, **3**, 211.
137. Makarenko, G. N. in *Boron and Refractory Borides*, Matkovich, V. I. (Ed.), Springer, Berlin, 1977, pp. 310–330.
138. Kuz'ma, Y. B., Telegus, V. S., and Marko, M. A. *Poroshkovaya Metallurgiya* 1972, **4(66)**, 215
139. Koval'chenko, M. S., Ochkas, L. F., and Vinokurov, V. B. *J. Less Common Met.* 1979, **67**, 297.
140. Klimenko, W. N. and Shunkowski, G. L. *Planseeber. Pulvermetall.* 1981, **2**, 405.
141. Okada, S., Satoh, M., and Agata, T. *Nippon Kagaku Kaishi* 1985, 685.
142. Liao, P. K. and Spear, K. E. *Binary Alloy Phase Diagrams Vol. 1*, ASM Internat., Metals Park, OH, 2nd Ed. 1990, 471–474.
143. Telegus, V. S. and Kuz'ma, Y. B. *J. Sov. Powder Metallurgy and Met. Ceram.*, 1968, **2**, 133.
144. Rudy, E. and Windisch, S. *Ternary Phase Equilibria in Transition Metal–Boron–Carbon–Silicon Systems*, Part I, Related Binary Systems, Technical Report No. AFML-TR-65–2, Wright Patterson Air Force Base, OH, 1966.

145. Yasinskaya, G. A. and Groisberg, M. S. *Sov. Powder Metall. Met. Ceram.* 1963, **2**, 457.
146. Kuz'ma, Yu. B., Svarichevskaya, S. I., and Telegus, V. S. *Soviet Powder Metall. Met. Ceram.* 1971, **10**, 478.
147. Kosterova, N. V. and Ordanyan, S. S. *Ivest. Akad. Nauk USSR – Neorg. Mater.* 1977, **13**, 1411.
148. Ariel, E, Barta, J. and Niedzwiedz, S. *J. Less-Common Met.* 1970, **29**, 199.
149. Ahn, D.-G., Kawasaki, A., and Watanabe, R. *Mater. Trans. JIM* 1996, **37**, 1078.
150. Pohl, A., Kitzler, P., Telle, R., and Aldinger, F. *Z. Metallkunde* 1994, **85**, 658.
151. Pohl, A., Telle, R., and Petzow, G. *Z. Metallkunde* 1995, **86**, 148.
152. Mitra, I. and Telle, R. in *Werkstoffwoche 1996: Symposium 7: Materialwissenschaftliche Grundlagen*, Aldinger, F. and Mughrabi, H. (Eds), DGB Informationsgesellschaft, Oberursel, 1996, pp. 521–525.
153. Mitra, I. and Telle, R. *J. Solid State Chem.* 1997, **133**, 25.
154. Schmalzried, C. and Telle, R. in *Proc. Werkstoffwoche 98: Band 7: Keramik/Simulation Keramik*, Heinrich, J., Ziegler, G., Hermel, W., and Riedel H. (Eds), Wiley-VCH, Weinheim, 1999, pp. 377–382.
155. Telle, R. in *Materials Science and Technology*, Cahn, R. W., Haasen, P., and Kramer, E. J. (Eds), Verlag Chemie, Weinheim, 1993, Vol. 11, Ch. 4, pp. 175–266.
156. Joly, A. C. *R. Acad. Sci.* 1883, **97**, 456.
157. Moisson, H., *C. R. Acad. Sci.* 1894, **118**, 556.
158. Ridgeway, R. R. *Trans. Electrochem. Soc.* 1934, **66**, 117; Norton Co. *US Patent 1897214*, 1933.
159. Lipp, A. *Tech. Rundschau* 1965, **57(14)**, 5; **57(28)**, 19; **57(33)**, 5.
160. Lipp, A. *Tech. Rundschau* 1966, **58(7)**, 3.
161. Dufek, G., Wruss, W., Vendl, A., and Kieffer, R. *Planseeber. Pulvermetall.* 1976, **24**, 280.
162. Gray, E. G. *Eur. Patent Appl. 1 152 428*, 1980.
163. Schwetz, K. A. and Lipp, A. in *Ullmanns Encyclopedia of Industrial Chemistry*, VCH, Weinheim, 1985, A4, pp. 295–307.
164. Thevenot, F. *J. Eur. Ceram. Soc.* 1990, **6**, 205.
165. Thevenot, F. in *The Physics and Chemistry of Carbides, Nitrides, and Borides*, Freer, R. (Ed.), NATO ASI Series E, Vol. 185, Kluwer, Dordrecht, The Netherlands, 1990, pp. 87–96.
166. McKinnon, I. M. and Reuben, B. G. *J. Electrochem. Soc.* 1975, **122**, 806.
167. Knudsen, A. K. *Adv. Ceram.* 1987, **21**, 237.
168. Walker, B. E., Rice, R. W., Becher, P. F., Bender, B. A., and Coblenz, W. S. *Ceram. Bull.* 1983, **62**, 916.
169. DeHoff, R. T., Rummel, R. A., LaBuff, H. P., and Rhines, F. N. in *Modern Development in Powder Metallurgy*, Hausner, H. H. (Ed.), Plenum, New York, 1966, pp. 310–330.
170. Greskovich, C. and Rosolowski, J. H. *J. Am. Ceram. Soc.* 1976, **59**, 336.
171. Prochazka, S. *Why Is It Difficult to Sinter Covalent Substances?* General Electric Corp. Res. and Dev. Center Technical Information Series, Report No. 89CRD025, Schenectady, NY, 1989.
172. Matje, P. and Schwetz, K. A. in *Proc. 2nd Int. Conf. on Ceramic Powder Processing Science*, Hausner, H., Messing, G. L., and Hirano, S. (Eds), Deutsche Keramische Gesellschaft, Köln, 1989, pp. 377–384.
173. Shaw, N. *J. Powder Met. Int.* 1989, **21**, 16; 31.
174. Grabchuk, B. L. and Kislyi, P. S. *Sov. Powder Metall. Met. Ceram.* 1976, **15**, 675.
175. Adlassnig, K. *Planseeber. Pulvermetall.* 1958, **6**, 92.
176. Grabchuk, B. L. and Kislyi, P. S. *Poroshkovaya Metallurgiya* 1974, **8**, 11.
177. Katz, J. D., Blake, R. D., and Petrovich, J. *Met. Powder Rep.* 1988, **43**, 835.
178. Dole, S. L. and Prochazka, S. in *Ceramic Engineering and Science Proc. 6(7/8)*, Smothers, W. J. (Ed.), American Ceramic Society, Westerville, OH, 1985, pp. 1151–1160.
179. Kislyi, P. S. and Grabchuk, B. L. in *Proc. 4th Eur. Symp. Powder Metallurgy*, Soc. Française Metall., Grenoble, 1975, p. 10–2–1.
180. Kriegesmann, J. *Ger. Patent DE 37 11 871 C2*, 1989.
181. Lange, R. G., Munir, Z. A., and Holt, J. B. *Mater. Sci. Res.* 1980, **13**, 311.
182. Kanno, Y., Kawase, K., and Nakano, K. *J. Ceram. Soc. Jpn.* 1987, **95**, 1137.
183. Glasson, D. R. and Jones, J. A. *J. Appl. Chem.* 1969, **19(5)**, 125.
184. Janes, S. and Nixdorf, J. *Ber. Deutsche. Keram. Ges.* 1966, **43**, 136.

185. Stibbs, D., Brown, C. G., and Thompson, R. *US Patent 3 749 571*, 1973.
186. Zakhariev, Z. and Radev, D. *J. Mater. Sci. Lett.* 1988, **7**, 695.
187. Prochazka, S. *U.S. Patent* 4005235, 1977.
188. Kriegesmann, J. in *Technische Keramische Werkstoffe*, 4th Suppl., Kriegesmann, J. (Ed.), Deutscher Wirtschaftsdienst, Köln, 1991.
189. Schwetz, K. A. and Vogt, G. *Ger. Patent 2 751 998*, 1977; *US Patent 4 195 066*, 1980.
190. Henney, J. W. and Jones, J. W. S. *Br. Patent Application 2014193A*, 1978.
191. Suzuki, H. and Hase, T. in Proc. Conf. Factors in Densification of Oxide and Nonoxide Ceramics, Somiya, S. and Saito, S. (Eds), Japan, 1979, p. 345.
192. Schwetz, K. A. and Grellner, W. *J. Less-Common Met.* 1981, **82**, 37.
193. Dole, S. L., Prochazka, S., and Doremus, R. H. *J. Am. Ceram. Soc.* 1989, **72**, 958.
194. Bougoin, M., Thevenot, F., Dubois, F., and Fantozzi, G. *J. Less-Common Met.* 1985, **114**, 257.
195. Matje, P., internal report, Elektroschmelzwerk Kempten GmbH, Germany, 1990.
196. Sigl, L. S. and Schwetz, K. A. *Powder Metall. Int.* 1991, **23**, 221.
197. Sigl, L. S. and Schwetz, K. A. *Euro-Ceramics II* 1991, **1**, 517.
198. Grabchuk, B. L. and Kislyi, P. S. *Poroshkovaya Metallurgiya* 1975, **7(151)**, 27. Engl. Transl.: *Sov. Powder Metall. Met. Ceram.* 1975, **14**, 538.
199. Oh, J. H., Orr, K. K., Lee, C. K., Kim, D. K., and Lee, J. K. *J. Korean Ceram. Soc.* 1985, **22**, 60.
200. Weaver, G. Q. *US Patent 4320204*, 1982.
201. Weaver, G. Q. *UK Patent GB 2093481 A*, 1982.
202. Borchert, W. and Kerler, A. R. *Metall* 1975, **29**, 993.
203. Schwetz, K. A., Reinmuth, K., and Lipp, A. *Sprechsaal* 1983, **116**, 1063.
204. Lörcher, R., Strecker, K., Riedel, R., Telle, R., and Petzow, G. in *Solid State Phenomena 8&9, Proc. Int. Conf. Sintering of Multiphase Metal and Ceramic Systems*, Upadhyaya, G. S. (Ed.), Sci-Tech Publications, Vaduz, Gower Publ. Co., Brookfield, VT, 1990, pp. 479–492.
205. Bougoin, M. and Thevenot, F. *J. Mater. Sci.* 1987, **22**, 109.
206. Halverson, D. C., Pyzik, A. J., Aksay, I. A., and Snowden, W. E. *J. Am. Ceram. Soc.* 1989, **72**, 775.
207. Sigl, L. *J. Europ. Ceram. Soc.* 1998, **18(11)**, 1521.
208. Williams, P. D. and Hawn, D. C. *J. Am. Ceram. Soc.* 1991, **74**, 1614.
209. Pyzik, A. J., Aksay, A., and Sarikaya, M. *Mater. Sci. Res.* 1987, **21**, 45.
210. Schwetz, K. A., Sigl, L. S., and Pfau, L. *J. Solid State Chem.* 1997, **133**, 68.
211. Kuzenkova, P. S., Kislyi, P. S., Grabchuk, B. L., and Bodnaruk, N. I. *J. Less-Common Met.* 1979, **67**, 217.
212. Ostapenko, I. T., Slezov, V. V., Tarasov, R. V, Kartsev, N. F., and Podtykan, V. P. *Poroshkovaya Metallurgiya* 1979, **197**, 38. Engl. Transl.: *Sov. Powder Met. Met. Ceram.* 1979, **19**, 312.
213. Brodhag, C., Bouchacourt, M., and Thevenot, F. in *Ceramic Powders*, Material Science Monographs 16, Vincenzini, P. (Ed.), Elsevier, Amsterdam, 1983, pp. 881–890.
214. Beauvy, M. and Angers, R. *Sci. Ceram.* 1980, **10**, 279.
215. Bouchacourt, M., Brodhag, C., and Thevenot, F. *Sci. Ceram.* 1981, **11**, 231.
216. Ekbom, L. B. and Amundin, C. O. *Sci. Ceram.* 1980, **10**, 237.
217. Champagne, B. and Angers, R. *J. Am. Ceram. Soc.* 1979, **62**, 149.
218. Telle, R. and Petzow, G. *Mater. Sci. Eng.* 1988, **A105/106**, 97.
219. Vasilos, T. and Dutta, S. K. *Ceram. Bull.* 1974, **53**, 453.
220. Furukawa, M. and Kitahira, T. *Nippon Tungsten Rev.* 1979, **12**, 55.
221. Larker, H. T., Hermansson, L., and Adlerborn, J. *Ind. Ceram.* 1988, **8**, 17.
222. Schwetz, K. A., Grellner, W., and Lipp, A. in *Science of Hard Materials*, Proc. 2nd Int. Conf. Science of Hard Materials, Almond, E. A., Brookes, C. A., and Warren, R. (Eds), Inst. Phys. Conf. Series, Elsevier, London, 1986, pp. 415–426.
223. Schwetz, K. A., "Boron Carbide, Boron Nitride, and Metal Borides", in *Ullmann's Encyclopedia of Industrial Chemistry*, 6th ed. on CD-ROM, Wiley-VCH, Weinheim, Germany, 1999.
224. Schwetz, K. A., de Groot, K., and Malkemper, W. *J. Less-Common Metals* 1981, **82**, 153.
225. Bouchacourt, M. and Thevenot, F. *J. Mater. Sci.* 1985, **20**, 1237.
226. Werheit, H. *Prog. Crystal Growth and Charact.* 1988, **16**, 179.

227. Wood, C., Emin, D., and Gray, P. E. *Phys. Rev. B* 1985, **31**, 6811.
228. Aselage, T. L. and Emin D. Proc. 10th Int. Symp. Boron, Borides, and Related Compounds, Albuquerque, NM, Boron-Rich Solids, *AIP Conf. Proc.* 1990, **231**, 177.
229. Heuberger, M., Telle, R., and Petzow, G. *Powder Metallurgy* 1992, **35**, 125.
230. Heuberger, M. *thesis*, University of Stuttgart, 1995.
231. Telle, R. in *Boron-Rich Solids*, Proc. 10th Int. Symp. Boron, Borides, and Related Compounds, Emin, D., Aselage, T. L., Switendick, A. C, Morosin, B., and Beekel, C. L. (Eds), AIP Conf. Proc. 231, American Institute of Physics, New York, 1991, pp. 553–560.
232. Telle, R. *Ceramics for High-Tech Applications*, Materials Science, European Concerted Action COST 503 Powder Metallurgy-Powder Based Materials, Vol. V, Valente, T. (Ed.), *The European Communities*, Brussels, 1997.
233. Telle, R. and Petzow, G. in *Proc. 9th Int. Symp. on Boron, Borides, and Related Compounds*, Werheit, H. (Ed.). University Press, Duisburg, 1987, pp. 234–245.
234. Efimenko, L. N., Lifshits, E. V., Ostapenko, I. T., Snezhko, I. A., and Shevyakova, E. P. *Sov. Powder Met. and Met. Ceram.* 1987, **26(3)**, 318.
235. Litz, L. M. and Mercuri, R. A. *J. Electrochemical Soc.* 1963, **110(8)**, 921.
236. Motzfeld, K. *Acta Chem. Scand.* 1964, **18**, 1596.
237. Hofmann, H. and Petzow, G. *J. Less-Common Met.* 1986, **117**, 121.
238. Telle, R. and Petzow, G. in *Horizons of Powder Metallurgy II*, Proc. 1986 Int. Powder Metall. Conf. Exhibit., Kaysser, W. A. and Huppmann, W. J. (Eds), Verlag Schmid, Freiburg, 1986, pp. 1155–1158.
239. Telle, R., Meyer, S., Petzow, G., and Franz, E. D. *Mater Sci. Eng.* 1988, **A105/106**, 125.
240. Kim, D. K. and Kim, C. H. *Adv. Ceram. Mater.* 1988, **3**, 52.
241. Nishiyama, K. and Umekawa, S. *Trans. Jpn. Soc. Ceram. Met.* 1985, **11(2)**, 53.
242. Rice, R. W., Richardson, G. Y., Kunetz, J. M., Schroeter, T., and McDonough, J. in *Proc. 10th Annu. Conf. Composites and Advanced Ceramic Materials*, Messier, D. R. (Ed.), American Ceramic Society, Westerville, OH, 1986, p. 737.
243. Richardson, G. Y., Rice, R. W., McDonough, W. J., Kunetz, J. M., and Schroeter, T. in *Proc. 10th Annu. Conf. Composites and Advanced Ceramic Materials*, Messier, D. R. (Ed.), American Ceramic Society, Westerville, OH, 1986, p. 760.
244. McCauley, J. W., Corbin, N. D., Resetar, T., and Wong, P. in *Proc. 10th Ann. Conf. Composites and Advanced Ceramic Materials*, Messier, D. R. (Ed.), American Ceramic Society, Westerville, OH, 1986, pp. 538–554.
245. Holt, B., Kingman, D. D., and Bianchi, G. M. *Mater. Sci. Eng.* 1985, **71**, 321.
246. McCauley, J. W. *Am. Ceram. Soc. Bull.* 1988, **67**, 1903.
247. Faber, K. T. and Evans, A. G. *Acta Metallurgica* 1983, **31(4)**, 565.
248. Selsing, J. *J. Am. Ceram. Soc.* 1961, **44**, 419.
249. Fendler, E., Babushkin, O., Lindbäck, T., Telle, R., and Petzow, G. in *Proc. Int.Symp. Ceramic Materials and Components for Engines*, Gothenberg, Sweden, June 10–12, 1991.
250. Fendler, E., Babushkin, O., Lindbäck, T., Telle, R., and Petzow, G. *J. Hard Mater.* 1993, **4**, 137.
251. Lönnberg, B. *J. Less-Common Met.* 1987, **141**, 145.
252. Rey, J. and Male, G. in *Proc. 9th Int. Symp. on Boron, Borides, and Related Compounds*, Werheit, H. (Ed.), University Press, Duisburg, 1987, pp. 419–420.
253. Ohya, Y. and Hoffmann, M. J., Petzow, G. *J. Am. Ceram. Soc.* 1992, **75**, 2479.
254. Sigl, L. S. and Schwetz, K. A. *Jap. J. Appl. Physics* 1994, **10**, 224.
255. Sigl, L. S. and Kleebe, H.-J. *J. Am. Ceram. Soc.* 1995, **78(9)**, 2374.
256. Thevenot, F. and Bouchacourt, M. *L'Industrie Ceramique* 1979, **732**, 655.
257. Binder, F. *Radex-Rundschau* 1975, **4**, 531.
258. Fitzer, E. *Arch. Eisenhüttenwesen* 1973, **44(9)**, 703.
259. Coble, R. L. and Hobbs, H. A. in *Investigation of Boride Compounds for Very High Temperature Applications*, NTIS Report AD 428006, Kaufman, L. and Clougherty, E. V. (Eds), Clearinghouse for Federal Scientific and Technical Information, Springfield, VA, 1973, pp. 82–120.
260. Kislyi, P. S. and Zaverukha, O. V. *Poroshkovaya. Metallurgiya* 1970, **7**, 32. Engl. Transl.: *Sov. Powder Metall. Met. Ceram.* 1970, **7**, 549.
261. Kislyi, P. S., Kuzenkova, M. A., and Zaveruha, O. V. *Phys. Sintering* 1972, **4**, 107.

262. Baumgartner, H. R. and Steiger, R. A. *J. Am. Ceram. Soc.* 1984, **67**, 207.
263. Baik, S. and Becher, P. *J. Am. Ceram. Soc.* 1987, **70**, 527.
264. Čech, R., Olivierus, P., and Seijbal, J. *Powder Metall.* 1965, **8(15)**, 142.
265. Rasskazov, N. I. in *Proc. 3rd Int. Powder Metallurgy Conf.* 2, Karlovy Vary, Czechoslovakia, 1970, pp. 228–238.
266. Krylov, Yu. I., Bronnikov, V. A., Krysina, V. G., and Prislavko, V. V. *Sov. Powder Metall. Met. Ceram.* 1976, **15**, 1000.
267. Munir, Z. A. *Am. Ceram. Soc. Bull.* 1988, **67**, 342.
268. Ouabdesselam, M. and Munir, Z. A. *J. Mater. Sci.* 1987, **22**, 1799.
269. Clougherty, E. V. and Pober, R. L. *Nucl. Met.* 1964, **10**, 422.
270. Samsonov, G. V., Goryachev, Y. M., and Kovenskaya, B. A. *Izv. Vuz Fiz.* 1972, **6**, 37.
271. Castaign, J. and Costa, P. in *Boron and Refractory Borides*, Matkovich, V. I. (Ed.), Springer, Berlin, Heidelberg, New York, 1977, pp. 390–412.
272. Lönnberg, B. *J. Less-Common Metals* 1988, **141**, 145.
273. Rahman, M. and Wang, C. C., Chen, W., Akbar, S. A. *J. Am. Ceram. Soc.* 1995, **78**, 1380.
274. Billehaug, K. and Øye, H. A. *Aluminium* 1980, **56**, 642.
275. Sørlie, M. and Øye, H. A. *Cathodes in Aluminium Electrolysis*, 2nd ed., Aluminium-Verlag, Düsseldorf, 1994.
276. Bannister, M. K. and Swain, M. V. *Ceram. Int.* 1989, **15**, 375.
277. Murata, Y., Julien, H. P., and Whitney, E. D. *Ceram. Bull.* 1967, **45**, 643.
278. Venables, J. D. *Philos. Mag.* 1967, **16**, 873.
279. Williams, W. S. *Trans AIME* 1966, **236**, 211.
280. Ramberg, J. R. and Williams, W. S. *J. Mater. Sci.* 1987, **22**, 1815.
281. Kang, E. S. and Kim, C. H. *J. Mater. Sci.* 1990, **25**, 580.
282. Ly Ngoc, D. *Dr. rer. nat Thesis*, University of Stuttgart, 1989.
283. Jimbou, R., Takahashi, K., and Matsushita, Y. *Adv. Ceram. Mater.* 1986, **1**, 341.
284. McMurtry, C. H., Böcker, W. D. G., Seshadri, S. G., Zanghi, J. S., and Garnier, J. E. *Am. Ceram. Soc. Bull.* 1986, **66**, 325.
285. Takahashi, K., Jimbou, R. *Commun. Am. Ceram. Soc.* 1987, C-369.
286. Janney, M. A. *Am. Ceram. Soc. Bull.* 1986, **64(5)**, 357.
287. Janney, M. A. *Am. Ceram. Soc. Bull.* 1987, **66(2)**, 322.
288. Cai, H., Gu, W.-H., and Faber, K. T. in *Proc. Am. Soc. Composites, 5th Techn. Conf. on Composite Materials*, 1990, pp. 892–901.
289. Faber, K. T., Gu, W.-H., Cai, H., Winholtz, R. A., and Magleg, D. J. in *Toughening Mechanisms in Quasi-Brittle Materials*, Shah, S. P. (Ed.), Kluwer, Dordrecht, The Netherlands, 1991, pp. 3–17.
290. de Mestral, F. and Thevenot, F. in *The Physics and Chemistry of Carbides, Nitrides, and Borides*, NATO ASI Series E, Vol. 185, Freer, R. (Ed.), Kluwer, Dordrecht, The Netherlands, 1990, pp. 457–482.
291. Phan-Tan-Luu, R., Mathieu, D., and Feneuille, D. *Méthodologie de la Recherche Expérimentale*, Fascicules de cours, L.P.R.A.I., Université d'Aix-Marseille, 1989.
292. Tanaka, H. and Iyi, N. *J. Am. Ceram. Soc.* 1995, **78**, 1223.
293. Watanabe, T. and Kouno, S. *Ceram. Bull.* 1982, **61**, 970.
294. Watanabe, T. *J. Am. Ceram. Soc.* 1977, **60**, 176.
295. Watanabe, T. *Am. Ceram. Soc. Bull.* 1980, **59**, 465.
296. Petzow, G. and Telle, R. in *Advanced Ceramics*, Somiya, S. (Ed.), Terra Scientific Publ. Co, Tokyo, 1987, pp. 131–144.
297. Watanabe, T. and Shobu, K. *Yogo-Kyokai-Shi* 1988, **96(7)**, 778.
298. Shobu, K. and Watanabe, T. *Yogo-Kyokai-Shi* 1987, **95(1)**, 991.
299. Telle, R., Meyer, S., Petzow, G., and Franz, E. D. *Mater. Sci. Eng.* 1988, **A105/106**, 125.
300. Watanabe, T. and Shobu, K. *J. Am. Ceram. Soc.* 1985, **68**, C34.
301. Shobu, K., Watanabe, T., Drennan, J., Hannink, R. H. J., and Swain, M. V. in *Proc. 4th Int. Conf. Sci. Techn. Zirconia Advanced Ceramics, Zirconia 86*, American Ceramic Society, Westerville, OH, 1987, pp. 1091–1099.
302. Swain, M. *J. Hard Mater.* 1991, **2**, 139.
303. Franz, E.-D., Tumback, M., and Telle, R. *Sprechsaal* 1992, **125**, 55.

304. Tumback, M., Franz, E.-D., and Telle, R. *Sprechsaal* 1992, **125**, 415.
305. Müller, R. *thesis*, University of Stuttgart, 1995.
306. McHale, A. E. and Scott, R. S. *J. Am. Ceram. Soc.* 1986, **69**, 827.
307. Takagi, K., Komai, M., Ando, T in *Sintering '87*, Vol. 2, Proc. 4th Inst. Symp. Science and Technology of Sintering, Somiya, S., Simada, M. (Eds.), Elsevier, Amsterdam, 1988, pp. 1296–1301.
308. McMeeking, R. M. *J. Am. Ceram. Soc.* 1986, **69**, C-301.
309. Ogata, T. *Toray Industries Internal Report*, private communication, 1989.
310. Watanabe, T., Shobu, K., Tani, E., and Nakanishi, T. in *Proc. 11th Int. Symp. Boron, Borides and Related Compounds*, JJAP Series 10, Uno, R. and Higashi, I. (Eds), Publication Office, Jap. J. Applied Physics, Tokyo, 1994, pp. 208–211.
311. Hamjian, H. J. and Lidman, W. G. *J. Am. Ceram. Soc.* 1952, **35**, 44.
312. Halverson, D. C., Pyzik, A. J., and Aksay, I. A. *Ceram. Eng. Sci. Proc.* 1985, **6**, 736.
313. Nishiyama, K. and Umekawa, S. in *Achievements in Composites in Japan and the United States*, Kobayashi, A. (Ed.), Proc. Vth Japan-US Conference on Composite Materials, Tokyo, 1990, pp. 371–378.
314. Hayes, F. H., Lukas, H. L., Effenberg, G., and Petzow, G. *Z. Metallkunde* 1989, **80**, 361.
315. Zupanic, F., Spaic, S., and Krizman, A. *Mater.Sci. Technol.* 1998, **14**, 1203.
316. Funke, V. F., Yudkovskii, S. I., Samsonov, and G. V. Russ. *J. Appl. Chem.* 1961, **34**, 973.
317. Tangermann, I. *Neue Hütte* 1963, **8**, 973.
318. Samsonov, G. V. *Metallov. i. Obrab.* 1958, **12**, 35.
319. Kieffer, R. and Benesovsky, F., *Hartstoffe*, Springer, Vienna, 1963.
320. Golczewski, J. and Aldinger, F., poster presentation at *Annu. Mtg. Germany Society for Materials (DGM)*, Göttingen, 1994.
321. Funke, V. F. and Yudkovskii, S. I. *Zhur. Fiz. Khim.* 1963, **37**, 1557; translation into English: *Russ. J. Phys. Chem.* 1963, **37**, 835.
322. Jüngling, T., Oberacker, R., Thümmler, F., Sigl, L. S., and Schwetz, K. A. *J. Hard Mater.* 1991, **2(3–4)**, 183.
323. Jüngling, T., Oberacker, R., Thümmler, F., and Sigl, L. S., in *Proc. Int. Conf. Advances in Hard Materials Production*, 1992, pp. 15/1–15/14.
324. Pastor, H., Allibert, C. H., Ottavi, M. L., Albajar, M., and Castro, F. *French Patent Application* No. 91 08030, 1991.
325. Pastor, H. et al., in *HMP'92 – Advances in Hard Materials Production*, Proc. 4th Int. Conf., MPR Publishing Services, Bellstone, UK, 1992, pp. 23/1–23–11.
326. Sánchez, J. M., Barandika, M. G., Gil-Sevillano, J., and Castro, F. *Scripta Metall. Mater.* 1992, **26**, 957.
327. Brynestad, J., Bamberger, C. E., Land, J. F., Fand inch, C. B. *J. Am. Ceram. Soc.* 1983, **66**, C215.
328. Ghetta, V., Gayraud, V., and Eustathopoulos, N. presented at *Sintering '91*, Vancouver, August, 1991.
329. Li, J. G. *J. Am. Ceram. Soc.* 1992, **75**, 3118.
330. Li, J. G. *Rare Met.* 1992, **11**, 177.
331. Savov, L., Heller, H.-P., and Janke, D. *Metall.* 1997, **51**, 475.
332. Panasyuk, D. and Umansky, J. *J. Less-Common Met.* 1986, **117**, 336.
333. Schwetz, K. A., Reinmuth, K., and Lipp, A. *Radex Rundschau* 1981, **3**, 568.
334. Janaf, Thermochemical Tables, 2nd ed., NBS, Washington, DC, 1971.
335. Kubaschewski, O. and Alcock, C. B. *Metallurgical Thermochemistry*, Int. Series on Materials Science and Technology, Vol. 24, Pergamon, Oxford, 1979.
336. Matuschka, A. V. *Borieren*, Carl Hanser Verlag, Munich, 1977.
337. Schwetz, K. A., Reinmuth, K., and Lipp, A. *Radex Rundschau* 1981, **3**, 568.
338. Sigl, L. S., Jüngling, T. Proc. 4th Int. Conf. Sci. Hard Materials (ICSHM4), Funchal, Madeira, Nov. 10–15, 1991, p. 321.
339. Ottavi., L., Chaix, J. M., Allibert, C., and Pastor, H. presented at *Sintering '91*, Vancouver, August, 1991.
340. Jüngling, T., *Dr.-Ing. Thesis*, University of Karlsruhe, Germany, 1992, IKM-Series 007.
341. Liao, P. K. and Spear, K. E. in *Binary Alloy Phase Diagrams*, Vol. 1, Massalski, T. B., et al. (Eds.), American Society of Metals, Metals Park, OH, 1986, p. 355.

342. Murray, J. L. in *Binary Alloy Phase Diagrams*, Vol. 1, Mas-salski, T. B., *et al.* (Eds.), American Society of Metals, Metals Park, OH, 1986, p. 1118.
343. Jüngling, T., Sigl, L. S., Oberacker, R., Thümmler, F., and Schwetz, K. A. *Int. J. Refractory Met. Hard Mater.* 1993/94, **12**, 71.
344. Angelini, P., Becher, P. F., Bentley, J., Brynestad, J., Ferber, M. K., Finch, C. B., and Sklad, P. S. in *Science of Hard Materials*, Proc. 2nd Int. Conf. on Science of Hard Materials, Almond, E. A., Brookes, C. A., and Warren, R. (Eds), Inst. Phys. Conf. Ser. 75, Adam Hilger, Bristol, 1986, pp. 1019–1032.
345. Sklad, S. and Yust, C. S. in *Proc. 1st Int. Conf. Science of Hard Materials*, Hagenmüller, R. and Thevenot, F. (Eds), Bordeaux, 1981.
346. Takatsu, S. and Ishimatsu, E. in *Proc. 10th Plansee-Seminar on Powder Metall*, Planseewerke, Reutte, 1981, Vol. 1, p. 535.
347. Finch, C. B., Cavin, O. B., and Becher, P. F. *J. Cryst. Growth* 1984, **67**, 556.
348. Nishiyama, K. and Umekawa, S. in *Proc. Composites '86: Recent Advances in Japan and in the United States*, Proc. 3rd Japan–US Conf. on Composite Materials, Tokyo, Kawata, K., Umekawa, S., Kobayahi, A. (Eds.), 1986, pp. 433–440.
349. Telle, R., Schmalzried, C., and Schunck, B, unpublished report to the Volkswagen-Foundation, 1999.
350. Johnson, W. B., Nagelberg, A. S., and Breval, E. *J. Am. Ceram. Soc.* 1991, **74**, 1093.
351. Newkirk, M. S., Urquhart, A. W., and Zwicker, H. R. *J. Mater. Res.* 1986, **1**, 81.
352. Antolin, S., Nagelberg, A. S. *J. Am. Ceram. Soc.* 1992, **75**, 447.
353. Nagelberg, A. S., Antolin, S., and Urquhart, A. W. *J. Am. Ceram. Soc.* 1992, **75(2)**, 455
354. Weinstein, J. in *Proc. Int. Symp. Advances in Processing and Characterization of Ceramic Metal Matrix Composites, CIM/ICM*; Vol. 17: Mostaghaci, H. (Ed.), Pergamon, Oxford, 1989, p. 132.
355. Lugscheider, E. and Eschnauer, H. in *Proc. 9th Int. Symp. on Boron, Borides, and Related Compounds*, Werheit, H. (Ed.), University Press, Duisburg, 1987, pp. 202–212.
356. Komai, M., Yamasaki, Y., Takagi, K., and Watanabe, T in *Advances in Powder Metallurgy and Particulate Materials 8*, Capus, J. M. and German, R. M. (Eds), Metal Powder Industries Federation, Princeton, NJ, 1992, pp. 81–88.
357. Rieger, W., Nowotny, H., and Benesowsky, F. *Monatsh. Chem.* 1966, **97**, 378.
358. Ozaki, S., Yamasaki, Y., Komai, M., and Takagi, K. in *Proc. 11th Int. Symp. Boron, Borides and Related Compounds*, JJAP Ser. 10, Uno, R., Higashi, I. (Eds.), Publication Office, Jap. J. Applied Physics, Tokyo, 1994, pp. 220–221.
359. Takagi, K. in *Proc. 11th Int. Symp. Boron, Borides and Related Compounds*, JJAP Series 10, Uno, R. and Higashi, I. (Eds), Publication Office, Jap. J. Applied Physics, Tokyo, 1994, pp. 200–203.
360. Sivaraman, K., German, R. M., and Takagi, K. in *Proc. Int. Conf. Powder Metallurgy PM'96*, Toronto, 1996, pp. 9/117–9/129.
361. Sivaraman, K., Griffo, A., German, R. M., and Takagi, K. in *Proc. Int. Conf. Powder Metallurgy PM'96*, Toronto, 1996, pp. 11/67–11/79.
362. Matsuo, S., Ozaki, S., Yamasaki, Y., Komai, M., and Takagi, K. in *Sintering Technology*, German, R. M., Messing, G. L., and Cornwall, R.G. (Eds), Marcel Dekker, New York, 1996, pp. 261–268.
363. Warren, R. *J. Mater. Sci.* 1968, **3**, 471.
364. Schmalzried, C. and Telle, R. in *Proc. Int. Conf. on Boron, Borides and Related Compounds (ISBB '99)*, Dinard, France, October 1999, to be published in *J. Solid State Chem.*
365. Takagi, K., Yamasaki, Y., and Komai, M. *J. Solid State Phys.* 1997, **133**, 243.
366. Sigl, L. S., Schwetz, K. A., and Dworak, U. *Int. J. Refractory and Hard Materials, 1994*, **12**, pp. 95–99.
367. Ramberg, J. R., Wolfe, C. F., and Williams, W. S. *J. Am. Ceram. Soc.* 1985, **68**, C78.
368. Freitag, B. and Mader, W. *J. Microsc.* 1999, **194**, 42.

8 The Hardness of Tungsten Carbide–Cobalt Hardmetal

S. Luyckx

8.1 Introduction

WC–Co hardmetal is a sintered material consisting of brittle tungsten carbide (WC) crystals bonded by a tough cobalt-based binder. Figure 1 shows the microstructure of a typical commercial WC–Co hardmetal.

WC–Co is used mostly on account of its hardness which is high for a metallic material, although other outstanding properties, such an extremely high Young modulus, a high thermal conductivity, and a low coefficient of thermal expansion, contribute to its success in a wide range of technical applications.

This chapter reviews the information available on the dependence of the hardness of WC–Co on microstructural parameters, composition and external conditions, such as temperature. It also reviews information available on the effect of microstructure and composition on the relationships between hardness and toughness and hardness and abrasive wear resistance.

The hardness of industrial hardmetal grades is characterized preferentially by the Rockwell A scale [1] in the USA [2] and by the Vickers scale [3] in Europe. Most research work, however, has been carried out, also in the USA, using the Vickers scale. Most of the results reported below were obtained using standard pyramidal Vickers indenters [3].

Figure 1. Microstructure of a typical WC–6 weight-% Co alloy.

Table 1. The hardness of three crystallographic faces of a WC single crystal [5].

WC crystal plane	Hardness, HV1
(0001)	2100
(1100)	1080
(1101)	1060

8.2 The Hardness of the Two Component Phases

8.2.1 The Hardness of Tungsten Carbide

WC is present in hardmetal in the form of single crystals. These have a hexagonal structure and are highly anisotropic [4]. Table 1 lists the Vickers hardness of three WC crystal faces and shows that the hardness can vary from face to face by up to 100% [5]. On account of the superior hardness of the (0001) faces, attempts have been made to produce WC–Co parts with a (0001) texture [6].

The literature contains contradictory results on the hardness of WC because it has been measured mostly on sections of hardmetal samples, where the grains are randomly oriented, rather than on single crystals.

An additional source of contradictory results is the dependence of hardness on the load applied to the indenter when microhardness is measured (and very low indenting loads must be used when measuring the hardness of WC grains in hardmetal samples, because the grains are typically between 1 and 10 µm in size). Figure 2 [7] is a

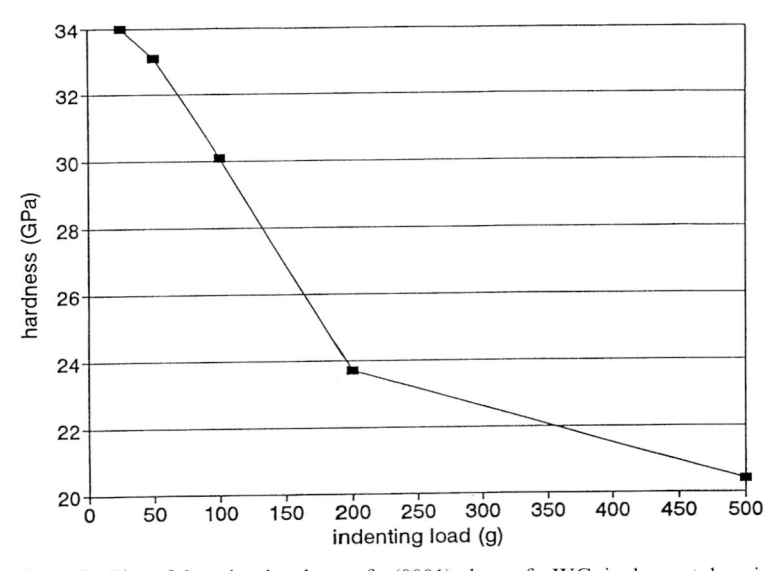

Figure 2. Plot of the microhardness of a (0001) plane of a WC single crystal against the indenting load.

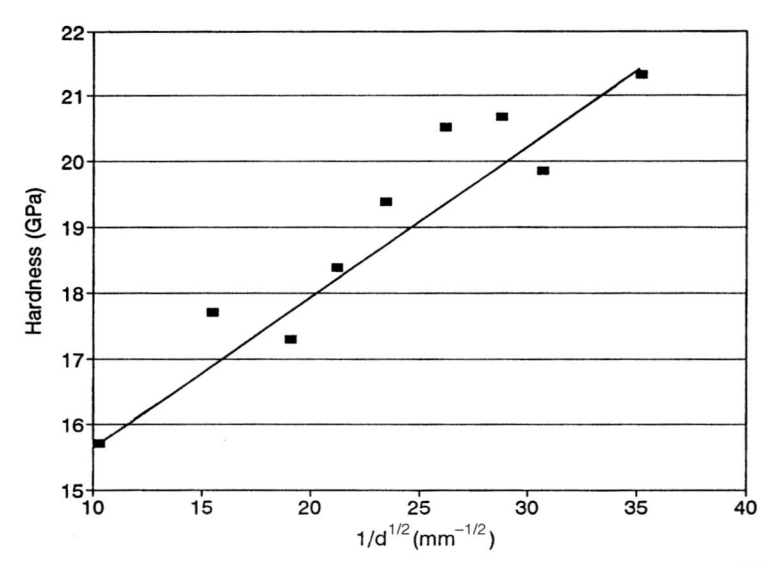

Figure 3. Plot of the hardness of hot-pressed polycrystalline WC against $d^{-1/2}$, where d is the mean size of the WC grains. Adapted from reference [9].

plot of the microhardness of the (0001) planes of WC crystals versus the indenting load. It shows that the microhardness of WC decreases with increasing load, which means that when measuring the microhardness of WC on sections of WC–Co samples one obtains different results at different loads.

The hardness of WC has also been measured on polycrystalline samples [8]. Figure 3 reports results from several investigators [9] which satisfy the following Hall–Petch relationship:

$$H_{WC} = H_{0WC} + K_{0WC} d^{-1/2}, \tag{1}$$

where H_{WC} = hardness of polycrystalline WC, d = mean grain size of the randomly oriented WC grains, H_{0WC} = average hardness of a WC crystal (over all possible crystallographic planes), and K_{0WC} = Hall–Petch coefficient, related to the ease of slip transfer across WC–WC grain boundaries.

From Fig. 3 it has been calculated that $H_{WC} = 1330 \, \text{kgf mm}^{-2}$ and $K_{0WC} = 24 \, \text{kgf mm}^{-3/2}$. It must be noted that the results reported in Fig. 3 were obtained at room temperature, thus the values of H_{0WC} and K_{0WC} are expected to be those given above only at room temperature.

8.2.2 The Hardness of Cobalt

The binder in WC–Co hardmetal is not pure cobalt but a solid solution of carbon and tungsten in cobalt. When cooling the material from the sintering temperature

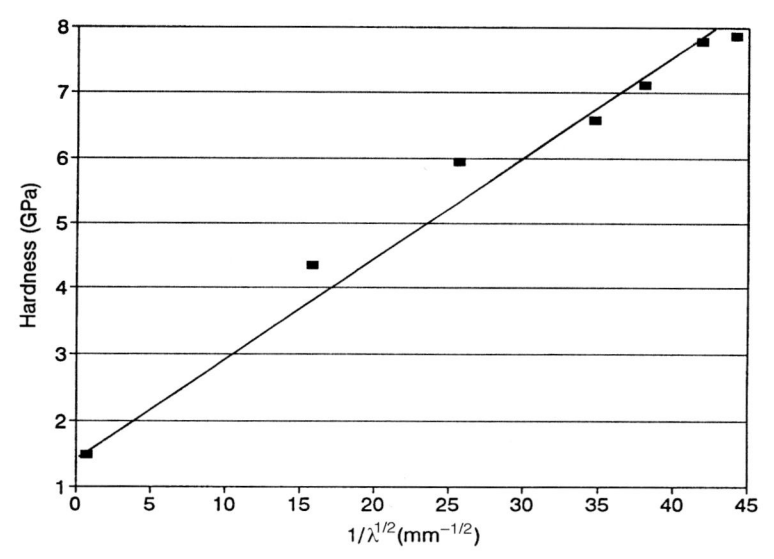

Figure 4. Plot of the hardness of the cobalt binder in WC–Co alloys against $\lambda^{-1/2}$, where λ is the mean width of the binder layers, or binder mean free path. Adapted from reference [9].

to room temperature only part of the cobalt transforms from the high temperature (above 417°C) f.c.c. structure to the low temperature h.c.p. structure, and the transformed regions are present as hexagonal lamellae in a predominantly cubic material [10].

The hardness of pure cobalt has been reported as being between 140 and 240 HV, depending on the method of sample preparation [11]. However, a solid solution of C and W in cobalt can be up to 100% harder than pure cobalt [12], depending on the composition. It has been difficult to measure directly the hardness of the binder in WC–Co because the width of the binder regions in the hardmetal is typically of the order of 0.1 μm.

However, Fig. 4 summarizes measurements by a number of investigators of the *in situ* hardness of the cobalt binder [9]. The hardness is plotted versus $\lambda^{-1/2}$, λ being the thickness of the cobalt layer where the hardness was measured, usually called 'the cobalt mean free path'. The results in Fig. 4 satisfy the following Hall–Petch type relationship

$$H_{Co} = H_{0Co} + K_{0Co}\lambda^{-1/2} \qquad (2)$$

where H_{Co} = the *in situ* hardness of the cobalt layers, λ = thickness of the layers, H_{0Co} = mean hardness of the binder and K_{0Co} = Hall–Petch coefficient.

From Fig. 4 it has been calculated that $H_{0Co} = 149\,\text{kgf}\,\text{mm}^{-2}$ and $K_{0Co} = 16\,\text{kgf}\,\text{mm}^{-3/2}$. K_{0Co} is related to the ease of slip transfer across Co–WC interfaces. The contribution of the Co–Co grain boundaries to the hardness of the material can be neglected because in WC–Co the size of the cobalt grains is of the order of 10^3 μm while the size of the WC grains is of the order of 1 μm. Thus the number of Co–Co boundaries is negligible compared to the number of Co–WC interfaces.

8.3 Factors Affecting the Hardness of WC–Co Hardmetal

8.3.1 Cobalt Content and Tungsten Carbide Grain Size

The two main factors affecting the hardness of WC–Co are the cobalt content and the tungsten carbide grain size. It is well established that the hardness of WC–Co decreases with increasing cobalt content at all carbide grain sizes and decreases with increasing grain size at all cobalt contents [13]. Figures 5 and 6 confirm this trend for the most comprehensive range of cobalt contents tested to date [14, 15].

As a result, by selecting appropriate combinations of cobalt contents and carbide grain sizes, the hardness of WC–Co can be varied from below 800 HV to more than 2000 HV. The most appropriate combination is determined by the properties which, besides hardness, are required for a specific application. The recent introduction of grain sizes of the order of 10^2 nm has extended the range of possible hardness well above 2000 HV [16]. However, nano-grade hardmetals are not included in the present review because they always contain grain refiners such as VC or Cr_2C_3, while this review is limited to two-phase WC–Co alloys.

Lee and Gurland [8] attempted to express the relationship that exists between the hardness of WC–Co, its microstructure and its composition by using a simple law of mixtures of the type

$$H_{\text{WC-Co}} = H_{\text{WC}} V_{\text{WC}} + H_{\text{Co}}(1 - V_{\text{WC}}) \tag{3}$$

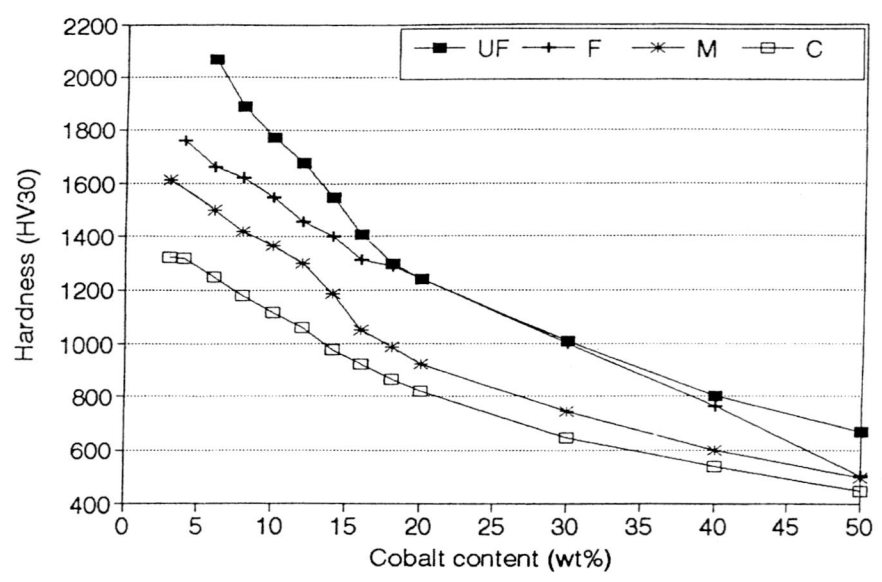

Figure 5. Plot of the hardness of WC–Co alloys against their cobalt content at four different carbide grain sizes. The grain sizes are indicated in the key where: UF = 0.7 μm; F = 1.0 μm; M = 3.0 μm; C = 5.0 μm. Reproduced by permission of the *Int. J. Refract. Met. & Hard Mater*.

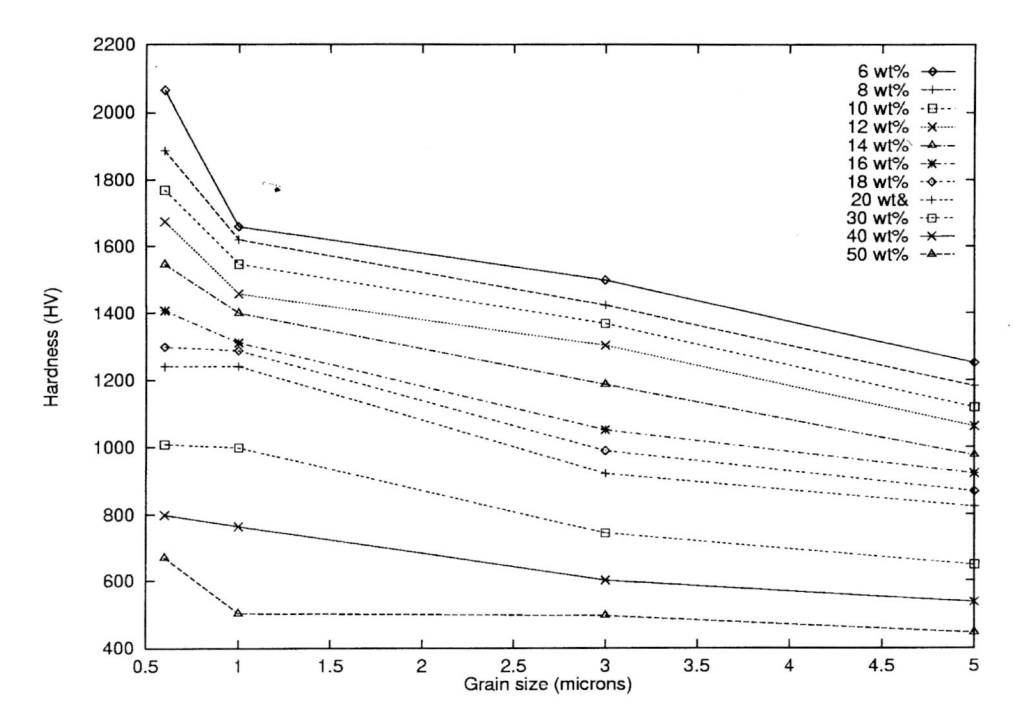

Figure 6. Plot of the hardness of WC–Co alloys against carbide grain size at the cobalt contents indicated in the key. Reproduced by permission of L. Makhele.

where $H_{\text{WC-Co}}$ = the hardness of the alloy, H_{WC} and H_{Co} = the hardness of the component phases as expressed by Eqs (1) and (2), and V_{WC} = volume fraction of the WC phase.

However, Lee and Gurland found that this simple law does not agree with the measured hardness values of WC–Co alloys. In order to reproduce the measured values, they had to introduce the concept of 'continuous carbide volume' or 'volume fraction of the carbide skeleton', V_c, which is defined as the volume fraction occupied by that part of the WC grains which are involved in forming infinitely long chains of connected particles while those parts which are surrounded by the cobalt binder do not contribute to V_c. Lee and Gurland derived that V_c is related to the volume fraction of the carbide phase, V_{WC}, via the contiguity, C:

$$V_c = CV_{\text{WC}} \tag{4}$$

where the contiguity, C, is defined as the ratio of the grain boundary area and the total surface area of the carbide phase.

By replacing V_{WC} in Eq. (3) with V_c, Lee and Gurland obtained the following equation:

$$H_{\text{WC-Co}} = CV_{\text{WC}}H_{\text{WC}} + (1 - CV_{\text{WC}})H_{\text{Co}} \tag{5}$$

which is in excellent agreement with experimental results (see Fig. 7).

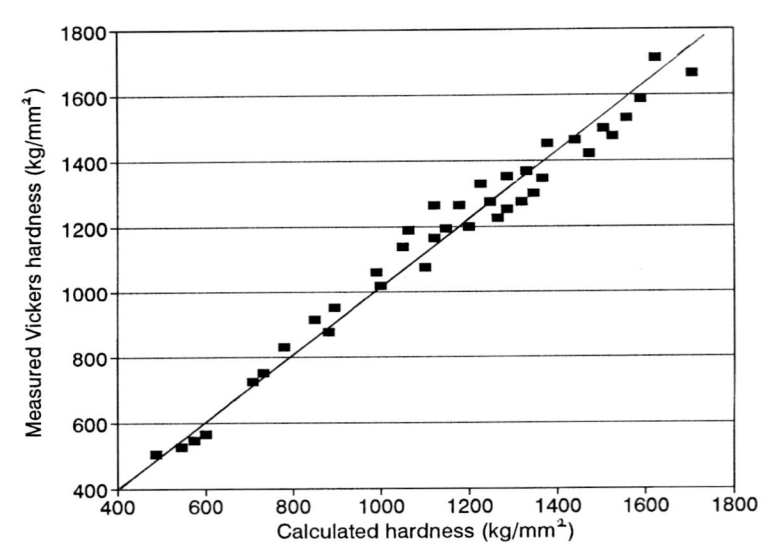

Figure 7. Plot of the hardness of WC–Co alloys, calculated from Eq. (5), against measured hardness. Adapted from reference [8].

By using Eq. (5) it is possible to calculate that in commercial WC–Co grades (i.e. grades with cobalt content ranging from about 6 to about 20 weight-%) the first term of the equation accounts for up to 80% of H_{WC-Co}, which means that the continuous carbide volume accounts for up to 80% of the hardness of the alloy.

8.3.2 Grain Size Distribution and Cobalt Mean Free Path

A wide carbide grain size distribution obviously leads to variations in the microhardness of the hardmetal, but it has not been established what effect it has on the macrohardness. However, if variations in grain size distribution lead to variations in the binder mean free path the effect on hardness can be substantial, as shown in Fig. 8 [17]. Figure 8 shows that at equal mean free path (and so equal binder hardness, according to Eq. (2)), finer grades have lower hardness than coarser grades.

8.3.3 Binder Composition and Carbon Content

Although the hardness of hardmetal is determined mostly by the carbide phase (see Section 8.3.1), the hardness of the binder does contribute to the overall hardness of the material. Therefore the composition of the binder, and specifically the amount of W dissolved in the binder, affects the overall hardness.

The composition of the binder depends on the total carbon content of the material since, according to Exner [18], there seems to exist a reciprocal relationship between

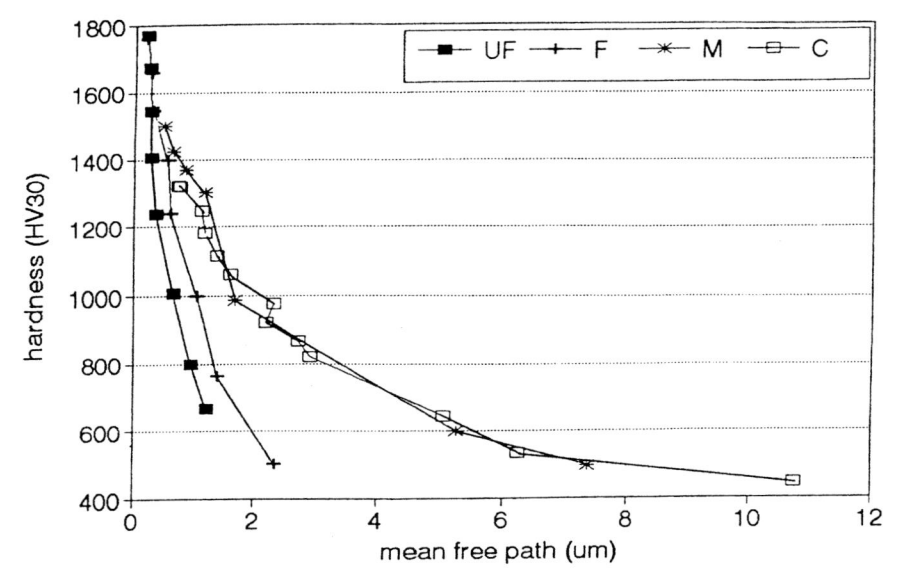

Figure 8. Plot of the hardness of WC–Co alloys of four different grain sizes against the binder mean free path. UF = 0.7 µm; F = 1.0 µm; M = 3.0 µm; C = 5.0 µm. Reproduced by permission of the *Int. J. of Refract. Met. & Hard Mater.*

the dissolved W and the dissolved C. The higher the total carbon, the lower is the amount of W dissolved in the binder and the lower is the hardness of the binder (Fig. 9) [19].

8.3.4 Porosity

The hardness of materials produced via powder metallurgy is known to decrease with increasing porosity level. However, hardmetals are almost pore-free materials and variations in their extremely low level of porosity does not affect their hardness appreciably.

8.3.5 Effect of Temperature

In most applications WC–Co is subjected to temperatures which can be as high as 900–1000°C (as in metal cutting). Therefore knowledge of the effect of temperature on hardness is essential for a correct selection of the grades suitable for a specific application.

The hardness of both WC and cobalt decreases with increasing temperature. Figure 10 shows the rate of hardness decrease with temperature for hot pressed polycrystalline WC and for WC single crystals [20] and Fig. 11 shows the rate of hardness decrease in cobalt [11].

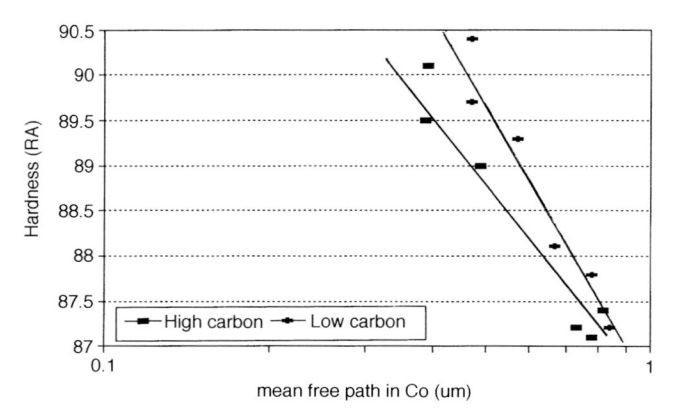

Figure 9. Plot of the hardness of WC–Co alloys of different carbon contents against the binder mean free path. Adapted from reference [19].

The most comprehensive work on the effect of temperature on the hardness of WC–Co has been carried out by Milman and co-workers [21] and has involved WC–Co grades ranging in grain size from 0.5 to 2.3 μm and in cobalt content from 6 to 15 weight-%. These grades have been tested at temperatures ranging from 20 to 900°C and the results are summarized in Figs 12 and 13. Figure 14 gives the same results in Hall–Petch coordinates, and shows that at all temperatures and cobalt contents a linear relationship exists between the hardness of WC–Co, H_{WC-Co}, and $d^{-1/2}$, d being the mean grain size of the WC grains.

The linear relationships shown in Fig. 14 can be explained by combining Eqs (1), (2), and (5) and by expressing λ, the mean free path in cobalt, in terms of d using the

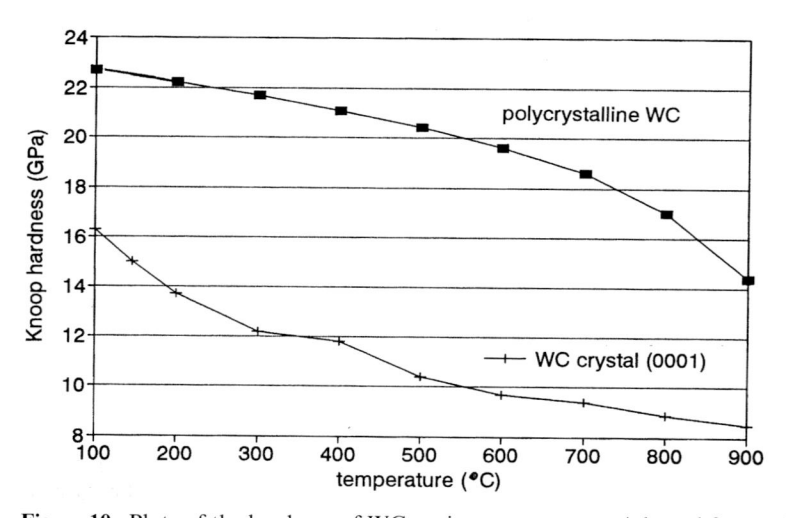

Figure 10. Plots of the hardness of WC against temperature. Adapted from reference [20].

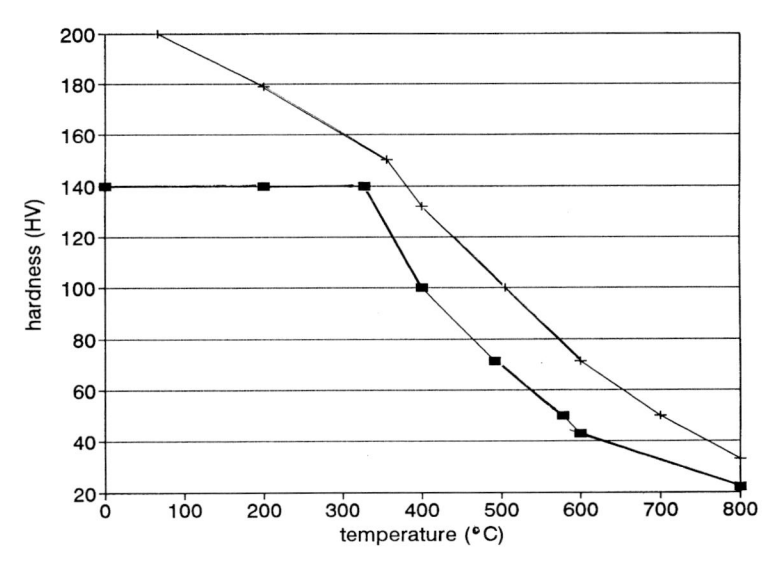

Figure 11. Band including possible values of the hardness of cobalt at various temperatures. Adapted from reference [11].

following equation due to Lee and Gurland [8]:

$$\lambda = \frac{1 - V_{WC}}{V_{WC}(1 - C)} d. \tag{6}$$

By combining the above equations one obtains

$$H_{WC-Co} = H_0 + K_y d^{-1/2} \tag{7}$$

where

$$H_0 = H_{0WC} C V_{WC} + H_{0Co}(1 - C V_{WC}) \tag{8}$$

and

$$K_y = [H_{0WC} C V_{WC} + K_{0Co}(1 - C V_{WC}) B^{-1/2}] d^{-1/2} \tag{9}$$

with

$$B = \frac{1 - V_{WC}}{V_{WC}(1 - C)}. \tag{10}$$

H_0 and K_y can be considered constant at constant carbide volume fraction, V_{WC}, since the contiguity, C, varies with cobalt content but does not vary appreciably with grain size [8]. Hence the relationship between H_{WC-Co} and $d^{-1/2}$ in Eq. (7) is

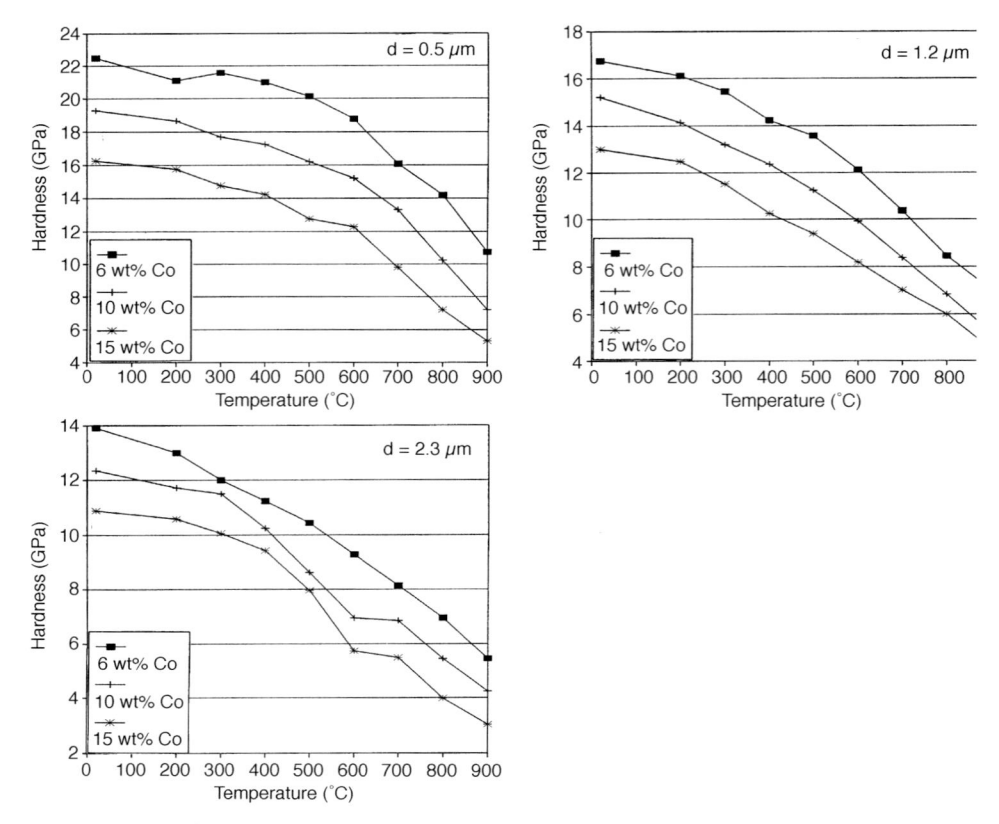

Figure 12. Plot of the hardness of WC–Co alloys against temperature. The cobalt content of the alloys and the grain sizes are indicated.

linear at any V_{WC}, in agreement with the results in Fig. 14. Figure 14 shows that the above relationship is linear also at all temperatures, which implies that the relationships between H_{WC} and d and H_{Co} and λ (see Eqs (1) and (2)) are linear at all temperatures.

Figures 15 and 16 show that H_0 in Eq. (7) decreases rapidly with increasing temperature up to about 600°C, while K_y decreases substantially with temperature only above 600°C. Since H_0 is a combination of the intrinsic hardness of WC and Co (see Eq. (8)) and K_y a combination of the Hall–Petch coefficients of Eqs (1) and (2) (see Eq. (9)), Figs 15 and 16 suggest that the softening of WC–Co with increasing temperature is due to the intrinsic softening of the component phases up to about 600°C but is controlled by the ease of slip transfer across grain boundaries and interfaces above that temperature.

Milman and co-workers' results [21] have established that the hardness of finer grained material decreases with increasing temperature at a lower rate than the hardness of coarser grained material and thus the advantage of using finer grained material increases with increasing temperature.

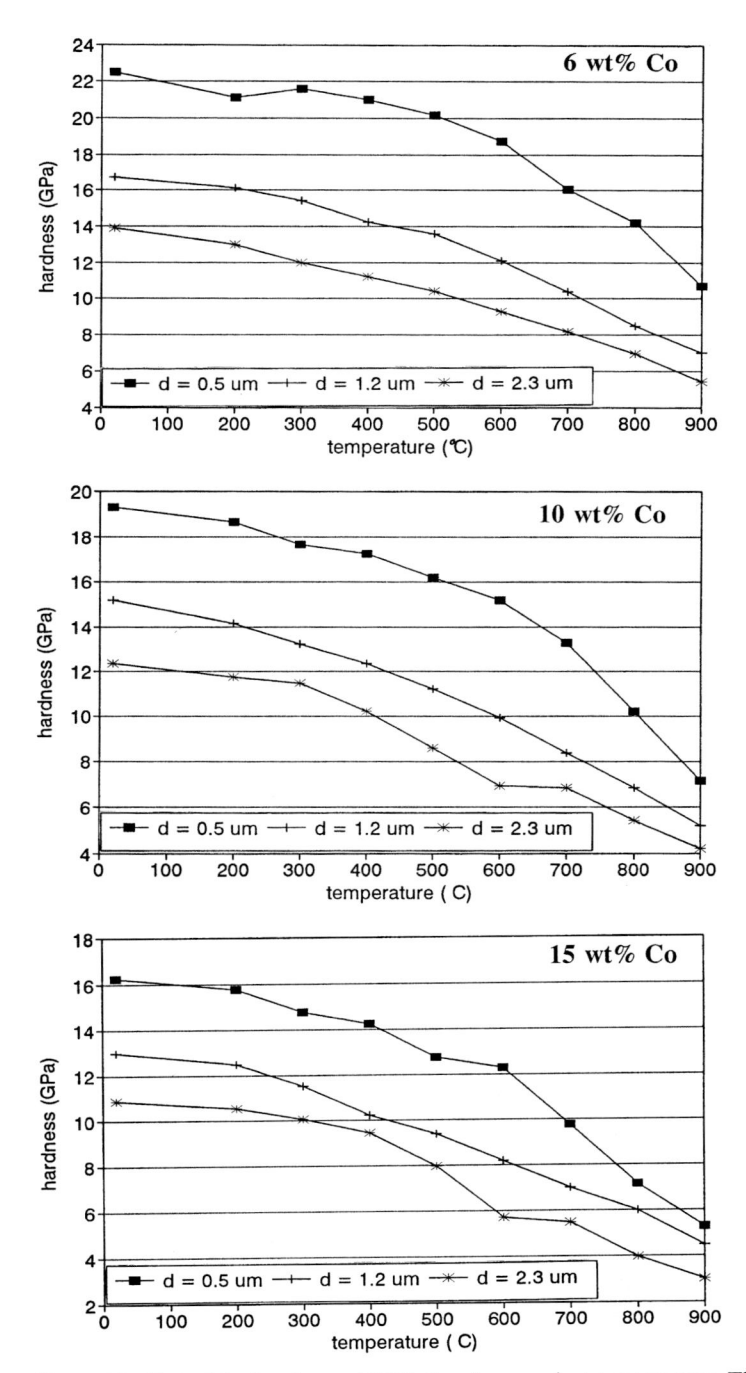

Figure 13. Plot of the hardness of WC–Co alloys against temperature. The grain size of the alloys and the cobalt contents are indicated.

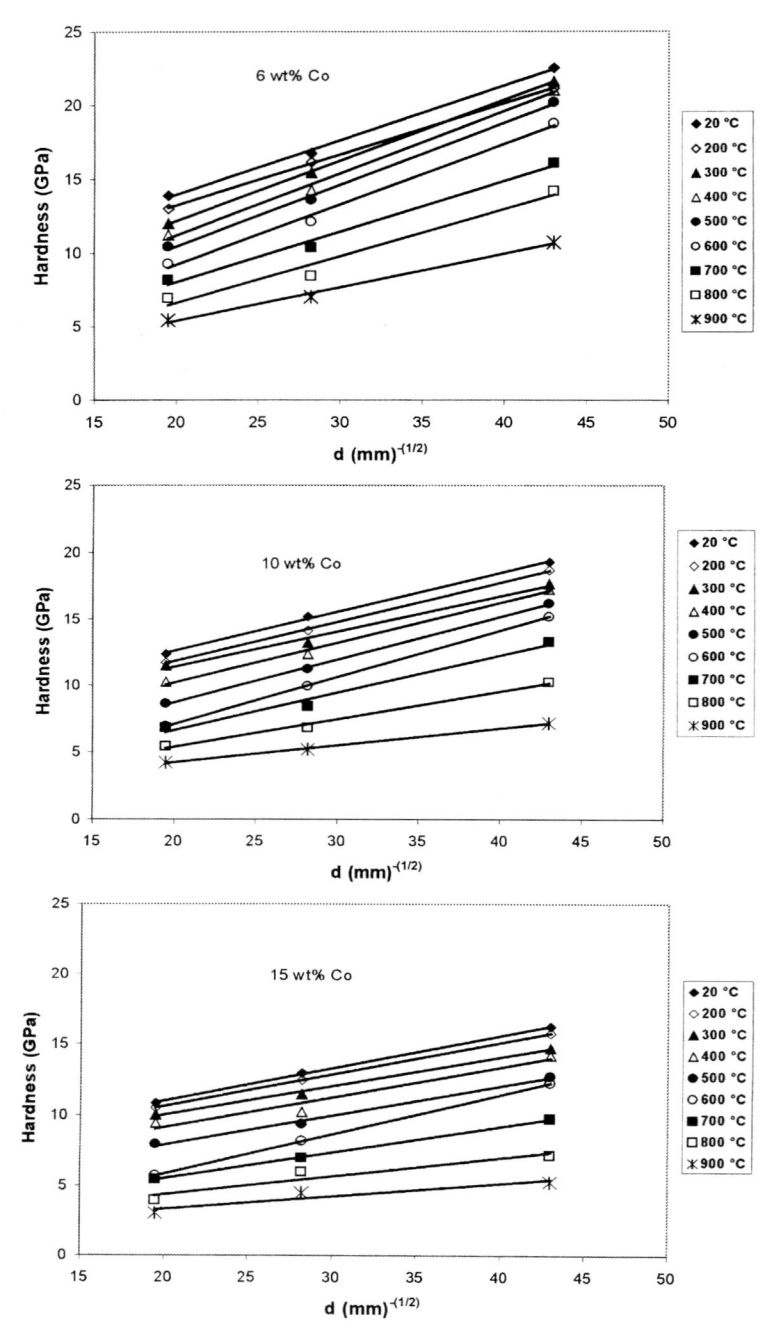

Figure 14. Plot of the hardness of WC–Co alloys against $d^{-1/2}$, where d is the carbide mean grain size, at temperatures in the range 20–900°C. The regression coefficients are all higher than 0.95 and mostly higher than 0.99.

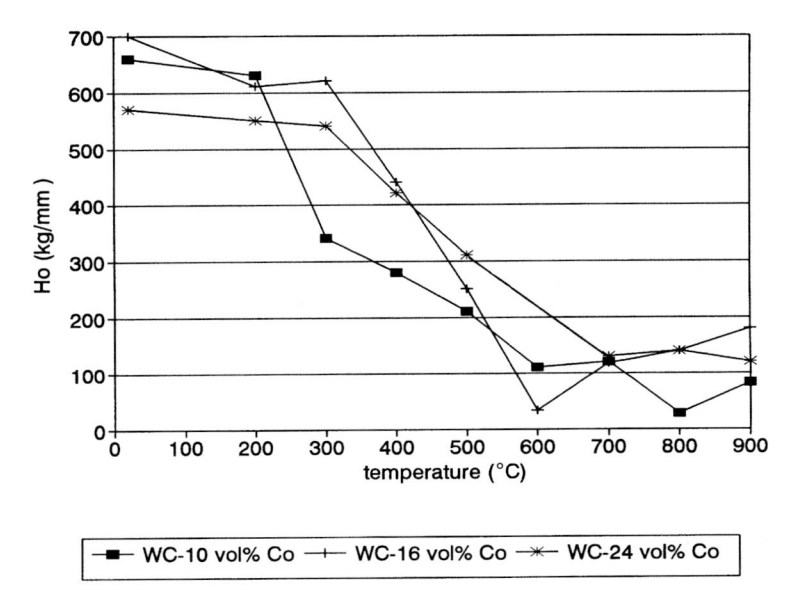

Figure 15. Plot of the parameter H_0 of Eq. (8) against temperature for WC–Co alloys of different cobalt content. The cobalt contents are indicated in the key. Reproduced by permission of the *Int. J. of Refract. Met. & Hard Mater.*

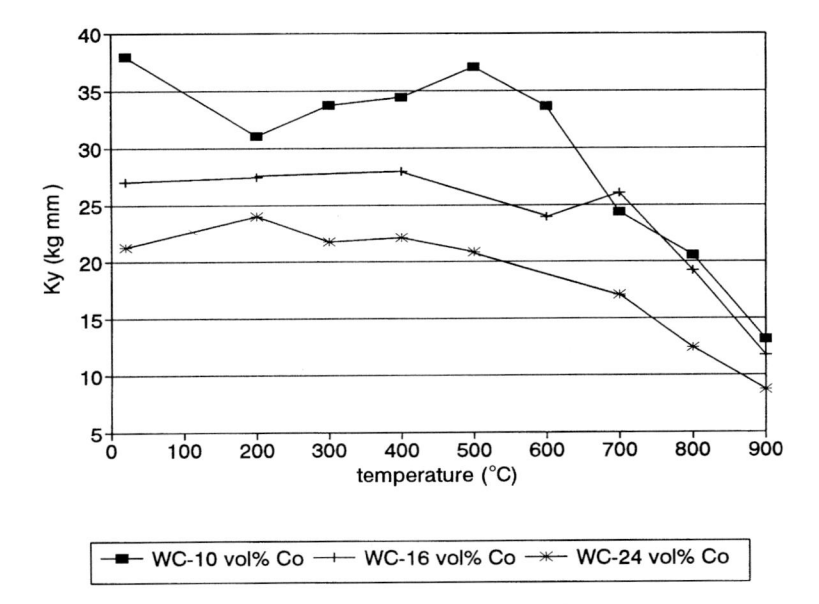

Figure 16. Plot of the parameter K_y of Eq. (8) against temperature for WC–Co alloys of different cobalt contents. The cobalt contents are indicated in the key. Reproduced by permission of the *Int. J. of Refract. Met. & Hard Mater.*

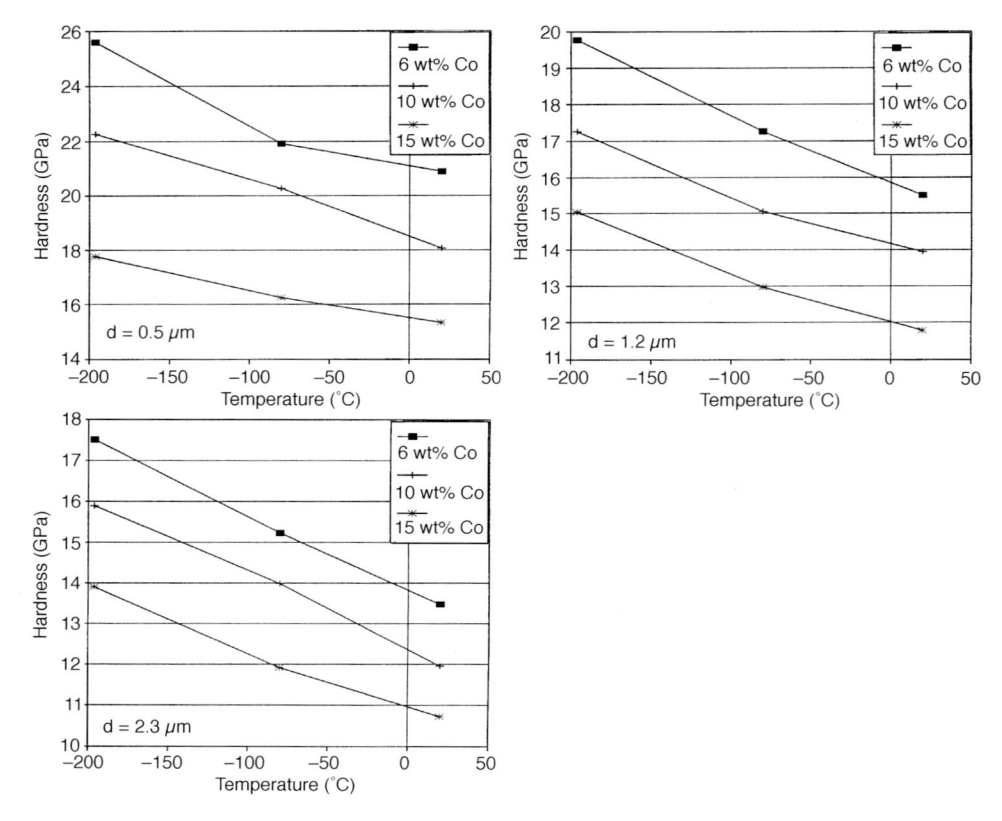

Figure 17. Plots of the hardness of WC–Co alloys of different cobalt contents against low temperatures. The grain size of the alloys is indicated in the keys.

Milman and co-workers tested WC–Co alloys of grain size ranging from 0.5 to 2.3 µm and cobalt content from 6 to 15 weight-% also at temperatures ranging from $-196°C$ to $20°C$. The results are shown in Figs 17 and 18 [21].

8.4 Relationships between Hardness and Other Hardmetal Properties

In most hardmetal applications it would be desirable to use the hardest possible grade. However, properties often have reciprocal relationships, i.e. an improvement in one leads to a deterioration in another. Therefore, in order to select the most appropriate grades for a specific application, it is desirable to know quantitatively the relationships between the various properties. So far, extensive work has been

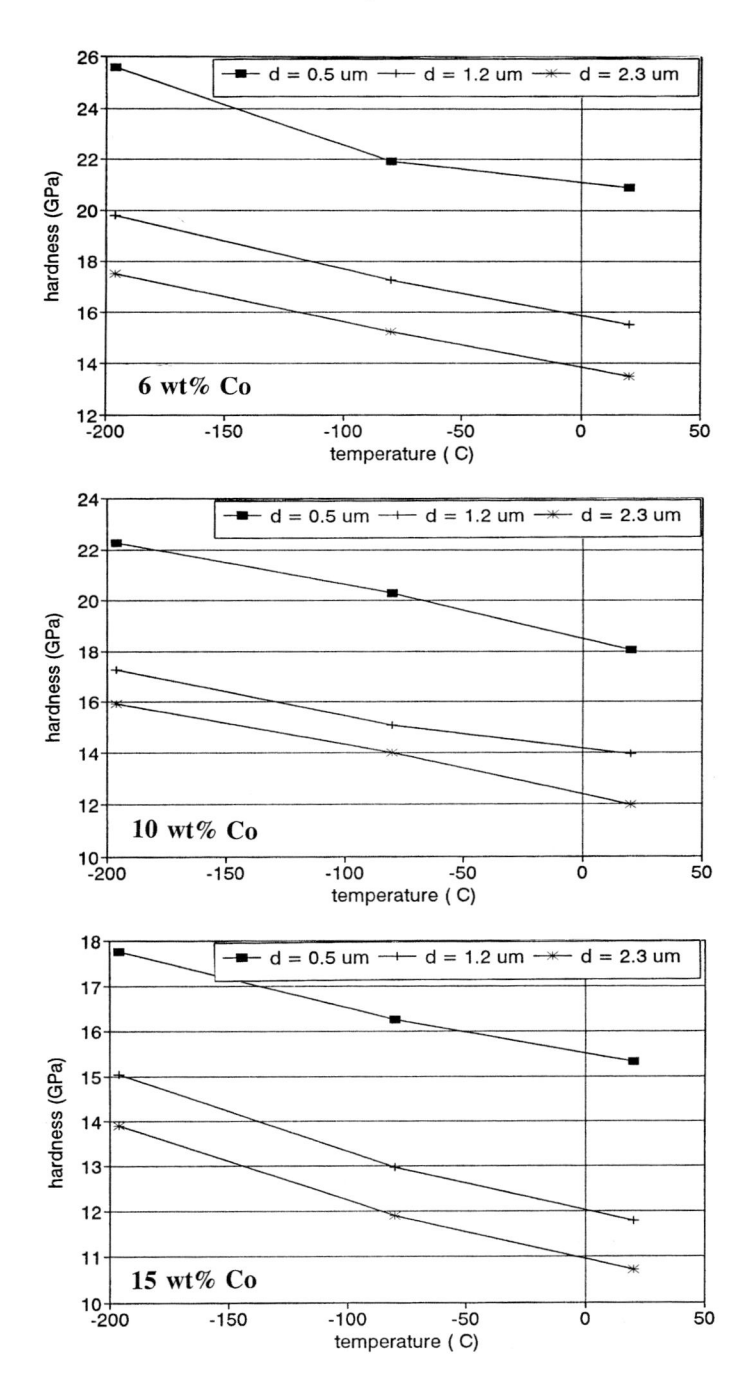

Figure 18. Plots of the hardness of WC–Co alloys of different grain size against temperature. The grain sizes and the cobalt content are indicated.

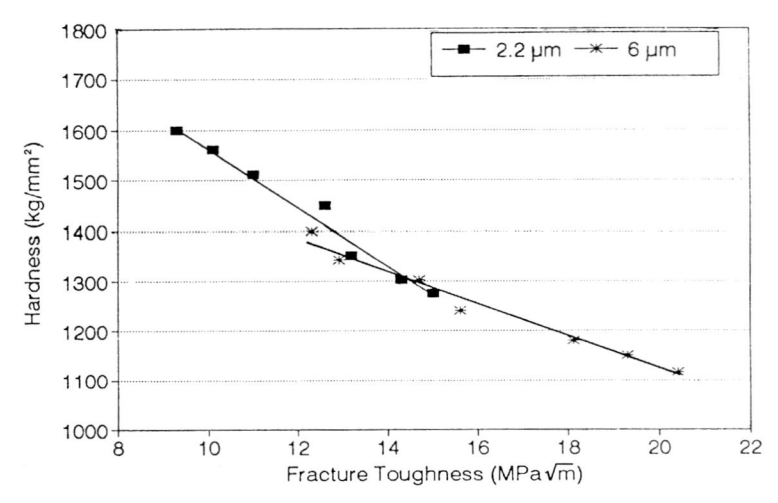

Figure 19. Plot of the hardness of WC–Co alloys of constant grain size against toughness. The grain sizes of the alloys are 2.2 μm and 6 μm. Adapted from reference [22].

done to establish the relationships that exist between hardness and toughness and hardness and abrasive wear resistance.

8.4.1 Relationship between Hardness and Toughness

It is well established that a reciprocal relationship exists between the hardness and the toughness of WC–Co. However, it has proved difficult to explain this relationship in terms of the material's microstructural parameters until it was observed that the relationship between the two properties is linear when the grain size is kept constant [22]. Figure 19 shows the particular case of alloys having grain size 2.2 and 6 μm, but the linearity of the relationship has been confirmed for a wide range of grain sizes [23].

Therefore it appears that, in general:

$$H_{\text{WC-Co}} = a(d)K_{\text{IC}} + b(d) \qquad (11)$$

where K_{IC} = toughness of WC–Co, and $a(d)$ and $b(d)$ = functions of the WC grain size.

8.4.2 Relationship between Hardness and Abrasive Wear Resistance

Figure 20 shows the results of a comprehensive investigation into the relationship between the hardness and the abrasive wear resistance of WC–Co, with the wear resistance being measured according to the ASTM Standard B611-85 for hardmetal,

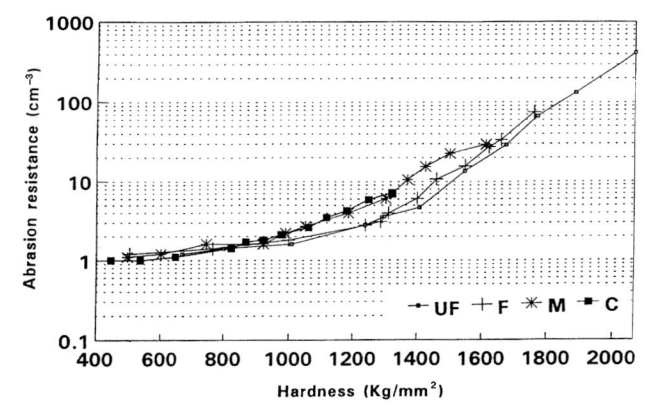

Figure 20. Plot of hardness against abrasive wear resistance for a wide range of WC–Co alloys. UF = 0.7 µm; F = 1.0 µm; M = 3.0 µm; C = 5.0 µm. Reproduced by permission of *Int. J. Refract. Met. & Hard Mater.*

using alumina particles as abradors. The alloys tested ranged in cobalt content from 3 to 50 weight-% and in grain size from 0.6 to 5.1 µm [17]. Figure 20 shows that up to a hardness of approximately 1000 HV there was a one-to-one correspondence between hardness and abrasion resistance, which suggests that the main wear mechanism was plastic deformation. Above 1000 HV the wear resistance of coarser grained alloys was higher than that of finer grained alloys of equal hardness, up to approximately 1600 HV. This suggests a wear process controlled by microfracture, since in that range of hardness coarser alloys are tougher than finer alloys. Therefore, the abrasive wear resistance of hardmetal increases with increasing hardness but, in the hardness range which is of interest to most applications, there is not a one-to-one correspondence between hardness and abrasive wear resistance since resistance to wear is determined also by other properties, such as toughness.

8.5 Conclusions

The knowledge of the factors affecting the hardness of WC–Co alloys has increased considerably in recent years, which has allowed a more precise selection of grades. However, much work remains to be done to quantify the relationships that exist between hardness and other properties which affect the performance of the material.

Acknowledgments

The author wishes to acknowledge the contribution of several colleagues to the research work carried out over the years and reported here. Professor Yu. V. Milman,

Mr D. G. F. O'Quigley and Mr I. T. Northrop must be mentioned in particular. The financial support of the Foundation for Research Development, THRIP and the Boart Longyear Research Centre, over the years, is also gratefully acknowledged.

References

1. ASTM Standard B 294-92.
2. J. J. Oakes, Teledyne Advanced Materials, Private Communication.
3. ISO Standard 3878.
4. S. Luyckx, F. R. N. Nabarro, Siu Wah Wai, and M. N. James, *Acta Metall. Mater.* 1992, **40**, 1623.
5. T. Takahashi and E. J. Freise, *Philos. Mag.* 1965, **12**, 1.
6. S. Luyckx and J. Katzourakis, *Mater. Sci. Technol.* 1991, **7**, 472.
7. S. Luyckx and L. C. Demanet, Unpublished Results.
8. H. C. Lee and J. Gurland, *Mater. Sci. Eng.* 1978, **33**, 125.
9. L. S. Sigl and H. E. Exner, *Mater. Sci. Eng.* 1989, **A108**, 121.
10. J. Freytag, PhD Thesis, University of Stuttgart, 1977.
11. W. Batteridge, Cobalt and its Alloys, Chichester: Ellis Horwood, 1982.
12. B. Roebuck and E. A. Almond, *Mater. Sci. Eng.* 1984, **66**, 179.
13. H. E. Exner and J. Gurland, *Powder Metall.* 1970, **13**, 13.
14. D. G. F. O'Quigley, MSc Thesis, University of the Witwatersrand, Johannesburg, South Africa, 1995.
15. L. Makhele, University of the Witwatersrand, Unpublished results.
16. D. F. Carroll, *14th International Plansee Seminar*, Plansee AG, Reutte, Austria, 1997, **2**, 168.
17. D. G. F. O'Quigley, S. Luyckx, and M. N. James, *Int. J. Refract. Met. Hard Mater.* 1997, **15**, 73.
18. H. E. Exner, *Int. Mater. Rev.* 1979, **4**, 149.
19. T. Sadahiro, *J. Jpn Soc. Powd. Powd. Met.* 1979, **26**, 33.
20. M. Lee, *Metall. Trans. A*, 1983, **14A**, 1625.
21. Yu. V. Milman, S. Luyckx, and I. T. Northrop, *Int. J. Refract. Met. Hard Mater.* in press.
22. D. G. F. O'Quigley, S. Luyckx, and M. N. James, *Mater. Sci. Eng.* 1996, **A209**, 228.
23. S. Luyckx, V. Richter, D. G. F. O'Quigley, and L. Makhele, *Proc. Int. Conf. Deformation and Fracture in Structural Materials*, Institute of Materials Research, Slovak Academy of Science, Kosice, Slovakia, 1996, **2**, 109.

9 Data Collection of Properties of Hard Materials*

G. Berg, C. Friedrich, E. Broszeit, and C. Berger

9.1 Introduction

Hard materials are used as thin hard coatings of some microns thickness for wear protection of tools and machine parts because of their high abrasive wear resistance. For the selection of the coating material the physical, mechanical, and technological properites of these coatings, required by the application, are decisive. The following data collection presents fundamental and available material properties for approximately 130 hard materials as a result of a literature search on carbides, nitrides, borides, silicides, and oxides.

9.2 Profile of Properties

Each technical application demands a special profile of properties concerning the material. In tribological applications the material selection of the system components plays a dominant role. The existing profile of material properties must, if possible, go beyond the demanded characteristics. For a functional selection of materials basic data must be available.

A 'profile' does not exist if only one property is known, it is built up from all relevant material characteristics, which could be called 'mechanical-technological', 'mechanical-tribological' or 'physical-chemical' for mechanical components. Only the combination and comparison of numerous characteristic data of one material make it possible to sketch a profile of properties (columns in the following table). First if there is a full characteristic profile for this material available, it makes sense to compare it with other materials in complex applications to recognize all positives and negatives (rows of the table).

Investigations of thin films show that the material properties are not constant. Responsible for this are different deposition technologies and procedures which result in different structure properties of the coatings, having various textures or various densities of defects. For this reason the evaluation of material properites by different references is of special interest (several data for one characterisic value).

---

* This contribution is the revised and translated version of the publication Datensammlung zu Hartstoffeigenschaften, published in Materialwissenschaft und Werkstofftechnik 1997, **28**, 57–76, © VCH Verlagsgesellschaft mbH, D-69451 Weinheim

The presented data are mainly of a physical-mechanical character. For applications technological/tribological data (i.e. load bearing capacity of hard coatings in different tribological systems under corrosive attack or under high temperature applications) are of special interest. For these uses, in practice physical data may only give first qualitative hints for decisions concerning a successful selection.

9.3 Organization and Contents of the Data Collection

The following data collection provides an overview of basic properties of hard materials found in the literature. The list is structured according to the chemical composition of the materials. To define 'hard materials' a minimum hardness of $HV = 1000$ is demanded (definition of Schedler [57]).

Values of bulk materials and coatings are listed together. A differentiation would be useful but is very difficult to achieve because basic information on design and structure of the layers is not available. For this reason multilayer and multiphase coatings like (Ti, Al)N or Ti(C, N) have not been taken into account.

Physical values like crystal structure and the Young modulus are cited, which may play a dominant role in mechanical/tribological loadings. Additionally values for microhardness and oxidation resistance are given. More than others these two values are a function of the test system, the test parameters, and the evaluation method. At this point very strong deviations appear (e.g. hardness of SiC: $HV = 1400 - 6000$). In many cases the boundary conditions of the investigations are not documented (e.g. no load specification for hardness testing). For this reason values for hardness and oxidation resistance have a qualitative character more than others.

This points out the necessity of further standardization efforts, to give the results of the system tests a transferrable and absolute character. In the field of thin films, the standardization is in progress, including special test procedures, but they are not yet established and for this reason not used and documented. Giving a guarantee of properties is only possible if firm regulations exist. For series investigations the realization of this demand is absolutely necessary and already realized for other test methods (e.g. macro hardness testing according to different DIN or ISO standards).

To reduce the size of the table, the given data are as far as possible compressed. One line in the data collection includes data from different authors. Mostly in one reference only single properties of the hard materials are given, so that only the collection of these properties leads to a profile of characteristic properties for a single material. The references are quoted in brackets.

The influence of the deposition procedure and exactness of the test methods on the profile of properties of hard coatings leads to very different values presented by the different authors. Especially the properties depending on crystal structure like thermal conductivity, electrical resistivity or the Young's modulus (see e.g. electrical conductivity for NbC or the Young modulus for TiN) differ very much.

Good correspondence is found for crystal structure, lattice parameters, density and melting point for one material.

The data collection (Table 1) does not claim to present all existing investigations on hard materials. The aim of the presented table is to give a representative overview on a special material. The number of quotaions gives a first impression on the number of published values (see e.g. TiN which is well known as layer for tribological applications). Looking on the five material groups, carbides and nitrides are dominant.

The database for the data collection comprises actual textbooks, theses, and publications in technical journals and from conferences in recent years. All in all approximately 3000 references were searched through. One result is that in new publications single phase materials are seldom studied. The quotation of older literature in these publications leads to the repetition of single values by different authors.

Acknowledgement

Many thanks to Mrs. R. Kurth for listing all the data on PC and thanks to Mr. U. Petzel as well as Mr. T. Eid for their help in increasing the data base.

Table 1. Data collection of properties of hard materials

Legend:

materials: carbides, nitrides, borides, silicides, oxides
α, β, γ, δ, ε, T1, T2: phases
HV: Vickers hardness, value after HV gives load in kp according to DIN
HK: Knoop hardness, value after HK gives load in kp according to DIN
HU: Universal hardness, value after HU gives load in N according to DIN

Crystal structure:

ortho: orthorhombic
hex: hexagonal
hcp: hexagonal close-packed
cub: cubic
cub-B1: cubic, NaCl-type
rhom: rhombohedral/trigonal

(mo) monocrystalline
(po) polycrystalline
(i) incongruent melting point (formation of liquified material and a second solid phase)
(per) peritectic melting point (several solid phases and liquified material in equilibrium)

fcc: face-centered cubic
bcc: body-centered cubic
tet: tetragonal

mono: monoclinic
tri: triclinic

(x) density as determined by X-rays, determined by lattice parameters and atomic mass
(d) decomposition
(p) "heat proofness"
(t) "thermal stability"

The crystal systems of single materials may be explained by different crystal structures e.g. hex/rhom for Al_2O_3

Carbides

No.	Symbol	Crystal structure		Lattice parameters (nm)		Density (g cm^{-3})		Melting point (°C)		Linear thermal expansion, α (10^{-6} K^{-1})		Thermal conductivity, λ (W m^{-1} K^{-1})		Electrical resistivity (10^{-6} Ωcm)		Enthalpy (kJ mol^{-1})		Young modulus (10^5 N mm^{-2})		Micro hardness (10 N mm^{-2})		Oxidation resistance (100°C)	
1	B_4C	rhom	[8]	0.5631/1.2144	[52]	2.52	[1]	2450	[1, 48, 57, 77]	6	[1]	27.63	[1]	10^6	[14]	72	[5]	4.5	[1]	3700	[1,33]	11–14	[25]
2		rhom	[31]	0.5804/1.2079	[55]	2.5	[7]	2470	[5]	4.5	[8]	29.3	[8]	5×10^5	[31]	40	[43]	4.48	[7]	3000	[7]	7–8 (P)	[35]
3		rhom	[43]	0.5599/1.2074	[55]	2.52	[14]	2350	[7]	7.32/8.33	[30]	10.42	[35]	4×10^8	[38]	71.6	[46]	4.4	[14]	4900–5000 HV	[8]	7	[43]
4		rhom	[52]	0.56003/1.2086	[55]	2.52	[25]	2350	[8]	4.5	[31]	29	[38]	10^5–10^7	[48]	57.7	[48]	2.96	[25]	2940 HK 0.05	[14]	8	[70]
5		hex	[112]		[112]	2.52	[31]	2450	[14]	4.5	[35]	28	[43]	4×10^8	[57]	73.6	[73]	4.41	[31]	3700 HV 0.05	[25]		
6						2.52	[35]	2450	[25]		[38]	35	[48]	10^6	[68]			4.5	[38]	3000–4000 HV	[31]		
7						2.52	[38]	2430	[30]		[43]	28	[57]	5×10^5	[77]			4.6	[43]	2800 HK 0.1	[33]		
8						2.52	[39]	2450	[31]	4.5	[48]	27.2	[106]	5×10^5	[97]			4.5	[48]	3700–4700 HV	[35]		
9						2.52	[43]	2450	[38]		[57]							4.48	[57]	4950 HV	[39]		
10						2.51	[48]	2447	[39]		[70]							4.2	[70]	2800 HK	[44]		
11						2.52	[52]	2447	[43]	4.5	[77]							4.41	[77]	3700 HV	[13, 44, 57, 68]		
12						2.52	[57]	2500	[55]	4.5	[97]							4.41	[97]	3500–4500 HV 0.1	[48]		
13						2.51	[70]	2450	[88]	4.5	[106]							2.9	[106]	2940 HK 0.1	[58]		
14						2.52	[77]	2350	[106]									4.4–4.7	[108]	3000–3500 HV	[58]		
15						2.5	[88]													4000–5500 HV 0.1	[70]		
16						2.52	[97]													3000–7000	[78]		
17						2.5	[106]													3000–4000 HV	[77]		
18						2.52	[108]													3500 HV 0.2	[88]		
19						2.45														3200 HV	[108]		
20	Cr_3C_2	ortho	[6]	1.146/0.552/0.2821	[6]	6.68	[6]	1900 (per)	[5]	10.3	[8]	18.8	[8]	75	[7]	88.8	[5]	4	[9]	1500–2000	[7]	12	[7]
21		ortho	[8]	1.147/0.5545/0.2830	[10]	6.68	[7]	1850	[7]	11.7	[9]	19	[9]	75	[9]	88	[6]	3.86	[10]	1300 HV	[8, 87]	11–14	[25]
22		ortho	[10]	1.147/0.5545/0.2830	[11]	6.68	[9]	1890	[10]	10.3	[10]			75	[40]	23	[11]	3.7	[40]	2100	[9]	11–12	[53]
23		ortho	[11]	1.147/0.553/0.282	[25]	6.68	[10]	1810	[9, 10, 40]	10.3	[40]			75	[75]	109.7	[46]	3.73	[57]	1300	[6, 10, 40, 76]		
24		ortho	[25]	1.147/0.5545/0.2839	[28]	6.68	[25]	1810	[76, 77, 87]	10.3	[57]			75	[77]	85.4	[102]	3.284	[74]	1000–1400	[11]		
25		ortho	[28]	1.1483/0.5531/0.2827	[52]	6.68	[39]	1895	[13]	10.3	[76]			75	[87]	94.2	[73]	3.7	[76]	2280 HV	[13]		
26		ortho	[46]	1.1488/0.5527/0.2829	[74]	6.66	[53]	1895	[25]	11.7	[77]			75	[106]			4	[77]	1300 HV 0.05	[25]		
27		ortho	[52]	1.14883/0.55273/0.28286	[112]	6.7	[57, 107]	1890	[39, 46, 90]	10.3	[87]							3.7	[87]	1350 HV	[39, 90]		
28		ortho	[57]			6.68	[74]	1895 (d)	[53]	10.3	[106]							3.73	[106]	2280 HV	[53]		
29		ortho	[74]			6.68	[77]	1895 (d)	[57]											2280	[57]		
30		ortho	[76]			6.68	[90, 106]	1850	[106]											1350	[59]		
31		ortho	[112]																	2150 HV	[77]		

#	Compound	Structure	Lattice	[ref]	Density	[ref]	Melting pt	[ref]	Thermal exp		[ref]	k	[ref]	E	[ref]	Hardness	[ref]	K1c	[ref]
32	Cr₇C₃	hex	1.4/0.45	[1]	6.9	[1]	1780 (d)	[1]						190.5	[1]	2200	[1]	7	[53]
33		hex	1.398/0.4523	[8]	6.85	[6]	1780	[46]						178	[6]	1600	[11]		
34		hex	1.401/0.4525	[10]	6.9	[52]	1600–1790	[102]						228.2	[46]	1900–2200	[53]		
35		hex	1.401/0.4532	[11]	6.7	[53]								162	[102]	1200–1600 HV	[58]		
36		hex	1.398/0.453	[25]										181.3	[73]	2200–2400 HV 0.05	[61]		
37		hex	1.4006/0.4532	[46, 76]												1910 HV 0.03	[64]		
38		rhom	0.7015/1.2153/0.4532	[52]												1200–1600 HV	[68]		
39		ortho	0.7014/1.2153/0.45320	[74]												1500 HV 0.025	[72]		
40		ortho		[112]												1200 HK 0.01	[127]		
41	Cr₂₃C₆	cub	1.06	[1]	7	[1]	1520	[1]						68.7	[1]	1650	[1]		
42		fcc	1.0638	[6]	6.95	[6]								68.6	[8]	1000	[11]		
43		fcc	1.0655	[10]	6.97	[52]								396.5	[73]	1800HV	[58]		
44		fcc	1.060–1.066	[11]												1800HV	[68]		
45		fcc	1.065	[25]															
46		fcc	1.0659	[52]															
47		fcc	1.066	[74, 78]															
48		cub	1.06599	[112]															
49	HfC	fcc	0.464	[8]	12.2	[7]	3830	[5]	6.3			6.73	[8]	209.7	[11]	2200 HV 2	[3]	12	[7]
50		fcc	0.464	[10]	12.67	[9]	3890	[7, 8, 9, 13]	6			6.6	[10]	205	[45]	2600	[7]	11–14	[25]
51		cub-B1	0.464	[11]	12.67 (x)	[10]	3928	[10]	29.31			6.78	[30]	219	[46]	2700 HV	[8, 13]		
52		cub-B1	0.464	[25]	12.2	[25]	3890	[25]	13			6.6	[32]	230.3	[73]	2800	[9]		
53		cub-B1	0.4646	[30]	12.7	[30]	3927	[30]				6.4	[38]			2700	[10, 57, 76]		
54		cub-B1	0.446	[52]	12.3	[32]	3950	[32]				6.9	[45]			2276 HK 0.1	[11]		
55		cub-B1	0.463765	[109]	~12	[38]	3890	[38]				6	[62]			2200	[32]		
56		fcc		[112]	12.7	[39]	3890	[39]				6.6	[76]			1932–2900 HV 0.05	[32]		
57		cub-B1		[76]	12.2	[52]	3387	[45]				6.7	[98]			1000 HV	[34]		
58		fcc		[98]	12	[90]	3830	[74]				6.59	[106]			2913 HV	[39]		
59		cub-B1		[109]	12.2	[106]	3890	[90]								3200	[45]		
60		cub		[112]	12.2	[108]	3890	[106]								1800–2500 HV	[58]		
61					12.7	[107]	3928	[108]								2600	[59]		
62							3890	[76]								1800–2500	[62]		
63							3890	[90]								2913 HV	[90]		
64							3900	[98, 106]								2750 HV	[98]		
65	MoC-γ	hex	0.29/0.284	[2]	8.7	[1]	2700	[1]								1600	[1]		
66		hex	0.2901/0.2785	[6]	8.4	[5]	2677	[5]								1500	[5]		
67		hex	0.2896/0.2809	[10]	8.65	[108]	2692	[6]								1800HV	[11]		
68		hex	0.2896/0.2809	[52]			2600	[10]								1950 HV	[73]		
69		hex	0.2932/1.097	[112]												49	[6]	5–8	[25]
70	MoC-β	fcc	0.427	[11]															
71		cub-B1	0.428	[25]															
72	Mo₂C-α	ortho	0.7244/0.6004/0.5199	[10]	8.9	[7]	2670	[7]				4.9/~8.2	[10]			1600	[7]	2.28	[10]
73		ortho	0.4733/0.6034/0.5206	[11]	9.06 (x)	[10]	2400	[13]				8.5/4.5/5.7	[28]			1500	[10]		
74		ortho	0.4732/0.6037/0.5204	[28, 112]	8.9	[39]	2410	[39]								1479–1800	[11]		
75																1950 HV	[13]		

Table 1. Continued

No.	Symbol	Crystal structure		Lattice parameters (nm)	Density (g cm^{-3})		Melting point (°C)		Linear thermal expansion, α (10^{-6} K^{-1})		Thermal conductivity, λ (W m^{-1} K^{-1})		Electrical resistivity (10^{-6} Ωcm)		Enthalpiy (kJ mol^{-1})		Young modulus (10^5 N mm^{-2})		Micro hardness (10 N mm^{-2})		Oxidation resistance (100°C)	
76	Mo$_2$C-β	hcp	[6]	0.2994/0.4722	8.9	[6]	2687	[6]	5.6/7.5	[28]			97.5	[6]	46	[11]	5.33	[57]	1500 HV U1410.05	[25]		
77		hex	[10, 112]	0.3005/0.4755	9.18	[10]	2690	[25]	7.8	[57]			97	[25]	46	[46]	5.4	[77]	1499 HV	[39]		
78		hex	[11]	0.3004/0.4722	9.2	[57]	2430 (per)	[46]	7.8–9.3	[77]			133	[57]	46.1	[73]	5.4	[97]	1800 HK	[44]		
79		hcp	[14]	0.3002/0.4724	9.18	[77]	2400 (d)	[57]	7.8–9.3	[97]			57	[77]			5.33	[106]	1500 HV	[44]		
80		hcp	[25]	0.414	8.9	[90]	2430	[73]	7.8	[106]			57	[97]					1950	[57]		
81		hex	[28, 46]	0.3012/0.47352	9.18	[97]	2517	[90]					71	[106]					1500 HV	[57, 68]		
82		hex	[52]		8.9	[106]	2410	[90]											1500	[59]		
83		hcp	[57]		8.9	[107]	2517	[97]											1499 HV	[90]		
84		fcc	[25]		8.9	[108]	2697	[106]											1660 HV	[77, 97]		
85	NbC	fcc	[1]	0.45	7.78	[1]	3490	[1]	6.65	[1]	14.24	[1]	35–74	[6]	139.8	[5]	3.4	[1]	2400	[1, 10, 57]	11	[7]
86		cub-B1	[6]	0.4471	7.51–7.82	[6]	3600	[5]	6.2	[6]	11.25	[6]	34	[7]	140.7	[6]	2.76–3.45	[6]	1950–2700	[6]	11–14	[25]
87		fcc	[8]	0.447	7.82	[7]	3477–3900	[6]	6.84	[8]	14.2	[8]	74	[25]	140.7	[11]	3.38	[10]	2000–2400	[7]		
88		cub-B1	[10]	0.4471	7.79 (x)	[10]	3500	[7]	6.6	[10]	14.24	[10]	51	[32]	138.2	[46]	3.45	[32]	2400 HV	8, 13		
89		cub-B1	[11]	0.4461	7.82	[25]	3480	[8]	5.2	[30]	14.24	[25]	35	[57]	140.7	[73]	3.38	[57]	2055–2400	[11]		
90		cub-B1	[25]	0.447	7.56	[39]	3600	[10]	6.6	[32]	14.24	[32]	19	[66]			3.19/3.5.10	[74]	2400 HV 0.05	[25]		
91		cub-B1	[30]	0.449	7.8	[57]	3500	[13]	6.7	[57]	14	[106]	19	[77]			5.8	[77]	1800	[32]		
92		cub-B1	[46]	0.447	7.6	[74]	3500	[25]	7.2	[77]			19	[97]			5.8	[97]	1961 HV	[39]		
93		cub-B1	[52]	0.44698	7.78	[77]	3500	[30]	7.2	[97]			35	[106]			3.38	[106]	1265	[58]		
94		fcc	[57]		7.56	[90]	3600	[32]	6.65	[106]									2000	[59]		
95		fcc	[74]		7.6	[106]	3480	[39]											2000 HV	[68]		
96		cub	[112]		7.9	[107]	3500	[57]											2400 HV 0.025	[72]		
97					7.9	[108]	3480	[73, 90]											1800 HV	[77]		
98							3613	[77]											1961 HV	[90]		
99							3613	[97]											1800 HV	[97]		
100							3480	[106]														
101	Nb$_2$C-α	ortho	[10, 64]	1.092/0.497/0.308	7.85 (x)	[52]	3090 (per)	[73]	4.1/11.0/7	[28]					186.3	[46]						
102		ortho	[112]	1.0920/0.4974/0.309											191.5	[73]						
103	Nb$_2$C-β	hex	[6]	0.540/0.4974	7.85 (x)	[6]	3100	[6]											1924–2322	[6]		
104		hex	[10, 112]																				
105		hex	[28]	0.3116/0.4958																		
106	Nb$_2$C-γ	hcp	[6]	0.3128/0.4974			~3500 (per)	[6]	6.6–7.0/8.7	[6]					167–195	[6]						
107		hcp	[10]	0.3127/0.4972					6.8/6.9	[28]												
108		hcp	[28]	0.3119/0.4959																		
109		hcp	[52]																			
110	SiC	α;hex	[22]	β;0.4360	3.2	[6]	2200	[6]	5.68	[1]	15.49	[1]	10^5	[9]	α 71.6	[6]	4.8	[1]	3500	[1, 33]	14–17	[25]
111		β;fcc	[22]	α;0.3–7.3/1–1.5	3.17	[22]	2700	[7]	5.3	[9]	63–155	[9]	10^5	[14]	β 73.3	[6]	3.9–4.1	[6]	1400 HV	[4]	13–14 (P)	[35]
112		α;hex	[31]	β;0.43	3.2	[7]	2760 (d)	[9]	4.8	[14]	42	[14]	10^5–10^{12}	[22]	80	[22]	4.8	[9]	2169–2428	[6]	α;16	[43]
113		β;cub	[31]	0.4358	3.22	[9]	2300 (d)	[14]	α-4.5	[22]	60	[22]	10^5	[31]	α 70.0	[43]	4.7	[14]	2600	[7, 45]	16	[70]

#	Phase	Structure	Ref	Lattice param	Ref		Ref		Ref		Ref		Ref		Ref		Ref	Ref
114		α:hex	[43]	α:0.3073/1.508	[112]	3.22	[14]	2700	[22]	β:6.0	[22]	15.49	[25]	>5×10⁶	[38]	α:4.40	[22]	2000–3000	[9]
115		β:cub	[43]			3.2	[22]	~2200	[25]	5.06	[30]	Oct-89	[35]	>10⁵	[67]	β:4.0	[22]	3500 HV	[13, 39, 57]
116		α:hex	[47]			3.2	[25]	~2500 (i)	[30]	5.3	[31]	59	[38]	105	[66]	4.8	[31]	2580 HK 0.05	[14]
117		β:cub	[47]			3.21	[31]	2760	[31]	4.4–5.3	[35]	α:60–150	[43]	10⁵	[77]	4.8	[38]	3000–4000	[22]
118		α:hex	[52]			3.2	[35]	2300	[38]	5	[38]	16	[57]	10⁵	[98]	4.8	[43]	3500–4000	[24]
119		β:cub	[52]			3.21	[38]	~2227	[39]	α:4.5–5.5	[43]	130	[67]	10⁸–10¹¹	[108]	4.8	[45]	3500 HV 0.05	[25]
120		α:hex	[76]			3.2	[39]	2827 (per)	[43]	5.3	[45]	155	[98]			4.8	[57]	2600 HV	[31, 44, 77]
121		β:cub	[76]			3.21	[43]	2760	[45]	5.7	[57]	90–200	[108]			4	[67]	2600 HV	[97, 98]
122		β:cub	[98]			3.22	[45]	2200	[57]	5	[67]	41.8	[106]			4	[70]	2500 HK 0.1	[33]
123		hex	[98]			3.2	[57]	2986 (d)	[73]	5	[70]					4.8	[77]	3000–3500 HV	[35]
124		α:hex	[112]			3.15	[67]	2760 (d)	[77]	5.3	[77]					4.8	[98]	2000–6000 HV 0.05	[38]
125						3.21	[70]	2180	[88]	5.3	[88]					4.8	[108]	2585 HK	[44]
126						3.22	[77]	2760	[97]	5.3	[97]					3.5–4.4		3000 HV 0.2	[88]
127						3.2	[88]	2760	[98]	3.5	[98]							2580 HK 0.1	[78]
128						3.22	[97]	2700	[106]	4.3–4.6	[106]							2500–4000 HV	[58]
129						3.22	[98]											3000	[58]
130						2.5–3.4	[108]											2500 HV 30	[67]
131						3.2	[106]											3000–3000 HV 0.1	[70]
132																		2400–3500 HV	[108]
133																			
134	TaC	fcc	[1]	0.44	[1]	14.5	[1, 8, 13, 57]	3780	[1]	6.29	[1]	22.19	[1]	20	[6]	2.9	[1]	1790	[1]
135		fcc	[6]	0.4454	[6]	14.65	[5]	3877	[7]	6.04	[6]	22.19	[6]	25	[7]	2.91	[6]	1490	[6]
136		cub-B1	[8]	0.4456	[10]	14.5 (x)	[7, 25, 38, 39]	3880	[8]	6.61	[8]	22	[8]	15	[14]	3.65	[10]	1800	[7, 59]
137		cub-B1	[10]	0.4455	[11]	14.5	[10, 32, 40, 76]	3983	[10]	6.3	[10]	22.19	[10]	30	[25]	5.5	[11]	1800 HV	[8]
138		cub-B1	[11]	0.4455	[14]	14.48	[14]	3980	[32]	4.86	[30]	22.19	[25]	42	[32]	2.91	[14]	2500	[10, 40, 76]
139		cub-B1	[25]	0.4456	[38]	14.5	[25]	3919	[30]	6.3	[38]	22.19	[32]	30	[38]	2.9	[32]	1547–1952	[11]
140		cub-B1	[30]	0.4456	[39]	14.3	[39]	4000	[46]	6.3	[46]	21	[38]	15	[40]	2.85	[38]	1790 HV	[13]
141		cub-B1	[46]	0.4456	[46]	14.5	[57]	3880	[53]	6.3	[53]	22.2	[98]	25	[57]	2.85	[40]	1550 HK 0.05	[14]
142		cub-B1	[52]	0.44	[52]	13.9	[74]	4000	[73]	6.3	[73]	22.2	[106]	15	[74]	3.04–5.34	[74]	1800 HV 0.05	[25]
143		fcc	[57]	0.445	[57]	14.48	[77]	3985	[77]	6.3	[77]			15	[76]	2.85	[76]	1600 HV	[32]
144		fcc	[74]	0.4454	[74]	14.5	[88]	3780	[88]	7.1	[88]			15	[97]	5.6	[77]	1000 HV	[34]
145		fcc	[98]	0.44547	[98]	14.3	[90]	3880	[90]	7.1	[97]			15	[98]	2.91	[80]	1599 HV	[39, 90]
146		cub-B1	[109]			14.5	[98]	3985	[98]	7.1	[98]			25	[87]	5.6	[97]	1790	[57]
147		cub	[112]			13.9	[106]	2985	[108]	6.3	[108]					5.6	[98]	1599	[80]
148						13.9	[108]	3983	[87]	7.1	[87]					2.85	[87]	1790 HV 0.2	[88]
149						14.5	[107]	3880	[106]	6.29	[106]					2.85	[106]	1550 HV	[77, 97, 98]
150																		2500 HV	[87]
151	Ta₂C-β	hex	[1, 112]	0.311/0.495	[1]	15.05 (x)	[1]	3400 (d)	[1]	6.2/4.8	[28]	20.93	[1]	40	[6]			1700	[1]
152		hex	[10]	0.3103/0.49378	[6]			3500 (per)	[73]			17–23.5	[8]					1000–1720	[6]
153		hex	[11]	0.3102/0.494														1000 HK	[11]
154		hcp	[14]	0.3106/0.4945															
155		hex	[25]	0.3091/0.493															
156		hex	[52]	0.31037/0.49394															
157				0.31042/0.4941															
158	Ta₂C-α	hex	[6]	0.31046/0.49444															
159	TiC	fcc	[1]	0.433	[1]	4.93	[1]	3067	[4]	7.42	[4]			52	[4]	4	[4]	3200 HV	[1, 13, 35, 44]
160		cub-B1	[3]	0.429–0.433	[6]	4.93	[4]	3150	[5]	7.4	[5]			68	[7]	3.22	[6]	3200 HV	[58, 68, 95]

Table 1. Continued

No.	Symbol	Crystal structure	Lattice parameters (nm)	Density (g cm^{-3})	Melting point (°C)	Linear thermal expansion, α (10^{-6} K^{-1})	Thermal conductivity, λ (W m^{-1} K^{-1})	Electrical resistivity (10^{-6} Ωcm)	Enthalpiy (kJ mol^{-1})	Young modulus (10^5 N mm^{-2})	Micro hardness (10 N mm^{-2})	Oxidation resistance (100° C)
161		cub-B1 [6]	0.4328 [10]	4.93 [6]	3170 [7]	7.61 [8]	21-32 [9]	52 [9]	239.7 [6]	4.7 [9]	2800 HV 2 [3]	31.4 (P) [35]
162		fcc [8]	0.4328 [11]	4.08 [7]	3180 [8]	8.0-8.6 [9]	20.93 [10]	52 [14]	184.6 [10]	2.59-4.62 [10]	2900 HV [4, 87]	~3 [53]
163		cub-B1 [10]	0.4319 [25]	4.93 [9]	3067 [9]	7.4 [9]	17.17 [25]	68.2 [25]	183.8 [46]	4.6 [14]	3200 [6, 33, 59, 69]	3 [92]
164		cub-B1 [11]	0.4328 [32]	4.91 (x) [10]	3067 [10]	8.74 [30]	20.93 [32]	52 [31]	184 [57]	4.7 [31]	2900 [7, 10, 40, 76]	
165		cub-B1 [25]	0.4327 [52]	4.91 [11]	3140 [13]	8.0-8.6 [31]	24-Mar [35]	45 [32]	184.2 [73]	3.22 [32]	2980-3800 HV [8]	
166		cub-B1 [30]	0.433 [57]	4.93 [14]	3070 [14]	7.4 [32]	21 [37]	70 [38]	183.6 [98]	3.22 [37]	2800-3500 [9]	
167		cub-B1 [31]	0.433 [59]	4.93 [25]	3140 [25]	7.4 [37]	29 [38]	52 [40]		3.2 [38]	1600-2800 HV 0.3 [11]	
168		cub-B1 [46]	0.4327 [74]	4.93 [31]	3067 [30, 31, 32, 40]	7.4 [38]	20 [57]	68 [57]		4.5 [40]	2800 HK 0.05 [14]	
169		cub-B1 [52]	0.43 [98]	4.94 [35]	3067 [45, 61, 76]	7.4 [40]	33 [67]	52 [61]		4.5 [45]	2800-3700 HV 0.05 [23]	
170		fcc [57]	0.431 [109]	4.9 [37]	3067 [77, 87]	7.4 [45]	21 [106]	52 [66]		4.5 [53]	2400 [24]	
171		fcc [67]	0.43274 [112]	4.93 [38]	3160 [37]	7.7 [57]		68 [67]		4.51 [57]	3200 HV 0.05 [25]	
172		fcc [74]		4.93 [39]	3140 [38]	7.42 [57]		52 [77]		4.58 [61]	2800 HV [31, 77, 97, 98]	
173		cub-B1 [76]		4.9 [45]	3147 [39]	6.7 [62]		52 [87]		4.58 [62]	2800 [32]	
174		fcc [98]		4.92 [53]	3150 [46]	7.74 [67]		52 [97]		4.51 [67]	2470 HK 0.1 [33]	
175		cub-B1 [109]		4.93 [57]	3160 [57]	7.4 [76]		52 [98]		~4 [69]	3200 HV 30 [37]	
176		cub [112]		4.93 [67]	3060 [62]	8.0-8.6 [87]		52 [92]		3.15-4.462 [74]	3000 HV [39, 71, 90]	
177				4.93 [74]	3070 [67]	7.4 [87]		68 [106]		4.5 [76]	2470 HK [44]	
178				4.93 [77]	3147 [90]	7.4 [90]				4.7 [77]	3900 [45]	
179				4.93 [90]	3150 [88]	6.7-7.6 [95]				3.5 [80]	3300-4000 HV [53]	
180				4.9 [88]	3067 [93]	8.8-8.6 [97]				4.5 [87]	3200 HV 0.1 [57]	
181				4.9 [93]	3067 [97]	8-8.6 [98]				3.02-5.86 [89]	3000-4000 HV 0.05 [59]	
182				4.93 [97]	3067 [98]	7.74 [106]				4.7 [93]	3400-5200 HV 0.05 [61, 92]	
183				4.93 [98]	3060 [92]					3.2 [95]	3400-5200 [62]	
184				4.92 [92]	3170 [106]					4.7 [97]	2200 HV 0.05 [64]	
185				4.08 [106]						4.7 [98]	1535 HV 0.05 [64]	
186				4.9 [107]						5 [92]	3000 HV 0.05 [67]	
187				4.93 [108]						4.51 [106]	3200 HV 0.025 [72]	
188											2988 HV [80]	
189											3200 HV 0.2 [88]	
190											2800-3700 HV 0.05 [93]	
191	VC	fcc [1]	0.42 [1]	5.5 [1]	2830 [1]	6.5 [8]	4.19 [6]	60-150 [6]	117.2 [1]	2.8 [1]	2950 [1, 57]	8-11 [25]
192		cub-B1 [6]	0.4173 [6]	5.36 [6]	2770 [5]	6.55 [30]	4.2 [8]	156 [25]	105.1 [5]	4.34 [8]	2340-2760 HK [6]	
193		fcc [8]	0.4166 [10]	5.25 (x) [7]	2750 [6]	7.2 [57]		60 [57]	101.9 [9]	4.22 [57]	2900 [7, 10, 58]	
194		cub-B1 [10]	0.416 [25]	5.4 [7]	2770 [7]	7.3 [77]		59 [66]	102.6 [11]	2.66-4.32 [74]	2800 HV [8, 44, 58, 68]	
195		cub-B1 [25]	0.4182 [52]	5.65 [10]	2830 [8]	7.2 [106]		59 [77]	100.9 [46]	4.3 [77]	2850-3000 [11]	
196		cub-B1 [30]	0.416 [74]	5.36 [25]	2648 [10]			59 [97]	100.9 [73]	4.3 [97]	2950 HV [13]	
197		cub-B1 [52]	0.415 [109]	5.36 [39]	2830 [13]			60 [106]		4.22 [106]	2800 HV 0.05 [25]	
198		fcc [57]		5.7 [57]	2830 [25]						2094 HV [39]	
199		fcc [74]		5.77 [74]	2734 [30]						2660 HK [44]	
200		cub-B1 [76]		5.41 [76]	2810 [39]						2000-3000 HV [58]	
201		cub-B1 [109]		5.36 [90]	2830 [57]						2600 HV 0.025 [72]	
202				5.41 [97]	2648 [77]						2600 [76]	

#	Phase	Struct	Ref	Lattice params	Ref	Val1	Ref	Density	Ref	Val2	Ref	Val3	Ref	Val4	Ref	Val5	Ref	Val6	Ref	Hardness	Ref	Val7	Ref
203						5.4	[106]	2810	[90]											2900 HV	[77, 97]		
204						5.8	[107]	2648	[97]											2094 HV	[90]		
205						5.77	[108]	2770	[106]														
206	V₂C-α	ortho	[10]	0.457/0.574/0.503																			
207		ortho	[28]	1.149/1.006/0.455																			
208		ortho	[112]	0.4577/0.5742/0.5037						12.6(6.9,3	[28]												
209	V₂C-β	hex	[6]	0.29043/0.45793		4.5	[6]	2165	[73]	8.2,9.7	[73]					69.1	[11]			2140	[11]		
210		hcp	[10,11]	0.2902/0.4577							[28]					147.4	[73]						
211		hcp	[14]	0.2902/0.4577																			
212		hex	[28,52]	0.2906/0.4597																			
213	WC	hex	[2]	0.29/0.28	[1]	15.7	[1]	2600	[1]	5.2-7.3	[1]	121.42	[1]	17	[4]	35.2	[1]	7.2	[1]	2080	[1,57]	8	[7]
214		hex	[6]	0.2906/0.2837	[4]	15.72	[4]	2776	[4]	5.2-7.3	[4]	29.3	[4]	53	[6]	40.5	[11]	7	[4]	2300 HV	[4]	5-8	[25]
215		hex	[8]	0.2906/0.2839	[6]	15.7	[6]	2785	[6]	6.2	[6]	29–120	[9]	22	[7]	38.1	[46]	7.2	[9]	2100	[7, 10, 32]	26 (P)	[35]
216		hex	[10]	0.2900/0.2831	[7]	15.7	[7]	2867	[7]	3.8–3.9	[7]	29.31	[10]	17	[9]	40.1	[73]	6.69	[10]	2000–2400 HV	[40, 45, 76]	5-8	[53]
217		hex	[11]	0.2906/0.2837	[8]	15.72	[9]	2730	[8]	5.0,4.2	[8]	29.31	[32]	17	[14]	35.2	[98]	7	[14]	2300	[8]		
218		hex	[14]	0.2906/0.2837	[9]	15.7	[10]	2867	[9]	4.6/4.83	[10]	29.33	[35]	53	[25]			7.27	[32]	2085	[9]		
219		hex	[25]	0.2906/0.2838	[10]	15.8 (x)	[14]	2776	[10]	5.0,4.2	[30]	121.6	[37]	19.2	[32]			7.07	[33]	2180 HV	[11]		
220		hex	[28]	0.291	[11]	15.7	[25]	2600	[13]	5.7–7.2	[32]	117	[38]	53	[33]			7.22	[37]	2350 HK 0.05	[14]		
221		hex	[30]	0.29062/0.28378	[14]	15.7	[28]	2780	[14]	5.2–7.3	[35]	90	[67]	20	[38]			7.3	[38]	2400 HV 0.05	[25]		
222		hex	[46]		[25]	15.77	[33]	2870 (d)	[25]	6.3	[37]	121.4	[98]	17	[40]			6.95	[40]	1800–2000 HV	[33]		
223		hex	[52]		[28]	15.7	[35]	2690–2880	[30]	5.2	[38]	12.1	[106]	22	[57]			6.95	[45]	1800 HK 0.1	[33]		
224		hex	[57]		[30]	15.7	[37]	2776	[32]	4.3	[40]			17	[66]			6.96	[57]	2400	[35]		
225		hex	[67]		[46]	15.7	[38]	2600	[33]	4.3	[45]			19	[67]			6.96	[67]	1600–2000 HV	[37]		
226		hex	[74]		[52]	15.7	[39]	2780	[35]	5.2	[57]			17	[77]			~7.2	[69]	2080 HV	[39, 80, 90]		
227		hex	[98]		[57]	15.7	[45]	2720	[37]	5.2,7.3	[67]			17	[76]			6.07–6.207	[74]	1780 HV	[44]		
228		hex	[112]		[67]	15.7	[53]	2776	[38]	4.3	[76]			17	[87]			6.95	[76]	1800 HK	[53, 95]		
229					[74]	15.7	[57]	2776	[39]	3.8–3.9	[77]			17	[97]			7.2	[77]	2400 HV	[58]		
230					[98]	15.7	[67]	2800 (d)	[45]	4.3	[87]			53	[106]			7.22	[80]	2000 HV	[59]		
231					[112]	15.63	[74]	2775	[53]	5.7–7.2	[95]							6.95	[87]	1213	[67]		
232						15.72	[77]	2785	[57]	3.8–3.9	[97]							7.2	[95]	1700 HV	[58]		
233						15.7	[88]	2600 (d)	[67]	3.8–3.9	[98]							7.2	[97]	1300–2000	[59]		
234						15.72	[90]	2776	[73]	7.3	[106]							6.96	[106]	1300–2000 HV 0.05	[67]		
235						15.72	[97]	2776	[76]											2400 HV 0.025	[68]		
236						15.7	[98]	2600	[77]											2500	[69]		
237						15.72	[106]	2720	[88]											2350 HV	[72]		
238						15.8	[107]	2776	[90]											2100 HV	[77, 97, 98]		
239						15.63	[108]	2776	[97]											2080 HV 0.2	[87]		
240								2776	[98]												[88]		
241								2860	[106]														
242	W₂C-α	hex	[10]	0.3001/0.4728	[6]	16.06	[6]	2800	[5]	1.2–11.4	[6]	29.31	[25]	81	[6]	38.1	[5]	4.28	[80]	2000	[7]	8	[7]
243		hex	[28]	0.299/0.472	[7]	17.15 (x)	[6]	2730–2880	[7]	6.6,7.7	[28]			80	[7]	26.4	[11]	4.3	[95]	1450	[11]	5-8	[53]
244	W₂C-β	hcp	[6]	0.3002/0.4755	[10]	16.06	[7]	2857	[25]	5.8	[62]			~80	[25]	26.4	[46]			3000 HV	[25, 80]		
245		hcp	[10]	0.298/0.471	[14]	17.2	[25]	2730	[53]	1.2–11.4	[95]					26.4	[73]			2000 HV	[53]		
246		hcp	[14]		[25]			2700 (d)	[62]											2000–2500 HV	[58]		
247		hcp	[25]	0.51809/0.47216	[28]			2730	[53]											1800	[59]		
248		hcp	[28]	0.29948/0.4726	[52]			2730	[62]														
249		hcp	[52]	0.416	[25]																		

Table 1. Continued

No.	Symbol	Crystal structure		Lattice parameters (nm)		Density (g cm^{-3})		Melting point (°C)		Linear thermal expansion, α (10^{-6} K^{-1})		Thermal conductivity, λ (W m^{-1} K^{-1})		Electrical resistivity (10^{-6} Ωcm)		Enthalpy (kJ mol^{-1})		Young modulus (10^5 N mm^{-2})		Micro hardness (10 N mm^{-2})		Oxidation resistance (100°C)	
250		fcc	[25]	0.47		6.5	[1]	2795	[73]											2000-2500	[62]		
251																				1900 HV	[95]		
252	ZrC	fcc	[1]	0.4689-0.476	[6]	6.51 (x)	[5]	3535	[1]	6.73	[1, 6]	20.93	[1]	63-156	[6]	263.8	[1]	3.8	[1]	2560	[1, 57]	12	[7]
253		cub-B1	[6, 76]	0.4698	[7]	6.51	[7]	3400	[5]	6.93	[8]	20.5	[8]	42	[7]	181.7	[5]	4	[9]	2600 HV	[6, 8]	11-14	[25]
254		fcc	[8]	0.4698	[10]	6.51	[6]	3532	[6]	7.0-7.4	[9]	12	[9]	42	[9]	196.8	[11]	4.07	[10]	2600	[7, 10, 32, 45]		
255		cub-B1	[10]	0.4685	[25]	6.59 (x)	[10]	3530	[7]	6.7	[10]	20.52	[25]	75	[25]	202.1	[46]	3.88	[32]	2500	[9]		
256		cub-B1	[11, 25]	0.4698	[32]	6.9	[25]	3530	[8]	8.3	[30]	20.52	[32]	43	[32]	196.8	[73]	3.9	[38]	2600-2900	[13]		
257		cub-B1	[30, 52]			6.6	[38]	3445	[9, 45, 77, 97, 98]	6.7	[32, 48]	19	[38]	50	[38]	184.6	[98]	4	[45]	2560 HV	[13, 77, 98]		
258		tet	[46]	0.4694	[52]	6.73	[39]	3420	[10]	5.7	[45]	20.9	[98]	42	[57]			3.324-4.02	[74]	2600 HV 0.05	[25]		
259		fcc	[57]	0.4693	[74]	6.73	[74, 97]	3530	[39, 57, 90]	7.0-7.4	[77]	20.5	[106]	42	[66]			4	[77]	2925 HV	[39, 90]		
260		fcc	[74]	0.47	[98]	6.63	[97, 98]	3443	[30]	7-7.4	[97]			42	[77]			4	[98]	2700	[59]		
261		fcc	[98]	0.467	[109]	6.63	[9, 45, 77]	3420	[32, 38, 46]	7-7.4	[98]			42	[97, 98]			4	[106]	1000-2800	[76]		
262		cub-B1	[109]	0.4693	[112]	6.73	[108]	3530	[13, 25, 106]	6.73	[106]			42	[106]			3.48					
263		cub	[112]			6.51	[106]																
264						6.7	[107]																

Nitrides

265	AlN	hex	[6]	0.311/0.498	[6]	3.05	[6]	>2200	[5, 106]	6	[38]	10	[38]	10^{11}	[38]	288.9	[5]	3.15	[43]	1200	[7]	13	[43]
266		hex	[43]	0.311/0.494	[47]	3.09	[7]	1900	[6]	5.4	[43]	110-170	[43]	10^{11}	[57]	290	[43]	~3	[57]	1230 HV	[39, 77, 97]		
267		hex	[52]	0.3110/0.4975	[52]	3.25	[38]	2230 (d)	[7]	4.8	[47]	30	[57]	10^{15}	[66]	318.6	[46]	3	[74]	2500 HV	[57]		
268		hex	[74]	0.311/0.4979	[74]	3.05	[39]	2300 (d)	[38]	7	[57]	165	[108]	10^{15}	[77]	318	[65]	3.437	[77]	1230 HK 1	[58]		
269		hex	[112]	0.31114/0.49792	[112]	3.26	[43]	2397	[39]	5.7	[77]			10^{15}	[97]	318.2	[73]	3.5	[38, 97]	1200	[59]		
270						3.26	[57]	2227 (d)	[43]	5.7	[97]			$10^{17}-10^{19}$	[108]					1200 HV	[108]		
271						3.26	[74]	2400	[47, 57]	5.3	[108]												
272						3.26	[77]	2517	[73]														
273						3.26	[97]	2250 (d)	[77, 97]														
274						3.09	[106]																
275						3.2	[107]																
276						3.25	[108]																
277	BN	hex	[1]	0.251/0.669	[6]	2.25	[1]	3000	[1]	3.8	[38]	284.7	[1]	3×10^{14}	[26]	252.5	[5]	0.9	[1]	4400 HV	[4]	10	[43]
278		hex	[30]	0.2504/0.6661	[52]	2.25	[38]	3000	[5]	-2.8/41.5	[30]	25	[38]	10^{16}	[38]	119.3	[6]	0.9	[38]	4700	[7]	11-14	[25]
279		hex	[39]	0.2504/0.6661	[78]	2.25	[39]	3000	[30]			25	[43]	10^{18}	[66]	225	[43]	3	[43]	4700 HV	[8]	12 (P, mo)	[35]
280		hex	[43]	0.362	[78]	2.27	[43]	3000 (d)	[38]	4.8	[43]	180-201	[9]	10^{18}	[9]	252.5	[46]	6.6	[9]	5000	[9]	14 (P, po)	[36]
281		hex	[46]	0.36158	[112]	2.34	[7]	2997	[39]	2.5-4.7	[9]	42	[9]	10^{18}	[14]	252.5	[73]	5.9	[14]	4700 HK 0.05	[14]	14 (t)	[36]
282		hex	[52]			3.48	[9]	2727	[43]	2.5-4.7	[35]	41.9	[35]	10^{20}	[26]			6.8	[26]	2100-3600	[26]	14	[43]
283		hex	[78]			3.48	[14]	2300 (d)	[43]			36	[36]	10^{18}	[36]			7.2	[36]	9000-9500 HV (mo)	[35]	14	[70]
284		cub	[9]			2.2	[25]	2730	[7]	3.5	[43]	200	[43]	10^{18}	[77]			8	[43]	7000-8000 HV (po)	[35]		
285		cub	[14]			3.44-3.49 (mo)	[35]	2730	[9]	1-6.0	[108]	42	[108]	$10^{17}-10^{19}$	[108]			6.8	[57]	4050 HK	[36]		
286		cub	[35]			3.3-3.4 (po)	[35]	2730	[25]			15-33	[108]					6.6	[77]	4700 HK	[36]		
												20	[106]										

#	Compound	Structure	[ref]	a	[ref]	b	[ref]	MP	[ref]	c	[ref]	d	[ref]	e	[ref]	f	[ref]	g	[ref]	Hardness	[ref]	extra	[ref]
287		cub	[36]	3.48	[39]															4000 HK	[57]		
288		cub	[39]	3.4	[36]															4700 HK 0.1	[58]		
289		cub	[43]	3.45	[39]															6000–8000 HK 0.1	[70]		
290		cub	[57]	3.48	[43, 88]															2000–4000	[76]		
291		cub	[58, 76]	3.45	[57]															5000 HV	[77]		
292		cub	[77, 78]	3.48	[70, 77]															8000 HV 0.2	[88]		
293		cub	[112]	2.34	[106]																	7–7.5	[125]
294				2.2	[107]																		
295				1.8	[108]																		
296	CrN	fcc	[1]	0.415	[1]	6.1		1050	[2]	2.3	[10]			640	[57]	118–124	[6]	4	[60]	1800–2100	[7]		
297		cub-B1	[3, 6]	0.4149	[3]	5.39–7.75	[6]	1450 (d)	[7]	2.3	[32]	11.72	[10]	640	[66]	123.1	[46]	3.236	[74]	1100	[10, 32, 59]		
298		cub-B1	[112]	0.414	[6]	7.7	[7]	1500	[10]	2.3	[60]	11.72	[32]	640	[77]	124.8	[102]	4	[77]	2000–2200 HV 0.05	[23]		
299		cub-B1	[10]	0.4149	[10]	6.1 (x)	[7]	1500 (d)	[32]	2.3	[97]			640	[97]	123.1	[73]	4	[97]	2600–2900 HK 0.05	[45]		
300		cub-B1	[25]	0.4148	[25]	6.1	[25]	1500 (d)	[57]	2.3	[81]					118.1	[111]	3.3	[81]	1090	[57]		
301		cub-B1	[46]	0.4149	[32]	5.9	[52]	1050	[60]	2.3	[110]							2.65–2.9	[104]	1800–2200 HV 0.05	[60]		
302		cub-B1	[52]	0.4148	[52]	6.1 (x)	[60]	1500	[77]	2.3	[111]							4	[110]	1965–2295 HV	[64]		
303		fcc	[57]	0.414	[74, 109]	6.1	[74]	1500	[111]	7.5 (850 C)–1040 C	[111]							4	[111]	1890 HV 0.01	[75]		
304		fcc	[60]	0.4149	[75]	5.9	[75]											2.45	[120]	1100 HV	[77]		
305		cub-B1	[74, 75]	0.416	[79]	6.12	[79]													2300 HV	[79]		
306		cub-B1	[76, 109]	0.415	[81]	6.12	[81]													1328–4140 HV 0.00005	[86]		
307		cub	[81]	0.4148	[111]	5.9	[97]													700–1100 HU 0.1	[103]		
308		cub	[104, 110]	0.414	[112]		[112]													680–780 HV 0.1	[104]		
309																				~1000–1700 HV 0.015	[115]		
310																				1300 HV 0.05	[118]		
311																				810 HV	[120]		
312																				1750–2650 HV 0.25	[125]		
313																				2600–2900 HK 0.05	[125]		
314	Cr$_2$N	hex	[25]	0.4760/0.4438	[25]	5.9	[74]	~1500	[25]	9.4	[110]					30.8	[111]	3.138	[74]	2250 HV 0.01	[75]		
315		hcp	[6, 52, 75]	0.4759/0.4438	[52]	5.9	[108]	1590	[52]	9.41	[111]					25.2	[111]	2.7		2500 HV	[79]		
316		hex	[3, 74, 76]	0.4811/0.4484	[74]	5.8	[107]		[74]											2100–2330 HK	[85]		
317		hex	[104]	0.4759/0.4438	[75]	6.51	[111]		[75]											1100 HU 0.1	[103]		
318		hex	[110, 112]	0.275/0.447	[111]		[111]													1570 HV 0.05	[111]		
319				0.48113/0.44841																		86.3–110.3	[6]
320	HfN	fcc	[8]	0.452	[7]	13.8	[7]	3310	[5]	6.9	[6]	11.3	[8]	26	[7]	369.4	[12]	3.33–4.8	[29]	1700–2000	[7]	114.3	[46, 73]
321		cub-B1	[10]	0.4526	[10]	13.8 (x)	[10]	3300	[7]	6.9	[8]	21.77	[10]	15	[32]	369	[17]	4.64	[40]	2000 HV	[9]	127.7 P	[102]
322		cub-B1	[25]	0.451	[25]	11.97	[74]	2700	[8]	6.9	[10]	21.77	[32]	33	[57]	369.3	[46]	4.64	[45]	1600	[10]		
323		fcc	[52]	0.4513	[52]	13.84	[98]	3387	[10]	6.6	[32]			56.5	[106]	300	[65]	3.8	[74]	1500–3500 HV	[18]		
324		fcc	[74]	0.4518	[29]	11.97	[81]	3300	[25]	6.6	[40]					369.3	[73]	4.64	[87]	1600–8590 HV	[20]		
325		cub-B1	[52]	0.452	[32]	13.8	[106]	3387	[32]	6.6	[45]							3.8	[81]	1600 HV 0.1	[32]		
326		cub-B1	[76]	0.4525	[52]	13.8	[107]	3928	[40]	6.6	[87]									2700	[40, 45]		
327		fcc	[98]	0.451	[74]			3330	[45]	6.9	[98]									1640	[57]		
328		fcc	[81]	0.452	[98]			3300	[46, 57]	6.9	[106]									1700	[59]		
329		cub	[112]	0.45253	[112]			3300	[62]	6.9	[81]									2000	[62]		
330								3928	[73]											1600–3500	[76]		
331								3928	[87]											2700 HV	[87]		
332								3200	[98]											2030	[98]		
333																				1300 HU 0.1	[54]		
334								3305	[106]											2240–2600 HK	[85]		

Table 1. Continued

No.	Symbol	Crystal structure	Lattice parameters (nm)		Density (g cm^{-3})		Melting point (°C)		Linear thermal expansion, α (10^{-6} K^{-1})		Thermal conductivity, λ (W m^{-1} K^{-1})		Electrical resistivity (10^{-6} Ωcm)		Enthalpy (kJ mol^{-1})		Young modulus (10^5 N mm^{-2})		Micro hardness (10 N mm^{-2})		Oxidation resistance (100°C)	
335	Mo$_2$N-γ	fcc	0.4139-0.4160	[6, 10]	9.46	[6]	700 (d)	[57]			17.88	[6]			70.3-81.6	[6]			630	[57]		
336		cub	0.4128	[112]			527	[73]							69.5	[46]						
337		fcc	0.4155-0.4160	[25]											69.5	[73]						
338		fcc	0.4163	[52]																		
339		fcc	0.4163	[112]																		
340	Mo$_2$N-β	tet	0.418-0.42/0.8-0.820	[6]	9.3	[6]																
341		tet	0.4210/0.8060	[76, 112]																		
342	NbN-δ	fcc	0.44	[1, 8]	8.3	[1]	2300 (d)	[1]	10.1	[6]	2.93-4.61	[6]	~200	[57]	237.8	[1]			1370-1422	[6]		
343		cub-B1	0.439	[3]	8.26-8.4	[6]	2300	[8]	10.1	[8]	3.74	[8]			237.8	[6]			1400 HV	[8]		
344		cub-B1	0.4392	[6]	8.4	[25]	2050	[25]											1400	[57]	5-8	[25]
345		fcc	0.4375	[57]			2630 (d)	[57]											1400	[59]		
346	NbN-ϵ	hex	0.2951/1.1271	[6]	8.4	[7]	2570	[7]	10.1	[10]	3.77	[10]	60	[32]	211.9	[5]	4.834	[74]	1400	[10]		
347		hex	0.2958/1.1272	[10]	7.3 (x)	[10]	2300	[9]	10.1	[32]	3.77	[32]	58	[66]	236.6	[12]	4.8	[77]	1400	[32]		
348		hex	0.2958/1.1272	[14, 46]	8	[74]	2204	[10, 32, 97]					58	[77]	234	[46]	4.8	[97]	1400 HV	[77]		
349		hex	0.2952/1.125	[74, 76]	8.43	[77]	2630	[46]					58	[97]	236.6	[73]						
350		hex-B1	0.2986/0.5548	[52]	8.43	[97]	2204 (d)	[77]														
351	NbN	hex	0.2980/1.1270	[112]	8.4	[106, 108]	2573	[106]	10.1	[106]			200	[106]								
352					7.3	[107]																
353	Nb$_2$N	hex	0.31/0.496	[1]	8.3	[1]	2430	[2]	3.26	[6]	4.27-8.71	[6]			255.8	[1]			2120	[1]		
354		hcp	0.3052/0.4964	[6, 52]	8.08-8.62	[6]	2400	[73]							255.8	[6]			1620-1820	[6]		
355		hex	0.3058/0.4961	[25]	8.33	[52]									253.3	[12, 73]						
356		ortho	0.3056/0.4948	[46]	8.31 (x)	[52]									248.7	[46]						
357	Si$_3$N$_4$	hex	α: 0.78/0.56	[22]	3.44	[1, 106]	1900	[1]	2.4	[1]	20-24	[9]	10^{18}	[9]	750.5	[6]	2.1	[9]	1410 HV	[4]	12-14	[43]
358		hex	α: 0.758/0.5623	[76]	3.44	[6]	1900	[6]	2.5	[9]	17-30	[22]	10^{18}	[14]	760	[22]	2.06	[14]	1700	[9]	16	[70]
359		α,β: hex	β: 0.76/0.30	[43]	3.44	[7]	1900	[7]	3	[14]	12	[38]	10^{16}-10^{18}	[22]	630	[43]	α: 3.20	[22]	1700 HK 0.05	[14]		
360		hex	α: 0.77541/0.56217	[76]	3.19	[9]	1900	[9]	2.7-3.5	[22]	10-15	[43]	10^{18}	[38]	745.2	[46]	β: 2.90	[22]	1700 HK	[22]		
361		α,β: hex	β: 0.76044/0.29075	[112]	3.18	[14]	1900 (d)	[14]	2.8	[38]	30	[57]	10^6	[57]			2.2	[38]	α: 2800-4000	[22]		
362					3.2	[22]	1900 (d)	[22]	2.7-3.3	[43]	32	[67]	10^{18}	[66]			1.6-3.0	[43]	β: 2000-3000	[22]		
363					3.2	[38]	1900 (d)	[38]	4.2	[47]	10-43	[108]	>10^{12}	[67]			~3	[57]	2300 HV 0.05	[38]		
364					3.44	[39]	1897	[39]	2.4	[57]			10^{18}	[77]			3.1	[67]	3340 HV	[39]		
365					3.19	[43]	1897 (d)	[43]	3.3	[67]			10^{18}	[97]			3	[70]	2000-4500 HK	[47]		
366					2.75-2.95	[47]	1900	[57]	3	[70]			10^{13}-10^{21}	[108]			2.1	[77]	1800 HV	[57]		
367					1.8-3.1	[57]	1900	[77]	~1.5	[76]							2.1	[97]	α: 1700 HK 0.1	[58]		
368					3.44	[52]	1900	[88]	2.5	[97]							1.7-3.0	[108]	2500-3000 HV	[58]		
369					3.15 (x)	[52]	1900	[106]	2.5	[108]									3300	[59]		
370					3.25	[57]			3.3										1500 HV 30	[67]		
371					3.2	[67]													2200-3200 HV 0.1	[70]		
372					3.21	[70]													500-3900	[76]		
373					3.19	[77, 97]													1720 HV	[77, 97]		
374					3.2	[88]													1400 HV 0.2	[88]		

#	Compound	Structure	Ref	Lattice param	Ref	Col1	Ref	Col2	Ref	Col3	Ref	Col4	Ref	Hardness	Ref	Col5	Ref
375										3.4	[107]						
376										2.4–3.44	[108]						
377	TaN	hcp	[1]	0.52/0.29	[1]					13.6–13.8	[1]	3000	[1]	225.7	[5]			800–2000 HV	[108]	
378		hex	[6]	0.5185/0.2908	[6]	128	[6]	8.58	[6]	14.35 (x)	[6]	3090	[5,25,88]	243.2	[6]	3240	[1]	5–8	[25]
379		hex	[8]	0.5185/0.2908	[10]	128	[25]	9.6	[8]	16.3	[7.74]	3360	[7]	251	[12]	1060	[6]		
380		hex	[10]	0.495/0.305	[25]	128	[32]	8.79	[10]	14.34 (x)	[10,52]	2090	[8]	252.5	[46]	1500–3000	[7]		
381		hcp	[25]	0.5185/0.2908	[32]	128	[57]	8.79	[32]	14.1	[25,107]	3093	[10]	252.5	[73]	1300 HV	[8]		
382		hex	[74,76]	0.5185/0.2908	[52]	130–250	[84]			13.6	[52]	3093	[32]			1000	[10,32,76]		
383		hex	[46,52]	0.5192/0.2908	[74]					13.8	[88]	2950	[46]			1060	[57]		
384		hcp	[57]	0.51918/0.29081	[112]					16.3	[106]	2930 (d)	[57]			1150–3000	[59]		
385		hex	[112]							13.8	[108]	3360	[106]			3230 HV 0.2	[88]		
386										16.3								
387	Ta₃N	hex	[2]	0.30/0.493	[2]	263	[6]	10.05	[6]	15.8	[2]	3000	[1]	270.9	[1]	3000	[1]		
388		hcp	[6,10]	0.3046/0.49187	[6]	190–250	[84]			15.46	[6]	2700	[46]	270.9	[6]	1220	[6]		
389		ortho	[46]	0.3048/0.4919	[10]					15.46	[52]	2727	[73]	270.5	[12]				
390		hcp	[14,52]	0.3048/0.4919	[52]					15.86 (x)	[52]			273	[46,73]				
391		hex	[112]	0.30445/0.49141	[112]													
392	TiN	fcc	[1]	0.423	[1]	25	[7]	29.31	[1]	5.21	[1]	2950	[1]	336.6	[1]	2450 HV	[1,98]	12	[7]
393		cub-B1	[3]	0.4238	[3]	25	[9]	28.9	[3]	5.21	[3]	2950	[5]	303.1	[5]	1800–2800	[7.62]	11–14	[25]
394		cub-B1	[6]	0.423	[6]	21.7	[25]	30	[8]	5.21	[6]	3220	[6]	336.2	[6]	2400 HV	[8.95]	~5	[53]
395		fcc	[8]	0.424	[8]	25–30	[27]	19.26	[9]	5.4	[7]	2950	[7]	338.1	[12]	2500–2800	[8]	5	[92]
396		cub-B1	[10]	0.424	[9]	75–500	[37]	19.26	[10]	5.39 (x)	[8]	2930	[8]	336	[17]	2000	[10,32,76]	~3.5	[113]
397		fcc	[23]	0.424	[10]	25	[32]	29.2	[32]	5.21	[9]	2950	[9]	336.6	[46]	2000–2500 HV 0.05	[19]	4.5	[122]
398		cub-B1	[25]	0.424	[15]	30	[38]	38	[37]	5.21	[10]	2949	[10]	337	[57]	1–6000 HV 0.001	[20]		
399		fcc	[41]	0.4245	[32]	25	[57]	30	[38]	5.21	[15]	2950	[25]	333	[65]	2000 HV 0.015	[21]		
400		cub-B1	[46]	0.423	[37]	17	[91]	19	[57]	5.43	[32]	2950	[27]	338.1	[73]	2200–2800 HV 0.05	[23]		
401		cub-B1	[52]	0.424	[38]	25	[66]	70	[91]	5.4	[37]	2930–2950	[29]	336.6	[98]	1770 HK 0.1	[25]		
402		fcc	[59]	0.4242	[39]	25	[77]			5.4	[38]	2949	[32]			2300–2600	[27]		
403		fcc	[60]	0.424	[45]	25	[87]			4.73	[39]	2950	[37]			1770 HK 0.1	[33]		
404		fcc	[74]	0.42	[57]	25	[91]			5.43	[45]	2950	[38]			2450 HV 30	[37]		
405		cub-B1	[76]	0.42	[60]	25	[92]			5.21	[53]	3205	[39]			1994 HV	[39]		
406		fcc	[81]	0.422	[74]	17	[106]			5.3	[57]	2900	[40]			2100	[40,59]		
407		fcc	[98]	0.42	[77]	22				5.21	[60]	2950	[45]			2200 HV	[41]		
408		fcc	[101]	0.42	[81]					5.2	[74]	2950	[46]			3300 HK 0.05	[42]		
409		cub-B1	[109]	0.424173	[98]					5.43	[77]	2950	[57]			2300	[45]		
410		cub	[112]		[109]					5.21	[88]	2950	[60]			1600–2000 HV	[47]		
411					[112]					5.4	[90]	2950	[61]			2000–2700 HV	[49,58]		
412										5.2	[91]	2949	[62]			1900–2400	[53]		
413										5.4	[93]	2950	[76]			2450 HV 0.1	[57]		
414										5.4	[96]	2949	[77]			1050 HV 0.1	[58]		
415										5.2	[97]	2950	[88]			2000–2400 HV 0.05	[59]		
416										5.43	[92]	2900	[87]			1700–2800 HV 0.05	[60]		
417										5.21	[81]	3205	[90]			1500–2700 HV 0.05	[61]		
418										5.21	[106]	2930	[91]			1725–1825 HV 0.02	[64]		
419										5.4	[107]	2950	[93]			1775–1935 HV	[64]		
420										5.22	[108]	2950	[98]			1800 HV	[71]		
421												2950	[92]			2160 HV	[80]		

Table 1. Continued

No.	Symbol	Crystal structure	Lattice parameters (nm)	Density (g cm⁻³)	Melting point (°C)	Linear thermal expansion, α (10⁻⁶ K⁻¹)	Thermal conductivity, λ (W m⁻¹ K⁻¹)	Electrical resistivity (10⁻⁶ Ωcm)	Enthalpy (kJ mol⁻¹)	Young modulus (10⁵ N mm⁻²)	Micro hardness (10 N mm⁻²)	Oxidation resistance (100°C)
422					2947 [106]						1666–3098 HV [82]	
423											1530–2730 HV 0.01 [86]	
424											2450 HV 0.2 [88]	
425											2100 HV [77,87]	
426											1994 HV [90]	
427											2300 HV 0.03 [91]	
428											2500–2800 HV 0.05 [93]	
429											2200 HV 0.05 [96]	
430											1800–2800 HV 0.05 [92]	
431											500–1200 HU 0.1 [54]	
432											2000–3350 HK [85]	
433											750–1200 HU 0.1 [103]; 1640 HV 0.05 [113]; 2225–2290 HK 0.01 [117]; 2200–2400 HV 0.01 [122]	
434	Ti₂N-ε	tet [6]	0.494/0.303 [101]								3000–4000 HV [41]	4 [122]
435		tet [41,112]	0.4452/0.3042 [112]								2400–2700 HV 0.01 [122]	5–8 [25]
436	VN	fcc [1,8]	0.41 [1]	6.13 [7,74]	2050 (d) [1]	8.1 [8]	11.3 [6]	85–100 [6]	147.8 [5]	4.6 [77]	1520 [6]	
437		cub-B1 [6,25]	0.4137 [6]	5.62 [6]	2050 [5,6,7,8,25,73]	8.1 [10]	11.3 [8]		217.3 [6]	4.6 [97]	1400–1600 [7]	
438		cub-B1 [46,52]	0.414 [3,10,32]	6.04 (x) [6]	2177 [10,32]	8.1 [32]	11.3 [10]		217.3 [46]		1500 HV [8]	
439		fcc [57,74]	0.4126 [25]	6 [10]	2350 [46]	9.2 [77]	11.3 [32]		217.3 [73]		1500 [10,32,59]	
440		cub-B1 [3,10]	0.4169 [52]	6.04 [25,52]	2350 (d) [57]	9.2 [97]		85 [97]			1520 [57]	
441		cub-B1 [76,109]	0.4139 [74]	6.11 [77,97]	2177 [77,97]	8.1 [106]		200 [106]			1560 HV [77,97]	
442		cub [112]	0.413 [109]	6.13 [108]	2050 [106]							
443			0.413916 [112]	6.13 [106]								
444				6.1 [107]								
445	WN-δ	hcp [6]	0.2893/0.2826 [10]								2500–4000 HV 0.05 [23]	
446		hex [10,52]	0.2893/0.2826 [52]									
447		hex [112]	0.2893/0.2826 [112]									
448	W₂N	fcc [6,52,57]	0.4126 [10,52,74]	12 [74]	700 (d) [57]			100–180 [83]				
449		cub [10,83]	0.4118 [25]									
450		fcc [25,74]	0.4128 [83]									
451		cub [112]	0.4126 [112]									
452	ZrN	fcc [1]	0.46 [1]	6.93 [6]	3000 [1]	6 [1]	16.75 [1]	13.6 [6]	365.5 [1]	5.1 [9]	2000 [1]	12 [7]
453		cub [3]	0.458 [6]	7.09 [7]	2980 [5,7,8]	7.9 [8]	10.9 [8]	21 [7]	336.2 [5]	5.1 [45]	1300–2000 [7]	11–14 [25]
454		fcc [8]	0.459 [8]	7.32 [9]	2980 [25,38]	7.2 [9]	28 [9]	21 [9]	365.5 [12]	3.93–4.60 [74]	1900 HV [8]	
455		cub-B1 [10]	0.4577 [10]	7.3 (x) [10]	2980 [39,90]	7.24 [10]	20.52 [10]	13.6 [25]	368.4 [46]	5.1 [77]	1600 [9,59]	
456		cub-B1 [25]	0.456 [25]	6.93 [25]	2982 [6,9,10]	7.24 [32]	20.52 [32]	21.1 [32]	366 [65]	5.1 [93]	1500 [10,32]	

No.	Compound	Struct.	[ref]	Lattice params	[ref]	ρ	[ref]	Tm	[ref]		[ref]		[ref]	E	[ref]		[ref]	Hardness	[ref]
457		cub-B1	[46]	0.4575	[29]														1510 HK 0.1	[25,33]
458		cub-B1	[52]	0.4577	[32]														1520 HV	[39,90]
459		fcc	[57]	0.4562	[52]														2300	[45]
460		fcc	[74]	0.4578	[74]														1520	[57]
461		cub-B1	[76]	0.463	[109]														1250 HV 0.1	[58]
462		cub-B1	[109]	0.457756	[112]														1500–2600	[76]
463		cub	[112]																1600 HV	[77,97]
464																			2500 HV 0.05	[93]
465																			2130 HV 0.05	[113]
466																				

Borides

No.	Compound	Struct.	[ref]	Lattice params	[ref]	ρ	[ref]	Tm	[ref]		[ref]	α	[ref]	E	[ref]		[ref]	Hardness	[ref]
467	AlB₂	hex	[52]	0.3006/0.3252	[52]	3.17 (x)	[52]	1975	[46]					67	[46]				
468		hex	[112]	0.30054/0.325276	[112]															
469	AlB₁₂	tet	[52]	0.1258/0.1020	[52]	2.57	[52]	2070 (per)	[5]			4.3		201	[46]				
470		tet	[112]	0.1030/0.1433	[52]	2.58	[77]	2070 (per)	[46]										
471				1.016/1.428	[112]														
472		mono	[52]	1.704/1.100/1.884	[52]			2150 (d)	[77]									14–18	[25]
473				0.8522/1.100/0.7393	[52]															
474	CrB	ortho	[6]	0.2969/0.7858/0.2932	[6]	6.05	[6]	1550	[6]		[25]			~75.4	[46]				
475		ortho	[25]	0.2969/0.7858/0.2932	[25,52]	6.05	[25,52]	1550	[25]					75.4	[73]				
476		ortho	[52]	0.2969/0.7858/0.2932	[52]	6.11 (x)	[52]	1550	[73]										
477		ortho	[112]	0.29663/0.78666/0.29322	[112]															
478	CrB₂	hex	[1]	0.279/0.307	[1,106]	5.6	[1]	2200	[1]	56	[7]			94.6	[5]	2.15		2250	[1,57]
479		hex	[6]	0.2969/0.3066	[6,107]	5.6	[5]	2170	[5]	21	[25]			129.8	[6]	2.15		1700	[6]
480		hex	[25]	0.2969/0.3066	[7]	5.6	[6]	1800–2300	[6]	56	[57]			~94.2	[46]	5.4		2200	[7]
481		hex	[52]	0.2969/0.3066	[25]	5.6	[25]	2150	[7]	18	[66]			94.2	[73]			1800 HV 0.1	[25]
482		hex	[57]	0.29730/0.30709	[57]	5.6	[57]	1850	[25,106]	18	[77]							1800	[33]
483		hex	[112]			5.58	[77]	2200	[57]	18	[97]							2250 HV	[77,97]
484						5.58	[97]	1850–1900	[73]	21	[106]							2100	[59]
485								2188	[77,97]										
486	Cr₅B₃	tet	[1]	0.55/1.06	[1]	6.1	[1,106]	1900	[1]					125.6	[1]				
487		tet	[6]	0.546/1.064	[6]	6.12	[6]	1960	[25]										
488		tet	[52]	0.546/1.064	[52]	6.14	[25]												
489		tet	[112]	0.54735/1.0115	[112]	6.12 (x)	[52]												
490	FeB	ortho	[52]	0.5506/0.2952/0.4061	[52]	6.3	[52]	1650	[5,46,73]					71.2		5,46,73		1400–1190	[7]
491		ortho	[109]	0.405/0.550/0.295	[109]			1550	[61]											
492		ortho	[112]	0.40587/0.55032/0.29474	[112]															
493	Fe₂B	tet	[52]	0.5109/0.4249	[52]	~7.0	[52]	1389	[46,61,73]					71.2		46,73			
494		tet	[109]	0.510/0.424	[109]															
495		tet	[112]	0.51103/0.42494	[112]															

Table 1. Continued

No.	Symbol	Crystal structure		Lattice parameters (nm)		Density (g cm^{-3})		Melting point (°C)		Linear thermal expansion, α (10^{-6} K^{-1})		Thermal conductivity, λ (W m^{-1} K^{-1})		Electrical resistivity (10^{-6} Ωcm)		Enthalpiy (kJ mol^{-1})		Young modulus (10^5 N mm^{-2})		Micro hardness (10 N mm^{-2})		Oxidation resistance (100°C)
496	HfB	cub-B1	[52]	0.462	[52]																	
497		ortho	[112]	0.6517/0.3218/0.4919	[112]																	
498	HfB$_2$	hex	[25]	0.3141/0.3470	[25]	11.01	[7]	3200	[5,63]	5.3	[57]	430	[98]	10	[25]	336.6	[5]			2800	[7]	11–17 [25]
499		hex	[52]	0.3141/0.3470	[52]	11.2	[25,107]	3380	[7]	5.7	[98]			15.8	[57]	336.2	[46]			2900	[57]	
500		hex	[57]			10.5	[52,108]	~3060	[25]	5.3	[106]			12	[106]	336.2	[73]			2900	[59]	
501		hex	[112]	0.314245/0.347602	[112]	11.2 (x)	[52]	3370	[46]											2850	[98]	
502						11	[57]	3250	[57,98]											4000–6000 HV 0.01	[51]	
503								2147	[73]													
504						10.96	[106]	3240	[106]													
505	LaB$_6$	cub	[109]	0.415	[109]	4.76	[90]	2530	[90]					28	[106]					2770 HV	[90]	
506		cub	[112]	0.41569	[112]	4.7	[106,107]	2200	[106]													
507						2.61	[108]															
508	LaB$_8$	cub	[52]	0.4156	[52]	4.73	[7]	2530	[5,39]	5.84	[30]			17	[7]					1400–2500	[7]	
509		cub	[55]	0.4145	[55]	4.76	[39]	2200	[7]	6.4	[77]			15	[77]					2770 HV	[39]	
510				0.4156		4.73	[77]	2770	[77]											2530 HV	[77]	
511	MoB-δ	tet	[25]	0.3110/1.695	[25]	8.3	[25]	~2500	[5]					50	[25]					1570 HV 0.1	[25]	
512		tet	[52]	0.3110/1.695	[52]	8.77 (x)	[52]	1930	[25]					50	[57]					1570	[33]	
513		tet	[112]	0.3105/1.697	[112]	8.3	[52]	2350	[57]											2500	[57]	
514				0.3108/1.697																		
515	MoB-β	ortho	[25, 52]	0.316/0.861/0.308	[25]	8	[57]													2500	[59]	
516		ortho	[57]	0.316/0.861/0.308	[52]	8.5																
517		ortho	[112]	0.316/0.844/0.308	[112]																	
518	MoB$_2$	hex	[1]	0.3/0.31	[1]	7.8	[1]	2100	[1,57]					45	[25]	96.3	[1]			1380 HV 0.1	[25]	11–14 [25]
519		hex	[25, 52]	0.305/0.3113	[25]	7.12	[7]	2300	[7]					30	[57]					3000	[57]	
520		hcp	[57]	0.306/0.310	[52]	8	[25]	2250	[25]													
521		hex	[112]	0.304/0.307	[112]																	
522	Mo$_2$B	tet	[1]	0.55/0.47	[1]	9.2	[1]	2140	[1]					40	[25]	106.8	[1]			2500	[1]	
523		tet	[25, 52]	0.5543/0.4735	[25, 52]	9.1	[25]	1850	[25]											1660 HV 0.1	[25]	
524		tet	[52, 109]	0.5547/0.4739	[112]	9.1	[52]													1660	[33]	
525		tet	[112]	0.554/0.474	[109]	9.31 (x)	[52]															
526	Mo$_2$B$_5$	hex	[1]	0.3/2.1	[1]	7.2	[1]	2300	[1]	8.6	[77]			18	[77]	209.3	[1]	6.7	[77]	3200	[1]	
527		hex	[25]	0.301/2.093	[25]	7.48 (x)	[52]	2140	[77]	8.6	[97]			18	[97]			6.7	[97]	2350 HV	[77]	
528		hex	[52]	0.301/2.093	[52]	7.45	[77,97]	2140	[97]	5	[106]			25	[106]					2350 HV	[97]	
529		hex	[112]	0.301174/2.09369	[112]	7.12	[106]	2100	[106]													
530						7.8	[107]															
531	Mo$_3$B$_2$	tet	[1]	0.60/0.31	[1]	9	[1]	2240	[1]							175.8	[1]			2300	[1]	

#	Compound	Structure	Lattice params	[ref]	Density	[ref]	Melting point	[ref]	α	[ref]	k	[ref]	ρ	[ref]	E	[ref]	Hardness	[ref]	[ref]
532	NbB	ortho	0.3298/0.8724/0.3166	[6, 25]	7.6												2200	[6]	
533		ortho	0.3292/0.8713/0.3165	[52]															
534		ortho	0.32973/0.87229/0.31883	[112]															
535	NbB$_2$	hex	0.31/0.33	[1]	6.8	[1]	3000	[1]	16.75	[1]			32	[25]	167.5	[6]	2600	[1, 6, 57, 59]	11-14 [25]
536		hex	0.3099/0.3271	[6]	7.2	[6]	3000	[6]	16.75–35.2	[6]			12	[57]			1300–2600	[7]	
537		hex	0.3086/0.3306	[25]		[77]	3036	[7, 57]		[106]			12	[77]			1800–1900	[28]	
538		hex	0.3096/0.3306	[52]	6.6	[97]	~2900	[25, 52]	16.75	[97]			12	[97]			2600 HV	[77]	
539		hex	0.311133/0.32743	[112]	7.21 (x)	[106]	3036	[52]	16.7	[106]			32	[106]					
540					6.98		3036	[77, 97]											
541					6.97		2900	[106, 108]											
542	SiB$_6$	ortho	1.4470/1.8350/0.9946	[112]	2.43	[1]	1950	[1]					10^7	[66]			1910	[1]	
543					2.47	[1]	1947	[6, 108]	3.3	[77]			10^7	[77]			2450–2800 HK 0.1	[39]	
544					2.43	[70]	1900	[39]	3.3	[97]			10^7	[97]			2300 HV 0.1	[58]	
545					2.43	[77]	1900	[70]									2400–2800 HV 0.1	[70]	
546					2.43		1900	[77]									2300 HV	[77, 97]	
547	TaB$_2$	hex	0.31/0.33	[1]	12.1 (x)	[1, 57]	3150	[1]	21.35	[1]			68	[25]	209.3	[1]	2200	[1, 57]	11-14 [25]
548		hex	0.3085/0.3249	[6]	12.58	[6]	3100	[6]	10.89	[6]			21	[57]	193.8	[5]	2615 HK 0.1	[6]	
549		hex	0.3078/0.3265	[25]	11.7	[7]	3040	[7]	10.9	[25]			14	[77]	209.3	[46]	2500	[28]	
550		hex	0.309803/0.322660	[52]	11.7	[25]	~3000	[25]		[106]			68	[106]	209.3	[73]	2100–2400	[47]	
551		hex		[57]	12.6 (x)	[46]	3100	[52]									2500 HV	[76]	
552		hex		[112]	12.58	[52]	3150	[57, 63]									2000–2700		
553					11.15	[77]	3037	[77]											
554					12.6	[106]	3000	[106, 108]											
555						[107]													
556	Ta$_2$B	tet	0.58/0.48	[1]	15.2	[1]	1920	[1]									2100 HV	[77]	
557		tet	0.5778/0.4864	[6, 25]	15.18 (x)	[6]											2200	[1]	
558		tet	0.5778/0.4846	[52]													2200	[6]	
559		tet	0.578/0.486	[109]													2430HV	[47]	
560		tet	0.5783/0.4866	[112]															
561	Ta$_3$B$_2$	tet	0.62/0.33	[1]	15	[1]	2120	[1]									2770	[1]	
562		tet	0.6184/0.3284	[6]	15.0 (x)	[6]											2770	[6]	
563		tet	0.6184/0.3286	[112]															
564	Ta$_3$B$_4$	ortho	0.329/1.40/0.313	[6, 25, 52]	13.5	[6, 52]											3350	[57]	
565		ortho	0.329/1.40/0.313	[112]	13.60 (x)	[6, 52]													
566		ortho																	
567	TiB$_2$	hex	0.3/0.32	[1]	4.5	[1, 107]	2900	[1, 25, 30]	25.96	[7]	6.39	[1]	9	[7]	150.7	[1]	3480	[1]	11-17 [25]
568		hex	0.3028/0.3228	[6]	4.38	[6, 106]	2900	[38, 57, 88]	26.13	[9]	7.8	[9]	7	[9]	324.1	[5]	3000–3400	[7]	13 [43]
569		hex	0.3026/0.3213	[25]	4.38	[25]	2920	[7]	26–39	[31]	5.6	[31]	7	[31]	280	[43]	2500–3500	[9]	
570		hex	0.3028/0.3228	[30]	4.5	[52]	3230	[9, 108]	26.13	[38]	4.84/9.83	[30]	10	[38]	324.1	[46]	3400 HV 0.05	[25]	
571		hex	0.303034/0.322953	[31]	4.4	[74]	3225	[25]	27	[40]	7.8	[38]	7	[40]	305	[48]	3000 HV	[31, 97, 98]	
572		hex	0.3	[43]	4.5	[98]	3225	[31]	65–120	[43]	8	[40]	7	[43]	324.1	[50]	2700 HK 0.1	[33]	

Table 1. Continued

No.	Symbol	Crystal structure	Lattice parameters (nm)		Density (g cm^{-3})		Melting point (°C)		Linear thermal expansion, α (10^{-6} K^{-1})		Thermal conductivity, λ (W m^{-1} K^{-1})		Electrical resistivity (10^{-6} Ωcm)		Enthalpy (kJ mol^{-1})		Young modulus (10^5 N mm^{-2})		Micro hardness (10 N mm^{-2})		Oxidation resistance (100 °C)	
573		hex	0.303034/0.322953	[52]	4.5	[38]	3225	[77, 97, 98]	8	[43]	24–56	[48]	9	[57]	324.1	[73]	4.8	[45]	3400			
574		hex	0.3018–0.3023,...	[57]	4.5	[39]	2980	[39, 40, 76]	7.8	[45]	26.9	[98]	7	[66]	150.7	[98]	5.6	[48]	3300 HV	[39, 90, 77]		
575		hex	.../0.319–0.322	[74]	4.52	[43]	2980	[87, 90]	6.6/8.6	[48]	26.1	[106]	7	[77]			3.74	[57]	3370	[40, 76]		
576		hex		[98]	4.5	[45]	2980	[40]	6.39	[57]			7	[87]			4.8	[76]	3000	[45]		
577		hex		[112]	4.52	[48]	3197	[43]	8	[76]			7	[97]			5.6	[77]	2900–3700 HV	[47]		
578					4.38	[52]	3325	[45]	7.8	[77]			7	[98]			4.8	[87]	2500–3000	[48]		
579					4.5	[57]	2790	[106]	8	[87]			15, 5	[106]			5.6	[97]	3480	[57]		
580					4.5	[74, 77]			7.8	[97]							5.6	[98]	3300	[59]		
581					4.5	[90, 98]			7.8	[98]									3370 HV	[87]		
582					4.4	[88]			5.9	[106]									3480 HV 0.2	[88]		
583																			1646–3115 HV 0.05	[123]		
584																			1843 HV 0.05	[119]		
585																			2200–6900 HV 0.01	[127]		
586	Ti$_2$B	tet	0.61/0.46	[1]			2200	[7]											2500	[1]		
587		tet	0.611/0.456	[25, 52]																		
588		tet	0.610/0.453	[52]																		
589	VB$_2$	hex	0.3/0.31	[1]	4.8	[1]	2400	[1, 5, 7]	5.3	[1]			16	[6]	203.9	[73]	5.1	[77]	2080	[1, 57]	13	[7]
590		hex	0.2998/0.3057	[6]	4.56	[6]	2450	[6]	5.3	[57]			16	[25]			5.1	[97]	2077	[6]	8–14	[25]
591		hex	0.2998/0.3057	[25]	5.1	[7, 107]	2100	[25]	7.6	[77]			38	[57]					1300–2100	[7]		
592		hex	0.2998/0.3057	[52]	5.1	[25, 108]	2400	[57]	7.6	[97]			6	[66]					2100	[59]		
593		hex	0.299761/0.305620	[57]	4.61	[52]	2747	[77, 97]	6.7	[106]			13	[77]					2150 HV	[77]		
594		hex		[112]	5.10 (x)	[52]							13	[97]					2120 HV	[97]		
595					5	[57]							16	[106]								
596					5.05	[77, 97]																
597					4.92	[106]																
598	WB	tet	0.31/1.7	[1]	15.5	[1]	2860	[1, 25]											3750	[1]		
599		tet	0.3115/1.693	[25]	15.3	[25]	2685–2920	[6]											3750	[57]		
600		δ: tet	0.3115/1.693	[52, 112]	15.3	[52]	2860 (d)	[57]														
601		β: ortho	0.319/0.840/0.307	[25]	16.0 (x)	[52]																
602		ortho	0.319/0.840/0.307	[52]	15.5	[57]																
603		ortho	0.311655/1.69101	[57]	15.7	[107]																
604					10.77	[108]																
605	W$_2$B	tet	0.56/0.47	[1]	16.5	[1]	2770	[1]	4.7	[106]			21–43	[106]					2350	[1]	8–14	[25]
606		tet	0.5564/0.4740	[25]	16	[25, 52]	2670–2780	[6]														
607		tet	0.5564/0.4740	[52]	16.72 (x)	[52]	2770	[25]														
608		tet	0.556/0.474	[109]	10.77	[106]	2900	[106]														
609		tet	0.5568/0.4744	[112]																		

#	Compound	Structure	Ref	Lattice parameter (nm)	Ref	Density (g/cm³)	Ref	Melting point (°C)	Ref	α (×10⁻⁶/K)	Ref	k (W/m·K)	Ref	Cp (J/mol·K)	Ref	Microhardness (HV)	Ref
610	WB$_2$																
611		hex	[1]	0.3, 1.39	[25, 52]	15.73		2920	[7]								[7]
612	W$_2$B$_5$	hex	[25, 52]	0.2982, 1.387	[25, 52]	13.1	[1]	2300	[1, 57]	7.8	[57]					2700	[1]
613		hcp	[57]	0.2982, 1.387	[52]	11	[52]	2980	[5]	7.8	[77]					2700	[57, 59]
614		hex	[112]	0.2982, 2.0715	[112]	13.1 (x)	[77, 97]	2200–2980	[6]		[97]					2700HV	[77]
						13.03		2365	[77, 97]								
615	ZrB																
616		fcc	[1]	0.47	[1]	6.5	[1]	3000	[6]							3600	[1]
617		cub-B1	[25]	0.468	[25]	5.7	[25]									3600	[57]
		cub-B1	[52]	0.465	[52]	6.7 (x)	[52]										
618	ZrB$_2$	hex	[1]	0.32, 0.35	[1]	6.1	[1]	3000	[1, 30, 106]	9.2	[6]	23.03	[1]	163.3	[1]	2200	[1, 33, 57]
619		hex	[6]	0.3170, 0.3533	[6]	5.64	[6, 7]	3040	[5, 39, 90]	7	[7]	23.03	[25]			2600	[7]
620		hex	[25]	0.3169, 0.3530	[25]	6.17	[25, 52]	2992	[7]	9.2	[25]	23	[38]			2200 HV 0.05	[25]
621		hex	[30]	0.3170, 0.3533	[30]	6.17	[39]	3250	[25, 38, 57, 88]	10	[38]	23	[98]			2252 HV	[90, 39]
622		hex	[52]	0.32	[47]	~6	[47]	2990	[46]	7	[47]	12.1	[106]			2200 HV	[47]
623		hex	[57]	0.316870, 0.353002	[52]	6.09 (x)	[52]	3200	[52]	6	[57]					965 HV 0.1	[58]
624		hex	[98]			6.11	[77, 97]	3245	[77, 97, 98]	6	[77]					2300	[59]
625		hex	[112]			6	[88]			6	[97]					2300–3000	[76]
626						6.17	[90]			6	[98]					2300 HV	[77, 97, 98]
627						6.1	[107]			9.2	[106]					2200 HV 0.2	[88]
628						6.085	[108]										
629						5.64	[106]										
630																	
631	ZrB$_{12}$	cub	[1]	0.74	[1]	3.65	[1]	2700	[1]			12.14	[25]	502.4	[1]	2500	[1]
632		fcc	[25]	0.7408	[25]	3.7	[25]									2500	[57]
633		cub	[52]	0.7408	[52]	3.7	[52]										
634		cub	[112]	0.7408	[112]	3.63 (x)	[52]										

Silicides

#	Compound	Structure	Ref	Lattice parameter (nm)	Ref	Density (g/cm³)	Ref	Melting point (°C)	Ref	α (×10⁻⁶/K)	Ref	k (W/m·K)	Ref	Cp (J/mol·K)	Ref	Microhardness (HV)	Ref
635	CrSi																
636		cub	[1]	0.462	[1]	5.38	[6]	1550	[1]					53.2	[46]	1000	[1]
637		cub	[6]	0.4629	[6]			1550	[6]					54.8	[73]	950–1050	[6]
638		cub	[25]	0.462	[25]			1475	[63]								
639		cub	[52]	0.4607–0.4629	[52]			1457 (per)	[73]								
		cub	[109]	0.462													
640	CrSi$_2$	hex	[1]	0.442/0.655	[1]	4.91	[6]	1630	[6]					100.5	[1]	1100	[1]
641		hex	[6]	0.4431/0.6364	[6, 52]	4.4	[25]	1457	[46]					80	[46]	880–110	[6]
642		hex	[25]	0.4422/0.6351	[25]	4.91	[25]	1570	[25]					80.1	[73]	1150 HV 0.1	[25]
643		hex	[28]	0.4427/0.6375	[28]	4.4	[107]	1457	[73]								
644		hex	[52]	0.4420/0.6349	[52]	5.5	[108]										
645		hex	[112]	0.4428/0.63691	[112]												
646	Cr$_3$Si	cub	[1]	0.455	[1]	6.52	[6]	1710	[1]					105.5	[1]	900–980	[6]
647		cub	[6]	0.4564	[6]	6.52	[25]	1770	[5]					105.5	[5]	1005	[6]
648		cub	[25]	0.4555	[25]	6.45	[52]	1750	[6]					92.1	[46]		
649		cub	[52]	0.4550–0.4564	[52]			1770	[46, 63, 73]					105.5	[73]		
650		cub	[112]	0.4558													

Table 1. Continued

No.	Symbol	Crystal structure		Lattice parameters (nm)		Density (g cm^{-3})		Melting point (°C)		Linear thermal expansion, α (10^{-6} K^{-1})		Thermal conductivity, λ (W m^{-1} K^{-1})		Electrical resistivity (10^{-6} Ωcm)		Enthalpy (kJ mol^{-1})		Young modulus (10^5 N mm^{-2})		Micro hardness (10 N mm^{-2})		Oxidation resistance (100°C)	
651	Cr$_5$Si$_2$					5.6	[6]	1510–1610	[6]					1420	[7]					1050–1200	[6]		
652						4.7	[7]	1550	[7]											1000	[7]		
653	Cr$_5$Si	tet	[1]	0.919/0.465		5.6	[1]	1560	[1]											1280	[1]		
654	Cr$_5$Si$_3$	tet	[6]	0.9170/0.4636		5.9	[6]	1647	[46]							211.4	[46]						
655		tet	[52]	0.9170/0.4636		5.6	[52]	1647	[73]							223.2	[73]						
656						5.9 (x)	[52]																
657	LaSi$_2$	tet	[30]	0.4281/1.375				~1520	[30]	10.78/17.05	[30]												
658		tet	[112]					>1500	[63]														
659	MoSi$_2$	tet	[1]	0.32/0.786		6.3	[1, 107]	2050	[1, 7, 22, 30]	8.4	[1]	221.9	[1]	21.	[7]	108.9	[1]			1290	[1]	17	[7]
660		tet	[22]	0.32/0.65		6.31	[7, 108]	2030	[25, 39, 73, 90]	8	[22]	50	[22]	20	[22]	131.9	[7, 22]	3.84	[1]	1300	[7, 22]	>17	[25]
661		tet	[25, 28]	0.320/0.786		6.2	[22]	2020	[50]	5.32/15.55	[30]			21.5	[25]	131.5	[25]	4.4	[22]	1290 HV 0.1	[25]	17	[100]
662		tet	[30, 52]	0.3205/0.7848		6.12	[25]	2190 (per)	[63]							131.8	[50]			1200 HV	[39, 90]		
663		tet	[46]	0.3203/0.7886		6	[39, 90]										[73]			900 HV 0.1	[58]		
664		tet	[109]	0.320/0.786			[109]																
665		tet	[112]	0.32047/0.78449			[112]																
666	Mo$_3$Si	cub	[1]	0.489		8.8	[1]	2150	[1]							100.5	[1]			1310	[1]		
667		cub	[25]	0.489		8.4	[25]	2050	[25]							116.4	[5]			1310 HV 0.1	[25]		
668		cub	[46]	0.489		8.4	[52]	2025	[73]							116.4	[46]						
669		cub	[52, 112]	0.489		8.97	[112]									116.5	[73]						
670	Mo$_5$Si$_3$	tet	[1, 112]	0.964/0.49		7.8	[1]	2100	[1]							280.5	[1]			1170	[1]		
671		cub	[46]	0.962/0.490				2180	[50]							310.2	[46, 50]						
672		tet	[52]	0.96483/0.49135				2190	[73]							309.8	[73]						
673	NbSi$_2$	hex	[1]	0.48/0.66		5.5	[1]	1950	[1]	8.4	[6]			6.3	[25]	50.2	[1]			700	[1]	8–11	[25]
674		hex	[6]	0.4785/0.6576		5.69 (x)	[6, 25]	1950	[7]							35.6–67.8	[6]			660–1320 HV	[6]		
675		hex	[25]	0.4797/0.6590		5.7	[7]	1950	[25]							138.2	[46]			1050 HV 0.1	[25]		
676		hex	[28, 52]	0.4795/0.6589		5.29	[25]	1930	[52]							138.2	[73]						
677		hex	[112]	0.4797/0.6592		5.37	[108]																
678						5.6	[107]																
679	Nb$_3$Si	hex	[1]	0.36/0.50		7.8	[1]	1950	[1]											550	[1]		
680		hex	[6]	0.359/0.446		8.01	[6]																
681	Nb$_5$Si$_3$	α: tet	[6]	α: 0.6570/1.1884	[6]	α: 7.09 (x)	[6]	2480	[5]	β: 7.3/4.6	[6]					452.2	[5]						
682		β: tet	[6]	β: 1.0010/0.5070	[6]	β: 7.20 (x)	[6]	2480	[63]							452.2	[73]						
683		α: tet	[52, 72]	α: 0.6583/1.1884	[52]																		
684		β: tet	[52, 71]	β: 1.0010/0.5070	[52]																		

#	Formula	Crystal	Lattice params	[refs]	ρ	[refs]	T_m	[refs]	α	[refs]			[refs]	E	[refs]		[refs]	H	[refs]	K_{IC}	[refs]
685		tet		[112]																	
686	TaSi$_2$	hex	0.4773/0.6552	[6][25]	9.2	[7][25][73]	2200	[7][25][73]	8.9/8.8	[6]		38	[6][7][25]	150.7	[6][46][73]			1410	[6][7][25][33]	11	[7]
687		hex	0.4773/0.6552	[25][28,52]	9.07 (x)	[6]	2400					46		119.3				1200		11–14	[25]
688		hex	0.4783/0.6567	[28,52]	8.83	[25]	2200					8.5		119.2				1560 HV 0.1			
689		hex	0.4781/0.6564	[46]	9.14	[52,108]												1560			
690		hex	0.478351/0.656980	[112]	8.4 (x)	[52]															
691					9.1	[107]															
692	Ta$_2$Si	tet	0.61/0.50	[1]	13.5	[1]	2450	[1]				124	[6]	83.7	[1]			1500	[1]		
693		tet	0.6157/0.5039	[6]	13.54 (x)	[6,52]								126.4	[6]						
694		tet	0.6157/0.5039	[52]	12.4	[52]								125.6	[46,73]						
695		tet	0.6157/0.5039	[112]																	
696	Ta$_5$Si$_3$	tet	T1: 0.9880/0.5060	[6]			2500	[63]				108	[6]	335.4	[6]						
697		tet	T2: 0.6516/1.1873	[6]			2500	[73]						334.9	[46]						
698		tet	T1: 0.9820/0.5010	[46]										334.9	[73]						
699		tet	T2: 0.6513/1.1864	[52]																	
700		tet	0.6516/1.1873	[112]																	
701	TiSi$_2$	ortho	0.8236/0.4773/0.8523	[6,25]	4.39	[90]	1520	[90]	11.5	[46]		18	[7]	134.4	[46]	2.556	[100]	892 HV	[35, 90]	11	[7]
702		ortho	0.8263/0.4800/0.8553	[28]	4	[7,108]	1470	[6]	–/–/11.3	[100]		123	[25]	134.3	[100]			700	[7]	8–11	[25]
703		ortho	0.8253/0.4783/0.8540	[52]	4.39	[25, 39]	1540	[7, 25]		[73]				134.4	[73]			870 HV 0.1	[25]		
704		ortho	0.8267/0.4800/0.8551	[100]	4.4	[107]	~1527	[39]										870 HV 1	[100]		
705		ortho	0.82687/0.85534/	[112]	4.07	[100]	~1550	[52]													
706			0.47983				1480	[100]			[100]										
707	Ti$_5$Si$_3$	hex	0.75/0.52	[1]	4.3	[1]	2120	[1,5,25]	7.03/15.61	[1]				577.8	[1]	1.56	[100]	986	[1]		
708		hex	0.7465/0.5162	[25]	4.32	[25]	2120	[52, 63]	9.2	[5]				579	[5]			986 HV 0.1	[25]		
709		hex	0.7465/0.5162	[30,52]			2130	[100]		[46]				579.9	[46]			968 HV 1	[100]		
710		hex	0.7448/0.5141	[100]			~2150	[30]		[100]				579.1	[100]						
711		hex	0.7444/0.5143	[112]						[73]				579.5	[73]						
712	VSi$_2$	hex	0.46/0.64	[1]	4.5	[1]	1650	[1,7,25]	11	[6]		9.5	[25]	95	[1]			960	[1]	16	[7]
713		hex	0.4562/0.6359	[6]	4.34–4.42	[6,25]	1680	[46]						125.6	[46]			890–960	[6]	14–18	[25]
714		hex	0.4573/0.6374	[25]	5.7	[7]	1750	[63]						125.6	[73]			1200–1320 HV	[6]		
715		hex	0.4571/0.6372	[28,52]	4.71	[25,107]												1090 HV 0.1	[25]		
716		hex	0.457230/0.63730	[112]	4.42	[108]															
717	V$_3$Si	cub	0.47	[1]	5.5	[1]	2060	[1]				97	[6]	154.9	[1]			1500	[1]		
718		cub	0.4712	[6]	5.33–5.67	[6,25]	2030 (d)	[6]						150.7	[46]			1430–1560	[6]		
719		cub	0.4721	[25, 52]	5.67	[52]	1935	[52]						150.7	[73]						
720		cub	0.47253	[112]																	
721	V$_5$Si$_3$	tet	T1: 0.9428/0.4750	[52]	9.5	[1]	2150	[5, 63]				12.5	[7]	462.2	[5, 73]	5.3	[22]	1090	[1]		
722		tet	0.94276/0.47555	[112]	9.4	[7, 108]	2010	[46]				10	[22]	462.6	[46]			1100	[7]		
723	WSi$_2$	tet	0.321/0.788	[1]			2165	[1, 39]	6.5		45			92.1	[1]						
724		tet	0.32/0.78	[6]			2170	[7]						92.9	[46]						

Table 1. Continued

No.	Symbol	Crystal structure		Lattice parameters (nm)		Density (g cm^{-3})		Melting point (°C)		Linear thermal expansion, α (10^{-6} K^{-1})		Thermal conductivity, λ (W m^{-1} K^{-1})		Electrical resistivity (10^{-6} Ωcm)		Enthalpy (kJ mol^{-1})		Young modulus (10^5 N mm^{-2})		Micro hardness (10 N mm^{-2})		Oxidation resistance (100°C)	
725		tet	[22]	0.3212/0.7880	[25]	9.8	[22]	2030–2200	[22]							92.8	[73]			1200			
726		tet	[25, 28]	0.3213/0.7829	[28]	9.3	[25, 107]	2150	[25, 46]					33.4	[25]					1090 HV 0.1	[22]		
727		tet	[46, 52]	0.3211/0.7868	[52]			2160	[73]											1074 HV	[25]		
728		tet	[112]	0.3211/0.7829	[112]			2210	[90]												[39]		
729	W$_5$Si$_3$	tet	[1]	0.561/0.496	[1]	14.56	[1]	2320	[1, 73]							125.6	[1]			770	[1]		
730		tet	[6]	0.964/0.497	[52]	13.06	[100]	2350	[99, 63]							134.4	[5]						
731		tet	[52]	0.9601/0.4972	[112]											135.2	[46]						
732		tet	[112]													134.6	[73]						
733	ZrSi	hex	[46, 52]	0.7005/1.2772	[52]	5.65 (x)	[52]	2090	[46]							147.8	[46]						
734		ortho	[52]	0.6982/0.3786/0.5302				2107 (per)	[73]							154.9	[73]						
735		ortho	[112]	0.6981/0.3785/0.5301																			
736	ZrSi$_2$	ortho	[6]	0.372/1.416/0.367	[6]	4.87	[6]	1700	[25]	9.7	[100]			161	[25]	159.4	[100]	2.348	[100]	1030 HV 0.1	[25]	8–11	[25]
737		ortho	[25, 28]	0.3698/1.4761/0.3665	[28]	4.88	[25, 108]	1520	[100]							151.1	[63]			942 HV 1	[100]		
738		ortho	[52, 100]	0.3721/1.468/0.3683	[100]	4.83	[52, 100]	1517 (per)	[73]							159.5	[73]						
739		ortho	[112]	0.36958/1.4751/0.36654	[112]	4.9	[107]																
740	Zr$_2$Si	tet	[1]	0.66/0.54	[1]	6	[1]	2110	[1, 46]							309.8	[1]			1230	[1]		
741		tet	[46, 52]	0.6612/0.5294	[52]	6.22	[52]	2107 (per)	[73]							339.1	[46]						
742		tet	[112]	0.6609/0.5298	[112]											208.5	[73]						
743	Zr$_5$Si$_3$	hex	[46]	0.7886/0.5551	[52]			2150	[46]							614.2	[46]						
744		hex	[52, 112]	0.7885/0.5558	[112]											576.1	[73]						

Oxides

No.	Symbol	Crystal structure		Lattice parameters (nm)		Density (g cm^{-3})		Melting point (°C)		Linear thermal expansion (10^{-6} K^{-1})		Thermal conductivity (W m^{-1} K^{-1})		Electrical resistivity (10^{-6} Ωcm)		Enthalpy (kJ mol^{-1})		Young modulus (10^5 N mm^{-2})		Micro hardness (10 N mm^{-2})		Oxidation resistance (100°C)	
745	Al$_2$O$_3$-α	hex	[8]	0.5127 - α = 55 16.7'	[52]	3.99	[9]	2043	[5]	8	[3]	30.1	[8]	10^{21}	[9]	1580.1	[5]	4	[7]	2100 HV 0.05	[3]	>17	[25]
746		hex	[30]	0.512	[57]	3.98	[16]	2045	[7]	8.6	[8]	27–36	[9]	10^{21}	[38]	1590	[43]	4	[9]	2200 HV	[4]	18	[43]
747		rhom	[43]	0.513 - α = 55.3°	[109]	3.9	[25]	2030	[8]	7.2–8.6	[9]	4.2–16.7	[16]	10^{21}	[40]	1678.5	[46]	3.6	[16]	2100 HV	[8, 77, 87, 98]	20	[70]
748		rhom	[46]	0.5544/0.9024	[112]	3.99	[38]	2047	[9]	7–9	[16]	25	[38]	10^{21}	[66]	1690	[57]	3.5	[25]	1800–2300	[9]		
749		rhom	[52]	0.47588/1.2992	[121]	3.8–3.9	[39]	2015	[16]	9.5/10	[30]	27	[43]	>10^{18}	[67]	1676.4	[73]	4.1	[38]	2500–3000 HV 0.05	[16, 92]		
750		hex	[57]			3.98	[43]	2050	[25]	8.3	[38]	25	[57]	10^{21}	[77]			4	[40]	2000 HK 0.1	[33]		
751		hex	[98]			4	[45]	2050	[30]	9	[40]	27	[63]	10^{21}	[87]			4	[43]	2800	[33, 69]		
752		rhom/hex	[109]			3.8	[47]	2050	[38]	8	[43]	28	[67]	10^{21}	[98]			4	[45]	1800	[34]		
753		hex	[112]			3.99	[52]	2047	[39]	7.7	[45]	30.1	[98]	10^{21}	[106]			8.5	[53]	2800 HV	[39]		
754		hex, rhom	[121]			3.96 (x)	[52]	2300	[40]	6.8	[53]	35	[106]	10^{21}	[121]			2.5	[53]	2100	[40]		
755						3.98	[53]	2047	[43]	α:8.5	[56]	27–36	[121]					5.2	[56]	2020 HK	[44]		
756						3.96	[56]	2040	[45]	7.2	[57]							4	[62]	1800 HV	[44]		
757						3.97	[57]	2050	[46]	8.4	[62]							4	[63]	2100	[45]		
758						3.95	[63]	2050	[53]	8	[63]							3.7	[67]	2050 HV	[53]		
759						3.99	[67]	2054	[56]	8.5	[67]							3.7	[69]	1200–1600 HV	[53]		
760						3.96	[70]	2015	[57]	9	[70]							4	[70]	2500–3000 HV	[53]		

#	Compound	Structure	Ref	Lattice params	Ref	ρ	Ref	T_m (°C)	Ref	Value	Ref	Value	Ref	Value	Ref	Hardness	Ref
761	Al$_2$O$_3$-γ	cub (spinell)	[52]	0.7859													
762		fcc	[52]	0.3958		3.98	[77]	2050	[77]	9	[76]			4	[76]	2000	[56, 76]
763		cub	[112]	0.79		3.9	[88]	2300	[88]	8.4	[77]			4	[77]	2100 HV 0.1	[57]
764		cub-B1	[109]	0.77–0.8		3.98	[98]	2047	[87]	9	[87]			3.8	[82]	2000–2300 HV	[58]
765						3.98	[92]	2300	[95]	8.0–8.6	[95]			4	[87]	2080 HK 0.1	[58]
766						4	[105]	2050	[98]	8.4	[98]			3.8	[95]	2000–2500 HV	[58]
767						3.99	[106]	2047	[105]					4	[98]	1200–1600 HV 0.1	[58]
768						3.9	[107]	2046	[106]	30	[106]			3.99	[92]	2300	[59]
769						3.9	[121]	2045	[121]	5.5–7.5	[121]			2.5	[106]	2000–2500	[62]
770														2.94		855–975 HV 0.02	[64]
771														4.00		2000 HV 30	[67]
772																1500–2200 HV 30	[69]
773																2000–3000 HV 0.1	[70]
774																2300 HV 0.2	[88]
775																2000 HV	[82]
776																2600–2800 HV	[95]
																1510 HK	[114]
777	BeO	hex	[43]	0.2699/0.4401	[52]	3	[52]	2450	[25]					3	[25]		
778		hex	[52]			3.03	[25]	2530	[39, 77]					3.9	[43]	2200	[121]
779						3.01	[39]	2570	[43]					3.9	[77]		
780						3.01	[43]	2567	[52]								
781						3.00 (x)	[45]	2580	[45]								
782							[77]	2550	[77]								
783	CrO$_2$	tet	[52]	0.441/0.291	[52]	4.8	[52]							10^{23}	[66]	1230–1490 HV	[39]
784		tet	[112]	0.4421/0.2916	[52]	4.90 (x)	[52]							10^{23}	[77]	1500 HV	[77]
785	CrO$_3$	ortho	[6]	0.573/0.852/0.474	[6]	2.81	[6]	170–198	[6]								
786		ortho	[52]	0.5743/0.8557/0.4789	[52]	2.7	[52]	185	[46]								
787		ortho	[112]	0.57494/0.8556/0.4796	[112]	2.82 (x)	[52]										
788	Cr$_2$O$_3$	rhom	[6]	0.536 – α=55°	[6]	5.21	[6]	2440	[6]	6.7	[53]	569	[5]	10^{13}	[6]	1000 HV	[34]
789		rhom	[52]	0.5361 – α=55°	[52]	5.2	[7]	2300	[7]	6.7	[60]	580	[43]	1.3×10^9	[106]	2915 HV	[53]
790		hex	[112]	0.495876/1.35942	[112]	5.21	[39]	2400	[39]	5.6	[106]	608.8	[46]			2300 HV	[53]
791						5.22	[46]	2343	[46]			599.1	[73]			1200–1700 HV	[53]
792						5.25 (x)	[52]	2343	[53]							1200–1700 HV 0.1	[58]
793						5.41	[53]	2708	[60]							2000	[59]
794						5.4	[60]		[106]							1200–1700 HV 0.05	[60]
795						5.21	[106]									1100 HV 0.02	[64]
796	Cr$_3$O	cub	[52]	0.4544	[52]							582.8	[46]				
797												582	[73]				
798	Cr$_3$O$_4$	tet	[52]	0.872/0.750	[52]							579.9	[46]				
799		tet	[112]	0.6145/0.755	[112]							589.9	[102]				
800												578.6	[73]				
801	HfO$_2$	mono	[43]	0.512	[52]	9.7	[7]	~2900	[5, 47]	10	[43]	1053.4	[5]	5×10^{15}	[106]	900	[59]
802																>17	[25]

Row 770 additional: 3 (value)

Table 1. Continued

No.	Symbol	Crystal structure		Lattice parameters (nm)		Density (g cm^{-3})		Melting point (°C)		Linear thermal expansion, α (10^{-6} K^{-1})		Thermal conductivity, λ (W m^{-1} K^{-1})		Electrical resistivity (10^{-6} Ωcm)		Enthalpy (kJ mol^{-1})		Young modulus (10^5 N mm^{-2})		Micro hardness (10 N mm^{-2})		Oxidation resistance (100°C)	
803		mono	[46]	0.51	[109]	9.68	[43]	2790	[7]	6.5	[77]					1010	[43]			780 HV	[77]		
804		cub	[109]	0.528/0.5181/0.5115	[112]	10.2	[77]	2897	[43]							1113.7	[46]					>17	[25]
805		mono	[112]			9.63	[106]	2810	[46]							1113.8	[73]					17	[43]
806						9.6	[107]	2790	[106]														
807	MgO	cub	[43]	0.4208	[52]	3.6	[7]	2850	[5]							568.6	[5]	3.2	[39]	745 HV	[39]		
808		cub-B1	[46]	0.42	[109]	3.5	[25, 107]	2800	[7, 25, 39, 53]			36	[43]	10^{12}	[66]	570	[43]	3.2	[53]	520 HV	[53]		
809		cub-B1	[52]	0.4213	[112]	3.65	[39]	2837	[43]	11.2		36	[106]	10^{12}	[77]	601.6	[46]	2.4	[77]	~400 HV	[58]		
810		cub-B1	[109]			3.58	[43, 53]	2825	[46, 73]					10^{18}	[106]	601.6	[73]		[106]	520 HV 0.1	[58]		
811		cub-B1	[112]			3.77	[77]	2827	[77]											750 HV	[77]		
812						3.58	[106, 108]	3073	[106]														
813	Nb$_2$O$_5$	mono	[112]	2.893/0.3827/1.758		4.5	[7]	1460-1520	[6]	1.76	[6]					1905.8	[6]			740	[6]		
814						4.47	[106]	1510	[7]							1903.3	[73]			700	[59]		
815						4.6	[107]	1512	[73]														
816								1785	[106]														
817	SiO					2.1	[7, 107]	1750	[7]							98.4	[46]						
818						2.1	[105]	1610	[46]														
819						2.13	[106, 108]	1705	[105]														
820								1975	[106]														
821	SiO$_2$	quartz:	[109]	quartz:		2.33	[6]	1703-1729	[6]	0.4	[106]	1.38	[106]	10^{22}	[66]	911	[46]	0.5-1.0	[82]	1130-1260	[6]		
822		trigonal		0.4093/0.5393		2.2	[7, 105]	1713	[7]	0.5-0.75	[108]	1.2-1.4	[108]	10^{21}	[106]	911.5	[73]	1.114	[105]	1200	[59]		
823				0.421/0.539		1.4-2.7	[47]	1722	[46]					10^{20}	8108]			0.65-0.75	[108]	1000-1200 HV	[82]		
824						2.1-2.2	[47]	1373-1473	[73]														
825		cristobalite:		cristobalite:		2.2	[106]	1713	[105]														
826		tet	[6]	0.70/0.69	[6]	2.6	[107]	1983	[106]														
827		cub	[109]	0.704	[109]	2.18	[108]																
828	Ta$_2$O$_5$-α	tet	[6]	0.3808/0.3567	[52]	8.37-9.48	[6]	1880	[5]	2	[6]					1953.1	[5]			660-1030	[99]		
829		tet	[52]			7.5	[7]	1887	[6]														
830						8.53 (x)	[52]	1880	[7]														
831						8.2	[99]																
832	Ta$_2$O$_5$-β	ortho	[6]	0.619/0.367/0.389	[6]	8.18-8.91	[6]	1785	[6]	2	[6]			10^{17}	[6]	2045.3	[6]						
833		ortho	[46]	0.6192/4.4019/0.3898	[52]	8.30 (x)	[52]	1900	[46]							2047.3	[46]						
834		ortho	[52]													2047.3	[73]						
835	Ta$_2$O$_5$	ortho	[112]	0.6198/4.029/0.3888	[112]	8.3	[105]	1880	[106]		[106]												
836						7.53	[106]	1880	[105]														
837						8.7	[107]																
838	ThO$_2$	fcc	[6]	0.5859	[6]	10.05 (x)	[6]	3250	[5]	10	[43]	10	[43]	10^{16}	[66]	1173.1	[5]	1.38	[6]	950 HV	[77]	>17	[25]
839		cub	[43]	0.5595	[52]	9.7	[7]	2997-3250	[6]	9.3	[77]	10	[106]	10^{16}	[77]	1170	[43]	1.4	[25]				

#	Compound	Structure	Lattice params	Ref	Density	Ref	Melting pt	Ref	Col	Ref	Col	Ref	Col	Ref	Hardness	Ref
840		fcc	0.521	[46]	9.7		2990	[7]								
841		fcc	0.5597	[52]	10		3050	[25]							2.4	[43]
842		cub		[109]	10		3217	[43]			2×10^{17}	[106]	1227.5	[46]	2.4	[77]
843		cub		[112]	9.9		3370	[46]					1227.2	[73]	1.48	[106]
844					10		3220	[73]								
845							3300	[77]								
846							3323	[106]								
847	TiO	cub-B1	0.417	[32]	4.88	[6]	1750	[6,7,32,105]	7.6	[106]						
848		cub	0.418	[52]	4.9	[7,105]	2020	[46]							1300	[32]
849		cub-B1	0.424	[52,76]			1700	[106]					520	[73]	1800	[59]
850		cub-B1	0.4177	[109]	4.93	[106]									1000	[76]
851	TiO$_2$	tet	0.4593/0.2959	[46]	4.19	[6]	1900	[6]	4.21–4.25	[6]			945.4	[46]	767–1000 HK	[6]
852		tet	0.449/0.2959	[52]	4.1	[7]	1855	[7]	9.0–9.4	[9]			945.4	[73]	700–1100	[9]
853		cub-B1	0.45933/0.29592	[25]	4.25	[9]	1867	[9]	9	[77]					1000 HV	[53]
854		tet		[109]	2.40–2.49	[47]	1860	[46]	9	[98]					600–1000 HV	[53]
855		tet		[112]	4.16	[53]	1840	[53]	9.2	[106]	1.2×10^{10}	[106]			600–1000 HV 0.1	[58]
856		tet			4.25	[77,98]	1867	[77]	8.8	[108]					1100 HV	[77]
857					4.2	[88]	1860	[88]							600–700	[82]
858					4.24	[106]	1860	[106]							1000 HV 0.2	[88]
859					4.2	[105,107]	1913	[105,107]							1150 HV	[98]
860					3.02–3.92	[124]		[124]								
861	Ti$_2$O$_3$	rhom	0.5454 · α=59°5′	[46]	4.6	[52,105]	2130	[5]					1433.1	[5]		
862		rhom	0.5139/1.3659	[52]	3.58 (x)	[52]	1842	[46]					1521.9	[46,73]		
863		rhom		[112]	4.05	[108]	1760	[105]								
864	Ti$_3$O$_5$	mono	0.9828/0.3776/0.9898	[46]			1780	[46]					2461	[46]		
865		mono		[112]									2460.8	[73]		
866	WO$_2$	mono	0.5560/0.4884/0.5546	[6,46,52]	12.1	[6]	1500–1600	[6]					587	[6]		
867		tet	0.4870/0.2776	[52]	11.05	[52]	1724 (per)	[6]					590.1	[46]		
868		tet	0.556/0.555	[109]	10.82 (x)	[52]	1724 (d)	[46]					590.1	[73]		
869		mono	0.55754/0.48995/0.55608	[112]												
870	WO$_3$	tri	0.382/0.748/0.728	[6]	7.16	[6]	1473	[6]					843.4	[46]		
871		mono	0.3835/0.7517/0.7285	[46,52]	7.16	[7,106]	1473	[7,106]			[106]		843.5	[73]		
872		tet	0.5250/0.3915	[52]			1472	[46]			0.6					
873		tri	0.7309/0.7522/0.7678	[112]												
874	Y$_2$O$_3$	bcc	1.0601	[46]	5	[7,105]	2465	[5]	8	[38]			1906.7	[46]	1.8	[38]
875		bcc	1.06	[52]	4.5	[38]	2415	[7]	8.1	[108]			1906.7	[73]		
876		cub	1.04041	[109]	5.03	[108]	2450	[38]	14	[38]						
877		cub		[112]	4.84	[106]	2704	[46]	8–12	[108]						
878							2420	[73]								
879							2410	[105]								
880							2683	[106]								

Table 1. Continued

No.	Symbol	Crystal structure		Lattice parameters (nm)		Density (g cm^{-3})		Melting point (°C)		Linear thermal expansion, α (10^{-6} K^{-1})		Thermal conductivity, λ (W m^{-1} K^{-1})		Electrical resistivity (10^{-6} Ωcm)		Enthalpiy (kJ mol^{-1})		Young modulus (10^5 N mm^{-2})		Micro hardness (10 N mm^{-2})		Oxidation resistance (100°C)	
881	ZrO$_2$	cub	[39]	0.511	[52]	5.6	[7, 105]	2750	[5]	7.5–10.5	[9]	0.7–2.4	[9]	10^{16}	[9]	1035	[5]	1.63	[6]	1200	[9]	>17	[25]
882		cub	[52]	0.521/0.5209/0.5375	[6]	5.7	[9, 88]	2687	[7]	10	[38]	2	[38]	10^{17}	[38]	1040	[43]	1.8	[9]	1200 HV	[39, 77]	17	[43]
883		mono	[39, 43]	0.5220/0.5271/0.5381	[52]	5.8	[25]	2677	[9, 43, 53, 77]	9–10.5	[43]	2	[43]	10^{16}	[66]	1101.5	[46]	1.7	[25]	1500 HV	[53]		
884		mono	[46, 52]	0.5084/0.5165	[52]	5.56	[38, 43]	2690	[25, 39]	10	[67]	2.9	[67]	10^{16}	[77]	1098.2	[73]	2.4	[38]	1000 HV	[58]		
885		tet	[52]	0.508	[109]	5.56 (mo)	[39]	2680	[38, 46]	10	[77]	1.7	[106]					2.05	[43]	1500 HK 0.1	[58]		
886		cub	[109]			6.27 (cub)	[39]	2700	[88]	17.2	[106]							2.1	[67]	1600	[59]		
887		mono	[6, 112]	0.51463/0.52135/0.5311	[112]	6.1	[53, 108]	2687	[106]									1.9	[77]	1300 HV 30	[67]		
888		tet	[112]	0.364/0.527		6.07	[67]	2700	[105]											400–1000	[76]		
889						5.76	[77]													1100 HV 0.2	[88]		
890						5.6	[106]													1500–1650	[116]		

References

1. R. Kieffer and F. Benesovsky, *Hartstoffe*. Springer, Vienna, 1968.
2. H. Holleck, *Binäre und ternäre Carbide und Nitride der Übergansmetalle und ihre Phasenbeziehungen*. Gebr. Borntrager Verlag, Berlin, Stuttgart, 1984.
3. U. König, Untersuchung von kathodenzerstäubten binären und ternären Hartstoffschichten zur Verschleißminderung von Hartmetallen. *Tribologie – Reibung, Verschleiß, Schmierung*. Band 9, Springer, Berlin 1985, pp. 282, 287, 296, 306, 311, 316.
4. H. Kolleck, Die Bedeutung von Phasengleichgewichten bei der Entwicklung verschleißfester Werkstoffe. *Chem. Z.* 1982, **106**, 216.
5. M. Rühle, Zum technischen Stand der Dispersionshärtung – Teil 1. *Metall*. 1982, **36**, 1281–1285.
6. *Gmelin: Handbuch der anorganischen Chemie*. Verlag Chemie: Weinheim 1960. Angaben in den jermeiligen Bände der Elemente (z.B. Cr, Ti, ...).
7. R. A. Haefer, *Oberflächen- und Dünnschichttechnologie Teil I: Beschichtung von Oberflächen*. Springer, Berlin, 1987, pp. 282–294.
8. H. Frey and G. Kienel, *Dünnschichttechnologie*. VDI-Verlag, Düsseldorf 1987, p. 121.
9. B. Rother and J. Vetter, *Plasma-Beschichtungsverfahren und Hartstoffschichten*. Deutscher Verlag für Grundstoffindustrie, Leipzig, 1992, p. 171–172.
10. L. E. Toth, *Transition Metal Carbides and Nitrides*. Academic, New York, 1971, pp. 5–7, 80–81, 92, 95–97 and 148–150.
11. E. K. Storms, *The Refractory Carbides*. Academic, New York, 1967.
12. H. L. Schick, *Thermodynamics of Certain Refractory Compounds. Vols. 1 and 2*, Academic, New York, 1966.
13. E. Hornbogen, *Werkstoffe: Aufbau und Eigenschaften von Keramik, Metallen, Polymer- und Verbundwerkstoffen*. Springer, Berlin, 1987, p. 286.
14. H.-J. Bargel and G. Schulze, *Werkstoffe*. CDI-Verlag, Düsseldorf, 1988, pp. 293–294.
15. V. Demarne and E. Bergmann, Verwendung von Hartstoff- und Hartmetallschichten für Schneidwerkzeuge. In *1. Int. PVD-Tagung an der TH Darmstadt am 15./16. März 1983*, Vol. 20, ISBN 3-88607-027-1, THD Schriftenreihe Wissenschaft und Technik 20, 1983, pp. 269–277.
16. K. H. Kloos et al. Herstellung und Eigenschaften gesputterter Aluminiumoxidschichten. In *2. Int. PVD-Tagung an der TH Darmstadt am 11./12. März 1986*, Vol. 30, THD Schriftenreihe Wissenschaft und Technik, 1986, p. 100.
17. H. Jehn et al. Beeinflussung der Schichteigenschaften mangetrongesputerter Nitridschichten durch Substrattemperatur und Sputteratmosphäre. In *2. Int. PVD-Tagung an der TH Darmstadt am 11./12. März 1986*, Vol. 30, THD Schriftenreihe Wissenschaft und Technik, 1986, p. 138.
18. J.-E. Sundgren et al. Structure and Properties of HfN Coatings grown by Reactive Sputtering. In *2. Int. PVD-Tagung an der TH Darmstadt am 11./12. März 1986*, Vol. 30, THD Schriftenreihe Wissenschaft und Technik, 1986, p. 185.
19. P. Johannsen, Beschichtete HSS- und Hartmetall-Werkzeuge in der Großserienfertigung. In *2. Int. PVD-Tagung an der TH Darmstadt am 11./12. März 1986*, Vol. 30, THD Schriftenreihe Wissenschaft und Technik, 1986, p. 192.
20. U. Kopacz and H. Jehn, Härte- und Haftfestigkeitsmessungen an Nitridschichten auf Schnellarbeitsstahl. In *2. Int. PVD-Tagung an der TH Darmstadt am 11./12. März 1986*, Vol. 30, THD Schriftenreihe Wissenschaft und Technik, 1986, pp. 217–218.
21. H. R. Stock and P. Mayr, Hartstoffbeschichtung mit dem Plasma-CVD-Verfahren. In *2. Int. PVD-Tagung an der TH Darmstadt am 11./12. März 1986*, Vol. 30, THD Schriftenreihe Wissenschaft und Technik, 1986, p. 250.
22. A. Petzold, *Physikalische Chemie der Silicate*. Deutscher Verlag für Grundstoffindustrie, Leipzig, 1991, p. 182.
23. D. Kammermeier, Charakterisierung von binären und ternären Hartstoffschichten anhand von Simulations- und Zerspanungsuntersuchungen. *Fortschritt-Ber. VDI Reihe 2 Nr. 271*, VDI-Verlag, Düsseldorf, 1992, p. 2.
24. R. F. Bunshah, *Deposition Technologies for Films and Coatings*. Noyes Publications, New Jersey, 1982, p. 149.

25. R. Kieffer and P. Schwarzkopf, *Hartstoffe und Hartmetalle.* Springer, Vienna, 1953.
26. S. u. Z. Marinkovic *et al. Boron Nitride Coatings.* Kernforschungsanlage Jülich GmbH, 1989.
27. S. u. Z. Marinkovic *et al. Titanium Nitride Coatings.* Kernforschungsanlage Jülich GmbH, 1988.
28. B. Lönnberg, Synthesis, Structure and Properties of some technologically important Carbides, Borides and Silicides. Acta Univ. Uppsala, 1988, pp. 17–21, 26 and 30.
29. T. Wolf, *Herstellung und Charakterisierung von TiN, ZrN und HfN.* Dissertation Universität Karlsruhe, 1982.
30. H. Bleckmann, *Röntgenographische Bestimmung der thermischen Ausdehnung an verschiedenen metallischen und nichtmetallischen Hartstoffen.* Dissertation RWTH Aachen, 1971.
31. G. Hilz, *Zum Einfluß innerer Grenzflächen auf Aufbau und Eigenschaften mehrphasiger Hartstoffschichten.* Kernforschungszentrum Karlsruhe GmbH, KfK 5022, 1992.
32. G. D. Brundiers, *Herstellung, Aufbau und Eigenschaften von Hafniumverbindungen im System Hf–C–N–O.* Kernforschungszentrum Karlsruhe GmbH, KfK 2161, 1975, p. 7.
33. P. Schwarzkopf and R. Kieffer, *Cemented Carbides.* Macmillan, New York, 1960, pp. 138 and 214.
34. A. R. Lansdown and A. L. Price, *Materials to Resist Wear.* Pergamon Press, Oxford, 1986, pp. 7 and 99.
35. R. Reinhold and S. Becker, Superharte Schneidwerkstoffe der Spanungstechnik. VEB Verlag Technik, Berlin, 1982, p. 17.
36. W.-E. Borys, *Vergleichsuntersuchung zum Einsatz hochharter polykristalliner Schneidstoffe beim Fräsen.* Dissertation Universität Hannover, 1984, p. 3.
37. W. König and H. U. Schemmel, *Untersuchung moderner Schneidstoffe – Beanspruchungsgerechte Anwendung sowie Verschleißursachen.* Forschungsbericht des Landes Nordrhein-Westfalen, Nr. 2472, 1975, p. 54.
38. F. Löffler, *Eigenschaften von keramischen Hartstoffschichten auf Silicium- und Aluminiumbasis.* Fortschritt-Ber. VDI Reihe 5 Nr. 180, VDI-Verlag, Düsseldorf, 1990, p. 23.
39. R.-J. Peters, *Beschichten mit Hartstoffen. VDI Technologiezentrum Physikalische Technologien.* VDI-Verlag, Düsseldorf, 1992, pp. 2–3.
40. H. Freller and H. P. Lorenz, Kriterien für die anwendungsbezogene Auswahl von Hartstoffschichten. VDI-Verlag, Düsseldorf, 1992, p. 22.
41. C. Ribeiro, Silbernes TiN durch Magnetronsputtern und seine Anwendungsbeispiele. VDI-Verlag, Düsseldorf, 1992, p. 124.
42. E. Ertürk and H.-J. Heuvel, Neue Schichtsysteme mit dem Arc-PVD-Verfahren. VDI-Verlag, Düsseldorf, 1992, p. 254.
43. F. Thümmler *et al.* Fortschritte mit neuen Werkstoffen: Keramik für den Maschinenbau. *Keram. Z.* 1988, **40**, 158–159.
44. K. Dettling, Verschleißfeste Werkstoffe für Extrusionswerkzeuge. *Keram. Z.* 1988, **40**, 802.
45. R. Riesenberg *et al. Mechanische Eigenschaften von Schichten.* Wissenschaftliche Schriftenreihe TU Chemnitz, Heft 9/1990, p. 53.
46. O. Kubaschewski *et al. Metallurgical Thermochemistry.* Pergamon, Oxford, 1979, pp. 268ff.
47. C. E. Morosanu, Thin films by chemical vapour deposition. *Thin Films Sci. Technol.* 1990, 7, pp. 225, 229, 381, 391 and 424.
48. R. Telle, Boride – eine neue Hartstoffgeneration? *Chemie in unserer Zeit* 1988, 22, 95.
49. P. Hedenqvist, Evaluation of vapour-deposited coatings for improved wear resistance. *Acta Univ. Uppsala,* 1991, p. 13.
50. T. C. Chou and T. G. Nieh, Anisotropic grain growth of d-NiMo. *Thin Solid Films* 1991, 219, 61.
51. G. Berg; E. Broszeit, C. Friedrich, W. Herr and K. H. Kloos, Grundlageneigenschaften und Verschleißverhalten von HfB_2- und Hf(B,N)-Schichtsystemen. *Mat.-wiss. u. Werkstofftech.* 1994, 25, 175–179.
52. W. B. Pearson, *Lattice Spacings and Structures of Metals and Alloys.* Pergamon, London, 1958, pp. 219ff.
53. H. Simon and M. Thoma, *Angewandte Oberflächentechnik für metallische Werkstoffe.* Hanser, München, 1985, pp. 103, 108 and 131.
54. G. Berg, C. Friedrich, E. Broszeit and K.-H. Kloos, Comparison of fundamental properties of r.f. sputtered Ti_xN-and Hf_xN coatings on steel substrates. *Surf. Coat. Technol.* 1995, 74–75, 135–142.

55. H. Bolmgren, Synthesis and structural studies of some refractory borides. *Acta Univ. Uppsala* 1992, p. 19.
56. N. P. Cheremisinoff, *Handbook of Ceramics and Composites*. Dekker, New York, 1990, p. 24.
57. W. Schedler, *Hartmetall für den Praktiker*. VDI-Verlag, Düsseldorf, 1988, pp. 3, 8–9, 12 and 207.
58. K.-H. Habig, *Verschleiß und Härte von Werkstoffen*. Hanser, München, 1980, pp. 125, 164, 211 and 263–268.
59. H.-D. Steffens and W. Brandl, *Moderne Beschichtungsverfahren*. Lehrstuhl für Werkstofftechnologie der Universität Dortmund, 1992, pp. 242 and 261.
60. J. H. Mittendorf, *Einsatz von Werkzeugbeschichtungen für ausgewählte Umformverfahren*. Dissertation RWTH Aachen D82, 1992, p. 123.
61. H. Benninghoff, Industriereife CVD-Verfahren. *Metalloberfläche* 1976, 30, 476.
62. E. Erben et al. CVD – ein modernes und vielseitiges Beschichtungsverfahren. *Metall* 1981, 35, 1256.
63. M. Rühle, Zum technischen Stand der Dispersionshärtung – Teil 2. *Metall* 1985, 39, 520, 522 and 528.
64. K.-H. Habig, Verschleiß-Schutzschichten – Entwicklungstendenzen zur Optimierung von Eigenschaften und Verfahren. *Metall* 1985, 39, 915.
65. H. Holleck, Neue Entwicklungen bei PVD-Hartstoffbeschichtungen. *Metall* 1989, 43, 622.
66. S. Hofmann, Charakterisierung keramischer Werkstoffe mit der Auger-Elektronenspektroskopie: Möglichkeiten und Grenzen. *Materialwissenschaft und Werkstofftechnik* 1990, 21, 99.
67. H. Kolaska and K. Dreyer, Hartmetalle, Cermets und Keramiken als verschleißbeständige Werkstoffe. *Metall* 1991, 45, 227 and 232.
68. W. Wahl, Verschleißfeste metallische Gußwerkstoffe. Verschleißfeste Werkstoffe. *Vorträge der VDI-Tagung Stuttgart 1973*, VDI-Berichte Nr. 194 (1973), p. 82.
69. G. Schumacher, Hartmetall für den Verschleißschutz. *Vorträge der VDI-Tagung Stuttgart 1973*, VDI-Berichte Nr. 194 (1973), p. 113.
70. E. Gugel, Nichtoxidkeramische Werkstoffe für die Verschleißtechnik. *Vorträge der VDI-Tagung Stuttgart 1973*, VDI-Berichte Nr. 194 (1973), p. 139.
71. B. Finnern, Konstitution, Verfahrenstechnik und Anwendung verschleißfester Diffusionsschichten. *Vorträge der VDI-Tagung Stuttgart 1973*, VDI-Berichte Nr. 194 (1973), p. 219.
72. K.-H. Kloos, *Werkstoffkunde IV*, Vorlesungsumdruck TH Darmstadt, 1994, pp. 2.4.2.5f.
73. I. Barin and O. Knacke, *Thermochemical Properties of Inorganic Substances*. Springer, Berlin, 1971 and 1973.
74. H. Jehn, G. Reiners et al. *Charakterisierung dünner Schichten. DIN-Fachbericht 39*. Beuth, Berlin, 1993, pp. 134, 155 and 171.
75. K. Kashigawi et al. Chromium nitride films by radio-frequency reactice ion-plating. *Vac. Sci. Technol.* 1986, **A4**, 210 and 214.
76. J.-E. Sundgren et al. A review of the present state of art in hard coatings grown from the vapor phase. *Vac. Sci. Technol.* 1986, **A4**, 2260ff.
77. H. Holleck, Material selection for hard coatings. *Vac. Sci. Technol.* 1986, **A4**, 2663.
78. W. Gissler et al. Mixed phase nanocrystalline boron nitride films: preparation and characterization. *Thin Solid Films* 1991, **199**, 115.
79. R. A. Andrievski et al. Structure and microhardness of TiN compositional and alloyed films. *Thin Solid Films* 1991, **205**, 173–174.
80. N. Zhang and V. Wang, Dislocations and hardness of hard coatings. *Thin Solid Films* 1992, **214**, 5.
81. C. Friedrich, G. Berg, E. Broszeit and K.-H. Kloos, X-ray diffractometry analysis of r.f. sputtered hard coatings based on nitrides of Ti, Cr, Hf. *Surf. Coat. Technol.* 1995, **74–75**, 279–285.
82. E. Vancoille et al. Mechanical properties of heat treated and worn PVD TiN, (Ti, Al)N, (Ti, Nb)N and Ti(C, N) coatings as measured by nanoindentation. *Thin Solid Films* 1993, **224**, 170 and 175.
83. S. D. Marcus and R. F. Foster, Characterization of low pressure chemically vapor-deposited tungsten nitride films. *Thin Solid Films* 1993, **236**, 331 and 332.

84. Sun, Xin et al. Properties of reactively sputter-deposited Ta-N thin films. *Thin Solid Films* 1993, **236**, 350.
85. W. Heinke, A. Leyland, A. Matthews, G. Berg, C. Friedrich and E. Broszeit, Evaluation of PVD nitride coatings, using impact, scratch and Rockwell C adhesion tests. *Thin Solid Films* 1995, **270**, 431–438.
86. A. Schröer, *Synthese und chemische, mikrostrukturelle und mechanische Charakterisierung von Hartstoffschichten auf Stählen*. Dissertation Heidelberg, 1992.
87. J. Berger *Beschichten mit Hartstoffen*. VDI-Verlag, Düsseldorf, 1992.
88. E. Dörre, Nichtmetallische Hartstoffe, in *Abrasion und Erosion*, H. Uetz (Ed.), Hanser, München, 1986, 433–437.
89. C. Biselli and L. Chollet, Young's modulus of TiN and TiC coatings, in *Mechanics of Coatings*. *Tribology Series 17* D. Dowson, M. Godet and C. M. Tayloer (Eds), Proccedings of the 16th Leeds-Lyon Symposium on Tribology, Lyon, Amsterdam, Elsevier, 1990.
90. P. Mayr and H. Vetters, *Gefüge, Oberflächenzustand und technologische Eigenschaften von Hartstoffschichten*. *HTM* 1993, **48**, 281–287.
91. R. Wild, Hartstoffschichten leisten Widerstand. *Metalloberfläche* 1990, **44**, 355–361.
92. R. Saß, and P. Thienel, Oberflächenbehandlungen von Spritzgießwerkzeugen. *Metalloberfläche* 1991, **45**, 313–322.
93. R. Rochotzki and J. Vetter, Festkörperreibung von PVD-Hartstoffschichten: TiN_x, ZrN_x, TiC_x, TiC_x/i-C. *Neue Hütte* 1989, **34**, 272–277.
94. R. S. Bonetti and M. Tobler, Amorphe diamantartige Kohlenstoffschichten. *Metalloberfläche* 1990, **44**, 209–211.
95. K. Keller and F. Koch, CVD-Beschichtung von Fließwerkzeugen. *VDI-Zeitschrift*. 1989, **131**, 42–50.
96. M. Fieber and W. König, Kontaktbedingungen beim Einsatz beschichteter Werkzeuge. *Tribologie + Schmierungstechnik* 1995, **42**, 135–140.
97. G. Kienel (Ed.): *Vakuumbeschichtung 4 und 5, Anwendungen Teil 1 und 2*. VDI-Verlag, Düsseldorf, 1993.
98. B. W. Matthes, *Abscheidung, anwendungsbezogene Prüfung und Optimierung von PVD-Hartstoffbeschichtungen des Systems Ti-B-N/Al_2O_3 aus der Hochfrequenz-Kathodenzerstäubung*. Dissertation, TH Darmstadt, 1993.
99. P. J. Martin et al. Properties of thin films of tantalum oxide deposited by ion-assisted deposition. *Thin Solid Films* 1994, **238**, 183–184.
100. R. Rosenkranz, *Strukturen, physikalische und mechanische Eigenschaften der hochschmelzenden Ti_5Si_3-, $TiSi_2$- und $ZrSi_2$-Phasen*. Fortschritt-Ber. VDI Reihe 5 Nr. 291, VDI-Verlag, Düsseldorf, 1993.
101. E. D'Anna et al. Synthesis of thin films of semiconductor and refractory metal nitrides by laser irradiation of solid samples in ambient gas. *Thin Solid Films* 1992, **218**, 224–225.
102. J. L. Davis et al. Thin metal films on polyether imide. *Thin Solid Films* 1992, **220**, 221.
103. G. Berg, C. Friedrich, E. Broszeit and C. Berger, Development of chromium nitride coatings substituting titanium nitride. *Surf. Coat. Technol.* 1996, **86–87**, 184–191.
104. C. Friedrich, G. Berg, E. Broszeit and C. Berger, Measurement of the hardness of hard coatings using a force indentation function. *Thin Solid Films* 1996, **290–291**, 216–220.
105. A. G. Balzers, *Coating materials and sources selector guide*. Company information. Balzers AG, FL-9496 Balzers, Fürstentum Liechtenstein.
106. A. G. Leybold, *Berechnungsgrundlagen für Dünnschicht-Technik*. Company information 01-017.9.70.61.02.21/KT 5.
107. Cerac Inc.: *Advanced special inorganic materials*. Company information 1995. Cerac Inc., PO Box 1178, Milwaukee, WI 53201-1178, USA.
108. Goodfellow: Firmenkatalog zu Metallen, Keramiken und anderen Materialien. Edition 1995/96. Deutsche Niederlassung: Goodfellow GmbH, Postfach 1343, D 61213 Bad Nauheim.
109. K. Schäfer and C. Synowitz, *Taschenbuch für Chemiker und Physiker. Bd. 3, 3. Aufl*. Springer, Berlin, 1970.
110. W. Herr, B. Matthes, E. Broszeit, M. Meyer and R. Suchentrunk, Influence of substrate material and deposition parameters on the structure, residual stress and adhesion of sputtered Cr_xN_y hard coatings. *Surf. Coat. Technol.*, 1993, **60**, 428–433.

111. M. Atzor, Aspekte des Magnetronsputterns zur Herstellung verschleiß- und korrosionsbeständiger Schichten auf Chrombasis. VDI-Fortschrittsbericht Reihe 5, Nr. 156, VDI, Düsseldorf, 1989.
112. International Center for Diffraction Data *Powder Diffraction File*, ASTM, Swarthmore, PA 19081, USA, CD-ROM 1992.
113. S. Luridiana and A. Miotello, Spectrographic study of oxide growth of arc evaporated TiN and ZrN coatings during hot air oxidation. *Thin Solid Films* 1996, **290–291**, 289–293.
114. P. J. Kelly, O. A. Abu-Zeid, R. D. Arnell and J. Tong, The deposition of oxide coatings by reactive unbalanced magnetron sputtering. *Surf. Coat. Technol.* 1996, **86–87**, 28–32.
115. C. Gautier, H. Moussaoui, F. Elstner and J. Machet, Comparative study of mechanical and structural properties of CrN films deposited by d.c. magnetron sputtering and vacuum arc evaporation. *Surf. Coat. Technol.* 1996, **86–87**, 254–262.
116. M. S. Wong, W. J. Chia, P. Yashar, J. M. Schneider, W. D. Sproul and S. A. Barnett, High-rate reactive d.c. magnetron sputtering of ZrO_x coatings. *Surf. Coat. Technol.* 1996, **86–87**, 381–387.
117. M. H. Staia, E. S. Puchi, D. B. Lewis, J. Cawley and D. Morel, Microstructural characterization of chemically vapor deposited TiN coatings. *Surf. Coat. Technol.* 1996, **86–87**, 432–437.
118. A. Igartua, J. Lauricirca, A. Aranzabe, T. Leyendecker, O. Lemmer, G. Erkens, M. Weck and G. Hanrath, Application of low temperature PVD coatings in rolling bearings: tribological tests and experiences with spindle bearing systems. *Surf. Coat. Technol.* 1996, **86–87**, 460–466.
119. S. Heck, T. Emmerich, I. Munder and J. Steinebrunner, Tribological behaviour of Ti–Al–B–N-based PVD coatings. *Surf. Coat. Technol.* 1996, **86–87**, 467–471.
120. K.-D. Bouzakis, N. Vidakis, T. Leyendecker, O. Lemmer, H.-G. Fuss and G. Erkens, Determination of fatigue behaviour of thin hard coatings using the impact test and FEM simulation. *Surf. Coat. Technol.* 1996, **86–87**, 467–471.
121. F. Fietzke, K. Goedicke and W. Hempel, The deposition of hard crystalline Al_2O_3 layers by means of bipolar pulsed magnetron sputtering. *Surf. Coat. Technol.* 1996, **86–87**, 657–663.
122. G. DErrico, R. Chiara and E. Guglielmi, PVD coatings of cermet inserts for milling applications. *Surf. Coat. Technol.* 1996, **86–87**, 735–738.
123. J. Steinebrunner, T. Emmerich, S. Heck, I. Munder and R. Steinbuch, A novel impact tester operating at elevated temperatures for characterising hard coatings. *Surf. Coat. Technol.* 1996, **86–87**, 748–752.
124. O. Anderson, C. R. Ottermann, R. Kuschnereit, P. Hess and K. Bange, Density and Youngs modulus of thin TiO_2 films. *Fresenius J Anal Chem* 1997, **358**, 315–318.
125. B. Navinsek, P. Panjan and I. Milosev, Industrial applications of CrN (PVD) coatings, depopited at high and low temperatures. *Surf. Coat. Technol.* 1997, **97**, 182–191.
126. J.-R. Park, Y. K. Song, K.-T. Rie and A. Gebauer, Hard coating by plasma-assited CVD on plasma nitrided stellite. *Surf. Coat. Technol.* 1998, **98**, 1329–1335.
127. E. Kelesoglu and C. Mitterer, Structure and properties of TiB_2 based coatings prepared by unbalanced DC magnetron sputtering. *Surf. Coat. Technol.* 1998, **98**, 148–1489.

Index

α-modification, silicon nitrides   753
α-SiAlON structure   755 ff, 768 ff
ab initio pseudopotential approach   256 f
Aboudi model   72
abrasive applications, diamond materials   528 ff
abrasive wear
– silicon nitrides   784
– tungsten carbide–cobalt hardmetal   962
abrasives   5 ff
– silicon carbides   688, 736
absorption bands, CVD diamond   577
absorption coefficient, diamond-like carbon   638
acetylene   629
Acheson process   683, 688 f
acid stability, boron carbides   855
acidic solutions
– CVD diamond   415
– silicon carbides   735
activated sintering   877
activation energy, self-propagating synthesis   357
active corrosion   141 ff
additives
– boron carbides   843
– cemented ternary borides   925
– diamond synthesis   498
adhesion
– diamond-like carbon films   640 f
– silicon carbides   722 f
adiabatic combustion temperature   323
agglomerates, alumina-based ceramics   654
aggregates
– diamond   380
– directed metal oxidation   307
aggressive environment   584
alkali halide gas   43
alkali stability, boron carbides   855
alkaline earth nitrosilicates   29
alkoxides
– polymer-ceramic transformations   447
– thin film processing   463
allenes   LI
allotropes   XLIV, 271–285, 485
– diamond synthesis   390 f
allylhydropolycarbosilanes   463
alumina   3, 292
– bulk components   69
– directed metal oxidation   291 f
– elastic moduli   71

– silicon carbides   695
alumina addition, titanium carbide SHS   362
alumina–aluminum DMO composites   314
alumina-based ceramics   184, 192
– tool applications   648–682
aluminothermic reduction   875
aluminum
– cemented borides   896
– diamond synthesis   498
aluminum borates   164
aluminum boride type structures   805 f
aluminum interlayers, diamond-like carbon films   641
aluminum kryolite melt   880
aluminum lithium alloys   289
aluminum–magnesium systems, directed metal oxidation   295
aluminum–magnesium–zinc system, directed metal oxidation   292
aluminum–nitride–silicon carbide system, self-propagating synthesis   354
aluminum nitrides   715
aluminum oxides   787
aluminum titanate   318
AMBORITE milling   559
ammonia, carbide synthesis   210
amorphous carbon   XLIX, 272
– films   623–647
amorphous composites, nanocrystalline   112
amorphous covalent ceramics (ACC)   446 ff
amorphous hard materials   36 ff
amorphous structures, carbon nitrides   261
amorphous zone, silicon carbides   690
anatase structures, titanium oxides   32
anisotropic growth, silicon nitrides   763
annealing, nanostructures   130
anthraxolite   376
antireflection coatings, CVD diamond   589
anvils, diamond synthesis   490, 565
apatite   XL
applications   477–995
– alumina-based ceramics   648–682
– borides   802, 933
– cemented borides   927 ff
– CVD diamond   410, 573–622
– diamond materials   479–572
– diamond/boron nitrides   527
– diamond-like carbon films   640 ff

- directed metal oxidation   316
- polycrystalline ultrahard materials   548 ff
- silicon carbides   736 ff
- silicon nitrides   751 ff, 782 ff, 792
- single crystal diamond   559 ff
- TM carbides/nitrides   202 ff, 238 ff
arc evaporation, diamond-like carbon films   627
Archimedian three connected nets   275
argon   43
argon sputtering
- boron carbides   856
- ncNiCr   111
armor, directed metal oxidation   317
armor modules   94, 97
Arrhenius behavior
- self-propagating synthesis   323
- silicon carbides   450
aspect ratio
- silicon carbides   695
- silicon nitrides   767
ASTM, silicon carbides   736
AT60 composite   674
atomic force microscope (AFM)   54, 405, 440
atomic hydrogen, diamond synthesis   507
attack modes, corrosion   141 f
attrition milled powder mixtures   916
Auger electron spectroscopy (AES), 211
augmented-plane wave (APW) calculations   15
austenitic phases, carbides   16
automated pressing, silicon carbides   704

β-modification, silicon nitrides   753
β-phases, silicon carbon nitrides   265
ballas   512
ballistic properties   94, 97 ff
band gaps, borides   803
barriers, coulombic   94
BC(8/C4/T4/T8) series, carbon allotropes   272 ff
bell jar reactor   398
belt devices, diamond synthesis   490
bending strength
- alumina-based ceramics   649
- boron carbides   852, 865
- silicon nitrides   772, 778
bending tests   73
benzene
- carbide synthesis   208
- diamond-like carbon films   627, 633
Berkovich model   86, 195
Berman–Simon line   374
beryllium oxides   30
bias voltage, diamond-like carbon films   624
bias-induced heteroepitaxy   404 f
binary carbides/nitrides   202 ff
binary systems, borides   813 ff

binders
- bulk components   68
- carbides   12
- cemented ternary borides   919
- polycrystalline diamond   516
- self-propagating synthesis   357
- titanium boride-iron composites   910
- tungsten carbide–cobalt hardmetal   952
- tungsten carbides   165
binding energies, boron carbide–silicon carbide ceramics   859
birefringence, silicon carbides   719
black body radiation   45
black composites, alumina   669
block sawing, diamond abrasives   543
blocking, corrosion   146
boehmite
- alumina-based ceramics   653, 661
- Vickers hardness   184
Boltzmann constant   45, 93, 322
bonded tools, diamond/boron nitrides   528
bonding   3 f, 198
- borides   803
- boron nitrides   420 f
- carbon allotropes   274
- carbon nitrides   258
- diamond-like carbon   630 ff
- TM carbides/nitrides   204 ff
borane-dimethylamine   60
boride-based hard materials   802–945
boride–zirconia composites   888 ff
borides
- bulk components   69
- crystal structures   8 f
- data collection   965–995
boron
- diamond synthesis   500
- silicon carbide-based materials   683
boron-based hard materials   LVIII
boron–boron interactions, borides   808
boron carbide-based cermets   895 ff
boron carbide-based composites   857 ff
boron carbide ceramics   837 ff
boron carbide process   875
boron carbide–silicon carbide ceramics   857
boron carbide–transition metal diboride ceramics   861
boron carbides   XL
- ballistic properties   94
- corrosion   161 ff, 176
- crystal structures   8 f
- silicon carbides   714
boron–carbon–aluminum system   820
boron–carbon–metal system   819
boron–carbon–silicon system   819, 822
boron–carbon system   813
boron–carbon–titanium system   823 ff

boron carbonitrides   LXII, 526
boron doping
– carbon allotropes   276
– diamond-like carbon   638
boron halides   855
boron nitrate   406
boron nitride/silicon carbide double layers   309
boron nitrides   XL–LXXI, 6f, 253 ff
– alumina-based ceramics   648
– bulk components   69
– corrosion   171 ff
– crystallization   510 ff
– cubic polycrystalline   479 f
– CVD diamond   597 ff
– high pressure melting   49
– nanostructures   104, 125 f
– self-propagating synthesis   340
– vaporphase deposition   420–445
boron-rich nitrides   526
boron–silicon system   815, 821
boron suboxides   XL, LXV, 526
boron–transition metal systems   830
boro-silica glass   860
borothermic reduction   876
boundary conditions, corrosion   142
boundary sliding, nanocomposites   121
Bragg angle
– crystal stuctures   36
– nanocomposites   119
breakdown
– corrosion   147
– spinel layers   300 f
Bridge method   853
Bridgmen anvils   490
Brillouin zone   276
Brinell hardness   XLI, 85, 107
brittle failure   67
brittleness   20, 74 ff
bromine reactions   855
bronzes   895
brookite structures   32
brown corundum   658
bubble formation, silicon carbides corrosion   160
bulk components, mechanical properties   68 ff
bulk modulus   XLII
– carbon allotropes   277
– carbon nitrides   263
– polycrystalline boron nitride   519
– silicon carbon nitrides   265
bulk properties, TM carbides/nitrides   203
Burgers vector   93, 110, 115

C_{60} see: fullerenes
calcite   XL
calcium silicate, melting temperatures   51
calcium sulfate coatings   307

calibration, diamond anvil cell   43
CALPHAD (calculation of phase diagrams)   213
carbide conversion   876
carbide nitride–boride–silicide composite-based
      hard materials   888
carbide-reinforced composite ceramics,
      alumina-based   669 ff
carbides
– bulk components   69
– corrosion   155 ff
– data collection   965–995
– superstoichiometric   127
– transition metal diboride cermets   916
– transition metals   12 ff
carbon
– corrosion   154 ff
– diamond-like   XLIX
– hydrothermal synthesis   374
– self-propagating synthesis   338
– silicon carbide-based materials   683
carbon allotropes   XLIV, 485
– diamond synthesis   390 f
– doping   271–285
carbon-based hard materials   XLIV ff
carbon black
– boron carbides   843
– directed metal oxidation   293
– titania–aluminum system   364
carbon black reduction   875
carbon–boron system   813
carbon clusters   470
carbon content, tungsten carbide–cobalt
      hardmetal   952
carbon dioxide, diamond synthesis   490
carbon dioxide laser heating technique   41–65
carbon filaments, silicon carbides   697
carbon films, diamond-like   623–647
carbon–hydrogen–oxygen system, hydrothermal
      synthesis   376 ff
carbon phase diagram   485
carbonado   512
carbonitrides   LIII, 202–270, 523
– corrosion   173 ff
– nanostructures   104 ff
carborundum   7, 683
carbothermic reduction
– boron oxides   838
– metal oxides   875
– silica   691
– silicon carbides   715
carburization rate, diamond synthesis   394
carrier gas hot extraction   211
cascade arc plasma jet   624
cast iron machining   660, 667, 670
cast self-propagating high temperature
      synthesis   342
catalysts, diamond synthesis   498 f

cathodic arc deposition 422
caustic alkalis 155
caustic oxidizing media 415
cavities 92
cemented borides 895 ff
cemented carbides 12, 238 f
cemented carbonitrides 238 f
cemented diamond compositions 512
cemented ternary borides 919 ff
centrifugal compaction 654
ceramic alumina 3
ceramic bodies processing 446–476
ceramically bonded silicon carbides 700, 721
cermets 202, 345
charge carrier density
– boron nitrides 438
– carbon nitrides 260, 266
charge coupled devices (CCD) 45
chemical analysis, TM carbides/nitrides 210 f
chemical bonding
– borides 803
– carbides 15
– nitrides 25
chemical compositions see: compositions
chemical properties
– boron carbides 851 f, 855 f
– silicon carbides 720
chemical vapor deposition (CVD)
– boron nitrides 421
– carbide synthesis 210
– diamond films 390–419
– hydrothermal synthesis 374, 386
– nanostructures 116 f
– silicon carbides 697, 707
– thin films 463
– transition metal borides 876
– transition metal carbides/nitrides 241
chemical vapor deposition (CVD) diamond XLVII, 484
– applications 573–622
chemical vapor infiltration (CVI) 310, 717
chevron-nothed threepoint bend beam 84
chlorides 208
chlorination 374
chlorine reactions 855
chromium
– diamond synthesis 498
– transition metal diboride cermets 915
chromium binder 923 f
chromium carbides 17, 208
chromium diborides 831
chromium–nitrogen system 219
cluster fragmentation, cemented ternary borides 922
clusters
– alumina-based ceramics 653
– carbon allotropes 272

– titanium boride–yttria stabilized zirconia composites 893
coagulation, alumina-based ceramics 654
coarse grain, CVD diamond 580
coarsening see: Ostwald ripening
coarser powders, polycrystalline diamond 513
coated abrasives 736
coatings 66, 70 f
– boron carbides 838
– boron nitrides 429
– diamond-like carbon films 640
– directed metal oxidation 307
– nanocomposites 121 f
– self-propagating synthesis 366
– TM carbides/nitrides 242 f
cobalt
– diamond synthesis 498
– hardness 948
– transition metal diboride cermets 915
cobalt hard metals 238
cobalt–iron–system 488
cobalt–tungsten–carbon system 516
coesite–stishovite phase boundary 55
cold compaction 513
cold isostatic pressing 651
colloid size range growth, diamond 504
color
– boron nitride 510
– diamond synthesis 498
– silicon carbides 690, 719 f
– TM carbides/nitrides 224
combustion analysis
– diamond-like carbon 631
– TM carbides/nitrides 211
combustion tubes
compact nitridation, silicon powders 749
compact tension test 84
compaction homogeneity, alumina-based ceramics 654
complex kinetics, corrosion 148 ff
composite ceramics, whisker-reinforced 669
composites
– aluminum–magnesium 300
– boron carbide-based 857 ff
– nanocrystalline 112
– silicon carbides 710
– transition metal carbides 880
compositions
– cemented ternary borides 923
– titanium boride-iron composites 901
compounds
– borides 809
– isoelectronic 6 f
– transition metals 203
compressive strength
– polycrystalline boron nitride 519
– silicon carbides 718

compressive stress, sapphire 186
concrete machining 540 ff
conducting target sputtering 430
conduction band, nanostructures 132
conductivity
– alumina-based ceramics 667
– boron carbides 854
– boron nitrides 421
– CVD diamond 412, 582
– data collection 965–995
– directed metal oxidation 289, 316
– polycrystalline boron nitride 519
– silicon carbides 721
– silicon nitrides 751
– TM carbides/nitrides 203, 225
CONFLAT vacuum flanges 606
continuous fiber-reinforced silicon carbide matrix composites (CMCs) 717 ff, 739
controlled atmosphere, thin film processing 464
conversion process
– self-propagating synthesis 328 f
– thin films 464
copper
– diamond synthesis 488
– directed metal oxidation 291, 305
cordierite, silicon carbides 695
core–mantle composite model 110
core drills 549
core-rim structures, transition metal diboride cermets 918
cores, silicon carbides 688
corner sharing
– carbides 19
– nitrides 26
– titanium oxides 33
corrosion 140–182
– directed metal oxidation 316
– silicon nitrides 749, 786 ff
corundum XL
– alumina-based ceramics 648 f, 653, 658
– crystal stuctures 30
– silicon carbide based materials 683
– single phase sintered 670
Coulomb barriers 94
Coulombic forces, carbon/silicon–nitirides 254
covalent bonding XL, 253
– borides 803
– boron carbides 841
– boron nitrides 420 f
– carbides 197
– hydrothermal synthesis 374
crack branching 113
crack bridging 96
– silicon carbides 695, 710 f
crack deflection
– boride–zirconia composites 890
– silicon carbides 710 f

– titanium boride–titanium carbide system 880
crack growth rate, silicon nitrides 772 f
crack propagation 67 f, 75 ff, 86
cracking, directed metal oxidation 291, 298
creep behavior 92 ff
– alumina-based ceramics 667
– directed metal oxidation 313
– nanocomposites 121
– silicon nitrides 751, 778
creep feed grinding 537
crosslinking, polymer-ceramic transformations 447
crushing
– boron carbides 838
– polycrystalline diamond 513
crystal structures 1–285
– borides 804
– boron carbides 854
– boron nitrides 420 f
– carbon allotropes 277
– carbonitrides 258
– cemented ternary borides 919
– data collection 965–995
– polycrystalline boron nitride 520
– silicon carbides 685 f
– silicon nitrides 753 ff
– TM carbides/nitrides 204 ff
– zirconium/titanium borides 879
crystallite size effect, hydrothermal synthesis 376, 382
crystallites
– alumina-based ceramics 655, 661
– boride–zirconia composites 889
– diamond synthesis 400, 503
– silicon nitrides 759
crystallization, diamond 485
cubic boron nitride XL–LXXI
 see also: boron nitride
cubic diamond
– hydrothermal synthesis 374
– synthesis 509
cubic structures
– borides 809
– boron nitrides 6, 420–445
– cobalt 949
– diamond 5 f
– silicon carbides 685
– silicon carbon nitrides 266
– TM carbides/nitrides 205
Cubitron 664
curing 448
cutting edge displacement (CED) 674, 679
cutting edges 658, 666
cutting tools
– alumina-based ceramics 648 ff
– carbonitrides 202 f
– cemented borides 927

- CVD diamond   411, 611 ff
- material selection   68 f
- polycrystalline boron nitride   519
- polycrystalline diamond   550
- silicon carbides   736
- silicon nitrides   784
- single crystal diamond   563
- temperature effects   199
- TM carbides/nitrides   241

cyclohexadiene   273
Czochralski growth   190

damage
- creep   94 f
- intergranular   199
- silicon nitrides   780 f
- tolerance   67

data collection, hard material properties   965–995
DBC50 milling   561
De Hoff diagram   840
Debye temperature   879
decomposition reactions, CO_2 laser heating   54 ff
deep grinding   537
defects
- borides   813
- boron carbides   842
- carbides   12
- ceramic tool materials   651, 662
- silicon nitrides   771 f

definded cutting edge   666
deformation   XLI
- indentation test   87
- modes   76, 94 f, 185
- polycrystalline diamond   514

dense materials, self-propagating synthesis   342 ff
dense shapes, silicon carbides   699
densification
- boron carbides   844
- silicon nitrides   755 ff
- titanium boride-iron composites   906 f
- transition metal borides   876 ff

density   XLII
- alumina-based ceramics   654
- boron carbides   854
- CVD diamond   601
- data collection   965–995
- diamond-like carbon   632
- polycrystalline boron nitride   519
- silicon carbides   695, 718, 721
- titanium boride-iron composites   910
- TM carbides/nitrides   203
- zirconium/titanium borides   879

density effect, self-propagating synthesis   339

density functional based tight-binding (DF-TB)   273, 276
density of states (DOS)
- carbon allotropes   279 f
- TM carbides/nitrides   206

deposited layers, TM carbides/nitrides   241
deposition mechanisms   66
deposition methods
- boron nitrides   421 f
  see also individual types (CVD, PVD, etc.)
deposition temperature, diamond-like carbon   633
design, superhard materials   109 ff
detector materials, CVD diamond   584
DIAFILM   575 ff, 583
diamond   XL–LXXI, 5 ff
- alumina-based ceramics   648 f
- carbon allotropes   271 ff
- classifications   498 f
- corrosion   154 ff
- hydrothermal synthesis   374–389

diamond abrasives, synthesis   491
diamond anvil cell (DAC)   508
- laser heating   41–65
diamond films   390–419
diamond incorporation   367
diamond materials   479–572
diamond single crystals   379
diamond synthesis   481 ff
- carbon nitrides   257
diamond-like carbon films (DLC)   623–647
diboride systems   831
dielectric loss, CVD diamond   601
dielectric properties, CVD diamond   583
dies, diamond synthesis   490
differential thermal analysis (DTA)   36, 151, 446
diffusion, directed metal oxidation   291
diffusion-controlled growth, silicon nitrides   763
diffusion-cooled CO_2 laser   596
diffusion layers, TM carbides/nitrides   246 f
diffusional techniques, carbide synthesis   210
diffusivity   229 ff
dilution effect, self-propagating synthesis   335
diodes   606
direct coagulation casting (DCC)   654
direct current plasma jet deposition   XLVIII
directed metal oxidation (DMO)   92, 289–321
dislocations   183, 188, 196
- boron carbides   842
- creep   94
- CVD diamond   412
- nanocomposites   121
- transition metal diboride cermets   918
displacements
- control   77
- Voigt bounds   71
dissolution, nanostructures   130

dissolution temperatures, diamond   380
distortions, CVD diamond   589, 595
doping
– alumina-based ceramics   661
– boron carbides   852
– diamond-like carbon   638
– directed metal oxidation   305
– sp^2 bonded carbon allotropes   271–285
double bonds, carbon allotropes   278
double layers, directed metal oxidation   309
double torsion test   84
dressers, CVD diamond   611, 616 f
dressing tools   562
drilling
– carbonitrides   202 f
– polycrystalline diamond   554
– polycrystalline ultrahard materials   549
dry grinding
– abrasives   532, 538
– silicon carbides   704
dry machining   614
dry powder processing   651
dry pressing   704
dry sawing   547
ductile binders   68
ductile-to-brittle transition
– self-propagating synthesis   356
– titanium carbide   235
duplex layer, directed metal oxidation   292, 298
duplex techniques, titanium nitride coatings   342

η-carbides   18
earth's lower mantle materials   51 f
edge sharing
– nitrides   27
– titanium oxides   33
edge toughness   722
Einstein equation   130
Ekasic T   722 ff
elastic modulus   71 ff, 81 ff
– directed metal oxidation   310
– SiAlON   460 f
– silicon carbides   718 f
– siliconaluminocarbosilanes   454, 457 f
elastic properties, TM carbides/nitrides   231 f
elastic recoil detection (ERD)   631
elastic recoil detection analysis (ERDA)   436
electric discharge grinding   613
electric resistivity
– boron carbides   854
– data collection   965–995
– zirconium/titanium borides   879
electrical conductivity   225, 316
electrical properties
– boron nitrides   441
– CVD diamond   413

– diamond-like carbon   637 f
– silicon carbides   719
electrochemical applications, CVD diamond   415
electro-corundum   658
electrolysis, transition metal borides   876
electron cyclotron heating (ECH)   598
electron cyclotron resonance (ECR)   422
electron energy loss spectroscopy (EELS)   631
electron mobility, CVD diamond   414
electron paramagnetic resonance (EPR)   379
electron probe microanalysis (EPMA)   211
electron spectroscopy for chemical analysis (ESCA)   725
electron transport, directed metal oxidation   290
electronic applications, CVD diamond   413
electronic properties, carbon allotropes   279
electronic speckle pattern interferometry (ESPI)   593
electroplated tools   529, 532
Elektroschmelzwerk Kempten (ESK) process   688 f
element synthesis, silicon carbides   692
energy-dispersive analysis of X-rays (EDX)   119
energy–volume relationship, carbon nitrides   264
engineering requirements   70
enstatite   488
enthalpy
– data collection   965–995
– self-propagating synthesis   323
enthalpy increase, diamond synthesis   505
epitaxial growth   291
equilibrium volumes   XLII
erosion tests   81, 89 f
Eshelby treatment   311
etching
– diamond synthesis   404
– silicon nitrides   770
– titanium boride–yttria stabilized zirconia composites   893
eutectic concentrations, titanium boride-iron composites   909
eutectic reactions, borides   823
evaporation
– carbide synthesis   210
– diamond-like carbon films   627
even/odd phases, carbon nitrides   259
evolutionary selection, diamond synthesis   401
experimentals
– CO_2 laser heating   44 f
– corrosion   150 ff
– diamond anvil cell   43 f
– self-propagating synthesis   331 ff
– SiAlON   460 f
– silicon carbide–water system   383
– siliconaluminocarbosilanes   452
– thin film processing   463
– ncTiN/aSi$_3$N$_4$ deposition   118

extended X-ray absorption fine structure spectroscopy (EXAFS)　264, 446
external corrosion　141 f
extinction coefficient, boron nitrides　440
extrusion, silicon carbides　704

failure　67, 94 ff
– cemented borides　929
fatigue, mechanical　95 f
fatigue strength, silicon nitrides　751
feldspar　XL
Fermi level
– borides　808
– carbon allotropes　274
– diamond-like carbon　637
– nanostructures　132
– TM carbides/nitrides　206
ferritic binders, cemented borides　929
ferrous material machining　557
fiber-reinforced directed metal oxidation composites　314
fiber-reinforced plastics　611
fibers
– directed metal oxidation　307 f
– silicon carbides　697, 718
Fick diffusion　142 ff
field-activated self-propagating high temperature synthesis　348 ff
figures-of-merit, CVD diamond　413
film adhesion　427
fine grain, CVD diamond　580
finish machining　613 f
flaking　70
flat faces, diamond synthesis　401
flatness
– CVD diamond windows　595
– diamond submundum　610
flaws　88, 184
– alumina-based ceramics　653, 662
– creep　95
– nanostructures　111 f
flexural strength　82
– boron carbide ceramics　870
– polyaluminocarbosilanes　456
– silicon carbides　714, 718, 721
– titanium boride-iron composites　913
fluorescence spectroscopy　48
fluorite　XL
flute grinding high-speed steel　541
Forschungszentrum Karlsruhe (FZK), CVD diamond　598
forsterite　488
Fourier heat transfer　322, 350
Fourier transform infrared spectroscopy (FT-IR)　36, 437, 464
– borides　811

fractional parameters, carbon/silicon–nitirides　255
fracture corrosion stress　158
fracture strength　81 f, 601
fracture stress
– CVD diamond　580
– nanostructures　105, 109
fracture toughness　67 f, 74, 83 f
– alumina-based ceramics　649, 659, 663
– borides　928, 934
– boron carbides　853 f, 865, 870
– CVD diamond　588
– directed metal oxidation　289
– polycrystalline boron nitride　519
– SiAlON　460 f
– silicon nitrides　751, 771
– titanium boride composites　890, 912
fragmentation, directed metal oxidation　291
Frank–Kasper polyhedron　18
Frenkel excitations　638
fretting wear　656
friction coefficients
– boron nitrides　440
– CVD diamond　618
– diamond-like carbon　640 f
– nanocomposites　127
– silicon carbides　722 f, 732
– silicon nitrides　783
full-width-at-half-maximum (FWHM)
– CVD diamond　575
– diamond synthesis　405
– silicon carbide–water system　386
fullerenes
– carbon allotropes　XLVI, LIII, 271 ff
– complexes　253 ff
– hydrothermal synthesis　374
furan coke　505
fused corundum　658
fused grinding materials　648 ff

γ–alumina　291
γ-point sampling, carbon allotropes　276
garnets, phase diagrams　56
gas corrosion
– boron carbides　163, 172
– silicon carbides　159
– silicon nitrides　167
gas fusion methods　735
gas phase diagnostics　395
gas pressure　624
gas pressure sintered silicon nitrides (GPSN)　752, 758
gas temperatures, diamond synthesis　392
gaseous nitridation　290
gasification, graphite　307
gaskets, diamond synthesis　490

GefStoffV 736
gelcasting 654
gemstons ruby 30
General Electric process 692
generalized gradient approxiamtion (GGA) 255
geometrical distortions, CVD diamond 595
germanium 488
gettering 498
GG25 cast iron 669
Gibbs–Thomson equation 840
Gibbs free energy
– diamond synthesis 407
– titanium boride-iron composites 902
Gibbs free enthalpy, nanocomposites 120, 124 f
girdle devices 490
glas-reinforced plastics 553
glassy phases 92
glow discharge deposition
– diamond-like carbon films 623
– nanocomposites 117 f
grades
– CVD diamond 574 ff
– silicon carbides 687
grain boundaries
– boron carbides 841
– corrosion 141
– deformation 183–201
– junctions 91
– magnesia 292
– nanostructures 110, 130
– silicon nitrides 751
grain boundaries interlocking 886
grain size 183–201
– alumina-based ceramics 654
– boron carbides 838, 871
– silicon nitrides 751
– tungsten carbides 950 f
grains
– alumina-based ceramics 664
– CVD diamond 575, 580
– polycrystalline diamond 514
granite block sawing 543
granules, alumina-based ceramics 651
graphite XLIV ff, 253 ff
– borides 813
– boron carbides 838
– cemented borides 895
– diamond films 390 ff
– diamond synthesis 486 ff
graphite–diamond transformation 407
graphite reduction, transition metal borides 875
graphitic components, directed metal oxidation 293
graphitic structures 259
graphitization
– diamond corrosion 155
– polycrystalline diamond 514

graphyne 272
gray cast iron 669
green bodies, polyaluminocarbosilanes 452, 458
green density
– alumina-based ceramics 660
– self-propagating synthesis 360
– silicon carbides 700
Griffith model 74 f, 110 f
grinding
– alumina-based ceramics 648 f, 658 ff
– boron nitride abrasives 530, 538
– CVD diamond 613
– diamond abrasives 530 ff
– dislocations 194, 199
– hardness XLI
– silicon carbides 704, 736
grits
– alumina-based ceramics 658, 665
– borides 802
group (IV–VI) elements 830
group IV carbide system 213
group IV nitride system 216
group IVB carbonitrides 202–252
group IVB transition metal carbide layers 241
group V carbide system 213
group VB nitride system 217
group VIB transition metal carbide system 213
group VIII elements
– borides 830
– diamond synthesis 498
growth
– boron nitrides 433 ff
– CVD diamond 580
– diamond synthesis 392, 407 ff, 492, 496
– diamond-like carbon films 625
– directed metal oxidation 300 ff
– patterns 147 ff
– silicon nitrides 760
gypsum XL
gyrotron tube windows 583, 597 f

H3, carbon allotropes 273 f
hafnium nitrides 231
halides 876
Hall–Petch relation 93
– nanostructures 110 ff
– TM carbides/nitrides 235
– tungsten carbide–cobalt hardmetal 948 ff
halogenated hydrocarbons 377
hard materials XLI–LXXI
– structures 1–285
hard tool ceramics 650
hardened steel machining
– alumina-based ceramics 667, 672 ff
– CVD diamond 618

hardness   XLI–LXXI, 79, 84 ff
- alumina-based ceramics   648–682
- boride–zirconia composites   891
- borides   802
- boron carbides   851, 854, 865
- boron nitrides   421, 440
- cemented ternary borides   925
- cobalt   948
- CVD diamond   410
- diamond-like carbon   632, 639
- polycrystalline boron nitride   519
- SiAlON   460 f
- silicon nitrides   751 ff, 775 f
- titanium boride-iron composites   910
- tungsten carbide–cobalt hardmetal   946–964
- tungsten carbides   947
- ultrahard materials   521
- zirconium/titanium borides   879
Hashin–Shtrikman model   72
health   735
heat resistance   607 f
heat sinks   318
heated filament-assisted chemical vapor deposisiton (HF-CVD)   XLVII
Hertzian contact damage   783
Herzian indentation test   84, 98
heteroepitaxial films   400, 404
heterostructures, nanocrystalline   114 ff
hexagonal–cubic transition   433 f
hexagonal graphite
- diamond synthesis   508
- hydrothermal synthesis   374
hexagonal structures
- boron nitrides   LVIII, 6, 420 ff
- cobalt   949
- diamond   5 ff
- silicon carbides   7, 685
- titanium borides   816
- TM carbides/nitrides   205
high frequency glow discharge   117 f
high power gyrotron tubes   593
high power infrared lasers   589
high pressure decomposition   54 ff
high pressure diamond synthesis   487 f
high pressure high temperature phase diagrams   54 ff
high pressure high temperature (HPHT) synthesis
- boron nitrides   420
- diamond   XLVII, 508
- hydrothermal   374, 380
high pressure laminates (HPL)   613
high pressure microwave source reactor (HPMS)   398
high pressure modifications, silicon nitrides   753
high pressure process, diamond synthesis   504
high pressure stability, CO_2 laser heating   54 ff
high purity materials   838

high resolution transmission electron microscopy (HR-TEM)   119
high speed finish machining   613
high speed steel (HSS)   541, 555
high temperature expansion, TM nitrides   230
high temperature hardness   934
high temperature properties
- silicon nitrides   777 ff
- polyaluminocarbosilanes   457 f
high temperature strength   313
highest occupied molecular orbitals (HOMO), carbon allotropes   279 ff
highly ordered pyrolytic graphite (HOPG)   416
hole mobility   414
hole transport   290
homogeneity
- alumina-based ceramics   654
- corrosion   141 f
Hooke law   71, 81
- nanostructures   104, 108, 129
Hopkinson bar test   97
host lattice, TM carbides/nitrides   205
hot compaction   513 f
hot corrosion   151
- boron carbides   163
- boron nitrides   172
- directed metal oxidation   316
- silicon carbides   157 f
- silicon nitrides   167 f
hot filament chemical vapor deposition (HF-CVD)   391 ff
hot isostatic pressing (HIP)
- boron carbides   847 f, 861
- self-propagating synthesis   345
- silicon carbides   707, 721
- silicon nitrides   166, 749 ff, 778 f
hot pressing
- boron carbides   847 f, 861
- silicon carbides   705, 721
- silicon nitrides   749 ff
- transition metal borides   877
hot target sputtering boron nitrides   431
hybridization   XLIV
- diamond-like carbon   630
- nanostructures   104
- silicon carbides   686
hydroboric acid layers   855
hydrocarbons
- CO_2 laser heating   60
- diamond synthesis   XLVIII, 489
- hydrothermal synthesis   377
hydrochloric acid   316
hydrogen, diamond synthesis   390 f, 395, 507
hydrogen-free amorphous carbon (ta-C)   627 f
hydrogenated amorphous carbon (a-C:H)   623, 630 ff

hydrothermal corrosion
- boron carbides   162
- silicon carbides   156
- silicon nitrides   166
hydrothermal synthesis, diamond   374–389

icosahedral structures
- borides   809, 815
- boron carbides   8
ignition, boron carbides   838
ignition temperature   322, 359
ilmenite structures   57
immersion tests   150
impurities
- CVD diamond   412
- diamond synthesis   501
incubation
- corrosion   150
- directed metal oxidation   295
indentation crack length method   910
indentation hardness   107
indentation size   183–201
indentation tests   80 f, 86
- hardness   XLI
- SiAlON   460 f
inductive plasma emission (ICP) spectroscopy   735
industrial applications
- diamond/boron nitrides   527
- TM carbides/nitrides   238 ff
inelastic deformation   183 ff
inert atmosphere pyrolysis   452 f
infrared lasers   589
infrared seekers   583
initiation process, directed metal oxidation   289, 295
injection molding
- boron carbides   846
- silicon carbides   704
injection molding machine parts   932
interface controlled growth, silicon nitrides   763
interfacial energy, diamond synthesis   505
intergranular damage   199
intergranular microcracks   91
interlaminar strength   718
interlocking grain boundaries   886
intermediate phases, titanium borides   816
internal corrosion   141 f
international thermonuclear experimental reactor (ITER)   598
interstitial compounds   15
interstitials, TM carbides/nitrides   205
ion–ion core repulsion, carbon allotropes   276
ion beam-assisted deposition (IBAD)   523, 627
ion energy distribution, diamond synthesis   406
ion irradiation, thin film processing   467 f

ion plating   624 f
ion sputtering   111
ionic bonding, boron nitrides   420
ionic diffusion   291
iron, diamond synthesis   498
iron binder   923
iron catalysis   411
iron–nickel–system, diamond synthesis   488
iron oxides/silicates, melting temperatures   51
isoelectronic compounds   6 f
isostatic pressing   704
isothermal microwave plasma chemical vapor deposition   397
isotopic labelling   409
isotropic compounds   809

Jagodzinski notation   7 f, 13
Japan Atomic Energy Research Institute (JAERI)   602

kimberlite   377, 488
kinetics
- corrosion   142 ff
- titanium boride-iron composites   906
kinked faces   401
Kjeldahl analysis   211
Knoop hardness   XL, 86, 193
- boron nitrides   440
- data collection   965–995
- nanostructures   107
- polycrystalline boron nitride   519
- SiAlON   460 f
- ultrahard materials   521
Koehler model   114
Koks model   110

laminate wood flooring manufacturing   613
Langmuir probe   438
Lanxide process   917
laser ablation
- carbon nitrides   257
- diamond-like carbon films   627
laser diode arrays (LDA)   606 ff
laser heating   41   65
laser-induced damage threshold (LIDT)   594
laser irradiation, carbide synthesis   210
laser synthesis, silicon carbides   693
lasers, CVD diamond   589
lattice defects   185, 188
lattice dislocations   93
lattice parameters
- boron nitrides   420 f
- carbon allotropes   274 f, 277
- silicon nitrides   753

- TM carbides/nitrides   203
- zirconium/titanium borides   879
- data collection   965–995
layers   241
- corrosion   141
Lee–Gurland equation   955
Lehoczky model   114
Lely process   708
lifetime diagrams, cemented borides   929 ff
light covalent ceramics   522
light emitting diodes   737
line spread function (LSF)   578
linear rate constant, corrosion   144
liquid media corrosion
- boron carbides   161
- boron nitirides   171
- silicon carbides   156
- silicon nitrides   166
liquid phase sintering (LPS)
- boron carbides   845
- cemented ternary borides   920
- polycrystalline diamond   513, 516
- silicon carbides   705, 709, 721, 738
- silicon nitrides   756
- titanium boride-iron composites   899
liquid polymer infiltration (LPI)   718 f
liquid shaping, alumina-based ceramics   651
liquid silicon infiltration (LSI)   719 f
lithium   289
lithium aluminosilicate (LAS)   315
lithium doping
- carbon allotropes   271
- directed metal oxidation   304
load control   76, 187 ff
local density approximation (LDA)   255
localized corrosion   141 f
long range order (LRO)   16
longer wavelength infrared (LWIR) band   584
lonsdaleite   XLII, 5
- carbon allotropes   273
- diamond synthesis   510
- hydrothermal synthesis   375
loss tangent, CVD diamond   588, 600
low angle scattering   578
low pressure solid-state source (LPSSS)
- diamond synthesis   490, 508
- hydrothermal synthesis   374
low pressure sustained growth, diamond   504
lubrication, silicon carbides   722 f

machining
- alumina-based ceramics   649 ff, 660
- diamond abrasives   540 ff
- laminate wood flooring   613
magic angle spinning nuclear magnetic resonance (MAS-NMR)   36

magnesia *see:* magnesium oxides
magnesiothermic reduction   875
magnesium
- diamond synthesis   488
- directed metal oxidation   289 ff, 304
magnesium oxides   779, 787 f
- crystal stuctures   31
- directed metal oxidation   292
magnesium oxides/silicates, melting temperatures   51
magnetron sputtering
- boron nitrides   430, 435 f
- carbon nitrides   524
main group element nitrides   24 f
main group element oxide ceramics   30 f
MAK (maximal zulässige Konzentration) value, silicon carbides   736
manganese   498
manufacturing
- alumina-based ceramics   650
- polycrystalline diamond   513
martensitic phases, carbides   16
mass selected ion beam deposition (MSIB)
- boron nitrides   422
- diamond-like carbon   632
mass spectroscopy (MS)   36, 446
mass transport, directed metal oxidation   290
material grades, CVD diamond   574 f
material properties   1–285
- boron carbides   851 ff
- boron nitrides   421
- cemented ternary borides   923
- CVD diamond   410, 574 ff
- data collection   965–995
- diamond-like carbon   632
- ncM_nN/Si_3N_4 composites   119 ff
- polycrystalline boron nitride   519
- silicon carbides   719 ff
- silicon nitrides   753 ff, 771 ff
- titanium boride-iron composites   909
- TM carbides/nitrides   224 ff
- transition metal borides   878
materials   477–995
- corrosion   154 ff
mean free path   949 f
meandering   113
   see also: cracking
measurement, corrosion   150 ff
measurement models   193 f
mechanical grade, CVD diamond   575
mechanical properties-microstructure relation   66–103
mechanical properties
- borides   802
- boron carbide ceramics   865
- boron nitrides   440
- cemented borides   923, 928

- data collection   965–995
- diamond-like carbon   639 f
- directed metal oxidation   310
- self-propagating synthesis   327 ff
- silicon carbides   720
- silicon nitrides   771 ff
- polyaluminocarbosilanes   456
- titanium boride–yttria stabilized zirconia composites   892
- transition metal borides   878

medium density fireboard (MDF)   553
melonophlogite   274
melting, directed metal oxidation   295
melting points   3 f
- borides   802, 814
- boron carbides   854
- cemented borides   897
- creep   94
- data collection   965–995
- diamond synthesis   498
- high pressure   49
- polycrystalline boron nitride   520
- self-propagating synthesis   331, 342
- silicon carbides   685
- TM carbides   202 ff, 224
- TM nitrides   224
- zirconium/titanium borides   879

melts corrosion
- boron carbides   163
- boron nitrides   172

metal-bonded tools   529, 534
metal borides, cemented   895 ff
metal–boron interactions   808
metal carbides
- cemented borides   895
- diamond-like carbon films   629
- transition metal borides   876

metal containing amorphous hydrocarbon   629 f, 634 f
metal hydrides   875
metal matrix composites (MMC)
- cemented borides   895
- CVD diamond   611
- polycrystalline diamond   553
- silicon carbide whiskers   696

metal oxidation, directed   289–321
metal oxide silicon field-effect transistor (MOSFET)   415
metallic binders, cemented ternary borides   919
metallic impurities, silicon carbides   735
metallic melts   855
metallic nanoclusters   132
metals corrosion
- silicon carbides   156
- silicon nitrides   167

metastable processes, diamond synthesis   504
methane   60, 208

- diamond synthesis   395, 406

methane hydrogen plasma sustained growth   504
methyl radicals   408
methylhydroxylsiloxanes   463
Meyer hardness   85
microanalysis, physical   211
microcracking
- boride–zirconia composites   890
- boron carbide ceramics   873
- coatings   70
- nanostructures   110, 113

microcrystalline diamonds   379
microhardness   XL, 84, 194
- carbides   215
- data collection   965–995
- light covalent ceramics   521
- TM carbides/nitrides   203, 234 f
- tungsten carbides   947
see also: Knoop hardness

micropores, alumina-based ceramics   653
microstructure-mechanical properties relation   66–103
microstructures
- α-SiAlON materials   768 ff
- alumina-based ceramics   656
- β–silicon nitrides   758
- boron carbides   844
- boron carbide-silicon carbide ceramics   857
- cemented carbonitrides   240
- diamond-like carbon   630 ff
- directed metal oxidation   302 ff
- Ekasic W   733
- nanocrystalline   128
- silicon carbides   706
- silicon nitrides   749
- titanium boride-iron composites   909 f
- tungsten carbide–cobalt hardmetal   946

microwave-plasma chemical vapor deposition (MWP-CVD)   391, 397
migration
- directed metal oxidation   304
- high pressure melting   49
- self-propagating synthesis   343

military applications, CVD diamond   584
milling
- cast iron   670
- single crystal diamond   559

mining bits, polycrystalline diamond   556
modeling procedures
- carbon/silicon–nitirides   254 ff
- heat resistance   607

modifications, silicon nitrides   753
modulation transfer function (MTF)   578
moduli
- carbon nitrides   263 ff
- elastic   71 ff, 81 ff
- polycrystalline boron nitride   519

1010 Index

- ultrahard materials   521
- zirconium/titanium borides   879
modulus of rupture (MOR) *see:* rupture modulus
Mohs hardness   XL, 80, 107
molding
- boron carbides   846
- silicon carbides   704
molding tools   932
molybdenum
- diamond-like carbon   634
- directed metal oxidation   291
- transition metal diboride cermets   915
molybdenum boride-type structures   807
molybdenum–iron–boron system   832
molybdenum–nitrogen system   219
molybdenum silicides   20 f, 330
monolithic ceramics   97
monolithic components processing   450
morphology
- boron nitride films   423 f
- diamond synthesis   495
Mott law   637
mounting techniques, CVD diamond   605
mullite
- directed metal oxidation   293
- silicon carbides   695
multi-anvil devices, diamond synthesis   491
multiblade sawing   544
multilayer coatings   242
multiphase hard materials   888
multiphase systems, transition metal diboride cermets   916
multiphonon absorptions bands, CVD diamond   577
multiple cracking   291

nanoceramics   709
nanocrystalline composites   116 ff
nanocrystalline diamond   379
nanocrystalline materials   110 f
nanoindentation   108
nanostructured superhard materials   104–139
nanotubes   271
natural occurrence, silicon carbides   684
natural polycrystalline diamond   512
Nd-YAG laser   41
near cubic structures   266
neck formation
- boron carbides   849
- cemented ternary borides   922
negative electron affinity (NEA)   415
Newkirk process   289 ff
Nicalon
- polymer-ceramic transformations   448
- reinforced lithium alumiosilicate   315
- silicon carbides   695

nickel   915
nickel binders   923 f
nickel–carbon phase diagram   491
nickel chromides   111
nickel doping   305
nickel–manganese system   488
nickel silicides   325
niobium
- diamond synthesis   498
- diamond-like carbon   634
niobium–nitrogen system   218
nitridation, aluminum   290
nitrides
- bulk components   69
- corrosion   166 ff
- crystal structures   23 ff
- data collection   965–995
- superlattices   114
nitridosilicates   29
nitrogen
- diamond anvil cell   43
- diamond synthesis   498 ff
- silicon carbide based materials   683
nitrogen doping
- carbon allotropes   271, 276
- diamond-like carbon   638
nitrogen pressure
- nitrides   212
- silicon nitrides   753
nitrogen-rich boron nitrides   LXII
noble gases   43
nodule spacing, directed metal oxidation   305
nomenclature, crystal structures   7 f
noncutting applications
- polycrystalline diamond   555
- single crystal diamond   564
nondestructive evaluation (NDE)   91
non-equilibrium thermodynamics, diamond synthesis   507
nongraphitic carbons   505
non-isothermal plasma deposition   400
notch-beam method   910
novolac type resins   843
nuclear magnetic resonance (NMR)   631
nuclear reaction analysis (NRA)   631
nucleation
- boron nitrides   433 ff
- CVD diamond   580
- diamond synthesis   400 ff, 492
- magnesia   292
- silicon nitrides   760
numerical simulations, heat resistance   607

occupational health   736
occurrence, silicon carbides   684
octahedral structures   205

odd/even phases, carbon nitrides 259
Ohm's law 719
onions,
– carbon 374
– diamond synthesis 507
optical applications, CVD diamond 575, 583 ff, 589
optical properties
– boron nitrides 440
– CVD diamond 575 f, 589
– diamond-like carbon 637 f
– silicon carbides 719
orbitals 279, 808
ordered defects 12
organic compounds 60 ff
oriented globular growth 512
ornamental layers 243
orthorhombic structures
– borides 10
– cemented ternary borides 920
– titanium borides 816
Ostwald–Volmer rules 489
Ostwald ripening
– nanocomposites 123, 129 f
– titanium boride-iron composites 911
Ostwald rule 686
output windows, gyrotron tubes 583
oversaturation, silicon nitrides 761 ff
oxidation
– boron carbides 163, 855, 858
– directed 289–321
oxidation resistance 965–995
oxide ceramics
– bulk components 69
– crystal structures 30 ff
oxides 965–995
oxyacetylene 391
oxycarbide composites 674
oxygen acetylene torch XLVIII
oxygen attack, corrosion 141
oxygen contamination, titanium boride-iron composites 900
oxygen pressures 159
oxynitride solution 760 ff

Palmqvist cracks 87
parabolic rate constant, corrosion 145
paracyanogen 104
Paris law 95
Parthé–Yvon rule 13
partial pressure, directed metal oxidation 291
particle size, self-propagating synthesis 338
passivation
– directed metal oxidation 291
– spinels 295 f
passive corrosion 141 f, 145 f

passive infrared windows 584
Pauling rules 33
PDA diamonds 533 f
Pearson symbol 4, 203
Peierls–Nabarro stress 110
Peierls barrier 359
Peierls stress 72
penetration depth, ballistic 79
periclase 302
periodic bond chains (PBC) 401
permittivity 598
perovskite oxides 59
perovskite structures 27
Petch law 92
petroleum coke 838
phase diagrams
– borides 812 ff
– carbonitride systems 221 f
– CO_2 laser heating 54 ff
– silicon carbide- aluminum nitride 716
– silicon carbides 685
– transition metal–nitrogen systems 216 ff
– transition metal carbides 15
– tungsten–carbon–cobalt system 239
phase equilibria, carbide systems 213 ff
phase properties, polycrystalline boron nitride 520
phase reactions, cemented ternary borides 920
phase segregation, nanocomposites 116, 125
phase transitions, laser heating 41–65
phases, silicon nitrides 755
phenolic resins 842
phosphorus doping 271
physical boundary conditions, corrosion 142 f
physical microanalysis, TM carbides/nitrides 211 ff
physical properties 66 f
– boron carbides 851 ff
– boron nitrides 421, 440
– cemented ternary borides 923
– CVD diamond 410 ff, 574
– data collection 965–995
– diamond-like carbon 632, 637 ff
– polycrystalline boron nitride 519
– silicon carbides 695, 719 ff
– silicon nitrides 753 ff, 771 ff
– titanium boride-iron composites 909
– transition metal borides 878 f
– transition metal carbides/nitrides 203, 224 ff
physical vapor deposition (PVD)
– boron nitrides 421
– carbide synthesis 210
– silicon carbides 708
– thin films 463
– TM carbides/nitrides 241
π–π bonding 274
pileup models 110

pitting 158 f, 786
Planck formula 45
plasma-assisted chemical vapor deposisiton (PA-CVD) XLVII
plasma decomposition carbonitrides LV
plasma deposition 624 f
plasma-enhanced chemical vapor deposition (PE-CVD) 523
plasma etching
– diamond synthesis 404
– silicon nitrides 770
– titanium boride–yttria stabilized zirconia composites 893
– titanium boride-iron composites 912
plasma vapor deposition 116 f
plastic deformation *see:* deformation
plastic indentation 108
platelets
– Ekasic W 733
– silicon carbides 696
platelets alignement 92
point defects
– boron carbides 842
– diamond synthesis 501
poisoning agents, directed metal oxidation 290, 307
Poisson ratio 72
– boron carbides 854, 865
– CVD diamond 588
– nanostructures 108
– polycrystalline boron nitride 519
– TM carbides/nitrides 232
– zirconium/titanium borides 879
polishing, diamond abrasives 544
polyaluminocarbosilanes (PAlC) 451 ff
polybenzene 274, 278
polyborosilizanes 174
polycarbosilane (PCS)
– boron carbides 845
– polymer-ceramic transformations 446 f
– silicon carbides 698
– thin film processing 463
polycrystalline boron nitrides 421, 512, 518 ff, 611
polycrystalline diamond 482 f, 512 ff, 611
polycrystalline grits 648
polycrystalline silicon carbides 685
polycrystalline superlattices 114
polycrystalline tungsten carbides 948
polyhedron structure, carbides 18
polymer to ceramic transformations 446–476
polymeric networks, carbon allotropes 272
polymer-pyrolysis derived (PP) fibers 697 f
polyphase ceramic composites 289 ff
polysilanes 446 f
polysiloxanes 446 f
polytitanocarbosilane (PTC) 447

Poole–Frenkel excitations 638
pore sizes, polycrystalline diamond 513
porosity 84, 91
– boron carbides 841
– residual 184
– self-propagating synthesis 339, 346
– polyaluminocarbosilanes 455
– tungsten carbide–cobalt hardmetal 953
potentials, cemented borides 927
powder processing
– alumina-based ceramics 651, 665 f
– carbides 12
– directed metal oxidation 293
– silicon carbides 691
precursors
– diamond XLVIII
– diamond-like carbon films 624
– polymer-ceramic transformations 447, 453 f
preforming, directed metal oxidation 307
preheating, self-propagating synthesis 366
preparation
– boron carbides 837 f
– carbonitrides LVI
– diamond synthesis 391 ff
– diamond-like carbon films 623
– SiAlON 460
– polyaluminocarbosilanes 452 ff
– transition metal borides 875
– transition metal carbides/nitrides 207 f
pressing 704
pressure, directed metal oxidation 291
pressure conditions, diamond anvil cell 43
pressure effects, self-propagating synthesis 347
pressure–temperature conditions, diamond films 390
pressure–temperature diagram, diamond synthesis 493
pressure–temperature regime, CO_2 laser 42
pressureless sintering
– boron carbides 841
– transition metal borides 876
process-related properties, alumina-based ceramics 654
processing 287–476
processing defects 91
processing zones, temperature effects 199
production
– silicon carbide nanoceramics 709
– silicon carbides 709, 717
production routes
– alumina-based ceramics 650
– boron carbides 838
– silicon carbides 688 ff
protective scales, corrosion 141
pseudo-hot-isostatic pressing 345 f
pseudopotential approach, carbon/silicon nitrides 256

purity
- boron carbides 838
- diamond synthesis 501
- silicon carbides 690
pyrolysis
- amorphous materials 36
- boron carbides 843
pyrope glass 57
pyrophyllite gaskets 490

QQC deposition 490, 508
quality control, silicon carbides 734 ff
quantum tunneling 291
quartz 377
Quasam 274
quaternary systems
- borides 818
- carbonitrides 222 ff
- diborides 831

R8, carbon allotropes 273 f
radiation effects, CVD diamond 605
radiofrequency diode sputtering 430
radiofrequency glow discharge deposition 623
rain drop impact, CVD diamond 584, 588
Raman spectra
- CVD diamond 412, 575
- diamond anvil cell 48
- diamond films 394
- diamond XLVIII
- silicon carbide–water system 383
- thin films 469 ff
- tungsten carbides 165
Ramsdell notation 7 f
random aggregates 512
rare earth hexaborides 11
rare earth nitridosilicates 29
reactants
- diamond synthesis 408
- self-propagating synthesis 342, 366
reaction-bonded aluminum oxide (RBAO) 290, 293 f
reaction-bonded mullite (RBM) 293
reaction-bonded silicon carbides 700, 721
reaction-bonded silicon nitride (RBSN) 290 ff, 749–801
reaction-boronizing sintering 920
reaction control, scale formation 145
reaction-sintered silicon carbide (RSSC) 290, 293
reactive atmosphere pyrolysis 460 f
reactive plasma techniques 210
reactor, microwave-plasma chemical vapor deposition 391
rebound hardness 107

rebound test 81
recession rate, corrosion 152
reciprocating grinding 538
recrystallization
- boron carbides 842
- silicon carbides 700, 721
REFEL process 290, 294
refractive index
- boron nitrides 421, 440
- CVD diamond 588, 599
- diamond-like carbon 633, 639
- silicon carbides 719
refractories, directed metal oxidation 289
refractory carbides 3
refractory grade silicon carbides 688
refractory materials, self-propagating synthesis 334
refractory metal wires 392
refractory metals 634
refractory nitrides 113
reinforced dies 490
reinforced grades, alumina-based ceramics 649, 668
relative stability, carbon/silicon nitrides 255, 263, 268
residual porosity 91, 184
residual stress 66
resin bond polishing head 545
resin-bonded tools, diamond/boron nitrides 529
resins, boron carbides 842 f
resistance curves 78, 89
resistivity
- boron nitrides 421
- data collection 965–995
- TM carbides/nitrides 203
- zirconium/titanium borides 879
Reuss–Voigt average 310
Reuss model 72
rhenium filaments 394
rhombohedral graphite 509
rhombohedral graphitic structures, carbon/silicon nitrides 256
rhombohedral structures XLIV
- boron nitrides LVIII, 420 f
- corundum 30
- silicon carbides 7, 685
rigidity 104
roadway sawing 546
Rockwell hardness XLI, 85
- titanium carbide reinforced alumina 674 f
- tungsten carbide–cobalt hardmetal 946
room temperature
- grain boundaries 195 ff
- silicon nitrides 771 f
- Young modulus 234
ruby fluorescence scale 43
rule-of-mixtures, elastic moduli 72

rupture modulus 72
– directed metal oxidation 310
– polyaluminocarbosilanes 456
rupture strength 926
Rutherford backscattering (RBS) 211, 395, 464
rutile structures
– nitrides 26
– titanium oxides 32

σ-bonds
– carbon allotropes 271
– diamond-like carbon 630
salt baths 210
salt melts corrosion
– silicon carbides 157
– silicon nitrides 167
sample preparation, diamond anvil cell 42
sand erosion 588
sapphire 30, 186 ff, 597 ff
saturation, silicon nitrides 761 ff
sawing 542
SC24, carbon allotropes 275
scales
– corrosion 141
– directed metal oxidation 305 f
scanning electron microscopy (SEM) 54
– thin film processing 464
– cemented carbonitrides 240
– cemented ternary borides 926
– diamond 379, 393, 404
– silicon nitrides 777 f
– titanium boride-iron composites 900
– titanium carbide–boron powder blend 862
– tungsten borides 885
scanning tunneling microscopy (STM) 636
Scherrer equation
– crystal stuctures 36 ff
– nanocomposites 119
Schrödinger equation 276
scratch hardness XLI, 80, 89, 107
secondary ion mass spectrometry (SIMS) 211
– boron nitrides 434
– diamond synthesis 396
– thin film processing 464
secondary neutron mass spectrometry (SNMS) 211
Seebeck coefficient 854
Seebeck elements 737
seeding, alumina-based ceramics 652, 661
selection criteria, diamond synthesis 401
self-propagating high-temperature synthesis (SHS) 322–373
self-propagating synthesis, boron carbides 839
self-scavenging reactions, diamond synthesis 408

self-sustaining high-temperature synthesis (SHS) 207
semi-empirical approaches, carbon/silicon nitirides 254 f
semiconductor properties
– blue diamond 500
– CVD diamond 414
– silicon carbides 719
SH1 composites, titanium carbide reinforced alumina 673 ff
shaping techniques, silicon carbides 699, 704 f
shear modulus
– boron carbide ceramics 865
– nanostructures 115
– titanium carbide 235
– zirconium/titanium borides 879
shear stress 105
shock wave production XLVII
shock wave tests
– hydrothermal synthesis 374
– sapphire 186
– methane 60
short range order (SRO), carbides 16
shrinkage
– boron carbides 841
– cemented ternary borides 921
– polycrystalline diamond 514
– silicon carbides 705
– polyaluminocarbosilanes 455
– titanium boride-iron composites 906 f
Si_3N_4, nanostructures 104
SiAlON see: silicon aluminum oxynitride
Sicalon 716
  see also: siliconcarbides
silica carbides, thin film processing 463
silica formers, corrosion 142
silica reactor tube 397
silicides 20 ff, 965–995
silicon, diamond synthesis 404
silicon aluminum oxynitride, pyrolysis 460 ff
silicon aluminum oxynitride-bonded silicon carbides 700
silicon based materials LXV
silicon borides 815
silicon carbide–titanium carbide–titanium diboride composites 883 f
silicon carbide Al_2OC 716
silicon carbide-based hard materials 683–748
silicon carbide–carbon systems 717 f
silicon carbide metal matrix composites 614
silicon carbide–silicon carbide systems 717 f
silicon carbides
– alumina-based ceramics 648, 668
– boron carbide ceramics 857
– corrosion 141, 156 ff
– crystal structures 6 f
– directed metal oxidation 308 ff

- fibers 446 f
- hydrothermal treatment 382 ff
- self-propagating synthesis 351
- temperature effects 198
- thin film processing 463
silicon carbon nitrides 264 ff
- corrosion 173 f
silicon carbon nitrogen systems 253–270
silicon containing systems, polymer-ceramic transformations 446 f
silicon nitride based hard materials 749–801
silicon nitrides XL, 253–270
- bulk components 69
- corrosion 166 ff
- crystal structures 24
- nanostructures 113 ff
silicon oxides 55
silicon sputtering 856
polyaluminocarbosilanes 452 ff
silicothermic reduction 875
single crystal data 188
single crystal diamond, applications 559 ff
single layers, self-propagating synthesis 366
single-phase ceramics 878
single-phase sintered corundum 670
single-point ignition, boron carbides 838
sintered grinding materials 648 f, 658 ff
sintered powder derived (SP) fibers 697 f
sintered reaction-bonded silicon nitride (SRBSN) 749–801
sintered silicon carbides 700 ff, 721
sintered silicon nitrides (SSN) 749–801
sintering
- alumina ceramics 184
- alumina-based ceramics 651
- borides 815, 828
- boron carbides 839 ff
- cemented ternary borides 920 ff
- polycrystalline diamond 512, 516
- titanium boride-iron composites 899
- transition metal borides 876 f
slide rings 737
sliding tests 785
sliding wear 723, 729 ff
slip casting 700, 704, 739
slip dislocations 95
small angle diffraction (SAD) pattern, silicon carbide–water system 385
small angle X-ray scattering (SAXS) 635
sodium 289 f
softening, nanostructures 111
sol-gel derived corundum 661
sol-gel technology 652 f
solid-state lasers 41
solid-state reactions, titanium boride-iron composites 909
solid-state sintering 703, 709, 721

solvents, diamond synthesis 498 f
Soret effect, self-propagating synthesis 343
sp^2–sp^2 bonding 274
sp^2 bonded carbon allotropes 271–285
sp^3 hybridization 104
space groups 4 ff
- borides 805, 815
- carbon nitrides 258
- cemented ternary borides 919
- silicon nitrides 753
spacing, directed metal oxidation 305
spalling
- coatings 70
- corrosion 147
sphalerite structure 173, 510
spin propagation 325, 336
spinels
- crystal stuctures 31
- directed metal oxidation 292, 295 ff, 305 f
- silicon nitrides 753
spray drying 651
springback, polyaluminocarbosilanes 452
sputtering
- boron carbides 856
- boron nitrides 429 ff
- carbon nitrides 524
- diamond-like carbon films 624 f
- hydrothermal synthesis 374
- ncNiCr 111
- thin films 463
ST12, carbon allotropes 273 f
stability
- alumina-based ceramics 667
- carbides 212
- carbon nitrides 255, 263, 268
- CO_2 laser heating 54 ff
- diamond 380
- perovskite oxides 59
- silicon nitrides 255, 268
stable processes, diamond synthesis 504
stacking faults 412
stacking sequences XLV
- borides 805
- boron nitrides 420
- diamond 5, 510
- nitrides 28
- silicides 20
- silicon carbides 7, 686
- TM carbides/nitrides 205
star of FZK 602
static properties, carbon allotropes 278
steel, elastic moduli 71
steel machining 660, 667 ff
stepped faces, diamond synthesis 401
stiffness 66
  see also: Young modulus
stishovite structures XL, 33, 55 526

Stone–Wales transformations 282
stone machining 540 ff
strain energy release rate (SERR) 76
strength
- alumina-based ceramics 649
- boron carbides 852, 865
- cemented ternary borides 926
- CVD diamond 580 ff
- directed metal oxidation 311
- mechanical 66 f, 72
- polycrystalline boron nitride 519
- silicon carbides 718
- silicon nitrides 751, 771
- titanium boride-zirconia composites 890
- titanium boride-iron composites 913
strength–grain size relation, diboride composites 887
stress
- CVD diamond 610
- residual 66 f
- silicon nitrides 776
stress intensity factor (SIF) 72, 77, 95
stress rupture 94
stress–strain characteristics, nanostructures 129
stress testing 186 ff
Stribeck test data 729 ff
strong tool ceramics 650
structural application, silicon nitrides 792
structural chemistry, hard materials 3–40
structures 1–285
- borides 804
- boron carbides 854
- carbon allotropes 277
- carbon nitrides 261
- cemented ternary borides 919
- cobalt 949
- diamond-like carbon 632
- silicon carbides 685 ff
- silicon carbon nitrides 265
- TM carbides/nitrides 204 ff
- zirconium/titanium borides 879
  see also: crystal structures
subgranular residual porosity 91
sublattice structures, borides 811
submicrometer cutting 666
submount heat resistance 607
suboxides, boron LXV, 526
substrate based CVD fibers 697
substrate bias voltage 624
Super FZK 601
superabrasive tools 479 ff
supercritical water 379
superequilibrium atomic hydrogen 507
superhard materials 253–270
- nanostructured 104–139
Superior Graphite Company, silicon carbides 692

superlattices 114
superplastic silicon carbide sintered body 710
supersaturation, diamond synthesis 492
superstoichiometric carbides 127
surface adsorption, diamond synthesis 493
surface chemistry, polycrystalline diamond 516
surface damage 67
surface graphitization, diamond corrosion 155
surface states 183–201
SYNDITE 025 553
synthesis 287–476
- boron carbides 838 f
- boron nitrides LX
- diamond XLVII, 504 ff
- polycrystalline diamond 512 f
- silicon carbides 692
- TM carbides/nitrides 207 f

talc gaskets 490
talcum XL
tantalum 498
tantalum carbides 12
tantalum filaments, diamond synthesis 394
tantalum–nickel–boron system 832
tantalum–nitrogen system 219
target materials, diamond-like carbon films 629
Tauc relation 638
technical applications
- borides 813 f
- cemented ternary borides 919 f
- TM carbides/nitrides 202 ff
technical demands, alumina-based ceramics 660, 667
technical grades, silicon carbides 687
technical scale production, boron carbides 838
technological properties 965–995
temperature strength 92
temperatures 183–201
- boron carbides 841
- CO_2 laser heating 45, 48
- CVD diamond 582
- diamond-like carbon 633
- dielectric constants 604
- self-propagating synthesis 325 f
- silicon carbides 688
- silicon nitrides 765
- tungsten carbide–cobalt hardmetal 953
tensile fracture, silicon carbides 710
tensile strength
- CVD diamond 588
- silicon carbides 697, 718
ternary composites, borides 888
ternary systems
- borides 818 ff
- boron–metals 824 ff, 836
- phase diagrams 56 f

ternary transition metal carbides 213
ternary transition metal nitrides 221
Tersoff potential 255, 261
testing 80 ff
tetragonal structures, cemented ternary
    borides 920
tetragonal zirconia 888
tetrahedral amorphous carbon 428
textural changes, melting temperatures 49, 54
textured films 400
thermal barriers, directed metal oxidation 318
thermal conductivity
– alumina-based ceramics 667
– boron carbides 854
– boron nitrides 421
– CVD diamond 412, 582
– data collection 965–995
– directed metal oxidation 289
– silicon carbides 721
– TM carbides/nitrides 225
thermal effects, CVD diamond 589
thermal emission, silicon carbides 46
thermal expansion coefficient (CTE)
– alumina 69
– boron carbides 854
– CVD diamond 588, 610
– data collection 965–995
– polycrystalline boron nitride 519
– silicon carbides 718, 721
– silicon nitrides 753
– TM carbides/nitrides 203, 228
– zirconium/titanium borides 879
thermal grade, CVD diamond 575
thermal management, laser diode arrays 606 ff
thermal migration, self-propagating
    synthesis 343
thermal properties, silicon carbides 720
thermal resistance, CVD diamond 607 f
thermal shock
– CVD diamond 587
– directed metal oxidation 313
thermal shock resistance
– silicon carbides 714
– silicon nitrides 749
thermal spikes, boron nitrides 427
thermal stress, CVD diamond 610
thermite process 838
thermochemistry, diamond synthesis 407 ff
thermodynamic stability, diamond 380, 504
thermodynamics
– hydrothermal synthesis 378
– self-propagating synthesis 349
– silicon carbide–water system 382
– TM carbides/nitrides 212 ff
thermoelastic effect, CVD diamond 589
thermogravimetric analysis (TGA) 36, 446
– corrosion 151

– directed metal oxidation 295
thermorefractive effect, CVD diamond 589
thin films processing 446, 463 ff
three dimensional carbon phases L
tight-binding
– carbon allotropes 273, 276
– carbon/silicon nitrides 255 f
time dependence, self-propagating synthesis 331
titanium
– diamond synthesis 498
– diamond-like carbon 634
– directed metal oxidation 305
titanium boride-iron composites 897
titanium boride–yttria-stabilized zirconia
    composites 892
titanium boride–zirconium boride system 879
titanium borides, physical properties 879
titanium–boron–nitrogen films 125
titanium–boron system 333, 816
titanium carbide–boron system 825
titanium carbide-reinforced alumina 673
titanium carbides XL, 12
– alumina-based ceramics 669
– corrosion 174
– self-propagating synthesis 324, 356
titanium–carbon–nitrogen system 222
titanium carbonitrides, corrosion 174
titanium–chromium mixed borides 833
titanium diboride–boron carbide system 880
titanium diboride–nickel, cobalt composites 914
titanium diboride–titanium carbide system 880
titanium diboride–transition metal boride
    composites 883
titanium diboride-based cermets 897 ff
titanium diborides XL, 831
– boron carbide ceramics 861
– corrosion 176
titanium dicarbides, silicon carbides 711
titanium interlayers, diamond-like carbon
    films 641
titanium–iron–boron system 827
titanium–molybdenum–carbon–nitrogen
    system 222
titanium–nickel–boron system 827
titanium nitrides
– corrosion 174
– diamond synthesis 500 f
titanium–nitrogen system 216, 323
titanium oxides, crystal stuctures 32
titanium silicides 330
titanium–tungsten–boron system 836
titanium–tungsten–carbon–nitrogen system 222
tools
– applications 648–682
– bonded 528
– superabrasive 479 f
topaz XL

toroidal anvils 490
toughness
– directed metal oxidation 311
– tungsten carbide–cobalt hardmetal 962
toxicology, silicon carbides 736
transformation processes, diamond synthesis 509
transformation toughening 92
transgranular fracture, silicon nitrides 774
transgranular microcracks 91
transition carbide systems 213
transition metal alloys 488, 498
transition metal boride ceramics 874 ff
transition metal borides, crystal structures 9
transition metal–boron system 817
transition metal carbides 202–252
– corrosion 174
– crystal structures 12 ff
transition metal diboride cermets 915 ff
transition metal diboride–silicon carbide composites 881 ff
transition metal diborides 711, 861 ff
transition metal doping 305
transition metal nitrides 15, 202–252
transition metal nitrides
– corrosion 174
– crystal structures 25 f
– nanostructures 113 ff
– wetting 131
transition metal–nitrogen systems 216 f
transition metal oxides 32 f
transition metal silicides 20 ff
transition pressures, carbon/silicon nitrides 255
transition temperatures, silicon carbides 159
transmission electron microscopy (TEM) 185, 446, 464
– CVD diamond 576 f
– diamond XLVIII
– diamond-like carbon 634
– silicon nitrides 763
– tungsten borides 808
transport properties, borides 803
transverse rupture strength
– cemented ternary borides 926
– directed metal oxidation 310
TRIBO 2000/-1, silicon carbides 729 ff
tribological properties
– boron nitrides 440
– diamond-like carbon films 639, 642 ff
– silicon carbides 723 ff
trigonal structures 7
– prismatic 205
triple layers
– directed metal oxidation 309
– self-propagating synthesis 366
triple point, boron nitride 511
tripling, tungsten silicides 21

truing tools 562
tungsten
– diamond-like carbon 634
– transition metal diboride cermets 915
tungsten carbide–cobalt hardmetal 946–964
tungsten carbide–nickel, bulk components 68
tungsten carbides 12, 512
– bulk components 68
– cobalt metals 238
– corrosion 165
– diamond-like carbon films 629
– elastic moduli 71
– grain size effects 197
– hardness 947
– self-propagating synthesis 324
– thermal expansion 228
– Young modulus 233
tungsten diborides 831
tungsten filaments
– diamond synthesis 392
– silicon carbides 697
tungsten nitrde–aSi_3N_4 system 124
tungsten silicides 20
turbostatic layers, boron nitride films 423
twins 183, 188, 196
– boron carbides 849
type I/II CVD diamond 574
type Ib/IIab diamonds 498
type IIa single crystal diamonds 606
Tyranno fibers 383, 697

ultrahard materials 479 ff, 521 ff
ultrahigh pressure anvils 565
ultraviolet spectroscopy (UPS) 211
undefined cutting edges 658
unit cell
– borides 804
– carbon nitrides 258
– crystal structures 8, 20

vacancies 130
vacancy complexes 272
vacancy ordering 205
vacuum arc evaporation 627
vacuum hot gas extraction 211
vacuum sintering, boron carbides 847
valence bonds 633
valence electron concentration (VEC) 206, 234
valence electrons 810
vanadium 634, 924
vanadium borides 808
vanadium carbides 68, 324
vanadium nitrogen system 217
vapor–liquid–solid process, silicon carbide whiskers 693

vapor–solid process, silicon carbide whiskers 693 f
vapor phase deposition 420–445, 692
vapor phase formation/condensation, silicon carbide whiskers 693
vapor phase reactions, titanium boride-iron composites 909
vented tube furnace 838
Vickers hardness XLI, 86, 183–201
– alumina-based ceramics 649, 655
– CVD diamond 588
– data collection 965–995
– nanostructures 105
– SiAlON 460 f
– titanium boride-titanium carbide system 880
– tungsten carbide–cobalt hardmetal 947
viscosity, boron carbides 846
vitrified bonded tools 529, 532
voids
– carbides 13
– nanostructures 110, 113
– polycrystalline diamond 513
– TM carbides/nitrides 205
Voigt bounds 71
volatile chlorides 208
Volmer equation 488

wafers, silicon carbides 708
Wagner model 159
Warren–Averbach analysis 119
water vapor corrosion 162
wear parts
– cemented borides 932
– CVD diamond 611, 617 f
wear properties 89 f
– boron nitrides 440
– CVD diamond 410
– diamond-like carbon films 643
– directed metal oxidation 314, 317
– silicon carbides 722 f
– transition metal borides 878
– transition metal carbides/nitrides 242
wear resistance
– alumina-based ceramics 655, 658, 667
– silicon nitrides 751, 782 ff
– tungsten carbide–cobalt hardmetal 962
Weibull modulus 73, 81
– CVD diamond 580 f, 590
– silicon nitrides 772, 783
weight gain
– boron carbides 856
– directed metal oxidation 297
weight loss
– silicon nitrides 789 f
– polyaluminocarbosilanes 455
wet grinding 532

wetting
– borides 803
– carbonitrides 202
– cemented borides 895
– diamond-like carbon films 644
– directed metal oxidation 307
– self-propagating synthesis 347
– titanium boride-iron composites 900
whiskers 92
– alumina-based ceramics 649, 668
– silicon carbide-boron carbide ceramics 857
– silicon carbides 693, 736
white corundum 658
wide band gap 441
Williams expansion 77
wire sawing 542
wood flooring manufacturing 613
work hardening model, nanostructures 110
working steels 658
wurtzite XLIV
– beryllium oxide 31
– boron nitrides LVIII, 6
– silicon carbides 685, 715
Wykoff parameters 255

X-ray diffraction (XRD) 91, 446
– boron carbides 163
– diamond 380, 394
– diamond-like carbon 634
– self-propagating synthesis 353
– silicon carbides 686
X-ray spectroscopy (XPS) 36, 211

yield stress XLI
Young modulus 71, 80 f, 93
– boron carbides 851, 854, 865
– boron nitrides 421
– cemented borides 928
– CVD diamond 581 f, 588
– data collection 965–995
– diamond-like carbon 639
– diboride composites 886
– directed metal oxidation 311
– nanostructures 110
– polycrystalline boron nitride 519
– silicon carbides 697, 721
– polyaluminocarbosilanes 458
– titanium boride-iron composites 910
– TM carbides/nitrides 203, 232
– zirconium/titanium borides 879
yttria-stabilized zirconia 892
yttrium–aluminum-garnet (YAG) 41
yttrium borides 805
yttrium oxides 787

Zeldovich criterion   336
zeta potential   846
Zhadanov notation   7 f
zinc
– diamond synthesis   488
– directed metal oxidation   289
zinc sulfide window   584
zincblende structures   XLIV
– boron nitrides   LVIII, 6, 420 ff, 510
– carbon nitrides   LVI, 258
– ultrahard materials   521
zirconia   92

zirconium
– diamond synthesis   498
– directed metal oxidation   305
zirconium borides   879
zirconium–boron system   817
– self-propagating synthesis   333
zirconium–carbon–nitrogen system   222
zirconium corundum   658
zirconium dioxide   35
zirconium oxide/aluminumoxide/silicon oxide layers   309
ZnSe windows   589 f